# GROUNDWATER ECOLOGY AND EVOLUTION

SECOND EDITION

# GROUNDWATER ECOLOGY AND EVOLUTION

## SECOND EDITION

*Edited by*

FLORIAN MALARD
*Univ Lyon, Université Claude Bernard Lyon 1, CNRS, ENTPE,
UMR 5023 LEHNA, Villeurbanne, France*

CHRISTIAN GRIEBLER
*University of Vienna, Department of Functional & Evolutionary Ecology,
Vienna, Austria*

SYLVIE RÉTAUX
*Paris-Saclay Institute of Neuroscience, Université Paris-Saclay and CNRS,
Saclay, France*

ELSEVIER

**ACADEMIC PRESS**
An imprint of Elsevier

Academic Press is an imprint of Elsevier
125 London Wall, London EC2Y 5AS, United Kingdom
525 B Street, Suite 1650, San Diego, CA 92101, United States
50 Hampshire Street, 5th Floor, Cambridge, MA 02139, United States
The Boulevard, Langford Lane, Kidlington, Oxford OX5 1GB, United Kingdom

**Notices**
Knowledge and best practice in this field are constantly changing. As new research and experience broaden our understanding, changes in research methods, professional practices, or medical treatment may become necessary.

Practitioners and researchers must always rely on their own experience and knowledge in evaluating and using any information, methods, compounds, or experiments described herein. In using such information or methods they should be mindful of their own safety and the safety of others, including parties for whom they have a professional responsibility.

To the fullest extent of the law, neither the Publisher nor the authors, contributors, or editors, assume any liability for any injury and/or damage to persons or property as a matter of products liability, negligence or otherwise, or from any use or operation of any methods, products, instructions, or ideas contained in the material herein.

ISBN: 978-0-12-819119-4

For information on all Academic Press publications visit our website at https://www.elsevier.com/books-and-journals

Publisher: Candice Janco
Acquisitions Editor: Maria Elekidou
Editorial Project Manager: Aleksandra Packowska
Production Project Manager: Bharatwaj Varatharajan
Cover Designer: Matthew Limbert

Typeset by TNQ Technologies

Front images credit: Marie Pavie, Krystel Saroul, Peter Pospisil, Robert Lepennec all rights reserved.
Image on the bottom left from Zagmajster et al. (2014), Global Ecology and Biogeography 23 (10), 1135—1145.

# Contents

List of contributors    xi
Preface    xv
Groundwater ecology and evolution: an
introduction    xvii

# I

## Setting the scene: groundwater as ecosystems

### 1. Hydrodynamics and geomorphology of groundwater environments

Luc Aquilina, Christine Stumpp, Daniele Tonina and
John M. Buffington

Introduction    3
The aquifer concept    5
Links to surface hydrology    13
Aquifer function    17
The chemical composition of groundwater    21
Chemical and nutrient fluxes in aquifers    24
Conclusion    27
Acknowledgments    28
References    28

### 2. Classifying groundwater ecosystems

Anne Robertson, Anton Brancelj, Heide Stein and
Hans Juergen Hahn

Introduction    39
Classification systems    41
Global scale    42
Continental scale    42
Landscape scale    44
Habitat/local scale    48
Conclusions    53
Glossary    54
Acknowledgments    55
References    55

### 3. Physical and biogeochemical processes of hyporheic exchange in alluvial rivers

Daniele Tonina and John M. Buffington

Introduction    61
The hyporheic zone    64
Predicting hyporheic exchange    65
The role of hyporheic flow on water quality    73
Conclusion    77
Acknowledgments    78
References    78

### 4. Ecological and evolutionary jargon in subterranean biology

David C. Culver, Tanja Pipan and Žiga Fišer

Introduction    89
Ecological classifications    90
Colonization and speciation    95
Morphological modification for subterranean life    99
Overall recommendations    103
Glossaries    104
Eco-Evo Glossary    104
Retired Speleobiological Glossary    105
Acknowledgments    106
References    106

# II

## Drivers and patterns of groundwater biodiversity

### 5. Groundwater biodiversity and constraints to biological distribution

Pierre Marmonier, Diana Maria Paola Galassi, Kathryn Korbel, Murray Close, Thibault Datry and Clemens Karwautz

Introduction   113
An overview of groundwater biodiversity   115
Physical constraints to biological distribution   122
Chemical constraints to biological distribution   125
Species interactions   128
The effect of the past: paleogeographic events and historical climates   130
Conclusion   132
Acknowledgments   133
References   133

### 6. Patterns and determinants of richness and composition of the groundwater fauna

Maja Zagmajster, Rodrigo Lopes Ferreira, William F. Humphreys, Matthew L. Niemiller and Florian Malard

Introduction   141
Patterns of species richness   143
Patterns of species composition   152
Toward a multifaceted approach to groundwater biodiversity patterns   156
Acknowledgments   159
References   159

### 7. Phylogenies reveal speciation dynamics: case studies from groundwater

Steven Cooper, Cene Fišer, Valerija Zakšek, Teo Delić, Špela Borko, Arnaud Faille and William Humphreys

Introduction   165
Single colonization versus multiple colonizations from surface ancestors   168
Speciation from subterranean ancestors   169
Speciation from subterranean ancestors: likely mechanisms   171
Drivers of subterranean diversity: the role of paleoclimatic and paleogeological events   173

Synthesis and future prospects   176
Acknowledgments   177
References   177

### 8. Dispersal and geographic range size in groundwater

Florian Malard, Erik Garcia Machado, Didier Casane, Steven Cooper, Cene Fišer and David Eme

Introduction   185
Evolution of dispersal   188
Range size   193
Groundwater landscape connectivity modulates dispersal   197
Conclusion   200
Acknowledgments   201
References   201

# III

## Roles of organisms in groundwater

### 9. Microbial diversity and processes in groundwater

Lucas Fillinger, Christian Griebler, Jennifer Hellal, Catherine Joulian and Louise Weaver

Introduction   211
Ecological processes determining microbial community diversity and composition   213
Microbial communities and biogeochemical cycles   217
Microbial attenuation of groundwater contaminants and bottlenecks   222
Resistance and resilience of groundwater microbial communities to perturbations   227
Outlook   230
Acknowledgments   230
References   231

### 10. Groundwater food webs

Michael Venarsky, Kevin S. Simon, Mattia Saccò, Clémentine François, Laurent Simon and Christian Griebler

Introduction   241
Basal energy dynamics in groundwater food webs   242

The role of habitat in groundwater food web
dynamics  245
The role of food web processes in groundwater
community dynamics  247
Trophic niche diversification in groundwater
ecosystems  248
Future directions  249
Acknowledgments  253
References  253

## 11. Role of invertebrates in groundwater ecosystem processes and services

Florian Mermillod-Blondin, Grant C. Hose, Kevin S. Simon,
Kathryn Korbel, Maria Avramov and Ross Vander Vorste

Introduction  263
Trophic actions of invertebrates  265
Ecosystem engineering activities by
invertebrates  269
Conceptual model of the role of invertebrates on
ecosystem processes and consequences for
ecosystem services  270
Environmental impacts on surface water–
groundwater interfaces and consequences for the
provision of ecosystem services by
invertebrates  273
Suggestions for future research directions  275
Acknowledgments  276
References  276

# IV

## Principles of evolution in groundwater

## 12. Voices from the underground: animal models for the study of trait evolution during groundwater colonization and adaptation

Sylvie Rétaux and William R. Jeffery

Introduction  285
Brief historical timeline  286
Groundwater model systems  287
Troglomorphic traits  289
Timeline of troglomorphic trait evolution  293
Evolutionary developmental biology of groundwater
organisms  293
Evolutionary genomics of groundwater
organisms  296
Conclusions  298

Acknowledgments  299
References  299

## 13. The olm (*Proteus anguinus*), a flagship groundwater species

Rok Kostanjšek, Valerija Zakšek, Lilijana Bizjak-Mali and
Peter Trontelj

Introduction  305
The historical rise to fame  306
Systematics and evolution  307
Molecular ecology and conservation
genetics  310
Morphology and sensory systems of a groundwater
top predator  313
Reproductive peculiarities  315
The overlooked part of groundwater ecology:
symbioses, pathogens and parasites  317
Conservation  320
Conclusive remarks on flagship species in
groundwater  322
Acknowledgments  324
References  324

## 14. The *Asellus aquaticus* species complex: an invertebrate model in subterranean evolution

Meredith Protas, Peter Trontelj, Simona Prevorčnik and Žiga Fišer

Introduction  329
Phylogeography and population structure  330
Phenotypic evolution of subterranean
populations  334
Raising and breeding in the laboratory  339
Genetic basis of subterranean-related traits  340
Evolutionary development (evo-devo)  342
Comparative transcriptomics  344
Conclusions and prospect  345
Acknowledgments  346
References  346

## 15. Developmental and genetic basis of troglomorphic traits in the teleost fish *Astyanax mexicanus*

Joshua B. Gross, Tyler E. Boggs, Sylvie Rétaux and Jorge Torres-Paz

The history of genetic and genomic studies of
troglomorphy in *Astyanax*  351
Developmental basis of troglomorphy in
*Astyanax*  357

Conclusions 366
Acknowledgments 366
References 366

## 16. Ecological and evolutionary perspectives on groundwater colonization by the amphipod crustacean *Gammarus minus*

Daniel W. Fong and David B. Carlini

Introduction 373
Ecological setting and morphological variation 374
Upstream colonization of subterranean waters by *Gammarus minus* 377
Impetus for colonizing cave streams 378
Multiple independent colonization of cave streams 380
Evolutionary perspectives 383
Melanin pigment loss and innate immunity 387
Future directions 388
Acknowledgments 389
References 389

## 17. Evolutionary genomics and transcriptomics in groundwater animals

Didier Casane, Nathanaelle Saclier, Maxime Policarpo, Clémentine François and Tristan Lefébure

Introduction 393
Evolution of genes and genome architecture 394
Evolution of gene expression in groundwater 405
Conclusion 410
Acknowledgments 410
References 410

# V

## Biological traits in groundwater

### 18. Dissolving morphological and behavioral traits of groundwater animals into a functional phenotype

Cene Fišer, Anton Brancelj, Masato Yoshizawa, Stefano Mammola and Žiga Fišer

Introduction 415
Habitat template 417
Morphological-behavioral functional phenotype 417
Synthesis and perspectives 430

Acknowledgments 432
References 432

## 19. Life histories in groundwater organisms

Michael Venarsky, Matthew L. Niemiller, Cene Fišer, Nathanaelle Saclier and Oana Teodora Moldovan

Introduction 439
A brief overview of life history evolution, life history traits, and life table variables 442
The current conceptual model of life history evolution in groundwater species 445
Support for the current conceptual model of life history evolution in groundwater species 446
Conclusions 451
Acknowledgments 452
References 452

## 20. Physiological tolerance and ecotoxicological constraints of groundwater fauna

Tiziana Di Lorenzo, Maria Avramov, Diana Maria Paola Galassi, Sanda Iepure, Stefano Mammola, Ana Sofia P.S. Reboleira and Frédéric Hervant

Introduction 457
Physiological tolerance of groundwater invertebrates to changing thermal conditions 458
Physiological tolerance of groundwater organisms to chemical stress 464
Physiological tolerance of groundwater organisms to light, food and oxygen variations: indications for ecotoxicological protocols 470
Conclusions 473
Acknowledgments 473
References 474

# VI

## Biodiversity and ecosystem management in groundwater

### 21. Global groundwater in the Anthropocene

Daniel Kretschmer, Alexander Wachholz and Robert Reinecke

Introduction 483
Groundwater availability and distribution 484
Frameworks for sustainable use of groundwater in the Anthropocene 489

Anthropogenic threats to groundwater   490
Outlook   494
Glossary   495
Acknowledgments   495
References   495

## 22. Assessing groundwater ecosystem health, status, and services

Grant C. Hose, Tiziana Di Lorenzo, Lucas Fillinger,
Diana Maria Paola Galassi, Christian Griebler, Hans Juergen Hahn,
Kim M. Handley, Kathryn Korbel, Ana Sofia Reboleira,
Tobias Siemensmeyer, Cornelia Spengler, Louise Weaver and
Alexander Weigand

Introduction   501
Assessing ecosystem health and condition   503
Indicators of ecosystem health and condition   508
Defining the reference condition for groundwater
  ecosystems   513
Combining indicators into summary indices   515
Predicting ecosystem health and condition   516
Future directions   517
Acknowledgments   518
References   519

## 23. Recent concepts and approaches for conserving groundwater biodiversity

Andrew J. Boulton, Maria Elina Bichuette, Kathryn Korbel,
Fabio Stoch, Matthew L. Niemiller, Grant C. Hose and
Simon Linke

Introduction   525
Past concepts and approaches in groundwater
  biodiversity conservation   527

Recent concepts and approaches in
  groundwater biodiversity
  conservation   531
Conclusion and future directions   543
Acknowledgments   545
References   545

## 24. Legal frameworks for the conservation and sustainable management of groundwater ecosystems

Christian Griebler, Hans Juergen Hahn, Stefano Mammola,
Matthew L. Niemiller, Louise Weaver, Mattia Saccò,
Maria Elina Bichuette and Grant C. Hose

Introduction   551
Conservation of groundwater ecosystems and species
  at risk   552
Why study, assess, and protect groundwater
  ecosystems?   553
Legal frameworks related to groundwater
  ecosystems   554
Current challenges and the future of groundwater
  conservation   563
Acknowledgments   566
References   566

**The ecological and evolutionary unity and
diversity of groundwater
ecosystems—conclusions and
perspective   573**
**Index   589**

# List of contributors

**Luc Aquilina**  Université Rennes 1- CNRS, UMR 6118 Géosciences Rennes, Rennes, France

**Maria Avramov**  Helmholtz Zentrum München, German Research Center for Environmental Health, Institute of Groundwater Ecology, Neuherberg, Germany

**Maria Elina Bichuette**  Laboratory of Subterranean Studies, Federal University of São Carlos, São Carlos, Brazil

**Lilijana Bizjak-Mali**  University of Ljubljana, Biotechnical Faculty, Department of Biology, Ljubljana, Slovenia

**Tyler E. Boggs**  Department of Biological Sciences, University of Cincinnati, Cincinnati, OH, United States

**Špela Borko**  University of Ljubljana, Biotechnical Faculty, Department of Biology, Ljubljana, Slovenia

**Andrew J. Boulton**  School of Environmental and Rural Science, University of New England, Armidale, NSW, Australia

**Anton Brancelj**  Université Paris-Saclay, CNRS, IRD, UMR Évolution, Génomes, Comportement et Écologie, Gif-sur-Yvette, France; Université de Paris, UFR Sciences du Vivant, Paris, France

**John M. Buffington**  Rocky Mountain Research Station, US Forest Service, Boise, ID, United States

**David B. Carlini**  Department of Biology, American University, Washington, DC, United States

**Didier Casane**  Université Paris-Saclay, CNRS, IRD, UMR Évolution, Génomes, Comportement et Écologie, Gif-sur-Yvette, France; Université Paris Cité, UFR Sciences du Vivant, Paris, France

**Murray Close**  Institute of Environmental Science and Research, Christchurch, Canterbury, New Zealand

**Steven Cooper**  South Australian Museum, Adelaide, SA, Australia; The University of Adelaide, School of Biological Sciences and Australian Centre for Evolution Biology and Biodiversity, Adelaide, SA, Australia

**David C. Culver**  Department of Environmental Science, American University, Washington, DC, United States

**Thibault Datry**  INRAE, UR-RiverLY, Lyon, France

**Teo Delić**  University of Ljubljana, Biotechnical Faculty, Department of Biology, Ljubljana, Slovenia

**Tiziana Di Lorenzo**  Research Institute on Terrestrial Ecosystems of the National Research Council of Italy (IRET-CNR), Florence, Italy; Emil Racovita Institute of Speleology, Cluj-Napoca, Romania; Centre for Ecology; Evolution and Environmental Changes (cE3c), Departamento de Biologia Animal, Faculdade de Ciências, Universidade de Lisboa, Lisbon, Portugal; National Biodiversity Future Center (NBFC), Palermo, Italy

**David Eme**  INRAE, UR-RiverLY, Lyon, France

**Arnaud Faille**  Stuttgart State Museum of Natural History, Stuttgart, Germany

**Rodrigo Lopes Ferreira**  Universidade Federal de Lavras (UFLA), Centro de Estudos em Biologia Subterrânea, Departamento de Ecologia e Conservação, Lavras, Minas Gerais, Brazil

**Lucas Fillinger**  University of Vienna, Department of Functional & Evolutionary Ecology, Vienna, Austria

**Cene Fišer** University of Ljubljana, Biotechnical Faculty, Department of Biology, Ljubljana, Slovenia

**Žiga Fišer** University of Ljubljana, Biotechnical Faculty, Department of Biology, Ljubljana, Slovenia

**Daniel W. Fong** Department of Biology, American University, Washington, DC, United States

**Clémentine François** Univ Lyon, Université Claude Bernard Lyon 1, CNRS, ENTPE, UMR 5023 LEHNA, Villeurbanne, France

**Diana Maria Paola Galassi** Department of Life, Health and Environmental Sciences, University of L'Aquila, L'Aquila, Italy

**Christian Griebler** University of Vienna, Department of Functional & Evolutionary Ecology, Vienna, Austria

**Joshua B. Gross** Department of Biological Sciences, University of Cincinnati, Cincinnati, OH, United States

**Hans Juergen Hahn** Institute for Environmental Sciences, University of Koblenz-Landau, Landau, Germany

**Kim M. Handley** School of Biological Sciences, The University of Auckland, Auckland, New Zealand

**Jennifer Hellal** BRGM, DEPA, Geomicrobiology and Environmental Monitoring Unit, Orléans, France

**Frédéric Hervant** Univ Lyon, Université Claude Bernard Lyon 1, CNRS, ENTPE, UMR 5023 LEHNA, Villeurbanne, France

**Grant C. Hose** School of Natural Sciences, Macquarie University, Sydney, Australia

**William F. Humphreys** University of Western Australia, School of Biological Sciences, Crawley, WA, Australia

**William Humphreys** Western Australian Museum, Welshpool DC, WA, Australia

**Sanda Iepure** Emil Racovita Institute of Speleology, Cluj-Napoca, Romania; Institutul Român de Ştiintă şi Tehnologie, Cluj-Napoca, Romania

**William R. Jeffery** Department of Biology, University of Maryland, College Park, MD, United States

**Catherine Joulian** BRGM, DEPA, Geomicrobiology and Environmental Monitoring Unit, Orléans, France

**Clemens Karwautz** University of Vienna, Department of Functional & Evolutionary Ecology, Vienna, Austria

**Kathryn Korbel** School of Natural Sciences, Macquarie University, Sydney, Australia

**Rok Kostanjšek** University of Ljubljana, Biotechnical Faculty, Department of Biology, Ljubljana, Slovenia

**Daniel Kretschmer** Institute of Environmental Science and Geography, University Potsdam, Potsdam, Germany

**Tristan Lefébure** Univ Lyon, Université Claude Bernard Lyon 1, CNRS, ENTPE, UMR 5023 LEHNA, Villeurbanne, France

**Simon Linke** CSIRO, Dutton Park, Brisbane, QLD, Australia

**Erik Garcia Machado** Institut de Biologie Intégrative et des Systèmes (IBIS), Pavillon Charles-Eugène-Marchand, Avenue de la Médecine, Université Laval Québec, Québec, Canada

**Florian Malard** Univ Lyon, Université Claude Bernard Lyon 1, CNRS, ENTPE, UMR 5023 LEHNA, Villeurbanne, France

**Stefano Mammola** Molecular Ecology Group (dark-MEG), Water Research Institute (IRSA), National Research Council (CNR), Verbania-Pallanza, Italy; University of Helsinki, Finnish Museum of Natural History (LUOMUS), Helsinki, Finland

**Pierre Marmonier** Univ Lyon, Université Claude Bernard Lyon 1, CNRS, ENTPE, UMR 5023 LEHNA, Villeurbanne, France

**Florian Mermillod-Blondin** Univ Lyon, Université Claude Bernard Lyon 1, CNRS, ENTPE, UMR 5023 LEHNA, Villeurbanne, France

**Oana Teodora Moldovan** Emil Racovitza Institute of Speleology, Cluj-Napoca, Romania

**Matthew L. Niemiller** Department of Biological Sciences, The University of Alabama in Huntsville, Huntsville, AL, United States

**Tanja Pipan** ZRC SAZU, Karst Research Institute, Postojna, Slovenia

**Maxime Policarpo** Université Paris-Saclay, CNRS, IRD, UMR Évolution, Génomes, Comportement et Écologie, Gif-sur-Yvette, France

**Simona Prevorčnik** University of Ljubljana, Biotechnical Faculty, Department of Biology, Ljubljana, Slovenia

**Meredith Protas** Dominican University of California, San Rafael, CA, United States

**Ana Sofia P.S. Reboleira** Natural History Museum of Denmark, University of Copenhagen, Copenhagen, Denmark; Centre for Ecology, Evolution and Environmental Changes (cE3c), Departamento de Biologia Animal, Faculdade de Ciências, Universidade de Lisboa, Lisbon, Portugal

**Ana Sofia Reboleira** Center for Ecology, Evolution and Environmental Changes (cE3c), Departamento de Biologia Animal, Faculdade de Ciências, Universidade de Lisboa, Lisbon Portugal; Natural History Museum of Denmark, University of Copenhagen, Copenhagen, Denmark

**Robert Reinecke** Institute of Environmental Science and Geography, University Potsdam, Potsdam, Germany

**Sylvie Rétaux** Paris-Saclay Institute of Neuroscience, Université Paris-Saclay and CNRS, Saclay, France

**Anne Robertson** School of Life & Health Sciences, University of Roehampton, London, United Kingdom

**Mattia Saccò** Subterranean Research and Groundwater Ecology (SuRGE) Group, Trace and Environmental DNA (TrEnD) Laboratory, School of Molecular and Life Sciences, Curtin University, Perth, WA, Australia

**Nathanaelle Saclier** ISEM, CNRS, Univ. Montpellier, IRD, EPHE, Montpellier, France; Univ Lyon, Université Claude Bernard Lyon 1, CNRS, ENTPE, UMR 5023 LEHNA, Villeurbanne, France

**Tobias Siemensmeyer** Institute for Environmental Sciences, University of Koblenz-Landau, Landau, Germany

**Kevin S. Simon** School of Environment, University of Auckland, Auckland, New Zealand

**Laurent Simon** Univ Lyon, Université Claude Bernard Lyon 1, CNRS, ENTPE, UMR 5023 LEHNA, Villeurbanne, France

**Cornelia Spengler** Institute for Environmental Sciences, University of Koblenz-Landau, Landau, Germany

**Heide Stein** Institute for Environmental Sciences, University of Koblenz-Landau, Landau, Germany

**Fabio Stoch** Evolutionary Biology & Ecology, Université libre de Bruxelles, Brussels, Belgium

**Christine Stumpp** University of Natural Resources and Life Sciences, Vienna, Department of Water, Atmosphere and Environment, Institute of Soil Physics and Rural Water Management, Vienna, Austria

**Daniele Tonina** Center for Ecohydraulics Research, University of Idaho, Boise, ID, United States

**Jorge Torres-Paz** Paris-Saclay Institute of Neuroscience, Université Paris-Saclay and CNRS, Saclay, France

**Peter Trontelj** University of Ljubljana, Biotechnical Faculty, Department of Biology, Ljubljana, Slovenia

**Michael Venarsky** Department of Biodiversity Conservation and Attractions, Kensington, WA, Australia; Australian Rivers Institute, Griffith University, Nathan, QLD, Australia

**Ross Vander Vorste** Department of Biology, University of Wisconsin - La Crosse, La Crosse, WI, United States

**Alexander Wachholz** Helmholtz Center for Environmental Research (UFZ), Department for Aquatic Ecosystem Analysis and Management, Magdeburg, Germany

**Louise Weaver** Institute of Environmental Science and Research (ESR) Christchurch, Canterbury, New Zealand

**Alexander Weigand** National Museum of Natural History Luxembourg, Luxembourg

**Masato Yoshizawa** University of Hawai'i at Mānoa, School of Life Sciences, Honolulu, HI, United States

**Maja Zagmajster** University of Ljubljana, Biotechnical Faculty, Department of Biology, Ljubljana, Slovenia

**Valerija Zakšek** University of Ljubljana, Biotechnical Faculty, Department of Biology, Ljubljana, Slovenia

# Preface

Since the first edition of *"Groundwater Ecology"* was published almost 3 decades ago, the knowledge of ecology and evolution of biodiversity in groundwater has grown tremendously. This overdue second edition does not replace the first one but is complementary to it. The first edition largely focused on case studies of groundwater ecosystems, while the second edition provides a much-needed synthesis of the current state of knowledge about the ecology and evolution of groundwater organisms. It has a stronger evolutionary emphasis, and the interplay of ecology and evolution provides the foundation for this second edition. Hence, its title is *"Groundwater Ecology and Evolution."*

This book covers the diversity of groundwater research conducted by ecologists and evolutionary biologists. This includes, but is not restricted to, the hydrogeological and hydrochemical attributes of groundwater habitats, the controls and patterns of groundwater biodiversity, the role of organisms in groundwater systems, the evolutionary processes and forces driving the acquisition of subterranean biological traits, and the way these traits are differently expressed among organisms. Finally, it covers the challenges and opportunities for conservation of groundwater biodiversity and management of groundwater ecosystems.

This book can be relished in its entirety or read "à la carte" because within the larger themes each chapter can stand alone. The contributors to each chapter, typically an international group of experts on a relevant topic, have successfully synthesized current research, analyzed controversies, identified knowledge gaps, and discussed future research avenues.

We like to express our gratitude to all contributors and reviewers who dedicated their time and efforts to the production of this book.

The editors: Florian Malard, Christian Griebler, and Sylvie Rétaux

# Groundwater ecology and evolution: an introduction

*Florian Malard[1], Christian Griebler[2] and Sylvie Rétaux[3]*

[1]Univ Lyon, Université Claude Bernard Lyon 1, CNRS, ENTPE, UMR 5023 LEHNA, Villeurbanne, France; [2]University of Vienna, Department of Functional & Evolutionary Ecology, Vienna, Austria; [3]Paris-Saclay Institute of Neuroscience, Université Paris-Saclay and CNRS, Saclay, France

## Rocks, water, and life

Groundwater occurs beneath the Earth's surface in void spaces of soil, sediment, and rock formations. It is contained in geological formations known as aquifers that can hold and transmit water. Water-bearing geological strata include consolidated rocks, such as limestone and granite, as well as unconsolidated sediments, such as sand and gravel. Recent estimates of the volume of groundwater in the upper 10 km of continental crust (43.9 million $km^3$) indicate that groundwater is the second largest reservoir of water globally, after the oceans (1.3 billion $km^3$), and ahead of ice sheets (30.158 million $km^3$) (Ferguson et al., 2021). However, only groundwater in the upper 1 km of the continental crust is likely to be fresh, representing an estimated volume of 15.9 million $km^3$ (Ferguson et al., 2021). At depths greater than 1 km, groundwater is essentially brackish to saline. The estimated volume of fresh groundwater is much higher than the 0.10 million $km^3$ of water in surface wetlands, large lakes, reservoirs, and rivers (Gleeson et al., 2016). Modern groundwater—the water in the first few hundred meters below ground that is less than 50 years old—represents a global volume of about 1.3 million $km^3$ (0.1–5.0 million $km^3$) (Gleeson et al., 2016). This volume dwarfs all other components of the active hydrologic cycle, namely (in order of decreasing importance) surface freshwater, soil water, atmosphere, and vegetation (Gleeson et al., 2016).

Groundwater hosts a high diversity of organisms including viruses, prokaryotes (bacteria and archaea), microeukaryotes (fungi and protists), and metazoans, including invertebrates, amphibians, and fishes (Euringer and Lueders, 2008; Karwautz and Griebler, 2022; Malard, 2022; Retter and Nawaz, 2022; Schweichhart et al., 2022). Groundwater ecosystems substantially contribute to the Earth's biodiversity and biomass. In fact, most of the global prokaryotic biomass is concentrated in the continental subsurface. Despite large uncertainties in the quantification of this biomass (21–62 gigatons), it represents about 15% of the total biomass in the biosphere (Bar-on et al., 2018, but see also Magnabosco et al., 2018). The subsurface vertical extent of metazoan life is probably less than that of prokaryotes (Pedersen, 2000; Fišer et al., 2014), although nematodes were recovered in South Africa from 0.9 to 3.6 kilometer deep fractures filled with 3000–12,000-year-old palaeometeoric water (Borgonie et al., 2011). An encyclopedia of the world fauna inhabiting subterranean waters published more than 35 years ago contains more than 5500 species of metazoans (Botosaneanu, 1986). From simple extrapolations of regional species richness data, Culver and Holsinger (1992) suggested that the world subterranean metazoan fauna could represent 50,000–100,000 species, among which one third would be aquatic species. Since the world surface freshwaters accommodate approximately 125,000 animal species (Balian et al., 2008), groundwater is expected to comprise a large fraction of the Earth's freshwater metazoan biodiversity. In Europe, the number of obligate groundwater crustaceans exceeds the number of surface-dwelling crustacean species, even though the description of groundwater species significantly lags behind that of surface species (Stoch and Galassi, 2010).

The groundwater organisms and their physical environment—water, rocks, and sediments—with which they interact constitute diverse groundwater ecosystems. Ecosystem functions, also referred to as ecosystem processes or ecological processes, correspond to the activities of organisms and the effects these activities have on their environment. In broad terms, ecosystem functions encompass the cycling of material and flow of energy. Ecosystem services are the benefits human populations obtain, directly or indirectly, from ecosystem functions (Haines-Young and Potschin, 2010). Most services provided by groundwater systems are mediated, if not entirely supported by groundwater organisms, involving microorganisms and metazoans (Boulton et al., 2008; Griebler and Avramov, 2015; Fenwick et al., 2018; Griebler et al., 2019). The capacity of aquifers to supply good-quality groundwater for various human uses without doubt depends on the activity of microorganisms and metazoans. There is compelling evidence from laboratory experiments showing that groundwater metazoans, through bioturbation and grazing of microbial biofilms, help maintain the effective porosity and hydraulic conductivity of unconsolidated sediments (Hose and Stumpp, 2019). Groundwater microorganisms exert a major control on the turnover of organic carbon and cycling of inorganic nutrients, thereby significantly contributing to the purification of groundwater along its paths from recharge to discharge areas (Griebler and Avramov, 2015). Groundwater organisms are also critical for the biodegradation of organic contaminants and elimination of pathogens, which in turn contributes to disease control (Sinton, 1984; Herman et al., 2001; Boulton et al., 2008; Griebler and Avramov, 2015).

Groundwater ecosystems are open systems that strongly interact with adjacent terrestrial and surface aquatic ecosystems. Discharge of groundwater at the land surface is crucial for the continued existence of many surface terrestrial and aquatic ecosystems, commonly referred to as groundwater-dependent ecosystems (Kløve et al., 2011a,b). Conversely, the functioning and biodiversity of groundwater ecosystems depend on their linkages with terrestrial and surface aquatic ecosystems. First, in the absence of light-driven primary production, groundwater food webs depend on the supply of organic carbon from surface environments. However, there is evidence that they can also be supported to various degrees by groundwater chemolithoautotrophs (Overholt et al., 2022). Second, groundwater metazoan communities are often composed of a mix of species that show different degrees of dependence to the subterranean environment. Specialist groundwater species strictly depend on groundwater; some of them, often referred to as obligate groundwater species or stygobionts, complete their entire life cycle in groundwater. Generalist species (stygophiles) exploit a wide range of resources from both groundwater and surface water. Third, a difficult-to-estimate, but probably significant, proportion of obligate groundwater species is derived directly from speciation events occurring during evolutionary transitions of surface aquatic species to groundwater. Many such transitions are currently ongoing in several taxa such as crustaceans and fishes (Malard, 2022). They provide ideal models to study adaptation to a novel environment because groundwater colonizers experience drastic and sudden environmental changes (i.e. darkness and food limitation) and evolve characteristic phenotypes.

## Research history in groundwater ecology and evolution

There is a long history of research on the ecology and evolution of groundwater organisms. It is marked by a series of influential events, which are briefly summarized in this paragraph (Fig. 1). The first mentions of blind and depigmented cavefish (Romero, 2001) and salamander (Aljancic, 1993) date back to the 16th and 17th century, respectively. However, biospeleology, the systematic study of organisms living in caves, began at the beginning of the 20th century with the launching of the "Biospeologica" international research program (Racovitza, 1907). Until the 1960s, ecological studies were mostly restricted to caves and the main theories on the evolution of cave animals, especially those explaining the loss of structures (e.g. eyes and pigmentation), were marked by Lamarckian ideas (Vandel, 1964). Groundwater ecology started to thrive in the 1970s with the definition of groundwater ecosystems and delineation of their physical boundaries. The ecosystem concept of groundwater, originally applied to carbonate rock aquifers (see review in Rouch, 1986), was then extended to unconsolidated sedimentary aquifers (Danielopol, 1989), and now serves as a basis for the study of the flow of energy and matter in groundwater systems (Simon, 2019). Groundwater ecology gained international recognition in the 1990s when the first book dedicated to the discipline was published (Gibert et al., 1994). At roughly the same time as the emergence of groundwater ecology, in the 1960s, American speleobiologists pushed forward neo-Darwinian hypotheses to explain the evolution of cave animals (Christiansen, 1961; Poulson, 1963; Barr, 1968). This endeavor culminated in the 1990s with the publication of a monograph on the importance of adaptations and natural selection in the cave amphipod *Gammarus minus* (Culver et al., 1995).

Since 2000s, research on the ecology and evolution of subterranean organisms has entered a new era dominated by hypothesis-testing and mechanistic explanations (Fig. 1). Now, research ranges from biodiversity to biogeography, from genes to mechanisms involved in adaptation and development of specific biological traits, from the inter- and intraspecific interactions to carbon and energy flow through food webs, and from physical–chemical and structural drivers to the role of individual organisms in groundwater ecosystem functioning and services. The 2000s scientific shift had several components. First, there was an increasing number of synthesis articles (see, for example, Jeffery, 2001; Gibert and Deharveng, 2002; Danielopol et al., 2004; Wondzell, 2011; Larned, 2012; Griebler et al., 2014; Torres-Paz et al., 2018; Mammola et al., 2020; Griebler et al., 2022), special issues (Gibert and Culver, 2009; Hancock et al., 2009; Gore et al., 2018; Kowalko et al., 2020; Griebler and Hose, 2022), and books dedicated to the ecology and evolution of groundwater organisms

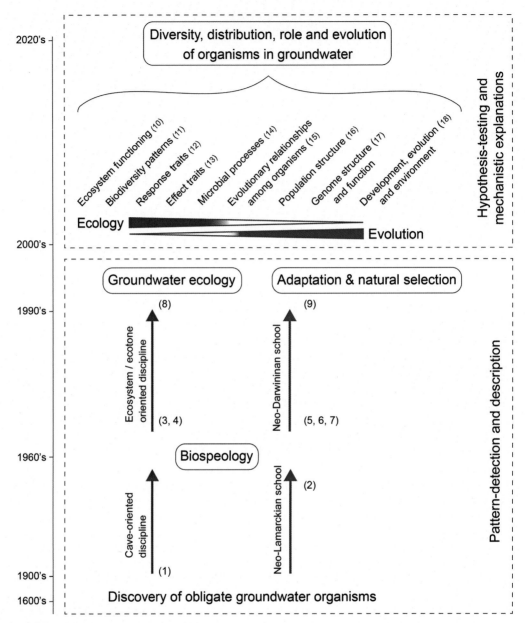

**FIGURE 1** A brief synopsis of research on the ecology and evolution of groundwater organisms. Numbers in parentheses refer to a nonexhaustive selection of landmark publications. (1) Racovitza (1907); (2) Vandel (1964); (3) Rouch (1986); (4) Danielopol (1989); (5) Christiansen (1961); (6) Poulson (1963); (7) Barr (1968); (8) Gibert et al. (1994); (9) Culver et al. (1995); (10) Baker et al. (2000); (11) Eme et al. (2018); (12) Hervant and Renault (2002); (13) Boulton et al. (2008); (14) Griebler and Lueders (2009); (15) Fišer et al. (2008); (16) Sbordoni et al. (2012); (17) Policarpo et al. (2021); (18) Torres-Paz et al. (2018).

(Chapelle, 2000; Jones and Mulholland, 2000; Wilkens et al., 2000; Griebler and Mösslacher, 2003; Romero, 2009; Culver and Pipan, 2014, 2019; Wilkens and Strecker, 2017; Moldovan et al., 2018; White et al., 2019). This period synthesized disparate results and concepts into a coherent theoretical eco-evolutionary framework for testing long-standing hypotheses and generating new ones. This theoretical framework largely benefited from the integration of concepts from several fields of research, including functional ecology (Calow, 1987), macroecology (Brown, 1995), and evolutionary developmental biology (Gilbert, 2003). Second, groundwater scientists have largely benefited from developments in biotechnology, molecular tools and data analysis for shedding new light on the functioning of groundwater food webs (Saccò et al., 2019), the relative importance of different environmental factors and evolutionary processes in shaping biodiversity patterns (Eme et al., 2018; Langille et al., 2021), and the mechanisms involved in the evolution of phenotypes (Protas et al., 2006). Third, increased connections between molecular genetics and ecology are now providing unprecedented opportunities for understanding the interplay between genes and ecological processes acting well above the organismic level. Finally, yet equally important, scientists are benefiting from the characteristic attributes of groundwater organisms and ecosystems to investigate general scientific questions that resonate well beyond the boundaries of groundwater ecology and evolution (Mammola et al., 2020). Subterranean organisms have been used as models to understand among-species variations in genome size (Lefébure et al., 2017) and rate of molecular evolution (Saclier et al., 2018), human diseases, such as degenerative eye diseases (Alunni et al., 2007), autism (Yoshizawa et al., 2018) and diabetes (Riddle et al., 2018), and the potential for life beyond Earth (Popa et al., 2012). A web of science search for the period 2019—22 indicates that approximately three percent of groundwater science articles are published in multidisciplinary journals. Contrary to the assumption that groundwater scientists contribute less than other scientists to broad-scope research questions (Griebler et al., 2014), this proportion is similar to that of other disciplines.

## Groundwater research in the Anthropocene

The Anthropocene refers to the time period—beginning potentially with the Great Acceleration in the mid-20th century (Steffen et al., 2015)—when the fast growing humanity started having a significant impact on the Earth's geology and ecosystems (Crutzen, 2002). Groundwater ecosystems have not escaped human impacts. The global extraction rate of groundwater (800—1000 km$^3$/year) exceeds that of oil by a factor of 20 (Velis et al., 2017). Groundwater is the primary source of drinking water for half of the world's population and provides $\geq$50% of global irrigation (Famiglietti, 2014; Velis et al., 2017). The water demand for all uses continues to increase worldwide and is predicted to increase by 20%—30% by 2050 (Boretti and Rosa, 2019).

In many areas of the world, groundwater extraction exceeds recharge from precipitation and surface water, thereby leading to groundwater depletion (Famiglietti, 2014). Moreover, excessive groundwater pumping has cascading effects on surface water ecosystems. De Graaf et al. (2019) estimated that environmentally critical stream flows will be reached by 2050 in approximately 42%—79% of the watersheds in which groundwater is extracted worldwide. The global groundwater crisis (Famiglietti, 2014) imposes trade-offs between various aspects of human development as stated in the United Nations Sustainable Development Goals (Griggs et al., 2013) and groundwater sustainability (Velis et al., 2017). Gleeson et al. (2020) defined groundwater sustainability as "maintaining long-term, dynamically stable storage and flow of high-quality groundwater using inclusive, equitable, and long-term governance and management."

Ensuring groundwater sustainability requires maintaining groundwater ecosystem functions and associated ecosystem services over time. Groundwater depletion and pollution are ecosystem disturbances (Griebler et al., 2019) and their consequences on ecosystem functions can potentially persist well after groundwater stores have been replenished and water quality has been restored (Rouch et al., 1993). Ecosystem responses to environmental disturbances depend on the traits of organisms, their ability to move (dispersal) and adapt. Pre- and postdisturbance communities may not be functionally equivalent, potentially leading to substantial changes in ecosystem functioning and connected services. Environmental, ecological, and evolutionary research provide the basis for integrating ecosystem function and resulting services into governance and management of groundwater resources, thereby generating joint benefits to people and biodiversity (Griebler et al., 2010, 2014; Devitt et al., 2019) (Fig. 2). However, groundwater ecosystem management is yet in its infancies: a web browser search for "groundwater management" returns about one million hits vs. only one thousand hits for "groundwater ecosystem management."

## Objective of the book

The objective of this book is to provide a synthesis of the current state of knowledge about the physics and biophysics of groundwater systems and the ecology and evolution of groundwater organisms. Thanks to the efforts of multiple investigators with wide-ranging expertise, this book brings together many facets of groundwater sciences ranging from hydrogeology to ecology to evolution of organisms. Bridging the gap between environmental, ecological, and evolutionary studies can foster the development of groundwater management practices that is needed to preserve the sustainability of groundwater ecosystems and their services to society.

## Audience

*Groundwater Ecology and Evolution*, second edition, is primarily intended for an audience of graduate students, postgraduate students, and academic researchers involved in the study of groundwater biodiversity, the function of groundwater ecosystems, and the evolution of groundwater organisms. This book not only represents an excellent resource for teaching groundwater ecology at the university level, but also provides fascinating case studies of evolution in caves for teaching evolutionary biology. Despite its focus on groundwater, this book is also highly relevant for general biologists. Groundwater ecosystems are excellent model systems to study general principles: in evolution, the mechanisms resulting in adaptation to a novel environment, in physiology, the mechanisms supporting life in extreme environments, and, in ecology, the importance of biodiversity for nutrient cycling at sedimentary interfaces. Groundwater is a crucial resource to humans. This book provides water managers

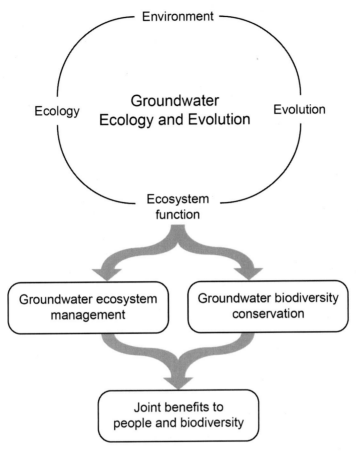

FIGURE 2    Groundwater research in the Anthropocene.

with key information on ecosystem management practices that contribute to maintaining groundwater sustainability in the face of future resource-use scenarios. We hope that the diverse readers will make frequent use of this book in basic science, applied science, and teaching.

## Structure and content of the book

This book is composed of 24 chapters grouped into six sections. The first section (Chapters 1–4) sets the environmental stage by describing the physical attributes of groundwater ecosystems. It also provides a glossary as part of Chapter 4 that revisits the specialized terminology used in subterranean biology. Chapter 1 describes the physical and hydrodynamic properties of aquifers, their linkages with surface water, groundwater age, the flow of groundwater, and transport of solutes and particulate matter. In addition, it provides insights into the chemical composition of groundwater, water–rock interactions, and chemical fluxes in aquifers. Chapter 2 focuses on the flow of water in the hyporheic zone of alluvial rivers, a functionally important transitional habitat between surface and ground waters. It describes operative definitions of the hyporheic zone, models for predicting hyporheic flow paths across different spatial scales, and the influence of hyporheic flow on water quality. Chapter 3 reviews existing classification systems for groundwater ecosystems at four hierarchical spatial scales—global, continental, landscape, and local—and emphasizes the importance of classification schemes for integrating groundwater ecosystems into the monitoring of freshwater aquatic ecosystems. Chapter 4 proposes to replace the specialized terminology used in subterranean biology with a more general ecological and evolutionary terminology. The specialized terminologies that are examined are those of ecological classifications of groundwater organisms, transitions of surface aquatic species to groundwater, and biological traits of obligate groundwater species.

The next four chapters in Section 2 review the controls and patterns of groundwater biodiversity. Chapter 5 provides an account of the taxonomic diversity of living forms in groundwater including viruses, prokaryotes, microeukaryotes, and metazoans. It documents the main constraints for biological distributions in groundwater, including physical and chemical constraints, biotic interactions, and the effects of past constraints linked to major paleogeographic events and historical climates. Chapter 6 gives an overview of the most striking features of patterns in species richness and taxonomic composition of the obligate groundwater fauna at regional to continental scales. It emphasizes patterns that are common enough among taxonomic groups and continents to be potentially recognized as "rules" and reports on the most probable combination of mechanisms shaping these patterns. The next two chapters focus on speciation and dispersal, two of the three processes that ultimately determine the number of species in a region. The third process, extinction, unfortunately is difficult to quantify in the absence of fossil records among groundwater species. Chapter 7 provides phylogenetic and phylogeographic studies of representative groundwater species-rich taxa and discusses the processes that have led to their speciation. It focuses on the evidence for speciation from surface aquatic ancestors *versus* subterranean speciation, where species evolve from groundwater ancestors within groundwater environments. Chapter 8 introduces dispersal in groundwater and highlights the factors controlling the eco-evolutionary dynamics of dispersal that are relevant for understanding species-range dynamics. It reports on the main barriers to dispersal of groundwater organisms and provides a case study showing how groundwater-landscape resistance controls the movement of organisms.

Section 3 contains three chapters that describe the roles of organisms in groundwater ecosystems. Chapter 9 provides an overview of the factors that determine the diversity and composition of microbial communities in groundwater, the resistance and resilience of microbial communities to disturbances, and their role in key biogeochemical processes including the cycling of organic matter and nutrients, and natural attenuation of contaminants. Chapter 10 begins with a discussion of basal energy dynamics in groundwater foodwebs that covers both organic matter dynamics as well as chemolithoautotrophic primary production. Then, it explores the role habitats play in structuring the inputs and processing of matter and energy and the influence of various food-web mechanisms on trophic-niche diversification and relative importance of bottom-up and top-down processes. Chapter 11 shows how invertebrates influence groundwater ecosystem function through their trophic interactions with biofilms and nontrophic actions that engineer the physical environment.

These trophic and engineering activities are modulated by habitat factors, in particular by the strength of surface water—groundwater exchanges, and they are sensitive to human-driven environmental disturbances.

Section 4 is devoted to developmental and evolutionary processes related to the acquisition of subterranean biological traits. The section starts with an introduction (Chapter 12) that gives an overview at the organismal level of recent research documenting the variety of troglomorphic traits specific to groundwater animals. It also introduces the fields and principles of evolutionary developmental biology (EvoDevo) and comparative genomics for the study of trait evolution in the dark. Chapters 13—16 give detailed presentations of emblematic model systems that have been the most studied so far and have brought most insights into the evolution of troglomorphy: the amphibian "olm" *Proteus anguinus* (Chapter 13), the isopod crustacean *Asellus aquaticus* (Chapter 14), the teleost fish *Astyanax mexicanus* (Chapter 15), and the amphipod crustacean *G. minus* (Chapter 16). The section ends with Chapter 17, which summarizes our current understanding of the evolution of gene sequences, gene expression, and genome architecture associated with surface to groundwater transition.

Section 5 summarizes the organismal-level research presented in the previous section by providing an overview of variation in morphological, behavioral, life-history, and physiological traits among groundwater organisms. Chapter 18 considers six general aspects of the morphological and behavioral phenotype, namely sensory input, locomotion, feeding, (micro)habitat choice, reproduction, and antipredator response. It introduces the "many-to-one relationship of form and function" principle, stating that enhancement of functional performance is possible either through morphological or behavioral changes, and proposes that this principle can explain variation in phenotypes among habitats, accounting for different trade-offs. Chapter 19 explores the current state of life-history research in groundwater fauna, including a brief review of life-history evolution, life-history traits, and life-table variables. The chapter outlines the current conceptual model of life-history evolution in groundwater species and reviews the support for this model, acknowledging that most inferences are from a few model taxa. Chapter 20 is devoted to the study of physiological tolerance of groundwater species. This includes the responses of organisms to light, food, and oxygen variations and their ability to cope with temperature changes and chemical contamination. Data are examined within the contexts of ecological risk-assessment of groundwater, global warming, and decreasing groundwater quality.

Section 6, the last section of the book, shows how knowledge derived from multiple research foci (Sections 1—5) can be used to manage groundwater biodiversity and ecosystem services in the face of future groundwater resource-use scenarios. Chapter 21 provides a global perspective of the crucial importance of groundwater resources, for both humans and ecosystems, and threats to these resources due to unsustainable use in the Anthropocene. Chapter 22 introduces current schemes for the assessment of groundwater ecosystem health, status, and services. These include conventional groundwater-assessment methods in the fields of community ecology, functional ecology, and ecotoxicology. It concludes with a discussion on future directions and knowledge gaps. Chapter 23 deals with recent concepts and approaches for conserving groundwater biodiversity, while Chapter 24 summarizes existing legal frameworks for the protection and conservation of groundwater organisms and ecosystems.

We finish the book with a conclusion chapter that identifies knowledge gaps, priorities in basic research, and challenges for the governance and management of groundwater ecosystems.

## Acknowledgments

We like to express our gratitude to all contributors and reviewers who dedicated their time and effort to the production of this book. This books represents the work of 78 contributors from 56 research laboratories in 17 countries. We like to thank Aleksandra Packowska, Editorial Project Manager, and Bharatwaj Varatharajan, Production Manager, for their continuous support throughout the realization of the book. We thank Björn Wissel for proofreading and English-language editing the General Introduction.

## References

Aljancic, M., Bulog, B., Kranjc, A., Habic, P., Josipovic, D., Sket, B., Skoberne, P., 1993. *Proteus*: The Mysterious Ruler of Karst Darkness. Speleo Projects. Caving Publications International, Ljubiana, Slovenia.

Alunni, A., Menuet, A., Candal, E., Pénigault, J.B., Jeffery, W.R., Rétaux, S., 2007. Developmental mechanisms for retinal degeneration in the blind cavefish *Astyanax mexicanus*. Journal of Comparative Neurology 505 (2), 221–233.

Baker, M.A., Dahm, C.N., Valett, H.M., 2000. Anoxia, anaerobic metabolism, and biogeochemistry of the stream-water-ground-water interface. In: Jones, J.B., Mulholland, P.J. (Eds.), Streams and Groundwaters. Academic Press, San Diego, California, pp. 259–283.

Balian, E.V., Segers, H., Lévêque, C., Martens, K., 2008. The freshwater animal diversity assessment: an overview of the results. Hydrobiologia 595, 627–637.

Bar-On, Y.M., Phillips, R., Milo, R., 2018. The biomass distribution on Earth. Proceedings of the National Academy of Sciences of the United States of America 115 (25), 6506–6511.

Barr, T.C., 1968. Cave ecology and the evolution of troglobites. Evolutionary Biology 2, 35–102.

Boretti, A., Rosa, L., 2019. Reassessing the projections of the world water development report. NPJ Clean Water 2, 15.

Borgonie, G., Garcia-Moyano, A., Litthauer, D., Bert, W., Bester, A., van Heerden, E., Möller, C., Erasmus, M., Onstott, T.C., 2011. Nematoda from the terrestrial deep subsurface of South Africa. Nature 474, 79–82.

Botosaneanu, L., 1986. Stygofauna Mundi: A Faunistic, Distributional, and Ecological Synthesis of the World Fauna Inhabiting Subterranean Waters. E.J. and Dr. W. Backhuys, Leiden, The Netherlands.

Boulton, A.J., Fenwick, G.D., Hancock, P.J., Harvey, M.S., 2008. Biodiversity, functional roles and ecosystem services of groundwater invertebrates. Invertebrate Systematics 22, 103–116.

Brown, J.H., 1995. Macroecology. The University of Chicago Press, Chicago.

Calow, P., 1987. Towards a definition of functional ecology. Functional Ecology 1, 57–61.

Chapelle, F.H., 2000. Ground-Water Microbiology and Geochemistry, second ed. Wiley.

Christiansen, K.A., 1961. Convergence and parallelism in cave entomobryinae. Evolution 15, 288–301.

Crutzen, P.J., 2002. Geology of mankind. Nature 415, 23.

Culver, D.C., Pipan, T., 2014. Shallow Subterranean Habitats. Ecology, Evolution, and Conservation. Oxford University Press, Oxford.

Culver, D.C., Pipan, T., 2019. The Biology of Caves and Other Subterranean Habitats, second ed. Oxford University Press, Oxford.

Culver, D.C., Kane, T.C., Fong, D.W., 1995. Adaptation and Natural Selection in Caves: The evolution of *Gammarus minus*. Harvard University Press, Cambridge.

Danielopol, D.L., 1989. Groundwater fauna associated with riverine aquifers. Journal of the North American Benthological Society 8 (1), 18–35.

Danielopol, D.L., Gibert, J., Griebler, C., Gunatilaka, A., Hahn, H.J., Messana, G., Notenboom, J., Sket, B., 2004. Incorporating ecological perspectives in European groundwater management policy. Environmental Conservation 31 (3), 185–189.

de Graaf, I.E.M., Gleeson, T., van Beek, L.P.H., Sutanudjaja, E.H., Bierkens, M.F.P., 2019. Environmental flow limits to global groundwater pumping. Nature 573, 90–94.

Devitt, T.J., Wright, A.M., Cannatella, D.C., Hillis, D.M., 2019. Species delimitation in endangered groundwater salamanders: Implications for aquifer management and biodiversity conservation. Proceedings of the National Academy of Sciences of the United States of America 116 (7), 2624–2633.

Eme, D., Zagmajster, M., Delić, T., Fišer, C., Flot, J.-F., Konecny-Dupré, L., Pálsson, S., Stoch, F., Zakšek, V., Douady, C.J., Malard, F., 2018. Do cryptic species matter in macroecology? Sequencing European groundwater crustaceans yields smaller ranges but does not challenge biodiversity determinants. Ecography 41, 424–436.

Euringer, K., Lueders, T., 2008. An optimised PCR/T-RFLP fingerprinting approach for the investigation of protistan communities in groundwater environments. Journal of Microbiological Methods 75 (2), 262–268.

Famiglietti, J.S., 2014. The global groundwater crisis. Nature Climate Change 4, 945–948.

Fenwick, G., Greenwood, M., Williams, E., Milne, J., Watene-Rawiri, E., 2018. Groundwater Ecosystems: Functions, Values, Impacts and Management. NIWA Report 2018184CH, Horizons Regional Council, Palmerston North, New Zealand.

Ferguson, G., McIntosh, J.C., Warr, O., Sherwood Lollar, B., Ballentine, C.J., Famiglietti, J.S., Kim, J.-H., Michalski, J.R., Mustard, J.F., Tarnas, J., McDonnell, J.J., 2021. Crustal groundwater volumes greater than previously thought. Geophysical Research Letters 48 e2021GL093549.

Fišer, C., Sket, B., Trontelj, P., 2008. A phylogenetic perspective on 160 years of troubled taxonomy of *Niphargus* (Crustacea: Amphipoda). Zoologica Scripta 37 (6), 665–680.

Fišer, C., Pipan, T., Culver, D.C., 2014. The vertical extent of groundwater metazoans: an ecological and evolutionary perspective. BioScience 64 (11), 971–979.

Gibert, J., Culver, D.C., 2009. Assessing and conserving groundwater biodiversity: an introduction. Freshwater Biology 54, 639–648.

Gibert, J., Deharveng, L., 2002. Subterranean ecosystems: a truncated functional biodiversity. BioScience 52 (6), 473–481.

Gibert, J., Danielopol, D.L., Stanford, J.A., 1994. Groundwater Ecology. Academic Press, San Diego, USA.

Gilbert, S.F., 2003. The morphogenesis of evolutionary developmental biology. International Journal of Developmental Biology 47 (7-8), 467–477.

Gleeson, T., Befus, K.M., Jasechko, S., Luijendijk, E., Cardenas, M.B., 2016. The global volume and distribution of modern groundwater. Nature Geoscience 9 (2), 161–167.

Gleeson, T., Cuthbert, M., Ferguson, G., Perrone, D., 2020. Global groundwater sustainability, resources, and systems in the Anthropocene. Annual Review of Earth and Planetary Sciences 48, 431–463.

Gore, A.V., Rohner, N., Rétaux, S., Jeffery, W.R., 2018. Seeing a bright future for a blind fish. Developmental Biology 441 (2), 207–208.

Griebler, C., Avramov, M., 2015. Groundwater ecosystem services: a review. Freshwater Science 34 (1), 355–367.

Griebler, C., Hose, G.C., 2022. Section introduction: groundwater sciences in limnology. In: Mehner, T., Tockner, K. (Eds.), Encyclopedia of Inland Waters, second ed. Elsevier, pp. 303–305.

Griebler, C., Lueders, T., 2009. Microbial biodiversity in groundwater ecosystems. Freshwater Biology 54 (4), 649–677.

Griebler, C., Stein, H., Kellermann, C., Berkhoff, S., Brielmann, H., Schmidt, S., Selesi, D., Steube, C., Fuchs, A., Hahn, H.J., 2010. Ecological assessment of groundwater ecosystems—vision or illusion? Ecological Engineering 36 (9), 1174–1190.

Griebler, C., Malard, F., Lefébure, T., 2014. Current developments in groundwater ecology - from biodiversity to ecosystem function and services. Current Opinion in Biotechnology 27, 159–167.

Griebler, C., Avramov, M., Hose, G., 2019. Groundwater ecosystems and their services: current status and potential risks. In: Schröter, M., Bonn, A., Klotz, S., Seppelt, R., Baessler, C. (Eds.), Atlas of Ecosystem Services. Springer, Cham, pp. 197–203.

Griebler, C., Fillinger, L., Karwautz, C., Hose, G.C., 2022. Knowledge gaps, obstacles, and research frontiers in groundwater microbial ecology. In: Mehner, T., Tockner, K. (Eds.), Encyclopedia of Inland Waters, second ed. Elsevier, pp. 611–624.

Griebler, C., Mösslacher, F., 2003. Grundwasser-Ökologie (Groundwater Ecology). UTB-Fakultas Verlag, Wien.

Griggs, D., Stafford-Smith, M., Gaffney, O., Rockström, J., Öhman, M.C., Shyamsundar, P., Steffen, W., Glaser, G., Kanie, N., Noble, I., 2013. Sustainable development goals for people and planet. Nature 495 (7441), 305–307.

Haines-Young, R., Potschin, M., 2010. The links between biodiversity, ecosystem services and human well-being. In: Raffaelli, D.G., Frid, C.L.J. (Eds.), Ecosystem Ecology: A New Synthesis. Cambridge University Press, British Ecological Society, pp. 110–139.

Hancock, P.J., Hunt, R.J., Boulton, A.J., 2009. Preface: hydrogeoecology, the interdisciplinary study of groundwater dependent ecosystems. Hydrogeology Journal 17, 1−3.

Hervant, F., Renault, D., 2002. Long-term fasting and realimentation in hypogean and epigean isopods: a proposed adaptive strategy for groundwater organisms. The Journal of Experimental Biology 205 (14), 2079−2087.

Culver, D.C., Holsinger, J.R., 1992. How many species of troglobites are there? NSS Bulletin 54, 79−80.

Herman, J.S., Culver, D.C., Salzman, J., 2001. Groundwater ecosystems and the service of water purification. Stanford Environmental Law Journal 20, 479−495.

Hose, G.C., Stumpp, C., 2019. Architects of the underworld: bioturbation by groundwater invertebrates influences aquifer hydraulic properties. Aquatic Sciences 81, 20.

Jeffery, W.R., 2001. Cavefish as a model system in evolutionary developmental biology. Developmental Biology 231, 1−12.

Jones, J.B., Mulholland, P.J., 2000. Streams and Ground Waters. Academic Press, San Diego, California.

Karwautz, C., Griebler, C., 2022. Microbial biodiversity in groundwater ecosystems. In: Mehner, T., Tockner, K. (Eds.), Encyclopedia of Inland Waters, Second ed. Elsevier, pp. 397−411.

Kløve, B., Ala-aho, P., Bertrand, G., Boukalova, Z., Ertürk, A., Goldscheider, N., Ilmonen, J., Karakaya, N., Kupfersberger, H., Kværner, J., Lundberg, A., Mileusnić, M., Moszczynska, A., Muotka, T., Preda, E., Rossi, P., Siergieiev, D., Šimek, J., Wachniew, P., Angheluta, V., Widerlund, A., 2011a. Groundwater dependent ecosystems. Part I: hydroecological status and trends. Environmental Science and Policy 14 (7), 770−781.

Kløve, B., Allan, A., Bertrand, G., Druzynska, E., Ertürk, A., Goldscheider, N., Henry, S., Karakaya, N., Karjalainen, T.P., Koundouri, P., Kupfersberger, H., Kværner, J., Lundberg, A., Muotka, T., Preda, E., Pulido-Velazquez, M., Schipper, P., 2011b. Groundwater dependent ecosystems. Part II. Ecosystem services and management in Europe under risk of climate change and land use intensification. Environmental Science and Policy 14 (7), 782−793.

Kowalko, J.E., Franz-Odendaal, T.A., Rohner, N., 2020. Introduction to the special issue—cavefish—adaptation to the dark. Journal of Experimental Zoology Part B: Molecular and Developmental Evolution 334, 393−396.

Langille, B.L., Hyde, J., Saint, K.M., Bradford, T.M., Stringer, D.N., Tierney, S.M., Humphreys, W.F., Austin, A.D., Cooper, S.J.B., 2021. Evidence for speciation underground in diving beetles (Dytiscidae) from a subterranean archipelago. Evolution 75 (1), 166−175.

Larned, Z.T., 2012. Phreatic groundwater ecosystems: research frontiers for freshwater ecology. Freshwater Biology 57, 885−906.

Lefébure, T., Morvan, C., Malard, F., François, C., Konecny-Dupré, L., Guéguen, L., Weiss-Gayet, M., Seguin-Orlando, A., Ermini, L., Der Sarkissian, C., Charrier, N.P., Eme, D., Mermillod-Blondin, F., Duret, L., Vieira, C., Orlando, L., Douady, C.J., 2017. Less effective selection leads to larger genomes. Genome Research 27, 1016−1028.

Magnabosco, C., Lin, L.-H., Dong, H., Bomberg, M., Ghiorse, W., Stan-Lotter, H., Pedersen, K., Kieft, T.L., van Heerden, E., Onstott, T.C., 2018. The biomass and biodiversity of the continental subsurface. Nature Geoscience 11, 707−717.

Malard, F., 2022. Groundwater Metazoans. In: Mehner, T., Tockner, K. (Eds.), Encyclopedia of Inland Waters, second ed. Elsevier, pp. 474−487.

Mammola, S., Amorim, I.R., Bichuette, M.E., Borges, P.A.V., Cheeptham, N., Cooper, S.J.B., Culver, D.C., Deharveng, L., Eme, D., Ferreira, R.L., Fišer, C., Fišer, Ž., Fong, D.W., Griebler, C., Jeffery, W.R., Jugovic, J., Kowalko, J.E., Lilley, T.M., Malard, F., Manenti, R., Martínez, A., Meierhofer, M.B., Niemiller, M.L., Northup, D.E., Pellegrini, T.G., Pipan, T., Protas, M., Reboleira, A.S.P.S., Venarsky, M.P., Wynne, J.J., Zagmajster, M., Cardoso, P., 2020. Fundamental research questions in subterranean biology. Biological Reviews 95, 1855−1872.

Moldovan, O.T., Kováč, L., Halse, S., 2018. Cave Ecology. Springer Nature, Cham, Switzerland.

Overholt, W.A., Trumbore, S., Xu, X., Bornemann, T.L.V., Probst, A.J., Krüger, M., Herrmann, M., Thamdrup, B., Bristow, L.A., Taubert, M., Schwab, V.F., Hölzer, M., Marz, M., Küsel, K., 2022. Carbon fixation rates in groundwater similar to those in oligotrophic marine systems. Nature Geoscience 15, 561−567.

Pedersen, K., 2000. Exploration of deep intraterrestrial microbial life: current perspectives. FEMS Microbiology Letters 185, 9−16.

Policarpo, M., Fumey, J., Lafargeas, P., Naquin, D., Thermes, C., Naville, M., Dechaud, C., Volff, J.-N., Cabau, C., Klopp, C., Møller, P.R., Bernatchez, L., García-Machado, E., Rétaux, S., Casane, D., 2021. Contrasting gene decay in subterranean vertebrates: insights from cavefishes and fossorial mammals. Molecular Biology and Evolution 38 (2), 589−605.

Popa, R., Smith, A.R., Popa, R., Boone, J., Fisk, M., 2012. Olivine-respiring bacteria isolated from the rock-ice interface in a lava-tube cave, a Mars analog environment. Astrobiology 12 (1), 9–18.

Poulson, T.L., 1963. Cave adaptation in amblyopsid fishes. The American Midland Naturalist 70, 257–290.

Protas, M.E., Hersey, C., Kochanek, D., Zhou, Y., Wilkens, H., Jeffery, W.R., Zon, L.I., Borowsky, R., Tabin, C.J., 2006. Genetic analysis of cavefish reveals molecular convergence in the evolution of albinism. Nature Genetics 38 (1), 107–111.

Racovitza, E.G., 1907. Essai sur les problèmes biospéologiques. Archives de Zoologie Expérimentale et Générale 6, 371–488.

Retter, A., Nawaz, A., 2022. The groundwater mycobiome: fungal diversity in terrestrial aquifers. In: Mehner, T., Tockner, K. (Eds.), Encyclopedia of Inland Waters, Second ed. Elsevier, pp. 385–396.

Riddle, M.R., Aspiras, A.C., Gaudenz, K., Peuß, R., Sung, J.Y., Martineau, B., Peavey, M., Box, A.C., Tabin, J.A., McGaugh, S., Borowsky, R., Tabin, C.J., Rohner, N., 2018. Insulin resistance in cavefish as an adaptation to a nutrient-limited environment. Nature 555, 647–651.

Romero, A., 2001. Scientists prefer them blind: the history of hypogean fish research. Environmental Biology of Fishes 62, 43–71.

Romero, A., 2009. Cave Biology. Life in Darkness. Cambridge University Press, Cambridge.

Rouch, R., 1986. Sur l'écologie des eaux souterraines dans le karst. Stygologia 2 (4), 352–398.

Rouch, R., Pitzalis, A., Descouens, A., 1993. Effets d'un pompage à gros débit sur le peuplement des Crustacés d'un aquifère karstique. Annales de Limnologie 29, 15–29.

Saccò, M., Blyth, A., Bateman, P.W., Hua, Q., Mazumder, D., White, N., Humphreys, W.F., Laini, A., Griebler, C., Grice, K., 2019. New light in the dark—a proposed multidisciplinary framework for studying functional ecology of groundwater fauna. Science of the Total Environment 662, 963–977.

Saclier, N., François, C.M., Konecny-Dupré, L., Lartillot, N., Guéguen, L., Duret, L., Malard, F., Douady, C.J., Lefébure, T., 2018. Life history traits impact the nuclear rate of substitution but not the mitochondrial rate in isopods. Molecular Biology and Evolution 35, 2900–2912.

Sbordoni, V., Allegrucci, G., Cesaroni, D., 2012. Population structure. In: White, W.B., Culver, D.C. (Eds.), Encyclopedia of Caves, Second edition. Academic/Elsevier Press, Amsterdam, the Netherlands, pp. 608–618.

Schweichhart, J.S., Pleyer, D., Winter, C., Retter, A., Griebler, C., 2022. Presence and role of prokaryotic viruses in groundwater environments. In: Mehner, T., Tockner, K. (Eds.), Encyclopedia of Inland Waters, Second ed. Elsevier, pp. 373–384.

Simon, K.S., 2019. Cave ecosystems. In: White, W.B., Culver, D.C., Pipan, T. (Eds.), Encyclopedia of Caves. Academic Press, London, UK, pp. 223–226.

Sinton, L.W., 1984. The macroinvertebrates in a sewage-polluted aquifer. Hydrobiologia 119, 161–169.

Steffen, W., Broadgate, W., Deutsch, L., Gaffney, O., Ludwig, C., 2015. The trajectory of the Anthropocene: the great acceleration. The Anthropocene Review 2 (1), 81–98.

Stoch, F., Galassi, D.M.P., 2010. Stygobiotic crustacean species richness: a question of numbers, a matter of scale. Hydrobiologia 653, 217–234.

Torres-Paz, J., Hyacinthe, C., Pierre, C., Rétaux, S., 2018. Towards an integrated approach to understand Mexican cavefish evolution. Biology Letters 14 (8), 20180101.

Vandel, A., 1964. Biospéologie: la biologie des animaux cavernicoles. Gauthier-Villars, Paris.

Velis, M., Conti, K.I., Biermann, F., 2017. Groundwater and human development: synergies and trade-offs within the context of the sustainable development goals. Sustainability Science 12, 1007–1017.

White, W.B., Culver, D.C., Pipan, T., 2019. Encyclopedia of Caves. Academic Press, London, UK.

Wilkens, H., Strecker, U., 2017. Evolution in the Dark. Darwin's Loss Without Selection. Springer, Cham.

Wilkens, H., Culver, D.C., Humphreys, W.F., 2000. Subterranean ecosystems. Ecosystems of the world, vol 30. Elsevier, Amsterdam.

Wondzell, S.M., 2011. The role of the hyporheic zone across stream networks. Hydrological Processes 25 (22), 3525–3532.

Yoshizawa, M., Settle, A., Hermosura, M.C., Tuttle, L.J., Cetraro, N., Passow, C.N., McGaugh, S.E., 2018. The evolution of a series of behavioral traits is associated with autism-risk genes in cavefish. BMC Evolutionary Biology 18, 89.

# Setting the scene: groundwater as ecosystems

# 1

# Hydrodynamics and geomorphology of groundwater environments

*Luc Aquilina[1], Christine Stumpp[2], Daniele Tonina[3] and John M. Buffington[4]*

[1]Université Rennes 1- CNRS, UMR 6118 Géosciences Rennes, Rennes, France; [2]University of Natural Resources and Life Sciences, Vienna, Department of Water, Atmosphere and Environment, Institute of Soil Physics and Rural Water Management, Vienna, Austria; [3]Center for Ecohydraulics Research, University of Idaho, Boise, ID, United States; [4]Rocky Mountain Research Station, US Forest Service, Boise, ID, United States

## Introduction

Within the global water cycle, the groundwater pool represents a substantial volume of water, containing approximately $8{,}000\text{-}23{,}000 \bullet 10^3$ km$^3$ (Abbott et al., 2019). Annual groundwater discharge to the ocean ($0.1-6.5 \bullet 10^3$ km$^3$/year) is two to three orders of magnitude smaller than oceanic evaporation ($350-510 \bullet 10^3$ km$^3$/year), which initiates the continental part of the water cycle through atmospheric condensation and landward precipitation ($88-120 \bullet 10^3$ km$^3$/year). The other segments of the water cycle are part of our daily life: the clouds in the sky, the polar ice seen from satellites and the snow, closer to us, the rivers we like to walk along, and the lakes where we go fishing and swimming. Conversely, the underground part of the water cycle remains poorly known to the general public. It is the invisible part of the water cycle. In fact, for many people, the notion of groundwater is generally associated with the idea of an underground lake or river. Instead, groundwater is a sponge-like hydrologic system that occupies an extensive network of voids in near-surface (crustal) rocks of the continents and ocean floor. While groundwater systems are understood conceptually, less is often known about the specific movement of water through the subsurface system and the associated annual water cycle.

Groundwater has long been difficult to comprehend in its entirety due to its subterranean nature and difficulty of access. Geologic maps offer clues for determining where groundwater may occur as a function of different rock types, sedimentary deposits, and surface topography, but the spatiotemporal extent of groundwater can only be measured from boreholes/wells and cave/spring systems, which are typically limited to a relatively small number of locations.

The underground hydrologic system is dynamic, made up of numerous flow paths (Fig. 1.1). The nature of these flows is complex, in response to the strong heterogeneity (spatial variability) of geological formations and, in turn, their ability to hold and transfer water (*permeability*). The soil (top layer of the sediment) constitutes the first compartment that controls flows feeding the groundwater system as a result of precipitation and snowmelt that percolate into the soil. Within a given groundwater system, we can find very fast flows, as well as areas with extremely slow flows. These variations can exist both on a regional scale and on a microscopic scale, with variations in flow controlling the physical and chemical interactions between water and rock. The physical control of water fluxes and the chemical control of elements are therefore intrinsically coupled.

Observed linkages between surface water and groundwater provide further information about the extent and function of the groundwater system, with recent studies emphasizing the need to evaluate river systems within the context of groundwater processes. For example, springs, which are a direct emergence of groundwater onto the landscape, can have important controls on headwater portions of the surface water system and downstream water quality (Peterson et al., 2001; Alexander et al., 2007; Meyer et al., 2007; Soulsby et al., 2007; Rhoades et al., 2021). Throughout their course, rivers also receive diffuse flows from the underground environment that modulate physical conditions. In turn, rivers drive complex exchanges of water between the surface and subsurface hydrologic systems. Consequently, surface water and groundwater are extremely dynamic, integrated systems.

Groundwater resources have become a subject of concern in the Anthropocene (the current geologic epoch in which humans have substantially altered physical and ecological processes (Crutzen, 2002), especially regarding climate change and pollution. Climate-driven increases in temperature and evapotranspiration may limit future groundwater volumes and the extent of groundwater flow. Weakening groundwater flows can have severe impacts on surface systems and may increase the length and occurrence of droughts. Drought, which occurs in nearly all regions, has affected more people worldwide in the last 40 years than any other natural hazard. The effects of water scarcity can manifest through environmental crises, as droughts may induce tipping points (Otto et al., 2020). As such, active research currently focuses on the effects of climate change on groundwater systems (Amanambu et al., 2020) given that a large number of human populations may have difficulty accessing drinking water under future climate scenarios. Human activities also can have major effects on the water quality of groundwater systems. Intensive agriculture can cause diffuse pollution of nitrate, creating extensive eutrophication (Vitousek et al., 1997; Tilman et al., 2001). Other pollutants entering groundwater systems, such as endocrine disruptors, nanoparticles, and microplastics, have become a major concern due to their impact on both human life and biodiversity (Kremen et al., 2002; Gallo et al., 2018).

Groundwater ecosystems also host a large variety of organisms that dwell in open spaces within the underground material, ranging from small pores or cracks to large voids and tunnels that are typically present in karst landscapes (e.g., limestone caverns) and lava tubes. The flows that traverse the underground environment also control the supply of nutrients accessible to living organisms. Porosity and flow, therefore, condition the subsurface living world and constitute an extremely diverse set of habitats in which physical, chemical, and biological systems are intimately interconnected.

The aim of this chapter is to describe the physical and chemical principles that characterize these underground environments. Specifically, we review the physical basis of aquifers (groundwater reservoirs), their hydrodynamics, and hydrogeological parameters (porosity and permeability) that collectively define different types of aquifers. We explore the relationships between groundwater and surface water and define how aquifers function in terms of (1) groundwater flow and the transport of solutes and particulate matter; (2) groundwater age, which affects ecosystem processes, physical and biological relations, and groundwater resources; and (3) modeling of the above processes. We also consider the chemical composition of groundwater and the origin of compounds and water—rock interactions that influence water quality. Finally, we discuss chemical and nutrient fluxes in aquifers and biogeochemical reactions, with a focus on oxygen and nitrogen.

## The aquifer concept

Most rocks, soils, and sediments near the surface of the earth have a certain degree of porosity caused by voids and fractures and, thus, are referred to as *porous media*. The ability of porous media to transmit water (*permeability*) depends on having connected pores. Permeable subsurface lithologies that contain extensive bodies of groundwater are termed *aquifers*. The upper surface of the aquifer is known as the *water table*, which separates saturated and unsaturated zones within geologic strata. In unconfined aquifers, the capillary rise may form a band of saturated sediment above the water table. This region is also known as a zone of tension saturation because tension forces pull the water upward into available pores, resulting in water pressures below atmospheric values in this zone. The capillary rise can extend above the water table from a few centimeters in sediments with large pores (e.g., clean gravel), up to several meters (e.g., 4–5m) in clay soils with small pores. Because the position of the water table varies with time due to seasonal and decadal changes in the supply and movement of groundwater, a variably saturated zone also can be defined (Fig. 1.1).

### Drivers of groundwater flow

The water content of the aquifer differs from the water flow, which is related to spatial gradients in the energy head. Water moves from high to low energy-head locations modulated by a conductivity coefficient describing the ease with which a given porous media transmits fluids. This phenomenon is described by Darcy's (1856) law

$$q = -K\frac{dh_T}{dl} \tag{1.1}$$

where $q$ is the flow per unit area (with dimensions of length, L, divided by time, T; L/T), $K$ is the hydraulic conductivity (L/T), $h_T$ is the total energy head defined as the sum of the hydraulic pressure head $h_p$, the elevation head $h_z$ (gravitational potential energy arising from elevation), and the velocity head $h_v$ (kinetic energy of the fluid velocity), all of which are expressed as the height of water (L), and $l$ is the distance (L) over which the change in $h_T$ is evaluated. Because interstitial flows through porous media tend to be slow, $h_v$ is typically

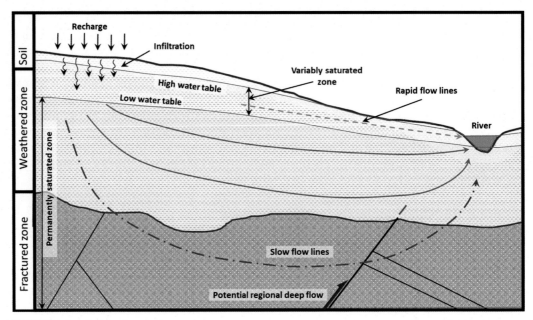

**FIGURE 1.1**  Cross-section of an aquifer. Blue lines with arrows show groundwater flow paths. Black diagonal lines portray main fractures.

negligible, such that $h_T$ is defined by the piezometric head $h_p + h_z$, which is simply the elevation of the water table measured by subtracting the depth to groundwater from the land surface. Consequently, it is the higher altitude of the water table within a landscape that creates groundwater motion, just like in a closed U-shaped tube with a difference in water level that is suddenly opened. Groundwater motion may be conceptually defined as successive flow lines that act as separate tubes (Fig. 1.1).

Figure 1.1 presents a cross-sectional view of a simple aquifer fully open to the atmosphere (*unconfined aquifer*), in which groundwater moves according to Darcy's law from the mountain top toward the river valley along three nested flow paths in the variably saturated, permanently saturated, and fractured zones, respectively. The variably saturated zone represents the seasonal variation of the water table due to competition between *recharge* and depletion of the aquifer. Recharge is mainly driven by precipitation percolating into the soil, but is modulated by vegetation. When precipitation infiltrates the soil, some water that is held against gravity in pores due to matric forces (i.e., adhesion of water to solid surfaces and the attraction of water molecules to one another) is accessible by plants and can be removed via evapotranspiration from this near-surface water reservoir. Particularly in summer and spring, plant demand progressively depletes the soil water content. Once the soil water content increases and gravitational forces exceed matric forces, water flows through the soil and the unsaturated zone down to the water table and into the aquifer. Over geologic time, the water within the aquifer weathers the bedrock to a certain depth, below which groundwater moves more slowly through fissures and fractures in the more competent (less weathered) parent bedrock (fractured zone). The rate of water movement (and thus its age) differs between the variably saturated, permanently saturated, and fractured (unweathered) zones due to differences in energy head, hydraulic conductivity, and the length of a given flow path (Fig. 1.1).

Aquifers may also be encountered below geological formations at depths of several hundred meters, particularly in sedimentary basins. Such geological formations also have limited zones of water inflow (recharge zones) and present extremely slow renewal rates. Although present at great depths, these aquifers represent active microbial ecosystems (e.g., Chapelle, 2001). The outflow of these systems also supports oases and specific groundwater-dependent ecosystems. Closer to the surface, aquifers may not be entirely open to the atmosphere, covered by clay-rich (low permeability) layers or geologic formations that cap and confine the aquifer. *Confined aquifers* can exhibit substantially different geochemical compositions and limited fluxes of nutrients, thus representing a different ecosystem within the aquifer compared to unconfined strata.

When aquifers are located at a great depth or close to a magmatic chamber in volcanic areas, temperature also becomes a driver of water motion and leads to the uprising and outflow of deep hot water (for example, geysers and geothermal vents). Indeed, volcanic areas are also the location of numerous thermal springs, which represent the outflow of hydrothermal convection cells. Thermal springs are frequent along mountain ranges, which induce large and deep hydrogeological loops of groundwater motion. In the deeper part of the loops, water encounters high temperatures and the upward movement of water is related to combined effects of thermal and head gradients. Such regional loops can also be present in nonmountainous areas, potentially with slight temperature anomalies (Fig. 1.2). This kind of hydrogeological situation is interesting as it induces mixing between deep and shallow groundwater with extremely different chemical compositions and biodiversity.

## Aquifer hydrodynamics

### Porosity

The total volume $V_t$ ($L^3$) of an aquifer can be divided into the volume of solids $V_s$ and the volume of voids $V_v$. The ratio of $V_v$ to $V_t$ defines the porosity $n$ (dimensionless) of the material, which is typically expressed as a percentage. There are several methods for determining porosity (Hao et al., 2008; Flint and Flint, 2018). Most commonly, it is determined by measuring the bulk density of the material and particle density of the solids. Other methods include obtaining a water-saturated sample of known volume and drying it in a laboratory oven (105°C). The difference in weight before and after drying (correcting for the temperature dependence of water density) gives information about the volume of water per total volume of the sample. For samples containing water that cannot be removed in a drying oven at those temperatures, the porosity can also be determined by sealing a sample of known volume with paraffin, placing it into the water, and measuring the displaced volume of water. The porosity gives the entire volume of the pore space, and thus gives information about the water volume potentially stored in an aquifer. However, a certain amount of pore space may contain entrapped air, rather than water, nor will all pores contribute to water flow due to, for example, dead-end pores, nonconnected pores, adhesively bound water, or hydrated water

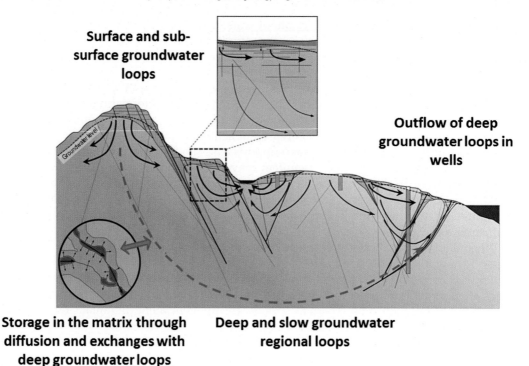

**Surface and sub-surface groundwater loops**

**Outflow of deep groundwater loops in wells**

**Storage in the matrix through diffusion and exchanges with deep groundwater loops**

**Deep and slow groundwater regional loops**

FIGURE 1.2    Different levels of mixing in aquifers. *Redrawn with permission from Fig. 1.14 of Roques (2013).*

of clay minerals. When considering only the amount of void volume contributing to the water flow, the volume ratio is denoted as an effective porosity $n_{eff}$. In addition, there is a distinction between primary porosity (i.e., that of the deposited sediment or parent rock material) and secondary porosity (i.e., that due to subsequent chemical or physical weathering). The latter is of particular importance in the weathered and fractured zones of bedrock aquifers, as well as in karst aquifers.

The value of porosity for unconsolidated material generally ranges from 25% to 70% (Freeze and Cherry, 1979) and is largely dependent on the size and shape of individual particles, and how well-sorted or uniform they are. Clay has higher porosity compared to sand or gravel, but is less permeable. For consolidated material, porosity values range from 0% to 50% (Freeze and Cherry, 1979) and are lower for crystalline rock or shale compared to fractured or karstic aquifers. When considering aquifers as habitat for organisms, the overall porosity is less important than the size of individual pores and the connectivity of the pore network. The pore size distribution defines whether aquifers are suitable habitats for biota, because they are restricted from actively moving or being transported through pores smaller than their own size (Fig. 1.3). Groundwater fauna is therefore mainly found in alluvial sediments with larger pores or in fractures or channels of fractured rocks or karst aquifers (Humphreys, 2009). However, it was found that some of these organisms not only use open voids, but are capable of modifying their environment by moving grains and digging through

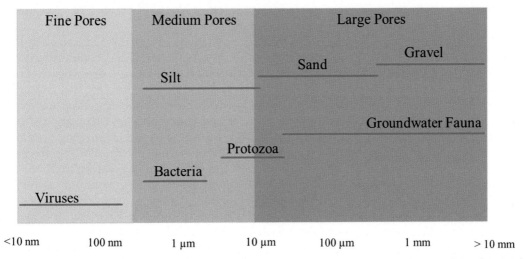

| Fine Pores | Medium Pores | Large Pores |

**FIGURE 1.3** Pore size classes (fine, medium, and large) as a function of different types of unconsolidated material (clay, silt, sand, and gravel) in comparison with size ranges for viruses, bacteria, protozoa, and fauna in aquifers. *Modified with permission from Krauss and Griebler (2011) and Matthess and Pekdeger (1981), with fauna data from Stein et al. (2012) and Thulin and Hahn (2008).*

porous sediment (Stumpp and Hose, 2017; Hose and Stumpp, 2019) or by moving into clay sediment (Korbel et al., 2019). For bacteria and viruses, most pores in unconsolidated material are wide enough for them to either be dispersed in water or attached to solid surfaces (Fig. 1.3).

### *Permeability and hydraulic conductivity*

The pore size distribution not only forms a habitat for biota, but also dictates how fast water flows through an aquifer for a given head gradient, as described by Darcy's law (1). The velocity of groundwater moving through pores and fractures, in turn, influences the energy needed for an organism to forage and live in such environments. The smaller the pore, the larger the flow resistance and the slower the flux. For uniform material and a unit head gradient, the fluid movement is proportional to the square of the mean pore diameter (Fetter, 2001). The proportional factor between the head gradient and the water flux is defined by the hydraulic conductivity $K$ in Eq. (1.1). It combines properties of the fluid (viscosity and density) and of the sediment/rock material. The inherent property of the porous medium alone is its permeability $k$ ($L^2$), which is mainly controlled by the size distribution of the voids. Therefore, unconsolidated fine materials like clay, glacial tills, or silt have smaller permeability values ($10^{-19}$–$10^{-13}$ m$^2$) compared to coarser materials like sand and gravel ($10^{-13}$–$10^{-7}$ m$^2$) (Freeze and Cherry, 1979; Gleeson et al., 2011). For consolidated rocks, permeability is generally low ($10^{-20}$–$10^{-12}$ m$^2$) due to low porosity, particularly in the absence of secondary porosity features (fractures, channels). If such features are present and augmented by weathering of the parent rock, permeability is larger ($10^{-15}$–$10^{-9}$ m$^2$) and depends on how well those features are connected (Worthington et al., 2016).

Permeability and connectivity of pores also may affect the bioenergetics of organisms in terms of nutrient availability and foraging distances.

Typically, the permeability and the hydraulic conductivity of a given medium are spatially variable (heterogeneous) depending on the structure, competence, and composition of the sediments/rocks. This is referred to as *continuous heterogeneity* because it describes the spatial variability of hydraulic properties within a given facies (e.g., a mixture of sand and gravel) due to the connectivity of porosity and fissures, which differs from *categorical heterogeneity* that describes changes in hydraulic conductivity among different facies (e.g., sand vs. gravel bodies). Permeability and hydraulic conductivity are often vector properties, leading to different behavior in the horizontal versus vertical directions (*anisotropy*). Both heterogeneity and anisotropy make it difficult to accurately measure the hydraulic conductivity in a representative elementary volume (REV, a volume of sediment large enough to capture the intrinsic process variability, but small enough to avoid combining variability among sediment types). Pumping tests or any other methods for determining hydraulic conductivity quantify the effective hydraulic conductivity, a lumped property controlled by the sediment/rock features having the largest hydraulic conductivity. Nevertheless, such values provide bulk information about the environment of the well during the pumping test. For scaling hydraulic conductivity and connectivity of specific hydrogeological features, regionalization methods can be used (Renard and Allard, 2013).

## Geologic types of aquifers

The above discussion of groundwater flow lines within aquifers is idealized, describing conditions that might exist in relatively homogeneous material, where hydraulic conductivity does not change spatially. In reality, any porous media has morphological and chemical variations that cause spatial changes in hydraulic conductivity, such that aquifers are intrinsically heterogeneous, but the degree of heterogeneity may vary from low to high. Furthermore, the geologic and geomorphic history of the landscape can have a strong influence on aquifer properties and function, with heterogeneous aquifers common in both karstic and fractured bedrock terrains.

Karstic aquifers are mainly carbonate rocks (e.g., limestone), which are progressively dissolved by carbon dioxide ($CO_2$) contained in water that originates from the soil due to organic matter degradation (White, 2002). Karst landscapes constitute 12%–15% of the continental surface. They are characterized by various dissolution features within the strata, including shafts and sinkholes, some of which may be primary locations for the inflow of surface water to the aquifer drainage system. Sinking streams are also a major feature of karst geomorphology and have attracted substantial attention due to the dramatic and *flashy* nature of flow in this part of the system (i.e., rapid flooding and drainage). The unsaturated zone of karstic systems also differs from more homogenous aquifers, typically characterized by dissolution features that may create specific local porosity in the first few meters of the karst surface. This zone of intense dissolution is termed the *epikarst* and may constitute a near-surface reservoir for the karst system, where water stored in the epikarst slowly infiltrates and recharges the underlying aquifer (Aquilina et al., 2006; Williams, 2008).

Karstic aquifers are characterized by a drainage system that traverses the entire carbonate formation from surface input to outflow, which may be characterized by a complex series of springs or a single dominant channel that controls a large part of the system. The drainage system is spectacular as it constitutes cavities that can be explored not only from a hydrogeological point of view but also for recreation and tourism. Karst aquifers are also important groundwater ecosystems because they can support larger groundwater fauna, such as fish (e.g., Hancock et al., 2005). Within the main channels of the karst drainage system (typically large caves), water flow is extremely rapid and produces spectacular floods that are often described as major characteristics of karstic systems. These open water flows are highly sensitive to human activities. Although the main channels of karst systems are typically flashy, matrix water within the carbonate aquifer contributes to the flow during the entire hydrological year, supplying water to outflow springs even when there is no surface-driven flooding in the main channels. Within the karst matrix, the transit time of water (i.e., the length of time spent traversing a given flow path) is often much longer than that of the primary drainage system of caverns and carbonate tunnels, providing strong mixing between rapid and slow flows (Long and Putnam, 2004; Bailly-Comte et al., 2011; Palcsu et al., 2021). Higher hydraulic head during flood events can cause a substantial flux of matrix water (Screaton et al., 2004), making many karstic systems flashy by nature, although the water that is flushed out during floods may have residence-times greater than assumed from the rapid pressure response (Kattan, 1997; Katz et al., 2001; Stuart et al., 2010; Han et al., 2015).

Fractured bedrock aquifers are also highly heterogeneous systems (Neuman, 2005). Hard-rock geologies represent about one-third of the continental surface and their aquifers thus comprise major water resources in many countries. In these systems, water circulates within the discontinuities of the rock (e.g., faults, fractures, and fissures). In contrast, the bedrock matrix itself has extremely low porosity and permeability and does not constitute a major water reservoir. Groundwater flow paths in fractured systems follow energy head gradients as presented in Fig. 1.1, but with more complex flow lines due to the heterogeneity of the medium. The formation of aquifers in hard-rock geologies requires intensive weathering of the bedrock over geologic time. Thus, fractured aquifers are often described as a weathered layer a few tens of meters thick, overlying the more competent parent bedrock. The weathered layer typically has a high clay content and is characterized as a hydrologically capacitive stratum, susceptible to human activities (Dewandel et al., 2006; Ayraud et al., 2008). The fractured part of the aquifer represents the transmissive part of the system and is more protected from human activities due to its deeper depth. The interface between the weathered and fractured zones is referred to as the *weathering front* and is a particularly reactive layer that is more intensively fractured than the underlying bedrock. Both layers represent fundamentally different systems, with differing hydrogeological properties and chemical compositions. Indeed, they also represent quite different ecosystems, with different microbial communities (Maamar et al., 2015).

Karstic and fractured systems present a high degree of heterogeneity between their low-permeability matrix material (competent bedrock) and the highly permeable karst conduits or hard-rock fractures. However, heterogeneity is manifested by different features over scales that vary by several orders of magnitude (Fig. 1.4A−C). For example, large regional faults may be present over several tens of kilometers along the surface of the landscape that transition to more localized fault networks at depth, with lengths of tens of meters. At smaller scales, fissures and cracks constitute discontinuities around faults or mineral boundaries. At even smaller scales, microscopic cracks and discontinuities in minerals create porosity

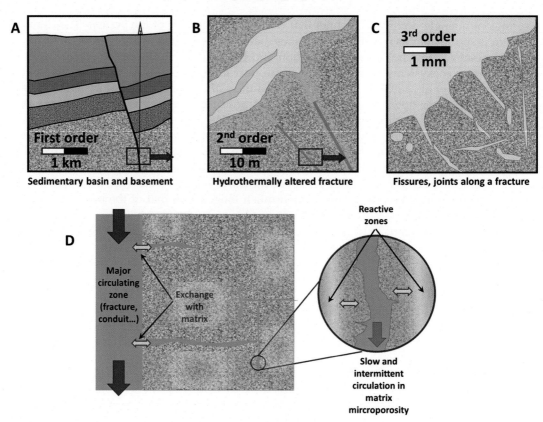

**FIGURE 1.4** Heterogeneity in aquifers. Panels (A–C) show successive scales of heterogeneity in a fractured aquifer, while panel (D) shows heterogeneity and hydrobiogeochemical functioning in peat.

within the rock. All scales of discontinuities contribute to the whole—water content of the aquifer, but with different properties. Large faults or fault zones may allow substantial fluid flow that is relatively more rapid, while the microporosity may act as a fluid reservoir. This affects both the hydrologic and chemical properties of the system. Even within relatively homogeneous aquifers, heterogeneity is also the rule, rather than the exception. Sandy aquifers contain both clay-rich and coarse facies that present local heterogeneity that respectively slows or accelerates water fluxes due to differences in grain size and hydraulic conductivity. These structures may also present distinct mineralogy and chemical reactivity. Heterogeneity is highly important for biogeochemical reactions and thus microbial ecosystems. Beyond the mean chemical characteristics of a given aquifer (e.g., chemical composition, pH, and redox), microsites within the aquifer may have very different conditions. For example, peatlands may be flushed by fresh water, resulting in overall oxic conditions, but intense sulfate reduction may occur in the microporosity of the peat, away from the main water flow (Fig. 1.4D). This allows resilience of peat and wetland systems that support relatively frequent water renewal while ensuring reducing functions such as denitrification (Racchetti et al., 2011).

# Links to surface hydrology

Aquifers are intimately linked to the surface hydrological system within watersheds. Rainfall and snowmelt percolate vertically through hillslope soils, helping to recharge the aquifer. However, most of this surface input moves laterally downslope and mainly supplies water to lakes and rivers in alluvial valleys (Fig. 1.1). The sediment in alluvial valleys is typically porous, allowing continued connection with the aquifer as the surface water flows down valley toward an ocean or terminal lake (endorheic or sink basin). Where the water table of the aquifer coincides with the water surface of rivers, subsurface flow is directed toward the river, since it represents the local topographic low point (Fig. 1.1). However, if the river flow is perched above the aquifer's water table, the river water will be driven into the streambed, recharging the aquifer through percolation and vertical head gradients; in extreme cases, rivers may become seasonally dry, disappearing into their streambeds (ephemeral streams). These two conditions, in which the river either receives groundwater or contributes river flow to the aquifer, are referred to as *gaining* versus *losing* conditions, respectively. Such phenomena may vary seasonally as the water table of the aquifer rises or falls (Fig. 1.1), and the process applies to any surface water body (i.e., rivers, lakes, and wetlands). These conditions can also vary with climate, such that water bodies in humid regions are typically gaining, while those in arid regions are frequently losing. Water bodies in karst terrain are typically losing systems, with surface water descending into the aquifer through a variety of surface fractures/inlets (e.g., sinkholes/swallets, ponors, and shafts) (e.g., Monroe, 1970; Taylor and Greene, 2008).

Because of the porous nature of alluvial sediments, the entire alluvial valley can be connected to the aquifer. For example, when rivers spill onto the valley floor during floods, water may pond for extensive periods of time, with some of this water percolating into the alluvial sediment and recharging the aquifer. Conversely, head gradients may cause the aquifer to direct water onto the floodplain via springs and seeps, which typically occur at the break in slope between hillslopes and the river valley or at topographic lows in the floodplain. Processes and rates of exchange between surface water and groundwater also may be influenced by geologic and geomorphic history in terms of how rugged the landscape is (topographic gradients) and the nature and stratigraphy of the bedrock geology and alluvial deposits. For example, volcanic eruptions can fill valleys with lava or ash, both of which have very different porosity and hydraulic conductivities. Similarly, glaciers create broad U-shaped valleys filled with sediments ranging from boulders to fine clay, resulting in different boundary conditions than valleys formed by faulting in the absence of glaciation. Extensive glacial advance can also erase topography and reorganize the direction and magnitude of surface runoff that occurs in river networks. Finally, human activity in river corridors (e.g., dams, diversions, levees, groundwater pumping, agriculture) can alter both surface and subsurface hydrological cycles and, in the long term, may significantly deplete aquifer resources and alter surface and subsurface ecosystems.

Another important linkage between surface and subsurface water occurs through *hyporheic exchange*, which is the movement of river water into and out of the alluvium within the river valley over relatively short time frames and short flow paths compared to deep groundwater movement in the aquifer (Fig. 1.5). This cycling of river water into and out of the alluvium

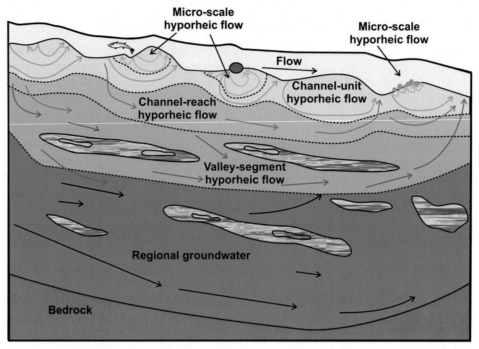

**FIGURE 1.5** Five scales of nested flow and exchange between surface and subsurface water for a longitudinal profile along a river valley: (A) microscale circulation (flow paths less than a channel width ($W$) in length) caused by local head variations at the streambed induced by objects in the flow (e.g., large particles, logs, and biotic mounds, such as salmon redds or root boles); (B) channel-unit scale due to streambed head variations around individual bed forms (e.g., dunes, bars, steps) or biotic structures (e.g., beaver dams, log jams), with flow paths up to several $W$; (C) channel-reach scale due to streambed head variations across a sequence of similar channel units (e.g., pool-riffle morphology *sensu* Montgomery & Buffington, 1997), with flow paths of tens of $W$; (D) valley-segment scale driven by head variations from spatial changes in valley confinement (floodplain width), alluvial depth, or underlying bedrock topography, with flow paths of hundreds to thousands of $W$ (Edwards, 1998; Baxter and Hauer, 2000; Dent et al., 2001; Malard et al., 2002); and (E) regional groundwater scale driven by head gradients due to overall basin slope and water-table slope, with flow paths $>10^3$ $W$. Horizontally shaded lenses are impervious clay layers that can alter the extent and direction of hyporheic exchange. *Modified from Alley et al. (1999) and Buffington and Tonina (2009b).*

allows mixing with the shallow groundwater, creating physical and biological gradients that structure both surface and subsurface ecosystems, creating complex biogeochemical cycles (Krause et al., 2011). Hyporheic exchange is driven by Darcy's law (1), but the energy head is controlled by conditions within the stream and, in particular, at the streambed surface. As such, the velocity head is no longer negligible when considering hyporheic flow.

In alluvial valleys, surface water and groundwater interact at different spatial and temporal scales that manifest as a hierarchy of nested circulation cells from near-surface hyporheic exchange to deep, regional, groundwater flow (Fig. 1.5). These scales are primarily formed because of different spatial patterns of energy head, some of which may vary temporally

(Tóth, 1963; Boano et al., 2014). Although each scale is idealized as a separate flow path, physical and chemical mixing between circulation cells occurs through numerous processes, including diffusion, advection, dispersion, divergence around heterogeneities of various scales (such as sand and clay lenses), and convergence in upwelling zones (Fig. 1.5).

At the regional scale, groundwater flow depends on (1) the overall water-table slope, (2) water levels in surface water bodies (e.g., rivers, lakes, and wetlands), and (3) characteristics of the porous media (thickness, hydraulic conductivity, heterogeneity) (Winter et al., 1998). At this scale, the water table typically follows topographic relief, but in a subdued way (Tóth, 1963; Wörman et al., 2006), and the river network can be modeled as a series of straight line segments, with energy head varying over length scales of several thousand channel widths (Guevara Ochoa et al., 2020). The water-surface elevation of other water bodies, such as lakes and wetlands, can be treated as constant or varying temporally. Groundwater flow is constrained and modulated by geological features as discussed above. Local perturbations of the energy head (for instance due to a single groundwater pumping well) may have negligible impacts at this scale. Streams can be divided into losing, gaining, or neutral segments. As the spatial scale becomes smaller, topographic features within the river corridor become progressively more important.

At the valley scale, circulation cells are driven not only by head gradients but by spatial changes in alluvial volume and hydraulic conductivity, which can be quantified by expanding the Darcy equation to account for spatial variation of those factors (Tonina and Buffington, 2009b). In particular, valley-scale circulation results from spatial changes in valley confinement (floodplain width), valley slope, geology (rock type), or underlying bedrock topography. Important features are bedrock knickpoints (Wondzell and Swanson, 1996, 1999; Baxter and Hauer, 2000) and large-scale changes in lithology and heterogeneity of alluvial deposits. For example, the frequency of elevation changes in the underlying bedrock topography may structure broad-scale variations in the depth of alluvium that in turn generate valley-scale circulation, with benthic animals congregating at upwelling locations due to relatively stable thermal and discharge regimes (Baxter and Hauer, 2000), although other physical factors may also influence habitat selection (Bean et al., 2014). Furthermore, at this scale, river hydraulics strongly influence the local connectivity between the river and its aquifer via relatively deep hyporheic exchange (Fig. 1.5). Rivers may now be modeled as a curvilinear bend moving through the floodplain, with changes in water-surface elevation and energy head dependent on spatial changes in channel roughness and geometry that can be predicted from one-dimensional hydraulic models (Fleckenstein et al., 2006; Lautz and Siegel, 2006). Sediment transport history also becomes important as buried paleochannels and stratigraphy of the streambed may form preferential hyporheic flow paths within the river valley (Stanford and Ward, 1993).

At the channel-reach scale, the river appears as a collection of repeating sequences of channel units (e.g., step-pool or pool-riffle couplets) (Fig. 1.5), with head gradients structured by the amplitude and wavelength of bed topography, and spatial changes in river depth and slope (Elliott and Brooks, 1997a,b; Kasahara and Wondzell, 2003; Buffington and Tonina, 2009). Hyporheic circulation at this scale is shallower than that of valley-segment scales but can be influenced by similar processes occurring at more local scales. These processes include changes in reach slope, mesoscale changes in the volume of alluvium, cross-valley head differences between the main channel and secondary channels (Kasahara and Wondzell, 2003), and flow through the floodplain (between meander bends (Wroblicky et al., 1998; Cardenas, 2009; Boano et al., 2010) or within buried paleochannels (Stanford and Ward, 1993). Reach-scale hyporheic circulation is also caused by irregularity amongst bed forms, with topographic low points driving larger-scale circulation and capturing hyporheic circulation of upstream channel units (Gooseff et al., 2006). Three-dimensional modeling of groundwater-surface water interaction at this scale may require two-dimensional hydraulic models of water-surface elevations within the stream to properly define both lateral and longitudinal gradients (Benjankar et al., 2016), but constant values also have been used (Storey et al., 2003).

At channel-unit and micro scales, hyporheic circulation is shallow and completely enveloped within larger-scale boundary conditions, because the physical domain is not large enough for interstitial flows to adjust to the local pore conditions, but instead is fully dominated by the boundary conditions (Tonina et al., 2016). Channel-unit circulation is associated with head variations around individual morphologic elements, such as pools and bars (Zarnetske et al., 2011), or biotic structures, such as beaver dams and log jams. In contrast, microscale circulation results from local, small-scale variations in channel characteristics (e.g., variation of the head around individual logs or clusters of streambed particles, or variation of hydraulic conductivity around a buried sand lens within otherwise coarse alluvium). At this scale, nesting and foraging activity of benthic animals also can alter bed topography, grain size, and porosity, thereby inducing and modulating hyporheic circulation (Ziebis et al., 1996; Statzner et al., 1999; Tonina and Buffington, 2009a; Pledger et al., 2017; Sansom et al., 2018). Similarly, aquatic and riparian vegetation can cause hydraulic pressure variations that drive hyporheic circulation and fine-sediment deposition that alters hydraulic conductivity (e.g., Magliozzi et al., 2018). At the microscale, pressure and velocity fluctuations due to turbulence may cause mass exchange between the river flow and near-surface pore water, resulting in hyporheic exchange (Blois et al., 2014; Roche et al., 2019). Because turbulence rapidly decreases within sediment interstices, turbulence exchange is generally limited to a near-surface layer, with thicknesses ranging from 2 to 10 times the median grain size of the streambed sediment (Packman et al., 2004; Detert et al., 2007; Tonina and Buffington, 2007). The turbulence-damping effect of the sediment increases with fine sediment, such that sand-bed streams may have shallow depths of turbulence-induced hyporheic exchange compared to gravel-bed rivers. However, sand-bed rivers are characterized by migrating bedforms at most discharges (Mohrig and Smith, 1996), with bed load transport causing a plug-flow mechanism of hyporheic exchange, known as turn-over (Elliott and Brooks, 1997a,b), in which surface water is trapped within sediment pores during bedform deposition, and released during erosion as the bedform migrates downstream. Consequently, the larger the bedform the deeper the exchange; similarly, the more active the bed load transport the faster the exchange. This mechanism may also occur in braided gravel-bed rivers due to their high rates of movement but is not expected to be important in other coarse-grained rivers, where bedforms are generally immobile except during bankfull flow or larger floods (Montgomery and Buffington, 1997).

# Aquifer function

## Flow and transport in aquifers

Streamlines in a fluid flow as described earlier indicate the main flow direction/path. Mixing of water from different flow paths occurs where those lines converge; for example, at groundwater discharge locations, such as wetlands, springs, and in the vicinity of a well when water is pumped. In addition, temporal variability of mixing processes may occur. For example, seasonality in groundwater recharge causes variability in the mixing of different water from the unsaturated and saturated zone throughout the year (Jasechko et al., 2014). Mixing water from different flow paths can be particularly relevant for geochemical and ecological processes. Water from individual flow paths may have distinct physical, chemical, and biological compositions. When different types of water with different physical and chemical compositions mix, other reactions might be triggered as well. Those mixing processes can create *hot spots* and *hot moments*, where reaction rates are larger compared to surrounding locations or to other time periods (McClain et al., 2003). Mixing is also relevant when pumping a fully screened well, as water from very different flow lines converges at the well (Tonina and Bellin, 2008). Thus, any water sampled at such locations represents the weighted average of fluxes contributing to the sample, which may have a different biogeochemical composition than would otherwise occur at that location in the absence of mixing.

For understanding the physical, chemical, and biological composition of groundwater, general transport processes are important in terms of how they influence the fate of reactive and nonreactive solutes or particles in groundwater. The general transport processes are advection, hydrodynamic dispersion, and diffusion. This applies to both reactive and nonreactive species, with the former also being affected by chemical reactions: dissolution, precipitation, and sorption. Advection is the entrainment of solutes/particles by the groundwater flux along flow paths. In contrast, hydrodynamic dispersion considers the entire distribution of fluid velocities within —microscopically and macroscopically- nonuniform media and the consequent spreading of solutes in both the main flow direction (longitudinal dispersion) and perpendicular to it (transverse dispersion). The ability of a given media to disperse solutes is described in terms of its *dispersivity*, which is an empirical parameter that increases with scale and flow distance (Gelhar et al., 1992).

Whereas hydrodynamic dispersion is a property of the flow field, and thus only relevant in a flowing system, diffusion results in the movement of solutes due to physical and chemical concentration gradients independent of fluid velocity. Because hydrodynamic dispersion and diffusion are often difficult to distinguish, their effects are often combined when modeling solute transport. Spreading due to diffusion becomes more important if fluid velocities are small and/or in the presence of immobile/stagnant water (Knorr and Blodau, 2009). Special behavior is shown by biporous systems, exhibiting large contrasts in permeability. Here, the transport of water is different from solutes, because the solutes diffuse from areas of high permeability into less permeable areas (or *vice versa* as a function of the concentration gradient). This process is of particular importance when considering the transport of pollutants, as these zones can act as long-term stores. Both, hydrodynamic dispersion and diffusion can strongly influence ecosystem services like the degradation of contaminants as they contribute to mass transfer, thus potentially causing mass transfer limitations at various scales (Meckenstock et al., 2015).

The processes described above may be limited by pore size. If particulate matter, either abiotic or biotic (e.g., viruses, bacteria, or groundwater fauna), is larger than a given pore size (Fig. 1.3), the particles simply cannot be transported through the porous media. Therefore, the particulate matter is filtered by the pores and/or is only transported in pores that are larger than the particle size. The former results in the retardation of particulate matter that may clog pores and further reduce transport, while the latter causes accelerated transport through larger pores with faster flow velocity. Thus, the average particle velocity (only transported through the larger pores) may be faster than the average velocity of all water in the system (flow through the entire pore system).

The main retardation process concerning reactive solutes or particles is sorption (i.e., their interaction with the solid phase due to chemical or physical processes). Sorption strongly depends on surface properties and/or properties of the reactant and surrounding water (e.g., ion strength). Sorption can be irreversible (attachment only) or reversible (attachment and detachment). Sorption is an important process for many nutrients and pollutants, but also for viruses and bacteria affecting their fate in groundwater (Tufenkji, 2007). Although pollutants might be immobilized for long periods due to sorption, this is only temporal retardation of transport, not the elimination of the pollutants. Complete elimination of pollutants occurs when they are sufficiently degraded; however, more toxic metabolites can be formed. Pollution can be particularly problematic in karst systems due to the rapid input of pollutants to the groundwater system via fissures, macropores, and tunnels, with relatively less filtration of particulate matter compared to granular porous media.

## Groundwater age

From the flow paths and transport properties along those paths, time scales of flow and transport can be assessed. Those time scales of flow and transport also influence ecosystem processes and water quality because the time since groundwater was recharged is relevant for biogeochemical processes (van der Velde et al., 2010), weathering of minerals (Maher, 2010), degradation of pollutants (Meckenstock et al., 2015), water availability in the critical zone (the area between the vegetation canopy and unweathered subsurface bedrock; Grant and Dietrich, 2017; Sprenger et al., 2019), renewal of groundwater (Jasechko, 2019; Moeck et al., 2020), and estimation of groundwater volume (Gleeson et al., 2011). Thus, those time scales are also used to assess both the intrinsic and specific vulnerability of groundwater bodies (Chatton et al., 2016; Wachniew et al., 2016; Jasechko et al., 2017). A particular concern for human use of aquifers is quantifying renewal times so that groundwater is not over-pumped and depleted relative to the time needed to replenish the aquifer; which is a common problem for aquifers containing very old groundwater, indicative of low renewal rates (le Gal La Salle et al., 2001; Favreau et al., 2002; Gonçalvès et al., 2013; Gardner and Heilweil, 2014).

Groundwater age is generally defined as the time elapsed since groundwater entered the subsurface (Bethke and Johnson, 2008). The mean water transit time, sometimes referred to as the turnover time, is defined as the ratio of the mobile water volume to the volumetric flow rate (Kreft and Zuber, 1978). It is the mean time between when water enters and leaves the system, also referred to as the *residence time*. As the mobile water volume is rarely known, tracers in combination with mathematical models are often used to determine the transit times (Maloszewski and Zuber, 1982). However, certain tracers only cover certain ranges of expected water ages (Newman et al., 2006; Suckow, 2014). Further, in the presence of immobile water or for nonideal conditions, the tracer transit time is different from the water transit time because the tracer diffuses into immobile zones and back, thus increasing its transit time relative to water flow (Maloszewski et al., 2004). This complication can be addressed through multi-tracer studies to better understand mixing processes, aquifer functioning, and groundwater age.

Understanding groundwater age and system behavior may also require knowing the distribution of transit and residence times. For example, very different distributions can have similar mean values (Wachniew et al., 2016), creating problems of equifinality. As such, one may need to know the entire distribution of transit and residence times to determine how potential pollutants are diluted or if preferential flow is of relevance. The estimation of mean transit times and transit-time distributions in groundwater has been a challenge for decades; an issue that has been summarized in several different reviews (Suckow, 2014; Turnadge and Smerdon, 2014; McCallum et al., 2015; Cartwright et al., 2017; Jasechko, 2019; Sprenger et al., 2019).

Groundwater ages range from days to weeks for near-surface flow in the variably saturated zone (Ayraud et al., 2008; Le Gal La Salle et al., 2012; Marçais et al., 2018) or during flood events, particularly in karstic systems (Delbart et al., 2014; Palcsu et al., 2021). Such low residence times may be quantified using a variety of tracers (e.g., anthropogenic gases, organic matter, tritium, stable isotopes, and dyes). Groundwater ages rapidly increase with depth, spanning several decades to hundreds of years for both unconfined and confined aquifers, while ages of millions of years have also been determined in sedimentary basins associated with marine environments (Gleeson et al., 2000). Saline and extremely old fluids were also discovered in deep crystalline formations, representing extremely slow regional flow (Bottomley et al., 1994; Aquilina et al., 1997; Greene et al., 2008; Bucher and Stober, 2010) that is controlled by microporosity (Fig. 1.3) (Waber and Smellie, 2008; Aquilina and Dreuzy, 2011; Aquilina et al., 2015). Radioactive or accumulative tracers ($^{14}$C, $^{4}$He, $^{36}$Cl, $^{39}$Ar, $^{40}$Ar, $^{81}$Kr, $^{129}$I) were used to characterize these old ages. Although relationships to surface and nutrient fluxes are extremely limited, these deep aquifers constitute ecosystems with specific microbial communities (Pedersen, 1997; Hallbeck and Pedersen, 2008).

## Modeling aquifers

Groundwater flow in aquifers is typically more complex than what is shown in Fig. 1.1. Complex aquifers require comprehensive definitions of the geologic features of each rock type within an aquifer to correctly parameterize spatial variation of hydraulic conductivity. However, the degree of heterogeneity and the morphologic distribution of different rock/sediment facies is usually unknown. Nevertheless, establishing a model of the aquifer remains a useful tool for analyzing factors such as groundwater flow patterns, human uses of the resource, and transport properties of nutrients and pollutants. Approaches for modeling groundwater can be broadly divided into spatially-distributed vs. lumped-parameter models. The former typically employs finite analyses, in which the aquifer is divided into small cells with distributed hydrogeological properties (e.g., MODFLOW; Langevin et al., 2017). These properties allow the flow and transport of material, along with their spatial variation, to be computed. Numerous other models exist that allow implementation of various hydrological and chemical processes, as well as coupling between surface and subsurface systems (e.g., PLFOTRAN (Hammond et al., 2012) and HydroGeoSphere (Therrien et al., 2010). Because the spatial distribution of hydraulic properties within the aquifer is typically unknown, they can be characterized as stochastic variables, producing a number of possible realizations. This is especially important for solute transport, where the fate of the solute is studied using a Monte Carlo approach in which many different realizations of hydraulic properties are generated, from which the solute transport and associated transformations are then studied statistically. "Black box" or lumped-parameter models have also been used to determine simple characteristics, such as groundwater flow or mean residence time (Małoszewski and Zuber, 1985). The simplest one is the piston-flow model, which describes the aquifer as a "tube" (Fig. 1.6A). The simple interpretation of groundwater "age" (discussed in the previous section) refers to this type of model. A more representative lumped-parameter model is the exponential approach, which integrates an infinity of flow lines along a vertical axis, with an exponentially decreasing proportion of flow lines with depth and age (Fig. 1.6B). The third type of lumped-parameter model, which characterizes conditions often observed in natural environments, is the binary mixing model (Fig. 1.6C). This model considers mixing between two water bodies, a modern component (which may be young recharge water or relatively recent groundwater) and an older component (which might be defined as containing no chlorofluorocarbons (CFCs, used as a tracer) or simply as older than the modern component).

A variety of models also have been developed for different scales of hyporheic exchange that elucidate different hydraulic environments and their consequences for fauna and water quality (further discussed by Tonina and Buffington, this volume). Although hyporheic and groundwater models employ similar mechanics related to Darcy's law, the coupling of the two is frequently uneven. For example, some studies combine complex hyporheic models with simplified groundwater flow (e.g., Boano et al., 2008; Marzadri et al., 2016), while others use complex groundwater models with simplified hyporheic flow (e.g., Therrien et al., 2010) that do not fully capture the various scales of flow and interaction illustrated in Fig. 1.5 without using high-resolution topography and a fine-scale numerical mesh. Recent modeling approaches use artificial intelligence, such as machine learning or neural networks, which are particularly promising for investigating heterogeneous and complex aquifers.

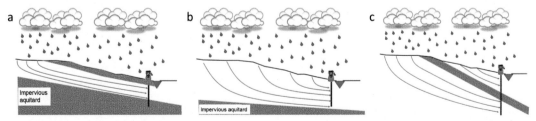

FIGURE 1.6    Lumped parameter models of aquifers: (A) piston flow model, (B) exponential model, and (C) binary mixing model. *Original drawing from Condate-Eau laboratory, used with permission.*

# The chemical composition of groundwater

## Origin of chemical compounds in surface waters

Groundwater derives its chemical composition initially from seawater through the integration of sea spray in vapor fluxes that give birth to clouds. Precipitation thus contains chemical elements that derive mainly from seawater. Seawater influence is maximized in coastal areas, with chloride (Cl) concentration (the major ion in seawater) ranging from 15 to 25 mg/L. A few kilometers inland, this concentration rapidly decreases toward <10 mg/L. Further inland, especially in mountainous areas, Cl concentration in precipitation is only a few mg/L.

For a chemical element $X$, the seawater contribution may be calculated in terms of the chloride concentration:

$$X_{sw} = (Clsw \times Xsp)/Clsp(x) \tag{1.2}$$

where the $sw$ and $sp$ subscripts refer to seawater and the sample, respectively, and $x$ is the proportion of the chemical component $X$ in the sample. A concentration higher than the marine contribution indicates another source, which might be derived either from mineral dust dissolution or from gaseous anthropogenic inputs. The composition of precipitation undergoes further chemical evolution in the soil. First, plant transpiration and, in high-temperature areas, evaporation, linked together as evapotranspiration (ETP), are responsible for the loss of a large amount of water to the atmosphere. In temperate areas, ETP leads to the loss of about half of the initial water content, thus doubling the chemical concentration. When ETP is higher, concentration may be multiplied by a factor of 3 to 4, further leading to evaporates and brines in specific environments. While ETP can concentrate all chemical elements, other processes in the soil can substantially modify the chemical composition. A major driver of the chemical reactivity of water in soils is related to the acidity of water. Precipitation contains dissolved $CO_2$ after equilibration with the atmosphere and contains about 0.5 mg/L bicarbonate ($HCO_3^-$) at 25°C. As $HCO_3^-$ concentration in groundwater is usually several tens of milligrams, the majority of $CO_2$ is related to organic matter degradation within the soil. Water acidity makes groundwater in the unsaturated zone or in the variably saturated zone highly reactive. The most easily soluble mineral that may account for this acidity is calcite ($CaCO_3$). Calcite is present in various proportions of sedimentary rocks, but is also present even in metamorphic or igneous rocks such as granites (White et al., 1999), thus making it ubiquitous at the earth's surface. This results in the dominance of $HCO_3^-$ and calcium ion ($Ca^{2+}$) concentrations in surface water and shallow groundwater.

A second chemical process that also strongly modifies the chemical composition of water that enters the unsaturated zone is exchange with clays. Clay minerals are made of layers with negative electrical charges. These negative charges are balanced by cations located in between the layers. These cations are likely to move out of the minerals if another cation is available to replace them. These cations constitute a pool contained in the soil that is termed the cation exchange capacity (CEC) of the soil. The CEC directly contributes to the cationic composition of water in the soil and strongly increases, with ETP, the overall concentration of cations in water.

## Water—rock interactions within aquifers

Precipitation enters the soil and migrates toward the saturated zone during recharge, increasing the total solute concentration in water, which progressively shifts the original composition from seawater ratios toward mineral-source ratios, especially for cations. Once at the top of the saturated zone, recharging water joins the groundwater flow lines and atmospheric exchanges stop, especially for gas exchange, which controls, for example, oxygen concentrations. While water moves downward, it crosses the weathered zone and reaches the transition to the unweathered zone. Water, which is chemically aggressive is the most reactive when it first encounters fresh minerals, which tend to dissolve. Mineral dissolution decreases the chemical reactivity and weathering appears as a front that slowly moves downward.

In competent bedrock, groundwater flows slowly along mineral facets in pores and microfractures of the aquifer, where water—rock interaction may also modify the groundwater composition. Most minerals are less reactive than calcite; however, the long groundwater residence time, from years to tens or even hundreds of years in some cases, allows for water and minerals to evolve toward equilibria. In hard-rock aquifers, water—rock interaction is dominated by silicate weathering of minerals such as alkali and plagioclase feldspars, which provide potassium, calcium, and sodium ions ($K^+$, $Ca^{2+}$, and $Na^+$) and silicon dioxide ($SiO_2$). Quartz is also a source of silica. In limestones, the dissolution of magnesium carbonates ($MgCO_3$) occurs with increasing residence time and follows calcite dissolution. Magnesium carbonate dissolution, in turn, induces calcite precipitation, providing typically decreasing ratios of Ca/Mg (Back et al., 1983). However, dissolution processes are fairly complex at the various spatial and temporal scales of the critical zone (Andrews et al., 2011; Brooks et al., 2015), even in carbonate aquifers. In addition to groundwater dissolution, carbon evolves during river transport toward the ocean (Hotchkiss et al., 2015; Ward et al., 2017). Weathering is also highly sensitive to human activities (Macpherson, 2009; Aquilina et al., 2012a).

Protons ($H^+$) promote mineral dissolution through acid-base reactions. Oxygen ($O_2$) also enhances mineral dissolution through redox reactions. It reacts with iron-rich minerals such as biotite. The spatial and vertical stratification of redox reactivity is similar to that of acid-base processes described above.

## Major water quality types

Surface water and groundwater are characterized by their chemical composition, with major components commonly used for the classification of different water types, such as calcium bicarbonate types or sodium chloride types. Piper ternary diagrams are used to plot groundwater composition for classifying different chemical types of water. Within Piper diagrams, water samples are represented as a point defined by the weight of chemical components (e.g., $Cl^-$, $SO_4^{2-}$, $HCO_3^-$) expressed as a percentage of the total concentration of anions or cations. The value of this approach is that it allows one to easily compare groundwater having different total concentrations. However, identifying slight seawater contamination (a few percent or less) is difficult in Piper diagrams. It is also difficult to identify in standard bivariate plots showing relative compositions because seawater is several orders of magnitude more concentrated than typical groundwater. Piper diagrams are more amenable to such analyses when seawater or any other water types are used as endmembers for comparison with groundwater samples. Piper diagrams are thus useful to determine water groups and the potential mixing of different water types. Statistical tools such as hierarchical classification more precisely help to define groups within water samples. Furthermore, as Piper diagrams only show element proportions, they are not adequate tools to elucidate or describe biogeochemical processes, which should be more precisely quantified in bivariate plots or other analyses that allow specific quantification of reactions. For example, bivariate plots of Ca, Mg, or Ca/Mg versus $HCO_3$ may quantify carbonate reactions (dissolution and/or precipitation, which might on some occasions be coupled), or Na—Cl, which may distinguish ETP enrichments with seawater Na/Cl molar ratios (0.8) from halite dissolution (1/1). More complex processes for trace elements may be solved using phase diagrams.

## Reduced environments

While acid-base reactions dominate cation processes, redox processes might influence both anion concentrations and trace elements, especially metals. Acid-base reactions represent the exchange of protons, while redox reactions represent an exchange of electrons. Redox reactions in groundwater environments require an abundance of electron acceptors, such as oxygen, nitrate, sulfate, and electron donors, such as natural organic matter. The element cycling of oxygen and nitrogen is specifically presented below. Sulfur represents a specific case as it is a constituent of many minerals in the environment, and depending on redox reactions, can bind or release metals, such as iron. Sulfur is present in both marine and continental environments, especially in the form of sulphate. It is introduced in marine sediments or lake sediments and might precipitate in reducing conditions below the seafloor or lake floor in the form of pyrite. Sulfur is also present in igneous rocks and metallic sulfur, such as pyrite, is ubiquitous in geological formations.

Metallic sulfur that has been formed in deep reducing environments, either sedimentary or igneous, is highly reactive in the surface environment, where it is rapidly oxidized by dissolved oxygen. Pyrite constitutes a source of iron, but several metals (e.g., As, Co, Zn, Au) are also associated with iron and may be released through pyrite dissolution. Furthermore, other metallic sulfur species may also be present locally and represent a major metal source.

Whether aquifers are oxic or anoxic depends on the availability of dissolved oxygen. If the reactivity of the system is low (e.g., as a consequence of lacking electron donors or slow reaction rates) or if enough dissolved oxygen is replenished (e.g., by gaseous exchange with the atmosphere or with new water entering the system), oxic conditions may prevail. In contrast, if dissolved oxygen is absent due to consumption or lack of replenishment, anoxic conditions establish. The common perception is that shallow subsurface environments are oxic, while confined and/or deep aquifers are reducing environments. However, even in unconfined and shallow water-table aquifers, reducing conditions can develop as a function of flow path length and slowing of groundwater flow velocity or due to the abundance of electron donors (i.e., related to the mineralogy of the geological formation of the aquifer). Reducing environments are characterized by low or negative redox potentials, also known as oxidation/reduction potentials that are measured (e.g., with an $E_h$ meter that measures electrical potential in millivolts) by low oxygen concentrations. Indeed, oxygen is the most well-known driver of redox potential although other components also contribute to the redox processes and control the redox potential (Stumm and Morgan, 1996). Groundwater exploitation and distribution result in mixing processes as explained above. Such mixing may induce reactive processes as reduced groundwater with no oxygen and high metal content mixes with surface groundwater with high oxygen or nitrate concentrations. These interactions produce metal precipitation, which has been identified as a major problem in groundwater well production (Williams and Oostrom, 2000; Burté et al., 2019).

## Chemical and nutrient fluxes in aquifers

### Biogeochemical reactions

Microorganisms obtain their energy from chemical reactions. Breaking nuclear bonds provides energy that is used to sustain various metabolic processes and bacterial growth. Several reactions are used depending on the redox environment, which is directly related to the nutrients available. A succession of microbial reactions has been described (Fig. 1.7) from gaining either the most or least energy out of a reaction. Oxygen reduction in combination to the oxidation of an organic or inorganic electron donor produces a high energy level. Nitrate reduction also produces a great deal of energy, although less than oxygen reduction, which partly explains that nitrate reduction appears only in low-oxygen environments. Sulfate

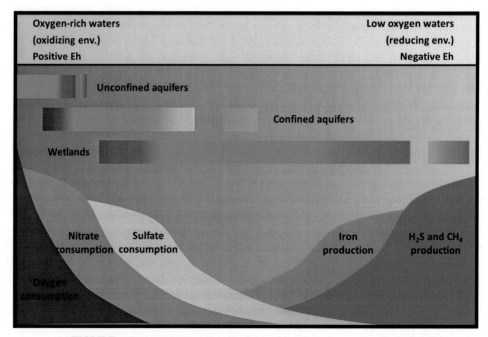

FIGURE 1.7   Succession of biogeochemical reactions in a redox gradient.

reduction to hydrogen sulfide ($H_2S$) is characteristic of highly reducing environments as well as methane production, which appears as the most reducing reaction. This succession may be observed and modeled in a variety of reduced environments, such as peatlands, along groundwater flow paths that can vary in length from a few centimeters to hundreds of meters (Hunter et al., 1998; Rezanezhad et al., 2016).

## Oxygen in groundwater

Oxygen is present in all surface water because of its interaction with atmospheric oxygen. However, oxygen supply may be limited to near-surface layers for stratified water bodies. Equilibria between water and the atmosphere for a given gas are described by Henry's law, which defines gas solubility:

$$P_E = X_E \times K_{HES} \tag{1.3}$$

where $P_E$ is the partial pressure of gas $E$ in the air equilibrated with the solution $S$, $X_E$ the molar fraction of $E$ in solution, and $K_{H\,E,S}$ the Henry's constant of $E$ in solution $S$ at a given temperature and pressure.

Solubility is specific to each gas and varies with both temperature and pressure. However, for a given temperature and pressure, an atmospheric increase leads to a dissolved concentration increase. This is the case for the $CO_2$ increase in the atmosphere during the 20th century, which has led to oceanic acidification. Gas solubility increases with decreasing temperature, and recharge in cold areas (below 10°C) allows oxygen concentrations equal to or slightly above 10 mg/L. This concentration decreases to 8−6 mg/L for recharge temperatures higher than 15−20°C.

Below the groundwater−atmosphere interface, oxygen is consumed by microorganisms through respiration. In unconfined aquifers, oxygen concentration decreases vertically over length scales of several tens of meters. Even though this represents a relatively shallow depth, the residence time may reach 10−20 yrs, which defines an oxygen consumption rate. The lower part of aquifers, with thickness greater than several tens of meters, may thus present hypoxic and reduced environments, with low to almost no oxygen available. While the above is generally true, the oxygen distribution also depends on the reactivity and availability of electron donors, as well as the distribution of flow lines. Recharge zones may be strongly oxic even below the surface, while confined areas with low oxygen concentration can be encountered at shallow depths (Hose et al., 2000; Röling et al., 2001; Maamar et al., 2015).

Beyond such stratification of oxygen concentrations, oxygen mixing zones may constitute hotspots of reactivity. This occurs in the hyporheic zone, where low-oxygen groundwater mixes with oxygenated river water within the streambed sediments. Within heterogeneous aquifers, fractures may transport fresh groundwater with high oxygen concentrations to depths of several tens or hundreds of meters. This leads to important mixing processes that may drive microbial reactions as observed for iron oxidizers (Bochet et al., 2020).

Peatlands and moors also constitute areas where oxygen distribution may be highly heterogeneous, with gas distributed throughout the whole environment (Rosenberry et al.,

2006). This distribution reflects the development of microsites of microbial activity within the peat (Nunes et al., 2015), with specific microbial habitats adapted to reduced conditions. These areas also constitute hot moments with periods of water renewal, oxygen and nutrient fluxes, and periods of reducing conditions. Following such periods, which are characterized by peat desiccation, peat rehumectation leads to the oxidation of organic matter and/or pyrite. Such oxidation induces acid pulses with high sulfate concentrations (Mitchell and Branfireun, 2005; Knorr and Blodau, 2009).

## Nitrogen and emerging contaminants in groundwater

While biological processes are key in nitrogen cycling in the environment, human activities have significantly modified the nitrogen cycle by converting huge amounts of atmospheric nitrogen to mineral fertilizers through the Haber-Bosch process (Erisman et al., 2011). Fertilizers, and with them the so-called green revolution, have led to intensive agriculture development in many parts of the world and an increase in both organic and inorganic nitrogen use (Ågren and Bosatta, 1988; Alvarez-Cobelas et al., 2008). The consequence of this development is massive nitrate pollution of surface waters that started in the 1970s and has increased through the end of the 20th century. Environmental regulations and public pressure have led to modifications of agricultural practices and decreasing nitrate concentrations in rivers during the last few decades, although the problem is still ongoing. According to the definition of planetary boundaries (i.e., environmental boundaries that define a habitable planet), nitrogen is one of the limits that anthropogenic activities have crossed (Steffen et al., 2015).

While agricultural activities have applied nitrogen fertilizers to the soil surface, N-compounds (organically bound N, ammonium, nitrite, and nitrate) have progressively entered recharge fluxes and have contaminated groundwater. Within such aquifers, a nitrogen reservoir exists that is a legacy of past agricultural practices that contributes to nitrate concentrations in rivers several years and decades after the application of surface fertilizers. This situation makes political decisions and environmental constraints challenging because it is difficult to disentangle current versus historic effects, and because time lags are often ignored in groundwater resources management (Aquilina et al., 2012b; Stumpp et al., 2016; Wachniew et al., 2016). Nitrate may accumulate at a slower rate in the lower parts of aquifers because of autotrophic denitrification reactions (Rivett et al., 2008). Once oxygen has been consumed, nitrate becomes the most energetically favorable electron acceptor, and various groups of microorganisms are able to reduce nitrate to dinitrogen or intermediate products, such as nitrite or nitrous oxide.

Beyond nitrogen, an extremely wide range of contaminants is related to human activities. In particular, a large number of inorganic and organic molecules have been introduced to the environment through human use of pesticides and pharmaceuticals. These contaminants are also transferred to groundwater through various soil processes (Jury and Wang, 2000; Arias-Estévez et al., 2008; Teuten et al., 2009). The extremely high number of introduced contaminants represents a major issue, as due to the generally low concentrations, many constituent molecules are not degraded by microbes nor easily detected by the analytical protocols available (Gavrilescu et al., 2015). Although pesticides are usually less concentrated in groundwater than in surface water, their degradation products (metabolites) are more prevalent in groundwater, making aquifers particularly vulnerable to such contaminants (Heberer, 2002; Lapworth et al., 2012). Other human activities have introduced a vast of new contaminants in the last few decades, termed *emerging contaminants*, such as endocrine-disrupting compounds, algal toxins, and nanoplastics. Organic contaminants and pathogens are also introduced to the aquatic environment through wastewater, which can lead to groundwater transfer of such contaminants (Nikolaou et al., 2007; Pandey et al., 2014; Bradford and Harvey, 2017).

# Conclusion

Groundwater represents the invisible part of the water cycle due to its subsurface nature. Aquifers occur in a broad range of geological formations, creating diverse physical properties and aquatic environments. Key physical characteristics of these geological formations are their ability to let groundwater flow (permeability), which in turn depends on the extent and connectivity of voids (porosity) within the geological formation, controlling aquifer function. Although the physical structure of aquifers varies between geological formations, most aquifers present a high degree of physical and chemical heterogeneity. This is particularly the case for fractured bedrock and karstic aquifers, which are characterized by two contrasting degrees of permeability: a highly-permeable system comprised of large fracture zones and tunnels, and a much less permeable system within the surrounding matrix of the geological formation. These hydraulic properties and boundary conditions define how fast water moves through the system and the age of the groundwater, which is an important variable for characterizing groundwater functioning and system dynamics.

Although groundwater aquifers are subsurface systems, they are intimately linked to surface water bodies, further enhancing the diversity of the hydrologic system. Depending on topography and local hydraulic gradients, groundwater may upwell into surface environments (springs, rivers, lakes, wetlands) in some locations, while in other parts of a basin, the aquifer will be recharged by surface water through processes of soil infiltration on hillslopes, hyporheic downwelling in rivers, and direct descent of water through surface fractures and inlets in karst terrain. The linkage between surface and subsurface systems and the hydraulic gradients that move water between the two allows for a broad range of physicochemical reactions and biogeochemical processes, emphasizing the need to better integrate disciplines of hydrogeology and ecology (Hancock et al., 2009).

Characterization of aquifer properties and heterogeneities is further used to parameterize aquifer models. Such models range from simple lumped-parameter approaches that represent the ensemble function of the aquifer to spatially-distributed models that reproduce the physical details of groundwater flow and transport. These models can provide important insight regarding system behavior. Quantifying the physical processes and characteristics of groundwater systems through such models also forms the basis for understanding the distribution and activity of living organisms in groundwater environments. In many aspects, the structural features of aquifers, through their control of groundwater flow, set the scene for the diversity and evolution of life in groundwater systems.

Aquifer dynamics and linkages to surface systems also control groundwater quality, which is influenced by processes occurring over multiple spatial and temporal scales. Close to the surface and over short time scales, groundwater quality is affected by the reactivity of rainwater percolating through various types of soil. Below the soil layer, the consumption of various electron acceptors (e.g., oxygen and nitrate) results in chemical reactions that progressively modify water quality through water—rock interactions. All of these processes depend on physicochemical conditions (acidity and redox conditions) that vary along groundwater flow lines. Biological processes additionally modulate chemical conditions, playing a critical role in the cycling of carbon, oxygen, and nutrients. Groundwater quality thus depends on abiotic processes related to soil/rock characteristics and biogeochemical reactions, as well as their interactions. Furthermore, human activities have strongly altered groundwater quality in many parts of the world, especially with the introduction of high nitrogen inputs to ecosystems, but also through the production and use of chemicals. The linkage between surface and subsurface systems, and the fact that aquifers are long-term reservoirs in which contaminants can concentrate, make groundwater quality a major concern, particularly in regions where communities rely on groundwater resources.

From the processes summarized in this chapter, it is clear that there is a need for better integration of hydrogeology, microbiology and ecology to facilitate "the study of interactions between the hydrology of all groundwater bodies and their ecological components" (e.g., Hancock et al., 2009). In this regard, a particularly important area for future research is the documentation of relationships between physical and biological parameters.

## Acknowledgments

We thank Jason Polk and Florian Malard for constructive comments that improved the chapter.

## References

Abbott, B.W., Bishop, K., Zarnetske, J.P., Minaudo, C., Chapin, F.S., Krause, S., Hannah, D.M., Conner, L., Ellison, D., Godsey, S.E., Plont, S., Marçais, J., Kolbe, T., Huebner, A., Frei, R.J., Hampton, T., Gu, S., Buhman, M., Sara Sayedi, S., Ursache, O., Chapin, M., Henderson, K.D., Pinay, G., 2019. Human domination of the global water cycle absent from depictions and perceptions. Nature Geoscience 12, 533—540. https://doi.org/10.1038/s41561-019-0374-y.

Ågren, G.I., Bosatta, E., 1988. Nitrogen saturation of terrestrial ecosystems. Environmental Pollution 54, 185—197. https://doi.org/10.1016/0269-7491(88)90111-X.

Alexander, R.B., Boyer, E.W., Smith, R.A., Schwarz, G.E., Moore, R.B., 2007. The role of headwater streams in downstream water quality. Journal of the American Water Resources Association 43, 41—59. https://doi.org/10.1111/j.1752-1688.2007.00005.x.

Alley, W.M., Reilly, T.E., Franke, O.L., 1999. Sustainability of ground-water resources. US Geological Survey, Circular 1186, Denver, CO.

Alvarez-Cobelas, M., Angeler, D.G., Sánchez-Carrillo, S., 2008. Export of nitrogen from catchments: a worldwide analysis. Environmental Pollution 156, 261—269. https://doi.org/10.1016/j.envpol.2008.02.016.

Amanambu, A.C., Obarein, O.A., Mossa, J., Li, L., Ayeni, S.S., Balogun, O., Oyebamiji, A., Ochege, F.U., 2020. Groundwater system and climate change: present status and future considerations. Journal of Hydrology 589, 125163. https://doi.org/10.1016/j.jhydrol.2020.125163.

Andrews, D.M., Lin, H., Zhu, Q., Jin, L., Brantley, S.L., 2011. Hot spots and hot moments of dissolved organic carbon export and soil organic carbon storage in the Shale Hills catchment. Vadose Zone Journal 10, 943—954. https://doi.org/10.2136/VZJ2010.0149.

Aquilina, L., Dreuzy, J.-R. De, 2011. Relationship of present saline fluid with paleomigration of basinal brines at the basement/sediment interface (Southeast basin—France). Applied Geochemistry 26, 1933—1945. https://doi.org/10.1016/j.apgeochem.2011.06.022.

Aquilina, L., Ladouche, B., Dörfliger, N., 2006. Water storage and transfer in the epikarst of karstic systems during high flow periods. Journal of Hydrology 327, 472—485. https://doi.org/10.1016/j.jhydrol.2005.11.054.

Aquilina, L., Pauwels, H., Genter, A., Fouillac, C., 1997. Water-rock interaction processes in the Triassic sandstone and the granitic basement of the Rhine Graben: geochemical investigation of a geothermal reservoir. Geochimica et Cosmochimica Acta 61, 4281–4295. https://doi.org/10.1016/S0016-7037(97)00243-3.

Aquilina, L., Poszwa, A., Walter, C., Vergnaud, V., Pierson-Wickmann, A.-C., Ruiz, L., 2012a. Long-term effects of high nitrogen loads on cation and carbon riverine export in agricultural catchments. Environmental Science and Technology 46 (17), 9447–9455. https://doi.org/10.1021/es301715t.

Aquilina, L., Vergnaud-Ayraud, V., Labasque, T., Bour, O., Molénat, J., Ruiz, L., de Montety, V., De Ridder, J., Roques, C., Longuevergne, L., 2012b. Nitrate dynamics in agricultural catchments deduced from groundwater dating and long-term nitrate monitoring in surface- and groundwaters. Science of the Total Environment 435–436, 167–178. https://doi.org/10.1016/j.scitotenv.2012.06.028.

Aquilina, L., Vergnaud-Ayraud, V., Les Landes, A.A., Pauwels, H., Davy, P., Pételet-Giraud, E., Labasque, T., Roques, C., Chatton, E., Bour, O., Ben Maamar, S., Dufresne, A., Khaska, M., La Salle, C.L.G., Barbecot, F., 2015. Impact of climate changes during the last 5 million years on groundwater in basement aquifers. Scientific Reports 5, 14132. https://doi.org/10.1038/srep14132.

Arias-Estévez, M., López-Periago, E., Martínez-Carballo, E., Simal-Gándara, J., Mejuto, J.C., García-Río, L., 2008. The mobility and degradation of pesticides in soils and the pollution of groundwater resources. Agriculture, Ecosystems & Environment 123, 247–260. https://doi.org/10.1016/J.AGEE.2007.07.011.

Ayraud, V., Aquilina, L., Labasque, T., Pauwels, H., Molenat, J., Pierson-Wickmann, A.C., Durand, V., Bour, O., Tarits, C., Le Corre, P., Fourre, E., Merot, P., Davy, P., 2008. Compartmentalization of physical and chemical properties in hard-rock aquifers deduced from chemical and groundwater age analyses. Applied Geochemistry 23, 2686–2707.

Back, W., Hanshaw, B.B., Plummer, L.N., Rahn, P.H., Rightmire, C.T., Rubin, M., 1983. Process and rate of dedolomitization: mass transfer and $^{14}C$ dating in a regional carbonate aquifer. The Geological Society of America Bulletin 94, 1415–1429. https://doi.org/10.1130/0016-7606(1983)94<1415:parodm>2.0.co;2.

Bailly-Comte, V., Martin, J.B., Screaton, E.J., 2011. Time variant cross correlation to assess residence time of water and implication for hydraulics of a sink-rise karst system. Water Resources Research 47, W05547. https://doi.org/10.1029/2010WR009613.

Baxter, C.V., Hauer, F.R., 2000. Geomorphology, hyporheic exchange, and selection of spawning habitat by bull trout (*Salvelinus confluentus*). Canadian Journal of Fisheries and Aquatic Sciences 57, 1470–1481.

Bean, J.R., Wilcox, A.C., Woessner, W.W., Muhlfeld, C.C., 2014. Multiscale hydrogeomorphic influences on bull trout (*Salvelinus confluentus*) spawning habitat. Canadian Journal of Fisheries and Aquatic Sciences 72, 514–526. https://doi.org/10.1139/cjfas-2013-0534.

Benjankar, R., Tonina, D., Marzadri, A., McKean, J., Isaak, D.J., 2016. Effects of habitat quality and ambient hyporheic flows on salmon spawning site selection. Journal of Geophysical Research: Biogeosciences 121, 1222–1235. https://doi.org/10.1002/2015JG003079.

Bethke, C.M., Johnson, T.M., 2008. Groundwater age and groundwater age dating. Annual Review of Earth and Planetary Sciences 36, 121–152. https://doi.org/10.1146/annurev.earth.36.031207.124210.

Blois, G., Best, J.L., Sambrook Smith, G.H., Hardy, R.J., 2014. Effect of bed permeability and hyporheic flow on turbulent flow over bed forms. Geophysical Research Letters 41, 6435–6442.

Boano, F., Camporeale, C., Revelli, R., 2010. A linear model for the coupled surface-subsurface flow in a meandering stream. Water Resources Research 46, W07535.

Boano, F., Harvey, J.W., Marion, A., Packman, A.I., Revelli, R., Ridolfi, L., Wörman, A., 2014. Hyporheic flow and transport processes: mechanisms, models, and biogeochemical implications. Reviews of Geophysics 52, 603–679.

Boano, F., Revelli, R., Ridolfi, L., 2008. Reduction of the hyporheic zone volume due to the stream-aquifer interaction. Geophysical Research Letters 35, L09401. https://doi.org/10.1029/2008GL033554.

Bochet, O., Bethencourt, L., Dufresne, A., Farasin, J., Pédrot, M., Labasque, T., Chatton, E., Lavenant, N., Petton, C., Abbott, B.W., Aquilina, L., Le Borgne, T., 2020. Iron-oxidizer hotspots formed by intermittent oxic–anoxic fluid mixing in fractured rocks. Nature Geoscience 13, 149–155. https://doi.org/10.1038/s41561-019-0509-1.

Bottomley, D.J., Gregoire, D.C., Raven, K.G., 1994. Saline groundwaters and brines in the Canadian shield: geochemical and isotopic evidence for a residual evaporite brine component. Geochimica et Cosmochimica Acta 5, 1483–1498.

Bradford, S.A., Harvey, R.W., 2017. Future research needs involving pathogens in groundwater. Hydrogeology Journal 25, 931–938. https://doi.org/10.1007/S10040-016-1501-0.

Brooks, P.D., Chorover, J., Fan, Y., Godsey, S.E., Maxwell, R.M., McNamara, J.P., Tague, C., 2015. Hydrological partitioning in the critical zone: recent advances and opportunities for developing transferable understanding of water cycle dynamics. Water Resources Research 51, 6973−6987. https://doi.org/10.1002/2015WR017039.

Bucher, K., Stober, I., 2010. Fluids in the upper continental crust. Geofluids 10, 241−253. https://doi.org/10.1111/j.1468-8123.2010.00279.x.

Buffington, J.M., Tonina, D., 2009. Hyporheic exchange in mountain rivers II: effects of channel morphology on mechanics, scales, and rates of exchange. Geography Compass 3 (3), 1038−1062. https://doi.org/10.1111/j.1749-8198.2009.00225.x.

Burté, L., Cravotta, C.A., Bethencourt, L., Farasin, J., Pédrot, M., Dufresne, A., Gérard, M.F., Baranger, C., Le Borgne, T., Aquilina, L., 2019. Kinetic study on clogging of a geothermal pumping well triggered by mixing-induced biogeochemical reactions. Environmental Science and Technology 53, 5848−5857. https://doi.org/10.1021/acs.est.9b00453.

Cardenas, M.B., 2009. A model for lateral hyporheic flow based on valley slope and channel sinuosity. Water Resources Research 45, 1−5. https://doi.org/10.1029/2008WR007442.

Cartwright, I., Cendón, D., Currell, M., Meredith, K., 2017. A review of radioactive isotopes and other residence time tracers in understanding groundwater recharge: possibilities, challenges, and limitations. Journal of Hydrology 555, 797−811. https://doi.org/10.1016/j.jhydrol.2017.10.053.

Chapelle, F.H., 2001. Ground-Water Microbiology and Geochemistry, 2nd. Wiley, New York, NY.

Chatton, E., Aquilina, L., Pételet-Giraud, E., Cary, L., Bertrand, G., Labasque, T., Hirata, R., Martins, V., Montenegro, S., Vergnaud, V., Aurouet, A., Kloppmann, W., Pauwels, 2016. Glacial recharge, salinisation and anthropogenic contamination in the coastal aquifers of Recife (Brazil). Science of the Total Environment 569-570, 1114−1125. https://doi.org/10.1016/j.scitotenv.2016.06.180.

Crutzen, P.J., 2002. Geology of mankind. Nature 415, 23. https://doi.org/10.1038/415023A.

Darcy, H., 1856. Les fontaines publiques de la ville de Dijon. Libraire des Corps Imperiaux des Ponts et Chaussées et des Mines, Paris.

Delbart, C., Barbecot, F., Valdes, D., Tognelli, A., Fourre, E., Purtschert, R., Couchoux, L., Jean-Baptiste, P., 2014. Investigation of young water inflow in karst aquifers using SF6-CFC-3H/He-85Kr-39Ar and stable isotope components. Applied Geochemistry 50, 164−176. https://doi.org/10.1016/J.APGEOCHEM.2014.01.011.

Dent, C.L., Grimm, N.B., Fisher, S.G., 2001. Multiscale effects of surface-subsurface exchange on stream water nutrient concentrations. Journal of the North American Benthological Society 20 (2), 162−181.

Detert, M., Klar, M., Wenka, T., Jirka, G.H., 2007. Pressure- and velocity-measurements above and within a porous gravel bed at the threshold of stability. In: Habersack, H., Piégay, H., Rinaldi, M. (Eds.), Gravel-Bed Rivers VI: From Process Understanding to River Restoration. Elsevier, Amsterdam, The Netherlands, pp. 85−105.

Dewandel, B., Lachassagne, P., Wyns, R., Maréchal, J.C., Krishnamurthy, N.S., 2006. A generalized 3-D geological and hydrogeological conceptual model of granite aquifers controlled by single or multiphase weathering. Journal of Hydrology 330, 260−284. https://doi.org/10.1016/j.jhydrol.2006.03.026.

Edwards, R.T., 1998. The hyporheic zone. In: Naiman, R.J., Bilby, R.E. (Eds.), River Ecology and Management: Lessons From the Pacific Coastal Ecoregion. Springer-Verlag, New York, NY, pp. 399−429.

Elliott, A., Brooks, N.H., 1997a. Transfer of nonsorbing solutes to a streambed with bed forms: laboratory experiments. Water Resources Research 33, 137−151.

Elliott, A., Brooks, N.H., 1997b. Transfer of nonsorbing solutes to a streambed with bed forms: theory. Water Resources Research 33, 123−136.

Erisman, J.W., van Grinsven, H., Grizzetti, B., Bouraoui, F., Powlson, D., Sutton, M.A., Bleeker, A., Reis, S., 2011. The European nitrogen problem in a global perspective. In: Sutton, M.A., Howard, C.M., Erisman, J.W., Billen, G., Bleeker, A., Grennfelt, P., van Grinsven, H., Grizzetti, B. (Eds.), The European Nitrogen Assessment: Sources, Effects, and Policy Perspectives. Cambridge University Press, https://doi.org/10.1017/cbo9780511976988.005.

Favreau, G., Leduc, C., Marlin, C., Dray, M., Taupin, J.D., Massault, M., le Gal La Salle, C., Babic, M., 2002. Estimate of recharge of a rising water table in semiarid Niger from 3H and 14C modeling. Ground Water 40, 144−151. https://doi.org/10.1111/J.1745-6584.2002.TB02499.X.

Fetter, C.W., 2001. Applied Hydrogeology, 4th. Prentice-Hall, Upper Saddle River, NJ.

Fleckenstein, J.H., Niswonger, R.G., Fogg, G.E., 2006. River-aquifer interactions, geologic heterogeneity, and low-flow management. Ground Water 44, 837−852.

Flint, L.E., Flint, A.L., 2018. Porosity. In: Dane, J.H., Topp, C.G. (Eds.), Methods of Soil Analysis. Soil Science Society of America, Inc., Madison, WI, pp. 241–254.

Freeze, R.A., Cherry, J.A., 1979. Groundwater. Prentice Hall, Englewood Cliffs, NJ.

Gallo, F., Fossi, C., Weber, R., Santillo, D., Sousa, J., Ingram, I., Nadal, A., Romano, D., 2018. Marine litter plastics and microplastics and their toxic chemicals components: the need for urgent preventive measures. Environmental Sciences Europe 30, 13. https://doi.org/10.1186/s12302-018-0139-z.

Gardner, P.M., Heilweil, V.M., 2014. A multiple-tracer approach to understanding regional groundwater flow in the Snake Valley area of the eastern Great Basin, USA. Applied Geochemistry 45, 33–49. https://doi.org/10.1016/J.APGEOCHEM.2014.02.010.

Gavrilescu, M., Demnerová, K., Aamand, J., Agathos, S., Fava, F., 2015. Emerging pollutants in the environment: present and future challenges in biomonitoring, ecological risks and bioremediation. New Biotech 32, 147–156. https://doi.org/10.1016/J.NBT.2014.01.001.

Gelhar, L.W., Welty, C., Rehfeldt, K.R., 1992. A critical review of data on field-scale dispersion in aquifers. Water Resources Research 28, 1955–1974. https://doi.org/10.1029/92WR00607.

Gleeson, S.A., Yardley, B.W.D., Boyce, A.J., Fallick, A.E., Munz, I.A., 2000. From basin to basement: the movement of surface fluids into the crust. Journal of Geochemical Exploration 69 (70), 527–531.

Gleeson, T., Smith, L., Moosdorf, N., Hartmann, J., Dürr, H.H., Manning, A.H., van Beek, L.P.H., Jellinek, A.M., 2011. Mapping permeability over the surface of the Earth. Geophysical Research Letters 38 (2), L02401. https://doi.org/10.1029/2010gl045565.

Gonçalvès, J., Petersen, J., Deschamps, P., Hamelin, B., Baba-Sy, O., 2013. Quantifying the modern recharge of the "fossil" Sahara aquifers. Geophysical Research Letters 40, 2673–2678. https://doi.org/10.1002/GRL.50478.

Gooseff, M.N., Anderson, J.K., Wondzell, S.M., LaNier, J., Haggerty, R., 2006. A modelling study of hyporheic exchange pattern and the sequence, size, and spacing of stream bedforms in mountain stream networks, Oregon, USA. Hydrological Processes 20, 2443–2457.

Grant, G.E., Dietrich, W.E., 2017. The frontier beneath our feet. Water Resources Research 53, 2605–2609. https://doi.org/10.1002/2017WR020835.

Greene, S., Battye, N., Clark, I., Kotzer, T., Bottomley, D., 2008. Canadian Shield brine from the Con Mine, Yellowknife, NT, Canada: noble gas evidence for an evaporated Palaeozoic seawater origin mixed with glacial meltwater and Holocene recharge. Geochimica et Cosmochimica Acta 72, 4008–4019. https://doi.org/10.1016/j.gca.2008.05.058.

Guevara Ochoa, C., Medina Sierra, A., Vives, L., Zimmermann, E., Bailey, R., 2020. Spatio-temporal patterns of the interaction between groundwater and surface water in plains. Hydrological Processes 34, 1371–1392. https://doi.org/10.1002/hyp.13615.

Hallbeck, L., Pedersen, K., 2008. Characterization of microbial processes in deep aquifers of the Fennoscandian Shield. Applied Geochemistry 23, 1796–1819. https://doi.org/10.1016/j.apgeochem.2008.02.012.

Hammond, G.E., Lichtner, P.C., Lu, C., Mills, R.T., 2012. PFLOTRAN: reactive flow and transport code for use on laptops to leadership-class supercomputers. In: Zhang, F., Yeh, G.T., Parker, J.C., Shi, X. (Eds.), Ground Water Reactive Transport Models. Bentham Science Publishers, Sharjah, United Arab Emirates, pp. 141–159.

Han, D., Cao, G., McCallum, J., Song, X., 2015. Residence times of groundwater and nitrate transport in coastal aquifer systems: Daweijia area, northeastern China. Science of the Total Environment 538, 539–554. https://doi.org/10.1016/J.SCITOTENV.2015.08.036.

Hancock, P.J., Hunt, R.J., Boulton, A.J., Hancock, P.J., Boulton, A.J., Hunt, R.J., 2009. Preface: hydrogeoecology, the interdisciplinary study of groundwater dependent ecosystems. Hydrogeology Journal 17, 1–3. https://doi.org/10.1007/s10040-008-0409-8.

Hancock, P.J., Boulton, A.J., Humphreys, W.F., 2005. Aquifers and hyporheic zones: towards an ecological understanding of groundwater. Hydrogeology Journal 13, 98–111. https://doi.org/10.1007/s10040-004-0421-6.

Hao, X., Ball, B.C., Culley, J.L.B., Carter, M.R., Parkin, G.W., 2008. Soil density and porosity. In: Carter, M.R., Gregorich, E.G. (Eds.), Soil Sampling and Methods of Analysis, second ed. Taylor & Francis Group, Boca Raton, FL, pp. 743–759.

Heberer, T., 2002. Occurrence, fate, and removal of pharmaceutical residues in the aquatic environment: a review of recent research data. Toxicology Letters 131, 5–17. https://doi.org/10.1016/S0378-4274(02)00041-3.

Hose, G.C., Stumpp, C., 2019. Architects of the underworld: bioturbation by groundwater invertebrates influences aquifer hydraulic properties. Aquatic Sciences 81, 20. https://doi.org/10.1007/s00027-018-0613-0.

Hose, L.D., Palmer, A.N., Palmer, M.V., Northup, D.E., Boston, P.J., DuChene, H.R., 2000. Microbiology and geochemistry in a hydrogen-sulphide-rich karst environment. Chemical Geology 169, 399–423. https://doi.org/10.1016/S0009-2541(00)00217-5.

Hotchkiss, E.R., Hall, R.O., Sponseller, R.A., Butman, D., Klaminder, J., Laudon, H., Rosvall, M., Karlsson, J., 2015. Sources of and processes controlling $CO_2$ emissions change with the size of streams and rivers. Nature Geoscience 8, 696–699. https://doi.org/10.1038/NGEO2507.

Humphreys, W.F., 2009. Hydrogeology and groundwater ecology: does each inform the other? Hydrogeology Journal 17, 5–21.

Hunter, K.S., Wang, Y., van Cappellen, P., 1998. Kinetic modeling of microbially-driven redox chemistry of subsurface environments: coupling transport, microbial metabolism and geochemistry. Journal of Hydrology 209, 53–80. https://doi.org/10.1016/S0022-1694(98)00157-7.

Jasechko, S., 2019. Global isotope hydrogeology–review. Reviews of Geophysics 57, 835–965. https://doi.org/10.1029/2018rg000627.

Jasechko, S., Birks, S.J., Gleeson, T., Wada, Y., Fawcett, P.J., Sharp, Z.D., McDonnell, J.J., Welker, J.M., 2014. The pronounced seasonality of global groundwater recharge. Water Resources Research 50, 8845–8867. https://doi.org/10.1002/2014wr015809.

Jasechko, S., Perrone, D., Befus, K.M., Bayani Cardenas, M., Ferguson, G., Gleeson, T., Luijendijk, E., McDonnell, J.J., Taylor, R.G., Wada, Y., Kirchner, J.W., 2017. Global aquifers dominated by fossil groundwaters but wells vulnerable to modern contamination. Nature Geoscience 10, 425–429. https://doi.org/10.1038/ngeo2943.

Jury, W.A., Wang, Z., 2000. Unresolved problems in vadose zone hydrology and contaminant transport. In: Faybishenko, B., Witherspoon, P.A., Benson, S.M. (Eds.), Dynamics of Fluids in Fractured Rock, Geophysical Monograph 122. American Geophysical Union, Washington, DC, pp. 67–72.

Kasahara, T., Wondzell, S.M., 2003. Geomorphic controls on hyporheic exchange flow in mountain streams. Water Resources Research 39, 1005.

Kattan, Z., 1997. Environmental isotope study of the major karst springs in Damascus limestone aquifer systems: case of the Figeh and Barada springs. Journal of Hydrology 193, 161–182. https://doi.org/10.1016/S0022-1694(96)03137-X.

Katz, B.G., Böhlke, J.K., Hornsby, H.D., 2001. Timescales for nitrate contamination of spring waters, northern Florida, USA. Chemical Geology 179, 167–186. https://doi.org/10.1016/S0009-2541(01)00321-7.

Knorr, K.H., Blodau, C., 2009. Impact of experimental drought and rewetting on redox transformations and methanogenesis in mesocosms of a northern fen soil. Soil Biology and Biochemistry 41, 1187–1198. https://doi.org/10.1016/j.soilbio.2009.02.030.

Korbel, K.L., Stephenson, S., Hose, G.C., 2019. Sediment size influences habitat selection and use by groundwater macrofauna and meiofauna. Aquatic Sciences 81, 39. https://doi.org/10.1007/s00027-019-0636-1.

Krause, S., Hannah, D.M., Fleckenstein, J.H., Heppell, C.M., Kaeser, D., Pickup, R., Pinay, G., Robertson, A.L., Wood, P.J., 2011. Inter-disciplinary perspectives on processes in the hyporheic zone. Ecohydrology 4 (4), 481–499. https://doi.org/10.1002/eco.176.

Krauss, S., Griebler, C., 2011. Pathogenic microorganisms and viruses in groundwater. Acatech Materialien Nr. 6. München, Germany.

Kreft, A., Zuber, A., 1978. On the physical meaning of the dispersion equation and its solution for different initial and boundary conditions. Chemical Engineering Science 33, 1471–1480.

Kremen, C., Williams, N.M., Thorp, R.W., 2002. Crop pollination from native bees at risk from agricultural intensification. Proceedings of the National Academy of Sciences of the United States of America 99 (26), 16812–16816. https://doi.org/10.1073pnas.262413599.

Langevin, C.D., Hughes, J.D., Banta, E.R., Niswonger, R.G., Panday, S., Provost, A.M., 2017. Documentation for the MODFLOW 6 groundwater flow model. US Geological Survey, Technical Methods, book 6, chap. A55. https://doi.org/10.3133/tm6A55.

Lapworth, D.J., Baran, N., Stuart, M.E., Ward, R.S., 2012. Emerging organic contaminants in groundwater: a review of sources, fate and occurrence. Environmental Pollution 163, 287–303. https://doi.org/10.1016/J.ENVPOL.2011.12.034.

Lautz, L., Siegel, D.I., 2006. Modeling surface and ground water mixing in the hyporheic zone using MODFLOW and MT3D. Advances in Water Resources 29, 1618–1633.

Le Gal La Salle, C., Aquilina, L., Fourre, E., Jean-Baptiste, P., Michelot, J.L., Roux, C., Bugai, D., Labasque, T., Simonucci, C., Van Meir, N., Noret, A., Bassot, S., Dapoigny, A., Baumier, D., Verdoux, P., Stammose, D., Lancelot, J., 2012. Groundwater residence time downgradient of Trench No. 22 at the Chernobyl Pilot Site: constraints on hydrogeological aquifer functioning. Applied Geochemistry 27, 1304–1319. https://doi.org/10.1016/j.apgeochem.2011.12.006.

Le Gal La Salle, C., Marlin, C., Leduc, C., Taupin, J.D., Massault, M., Favreau, G., 2001. Renewal rate estimation of groundwater based on radioactive tracers ($^3$H, $^{14}$C) in an unconfined aquifer in a semi-arid area, Iullemeden Basin, Niger. Journal of Hydrology 254, 145–156. https://doi.org/10.1016/S0022-1694(01)00491-7.

Long, A.J., Putnam, L.D., 2004. Linear model describing three components of flow in karst aquifers using $^{18}$O data. Journal of Hydrology 296, 254–270. https://doi.org/10.1016/J.JHYDROL.2004.03.023.

Maamar, S.B., Aquilina, L., Quaiser, A., Pauwels, H., Michon-Coudouel, S., Vergnaud-Ayraud, V., Labasque, T., Roques, C., Abbott, B.W., Dufresne, A., 2015. Groundwater isolation governs chemistry and microbial community structure along hydrologic flowpaths. Frontiers in Microbiology 6, 1457. https://doi.org/10.3389/fmicb.2015.01457.

Macpherson, G.L., 2009. $CO_2$ distribution in groundwater and the impact of groundwater extraction on the global C cycle. Chemical Geology 264, 328–336. https://doi.org/10.1016/J.CHEMGEO.2009.03.018.

Magliozzi, C., Grabowski, R.C., Packman, A.I., Krause, S., 2018. Toward a conceptual framework of hyporheic exchange across spatial scales. Hydrology and Earth System Sciences 22, 6163–6185. https://doi.org/10.5194/hess-22-6163-2018.

Maher, K., 2010. The dependence of chemical weathering rates on fluid residence time. Earth and Planetary Science Letters 294, 101–110. https://doi.org/10.1016/j.epsl.2010.03.010.

Malard, F., Tockner, K., Dole-Olivier, M.-J., Ward, J.V., 2002. A landscape perspective of surface–subsurface hydrological exchanges in river corridors. Freshwater Biology 47 (4), 621–640. https://doi.org/10.1046/j.1365-2427.2002.00906.x.

Maloszewski, P., Stichler, W., Zuber, A., 2004. Interpretation of environmental tracers in groundwater systems with stagnant water zones. Isotopes in Environmental and Health Studies 40, 21–33.

Małoszewski, P., Zuber, A., 1985. On the theory of tracer experiments in fissured rocks with a porous matrix. Journal of Hydrology 79, 333–358. https://doi.org/10.1016/0022-1694(85)90064-2.

Maloszewski, P., Zuber, A., 1982. Determining the turnover time of groundwater systems with the aid of environmental tracers : 1. Models and their applicability. Journal of Hydrology 57, 207–231.

Marçais, J., Gauvain, A., Labasque, T., Abbott, B.W., Pinay, G., Aquilina, L., Chabaux, F., Viville, D., de Dreuzy, J.R., 2018. Dating groundwater with dissolved silica and CFC concentrations in crystalline aquifers. Science of the Total Environment 636, 260–272. https://doi.org/10.1016/j.scitotenv.2018.04.196.

Marzadri, A., Tonina, D., Bellin, A., Valli, A., 2016. Mixing interfaces, fluxes, residence times and redox conditions of the hyporheic zones induced by dune-like bedforms and ambient groundwater flow. Advances in Water Resources 88, 139–151. https://doi.org/10.1016/j.advwatres.2015.12.014.

Matthess, G., Pekdeger, A., 1981. Concepts of a survival and transport model of pathogenic bacteria and viruses in groundwater. The Science of the Total Environment 21, 149–159. https://doi.org/10.1016/0048-9697(81)90148-0.

McCallum, J.L., Cook, P.G., Simmons, C.T., 2015. Limitations of the use of environmental tracers to infer groundwater age. Groundwater 53, 56–70. https://doi.org/10.1111/gwat.12237.

McClain, M.E., Boyer, E.W., Dent, C.L., Gergel, S.E., Grimm, N.B., Groffman, P.M., Hart, S.C., Harvey, J.W., Johnston, C.A., Mayorga, E., McDowell, W.H., Pinay, G., 2003. Biogeochemical hot spots and hot moments at the interface of terrestrial and aquatic ecosystems. Ecosystems 6, 301–312. https://doi.org/10.1007/s10021-003-0161-9.

Meckenstock, R.U., Elsner, M., Griebler, C., Lueders, T., Stumpp, C., Aamand, J., Agathos, S.N., Albrechtsen, H.J., Bastiaens, L., Bjerg, P.L., Boon, N., Dejonghe, W., Huang, W.E., Schmidt, S.I., Smolders, E., Sørensen, S.R., Springael, D., Van Breukelen, B.M., 2015. Biodegradation: updating the concepts of control for microbial cleanup in contaminated aquifers. Environmental Science and Technology 49, 7073–7081. https://doi.org/10.1021/acs.est.5b00715.

Meyer, J.L., Strayer, D.L., Wallace, J.B., Eggert, S.L., Helfman, G.S., Leonard, N.E., 2007. The contribution of headwater streams to biodiversity in river networks. Journal of the American Water Resources Association 43 (1), 86–103. https://doi.org/10.1111/j.1752-1688.2007.00008.x.

Mitchell, C.P.J., Branfireun, B.A., 2005. Hydrogeomorphic controls on reduction-oxidation conditions across boreal upland-peatland interfaces. Ecosystems 8, 731–747. https://doi.org/10.1007/s10021-005-1792-9.

Moeck, C., Grech-Cumbo, N., Podgorski, J., Bretzler, A., Gurdak, J.J., Berg, M., Schirmer, M., 2020. A global-scale dataset of direct natural groundwater recharge rates: a review of variables, processes and relationships. Science of the Total Environment 717, 137042. https://doi.org/10.1016/j.scitotenv.2020.137042.

Mohrig, D., Smith, J.D., 1996. Predicting the migration rates of subaqueous dunes. Water Resources Research 32, 3207–3217. https://doi.org/10.1029/96WR01129.

Monroe, W.H., 1970. A glossary of karst terminology. US Geological Survey, Water-Supply Paper 1899-K, Washington, DC.

Montgomery, D.R., Buffington, J.M., 1997. Channel-reach morphology in mountain drainage basins. Geological Society of America Bulletin 109 (5), 596–611.

Neuman, S.P., 2005. Trends, prospects and challenges in quantifying flow and transport through fractured rocks. Hydrogeology Journal 13, 124–147. https://doi.org/10.1007/s10040-004-0397-2.

Newman, B.D., Wilcox, B.P., Archer, S.R., Breshears, D.D., Dahm, C.N., Duffy, C.D., McDowell, N.G., Phillips, F.M., Scanlon, B.R., Vivoni, E.R., 2006. Ecohydrology of water-limited environments: a scientific vision. Water Resources Research 42, W06302.

Nikolaou, A., Meric, S., Fatta, D., 2007. Occurrence patterns of pharmaceuticals in water and wastewater environments. Analytical and Bioanalytical Chemistry 387, 1225–1234. https://doi.org/10.1007/S00216-006-1035-8.

Nunes, F.L.D., Aquilina, L., De Ridder, J., Francez, A.J., Quaiser, A., Caudal, J.P., Vandenkoornhuyse, P., Dufresne, A., 2015. Time-scales of hydrological forcing on the geochemistry and bacterial community structure of temperate peat soils. Scientific Reports 5 (14612). https://doi.org/10.1038/srep14612.

Otto, I.M., Donges, J.F., Cremades, R., Bhowmik, A., Hewitt, R.J., Lucht, W., Rockström, J., Allerberger, F., McCaffrey, M., Doe, S.S.P., Lenferna, A., Morán, N., van Vuuren, D.P., Schellnhuber, H.J., 2020. Social tipping dynamics for stabilizing Earth's climate by 2050. Proceedings of the National Academy of Sciences of the United States of America 117, 2354–2365. https://doi.org/10.1073/pnas.1900577117.

Packman, A.I., Salehin, M., Zaramella, M., 2004. Hyporheic exchange with gravel beds: basic hydrodynamic interactions and bedform-induced advective flows. Journal of Hydraulic Engineering 130 (7), 647–656.

Palcsu, L., Gessert, A., Túri, M., Kovács, A., Futó, I., Orsovszki, J., Puskás-Preszner, A., Temovski, M., Koltai, G., 2021. Long-term time series of environmental tracers reveal recharge and discharge conditions in shallow karst aquifers in Hungary and Slovakia. Journal of Hydrology: Regional Studies 36, 100858. https://doi.org/10.1016/J.EJRH.2021.100858.

Pandey, P.K., Kass, P.H., Soupir, M.L., Biswas, S., Singh, V.P., 2014. Contamination of water resources by pathogenic bacteria. AMB Express 4, 1–16. https://doi.org/10.1186/S13568-014-0051-X.

Pedersen, K., 1997. Microbial life in deep granitic rock. FEMS Microbiology Reviews 20, 399–414. https://doi.org/10.1016/S0168-6445(97)00022-3.

Peterson, B.J., Wollheim, W.M., Mulholland, P.J., Webster, J.R., Meyer, J.L., Tank, J.L., Marti, E., Bowden, W.B., Valett, H.M., Hershey, A.E., McDowell, W.H., Dodds, W.K., Hamilton, S.K., Gregory, S., Morrall, D.D., 2001. Control of nitrogen export from watersheds by headwater streams. Science 292, 86–90. https://doi.org/10.1126/science.1056874.

Pledger, A.G., Rice, S.P., Millett, J., 2017. Foraging fish as zoogeomorphic agents: an assessment of fish impacts at patch, barform, and reach scales. Journal of Geophysical Research: Earth Surface 122, 2105–2123. https://doi.org/10.1002/2017JF004362.

Racchetti, E., Bartoli, M., Soana, E., Longhi, D., Christian, R.R., Pinardi, M., Viaroli, P., 2011. Influence of hydrological connectivity of riverine wetlands on nitrogen removal via denitrification. Biogeochemistry 103, 335–354. https://doi.org/10.1007/s10533-010-9477-7.

Renard, P., Allard, D., 2013. Connectivity metrics for subsurface flow and transport. Advances in Water Resources 51, 168–196. https://doi.org/10.1016/j.advwatres.2011.12.001.

Rezanezhad, F., Price, J.S., Quinton, W.L., Lennartz, B., Milojevic, T., van Cappellen, P., 2016. Structure of peat soils and implications for water storage, flow and solute transport: a review update for geochemists. Chemical Geology 429, 75–84. https://doi.org/10.1016/J.CHEMGEO.2016.03.010.

Rhoades, C.C., Fegel, T.S., Covino, T.P., Dwire, K.A., Elder, K., 2021. Sources of variability in springwater chemistry in Fool Creek, a high-elevation catchment of the Rocky Mountains, Colorado, USA. Hydrological Processes 35, e14089. https://doi.org/10.1002/hyp.14089.

Rivett, M.O., Buss, S.R., Morgan, P., Smith, J.W.N., Bemment, C.D., 2008. Nitrate attenuation in groundwater: a review of biogeochemical controlling processes. Water Research 42, 4215–4232. https://doi.org/10.1016/J.WATRES.2008.07.020.

Roche, K.R., Li, A., Bolster, D., Wagner, G.J., Packman, A.I., 2019. Effects of turbulent hyporheic mixing on reach-scale transport. Water Resources Research 55, 3780–3795. https://doi.org/10.1029/2018WR023421.

Röling, W.F.M., Van Breukelen, B.M., Braster, M., Lin, B., Van Verseveld, H.W., 2001. Relationships between microbial community structure and hydrochemistry in a landfill leachate-polluted aquifer. Applied and Environmental Microbiology 67, 4619–4629. https://doi.org/10.1128/AEM.67.10.4619-4629.2001.

Roques, C., 2013. Hydrogéologie des zones de faille du socle cristallin: implications en terme de ressources en eau pour le Massif Armoricain. PhD thesis. Université Rennes 1. URL. http://www.theses.fr/2013REN1S138 (accessed 10.06.2022).

Rosenberry, D.O., Glaser, P.H., Siegel, D.I., 2006. The hydrology of northern peatlands as affected by biogenic gas: current developments and research needs. Hydrological Processes 20, 3601–3610. https://doi.org/10.1002/hyp.6377.

Sansom, B.J., Atkinson, J.F., Bennett, S.J., 2018. Modulation of near-bed hydrodynamics by freshwater mussels in an experimental channel. Hydrobiologia 810, 449–463. https://doi.org/10.1007/s10750-017-3172-9.

Screaton, E., Martin, J.B., Ginn, B., Smith, L., 2004. Conduit properties and karstification in the unconfined Floridan Aquifer. Ground Water 42, 338–346. https://doi.org/10.1111/J.1745-6584.2004.TB02682.X.

Soulsby, C., Tetzlaff, D., van den Bedem, N., Malcolm, I.A., Bacon, P.J., Youngson, A.F., 2007. Inferring groundwater influences on surface water in montane catchments from hydrochemical surveys of springs and streamwaters. Journal of Hydrology 333, 199–213. https://doi.org/10.1016/j.jhydrol.2006.08.016.

Sprenger, M., Stumpp, C., Weiler, M., Aeschbach, W., Allen, S.T., Benettin, P., Dubbert, M., Hartmann, A., Hrachowitz, M., Kirchner, J.W., McDonnell, J.J., Orlowski, N., Penna, D., Pfahl, S., Rinderer, M., Rodriguez, N., Schmidt, M., Werner, C., 2019. The demographics of water: a review of water ages in the critical zone. Reviews of Geophysics 57, 800–834. https://doi.org/10.1029/2018rg000633.

Stanford, J.A., Ward, J.V., 1993. An ecosystem perspective of alluvial rivers: connectivity and the hyporheic corridor. Journal of the North American Benthological Society 12, 48–60.

Statzner, B., Arens, M.-F., Champagne, J.-Y., Morel, R., Herouin, R., 1999. Silk-producing steam insects and gravel erosion: significant biological effects on critical shear stress. Water Resources Research 35 (11), 3495–3506.

Steffen, W., Richardson, K., Rockström, J., Cornell, S.E., Fetzer, I., Bennett, E.M., Biggs, R., Carpenter, S.R., De Vries, W., De Wit, C.A., Folke, C., Gerten, D., Heinke, J., Mace, G.M., Persson, L.M., Ramanathan, V., Reyers, B., Sörlin, S., 2015. Planetary boundaries: guiding human development on a changing planet. Science 347 (6223). https://doi.org/10.1126/science.1259855.

Stein, H., Griebler, C., Berkhoff, S., Matzke, D., Fuchs, A., Hahn, H.J., 2012. Stygoregions - a promising approach to a bioregional classification of groundwater systems. Scientific Reports 2, 673. https://doi.org/10.1038/srep00673.

Storey, R.G., Howard, K.W.F., Williams, D.D., 2003. Factors controlling riffle-scale hyporheic exchange flows and their seasonal changes in gaining stream: a three-dimensional groundwater model. Water Resources Research 39, 17. https://doi.org/10.1029/2002WR001367.

Stuart, M.E., Maurice, L., Heaton, T.H.E., Sapiano, M., Micallef Sultana, M., Gooddy, D.C., Chilton, P.J., 2010. Groundwater residence time and movement in the Maltese islands — a geochemical approach. Applied Geochemistry 25, 609–620. https://doi.org/10.1016/j.apgeochem.2009.12.010.

Stumm, W., Morgan, J.J., 1996. Aquatic Chemistry: Chemical Equilibria and Rates in Natural Waters, 3rd. Wiley, New York, NY.

Stumpp, C., Hose, G.C., 2017. Groundwater amphipods alter aquifer sediment structure. Hydrological Processes 31, 3452–3454. https://doi.org/10.1002/hyp.11252.

Stumpp, C., Żurek, A.J., Wachniew, P., Gargini, A., Gemitzi, A., Filippini, M., Witczak, S., 2016. A decision tree tool supporting the assessment of groundwater vulnerability. Environmental Earth Sciences 75, 1057. https://doi.org/10.1007/s12665-016-5859-z.

Suckow, A., 2014. The age of groundwater — definitions, models and why we do not need this term. Applied Geochemistry 50, 222–230. https://doi.org/10.1016/j.apgeochem.2014.04.016.

Taylor, C.J., Greene, E.A., 2008. Hydrogeologic characterization and methods used in the investigation of karst hydrology. In: Rosenberry, D.O., LaBaugh, J.W. (Eds.), Field Techniques for Estimating Water Fluxes Between Surface Water and Ground Water. US Geological Survey, Techniques and Methods 4-D2, pp. 75–114.

Teuten, E.L., Saquing, J.M., Knappe, D.R.U., Barlaz, M.A., Jonsson, S., Björn, A., Rowland, S.J., Thompson, R.C., Galloway, T.S., Yamashita, R., Ochi, D., Watanuki, Y., Moore, C., Viet, P.H., Tana, T.S., Prudente, M., Boonyatumanond, R., Zakaria, M.P., Akkhavong, K., Ogata, Y., Hirai, H., Iwasa, S., Mizukawa, K., Hagino, Y., Imamura, A., Saha, M., Takada, H., 2009. Transport and release of chemicals from plastics to the environment and to wildlife. Philosophical Transactions of the Royal Society B: Biological Sciences 364, 2027–2045. https://doi.org/10.1098/RSTB.2008.0284.

Therrien, R., McLaren, R.G., Sudicky, E.A., Panday, S.M., 2010. HydroGeoSphere: a three-dimensional numerical model describing fully-integrated subsurface and surface flow and solute transport. Groundwater Simulations Group, URL https://www.ggl.ulaval.ca/fileadmin/ggl/documents/rtherrien/hydrogeosphere.pdf.

Thulin, B., Hahn, H.J., 2008. Ecology and living conditions of groundwater fauna. Technical Report TR-08-06. Swedish Nuclear Fuel and Waste Management CO, Stockholm.

Tilman, D., Fargione, J., Wolff, B., D'Antonio, C., Dobson, A., Howarth, R., Schindler, D., Schlesinger, W.H., Simberloff, D., Swackhamer, D., 2001. Forecasting agriculturally driven global environmental change. Science 292, 281–284. https://doi.org/10.1126/science.1057544.

Tonina, D., Bellin, A., 2008. Effects of pore-scale dispersion, degree of heterogeneity, sampling size, and source volume on the concentration moments of conservative solutes in heterogeneous formations. Advances in Water Resources 31, 339–354. https://doi.org/10.1016/j.advwatres.2007.08.009.

Tonina, D., Buffington, J.M., 2007. Hyporheic exchange in gravel-bed rivers with pool-riffle morphology: laboratory experiments and three-dimensional modeling. Water Resources Research 43, W01421. https://doi.org/10.1029/2005WR004328.

Tonina, D., Buffington, J.M., 2009a. A three-dimensional model for analyzing the effects of salmon redds on hyporheic exchange and egg-pocket habitat. Canadian Journal of Fisheries and Aquatic Science 66, 2153–2157. https://doi.org/10.1139/F09-146.

Tonina, D., Buffington, J.M., 2009b. Hyporheic exchange in mountain rivers I: Mechanics and environmental effects. Geography Compass 3 (3), 1063–1086. https://doi.org/10.1111/j.1749-8198.2009.00226.x.

Tonina, D., de Barros, F.P.J., Marzadri, A., Bellin, A., 2016. Does streambed heterogeneity matter for hyporheic residence time distribution in sand-bedded streams? Advances in Water Resources 96, 120–126. https://doi.org/10.1016/j.advwatres.2016.07.009.

Tóth, J., 1963. A theoretical analysis of groundwater flow in small drainage basins. Journal of Geophysical Research 68, 4795–4812.

Tufenkji, N., 2007. Modeling microbial transport in porous media: traditional approaches and recent developments. Advances in Water Resources 30, 1455–1469.

Turnadge, C., Smerdon, B.D., 2014. A review of methods for modelling environmental tracers in groundwater: advantages of tracer concentration simulation. Journal of Hydrology 519, 3674–3689. https://doi.org/10.1016/j.jhydrol.2014.10.056.

van der Velde, Y., de Rooij, G.H., Rozemeijer, J.C., van Geer, F.C., Broers, H.P., 2010. Nitrate response of a lowland catchment: on the relation between stream concentration and travel time distribution dynamics. Water Resources Research 46, W11534. https://doi.org/10.1029/2010WR009105.

Vitousek, P.M., Aber, J.D., Howarth, R.W., Likens, G.E., Matson, P.A., Schindler, D.W., Schlesinger, W.H., Tilman, D.G., 1997. Human alteration of the global nitrogen cycle: sources and consequences. Ecological Applications 7, 737–750. https://doi.org/10.2307/2269431.

Waber, H.N.N., Smellie, J.a.T.A.T., 2008. Characterisation of pore water in crystalline rocks. Applied Geochemistry 23, 1834–1861. https://doi.org/10.1016/j.apgeochem.2008.02.007.

Wachniew, P., Zurek, A., Stumpp, C., Gemitzi, A., Gargini, A., Filippini, M., Rozanski, K., Meeks, J., Kvaerner, J., Witczak, S., 2016. Toward operational methods for the assessment of intrinsic groundwater vulnerability: a review. Critical Reviews in Environmental Science and Technology 46, 827–884. https://doi.org/10.1080/10643389.2016.1160816.

Ward, N.D., Bianchi, T.S., Medeiros, P.M., Seidel, M., Richey, J.E., Keil, R.G., Sawakuchi, H.O., 2017. Where carbon goes when water flows: carbon cycling across the aquatic continuum. Frontiers in Marine Science 4, 7. https://doi.org/10.3389/FMARS.2017.00007/PDF.

White, A.F., Bullen, T.D., Vivit, D.V., Schultz, M.S., Clow, D.W., 1999. The role of disseminated calcite in the chemical weathering of granitoid rocks. Geochimica et Cosmochimica Acta 63 (13/14), 1939−1953.

White, W.B., 2002. Karst hydrology: recent developments and open questions. Engineering Geology 65, 85−105. https://doi.org/10.1016/S0013-7952(01)00116-8.

Williams, M.D., Oostrom, M., 2000. Oxygenation of anoxic water in a fluctuating water table system: an experimental and numerical study. Journal of Hydrology 230, 70−85. https://doi.org/10.1016/S0022-1694(00)00172-4.

Williams, P.W., 2008. The role of the epikarst in karst and cave hydrogeology: a review. International Journal of Speleology 37 (1), 1−10. https://doi.org/10.5038/1827-806X.37.1.1.

Winter, T.C., Harvey, J.W., Franke, O.L., Alley, W.M., 1998. Ground water and surface water: a single resource. US Geological Survey, Circular 1139, Denver, CO.

Wondzell, S.M., Swanson, F.J., 1996. Seasonal and storm dynamics of the hyporheic zone of a $4^{th}$ order mountain stream. 1: Hydrologic processes. Journal of the North American Benthological Society 15, 3−19.

Wondzell, S.M., Swanson, F.J., 1999. Floods, channel change, and the hyporheic zone. Water Resources Research 35, 555−567.

Wörman, A., Packman, A.I., Marklund, L., Harvey, J.W., 2006. Exact three-dimensional spectral solution to surface-groundwater interaction with arbitrary surface topography. Geophysical Research Letters 33, L07402.

Worthington, S.R.H., Davies, G.J., Alexander, E.C., 2016. Enhancement of bedrock permeability by weathering. Earth-Science Reviews 160, 188−202. https://doi.org/10.1016/j.earscirev.2016.07.002.

Wroblicky, G.J., Campana, M.E., Valett, H.M., Dahm, C.N., 1998. Seasonal variation in surface-subsurface water exchange and lateral hyporheic area of two stream-aquifer systems. Water Resources Research 34, 317−328.

Zarnetske, J.P., Haggerty, R., Wondzell, S.M., Baker, M.A., 2011. Dynamics of nitrate production and removal as a function of residence time in the hyporheic zone. Journal of Geophysical Research: Biogeosciences 116, G01025.

Ziebis, W., Forster, S., Huettel, M., Jørgensen, B.B., 1996. Complex burrows of mud shrimp *Callianassa truncata* and their geochemical impact in the sea bed. Nature 382 (6592), 619−622.

# Classifying groundwater ecosystems

*Anne Robertson[1], Anton Brancelj[2,3], Heide Stein[4] and Hans Juergen Hahn[4]*

[1]School of Life & Health Sciences, University of Roehampton, London, United Kingdom;
[2]Université Paris-Saclay, CNRS, IRD, UMR Évolution, Génomes, Comportement et Écologie, Gif-sur-Yvette, France; [3]Université de Paris, UFR Sciences du Vivant, Paris, France; [4]Institute for Environmental Sciences, University of Koblenz-Landau, Landau, Germany

## Introduction

There is a very strong human imperative, reaching back to the Greek philosophers Aristotle and Theophrastus, to bring order to the apparent chaos of the natural world by developing classification systems. Many systems exist, for example, large scale terrestrial ecosystems are classified by identifying communities that develop in response to abiotic drivers such as temperature and precipitation (biomes) or by delineating biogeographic regions where a given region is home to similar species, which evolved from common ancestors. In surface freshwaters, ecosystems may be organized into ecoregions — large areas that encompass freshwater systems with a distinct assemblage of communities, determined by climate and geology (Illies, 1978; Abell et al., 2008), or at smaller scales based on catchment or stream order. Such classification systems are of great importance because they provide underpinning data that supports management and conservation decisions (Markovic et al., 2017; Lopes-Lima et al., 2018). This is particularly important in freshwater ecosystems because they are among the most threatened on the planet facing unprecedented pressures resulting from increasing human populations and socioeconomic development (Dudgeon et al., 2006; Vorosmarty et al., 2010).

Groundwaters (GWs) are an essential part of the hydrological continuum containing 95% of global unfrozen freshwaters and sustaining many surface ecosystems including wetlands, baseflow fed rivers and groundwater dependent terrestrial vegetation. Groundwaters are home to unique and diverse communities comprising microbes, invertebrates and a few vertebrates that play a potentially pivotal role in the maintenance of GW quality (Mermillod-Blondin et al., this volume; Griebler and Avramov, 2014; Griebler et al., 2019). They supply essential ecosystem services to humans; globally 2.5 billion people exclusively depend on GW to meet their daily freshwater needs (Connor, 2015). Nevertheless, many aquifers are threatened by overexploitation, pollution or climate change (e.g., Retter et al., 2021) and despite the hydrological, ecological, economic, and social importance of GW, research on their conservation lags far behind that on surface waters (Boulton, 2020; Griebler et al., this volume), in part because in most countries GW is considered by law as a resource, but not as an ecosystem (Hahn et al., 2018). In the same way as classification systems for surface waters underpin their management, effective GW ecosystem classification systems could support both conservation efforts for, and resource protection of GWs.

The development of classification systems that adequately characterize GW ecosystems has been painfully slow and is still incomplete. There are several main explanations for why this is so; (1) global scale distribution patterns of GW organisms are still elusive because of taxonomic and geographic biases (Mammola et al., 2019), (2) access to the GW biotopes is limited and (3) it is difficult to describe and map biodiversity in an environment that has never been physically explored and mapped — the so-called Racovitsa's impediment (Ficetola et al., 2019). An added complication is that the distribution of GW fauna is very patchy and a considerable and repeated sampling effort over a large area is necessary to assess the true biodiversity of a given region (Gibert and Deharveng, 2002; Ferreira et al., 2007). Furthermore, research funding, particularly in Europe, remains limited due to a lack of inclusion of GW ecosystems in the relevant legislation.

In contrast, hydrogeological classification schemes characterizing the productivity of units in terms of the yield of water and its movement through the aquifer are well developed (Struckmeier and Margat, 1995) and are used to support economic development, planning and management. Typical components include the permeability of the matrix, the pore size and hydrological exchange with surface waters, features that are also relevant to ecological classifications (Hahn, 2009; Weitowitz et al., 2017). However, hydrogeological classification schemes alone cannot fully characterize groundwater ecosystems because they do not consider the distribution of groundwater species, which is influenced by factors depending on the spatial scale (see later in this chapter). Another approach, has been to classify GWs in terms of their vulnerability to pollution (Sinreich, 2009) using hydrogeological or ecological indices such as DRASTIC (Ferreira and Oliviera, 2004) or AQUALIFE (Strona et al., 2019).

Intrinsic vulnerability is a hydrogeological concept that has been widely used in risk assessment and to inform management strategies. It describes the sensitivity of an aquifer to pollution, based on intrinsic characteristics such as hydrological, geological, hydrogeological and soil properties (e.g., https://www.bgs.ac.uk/products/hydrogeology/Groundwater Vulnerability.html; Ferreira and Oliviera, 2004; Heiss et al., 2020). Since hydrological exchange is one of the key factors shaping GW communities, it is highly probable that the intrinsic vulnerability of an aquifer is also relevant to GW ecosystems, however, these schemes do not include a consideration of species distributions and so their usefulness in GW ecosystem classification and assessment is restricted.

Classification systems developed for surface freshwaters have limited use for groundwater ecosystems. The bioregions delineated for surface freshwaters (Illies, 1978; Abell et al., 2008) are not necessarily the same for GW ecosystems. Groundwater habitats are often fragmented and the poor dispersal capacity of many GW organisms (Stoch and Galassi, 2010) means that their present day distributions are still heavily influenced by past glaciations and palaeodrainages (Strayer, 1994; Humphreys, 1999; Robertson et al., 2009). Similarly, surface freshwaters are often classified at the catchment scale, which do not necessarily map onto aquifers. Thus, attempts to model GW ecosystem classifications on the longitudinal zonation observed in surface freshwaters have had limited success (Husmann, 1966, 1967; Mösslacher, 2003).

It is clear that dedicated classification schemes for groundwater ecosystems are needed, particularly as GW ecosystems are in intensive hydrological connection with surface waters and at the same time many surface ecosystems are strongly dependent on GW (so-called groundwater dependent ecosystems — GWDEs). It is vital to embrace this connectivity and set GW ecosystems in the wider context of aquatic and terrestrial ecosystems to support an integrated approach to GW management (Jakeman et al., 2016). In this chapter we describe and evaluate existing GW classification systems for aquifer ecosystems, distinguishing four main spatial scales: (1) global, (2) continental (3) landscape and (4) local/habitat (Fig. 2.1).

Finally, we identify the main shortfalls that restrict our understanding and development of GW ecosystem classifications and propose a future road map to address these.

## Classification systems

For many years, GW was considered as a single habitat (Illies, 1978) and it was not until 1986 that the first comprehensive list of different subterranean habitats was published (Botosaneanu, 1986). To date, researchers have employed a variety of approaches to develop GW classification systems across different spatial and functional scales (Husmann, 1967; Illies, 1978; Botosaneanu, 1986; Gibert et al., 1997; Stoch et al., 2004; Hahn, 2009; Stoch and Galassi, 2010; Stein et al., 2012; Cornu et al., 2013; Weitowitz et al., 2017). The attributes

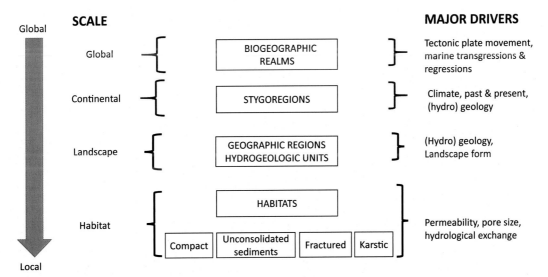

**FIGURE 2.1**   Classifying groundwaters ecosystems from global to local scales.

used to classify GW ecosystems vary across spatial scales. Thus at the global and continental scale, species are used to delimit biogeographic realms and stygoregions. At the landscape scale, hydrogeological parameters such as permeability, pore size and hydrological exchange with surface water are used to delimit hydrogeological units, whereas geographic regions are distinguished by common natural features such as landscape form and geology. Aquifer types, differing in the size and density of openings in the matrix and their degree of connectivity, are used at the local/habitat scale.

Approaches such as those of the AQUALIFE project (Strona et al., 2019), which developed a set of procedures and indicators for the ecological assessment of GWDEs based on biodiversity and it's threats, may be used to complement classification systems for GW ecosystems.

## Global scale

Here, the movement of tectonic plates, climate changes, including glaciation, marine transgressions and regressions and hydrography shape the structure of communities and species occurrence (Fig. 2.1).

Areas are categorized in terms of the distribution of communities and their taxonomic relationships. This approach has a long history stretching back to the early 1800s. The delineation of these realms is intimately entwined with plate tectonics (Trewick, 2017). Plate movements have resulted in the isolation and subsequent evolution of taxa (vicariance speciation) and the formation of distinctive groupings of terrestrial taxa in the different realms. Although the applicability of these realms to surface freshwater taxa is unresolved (Abell et al., 2008), they are much more likely to be relevant for GW because of the very poor dispersal abilities of subterranean species and their close connectivity with terrestrial ecosystems (Stoch and Galassi, 2010). Ancient marine transgressions and regressions may also explain present day distributions of GW fauna (Humphreys, 2000). Udvardy (1975) and Pielou (1979) identified eight terrestrial biogeographic realms that are today broadly accepted. These are: (1) Nearctic, (2) Palaearctic, (3) Africotropical, (4) Indomalayan, (5) Oceania, (6) Australian, (7) Antarctic and (8) Neotropical.

Botosaneanu (1986) developed a spatially explicit classification system for GW fauna, using the terms zones, provinces and districts based on published work and expert knowledge that partially deviates from the terrestrial realms listed above. He recognized nine zones. These are: (1) Periponto- Caspi- Mediterranean, (2) Western and Central Europe, (3) Palaearctic, (4) non-Palaearctic Africa, islands of the Indian Ocean, islands of the Red Sea, (5) non-Palaearctic Asia, westward from the Wallace line, (6) Indonesia eastward from the Wallace line, New Guinea, Australia, Tasmania, New Zealand, other islands of the Pacific Ocean, (7) Pericaribbean & Mexican zone, (8) North America, north of Mexico and (9) Neotropical zone.

## Continental scale

At this scale most classification systems are either focused on surface waters or on the terrestrial ecosystem. Delineated areas are often referred to as bioregions or ecoregions. For example, the limnetic surface water ecoregions of the Water framework Directive (WFD) are based on Illies (1978), while the widely accepted ecoregions of Olson et al. (2001) are terrestrial based. The distribution patterns of GW fauna at this level are clearly different from those found in surface ecosystems. Climate changes over geological time are still powerful predictors of present day subterranean community presence and composition (Fig. 2.1) because stygobites may be extirpated by glaciations or aridity and subsequent recolonization is extremely slow (Stoch and Galassi, 2010). North of the Alps, GW communities are characterized by a decreasing gradient of species richness and endemicity in a northerly direction (Thienemann, 1950; Gibert et al., 2009; Martin et al., 2009; Stoch and Galassi, 2010; Stein et al., 2012). Similar patterns have been shown in the British Isles (Robertson et al., 2009) and the United States (Strayer, 1994). In contrast, in southern Europe (Italy; the former Yugoslavia i.e., the Balkans), which was unaffected by glaciation, old tertiary fauna (i.e., from the period 66 million to 2.6 million years ago before the onset of the quaternary glaciations), have survived in small isolated refuges, leading to a high level of endemism (Stoch and Galassi, 2010). This finding is supported by a detailed analysis of diversity patterns in the European groundwater fauna, which found that highest species richness occurred at intermediate latitudes (Zagmajster et al., 2014).

The term "stygoregions," used in the sense of eco- or bioregions for GW was first coined by Gibert et al. (2009) and has been examined by Ferreira et al. (2005), Hahn (2009); Stoch and Galassi (2010) and Stein et al. (2012). To date stygoregions have been mostly (but not entirely) used as a classification system at the European level. The degree of internal fragmentation and the size of stygoregions can vary strongly, depending on the history of the region over geological time. This means that on a worldwide scale, stygoregions have to be defined individually. However, the observed patterns must be interpreted carefully because they may also be influenced by the intensity of stygofauna research activity. Here we give two stygoregion examples.

Analysis of a comprehensive groundwater dataset for Central Europe demonstrated that groundwater communities assembled into four major clusters (stygoregions) that differed from the surface water ecoregions identified for this area (Stein et al., 2012, Fig. 2.2).

(1) The Northern Lowlands comprising groundwater systems that had been heavily influenced by Pleistocene ice shields.
(2) The Central Uplands encompassing the Central Mountain Ranges and the adjacent sub-mountainous forelands. Although not covered by ice shields these areas were strongly affected by permafrost soils and low precipitation.
(3) The South-Western Uplands were unaffected by the last glaciation.
(4) The Southern Uplands and Northern Alps comprise those areas that were covered by the local/regional Pleistocene ice shields of the Alps and the Black Forest.

South of the Alps in Italy, Stoch and Galassi (2010) described a highly fragmented and comparatively smaller stygoregion in the eastern Alpine area of Italy comprising eight karstic areas from the Italian-Slovenian border in the east to Piedmont in the west, most of which was ice-free during the last glacial maximum. The groundwater communities in this stygoregion are characterized by a high degree of endemicity. The Dinaric region in the Balkans is similarly highly fragmented and rich in stygobionts. This area was not glaciated in the Pleistocene and was connected to the Tethys Ocean for about 30 million years (Sket, 2005).

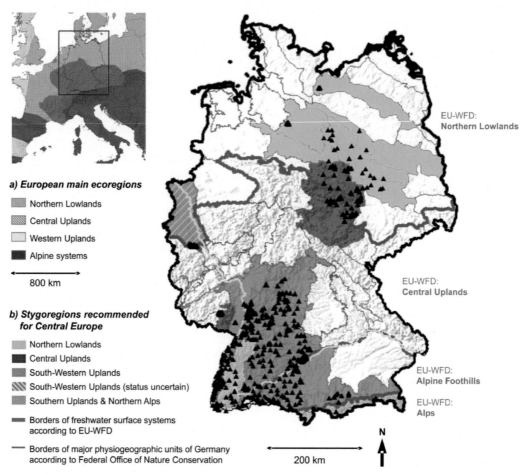

**FIGURE 2.2** (A) European main ecoregions and (B) a topographic map of Germany depicting the proposed stygoregions (colored), which were delineated according to invertebrate distribution patterns found in groundwater. The white areas comprise major physiographic units where no data was available. The affiliation of the Lower Rhine Valley is under debate, indicated by the blue-white hatching. EU-WFD is the European Union Water Framework Directive. Topographic map/GIS: http://www.eea.europa.eu/legal/copyright. *Credit: Stein et al. (2012).*

## Landscape scale

Landscapes are the visible characteristic features of an area of land and its landforms and are frequently heavily modified by humans. The Alps, for example, are a landscape, as well as the Blackforest; the Paris Basin, the Great Plains, the Carpats, the Dinarids and the Darling Riverine Plains. Landscapes integrate natural and man-made features with their ecosystems (Schubert, 1985). To classify GW habitats and ecosystems at the landscape scale it is necessary to select appropriate landscape units. Typical landscape ecological factors are (hydro)geology, landscape form, climate, soils, vegetation, landuse and landscape history (Fig. 2.1). All these factors influence the general exchange processes between precipitation, surface water and GW, in effect the landscape water regime ("Landschaftswasserhaushalt") (Wohlrab et al., 1992), which describes how precipitation interacts with surface water flow, GW recharge/discharge, surface run-off and evapotranspiration.

The landscape water regime is strongly affected by the heterogeneity and the particularities of a catchment or a hydrographical system within a landscape. For example, in forests on sandy soils with an underlying fractured aquifer, there is strong connectivity with the surface resulting in high groundwater recharge rates and low surface run-off. In contrast, areas used for agriculture with compacted clay soils are characterized by a strong surface run-off and a low groundwater recharge rate. On steep slopes groundwater recharge is lower than on smooth slopes or plains. Karst springs behave differently to alluvial springs (Hölting, 2013), and the interactions between streams and groundwater are regulated by hydromorphological and hydrogeological features both on a local and a regional scale (Boulton et al., 1997).

At the landscape scale, groundwater-surface water interactions are one of the key factors regulating the fluxes of energy and matter and thus directly shaping the ecosystems themselves (Datry, 2005; Hahn, 2006; Dole-Olivier et al., 2009). Groundwater-surface water interactions and connectivity are strongly related to another key factor — aquifer geology—which provides the physical characteristics of a habitat and determines permeability and hydrochemistry (Datry et al., 2005; Hahn, 2006). Generally, rocks with higher permeability are thought to provide higher levels of oxygen (due to faster groundwater replenishment) and organic matter (detritus) than less permeable rocks and sediments (Hahn, 2006; Bork et al., 2009). However, an enhanced hydrological exchange also increases the risk of the intrusion of pollutants. Thus, a stronger hydrological exchange is often linked to a greater vulnerability of the aquifer to pollutants (Di Lorenzo et al., 2019; Strona et al., 2019).

At the landscape scale, two main systems are used for the classification of GW communities and ecosystems: (1) hydrogeological units and (2) geographic regions (Fig. 2.1). Hydrogeological units depict aquifer properties and are widely used to foster the planning, protection and monitoring of groundwater resources (Gogu et al., 2001). Maps of these hydrogeological units are available for most of the World e.g., for Europe see http://www.europe-geology.eu/. They also provide key information for inferring the distribution of distinct GW habitats for GW communities because they integrate ecologically important hydrogeological features such as GW flow type, permeability, pore size and hydrological exchange with surface water (Dole-Olivier et al., 2009; Hahn, 2009). Cornu et al. (2013) used the International Hydrogeological Map of Europe (IHME) to classify GW habitats in Europe (Fig. 2.3).

The classification was based on three criteria: (1) GW flow type, divided into two categories: (a) consolidated and (b) unconsolidated rocks, because each of them provides distinct habitats colonized by different species assemblages (Gibert et al., 1994; Malard et al., 2009). (2) Permeability, divided into three classes — (a) high, (b) moderate and (c) low — according to the information provided in the IHME and (3) void size divided into: (a) large and (b) small, because it determines the range of species having different body sizes that can co-occur in an aquifer. A final category of nonaquiferous rock was added giving a total of 13 habitat categories. Weitowitz et al. (2017) developed a new national-scale typology for GW ecosystems in England and Wales. Initially bedrock was separated into karstic, porous and fractured rock following the approach of some other authors (Galassi et al., 2009; Hahn and Fuchs, 2009; Malard et al., 2009; Martin et al., 2009). Then, further subdivisions were made based

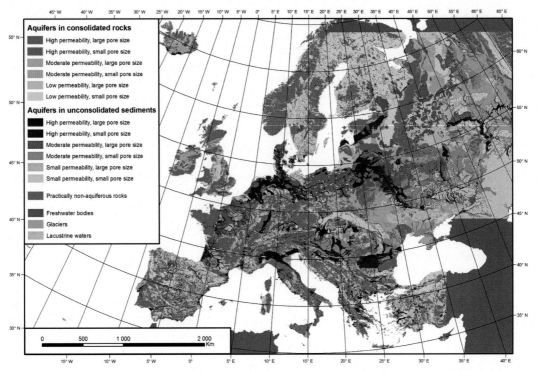

**FIGURE 2.3**  The groundwater habitat map of Europe (Lambert azimuthal equal-area projection). *Credit: Cornuet al. (2013).*

on differences in hydrogeological features (e.g., karstification, fractures, and pore space sizes), which affect the available habitat space and water chemistry to produce a higher resolution typology of 11 geo-habitats that differed significantly in their hydrogeological and hydrochemical characteristics. To assess the quality of geo-habitats (in terms of their suitability for sustaining GW communities), seven parameters (the concentrations of dissolved oxygen, dissolved organic carbon, nitrates, and calcium, transmissivity, cave development and physical habitat space) were used to compute a final geohabitat quality score (Weitowitz et al., 2017). Low quality habitats were dominant in Wales, northern and south-west England (Fig. 2.4). High quality, and some medium quality, habitats provided highly permeable corridors connecting southern to northern England. Medium quality habitats covered small geographical areas and were spatially patchy, particularly in Wales and southern England. Johns et al. (2015) classified groundwater communities in Southwest England using a combination of hydrogeological units and hydrochemistry. They found that hydrogeological units explained a greater proportion of the variance in the community than water chemistry but that communities also differed with location in the study area.

However, hydrogeological units have a major weakness because they disregard landscape history and biogeographical aspects. Weitowitz et al. (2017) had good results because they worked in a relatively small area (England and Wales). Cornu et al. (2013) worked on a larger scale and their predictive model classified the Northern European Lowlands as good habitats

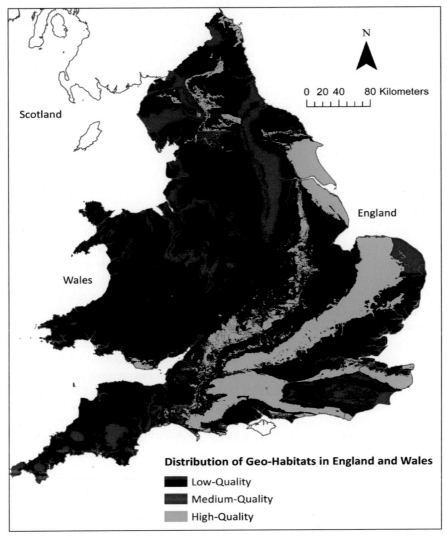

**FIGURE 2.4** Distribution map of groundwater habitats in England and Wales grouped by their habitat quality scores calculated from abiotic parameters important to groundwater ecosystems. Contains British Geological Survey materials © NERC 2017. Contains Ordnance Survey data © Crown Copyright and database rights 2017. *Credit: Weitowitz et al., 2017.*

for groundwater fauna whereas in reality these aquifers harbor one of the most impoverished groundwater fauna in Europe (Stein et al., 2012). Furthermore this model suggested that igneous rocks were unsuitable habitats but in the UK, these rocks support significant faunal assemblages (Johns et al., 2015). To solve this problem, Cornu et al. (2013) recommended the inclusion of hydrochemistry (oxygen) and biogeographical aspects (glaciation, latitude) into future iterations of the model, bringing it closer to the geographic region approach.

In the German speaking part of Europe, geographic regions ("Naturraum") are used as the basis for biotope protection and species conservation in surface ecosystems (BfN, 2008). They are distinguished by common natural features such as landscape form, geography, geology, landuse, climate and landscape history. Naturraums, like the Blackforest or the Upper Rhine valley, represent the regional expression of ecoregions. Hahn and Fuchs (2009) explored this approach for GW ecosystems and found that a combination of permeability and natural geographic regions ("georegs") gave the best explanation of observed GW community composition and distribution in Southwestern Germany. However, there are few supranational systems of geographic regions available, and often they are not comparable because different spatial scales are used.

## Habitat/local scale

At the local scale (i.e.,<1 km), a four aquifer type classification system is widely used (Fig. 2.1): (A) compact/aquitard rocks, (B) unconsolidated sediments, (C) fractured rocks, and (D) karstic rocks. Three types of physical structures are available as groundwater habitats within these aquifers: (1) pore spaces, (2) fractures, and (3) karstic voids/caves. The size and density of openings in the matrix and their degree of connectivity, together with the geology, determine the amount of living space available for groundwater animals and the ambient water chemistry (Goldscheider et al., 2006) as well as permeability. Pore spaces in unconsolidated sediments (alluvium) are generally larger in coarse-grained aquifers, providing greater living space (Dole-Olivier et al., 2009; Hahn, 2009; Hahn and Fuchs, 2009; Malard et al., 2009). In fractured rocks (like sandstones), the habitat is determined by fracture size and permeability (Hahn and Matzke, 2005; Hahn and Fuchs, 2009). In karstic rocks (limestone or dolomite), water action transforms fractures into large voids and cave systems with rapid water flow and high connectivity with surface waters (Danielopol et al., 2004; Malard et al., 2009; Robertson et al., 2009). In general, karstic aquifers have the highest permeability and compact aquifers the lowest; fractured and unconsolidated aquifers have intermediate permeability. However, there is high intra-aquifer type variability (Weitowitz et al., 2017). Furthermore, a single aquifer may fall into more than one aquifer type. For example, mixed unconsolidated sediment and fractured aquifers are common where rocks have locally variable degrees of consolidation (some sandstones) or locally variable degrees of weathering (granite) (Maurice and Bloomfield, 2012). With the exception of isolated aquifers such as calcretes and mound springs, there are likely to be connections between aquifers and each aquifer type is unlikely to be a discrete habitat harboring a distinctive fauna. There is often a tendency to habitat generality in stygobites leading to a broad overlap of taxa between aquifer types, although richness and abundance differ. For example, Eberhard et al. (2005) found that, except for large bodied taxa such as fish that require large voids, stygofauna from the Pilbara and adjacent areas are found wherever GW environments provide suitable living space whether these be in unconsolidated sediment, karstic or fractured rock aquifers. Nevertheless, it is possible that a given aquifer type could represent a 'sink' for some stygobiont species that have drifted from a neighboring aquifer type with 'source' populations (*sensu* Dunning et al., 1992).

The majority of research effort has concentrated on karstic aquifers and we focus on them here to exemplify the diversity of GW habitats. Compact, unconsolidated sediment and fractured aquifers and ecotone GW biotopes are described briefly. For an ecological comparison of the main aquifer types see Hahn and Fuchs (2009).

## Compact aquifers

Compact aquifers (sometimes known as aquitards), such as loess and silted glacial moraines, are characterized by a reduced permeability and very small pore sizes. Thus, in an ecological sense, they are stressful habitats because there is limited living space and low hydrological exchange (Hahn and Fuchs, 2009). A depleted fauna is typical of compact aquifers although ecological information is scarce; diversity is low and metazoans occur only in around 50% of all samples (Hahn and Fuchs, 2009). As pore spaces are small it is likely that stygobiont populations live in flow channels that also serve as preferential flow paths. In many cases compact aquifers are likely to be sink habitats for stygobites that have drifted from nearby source habitats such as the hyporheic zones of rivers. Yet, permanent populations of stygobites are found in some compact aquifers such as the loess aquifers of Southern Germany (Hahn and Fuchs, 2009) and the mudstones of Southwestern England (Johns et al., 2015).

## Unconsolidated sediment aquifers

These comprise alluvial deposits from past (i.e., palaeo-channels formed by the migration of a rivers course in the past), and present day rivers and standing water bodies. Unconsolidated sediment habitats associated with palaeo-rivers are often present in semidesert and desert areas, where groundwater is a few meters below the surface and fed exclusively by precipitation or temporary streams, e.g., wadis or gueltas in the Sahara or Western Australia (Karanovic, 2004, 2006; Brancelj, 2015b). Sometimes these habitats are associated with karstic calcretes, which are formed as a result of the precipitation of calcium carbonate (Humphreys, 1999; Sacco et al., 2019).

In unconsolidated sediment aquifers, water is stored and transported through the voids between the sediment particles. The size of these voids, that also comprise the living space for GW fauna, is strongly dependent on sediment size and sorting. In most unconsolidated sediment aquifers, living space is in the mm scale and only exceptionally in cm. Unconsolidated sediment habitats have relatively stable physical and chemical characteristics compared to fractured or karstic habitats.

Unconsolidated sediment aquifers are globally distributed and are diverse habitats in terms of pore size; ranging from glacio-fluvial deposits (large pore size) to glacial-morainic deposits (small pore size). They can be extensive, for example, in the Northern European lowlands, sandy deposits of glacial origin cover ten thousands of square kilometers, forming large unconsolidated sediment aquifers. Here pore spaces are small, resulting in weak hydrological exchange with surface water, low oxygen concentrations and impoverished faunal communities (Stein et al., 2012). However, depending on grain size and oxygen concentrations, stygobite diversity in unconsolidated sediment aquifers can be considerable. Hancock and Boulton (2008) recorded 87 groundwater taxa from alluvial aquifers in four regions of Eastern Australia, Eberhard et al. (2009) found 78 stygobite species in alluvial aquifers of the Pilbara region, Western Australia. Also, 35 stygobite taxa were found in the shallow alluvial aquifer of the Danube floodplains near Vienna, Austria (Danielopol and Pospisil, 2001) and the Jons Plain near the city of Lyon, France, contains approximately 40 stygobite species (Dole-Olivier et al., 1994). In the state of Baden-Württemberg, Southern Germany, Hahn and Fuchs (2009) found 94 taxa in samples from unconsolidated sediment aquifers including 59 stygobite species; these samples were even more diverse than those from nearby karst aquifers.

## Fractured aquifers

These are the most common type of aquifer in consolidated rock but they have been little sampled. Water is transported along secondary openings in the rock such as joints, fractures and bedding planes that also provide living space for GW fauna. Sandstones are important examples of fractured aquifers that comprise consolidated porous material where the grains are cemented together. Permeability is low to moderate and these aquifers provide good habitats for groundwater communities (Humphreys, 2009) that can exhibit levels of diversity similar to karst habitats. In Southwest Germany, Hahn and Fuchs (2009) found 26 stygobitic species in a sandstone aquifer and 28 species in a nearby karst aquifer. Man-made fractures in rocks (e.g., mines) may also form good habitats for groundwater fauna (Knight, 2001).

## Karst aquifers

Karstic aquifers are an intensive fractured geological formation in limestone or dolomite with the highest habitat and fauna diversity. Over the last 3 decades, studies have classified karstic aquifers into four distinct environments, organized in a vertical orientation, each with specific hydrological, energetic and faunal parameters including increasing α diversity and abundance/biomass: (1) epikarstic zone; (2) vadose zone; (3) epiphreatic zone, (4) phreatic zone. Furthermore, karstic springs are an important habitat (Brancelj, 2002; Sket, 2004; Bakalowicz, 2005; Brancelj and Culver, 2005; White and Culver, 2005; Ravbar, 2007; Culver and Pipan, 2009).

Epikarstic habitats (Table 2.1) are the thin zone at the uppermost surface of the karst, extending to only a few meters in depth. This zone typically has enhanced porosity and permeability compared to the bulk rock mass below, is very heterogeneous and supports a perched aquifer, where vertical water flow prevails. Its flow is fed exclusively by precipitation and thus it is very dynamic at hourly/daily/seasonal scales (Bakalowicz, 2005; Brancelj, 2015a). Water in the epikarst is collected from a relatively small surface area to a specific drip-point, thus they represent "islands" from an ecological point of view (Brancelj and Culver, 2005). Water from the epikarst is discharged either into galleries where it collects in small pools from ml to tens of liters in volume, or into voids that are connected to the vadose habitat. Living space in the epikarst is small cracks/voids, normally in mm scale. It is a source habitat for very specialized aquatic fauna, dominated by Copepoda, which may accidently drift downwards to the vadose zone (Brancelj, 2002; Pipan et al., 2006; Pipan and Culver, 2022).

Vadose habitats (Table 2.1) are located just below the epikarst. Here, water flow through voids/channels is better defined with a more stable discharge, even though some trickles can dry-out during periods of low precipitation. This water can form pools of drip water, with a volume of several hundreds of liters, or permanent subterranean streams with rifles and

**TABLE 2.1**   Subterranean karst habitats.

| Habitat type | Main source of water | Source or SINK stygobite habitat | Biodiversity potential |
|---|---|---|---|
| Epikarst zone — small fissures, issuing as drips from cave ceilings; connected with local precipitation | Exclusively infiltrating precipitation | Source habitat | Very high; microendemics; poorly studied |
| Vadose zone — water from epikarst typically collected in pools, streams in a fossil/dry cave zone | Infiltrating precipitation and temporary sinking streams | Source habitat; but sink habitat for fauna from the epikarst | Very high |
| Epiphreatic zone — Pools, streams and rivers; zone with seasonal oscillations of water level (floods) | Infiltrating precipitation and sinking streams/rivers and phreatic aquifer | Source habitat; but sink habitat for fauna from the epikarst and also from the vadose zone | Very high |
| Phreatic zone — saturated zone; including anchialine caves and cenotes | Infiltrating precipitation and sinking streams/rivers and/or sea | Source habitat; but sink habitat for fauna from the epikarst and also from the vadose zone | Very high |
| Springs — temporary or permanent | Infiltrating precipitation and/or sinking streams/rivers and phreatic water (either karst or adjacent porous) | Sink habitat for fauna drifting from the karst or adjacent aquifers | Very high; normally as drift |

pools, associated with compact sinter forms such as stalactites and stalagmites, and flow-stones. Here, living space for groundwater animals ranges from millimeters to several meters in size. The habitat is still controlled by local/daily precipitation but with some time delay and a dampening of extreme values.

Epiphreatic habitats (Table 2.1) are well researched because they are relatively easy to access. An important source of water for these habitats are sinking rivers, which drain from near-by permeable or impermeable slopes into the rivers entering the caves. These habitats can oscillate in their vertical extent over several tens of meters (Mihevc, 2004) as a result of differences in the infiltration rate from the surface and in the discharge ability of channels of subterranean rivers. The living space for stygobites, that include fauna both from the vadose and the phreatic zones, is unlimited and characterized by fine sediment deposition, sometimes intermixed with sinter deposits. The habitat is characterized by seasonal inputs of allochthonous food sources during floods when connectivity with sinking rivers is high. Stygobites from the phreatic and vadose zones are trapped in pools and dead-arms of subterranean rivers during low water levels when these habitats are replenished with water from the vadose zone. When water levels rise phreatic waters also contribute and the epiphreatic habitat is actually an ecotone between the unsaturated (vadose) and saturated (phreatic) zones.

Karstic phreatic habitats (Table 2.1) are accessible only: (1) during the lowest water level in the caves, (2) by human divers/remote-control vehicles/wells and boreholes or (3) by deep-feeding springs. Water quality and discharge as well as composition of fauna and biodiversity are stable over long-time periods. Living space for stygobites is unlimited as channels can be up to several meters in diameter and kilometers in length. Food resources are limited and rather constant except in cases where local surface pollution/eutrophication is present, or phreatic habitats are connected to sinking river channels (Di Lorenzo et al., 2019; Strona et al., 2019). Many karstic phreatic habitats are under heavy water extraction pressure, meaning that they are vulnerable and those close to the coast are already subject to salt water intrusion (Kresic, 2012). Some phreatic habitats in karstic caves close to the sea are actually a combination of freshwater and marine environments, so-called "anchialine caves". Here voids or channels connect inland and sea aquifers and a water density gradient is established (chemocline), with less dense freshwater positioned above sea water of greater density. The largest example of this kind of habitat are "cenotas" in the Yucatan (Mexico) peninsula (Beddows, 2004).

## Ecotonal groundwater habitats

Ecotones are dynamic interfaces between different habitats with significant ecological gradients and often inhabit a very diverse fauna and flora (Décamps and Naiman, 1990; Risser, 1990; Gibert, 1991). The most common groundwater ecotones are: (1) the capillary fringe above the GW table, incl. GWDEs (see glossary), (2) the hyporheic zone, (3) interstitial habitats connected to standing water bodies and (4) springs.

(1) The capillary fringe above the GW table in alluvial sediments is the interface between saturated and unsaturated soils and sediments. Food, oxygen supply and density of fauna populations are highest near the GW table and heavily influenced by surface water intrusion (Schmidt and Hahn, 2012). It is assumed that the capillary fringe is where the majority of GW self-purification processes take place, and that there is a gradual transition from terrestrial to aquatic fauna between the saturated and unsaturated zone (Fiers and Ghenne, 2000), but few studies have focused on this ecotone.

(2) The hyporheic zone is an ecotone between the surface river and subsurface (phreatic) ecosystems. It comprises the saturated interstitial areas beneath the stream bed and into the stream banks that contain some proportion of channel water (White, 1993). It is a dynamic area, where intensive exchange between surface and subsurface water occurs (Williams et al., 2010; Cardenas, 2015, Tonina et al., this volume) controlled by advection. Surface river water enters the hyporheic habitat in down-welling zones, where hydrostatic pressure is lower than in the river, bringing with it food and oxygen. Water in the hyporheic habitat moves along flow paths in parallel with the surface and, with increased residence time below the surface, hyporheic water generally becomes less oxygenated, with reducing biogeochemical processes initiated (Malard et al., 2002). However, these processes vary across cm scales creating 'hot spots' and 'hot moments' of biogeochemical activity in space and time (Krause et al., 2011). Eventually some of this water returns to the surface in upwelling zones that may be rich in nutrients but low in oxygen. Groundwater from the aquifer below also exchanges with hyporheic water. The hyporheic community (hyporheos) is a distinct entity, different from that found in surface waters (Peralta-Maraver et al., 2018) and in groundwaters, but with elements of both these adjacent ecosystems (Williams et al., 2010). Stygobites are frequently found in upwelling areas of hyporheic zones whereas surface taxa are found in downwelling zones.

(3) Interstitial habitat connected to standing water bodies. These can be freshwater, brackish or saline, and, in principle, have the same structure as hyporheic zones associated with rivers although they are not as extensive. They are referred to as the wetted or epiphreatic zone, where waves or water level oscillations contribute to the exchange of surface and subsurface water. Below this zone fine sediments (mineral or organic) accumulate, reducing water exchange between surface and subsurface, a process that occurs largely by diffusion. Thus the habitat often becomes anoxic. Interstitial pore sizes are small in these low energy environments and the resident communities typically comprise organisms in the meiofaunal size range such as Gastrotricha, Rotifera and Copepoda (Enckell, 1968). These organisms can be very abundant. For example, in a classic study on Mirror Lake, USA, Strayer (1985) found on average 1.2 million meiobenthic organisms/$m^2$ of lake bottom mostly in the top 1 cm of sediment. As for riverine hyporheic zones this habitat can extend lateral to the surface lake onto lake/sea beaches, in some cases as much as 30 m inland (Whitman et al., 1994).

(4) Springs are ecotonal "windows" between subterranean and surface habitats where surface and sub-surface species co-exist. Spring hydrology is complex and highly dynamic, with alternating proportions of waters of different origin. In springs, deeper and shallower groundwater mix with interflow and surface water. Some springs are permanent and some temporary and their discharge and duration are determined by local precipitation. Thus fauna is diverse comprising a mixture of stygobites and nonstygobites. The former drift into springs from all parts of the aquifer system and thus springs are sink habitats for these species. However, some stygobiontic taxa like *Niphargus* actively migrate between GW and springs (Kureck, 1967). Temporary springs are active only when groundwater levels are high enough to discharge water, i.e., after intensive or long-lasting precipitation and are populated with stygofauna. Permanent springs are fed from the phreatic habitat and thus the stygofauna found originate from there (Mori et al., 2015). As stygobites are related to the origin of groundwater their use as hydrological biotracers has been advocated (Mori et al., 2015; Stoch et al., 2016; van den Berg-Stein et al., 2019; Brancelj et al., 2020).

## Conclusions

Mammola et al. (2020) identified the top 50 unanswered fundamental research questions in subterranean biology. Twelve are centered on conservation and include questions such as 'How does climate change effect subterranean-adapted organisms?', 'How can we evaluate the ecological status of subterranean ecosystems?', 'What would be the best protocols to quantify long term changes in the distribution and abundance of subterranean invertebrates?'

Clearly, the further development of classification systems for GW ecosystems across a range of scales is a prerequisite to obtaining answers to these questions and, indeed, to the overall monitoring, management and conservation of these vulnerable systems. Our understanding of how to classify GW ecosystems has developed exponentially since the first edition of 'Groundwater Ecology' (Gibert et al., 1994) but much remains to be done to reach our goal of effective systems for classifying GW ecosystems. Comprehensive ecological and organismal data are scarce for GWs and patchily distributed. One approach must be to 'fill the gaps' in the existing datasets but this is a long term aim. In the meantime, we believe the most fruitful approach to developing fit for purpose classification systems is to follow the accepted classifications of biogeography and landscape ecology, but with GW related speci-fications. We advocate further testing of the stygoregions approach, particularly outside of Europe and additional validation of the hierarchical approach proposed here. Inevitably the classification system that is most useful will depend on the question that is asked making it sensible to continue to develop classification systems across a range of scales. Regardless, classification systems are important to integrate GW ecosystems into the monitoring of fresh-water aquatic ecosystems to improve safeguarding of their biodiversity, health and the ecosystem services they deliver.

## Glossary

| Key words | Meaning |
|---|---|
| Aquifer | Any underground water body providing enough water for human use |
| Alluvial aquifer | Aquifer comprising unconsolidated material deposited by running water, typically occurring adjacent to present-day rivers and in paleo-channels. |
| Bioregions/Ecoregions | Geographical units with characteristic flora, fauna and ecosystems. |
| Catchment | The area of land from which water flows into running or standing water bodies, including aquifers. |
| Connectivity | Indicates degree of hydrological connections between water bodies. Allows organisms to migrate. |
| Drift | Passive transport of organisms by water flow. Organisms enter drift by accident/randomly, as part of their life cycle or during extreme events (i.e., floods). |
| Ecosystem services | The direct and indirect contributions of ecosystems to human wellbeing. |
| Groundwater dependent ecosystems (GWDEs) | Surface-water ecosystems receiving water from aquifers (wetlands, rivers, springs). |
| Fractured aquifers | Aquifers in which water moves through fractures within the rock. |
| Hydrogeological unit | Any soil or rock unit or zone, which by virtue of its hydraulic properties has a distinct influence on the storage or movement of groundwater (see aquifer). |
| Hydrological exchange/ hydrological exchange flows | The exchange of water between surface/subsurface water bodies or between aquifers (see connectivity). |
| Karstic aquifers | Aquifers in which fractures in limestone or dolomite are enlarged by dissolution and/or erosion to form larger voids. These may be fissures, conduits and caves. |
| Marine transgression (regression) | Event during which sea level oscillates relative to the land leading to inundation by seawater or exposure of previously inundated land. |
| Palaeochannels | A system of drainage channels produced by ancient rivers. |

—cont'd

| Key words | Meaning |
| --- | --- |
| Perched aquifer | An aquifer, which is situated between the main aquifer and the infiltration zone. It is disconnected from the main aquifer below the infiltration zone. |
| Permeability | Whether and how water can flow through a rock |
| Pore space | The amount and size of pores, or open space, within soil, sand or rock. |
| Riparian zone | Transitional areas between land and fresh/brackish/saline water bodies such as streams, rivers, lakes, sea and wetlands. |
| Source/sink habitats | Source habitats support viable populations of stygobites for long (geological) periods. Sink habitats are places where stygobites are regularly/accidently transported from source habitats, but cannot reproduce or establish persistent populations and die within a generation. |
| Stygobites | Animals that live their whole lives in groundwater and cannot survive in surface waters. |
| Tectonic plates | Massive slabs of solid rock varying from a few hundreds to thousands of kilometers across that move on the surface of the earth. |
| Unconsolidated sediment aquifers | Aquifer in which water is stored and transmitted in pores between sand and gravel grains. |
| Vulnerability, intrinsic | The risk of an aquifer to be affected by surface water intrusion (and contaminants), mainly depending on geology and soils. |

# Acknowledgments

We thank P. Marmonier, C. Griebler and F. Malard for constructive comments that improved this chapter.

# References

Abell, R., Thieme, M.L., Revenga, C., Bryer, M., Kottelat, M., Bogutskaya, N., Coad, B., Mandrak, N., Contreras Balderas, S., Bussing, W., Stiassny, M.L.J., Skelton, P., Allen, G.R., Unmack, P., Naseka, A., Ng, R., Sindorf, N., Robertson, J., Armijo, E., Higgins, J.V., Heibel, T.J., Wikramanayake, E., Olson, D., López, H.L., Reis, R.E., Lundberg, J.G., Sabaj Pérez, M.H., Petry, P., 2008. Freshwater ecoregions of the world: a new map of biogeographic units for freshwater biodiversity conservation. BioScience 58 (5), 403—414.

Bakalowicz, M., 2005. Epikarst. In: Culver, D.C., White, W.B. (Eds.), Encyclopedia of Caves. Academic Press, Boston, MA, pp. 220—223.

Beddows, P.A., 2004. Yucatan phreas, Mexico. In: Gunn, J. (Ed.), Encyclopedia of Caves and Karst Science. Fitzroy Dearborn, New York, NY, pp. 786—788.

Bork, J., Berkhoff, S., Bork, S., Hahn, H.-J., 2009. Using subsurface metazoan fauna to indicate groundwater—surface water interactions in the Nakdong River floodplain, South Korea. Hydrogeology Journal 17, 61—75.

Botosaneanu, L., 1986. Stygofauna Mundi. A Faunistic, Distributional and Ecological Synthesis of the World Fauna Inhabiting Subterranean Waters (Including the Marine Interstitial). Brill, Backhuys, Leiden.

Boulton, A.J., 2020. Conservation of groundwaters and their dependent ecosystems: integrating molecular taxonomy, systematic reserve planning and cultural values. Aquatic Conservation: Marine and Freshwater Ecosystems 30, 1—7.

Boulton, A.J., Scarsbrook, M.R., Quinn, J.M., Burrell, G.P., 1997. Land-use effects on the hyporheic ecology of five small streams near Hamilton, New Zealand. New Zealand Journal of Marine & Freshwater Research 31 (5), 609—622.

Brancelj, A., 2002. Microdistribution and high diversity of Copepoda (Crustacea) in and small cave in central Slovenia. Hydrobiologia 477, 59–72.

Brancelj, A., 2015a. Jama Velika Pasica — zgodovina, okolje in življenje v njej/The Velika Pasica Cave — The History, Environment and Life in it. ZRC Publishing and National Institute of Biology, Ljubljana, p. 110.

Brancelj, A., 2015b. Two new stygobiotic copepod species from the Tibesti area (Northern Chad) and a re-description of *Pilocamptus schroederi* (van Douwe, 1915). Zootaxa 3994 (4), 531–555.

Brancelj, A., Culver, D.C., 2005. Epikarstic communities. In: Culver, D.C., White, W.B. (Eds.), Encyclopedia of Caves. Academic Press, Boston, MA, pp. 223–229.

Brancelj, A., Mori, N., Treu, F., Stoch, F., 2020. The groundwater fauna of the Classical Karst: hydrogeological indicators and descriptors. Aquatic Ecology 54, 205–224.

Cardenas, M.B., 2015. Hyporheic zone hydrologic science: a historical account of its emergence and a prospectus. Water Resources Research 51, 3601–3616.

Connor, R., 2015. The United Nations World Water Development Report 2015: Water for a Sustainable World. UNESCO Publishing.

Cornu, J.-F., Eme, D., Malard, F., 2013. The distribution of groundwater habitats in Europe. Hydrogeology Journal 21, 949–960.

Culver, D.C., Pipan, T., 2009. The Biology of Caves and Other Subterranean Habitats. Oxford University Press, New York, NY, p. 254.

Danielopol, D.L., Pospisil, P., 2001. Hidden biodiversity in the groundwater of the Danube flood plain national park (Austria). Biodiversity & Conservation 10, 1711–1721.

Danielopol, D.L., Gibert, J., Griebler, C., Gunatilaka, A., Hahn, H.-J., Messana, G., Notenboom, J., Sket, B., 2004. Incorporating ecological perspectives in European groundwater management policy. Environmental Conservation 31, 1–5.

Datry, T., Malard, F., Gibert, J., 2005. Response of invertebrate assemblages to increased groundwater recharge rates in a phreatic aquifer. Journal of the North American Benthoogical Society 24, 461–477.

Décamps, H., Naiman, R.J., 1990. Towards an ecotone perspective. In: Naiman, R.J., Décamps, H. (Eds.), Ecology and Management of Aquatic-Terrestrial Ecotones. Man and the Biosphere Series. UNESCO Paris & Parthenon Publishing, Carnforth, pp. 1–5.

Di Lorenzo, T., Murolo, A., Fiasca, B., Di Camillo, A.T., Di Cicco, M., Galassi, D.M.P., 2019. Potential of a trait-based approach in the characterization of an N-contaminated alluvial aquifer. Water 11, 2553.

Dole-Olivier, M.-J., Marmonier, P., Creuzé des Châtelliers, M., Martin, D., 1994. Interstitial fauna associated with the alluvial floodplains of the Rhône River (France). In: Gibert, J., Danielopol, D.L., Stanford, J.A. (Eds.), Groundwater Ecology. Academic Press, San Diego, CA, pp. 313–346.

Dole-Olivier, M.-J., Malard, F., Martin, D., Lefébure, T., Gibert, J., 2009. Relationships between environmental variables and groundwater biodiversity at the regional scale. Freshwater Biology 54, 797–813.

Dudgeon, D., Arthington, A.H., Gessner, M.O., Kawabata, Z.-I., Knowler, D.J., Leveque, C., Naiman, R.J., Prieur-Richard, A.-H., Soto, D., Melanie, L.J., Stiassny, M.L.J., Sullivan, C.A., 2006. Freshwater biodiversity: importance, threats, status and conservation challenges. Biological Reviews 81, 163–182.

Dunning, J.B., Danielson, B.J., Pulliam, H.R., 1992. Ecological processes that affect populations in complex landscapes. Oikos 65, 169–175.

Eberhard, S.M., Halse, S.A., Humphreys, W.F., 2005. Stygofauna in the Pilbara region, north-west Western Australia: a review. Journal of the Royal Society of Western Australia 88, 167–176.

Eberhard, S.M., Halse, S.A., Williams, M.R., Scanlon, M.D., Cocking, J., Barron, H.J., 2009. Exploring the relationship between sampling efficiency and short-range endemism for groundwater fauna in the Pilbara region, Western Australia. Freshwater Biology 54, 885–901.

Enckell, P.H., 1968. Oxygen availability and microdistribution of interstitial mesofauna in Swedish fresh-water sandy beaches. Oikos 19 (2), 271–291.

Ferreira, J.P.L., Oliveira, M.M., 2004. Groundwater vulnerability assessment in Portugal. Geofísica Internacional 43 (4), 541–550.

Ferreira, D., Malard, F., Dole-Olivier, M.-J., Gibert, J., 2005. Hierarchical patterns of obligate groundwater biodiversity in France. In: Gibert, J. (Ed.), Proceedings of an International Symposium on World Subterranean Biodiversity. University Claude Bernard, Lyon, Lyon, France, pp. 5–78.

Ferreira, D., Malard, F., Dole-Olivier, M.-J., Gibert, J., 2007. Obligate groundwater fauna of France: diversity patterns and conservation implications. Biodiversity & Conservation 16, 567–596.

Ficetola, G.F., Canedoli, C., Stoch, F., 2019. The Racovitzan Impediment and the hidden diversity of unexplored environments. Conservation Biology 33, 214–216.

Fiers, F., Ghenne, V., 2000. Cryprozoic copepods from Belgium. diversity and biogeographic implications. Belgian Journal of Zoology 130 (1), 11–19.

Galassi, D.M.P., Stoch, F., Fiasca, B., Di Lorenzo, T., Gattone, E., 2009. Groundwater biodiversity patterns in the Lessinian Massif of northern Italy. Freshwater Biology 54, 830–847.

German Federal Agency for Nature Conservation (Bundesamt für Naturschutz, BfN), 2008. Daten zur Natur 2008. Bundesamt für Naturschutz. Münster (Landwirtschaftsverlag):10–11.

Gibert, J., 1991. Groundwater systems and their boundaries: conceptual framework and prospects in groundwater ecology. Verhandlung der Internatational Verein der Limnologie 24, 1605–1608.

Gibert, J., Deharveng, L., 2002. Subterranean ecosystems: a truncated functional biodiversity. BioScience 52 (6), 473–481.

Gibert, J., Danielopol, D.L., Stanford, J.A. (Eds.), 1994. Groundwater Ecology. Academic Press, San Diego, CA.

Gibert, J., Mathieu, J., Fournier, F. (Eds.), 1997. Groundwater/Surface Ecotones: Biological and Hydrological Interactions and Management Options. Cambridge University Press, Cambridge.

Gibert, J., Culver, D.C., Dole-Olivier, M., Malard, F., Christman, M.C., Deharveng, L., 2009. Assessing and conserving groundwater biodiversity: synthesis and perspectives. Freshwater Biology 54, 930–941.

Gogu, R.C., Carabin, G., Hallet, V., Peters, V., Dassargues, A., 2001. GIS-based hydrogeological databases and groundwater modeling. Hydrogeology Journal 9, 555–569.

Goldscheider, N., Hunkeler, D., Rossi, P., 2006. Microbial biocenoses in pristine aquifers and an assessment of investigation methods. Hydrogeology Journal 14, 926–941.

Griebler, C., Avramov, M., 2014. Groundwater ecosystem services: a review. Freshwater Science 34 (1), 355–367.

Griebler, C., Avramov, M., Hose, G., 2019. Groundwater ecosystems and their services: current status and potential risks. In: Schröter, M., Bonn, A., Klotz, S., Seppelt, R., Baessler, C. (Eds.), Atlas of Ecosystem Services. Springer, Cham, pp. 197–203.

Hahn, H.-J., 2006. The GW-fauna-index: a first approach to a quantitative ecological assessment of groundwater habitats. Limnologia 36, 119–139.

Hahn, H.-J., 2009. A proposal for an extended typology of groundwater habitats. Hydrogeology Journal 17, 77–81.

Hahn, H.-J., Fuchs, A., 2009. Distribution patterns of groundwater communities across aquifer types in southwestern Germany. Freshwater Biology 54, 848–860.

Hahn, H.-J., Matzke, D., 2005. A comparison of stygofauna communities inside and outside groundwater bores. Limnologica 35, 31–44.

Hahn, H.-J., Schweer, C., Griebler, C., 2018. Grundwasserökosysteme im Recht? – Eine kritische Betrachtung zur rechtlichen Stellung von Grundwasserökosystemen. Grundwasser. Bd. 23. H. 3. Springer-Verlag GmbH, Heidelberg, pp. 209–218.

Hancock, P.J., Boulton, A.J., 2008. Stygofauna biodiversity and endemism in four alluvial aquifers in eastern Australia. Invertebrate Systematics 22, 117–126.

Heiss, L., Bouchaou, L., Tadoumant, S., Reichert, B., 2020. Index-based groundwater vulnerability and water quality assessment in the arid region of Tata City (Morroco). Groundwater for Sustainable Development 10, 100344.

Hölting, B., Coldewey, G., 2013. Einführung in die Allgemeine und Angewandte Hydrogeologie. Elsevier, München. Spektrum Akad. Verlag.

Humphreys, W.F., 1999. Relict stygofaunas living in sea salt, karst and calcrete habitats in arid northwestern Australia contain many ancient lineages. In: Ponder, W., Lamney, D. (Eds.), The Other 99%: The Conservation and Biodiversity of Invertebrates. Transactions of the Royal Zoological Society of New South Wales, pp. 219–227.

Humphreys, W.F., 2000. The hypogean fauna of the cape range peninsula and Barrow island, northwestern Australia. In: Wilkens, H., Culver, D.C., Humphreys, W.F. (Eds.), Subterranean Ecosystems. Ecosystems of the World Volume 30. Elsevier, Amsterdam, pp. 581–602.

Humphreys, W.F., 2009. Hydrogeology and groundwater ecology: does each inform the other? Hydrogeology 17, 5–21.

Husmann, S., 1966. Versuch einer/Skologischen Gliederung des interstitiellen Grundwassers in Lebensbereiche eigener Pragung. Archiv fur Hydrobiologie 62, 231–268.

Husmann, S., 1967. Klassifizierungmariner, brackiger, und limnischer Grundwasserbiotope. Helgoländer wisennschaftliche Meeresuntersuchungen 16, 271–278.

I. Setting the scene: groundwater as ecosystems

Illies, J. (Ed.), 1978. Limnofauna Europaea. Gustav Fischer Verlag, New York, NY.

Jakeman, A.J., Barreteau, O., Hunt, R.J., Rinaudo, J.-D., Ross, A. (Eds.), 2016. Integrated Groundwater Management Concepts, Approaches and Challenges. Springer International Publishing, Cham.

Johns, T., Jones, I., Maurice, L., Wood, P., Robertson, A.L., 2015. Regional scale drivers of groundwater faunal distributions. Freshwater Science 34, 316–328.

Karanovic, T., 2004. Subterranean Copepoda from arid western Australia. Crustaceana Monographs 3, 1–366.

Karanovic, T., 2006. Subterranean copepods (Crustacea, Copepoda) from the Pilbara region in western Australia. Records of the Western Australian Museum 70, 1–239. Supplement.

Knight, L., 2001. The occurrence of *Niphargus glenniei* (Crustacea: Amphipoda: Niphargidae) in west Cornwall. Cave and Karst Science 28, 43–44.

Krause, S., Hannah, D.M., Fleckenstein, J.H., Heppell, C.M., Kaeser, D., Pickup, R., Pinay, G., Robertson, A.L., Wood, P.J., 2011. Inter-disciplinary perspectives on processes in the hyporheic zone. Ecohydrology 4, 481–499.

Kresic, N., 2012. Water in Karst: Management, Vulnerability, and Restoration, first ed. McGraw-Hill Education, New York, NY, p. 736.

Kureck, A., 1967. Über die tagesperiodische Ausdrift von *Niphargus aquilex schellenbergi* Karaman aus Quellen. Zeitschrift für Morphologie und Ökologie der Tiere 58, 247–262.

Lopes-Lima, M., Burlakova, L.E., Karatayev, A.Y., Mehler, K., Seddon, M., Sousa, R., 2018. Conservation of freshwater bivalves at the global scale: diversity, threats and research needs. Hydrobiologia 810, 1–14.

Malard, F., Tockner, K., Dole-Olivier, M.J., Ward, J.V., 2002. A landscape perspective of surface–subsurface hydrological exchanges in river corridors. Freshwater Biology 47 (4), 621–640.

Malard, F., Boutin, C., Camacho, A.I., Ferreira, D., Michel, G., Sket, B., Stoch, F., 2009. Diversity patterns of stygobiotic crustaceans across multiple spatial scales in Europe. Freshwater Biology 54, 756–776.

Mammola, S., Cardoso, P., Culver, D.C., Deharveng, L., Ferreira, R.L., Fišer, C., Galassi, D.M.P., Griebler, C., Halse, S., Humphreys, W.F., Isaia, M., Malard, F., Martinez, A., Moldovan, O.T., Niemiller, M.L., Pavek, M., Reboleira, A.S.P.S., Souza-Silva, M., Teeling, E.C., Wynne, J.J., Zagmajster, M., 2019. Scientists' warning on the conservation of subterranean ecosystems. BioScience 69, 641–650.

Mammola, S., Amorim, I.R., Bichuette, M.E., Borges, P.A.V., Cheeptham, N., Cooper, S.J.B., Culver, D.C., Deharveng, L., Eme, D., Ferreira, R.L., Fišer, C., Fišer, Ž., Fong, D.W., Griebler, C., Jeffery, W.R., Jugovic, J., Kowalko, J.E., Lilley, T.M., Malard, F., Manenti, R., Martínez, A., Meierhofer, M.B., Niemiller, M.L., Northup, D.E., Pellegrini, T.G., Pipan, T., Protas, M., Reboleira, A.S.P.S., Venarsky, M.P., Wynne, J.J., Zagmajster, M., Cardoso, P., 2020. Fundamental research questions in subterranean biology. Biological Reviews 95, 1855–1872.

Markovic, D., Carizo, S.F., Kärcher, O., Walz, A., David, J.N.W., 2017. Vulnerability of European freshwater catchments to climate change. Global Change Biology 23, 3567–3580.

Martin, P., De Broyer, C., Fiers, F., Michel, G., Sablon, R., Wouters, K., 2009. Biodiversity of Belgian groundwater fauna in relation to environmental conditions. Freshwater Biology 54, 814–829.

Maurice, L., Bloomfield, J., 2012. Stygobitic invertebrates in groundwater: a review from a hydrogeological perspective. Freshwater reviews 5 (1), 51–71.

Mihevc, A., 2004. Škocjanke Jama, Slovenia. In: Gunn, J. (Ed.), Encyclopedia of Caves and Karst Science. Fitzroy Dearborn, New York, NY, pp. 653–655.

Mori, N., Kanduč, T., Opalički Slabe, M., Brancelj, A., 2015. Groundwater drift as a tracer for identifying sources of spring discharge. Ground Water 53, 123–132.

Mösslacher, F., 2003. Evolution, Adaption und Verbreitung. In: Griebler, C., Mösslacher, F. (Eds.), Grundwasserökologie. Facultas Universitätsverlag, Vienna, pp. 209–251.

Olson, D.M., Dinerstein, E., Wikramanayake, E.D., Burgess, N.D., Powell, G.V.N., Underwood, E.C., D amico, J.A., Itoua, I., Strand, H.E., Morrison, J.C., Loucks, C.J., Allnutt, T.F., Ricketts, T.H., Kura, Y., Lamoreux, J.F., Wettengel, W.W., Hedao, P., Kassem, K.R., 2001. Terrestrial ecoregions of the world: a new map of life on Earth: a new global map of terrestrial ecoregions provides an innovative tool for conserving biodiversity. BioScience 51 (11), 933–938.

Peralta-Maraver, I., Galloway, J., Posselt, M., Arnon, S., Reiss, J., Lewandowski, J., Robertson, A.L., 2018. Environmental filtering and community delineation in the streambed ecotone. Scientific Reports 8, 1–11.

Pielou, E.C., 1979. Biogeography. John Wiley and Sons, New York, NY.

Pipan, T., Culver, D.C., 2022. Epikarst: an important aquatic and terrestrial habitat. In: Mehner, T., Tockner, K. (Eds.), Encyclopedia of Inland Waters, second ed. Elsevier, pp. 437–448.

Pipan, T., Blejec, A., Brancelj, A., 2006. Multivariate analysis of copepod assemblages in epikarstic waters of some Slovenian caves. Hydrobiologia 559, 213—223.

Ravbar, N., 2007. The Protection of Karst Waters. A Comprehensive Slovene Approach to vulnerability and Contamination Risk Mapping. Založba ZRC, Ljubljana, p. 254.

Retter, A., Karwautz, C., Griebler, C., 2021. Groundwater microbial communities in times of climate change. Current Issues in Molecular Biology 41, 509—538.

Risser, P.G., 1990. The ecological importance of land-water ecotones. In: Naiman, R.J., Décamps, H. (Eds.), Ecology and Management of Aquatic-Terrestrial Ecotones. Man and the Biosphere Series. UNESCO Paris & Parthenon Puplishing, Carnforth, pp. 7—21.

Robertson, A.L., Smith, J.W.N., Johns, T., Proudlove, G.S., 2009. The distribution and diversity of stygobites in Great Britain: an analysis to inform groundwater management. The Quarterly Journal of Engineering Geology and Hydrogeology 42, 359—368.

Saccò, M., Blyth, A., Bateman, P.W., Hua, Q., Mazumder, D., White, N., Humphreys, W.F., Laini, A., Griebler, C., Grice, K., 2019. New light in the dark — a proposed multidisciplinary framework for studying functional ecology of groundwater fauna. Science of the Total Environment 662, 963—977.

Schmidt, S.I., Hahn, H.-J., 2012. What is groundwater and what does this mean to fauna? An opinion. Limnologica 42, 1—6.

Schubert, R., 1985. Bioindikation in Terrestrischen Okosystemen. VEB Gustav Fischer Verlag, Jena, 327 pp.

Sinreich, M., 2009. Konzept der Vulnerabilität im Grundwasserschutz — Anwendung auf die Verhältnisse der Schweiz. Gas, Wasser, Abwasser 2/09, 109—217.

Sket, B., 2004. Subterranean habitats. In: Gunn, J. (Ed.), Encyclopedia of Caves and Karst Science. Fitzroy Dearborn, New York, NY, pp. 709—713.

Sket, B., 2005. Dinaric karst, diversity in. In: Culver, D.C., White, W.B. (Eds.), Encyclopedia of Caves. Academic Press, Boston, MA, pp. 158—165.

Stein, H., Griebler, C., Berkhoff, S., Matzke, D., Fuchs, A., Hahn, H.-J., 2012. Stygoregions — a promising approach to a bioregional classification of groundwater systems. Scientific Reports 2, 1—9.

Stoch, F., Galassi, D.M.P., 2010. Stygobiotic crustacean species richness: a question of numbers, a matter of scale. Hydrobiologia 653, 217—234.

Stoch, F., Malard, F., Castellarini, F., Dole-Olivier, M.-J., Gibert, J., 2004. PASCALIS D8 deliverable for workpackage 7: statistical analyses and identification of indicators. In: European Project Protocols for the Assessment and Conservation of Aquatic Life in the Subsurface (Contract No. EVK-CT-2001-00121), pp. 1—156.

Stoch, F., Fiasca, B., Di Lorenzo, T., Porfirio, S., Petitta, M., Galassi, D.M.P., 2016. Exploring copepod distribution patterns at three nested spatial scales in a spring system: habitat partitioning and potential for hydrological bioindication. Journal of Limnology 75 (1), 1—13.

Strayer, D.L., 1985. The benthic micrometazoans of Mirror Lake, New Hampshire. Archiv für Hydrobiologie Supplement 72, 287—426.

Strayer, D.L., 1994. Limits to biological distribution in groundwater. In: Gibert, J., Danielopol, D.L., Stanford, J.L. (Eds.), Groundwater Ecology. Academic Press, San Diego, CA, pp. 287—310.

Strona, G., Fattorini, S., Fiasca, B., Di Lorenzo, T., Di Cicco, M., Lorenzetti, W., Boccacci, F., Galassi, D.M.P., 2019. AQUALIFE software: a new tool for a standardized ecological assessment of groundwater dependent ecosystems. Water 11, 2574.

Struckmeier, W.F., Margat, J., 1995. Hydrogeological Map. A Guide and a Standard Legend, International Association of Hydrogeologists. Publication 17, Heise, Hannover, 177 pp.

Thienemann, A., 1950. Verbreitungsgeschichte der Suesswassertierwelt Europas. Die Binnengewasser. Band XVIII. Schweizerbart'sche Verlagsbuchhandlung, Stuttgart.

Trewick, S., 2017. Plate tectonics in biogeography. In: Richardson, D., Castree, N., Goodchild, M.F., Kobayashi, A.W., Liu, W., Marston, R.A. (Eds.), The International Encyclopedia of Geography. John Wiley & Sons, Ltd, pp. 1—9.

Udvardy, M.D.F., 1975. A Classification of the Biogeographical Provinces of the World. International Union of Conservation of Nature and Natural Resources. IUCN Occasional Paper no. 18, Morges (Switzerland).

Van den Berg-Stein, S., Thielsch, A., Schenk, K., Hahn, H.J., 2019. StygoTracing — Ein biologisches Tracerverfahren für Grund- und Trinkwasser. Erste Erfahrungen mit biologischen Tracern im Grund- und Trinkwasser auf unterschiedlichen räumlichen Skalen. Korrespondenz Wasserwirtschaft 12, 4.

Vorosmarty, V.C.J., McIntyre, P.B., Gessner, M.O., Dudgeon, D., Prusevich, A., Green, P., Glidden, S., Bunn, S.E., Sullivan, C.A., Reidy Liermann, C., Davies, P.M., 2010. Global threats to human water security and river biodiversity. Nature 467, 555–561.

Weitowitz, D., Maurice, L., Lewis, M., Bloomfield, J., Reiss, J., Robertson, A.L., 2017. Defining geo-habitats for groundwater ecosystem assessments: an example from England and Wales (UK). Hydrogeology Journal 25, 2453–2466.

White, D.S., 1993. Perspectives on defining and delineating hyporheic zones. Journal of the North American Benthological Society 12, 61–69.

White, W.B., Culver, D.C., 2005. Cave, definition of. In: Culver, D.C., White, W.B. (Eds.), Encyclopedia of Caves. Academic Press, Boston, MA, pp. 223–229.

Whitman, R.L., Andrzejewski, M.C., Kennedy, K.J., Sobat, T.A., 1994. Composition, spatial-temporal distribution and environmental factors influencing the interstitial beach meiofauna of southern Lake Michigan. SIL Proceedings 25 (3), 1389–1397.

Williams, D.D., Febria, C.M., Wong, J.C.Y., 2010. Ecotonal and other properties of the hyporheic zone. Fundamental and Applied Limnology 176, 349–364.

Wohlrab, B., Ernstberger, H., Meuser, A., 1992. Landschaftswasserhaushalt Wasserkreislauf und Gewässer im ländlichen Raum. Veränderungen durch Bodennutzung, Wasserbau und Kulturtechnik, Parey, p. 352.

Zagmajster, M., Eme, D., Fišer, C., Galassi, D., Marmonier, P., Stoch, F., Cornu, J.F., Malard, F., 2014. Geographic variation in range size and beta diversity of groundwater crustaceans: insights from habitats with low thermal seasonality. Global Ecology and Biogeography 23, 1135–1145.

# Physical and biogeochemical processes of hyporheic exchange in alluvial rivers

*Daniele Tonina[1] and John M. Buffington[2]*

[1]Center for Ecohydraulics Research, University of Idaho, Boise, ID, United States;
[2]Rocky Mountain Research Station, US Forest Service, Boise, ID, United States

## Introduction

Riverine water flows both above the streambed and within the porous sediment surrounding rivers (Nagaoka and Ohgaki, 1990; Mendoza and Zhou, 1992), ubiquitously entering and exiting the permeable streambed and banks as the flow moves downstream (Boano et al., 2014). This interstitial movement of riverine water is called *hyporheic flow* and forms a saturated volume of sediment surrounding alluvial rivers that is termed the *hyporheic zone* (Fig. 3.1A) (Gooseff, 2010). The extent of the hyporheic zone can be defined through biological, geochemical, or hydraulic means and is a transitional area between surface and ground waters. Hyporheic water is primarily stream water (Vaux, 1968) with limited mixing with the ambient groundwater (Hester et al., 2013, 2017). Stream water enters bed and bank sediments in downwelling areas and emerges downstream in upwelling areas (Edwards, 1998) (Fig. 3.1). The downwelling and upwelling fluxes are collectively referred to as *hyporheic exchange,* which stems from a variety of processes (Fig. 3.1C), including (1) spatial variation of hydraulic head along the boundaries of the channel due to flow—topography interactions and changes in channel dimensions, shape, and slope (Thibodeaux and Boyle, 1987; Harvey and Bencala, 1993; Elliott and Brooks, 1997; Boano et al., 2006; Tonina and Buffington, 2007) (Fig. 3.1C1), (2) spatial changes in the lateral and vertical extent of alluvium (Stanford and Ward, 1993; Baxter and Hauer, 2000; Malcolm et al., 2005) (Fig. 3.1C2 and 1C3), (3) spatial changes in the hydraulic conductivity of alluvium resulting from variation in grain size and depositional history (Salehin et al., 2004; Ward et al., 2011; Herzog et al., 2016) (Fig. 3.1C5), (4) flow turbulence that causes pressure fluctuations along the boundaries of the channel and diffusion of momentum into the near-surface sediment (Blois et al., 2014; Grant et al., 2018; Voermans et al., 2018; Roche et al., 2019; Rousseau and Ancey, 2020; Shen et al., 2020)

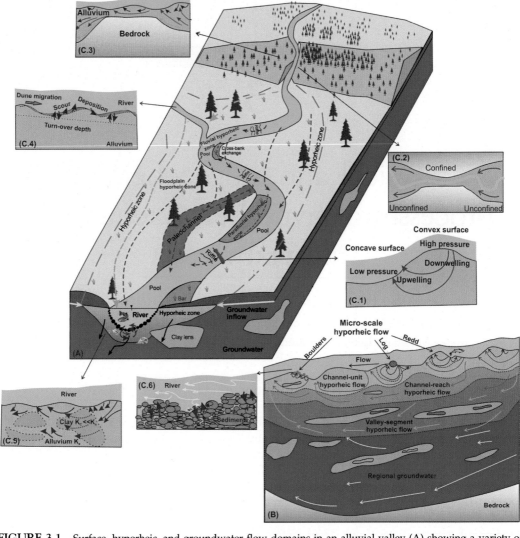

**FIGURE 3.1** Surface, hyporheic, and groundwater flow domains in an alluvial valley (A) showing a variety of hyporheic flow paths that move river water into the surrounding porous alluvium and back again (dashed lines from red to orange illustrating micro to valley hyporheic flows), a process termed *hyporheic exchange*. The spatial extent of this exchange defines the *hyporheic zone*, which can be subdivided into hyporheic flow within the channel (*fluvial hyporheic zone*), below exposed bars surfaces (*parafluvial hyporheic zone*), and through buried paleochannels and between meander bends (*floodplain hyporheic zone*) (A). Hyporheic flow is characterized by nested circulation cells (B, red to orange arrows) driven by successive scales of topography, ranging from small-scale objects in the streamflow (e.g., logs, boulders, biotic mounds), to individual meso-scale bedforms, to reach-scale repeating sequences of bedforms (e.g., pool-riffle morphology), and to valley-scale undulations in subsurface bedrock topography, with correspondingly greater mean residence times (from seconds to years) of exchange for longer flow paths. Hyporheic exchange is distinguished from far-field inflow of groundwater (A and B, yellow arrows) and from one-way outflow of river water (i.e., flow paths that do not circulate water from the river and back again). Horizontal and vertical components of regional groundwater flow further modulate the extent of the hyporheic zone (A and B, yellow arrows). Major mechanisms inducing hyporheic exchange are spatial variation of hydraulic head along the boundaries of the channel due to flow−topography interactions (C1), spatial variation in alluvial area due to lateral changes in valley confinement (C2) and vertical changes in alluvial depth driven by underlying bedrock topography (C3), sediment transport and bedform migration (*turnover exchange*) (C4), spatial variation in hydraulic conductivity of the sediment due to changes in grain size and depositional history (C5), and near-bed turbulence causing pressure fluctuations and diffusion of momentum into the near-surface sediment (C6). *Modified from Buffington and Tonina (2009) and Tonina and Buffington (2009), based on drawings by White (1993), Gibert et al. (1994), Harvey et al. (1996), Edwards (1998), Alley et al. (1999), Baxter and Hauer (2000), Dent et al. (2001), Malard et al. (2002), and Stanford (2006).*

(Fig. 3.1C6), (5) density differences between surface and pore waters due to differences in water temperature and chemistry (Boano et al., 2009), and (6) bedload transport that induces bedform migration in sand-bed rivers (Wolke et al., 2019) such that pore water is released from the eroded sediment and stream water is entrained within the deposited sediment; a phenomenon termed *turnover exchange* (Elliott and Brooks, 1997) (Fig. 3.1C4). The relative importance of these mechanisms of hyporheic exchange depends on stream geometry, discharge, and channel type, which vary along the stream network primarily as a function of slope and drainage area (Buffington and Tonina, 2009; Wondzell and Gooseff, 2013; Magliozzi et al., 2018). Furthermore, hyporheic flow typically exhibits nested scales of circulation cells driven by successive scales of topography, ranging from small-scale objects in the streamflow (e.g., boulders, logs, biotic mounds), to individual meso-scale bedforms (e.g., dunes), to reach-scale sequences of repeating bedforms (e.g., pool-riffle or step-pool morphology), and finally to valley-scale undulations of subsurface bedrock topography (Edwards, 1998; Baxter and Hauer, 2000; Dent et al., 2001; Malard et al., 2002; Poole et al., 2008; Buffington and Tonina, 2009; Stonedahl et al., 2013) (Fig. 3.1B). Similarly, nested scales of circulation occur in plan-view due to hyporheic exchange within the wetted channel (*fluvial hyporheic zone*), below exposed bars surfaces (*parafluvial hyporheic zone*), and through buried paleochannels and between meander bends (*floodplain hyporheic zone*) (Fig. 3.1A). These nested circulation cells are conceptualized in vertical and horizontal domains, but are actually complex three-dimensional flow fields. The vertical (basal) and horizontal (underflow) components of ambient groundwater flow further modulate hyporheic flows, their width and depth, *residence time* (the time stream water spends in the sediment), and discharge (Cardenas and Wilson, 2007b; Boano et al., 2008; Cardenas, 2009a; Fox et al., 2014) (Fig. 3.1A and B). Thus, both vertical and horizontal extents of the hyporheic zone vary spatially due to changes in stream and valley size, channel morphology, sediment characteristics, and temporally due to changes in water-table height, groundwater flow, fluvial discharge, sediment transport, and evolution of channel topography (Fig. 3.1) (Stanford and Ward, 1993; Malard et al., 2002; Stanford, 2006; Buffington and Tonina, 2009; Magliozzi et al., 2018). Hyporheic exchange can be predicted for different channel morphologies and across a range of spatial scales, each of which is characterized by different rates and magnitudes of exchange.

Hyporheic exchange also influences water quality by bringing surface water laden with solutes and suspended particles into sediment interstices (Ren and Packman, 2004a; Harvey et al., 2012; Drummond et al., 2014), where reactive solutes undergo biogeochemical transformations due to biofilms attached to streambed particles or are taken up by organisms dwelling within particle interstices (Janssen et al., 2005) (Fig. 3.2). Products of such transformations are then carried away by hyporheic flows (Bott et al., 1984; Triska et al., 1993) and brought to the surface water by upwelling fluxes (Triska et al., 1989a; Nagaoka and Ohgaki, 1990; Mulholland et al., 2008). These transformations depend on reaction time, water temperature, solute concentrations, flow velocity, and length of the hyporheic flow path (Findlay et al., 1993; Ocampo et al., 2006), all of which are stream-size dependent. In small and intermediate streams (<30 m wide), population densities of microorganisms are higher in the hyporheic zone than in the water column, such that most microbially mediated transformations occur in the hyporheic zone rather than in the water column (Wuhrmann, 1972; Master et al., 2005; Marzadri et al., 2017). In this chapter, we describe operative definitions of the hyporheic zone, models for predicting hyporheic exchange in different channel types and across different spatial scales, and finally the role of hyporheic exchange in water quality.

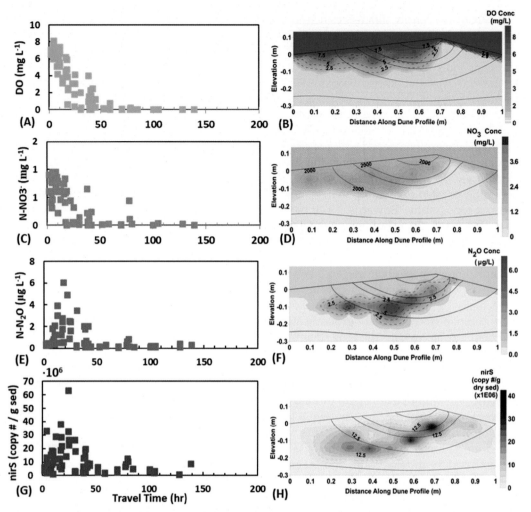

**FIGURE 3.2** Trends of dissolved oxygen (DO) (A), nitrate (NO₃⁻) (C) and nitrous oxide (N₂O) concentrations (E) and abundance of microbial nitrate reductase genes (nirS) associated with denitrifying environments (G) as a function of travel time within the hyporheic zone induced by dune bedforms (9 cm high × 100 cm long) in a laboratory flume. Spatial distributions of each constituent within the dune are respectively shown in (B), (D), (F), and (H), with surface flow from left to right. Solid black lines in (B), (D), (F), and (H) show subsurface hyporheic flow and groundwater flow (lowest flow line). *Modified from Quick et al. (2016) and Reeder et al. (2018a, 2019).*

# The hyporheic zone

The hyporheic zone has been traditionally identified via three operational approaches: biological, geochemical, and hydraulic. The hyporheic zone was first studied by biologists because of its function in carrying oxygen-rich stream waters to salmonid embryos incubating in streambed nests called redds (Stuart, 1954). Subsequently, the identification of a characteristic group of macroinvertebrates, termed *hyporheos*, which were neither benthos nor groundwater fauna (Orghidan, 1959), led to the identification of the hyporheic zone as an ecotone whose extent was delineated by the presence or absence of hyporheos (Edwards, 1998).

In contrast, the geochemical method for defining the hyporheic zone uses the chemical signature of surface water, which is different from that of the ambient groundwater. The approach defines the hyporheic zone as the volume of sediment containing an arbitrary amount of surface water (e.g., 10%) (Triska et al., 1989a; Singh et al., 2020). The selected constituent for geochemical differentiation could be a natural component of the fluid, such as pH, radon (Cranswick et al., 2014), or water temperature (Hatch et al., 2006; Constantz, 2008), or it could be an introduced tracer, such as fluorescein, rhodamine, and various types of chlorides, added to the surface flow (Bencala and Walters, 1983; Castro and Hornberger, 1991; Gooseff and McGlynn, 2005; Ward et al., 2010). Recently, the use of "smart" tracers, which are biologically sensitive, like resazurin that irreversibly transforms to resorufin in the presence of aerobic microbial communities (Haggerty et al. 2009; Argerich et al., 2011; González-Pinzón et al., 2012) help in quantifying the aerobically reactive portion of the hyporheic zone (Briggs et al., 2013; Knapp et al., 2017). Selected constituent concentrations can be measured within the sediments with an array of sampling devices placed within the channel and the surrounding floodplain (Wondzell and Swanson, 1996; Gariglio et al., 2013) from which three-dimensional maps of the hyporheic zone can be generated (Harvey et al., 1996).

Finally, the hydraulic method delineates the hyporheic zone as the assemblage of all downwelling flowlines that reemerge in downstream upwelling areas over a given spatial scale of analysis and are predicted with coupled surface-subsurface hydraulic models (Cardenas and Zlotnik, 2003; Lautz and Siegel, 2006) or quantified from measured piezometric heads in field and flume studies (Winter et al., 1998; Storey et al., 2003).

Whereas the three methods can provide different boundaries due to spatial and temporal changes in physicochemical conditions (e.g., discharge, solute concentration, and water table levels), they all identify the unique characteristic of the hyporheic zone as a transitional area between surface and ground waters. In contrast to early studies, which suggested that the hyporheic zone is a mixing area between surface and subsurface water (i.e., groundwater) (e.g., Edwards, 1998), recent investigations highlighted that the hyporheic zone is surface water only, whose physicochemical signature changes with residence time within the alluvium (Gooseff, 2010), and with limited mixing with the surrounding groundwater (Hester et al., 2013, 2017).

## Predicting hyporheic exchange

Any object, e.g., ripples (Marion et al., 2002), dunes (Elliott and Brooks, 1997), protruding particles (Hutchinson and Webster, 1998; Dudunake et al., 2020), animal and plant mounds (Cooper, 1965; Ziebis et al., 1996; Yuan et al., 2021; Moreto et al., 2022), boulder steps (Hassan et al., 2015), logs (Sawyer et al., 2011), and beaver dams (Briggs et al., 2012), placed in the streamflow will induce a pressure distribution from spatial changes in flow depth and bed elevation (piezometric head) and/or flow velocity (dynamic head), driving local hyporheic downwelling and upwelling, referred to as *pumping exchange*. Similarly, changes in flow direction within channel bends (Cardenas and Zlotnik, 2003) and other sinuous flow paths (Boano et al., 2010a) can induce pressure distributions that form hyporheic flows across the streambed and banks. These mechanisms may be superimposed on one another, i.e., a set of ripples could be on top of a dune within a meandering channel (Stonedahl et al., 2010), forming a nested series of hyporheic flow paths with different spatial scales (Fig. 3.1B). Hyporheic exchange models have been developed for select bedforms and macro-roughness elements at the bedform (Elliott and Brooks, 1997), reach (Wörman et al., 2002; Lautz and Siegel, 2006; Gooseff et al., 2007; Zarnetske et al., 2008), and landscape (Wörman et al., 2006; Stewart et al., 2011; Gomez-Velez and Harvey, 2014) scales.

## Bedform scale

At the bedform scale, Elliott and Brooks (1997) proposed the first Darcy-based "pumping" model for predicting pressure-driven hyporheic exchange in sand-bedded channels with two-dimensional dune-like bedforms. The near-bed head, $h$, was modeled as a regular sinusoidal distribution reflecting periodic drops in pressure due to bedform drag and turbulent energy dissipation along the length of the bedform (Vittal et al., 1977; Shen et al., 1990)

$$h = h_m \sin\left(\frac{2\pi}{\lambda} x\right) \tag{3.1}$$

$$h_m = 0.28 \frac{U^2}{2g} \begin{cases} (0.34 Y^*)^{-3/8}, & Y^{*-1} \le 0.34 \\ (0.34 Y^*)^{-3/2}, & Y^{*-1} > 0.34 \end{cases} \tag{3.2}$$

where $h_m$ is the amplitude of near-bed head variation over the bedform, $x$ is the longitudinal distance, $U$ is the mean flow velocity, $g$ is the gravitational acceleration, $Y^*$ is the dimensionless water depth, defined as the ratio between $Y_0$, the mean hydraulic flow depth, and $\Delta$, the bedform amplitude, and $\lambda$ is the bedform length. The constant 0.28 is also a function of $Y^*$ (Vittal et al., 1977; Fox et al., 2014), which may lead to a certain degree of error when applied beyond the experimental conditions of Shen et al. (1990). The spatially averaged hyporheic flux, $\bar{q}$, through a dune with homogeneous hydraulic conductivity, $K_c$, and a basal (vertical) groundwater flow, $v_{gw}$, at a given alluvial depth, $d_b$, can be quantified as (Boano et al., 2008; Marzadri et al., 2016)

$$\begin{cases} \bar{q}_{H,G} = \dfrac{u_m}{\pi} \sqrt{1 - \dfrac{v_{gw}^2}{u_m^2}} - \dfrac{v_{gw}}{\pi} \arccos\left(\dfrac{v_{gw}}{u_m}\right) \\[4mm] \bar{q}_{H,L} = \dfrac{u_m}{\pi} \sqrt{1 - \dfrac{v_{gw}^2}{u_m^2}} - \dfrac{v_{gw}}{\pi} \arccos\left(\dfrac{v_{gw}}{u_m}\right) + v_{gw} \end{cases} \tag{3.3}$$

$$u_m = K_C \lambda h_m \tanh(\lambda d_b) \tag{3.4}$$

where $\bar{q}_{H,G}$ and $\bar{q}_{H,L}$ are the mean hyporheic fluxes for gaining and loosing reaches, respectively, and $u_m$ is the maximum downwelling velocity. Analytical solutions of the above equations demonstrate that the groundwater underflow (horizontal) velocity, $u_{gw}$, does not affect the magnitude of hyporheic exchange, but does influence its extent and residence time (Boano et al., 2008; Fox et al., 2014; Marzadri et al., 2016). However, the vertical component of groundwater flow, $v_{gw}$, strongly impacts the hyporheic exchange, such that the hyporheic zone is fully suppressed for $v_{gw} > u_m$ in Eq. (3.3) (Marzadri et al., 2016). Predictions from these equations have been confirmed by flume experiments, which mimicked ambient groundwater flows (Fox et al., 2014). The residence time distribution $\overline{R}$ of hyporheic flow weighted over the downwelling flux across a dune can be estimated by the following implicit equation as a function of time, $t$, and sediment porosity, $n$, for the case of $v_{gw} = 0$

$$\frac{4\pi^2 K_C h_m t}{\lambda^2 n} = \frac{2a \cos(\overline{R}) \tanh\left(\frac{2\pi}{\lambda} d_b\right)}{\overline{R}} \tag{3.5}$$

and semianalytical solutions, which account for ambient groundwater flow, are also available (Bottacin-Busolin and Marion, 2010; Marzadri et al., 2016). In all cases, a lognormal distribution provides a good approximation of $\overline{R}$ (Wörman et al., 2002; Marzadri et al., 2016), although it has been suggested that its accuracy is less reliable in describing the tail of the distribution (Haggerty et al., 2002; Cardenas, 2007) and other distributions have been proposed (Grant et al., 2020). The vertical depth of hyporheic exchange, $d_h$, when not confined by a finite depth of alluvium, $d_b$, scales with bedform length, with $d_h \sim 1\,\lambda$ (Wörman et al., 2002), and also depends on the Reynolds number calculated from the mean velocity at the bedform crest, $\Delta$ (Cardenas and Wilson, 2007b). The above results for two-dimensional dunes can be used to approximate those of three-dimensional dunes, especially at high Reynolds numbers (Chen et al., 2015, 2018).

Typically, dunes are not stationary, but migrate at a downstream velocity, $U_b$, causing progressive turnover of the sediment, which simultaneously releases and entraps hyporheic fluid (Fig. 3.2D). The dimensionless turnover number is (Elliott and Brooks, 1997)

$$T_{over} = \frac{n U_b - u_{gw}}{u_m} \tag{3.6}$$

and can be used to estimate the relative dominance of pumping exchange ($T_{over}<0.5$) vs. turnover exchange ($T_{over}>7$) (Elliott and Brooks, 1997), where $U_b$ is multiplied by sediment porosity, $n$, to create a pore-water velocity of the same type as the Darcy velocities $u_{gw}$ and $u_m$. For turnover exchange ($T_{over}>7$), the mean downwelling flux can be quantified as

$$\overline{q} = \frac{Y_0}{\lambda} n U_b \tag{3.7}$$

The maximum $d_h$ for turnover exchange was originally suggested to be limited by the amplitude of the largest migrating dune (Elliott, 1990), such that rapid dune migration reduces the depth of hyporheic exchange, which scales with $\Delta$, compared to the case of pumping exchange, where $d_h$ scales with $\lambda$. Subsequently, it has been proposed that turnover is always impacted by pumping, which advectively moves solute from the well-mixed turnover region to the subsurface sediment, effectively increasing the depth of hyporheic exchange (Packman and Brooks, 2001).

The original Elliott and Brooks model has been expanded for unsteady surface discharge (Boano et al., 2007b; 2010b), for horizontally stratified streambeds having different hydraulic conductivity (Marion et al., 2008a), and for the effects of colloid deposition within the hyporheic zone (Packman et al., 2000; Ren and Packman, 2004b, 2007). In addition to these analytical solutions, numerical simulations also have been used by coupling surface and groundwater models in fluvial (Cardenas and Wilson, 2007a; Janssen et al., 2012) and marine (Cardenas et al., 2008) dunes. Dune morphology quickly adapts to the applied discharge as sand is typically mobile, even at low flows. This suggests that analysis of hyporheic exchange in sand-bed rivers should account for the dependence of dune size on discharge (Boano et al., 2013). For instance, dune height and length in equilibrium with the stream flow can be assumed equal to 0.167 and 6 times $Y_0$, respectively (Yalin, 1964). Furthermore, sand-bed rivers can exhibit a broad range of bed topography that varies systematically with discharge and bed load transport rate, transitioning from plane-bed to ripples to dunes and antidunes (e.g., Middleton and Southard, 1984), creating complex spatiotemporal changes in hyporheic exchange.

Most solutions for hyporheic flow through dunes have been quantified assuming that hydraulic conductivity, $K_c$, is homogenous (same value throughout the domain) and isotropic (same value in all directions). For such conditions, $K_c$ can be estimated as a function of a reference grain size, for instance the median grain size $d_{50}$ (m) (Salarashayeri and Siosemarde, 2012; Gomez-Velez et al., 2015),

$$K_c = 119.06 d_{50}^{1.62} \qquad (3.8)$$

Homogeneous and isotropic $K_c$ values yield regular, smoothly varying hyporheic flow paths. In contrast, sediment heterogeneity (i.e., spatially variable $K_c$) leads to a seemingly erratic distribution of the interstitial flow field (Hester et al., 2013) and generates hyporheic exchange fluxes due to conservation of mass, causing the velocity and direction of subsurface flow to diverge due to spatial changes in sediment porosity (Vaux, 1968; Ward et al., 2011). The phenomenon can manifest at two distinct scales: (1) that of a single textural facies (continuous heterogeneity) (Hester et al., 2013), which is the intrinsic heterogeneity of a given porous material (e.g., sand) due to random spatial changes in grain size and porosity within a given medium and (2) that of multiple, juxtaposing, textural facies at larger spatial scales (categorical heterogeneity) (Winter and Tartakovsky, 2002; Riva et al., 2006; Pryshlak et al., 2015), forming a composite porous medium (e.g., sand and gravel layers, interspersed with clay lenses).

Recent investigations for typical degrees of continuous heterogeneity of hydraulic conductivity in sand-bed streams (standard deviation of the log transformed $K_c$, i.e., $\sigma_Y^2 < 0.6$, where $Y = ln(K_c)$), show that hyporheic exchange statistics at the bedform scale (e.g., residence time and mean fluxes) can be reasonably approximated using homogeneous $K_c$ values (Tonina et al., 2016). Heterogeneity causes the hyporheic zone to be compressed toward the stream-water interface and an increase in tortuosity of hyporheic flow lines. The former effect reduces the length of the long flow lines (and their residence times), while the latter effect increases the length (and thus the residence times) of the short flow lines with respect to the case of homogeneous hydraulic conductivity. Because the hyporheic zone is strongly dominated by boundary conditions, especially the pressure profile at the water–sediment interface, heterogeneity may reduce the ensemble means of the hyporheic flow (e.g., residence time and fluxes) (Tonina et al., 2016) or increase them (Laube et al., 2018), depending on surface morphodynamic conditions and the consequent pressure distribution at the water–sediment interface. Anisotropy tends to reduce the uncertainty around the ensemble value, namely the difference in flow paths among all the possible heterogeneous realizations of the $K_c$-field. Investigations of categorical heterogeneity (flow through multiple textural facies) show similar results of compression of the hyporheic zone, but in general report a net result of reducing the hyporheic residence time and increasing the hyporheic discharge with increasing heterogeneity of hydraulic conductivity (Sawyer and Cardenas, 2009; Gomez-Velez et al., 2014; Pryshlak et al., 2015; Laube et al., 2018).

Modeling of hyporheic exchange in gravel-bed rivers having a pool-riffle morphology (e.g., Montgomery and Buffington 1997) is more complex than dunes because the bedforms are three-dimensional and may be fully (Marzadri et al., 2010; Trauth et al., 2013) or partially (Tonina and Buffington, 2007; Monofy and Boano, 2021) submerged. For the fully-submerged case, semianalytical models show that the mean and variance of the residence time distribution ($\mu_t$ and $\sigma_t$, respectively) depend on streambed morphology and surface hydrology, such that (Marzadri et al., 2010)

$$\mu_t^* = 1.39\ Y^{*0.6};\ R^2 = 0.96$$
$$\mu_t = \frac{\lambda \mu_t^*}{K_C s_i C_z} \tag{3.9}$$

$$\sigma_t^{2*} = 2.07\ Y^{*0.89};\ R^2 = 0.91$$
$$\sigma_t^2 = \left(\frac{\lambda}{K_C s_i C_z}\right)^2 \sigma_t^{2*} \tag{3.10}$$

where $\mu_t^*$ and $\sigma_t^{2*}$ are the dimensionless mean and variance of the residence time distribution, respectively, $s_i$ is the streambed slope, and $C_z$ is the dimensionless Chézy number

$$C_z = 6 + 2.5 \ln\left(\frac{1}{2.5\ d_s}\right) \tag{3.11}$$

with $d_s$ ($=d_{50}/Y_0$) being the relative submergence of the sediment. The full hyporheic residence time can be approximated from the mean and variance, assuming a lognormal distribution (Marzadri et al., 2010; Tonina and Buffington, 2011). Similarly, the spatially averaged downwelling flow is estimated as (Tonina, 2012)

$$\overline{q}^* = 41.108\ Y^{*-0.732};\ R^2 = 0.96$$
$$\overline{q} = \frac{\overline{q}^* K_C}{C_z^4} \tag{3.12}$$

Semiempirical relationships, which account for the role of ambient groundwater (both basal (vertical) and underflow (horizontal)) are also available for fully submerged pool-riffle bedforms (Trauth et al., 2013).

For the case of partially submerged bars in gravel pool-riffle channels, the prediction of hyporheic exchange is based on empirical regressions (Tonina and Buffington, 2011; Trauth et al., 2013; Huang and Chui, 2018; Monofy and Boano, 2021), using a set of dimensional numbers such as those suggested by Tonina and Buffington (2011).

$$\Pi'_1 = \frac{\rho U Y_0}{\mu}; \ \Pi'_2 = \frac{U}{\sqrt{g Y_0}}; \ \Pi'_3 = \frac{\Delta}{Y_0}; \ \Pi'_4 = \frac{\lambda}{Y_0}; \ \Pi'_5 = s_i; \tag{3.13}$$

$$\Pi_1 = \frac{\rho U \sqrt{K}}{\mu}; \ \Pi_2 = \frac{U^2}{g\sqrt{K}}; \ \Pi_3 = \frac{Y_0}{\sqrt{K}}; \ \Pi_4 = s_i; \ \Pi_5 = \frac{\Delta}{\lambda}; \ \Pi_6 = \frac{d_b}{\sqrt{K}}; \tag{3.14}$$

where $\rho$ is the fluid density, $\mu$ is the dynamic fluid viscosity, and $K$ is the sediment permeability. The hyporheic exchange depth, $d_h$, and the mean, $\mu_t$, and standard deviation, $\sigma_t$, of the hyporheic residence time distribution are then quantified as

$$\frac{d_h}{Y_0} = \exp\left( \sum_{i=1}^{5} A_i \ln(\Pi'_i) \right) \tag{3.15}$$

$$\frac{\mu_t U}{\sqrt{K}} = \exp\left( \sum_{i=1}^{6} B_i \ln(\Pi_i) \right) \tag{3.16}$$

$$\frac{\sigma_t U}{\sqrt{K}} = \exp\left( \sum_{i=1}^{6} B'_i \ln(\Pi_i) \right) \tag{3.17}$$

and the regression coefficients are reported in Table 3.1.

In contrast to dunes, pool-riffle sequences form at high flows near bankfull stage (e.g., Montgomery and Buffington, 1997, 1998), when gravel substrates are set in motion (Tubino, 1991). Thus, for applications with discharges lower than bankfull, the streambed morphology

**TABLE 3.1** Empirically determined parameters for, Eqs. (3.15)–(3.17), respectively, as proposed by Tonina and Buffington (2011).

| i | | 1 | 2 | 3 | 4 | 5 | 6 |
|---|---|---|---|---|---|---|---|
| Mean hyporheic depth (3.15) | $A_i$ | 0.152 | −0.058 | −0.509 | 0.074 | 0.906 | |
| Mean hyporheic residence time (3.16) | $B_i$ | −0.682 | 0.387 | 1.619 | 0.314 | −1.339 | 0.407 |
| Standard deviation of the hyporheic residence time (3.17) | $B'_i$ | −0.533 | 0.652 | 1.369 | 0.098 | −1.066 | 0.456 |

in pool-riffle channels can be assumed stationary, while surface hydraulics change with discharge. Bedform length and amplitude can be predicted from bankfull width, $W_b$, and depth, $Y_{0b}$, such that: $\lambda = 6.5W_b$, and $\Delta = Y_{0b}/(0.18d_s^{0.45}\beta^{1.45})$, where $\beta$ is the half-channel aspect ratio $\beta = Y_0/(2W)$ and valid for $2<\beta<35$ (Ikeda, 1984). Mean hydraulic width, $W$, depth, $D$, and velocity, $U$, can be estimated from discharge, $Q$, using hydraulic geometry relations: $W = 12.936Q^{0.423}$ ($R^2 = 0.82$), $D = 0.409Q^{0.294}$ ($R^2 = 0.62$) and $U = 0.196Q^{0.285}$ ($R^2 = 0.49$) (e.g., Raymond et al., 2012) for cases where only discharge is known as applied by Marzadri et al. (2017). In addition to the bar and pool topography of these channels, the protruding riffle volume above the mean streambed elevation may also induce hyporheic exchange. This mechanism is particularly important in low-gradient streams (Bray and Dunne, 2017), which typically have larger riffle amplitudes than steep streams (Marzadri et al., 2010).

The above investigations typically assume homogenous hydraulic conductivities, but gravel-bed rivers tend to have a complex stratigraphy (Dai et al., 2003; Ritzi et al., 2004) that produces chiefly categorical heterogeneity, which has been shown to have important implications for hyporheic exchange (Zhou et al., 2014; Pryshlak et al., 2015) in terms of reducing exchange depths and residence times, but increasing fluxes.

In addition to models developed for dune and pool-riffle channels, semiempirical solutions have been proposed for other in-channel features, such as channel-spanning logs (Sawyer et al., 2011), steps (Hester and Doyle, 2008), and fully-submerged semihemispherical features (e.g., boulders and cobbles) (Hutchinson and Webster, 1998). For sinuous channels, predictive models have been proposed for intrameandering flows (those due to pressure differences between meander bends) (Cardenas, 2009b; Boano et al., 2010a; Kiel and Cardenas, 2014), which typically have much longer residence times and flow paths than those induced by in-channel bedforms. However, predictive models are not yet available for reach-scale hyporheic flows induced by bedforms and cobble/boulder substrate in cascade, step-pool, and plane-bed streams (*sensu* Montgomery and Buffington, 1997), although the mechanics of hyporheic exchange in such channels are understood conceptually (Buffington and Tonina, 2009).

## Reach and landscape scales

Reach-scale hyporheic models were originally developed to improve prediction of in-stream solute transport by accounting for transient storage zones, which are comprised of surface storage zones (e.g., backwaters, eddies, bank alcoves) and subsurface storage (the hyporheic zone), both of which are characterized by slower velocities than the mainstream flow, causing solutes to be retained and then released at later times. These surface and subsurface transient storage zones form concentration curves with long tails that traditional advection-dispersion-reaction equations, ADRE, do not model well. The One-dimensional Transport with Inflow and Storage (OTIS) program was the first to model this phenomenon (Runkel, 1998). OTIS models the surface transport with an ADRE that exchanges solute concentration with a fixed volume of transient storage using a first-order transfer function characterized by two parameters (storage area and exchange rate), plus two additional parameters for the stream water (cross-sectional area and an effective diffusion coefficient) (Runkel, 1998). The four parameters are typically calibrated with measured concentration breakthrough curves derived from surface tracer experiments (Harvey and Wagner, 2000) because of the lack of relationships between stream hydromorphological characteristics and storage zone parameters. Calibration is difficult because the problem suffers from equifinality, as there are four unknowns but only two equations (Mrokowska and Osuch, 2011). OTIS has been applied for simulating hyporheic exchange by assuming that most transient storage chiefly occurs in the hyporheic zone, with negligible storage in surface (in-channel) reservoirs (Bencala and Walters, 1983; Harvey et al., 1996; Harvey and Gorelick, 2000). This has been argued to be true in small steep streams (Mulholland et al., 1997), where most studies using OTIS have been conducted (Gooseff et al., 2006), as opposed to low-gradient channels that commonly have extensive surface storage zones (e.g., pools and backwaters), particularly in meandering floodplain rivers or forest channels with abundant wood debris (Montgomery and Buffington, 1997, 1998). Therefore, the transient storage approach is best applied to streams with few surface storage zones; models with multiple transient storage zones are available, but they are difficult to calibrate (Choi et al., 2000; Stewart et al., 2011). Transient storage models such as OTIS typically result in an exponential residence time distribution (transfer function between the surface flow and the hyporheic zone), which has been shown to poorly capture long exchange paths (i.e., the tail of the residence time distribution) (Zaramella et al., 2003). Nevertheless, transient storage models are a useful first-order approximation that captures the bulk of the hyporheic exchange because most of the exchange occurs in the shallow, near-surface portion of the hyporheic zone and is characterized by rapid exchange along short flow paths that are strongly coupled to surface hydraulics (Harvey and Wagner, 2000; Zaramella et al., 2003; Marzadri et al., 2010).

Other researchers have proposed different forms for the transfer function, which models the interaction between the stream water and the hyporheic zone. The transfer function is typically associated with hyporheic exchange driven by topographic features. For instance, Wörman et al. (2002) modeled the transfer function assuming pumping exchange driven by dune-like bedforms, which generates a lognormal residence time distribution. Other approaches, such as the Solute Transport and Multirate Mass Transfer-Linear Coordinates (STAMMT-L) model (Haggerty and Reeves, 2002) and the Solute Transport in Rivers (STIR) routine (Marion et al., 2008b) select different forms of the hyporheic residence time distribution. Boano et al. (2007a) adopted a continuous time random walk (CTRW) theory, which has been further developed into a double domain approach (one domain for the surface and one for the subsurface processes) (Sherman et al., 2019) for modeling surface and subsurface exchange. These methods have been successfully applied at both reach and valley-segment scales. Stonedahl et al. (2010) explored the role of different topographic scales (e.g., ripples, dunes, and planimetric features such as meanders) on hyporheic exchange as the domain increases from micro to channel-reach scales. They suggested a superposition of the effects of bedform-induced hyporheic exchange and showed the influence of local and large-scale topography on hyporheic exchange. They assumed that dynamic head variation dominates at the micro scale, which they modeled as dune-like bedforms, while piezometric head dominates at the channel-reach scale due to water-surface variations.

Numerical models of hyporheic exchange have typically been applied at channel-unit (individual bedforms) and channel-reach (sequences of bedforms) scales, but Wörman et al. (2006) proposed a generalized three-dimensional model that can be applied at landscape scales. Although, they assumed a sinusoidal streambed pressure distribution throughout the entire river network (which is only suited to dune-ripple channels) and an energy profile approximated by the land topography (rather than that of water-surface elevations), their study is an important attempt to model hyporheic exchange at broad scales using digital elevation models. More recently, Gomez-Velez and Harvey (2014) developed a quantitative landscape perspective of hyporheic exchange along river networks using the Networks with EXchange and Subsurface Storage (NEXSS) model. The approach employs available hyporheic models to quantify mean values of exchange at reach scales resulting from multiple topographic features. Marzadri et al. (2017) proposed a similar method, but their model focusses on the primary topographic feature in a given reach (e.g., dune or pool-riffle bedforms).

## The role of hyporheic flow on water quality

Field observations (Triska et al., 1993; Argerich et al., 2011; Zarnetske et al., 2011) and large-scale flume experiments (Quick et al., 2016; Kaufman et al., 2017; De Falco et al., 2018) have quantified the role of hyporheic fluxes on reactive solutes and on water temperature as this regulates biogeochemical reactions. Hyporheic exchange affects both the chemical (Fig. 3.2) (Zarnetske et al., 2011) and temperature (Wu et al., 2020) regimes of pore water, as well as affecting surface-water characteristics (Triska et al., 1989b; Harvey et al., 2013). Similarly, numerical models that couple hydromorphological and biogeochemical processes using ADRE (Sheibley et al., 2003; Cardenas et al., 2008; Fanelli and Lautz, 2008; Marzadri et al., 2012; Zarnetske et al., 2012) support the importance of the hyporheic zone on reactive solute transformations for both surface and pore waters (Fig. 3.3). These numerical models, which typically solve the groundwater flow based on homogenous hydraulic conductivity and Darcy's flow velocity, are chiefly run at the bedform (Cardenas et al., 2008; Trauth et al., 2013) or stream reach (Fleckenstein et al., 2006; Lautz and Siegel, 2006) scales, where sufficient information about boundary conditions (e.g., streambed topography, discharge, and water stage) are available (e.g., Fig. 3.3A). Results typically include detailed mapping of the hyporheic flow field (magnitude and direction of flow, Fig. 3.3B) and hyporheic residence time (Fig. 3.3C), which have important implications for water quality and locations of upwelling and downwelling fluxes at the streambed surface. Adoption of a Lagrangian reference system, where transport is modeled along flowlines (assuming negligible transversal exchange), simplifies the problem from three dimensional to one dimensional, replacing the horizontal ($x, y$) and vertical ($z$) Cartesian coordinates with travel time, $\tau$, which is the time that a particle spends traversing the hyporheic zone along a given flowline between downwelling into the subsurface and upwelling back into the stream. This Lagrangian approach allows the prediction of hyporheic water quality, including temperature (Fig. 3.3D) and reactive solute concentrations (Fig. 3.3E) along flowlines based on $\tau$ only. Numerical modeling supported by accurate boundary conditions has been shown to provide good predictions of hyporheic processes, including mapping of hyporheic velocity and concentrations of reaction solutes (e.g., Janssen et al., 2012; Cardenas et al., 2016; Reeder et al., 2019; Trauth et al., 2014), although such approaches are limited to small well-characterized domains and well-constrained field applications (Marzadri et al., 2013a; Tonina et al., 2015).

Heterogeneity of streambed sediment can further strengthen the hyporheic zone as a hot spot for biogeochemical transformations, as recent numerical simulations have shown that mixtures of sands and silts have a high capacity to remove nitrate (Pescimoro et al., 2019).

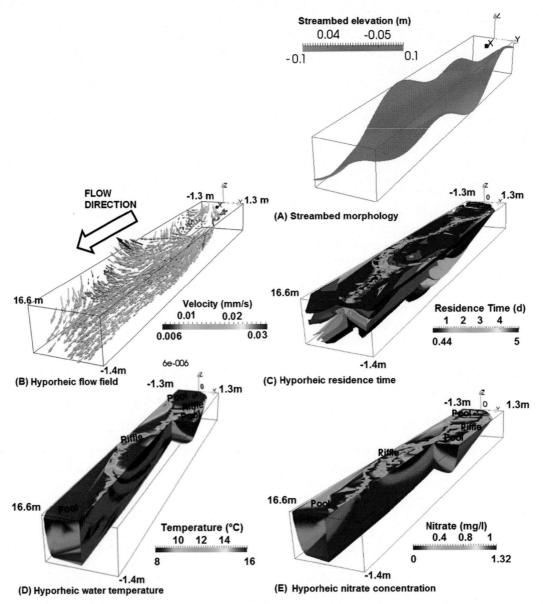

**FIGURE 3.3**   Pool-riffle morphology (A) induced hyporheic flow field (magnitude and direction of flow velocity) (B), residence time (shown as isotemporal curves) (C), water temperature (D), and nitrate concentration (E) simulated with a Lagrangian advection dispersion and reaction (ADRE) model presented by Marzadri et al. (2010, 2011, 2012, 2013a, 2013b). Hyporheic exchange was induced by a fully-submerged pool-riffle morphology (A), with bedform length of 17 m and amplitude of 0.2 m, for a mean streamflow depth of 0.1 m, discharge of 0.18 m$^3$/s, and a sediment thickness of 2.6 m. In-stream concentrations of dissolved oxygen were at saturation and that of nitrates at 1.32 mg/L, while stream water temperature changed sinusoidally through the day, with an amplitude of 8°C and a mean temperature of 12°C. Input parameters and model description are further detailed in Marzadri et al. (2010, 2011, 2012, 2013a, 2013b)

This mechanism could be used to enhance stream nutrient removal in systems where nutrients are problematic (Herzog et al., 2016). Furthermore, the spatial distribution of reaction rate constants may structure biological heterogeneity and the subsequent occurrence of biogeochemical hotspots within the hyporheic zone (Reeder et al., 2018a). However, the importance of the hyporheic zone on surface water quality is expected to be stream-size dependent (Wondzell, 2011), with hyporheic processes dominating biogeochemical reactions in small streams (Naegeli and Uehlinger, 1997; Marzadri et al., 2017) (channel widths, $W$, less than 10 m), but having a lesser effect in intermediate (10 m $< W <$ 100 m) streams, and a negligible influence in larger rivers ($W >$ 100 m) (Marzadri et al., 2022). For instance, recent studies on nitrous oxide emissions from streams and rivers show that the hyporheic zone is the main source of $N_2O$ in small streams, whereas the *benthic zone* (stream bottom and shallow subsurface sediment reached by sunlight) becomes important for larger streams (Marzadri et al., 2017, 2022). For large rivers ($W >$ 100 m), such as the upper Mississippi, biogeochemical reactions within the water column are the main source of $N_2O$ (Marzadri et al., 2020). Four physical causes have been suggested for the reduced importance of the hyporheic zone relative to instream processes in controlling biogeochemical reactions with increasing river size: (1) lower rates of exchange between surface and subsurface flows due to systematic decreases in hydraulic conductivity as one moves downslope from small streams with gravel beds (high permeability) to large rivers with sand and gravel mixtures or sand and silt substrates (low permeability); (2) the above reduction in exchange combined with systematic increases in surface discharge as watercourses become larger causes a smaller ratio of hyporheic-to-instream discharge (Wondzell, 2011), (3) lower reaction-rate constants within streambed sediments, potentially due to low interstitial velocities (Reeder et al., 2018b), and (4) increases in the density and diversity of microbial communities and biofilms in the water column (Liu et al., 2013; Reisinger et al., 2015) due to increases in suspended sediment, which provide the supportive matrix for these organisms (Xia et al., 2009).

Numerical models at bedform and reach scales have highlighted the role of stream topography, stream flow, groundwater flow, and hyporheic fluxes in controlling biogeochemical reactions (Fig. 3.3). However, application of these models at river-network scales is unfeasible. Recent studies, rooted in the Lagrangian approach (Ocampo et al., 2006; Marzadri et al., 2012; Zarnetske et al., 2012; Hampton et al., 2020), suggest that biogeochemical processes may scale with the Damköhler number, $Da$, defined as the ratio between the characteristic time scales of fluid advection and biogeochemical transformation. At the scale of individual flowlines, the advective time is simply the travel time, $\tau$, of the fluid along the flowline (Marzadri et al., 2012). However, at bedform scales, hyporheic flow has a distribution of residence times, which can be indexed by the median value, $\tau_{50}$, to describe overall response to the bedform. Thus, at bedform and reach scales, $Da$ is defined as the ratio between the corresponding $\tau_{50}$ and the time scale of a given biogeochemical process, $\tau_{lim}$, expressed as the inverse of the reaction rate constant, $K_{re}$

$$Da = \frac{\tau_{50}}{\tau_{lim}} = \tau_{50} K_{re} \qquad (3.18)$$

This approach has been shown to be effective for nitrous oxide, $N_2O$, emissions in streams, with potential extension to other biogeochemical reactions (Zarnetske et al., 2011; Marzadri et

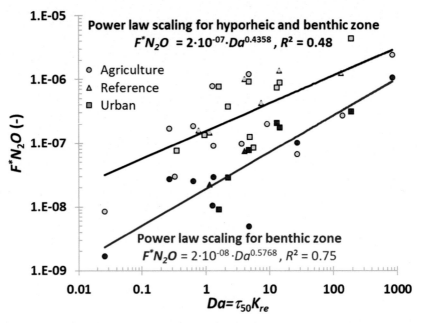

**FIGURE 3.4** Reach-scale dimensionless $N_2O$ fluxes, $F^*N_2O$, generated from the benthic (red) and combined benthic–hyporheic (gray) zones of headwater streams (channel widths less than 10 m) with pool-riffle or dune morphologies as a function of Damköhler number, $Da$, where $\tau_{50}$ is the median residence time of the hyporheic flow and $K_{re}$ is the reaction rate constant of denitrification, defined as the uptake rate normalized by the mean flow depth of the channel. $F^*N_2O$ is defined as the $N_2O$ flux per unit area scaled by the total in-stream flux of a given dissolved inorganic nitrogen species ($NO_3^-$ and $NH_4^+$). Data are from the Kalamazoo River (Michigan) (Beaulieu et al., 2008, 2009) and the second Lotic Intersite Nitrogen eXperiment (LINXII), which includes sites throughout the United States and Puerto Rico, spanning a broad range of land use, land cover, biomes, climatic zones, and channel morphologies (Mulholland et al., 2008). Following Mulholland et al.'s (2008) classification, reference streams had more than 85% native vegetation, while agricultural and urban streams drained watersheds with varying proportions of those land-cover types, respectively. *Modified from Marzadri et al. (2017).*

al., 2012, 2022). In particular, analysis of data from headwater streams shows that the reach-scale dimensionless flux of $N_2O$, $F^*N_2O$, exhibits a power-law scaling with $Da$ that is independent of land use, ecohydromorphology, or climate (Marzadri et al., 2017) (Fig. 3.4). The study showed that the production of $N_2O$ from the benthic layer (red curve in Fig. 3.4) was almost an order of magnitude smaller than that from the combined emissions of the benthic and hyporheic zones in these small streams (black curve in Fig. 3.4). The two empirical power laws were then applied to a set of more than 400 stream reaches worldwide, providing good predictive power (Marzadri et al., 2017).

The above analyses also show that not only are reactant loads (or concentrations) important, but stream morphology and hydraulics can regulate the amount of products generated (Marzadri et al., 2014). This is probably due to the fact that hyporheic exchange depends on stream hydromorphology (e.g., Buffington and Tonina 2009), which, in turn, may specialize the distribution of microbial communities and functioning (Quick et al., 2016). For example, microbes having the nirS gene, associated with denitrifying enzymes, cluster at the interface between the aerobic–anaerobic zone, the position of which is controlled by both flow hydraulics and biogeochemical reaction rates (cf. Fig. 3.2G and H). Overall, local-scale processes within the hyporheic zone, benthic zone, and surface water column may have important effects at the global scale, because they may regulate riverine production of $N_2O$ and, thus, should be accounted for in predicting the impact of anthropogenic activities on biogeochemical processes in riverine systems (e.g., Marzadri et al., 2021).

## Conclusion

Surface water of alluvial rivers ubiquitously flows into and out of the interstitial spaces of the surrounding alluvium, creating a highly connected riverine ecosystem that extends well beyond the visible surface flow of rivers (Poole et al., 2008; Hauer et al., 2016). The extent of these flows depends on local conditions across multiple scales, including bed-surface grain size and heterogeneity, presence of microroughness elements (e.g., logs, boulders, biotic mounds), morphological characteristics of the stream (e.g., meso-scale variation of bed topography, flow depth, width, velocity), reach- to valley-scale morphology (macro-scale variation of bedform sequences, slope, confinement, and alluvial volume), associated geologic constraints, and the position of the groundwater table (Krause et al., 2022). These different scales of physical conditions and constraints form a hierarchy of nested hyporheic cells, with larger scales enveloping smaller ones (Baxter and Hauer, 2000; Malard et al., 2002; Poole et al., 2008; Buffington and Tonina, 2009; Stonedahl et al., 2013). As the size of hyporheic cells increases, so does the residence time of stream water traversing the hyporheic zone, while the average interstitial flow velocity is reduced because pressure gradients become smaller over longer distances between two points. The net result is a set of nested transient storage zones, where biogeochemical reactions transform reactive solutes, with resultant products upwelling into the surface streamflow. These physical and chemical gradients have important implications for water quality in both surface and pore waters. However, the importance of hyporheic exchange on surface water quality is expected to be stream-size dependent, because as stream size increases, the ratio between hyporheic discharge and surface discharge decreases (Wondzell, 2011), diluting hyporheic products within the surface water.

The fading of the importance of hyporheic processes on surface water quality in larger rivers is still under investigation, but the hyporheic zone is expected to remain an important and key element of river corridors even in large rivers as it may provide localized thermal, chemical, and hydraulic refugia for aquatic species, particularly in upwelling zones (Baxter and Hauer, 2000; Malard et al., 2002; Stanford et al., 2005; see review by Wondzell, 2011). In addition, the hyporheic zone has become a key element in stream and river restoration (Hester and Gooseff, 2010; Ward et al., 2011; Herzog et al., 2016), particularly where engineering solutions are required to better design this environment. Overall, a broad range of investigations are exploring the impact of morphological and biological heterogeneities on hyporheic conditions and how hyporheic processes respond to dynamic changes of solutes (Marzadri et al., 2013b; Kaufman et al., 2017) and surface and groundwater hydraulics (Krause et al., 2022). Furthermore, local hyporheic processes are being upscaled to more broadly predict water quality within riverine systems at watershed and regional scales (e.g., to address nutrient removal and greenhouse gas emissions) (Stewart et al. 2011; Gomez-Velez and Harvey, 2014; Marzadri et al., 2017, 2021). These investigations may be supported by novel methodologies (1) in the laboratory, e.g., non-invasive approaches like optical measurements (Cardenas et al., 2016; Reeder et al., 2018b) and image analysis coupled with refractive index matching, RIM, which allows mapping of the flow field through solid grains (Voermans et al., 2018; Moreto et al., 2022) with biological activities (Rubol et al., 2018) and (2) in the field, including optical measurements (Vieweg et al., 2013; Reeder et al., 2019) and improved data analysis (e.g., water temperature time series (van Kampen et al., 2022)) and (3) data assimilation and distribution via machine learning (Marzadri et al., 2021) and numerical simulations (Chowdhury et al., 2020; Shen et al., 2020; Yuan et al., 2021).

## Acknowledgments

We thank Annika Quick, William Jeff Reeder and Alessandra Marzadri for providing data and helping with Figs. 3.2–3.4. Data were collected and analyzed with support from NSF grants 1141690, 1141752, 1344661, 1344602 and IIA-1301792 D.T. acknowledges support from the USDA National Institute of Food and Agriculture, Hatch project 1012806. We thank F. Boano for constructive comments that improved this chapter.

## References

Alley, W.M., Reilly, T.E., Franke, O.L., 1999. Sustainability of Ground-Water Resources. US Geological Survey, Circular 1186, Denver, CO.

Argerich, A., Haggerty, R., Martí, E., Sabater, F., Zarnetske, J., 2011. Quantification of metabolically active transient storage (MATS) in two reaches with contrasting transient storage and ecosystem respiration. Journal of Geophysical Research: Biogeosciences 116, G03034.

Baxter, C.V., Hauer, F.R., 2000. Geomorphology, hyporheic exchange, and selection of spawning habitat by bull trout (*Salvelinus confluentus*). Canadian Journal of Fisheries and Aquatic Sciences 57 (7), 1470–1481. https://doi.org/10.1139/f00-056.

Beaulieu, J.J., Arango, C.P., Hamilton, S.K., Tank, J.L., 2008. The production and emission of nitrous oxide from headwater streams in the Midwestern United States. Global Change Biology 14 (4), 878–894. https://doi.org/10.1111/j.1365-2486.2007.01485.x.

Beaulieu, J.J., Arango, C.P., Tank, J.L., 2009. The effects of season and agriculture on nitrous oxide production in headwater streams. Journal of Environment Quality 38 (2), 637–646. https://doi.org/10.2134/jeq2008.0003.

Bencala, K.E., Walters, R.A., 1983. Simulation of solute transport in a mountain pool-and-riffle stream: a transient storage model. Water Resources Research 19 (3), 718–724.

Blois, G., Best, J.L., Sambrook Smith, G.H., Hardy, R.J., 2014. Effect of bed permeability and hyporheic flow on turbulent flow over bed forms. Geophysical Research Letters 41 (18), 6435–6442.

Boano, F., Camporeale, C., Revelli, R., 2010a. A linear model for the coupled surface-subsurface flow in a meandering stream. Water Resources Research 46, W07535.

Boano, F., Camporeale, C., Revelli, R., Ridolfi, L., 2006. Sinuosity-driven hyporheic exchange in meandering rivers. Geophysical Research Letters 33, L18406. https://doi.org/10.1029/2006GL027630.

Boano, F., Harvey, J.W., Marion, A., Packman, A.I., Revelli, R., Ridolfi, L., Wörman, A., 2014. Hyporheic flow and transport processes: mechanisms, models, and biogeochemical implications. Reviews of Geophysics 52 (4), 603–679. https://doi.org/10.1002/2012RG000417.

Boano, F., Packman, A.I., Cortis, A., Revelli, R., Ridolfi, L., 2007a. A continuous time random walk approach to the stream transport of solutes. Water Resources Research 43, W10425.

Boano, F., Poggi, D., Revelli, R., Ridolfi, L., 2009. Gravity-driven water exchange between streams and hyporheic zones. Geophysical Research Letters 36, L20402.

Boano, F., Revelli, R., Ridolfi, L., 2007b. Bedform-induced hyporheic exchange with unsteady flows. Advances in Water Resources 30, 148–156.

Boano, F., Revelli, R., Ridolfi, L., 2008. Reduction of the hyporheic zone volume due to the stream-aquifer interaction. Geophysical Research Letters 35, L09401.

Boano, F., Revelli, R., Ridolfi, L., 2010b. Effect of streamflow stochasticity on bedform-driven hyporheic exchange. Advances in Water Resources 33, 1367–1374.

Boano, F., Revelli, R., Ridolfi, L., 2013. Modeling hyporheic exchange with unsteady stream discharge and bedform dynamics. Water Resources Research 49, 4089–4099. https://doi.org/10.1002/wrcr.20322.

Bott, T.L., Kaplan, L.A., Kuserk, F.T., 1984. Benthic bacterial biomass supported by stream water dissolved organic matter. Microbial Ecology 10, 335–344.

Bottacin-Busolin, A., Marion, A., 2010. Combined role of advective pumping and mechanical dispersion on time scales of bed form—induced hyporheic exchange. Water Resources Research 46, W08518. https://doi.org/10.1029/2009WR008892.

Bray, E.N., Dunne, T., 2017. Subsurface flow in lowland river gravel bars. Water Resources Research 53 (9), 7773–7797.

Briggs, M.A., Lautz, L.K., Hare, D.K., González-Pinzón, R., 2013. Relating hyporheic fluxes, residence times, and redox-sensitive biogeochemical processes upstream of beaver dams. Freshwater Science 32 (2), 622–641. https://doi.org/10.1899/12-110.1.

Briggs, M.A., Lautz, L.K., McKenzie, J.M., Gordon, R.P., Hare, D.K., 2012. Using high-resolution distributed temperature sensing to quantify spatial and temporal variability in vertical hyporheic flux. Water Resources Research 48, W02527.

Buffington, J.M., Tonina, D., 2009. Hyporheic exchange in mountain rivers II: effects of channel morphology on mechanics, scales, and rates of exchange. Geography Compass 3 (3), 1038–1062. https://doi.org/10.1111/j.1749-8198.2009.00225.x.

Cardenas, M.B., 2007. Potential contribution of topography-driven regional groundwater flow to fractal stream chemistry: residence time distribution analysis of Tóth flow. Geophysical Research Letters 34, L05403.

Cardenas, M.B., 2009a. Stream-aquifer interactions and hyporheic exchange in gaining and losing sinuous streams. Water Resources Research 45, W06429.

Cardenas, M.B., 2009b. A model for lateral hyporheic flow based on valley slope and channel sinuosity. Water Resources Research 45 (1), W01501. https://doi.org/10.1029/2008WR007442.

Cardenas, M.B., Wilson, J.L., 2007a. Dunes, turbulent eddies, and interfacial exchange with permeable sediments. Water Resources Research 43, W08412.

Cardenas, M.B., Wilson, J.L., 2007b. Exchange across a sediment—water interface with ambient groundwater discharge. Journal of Hydrology 346, 69–80.

Cardenas, M.B., Zlotnik, V.A., 2003. Three-dimensional model of modern channel bend deposits. Water Resources Research 39 (6), 1141. https://doi.org/10.1029/2002WR001383.

Cardenas, M.B., Cook, P.L.M., Jiang, H., Traykovski, P., 2008. Constraining denitrification in permeable wave-influenced marine sediment using linked hydrodynamic and biogeochemical modeling. Earth and Planetary Science Letters 275 (1–2), 127–137.

Cardenas, M.B., Ford, A.E., Kaufman, M.H., Kessler, A.J., Cook, P.L.M., 2016. Hyporheic flow and dissolved oxygen distribution in fish nests: the effects of open channel velocity, permeability patterns, and groundwater upwelling. Journal of Geophysical Research: Biogeosciences 121, 3113–3130.

Castro, N.M., Hornberger, G.M., 1991. Surface-subsurface water interaction in an alluviated mountain stream channel. Water Resources Research 27 (7), 1613–1621.

Chen, X., Cardenas, M.B., Chen, L., 2015. Three-dimensional versus two-dimensional bed form-induced hyporheic exchange. Water Resources Research 51, 2923–2936. https://doi.org/10.1002/2014WR016848.

Chen, X., Cardenas, M.B., Chen, L., 2018. Hyporheic exchange driven by three-dimensional sandy bed forms: sensitivity to and prediction from bed form geometry. Water Resources Research 54 (6), 4131–4149. https://doi.org/10.1029/2018WR022663.

Choi, J., Harvey, J.W., Conklin, M.H., 2000. Characterizing multiple timescales of stream and storage zone interaction that affect solute fate and transport in streams. Water Resources Research 36 (6), 1511–1518.

Chowdhury, R.S., Zarnetske, J.P., Phanikumar, M.S., Briggs, M.A., Day-Lewis, F.D., Singha, K., 2020. Formation criteria for hyporheic anoxic microzones: assessing interactions of hydraulics, nutrients, and biofilms. Water Resources Research 56 (3), e2019WR025971. https://doi.org/10.1029/2019WR025971.

Constantz, J., 2008. Heat as a tracer to determine streambed water exchanges. Water Resources Research 44, W00D10.

Cooper, A.C., 1965. The effect of transported stream sediments on the survival of sockeye and pink salmon eggs and alevin. International Pacific Salmon Fisheries Commission Bulletin 18, New Westminster, BC, Canada.

Cranswick, R.H., Cook, P.G., Lamontagne, S., 2014. Hyporheic zone exchange fluxes and residence times inferred from riverbed temperature and radon data. Journal of Hydrology 519, 1870–1881. https://doi.org/10.1016/j.jhydrol.2014.09.059.

Dai, Z., Ritzi, R.W., Dominic, D.F., Rubin, Y.N., 2003. Estimating spatial correlation structure for permeability in sediments with hierarchical organization. In: Mishra, S. (Ed.), Probabilistic Approaches to Groundwater Modeling Symposium 2003. American Society of Civil Engineers, New York, NY.

Dent, C.L., Grimm, N.B., Fisher, S.G., 2001. Multiscale effects of surface-subsurface exchange on stream water nutrient concentrations. Journal of the North American Benthological Society 20 (2), 162–181.

Drummond, J.D., Aubeneau, A., Packman, A.I., 2014. Stochastic modeling of fine particulate organic carbon dynamics in rivers. Water Resources Research 50 (5), 4341–4356.

Dudunake, T., Tonina, D., Reeder, W.J., Monsalve, A., 2020. Local and reach-scale hyporheic flow response from boulder-induced geomorphic changes. Water Resources Research 56 (10). https://doi.org/10.1029/2020WR027719.

Edwards, R.T., 1998. The hyporheic zone. In: Naiman, R.J., Bilby, R.E. (Eds.), River Ecology and Management: Lessons From the Pacific Coastal Ecoregion. Springer-Verlag, New York, NY, pp. 399–429.

Elliott, A.H., 1990. Transfer of Solutes Into and Out of Streambeds, Rep. KH-R-52. W.M. Keck Laboratory of Hydraulics and Water Resources, California Institute of Technology, Pasadena, CA.

Elliott, A.H., Brooks, N.H., 1997. Transfer of nonsorbing solutes to a streambed with bed forms: laboratory experiments. Water Resources Research 33 (1), 137–151.

De Falco, N., Boano, F., Bogler, A., Bar-Zeev, E., Arnon, S., 2018. Influence of stream-subsurface exchange flux and bacterial biofilms on oxygen consumption under nutrient-rich conditions. Journal of Geophysical Research: Biogeosciences 123 (7), 2021–2034.

Fanelli, R.M., Lautz, L.K., 2008. Patterns of water, heat and solute flux through streambeds around small dams. Ground Water 46 (5), 671–687.

Findlay, S., Strayer, W., Goumbala, C., Gould, K., 1993. Metabolism of streamwater dissolved organic carbon in the shallow hyporheic zone. Limnology & Oceanography 38, 1493–1499.

Fleckenstein, J.H., Niswonger, R.G., Fogg, G.E., 2006. River-aquifer interactions, geologic heterogeneity, and low-flow management. Ground Water 44, 837–852.

Fox, A., Boano, F., Arnon, S., 2014. Impact of losing and gaining streamflow conditions on hyporheic exchange fluxes induced by dune-shaped bed forms. Water Resources Research 50, 1895–1907.

Gariglio, F.P., Tonina, D., Luce, C.H., 2013. Spatio-temporal variability of hyporheic exchange through a pool-riffle-pool sequence. Water Resources Research 49 (11), 7185–7204. https://doi.org/10.1002/wrcr.20419.

Gibert, J., Stanford, J.A., Dole-Olivier, M.-J., Ward, J.V., 1994. Basic attributes of groundwater ecosystems and prospects for research. In: Gilbert, J., Danielpol, D.L., Standford, J.A. (Eds.), Groundwater Ecology. Academic Press, San Diego, CA, pp. 8–40.

Gomez-Velez, J.D., Harvey, J.W., 2014. A hydrogeomorphic river network model predicts where and why hyporheic exchange is important in large basins. Geophysical Research Letters 41 (18), 6403–6412. https://doi.org/10.1002/2014GL061099.

Gomez-Velez, J.D., Harvey, J.W., Cardenas, M.B., Kiel, B., 2015. Denitrification in the Mississippi River network controlled by flow through river bedforms. Nature Geoscience 8 (October), 1–8. https://doi.org/10.1038/ngeo2567.

Gomez-Velez, J.D., Krause, S., Wilson, J.L., 2014. Effect of low-permeability layers on spatial patterns of hyporheic exchange and groundwater upwelling. Water Resources Research 50, 5196–5215.

González-Pinzón, R., Haggerty, R., Myrold, D.D., 2012. Measuring aerobic respiration in stream ecosystems using the resazurin-resorufin system. Journal of Geophysical Research 117, G00N06.

Gooseff, M.N., 2010. Defining hyporheic zones- Advancing our conceptual and operational definitions of where stream water and groundwater meet. Geography Compass 4 (8), 945–955.

Gooseff, M.N., McGlynn, B.L., 2005. A stream tracer technique employing ionic tracers and specific conductance data applied to the Maimai catchment, New Zealand. Hydrological Processes 19, 2491–2506. https://doi.org/10.1002/hyp.5685.

Gooseff, M.N., Anderson, J.K., Wondzell, S.M., LaNier, J., Haggerty, R., 2006. A modelling study of hyporheic exchange pattern and the sequence, size, and spacing of stream bedforms in mountain stream networks, Oregon, USA. Hydrological Processes 20 (11), 2443–2457.

Gooseff, M.N., Hall, R.O.J., Tank, J.L., 2007. Relating transient storage to channel complexity in streams of varying land use in Jackson Hole, Wyoming. Water Resources Research 43, W01417.

Grant, S.B., Gomez-Velez, J.D., Ghisalberti, M., 2018. Modeling the effects of turbulence on hyporheic exchange and local-to-global nutrient processing in streams. Water Resources Research 54 (9), 5883–5889. https://doi.org/10.1029/2018WR023078.

Grant, S.B., Monofy, A., Boano, F., Gomez-Velez, J.D., Guymer, I., Harvey, J., Ghisalberti, M., 2020. Unifying advective and diffusive descriptions of bedform pumping in the benthic biolayer of streams. Water Resources Research 56 (11), e2020WR027967. https://doi.org/10.1029/2020WR027967.

Haggerty, R., Wondzell, S.M., Johnson, M.A., 2002. Power-law residence time distribution in hyporheic zone of a 2nd-order mountain stream. Geophysical Research Letters 29 (13), 4.

Haggerty, R., Reeves, P., 2002. STAMM-L Version 1.0 User's Manual. Sandia National Laboratory, Albuquerque, NM.

Haggerty, R., Martí, E., Argerich, A., von Schiller, D., Grimm, N.B., 2009. Resazurin as a "smart" tracer for quantifying metabolically active transient storage in stream ecosystems. Journal of Geophysical Research 114 (G3), G03014. https://doi.org/10.1029/2008JG000942.

Harvey, A.M., Wagner, B.J., 2000. Quantifying hydrologic interactions between streams and their subsurface hyporheic zones. In: Jones, J.B., Mulholland, P.J. (Eds.), Streams and Ground Waters. Academic Press, San Diego, CA, pp. 3–44.

Hampton, T.B., Zarnetske, J.P., Briggs, M.A., MahmoodPoor Dehkordy, F., Singha, K., Day-Lewis, F.D., Harvey, J.W., Chowdhury, S.R., Lane, J.W., 2020. Experimental shifts of hydrologic residence time in a sandy urban stream sediment–water interface alter nitrate removal and nitrous oxide fluxes. Biogeochemistry 149 (2), 195–219. https://doi.org/10.1007/s10533-020-00674-7.

Harvey, C., Gorelick, S.M., 2000. Rate-limited mass transfer or macrodispersion: which dominates plume evolution at the Macrodispersion Experiment (MADE) site? Water Resources Research 36 (3), 637–650.

Harvey, J.W., Bencala, K.E., 1993. The effect of streambed topography on surface-subsurface water exchange in mountain catchments. Water Resources Research 29 (1), 89–98. https://doi.org/10.1029/92WR01960.

Harvey, J.W., Böhlke, J.K., Voytek, A.M., Scott, D., Tobias, C.R., 2013. Hyporheic zone denitrification: controls on effective reaction depth and contribution to whole-stream mass balance. Water Resources Research 49 (10), 6298–6316.

Harvey, J.W., Drummond, J.D., Martin, R.L., McPhillips, L.E., Packman, A.I., Jerolmack, D.J., Stonedahl, S.H., Aubeneau, A.F., Sawyer, A.H., Larsen, L.G., Tobias, C.R., 2012. Hydrogeomorphology of the hyporheic zone: stream solute and fine particle interactions with a dynamic streambed. Journal of Geophysical Research: Biogeosciences 117, G00N11.

Harvey, J.W., Wagner, B.J., Bencala, K.E., 1996. Evaluating the reliability of the stream tracer approach to characterize stream-subsurface water exchange. Water Resources Research 32 (8), 2441–2451.

Hassan, M.A., Tonina, D., Beckie, R.D., Kinnear, M., 2015. The effects of discharge and slope on hyporheic flow in step-pool morphologies. Hydrological Processes 29 (3), 419–433. https://doi.org/10.1002/hyp.10155.

Hatch, C., Fisher, A.T., Revenaugh, J.S., Constantz, J., Ruehl, C., 2006. Quantifying surface water–groundwater interactions using time series analysis of streambed thermal records: method development. Water Resources Research 42, W10410.

Hauer, F.R., Locke, H., Dreitz, V.J., Hebblewhite, M., Lowe, W., Muhfeld, C., Nelson, C., Proctor, M., Rood, S., 2016. Gravel-bed river floodplains are the ecological nexus of glaciated mountain landscapes. Science Advances 2 (6), 2:e1600026. https://doi.org/10.1126/sciadv.1600026.

Herzog, S.P., Higgins, C.P., McCray, J.E., 2016. Engineered streambeds for induced hyporheic flow: enhanced removal of nutrients, pathogens, and metals from urban streams. Journal of Environmental Engineering 142 (1), 1–10. https://doi.org/10.1061/(ASCE)EE.1943-7870.0001012.

Hester, E.T., Cardenas, M.B., Haggerty, R., Apte, S.V., 2017. The importance and challenge of hyporheic mixing. Water Resources Research 53 (5), 3565–3575.

Hester, E.T., Doyle, M.W., 2008. In-stream geomorphic structures as drivers of hyporheic exchange. Water Resources Research 44 (3), W03417. https://doi.org/10.1029/2006WR005810.

Hester, E.T., Gooseff, M.N., 2010. Moving beyond the banks: hyporheic restoration is fundamental to restoring ecological services and functions of streams. Environmental Science and Technology 44 (5), 1521−1525. https://doi.org/10.1021/es902988n.

Hester, E.T., Young, K.I., Widdowson, M.A., 2013. Mixing of surface and groundwater induced by riverbed dunes: implications for hyporheic zone definitions and pollutant reactions. Water Resources Research 49, 5221−5237.

Huang, P., Chui, T.F.M., 2018. Empirical equations to predict the characteristics of hyporheic exchange in a pool-riffle sequence. Ground Water 56 (6), 947−958. https://doi.org/10.1111/gwat.12641.

Hutchinson, P.A., Webster, I.T., 1998. Solute uptake in aquatic sediment due to current-obstacle interactions. Journal of Environmental Engineering 124 (5), 419−426. https://doi.org/10.1061/(ASCE)0733-9372(1998)124:5(419).

Ikeda, S., 1984. Prediction of alternate bar wavelength and height. Journal of Hydraulic Engineering 110, 371−386.

Janssen, F., Cardenas, M.B., Sawyer, A.H., Dammrich, T., Krietsch, J., Beer, D., 2012. A comparative experimental and multiphysics computational fluid dynamics study of coupled surface−subsurface flow in bed forms. Water Resources Research 48 (8), W08514.

Janssen, F., Huettel, M., Witte, U., Faerber, P., Huettel, M., Meyer, V., Witte, U., 2005. Pore-water advection and solute fluxes in permeable marine sediments (I): calibration and performance of the novel benthic chamber system Sandy. Limnology & Oceanography 50 (3), 768−778. https://doi.org/10.4319/lo.2005.50.3.0779.

Kaufman, M.H., Cardenas, M.B., Buttles, J., Kessler, A.J., Cook, P.L.M., 2017. Hyporheic hot moments: dissolved oxygen dynamics in the hyporheic zone in response to surface flow perturbations. Water Resources Research 53, 6642−6662.

Kiel, B.A., Cardenas, M.B., 2014. Lateral hyporheic exchange throughout the Mississippi River network. Nature Geoscience 7 (6), 413−417. https://doi.org/10.1038/ngeo2157.

Knapp, J.L.A., González-Pinzón, R., Drummond, J.D., Larsen, L.G., Cirpka, O.A., Harvey, J.W., 2017. Tracer-based characterization of hyporheic exchange and benthic biolayers in streams. Water Resources Research 53 (2), 1575−1594. https://doi.org/10.1002/2016WR019393.

Krause, S., Abbott, B.W., Baranov, V., Bernal, S., Blaen, P., Datry, T., Drummond, J., Fleckenstein, J.H., Velez, J.G., Hannah, D.M., Knapp, J.L.A., Kurz, M., Lewandowski, J., Martí, E., Mendoza-Lera, C., Milner, A., Packman, A., Pinay, G., Ward, A.S., Zarnetzke, J.P., 2022. Organizational principles of hyporheic exchange flow and biogeochemical cycling in river networks across scales. Water Resources Research 58 (3), e2021WR029771. https://doi.org/10.1029/2021WR029771.

Laube, G., Schmidt, C., Fleckenstein, J.H., 2018. The systematic effect of streambed conductivity heterogeneity on hyporheic flux and residence time. Advances in Water Resources 122, 60−69. https://doi.org/10.1016/j.advwatres.2018.10.003.

Lautz, L.K., Siegel, D.I., 2006. Modeling surface and ground water mixing in the hyporheic zone using MODFLOW and MT3D. Advances in Water Resources 29 (11), 1618−1633.

Liu, T., Xia, X., Liu, S., Mou, X., Qiu, Y., 2013. Acceleration of denitrification in turbid rivers due to denitrification occurring on suspended sediment in oxic waters. Environmental Science and Technology 47 (9), 4053−4061. https://doi.org/10.1021/es304504m.

Magliozzi, C., Grabowski, R.C., Packman, A.I., Krause, S., 2018. Toward a conceptual framework of hyporheic exchange across spatial scales. Hydrology and Earth System Sciences 22 (12), 6163−6185. https://doi.org/10.5194/hess-22-6163-2018.

Malard, F., Tockner, K., Dole-Olivier, M.-J.J., Ward, J.V., 2002. A landscape perspective of surface−subsurface hydrological exchanges in river corridors. Freshwater Biology 47 (4), 621−640. https://doi.org/10.1046/j.1365-2427.2002.00906.x.

Malcolm, I.A., Soulsby, C., Youngson, A.F., Hannah, D.M., 2005. Catchment-scale controls on groundwater-surface water interactions in the hyporheic zone: implications for salmon embryo survival. River Research and Applications 21, 977−989.

Marion, A., Bellinello, M., Guymer, I., Packman, A.I., 2002. Effect of bed form geometry on the penetration of nonreactive solutes into a streambed. Water Resources Research 38 (10), 1209.

Marion, A., Packman, A.I., Zaramella, M., Bottacin-Busolin, A., 2008a. Hyporheic flows in stratified beds. Water Resources Research 44, W09433.

Marion, A., Zaramella, M., Bottacin-Busolin, A., 2008b. Solute transport in rivers with multiple storage zones: the STIR model. Water Resources Research 44, W10406.

Marzadri, A., Bellin, A., Tank, J.L., Tonina, D., 2022. Predicting nitrous oxide emissions through riverine networks. Science of The Total Environment 843, 156844. https://doi.org/10.1016/j.scitotenv.2022.156844.

Marzadri, A., Dee, M.M., Tonina, D., Bellin, A., Tank, J.L., 2014. A hydrologic model demonstrates nitrous oxide emissions depend on streambed morphology. Geophysical Research Letters 41 (15), 5484–5491. https://doi.org/10.1002/2014GL060732.

Marzadri, A., Dee, M.M., Tonina, D., Bellin, A., Tank, J.L., 2017. Role of surface and subsurface processes in scaling $N_2O$ emissions along riverine networks. Proceedings of the National Academy of Sciences 114 (17), 4330–4335. https://doi.org/10.1073/pnas.1617454114.

Marzadri, A., Tonina, D., Bellin, A., 2011. A semianalytical three-dimensional process-based model for hyporheic nitrogen dynamics in gravel bed rivers. Water Resources Research 47 (11), W11518. https://doi.org/10.1029/2011WR010583.

Marzadri, A., Tonina, D., Bellin, A., 2012. Morphodynamic controls on redox conditions and on nitrogen dynamics within the hyporheic zone: application to gravel bed rivers with alternate-bar morphology. Journal of Geophysical Research: Biogeosciences 117 (3). https://doi.org/10.1029/2012JG001966.

Marzadri, A., Tonina, D., Bellin, A., 2013a. Effects of stream morphodynamics on hyporheic zone thermal regime. Water Resources Research 49 (4), 2287–2302. https://doi.org/10.1002/wrcr.20199.

Marzadri, A., Tonina, D., Bellin, A., 2013b. Quantifying the importance of daily stream water temperature fluctuations on the hyporheic thermal regime: implication for dissolved oxygen dynamics. Journal of Hydrology 507, 241–248. https://doi.org/10.1016/j.jhydrol.2013.10.030.

Marzadri, A., Tonina, D., Bellin, A., 2020. Power law scaling model predicts $N_2O$ emissions along the Upper Mississippi River basin. Science of the Total Environment 732, 138390. https://doi.org/10.1016/j.scitotenv.2020.138390.

Marzadri, A., Tonina, D., Bellin, A., Valli, A., 2016. Mixing interfaces, fluxes, residence times and redox conditions of the hyporheic zones induced by dune-like bedforms and ambient groundwater flow. Advances in Water Resources 88, 139–151. https://doi.org/10.1016/j.advwatres.2015.12.014.

Marzadri, A., Tonina, D., Bellin, A., Vignoli, G., Tubino, M., 2010. Semianalytical analysis of hyporheic flow induced by alternate bars. Water Resources Research 46 (7), W07531. https://doi.org/10.1029/2009WR008285.

Marzadri, A., Amatulli, G., Tonina, D., Bellin, A., Shen, L.Q., Allen, G.H., Raymond, P.A., 2021. Global riverine nitrous oxide emissions: the role of small streams and large rivers. Science of the Total Environment 776, 145148. https://doi.org/10.1016/j.scitotenv.2021.145148.

Master, Y., Shavit, U., Shaviv, A., 2005. Modified isotope pairing technique to study N transformations in polluted aquatic systems: theory. Environmental Science and Technology 39, 1749–1756.

Mendoza, C., Zhou, D., 1992. Effects of porous bed on turbulent stream flow above bed. Journal of Hydraulic Engineering 118 (9), 1222–1240.

Middleton, G.V., Southard, J.B., 1984. Mechanics of Sediment Movement. Society for Economic Paleontologists and Mineralogists Short Course No. 3, Tulsa, OK.

Monofy, A., Boano, F., 2021. The effect of streamflow, ambient groundwater, and sediment anisotropy on hyporheic zone characteristics in alternate bars. Water Resources Research 57, 1–22. https://doi.org/10.1029/2019WR025069.

Montgomery, D.R., Buffington, J.M., 1997. Channel-reach morphology in mountain drainage basins. Geological Society of America Bulletin 109, 596–611.

Montgomery, D.R., Buffington, J.M., 1998. Channel processes, classification, and response. In: Naiman, R.J., Bilby, R. (Eds.), River Ecology and Management. Springer-Verlag, New York, NY, pp. 13–42.

Moreto, J.R., Reeder, W.J., Budwig, R., Tonina, D., Liu, X., 2022. Experimentally mapping water surface elevation, velocity, and pressure fields of an open channel flow around a stalk. Geophysical Research Letters 49 (7), 1–10. https://doi.org/10.1029/2021gl096835.

Mrokowska, M.M., Osuch, M., 2011. Assessing validity of the dead zone model to characterize transport of contaminants in the River Wkra. In: Rowinski, P. (Ed.), Experimental Methods in Hydraulic Research, Geoplanet: Earth and Planetary Sciences, vol 1. Springer, Berlin, Heidelberg, pp. 235–246.

Mulholland, P.J., Helton, A.M., Poole, G.C., Hall, R.O., Hamilton, S.K., Peterson, B.J., Tank, J.L., Ashkenas, L.R., Cooper, L.W., Dahm, C.N., et al., 2008. Stream denitrification across biomes and its response to anthropogenic nitrate loading. Nature 452 (7184), 202–205. https://doi.org/10.1038/nature06686.

Mulholland, P.J., Marzorf, E.R., Webster, J.R., Hart, D.D., Hendricks, S.P., 1997. Evidence that hyporheic zones increase heterotrophic metabolism and phosphorus uptake in forest streams. Limnology & Oceanography 42, 443–451.

Naegeli, M.W., Uehlinger, U., 1997. Contribution of the hyporheic zone to ecosystem metabolism in a prealpine gravel-bed-river. Journal of the North American Benthological Society 16 (4), 794–804.

Nagaoka, H., Ohgaki, S., 1990. Mass transfer mechanism in a porous riverbed. Water Research 24 (4), 417–425.

Ocampo, C.J., Oldham, C.E., Sivapalan, M., 2006. Nitrate attenuation in agricultural catchments: shifting balances between transport and reaction. Water Resources Research 42 (W01408), 16.

Orghidan, T., 1959. Ein neuer lebensraum des unterirdischen wassers: der hyporheische biotop. Archiv für Hydrobiologie 55, 392–414.

Packman, A.I., Brooks, N.H., 2001. Hyporheic exchange of solutes and colloids with moving bed forms. Water Resources Research 37 (10), 2591–2605. https://doi.org/10.1029/2001WR000477.

Packman, A.I., Brooks, N.H., Morgan, J.J., 2000. A physicochemical model for colloid exchange between a stream and a sand streambed with bed forms. Water Resources Research 36 (8), 2351–2361.

Pescimoro, E., Boano, F., Sawyer, A.H., Soltanian, M.R., 2019. Modeling influence of sediment heterogeneity on nutrient cycling in streambeds. Water Resources Research 55 (5), 4082–4095. https://doi.org/10.1029/2018WR024221.

Poole, G.C., O'Daniel, S.J., Jones, K.L., Woessner, W.W., Bernhardt, E.S., Helton, A.M., Stanford, J.A., Boer, B.R., Beechie, T.J., 2008. Hydrologic spiralling: the role of multiple interactive flow paths in stream ecosystems. River Research and Applications 24, 1018–1031.

Pryshlak, T.T., Sawyer, A.H., Stonedahl, S.H., Soltanian, M.R., 2015. Multiscale hyporheic exchange through strongly heterogeneous sediments. Water Resources Research 51 (11), 9127–9140.

Quick, A.M., Reeder, W.J., Farrell, T.B., Tonina, D., Feris, K.P., Benner, S.G., 2016. Controls on nitrous oxide emissions from the hyporheic zones of streams. Environmental Science and Technology 50 (21), 11491–11500. https://doi.org/10.1021/acs.est.6b02680.

Raymond, P.A., Zappa, C.J., Butman, D., Bott, T.L., Potter, J., Mulholland, P.J., Laursen, A.E., McDowell, W.H., Newbold, D., 2012. Scaling the gas transfer velocity and hydraulic geometry in streams and small rivers. Limnology & Oceanography 2 (1), 41–53. https://doi.org/10.1215/21573689-1597669.

Reeder, W.J., Quick, A.M., Farrell, T.B., Benner, S.G., Feris, K.P., Basham, W.J.R., Marzadri, A., Tonina, D., Huber, C., 2019. A novel fiber optic system to map dissolved oxygen concentrations continuously within submerged sediments. Journal of Applied Water Engineering and Research 7 (3), 1–12. https://doi.org/10.1080/23249676.2019.1611495.

Reeder, W.J., Quick, A.M., Farrell, T.B., Benner, S.G., Feris, K.P., Marzadri, A., Tonina, D., 2018a. Hyporheic source and sink of nitrous oxide. Water Resources Research 54 (7), 5001–5016. https://doi.org/10.1029/2018WR022564.

Reeder, W.J., Quick, A.M., Farrell, T.B., Benner, S.G., Feris, K.P., Tonina, D., 2018b. Spatial and temporal dynamics of dissolved oxygen concentrations and bioactivity in the hyporheic zone. Water Resources Research 54 (3), 2112–2128. https://doi.org/10.1002/2017WR021388.

Reisinger, A.J., Tank, J.L., Rosi-Marshall, E.J., Hall, R.O., Baker, M.A., 2015. The varying role of water column nutrient uptake along river continua in contrasting landscapes. Biogeochemistry 125 (1), 115–131. https://doi.org/10.1007/s10533-015-0118-z.

Ren, J., Packman, A.I., 2004a. Modeling of simultaneous exchange of colloids and sorbing contaminants between streams and streambeds. Environmental Science and Technology 38 (10), 2901–2911.

Ren, J., Packman, A.I., 2004b. Coupled stream-subsurface exchange of colloidal hematite and dissolved zinc, copper and phosphate. Environmental Science and Technology 39 (17), 6387–6394.

Ren, J., Packman, A.I., 2007. Changes in fine sediment size distributions due to interactions with streambed sediments. Sedimentary Geology 202, 529–537.

Ritzi, R.W., Dai, Z., Dominic, D.F., Rubin, Y.N., 2004. Spatial correlation of permeability in cross-stratified sediment with hierarchical architecture. Water Resources Research 40, W03513.

Riva, M., Guadagnini, L., Guadagnini, A., Ptak, T., Martac, E., 2006. Probabilistic study of well capture zones distribution at the Lauswiesen field site. Journal of Contaminant Hydrology 88 (1), 92–118.

Roche, K.R., Li, A., Bolster, D., Wagner, G.J., Packman, A.I., 2019. Effects of turbulent hyporheic mixing on reach-scale transport. Water Resources Research 55 (5), 3780–3795. https://doi.org/10.1029/2018WR023421.

Rousseau, G., Ancey, C., 2020. Scanning PIV of turbulent flows over and through rough porous beds using refractive index matching. Experiments in Fluids 61 (8), 1—24. https://doi.org/10.1007/s00348-020-02990-y.

Rubol, S., Tonina, D., Vincent, L., Sohm, J.A., Basham, W., Budwig, R., Savalia, P., Kanso, E., Capone, D.G., Nealson, K.H., 2018. Seeing through porous media: an experimental study for unveiling interstitial flows. Hydrological Processes 32 (3), 402—407. https://doi.org/10.1002/hyp.11425.

Runkel, R.L., 1998. One-dimensional Transport with Inflow and Storage (OTIS): A Solute Transport Model for Streams and Rivers. US Geological Survey, Water-Resources Investigations Report 98-4018, Denver, CO.

Salarashayeri, A.F., Siosemarde, M., 2012. Prediction of soil hydraulic conductivity from particle-size distribution. International Journal of Geological and Environmental Engineering 61, 454—458.

Salehin, M., Packman, A.I., Paradis, M., 2004. Hyporheic exchange with heterogeneous streambeds: laboratory experiments and modeling. Water Resources Research 40 (11), W11504.

Sawyer, A.H., Cardenas, M.B., 2009. Hyporheic flow and residence time distributions in heterogeneous cross-bedded sediment. Water Resources Research 45, W08406.

Sawyer, A.H., Cardenas, M.B., Buttles, J., 2011. Hyporheic exchange due to channel-spanning logs. Water Resources Research 47, W08502.

Sheibley, R.W., Jackman, A.P., Duff, J.H., Triska, F.J., 2003. Numerical modeling of coupled nitrification—denitrification in sediment perfusion cores from the hyporheic zone of the Shingobee River, MN. Advances in Water Resources 26, 977—987.

Shen, H.V., Fehlman, H.M., Mendoza, C., 1990. Bed form resistance in open channel flows. Journal of Hydraulic Engineering 116 (6), 799—815.

Shen, G., Yuan, J., Phanikumar, M.S., 2020. Direct numerical simulations of turbulence and hyporheic mixing near sediment-water interfaces. Journal of Fluid Mechanics 892 (A20). https://doi.org/10.1017/jfm.2020.173.

Sherman, T., Roche, K.R., Richter, D.H., Packman, A.I., Bolster, D., 2019. A dual domain stochastic Lagrangian model for predicting transport in open channels with hyporheic exchange. Advances in Water Resources 125, 57—67. https://doi.org/10.1016/j.advwatres.2019.01.007.

Singh, T., Gomez-Velez, J.D., Wu, L., Wörman, A., Hannah, D.M., Krause, S., 2020. Effects of successive peak flow events on hyporheic exchange and residence times. Water Resources Research 56 (8), e2020WR027113. https://doi.org/10.1029/2020WR027113.

Stanford, J.A., 2006. Landscapes and riverscapes. In: Hauer, F.R., Lamberti, G.A. (Eds.), Methods in Stream Ecology. Academic Press, San Diego, CA, pp. 3—21.

Stanford, J.A., Ward, J.V., 1993. An ecosystem perspective of alluvial rivers: connectivity and the hyporheic corridor. Journal of the North American Benthological Society 12 (1), 48—60.

Stanford, J.A., Lorang, M.S., Hauer, F.R., 2005. The shifting habitat mosaic of river ecosystems. Verhandlungen Internationale Vereinigung für Theoretische und Angewandte Limnologie 29 (1), 123—136. https://doi.org/10.1080/03680770.2005.11901979.

Stewart, R.J., Wollheim, W.M., Gooseff, M.N., Briggs, M.A., Jacobs, J.M., Peterson, B.J., Hopkinson, C.S., 2011. Separation of river network—scale nitrogen removal among the main channel and two transient storage compartments. Water Resources Research 47, W00J10.

Stonedahl, S.H., Harvey, J.W., Packman, A.I., 2013. Interaction between hyporheic flow produced by stream meanders, bars and dunes. Water Resources Research 49, 5450—5461.

Stonedahl, S.H., Harvey, J.W., Wörman, A., Salehin, M., Packman, A.I., 2010. A multi-scale model for integrating hyporheic exchange from ripples to meanders. Water Resources Research 46, W12539.

Storey, R.G., Howard, K.W.F., Williams, D.D., 2003. Factor controlling riffle-scale hyporheic exchange flows and their seasonal changes in gaining stream: a three-dimensional groundwater model. Water Resources Research 39 (2), 1034. https://doi.org/10.1029/2002WR001367.

Stuart, T.A., 1954. Spawning sites of trout. Nature 173, 354.

Thibodeaux, L.J., Boyle, J.D., 1987. Bedform-generated convective transport in bottom sediment. Nature 325 (22), 341—343.

Tonina, D., 2012. Surface water and streambed sediment interaction: the hyporheic exchange. In: Gualtieri, C., Mihailovic, D.T. (Eds.), Fluid Mechanics of Environmental Interfaces. CRC Press, Taylor and Francis Group, London, pp. 255—294.

Tonina, D., Buffington, J.M., 2007. Hyporheic exchange in gravel bed rivers with pool-riffle morphology: laboratory experiments and three-dimensional modeling. Water Resources Research 43, W01421. https://doi.org/10.1029/2005WR004328.

Tonina, D., Buffington, J.M., 2009. Hyporheic exchange in mountain rivers I: mechanics and environmental effects. Geography Compass 3 (3), 1063–1086. https://doi.org/10.1111/j.1749-8198.2009.00226.x.

Tonina, D., Buffington, J.M., 2011. Effects of stream discharge, alluvial depth and bar amplitude on hyporheic flow in pool-riffle channels. Water Resources Research 47 (8), W08508. https://doi.org/10.1029/2010WR009140.

Tonina, D., de Barros, F.P.J., Marzadri, A., Bellin, A., 2016. Does streambed heterogeneity matter for hyporheic residence time distribution in sand-bedded streams? Advances in Water Resources 96, 120–126. https://doi.org/10.1016/j.advwatres.2016.07.009.

Tonina, D., Marzadri, A., Bellin, A., 2015. Benthic uptake rate due to hyporheic exchange: the effects of streambed morphology for constant and sinusoidally varying nutrient loads. Water (Switzerland) 7 (2), 398–419. https://doi.org/10.3390/w7020398.

Trauth, N., Schmidt, C., Maier, U., Vieweg, M., Fleckenstein, J.H., 2013. Coupled 3-D stream flow and hyporheic flow model under varying stream and ambient groundwater flow conditions in a pool-riffle system. Water Resources Research 49 (9), 5834–5850. https://doi.org/10.1002/wrcr.20442.

Trauth, N., Schmidt, J.C., Vieweg, M., Maier, U., Fleckenstein, J.H., 2014. Hyporheic transport and biogeochemical reactions in pool-riffle systems under varying ambient groundwater flow conditions. Journal of Geophysical Research: Biogeosciences 119, 910–928.

Triska, F.J., Duff, J.H., Avanzino, R.J., 1993. Patterns of hydrological exchange and nutrient transformation in the hyporheic zone of a gravel bottom stream: examining terrestrial-aquatic linkages. Freshwater Biology 29, 259–274.

Triska, F.J., Kennedy, V.C., Avanzino, R.J., Zellweger, G.W., Bencala, K.E., 1989a. Retention and transport of nutrients in a third-order stream: channel processes. Ecology 70, 1894–1905.

Triska, F.J., Kennedy, V.C., Avanzino, R.J., Zellweger, G.W., Bencala, K.E., 1989b. Retention and transport of nutrients in a third-order stream in Northwestern California: hyporheic processes. Ecology 70 (6), 1893–1905.

Tubino, M., 1991. Growth of alternate bars in unsteady flow. Water Resources Research 27 (1), 37–52.

van Kampen, R., Schneidewind, U., Anibas, C., Bertagnoli, A., Tonina, D., Vandersteen, G., Luce, C., Krause, S., van Berkel, M., 2022. LPMLEn—A frequency domain method to estimate vertical streambed fluxes and sediment thermal properties in semi-infinite and bounded domains. Water Resources Research 58 (3), e2021WR030886. https://doi.org/10.1029/2021WR030886.

Vaux, W.G., 1968. Intragravel flow and interchange of water in a streambed. Fishery Bulletin 66 (3), 479–489.

Vieweg, M., Trauth, N., Fleckenstein, J.H., Schmidt, C., 2013. Robust optode-based method for measuring in situ oxygen profiles in gravelly streambeds. Environmental Science & Technology 47 (17), 9858–9865. https://doi.org/10.1021/es401040w.

Vittal, N., Ranga Raju, K.G., Garde, R.J., 1977. Resistance of two-dimensional triangular roughness. Journal of Hydraulic Research 15 (1), 19–36.

Voermans, J.J., Ghisalberti, M., & Ivey, G.N., 2018. The hydrodynamic response of the sediment-water interface to coherent turbulent motions. Geophysical Research Letters 45(19), 10,520–10,527. https://doi.org/10.1029/2018GL079850

Ward, A.S., Gooseff, M.N., Johnson, P.A., 2011. How can subsurface modifications to hydraulic conductivity be designed as stream restoration structures? Analysis of Vaux's conceptual models to enhance hyporheic exchange. Water Resources Research 47 (8), W09418.

Ward, A.S., Gooseff, M.N., Singha, K., 2010. Imaging hyporheic zone solute transport using electrical resistivity. Hydrological Processes 24, 948–953.

White, D.S., 1993. Perspective on defining and delineating hyporheic zones. Journal of the North American Benthological Society 12 (1), 61–69.

Winter, C.L., Tartakovsky, D.M., 2002. Groundwater flow in heterogeneous composite aquifers. Water Resources Research 38 (8).

Winter, T.C., Harvey, J.W., Franke, O.L., Alley, W.M., 1998. Ground Water and Surface Water: A Single Resource. U.S. Geological Survey, Circular 1139, Denver, CO.

Wolke, P., Teitelbaum, Y., Deng, C., Lewandowski, J., Arnon, S., 2019. Impact of bed form celerity on oxygen dynamics in the hyporheic zone. Water 12 (1), 62. https://doi.org/10.3390/w12010062.

Wondzell, S.M., 2011. The role of the hyporheic zone across stream networks. Hydrological Processes 25 (22), 3525–3532.

Wondzell, S.M., Gooseff, M.N., 2013. Geomorphic controls on hyporheic exchange across scales: watersheds to particles. In: Shroder (Ed.-in-chief), J.F., Wohl (vol. Ed.), E. (Eds.), Treatise on Geomorphology, 9. Fluvial Geomorphology. Academic Press, San Diego, CA, pp. 203–218. Fluvial Geomorphology.

Wondzell, S.M., Swanson, F.J., 1996. Seasonal and storm dynamics of the hyporheic zone of a 4th order mountain stream. 1: hydrologic processes. Journal of the North American Benthological Society 15, 3–19.

Wörman, A., Packman, A.I., Johansson, H., Jonsson, K., 2002. Effect of flow-induced exchange in hyporheic zones on longitudinal transport of solutes in streams and rivers. Water Resources Research 38 (1), 2–15.

Wörman, A., Packman, A.I., Marklund, L., Harvey, J.W., 2006. Exact three-dimensional spectral solution to surface-groundwater interaction with arbitrary surface topography. Geophysical Research Letters 33, L07402.

Wu, L., Gomez-Velez, J.D., Krause, S., Singh, T., Wörman, A., Lewandowski, J., 2020. Impact of flow alteration and temperature variability on hyporheic exchange. Water Resources Research 56 (3), e2019WR026225. https://doi.org/10.1029/2019wr026225.

Wuhrmann, K., 1972. River purification. In: Mitchell, R. (Ed.), Water Pollution Microbiology. Wiley-Interscience, New York, NY, pp. 119–151.

Xia, X., Yang, Z., Zhang, X., 2009. Effect of suspended-sediment concentration on nitrification in river water: importance of suspended sediment − water interface. Environmental Science and Technology 43 (10), 3681–3687. https://doi.org/10.1021/es8036675.

Yalin, M.S., 1964. Geometrical properties of sand waves. Journal of the Hydraulics Division 90, 105–119.

Yuan, Y, Chen, X., Cardenas, M.B., Liu, X., Chen, L., 2021. Hyporheic exchange driven by submerged rigid vegetation: a modeling study. Water Resources Research 57 (6), e2019WR026675. https://doi.org/10.1029/2019WR026675.

Zaramella, M., Packman, A.I., Marion, A., 2003. Application of the transient storage model to analyze advective hyporheic exchange with deep and shallow sediment beds. Water Resources Research 39 (7), 1198.

Zarnetske, J.P., Gooseff, M.N., Bowden Morgan, J., Greenwald, M.J., Brosten, T.R., Bradford, J.H., McNamara, J.P., 2008. Influence of morphology and permafrost dynamics on hyporheic exchange in arctic headwater streams under warming climate conditions. Geophysical Research Letters 35, L02501.

Zarnetske, J.P., Haggerty, R., Wondzell, S.M., Baker, M.A., 2011. Dynamics of nitrate production and removal as a function of residence time in the hyporheic zone. Journal of Geophysical Research: Biogeosciences 116, G01025.

Zarnetske, J.P., Haggerty, R., Wondzell, S.M., Bokil, V.A., González-Pinzón, R., 2012. Coupled transport and reaction kinetics control the nitrate source-sink function of hyporheic zones. Water Resources Research 48, W11508.

Zhou, Y., Ritzi, R.W., Soltanian, M.R., Dominic, D.F., 2014. The influence of streambed heterogeneity on hyporheic flow in gravelly rivers. Ground Water 52 (2), 206–216.

Ziebis, W., Forster, S., Huettel, M., Jørgensen, B.B., Jorgensen, B.B., 1996. Complex burrows of mud shrimp *Callianassa truncata* and their geochemical impact in the sea bed. Nature 382 (6592), 619–622.

# 4

# Ecological and evolutionary jargon in subterranean biology

*David C. Culver[1], Tanja Pipan[2] and Žiga Fišer[3]*

[1]Department of Environmental Science, American University, Washington, DC, United States;
[2]ZRC SAZU, Karst Research Institute, Postojna, Slovenia; [3]University of Ljubljana,
Biotechnical Faculty, Department of Biology, Ljubljana, Slovenia

## Introduction

Just as lack of eyes and pigment and the increase in extra-optic sensory structures are the hallmarks of subterranean organisms, the terminology associated with subterranean life (troglobiont, stygobiont, etc.) is the hallmark of papers about subterranean life. With the exception of the topic of how subterranean species came to lose their eyes, discussions about the ecological classification of cave organisms in particular and subterranean organisms in general are probably as frequent as any other topic about subterranean life (e.g., Racovitza, 1907; Barr, 1968; Sket, 2008; Giachino and Vailati, 2017; Trajano and de Carvalho, 2017; Culver and Pipan, 2019a). Next in frequency, and perhaps surpassing discussions of ecological classification, is the question of how organisms came to be in subterranean habitats and how they managed to evolve at all (e.g., Racovitza, 1907; Howarth, 1980; Coineau and Boutin, 1992; Desutter-Grandcolas and Grandcolas, 1996; Culver and Pipan, 2019b).

Almost simultaneously with the discovery of eyeless and pigmentless cave animals, ecological classification of cave dwellers began. In 1849, the Danish biologist Jørgen Schiødte proposed the first such classificatory scheme, based on the amount of light and the nature of the walls in the preferred habitat. The question arises why Schiødte and many who followed him found it necessary to have a classificatory scheme. If all cave (and by extension all subterranean) animals were without eyes and pigment, then presumably only one word would be necessary for all these species, e.g., cavernicole. However, the nearly universal presence of at least partially eyed and/or pigmented species in subterranean habitats (Pipan and Culver, 2012; Culver and Pipan, 2015) makes such a simple solution unrealistic and impractical.

Classification schemes like Schiødte's and others we take up below, do have the advantage of providing a common vocabulary for the field of speleobiological researchers, who come from a wide variety of language groups and for whom English is typically not their first language. Disadvantages include an elaborate system that is nearly impenetrable to most readers and nonspecialists, imprecise and/or conflicting definitions, and equivalence of many terms to standard ecological and evolutionary terminology, a terminology many more readers are familiar with. Even though the speleobiological jargon can be fit into modern evolutionary thinking, its origins are often non-Darwinian and even anti-Darwinian (anti-selectionist) (Romero, 2009). This leads to the perception that subterranean life is somehow subject to rules distinct from standard evolutionary and ecological processes.

The terminology we examine is that of the extent of dependence on the subterranean environment (e.g., troglobionts), the mode of colonization and speciation (e.g., the Adaptive Shift Hypothesis), and the extent of modification for subterranean life (e.g., troglomorphy). Epistemologically, these are all hypotheses about the evolution of subterranean life. While we argue that there are far too many specialized terms for the evolution of subterranean life, it is probably true that there are not enough terms to describe subterranean habitats, given their unfamiliarity. The description of aquatic subterranean habitats themselves is covered in the chapter by Robertson et al. (this volume).

Our goals are to (1) place the speleobiological terminology in a modern neo-Darwinian context, (2) suggest the minimal and judicious use of jargon, and (3) provide a glossary of ecological and evolutionary terms used in the study of subterranean aquatic systems.

## Ecological classifications

### Schiødte's classification and two possible explanations

Schiødte (1849) proposed four categories: (1) shadow animals, (2) twilight animals, (3) obscure (=aphotic) area animals, and (4) obscure area with flowstone animals. This categorization is based on the frequently described zonation of caves ranging from the entrance zone (shadow animals), to the twilight zone, to the zone of constant darkness (obscure area). Although he did not describe it as such, his categorization is a hypothesis where the distribution of organisms is in equilibrium with the environment. In spite of this very fundamental division of the cave habitat (see Howarth and Moldovan, 2018), there is little real evidence that cave dwellers show a zonation from eyed entrance animals to eyeless deep cave animals. Descriptions of zonation in caves are limited to terrestrial examples and are often nonquantitative (see Mohr and Poulson, 1966). In those few cases where quantitative data are available, eyed and pigmented individuals occur throughout the cave and eyeless and depigmented animals occur throughout most of the cave (Howarth and Stone, 1990; Tobin et al., 2013). The study of Tobin et al. (2013) provides a quantitative estimate of the importance of zonation in the distribution of species throughout the year—they found it accounted for 55% of the variation in species abundance. Kozel et al. (2019) provide a counterexample —in Zguba jama in Slovenia, many eyeless, depigmented species were more common in shallower area of the cave. In other cases, zonation was completely lacking, notably in a large cave in Quintana Roo, Mexico, one with a minimal dark zone (Mejía-Ortíz et al., 2018).

Equivalent patterns exist in the hyporheic zone. Gibert et al. (1990) showed the vertical zonation, with overlap of different species groups in the hyporheic zone of the Rhône River (Fig. 4.1A). Ward and Voelz (1997) provide data on the occurrence of different species in the

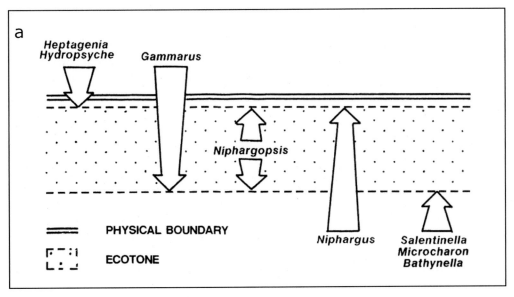

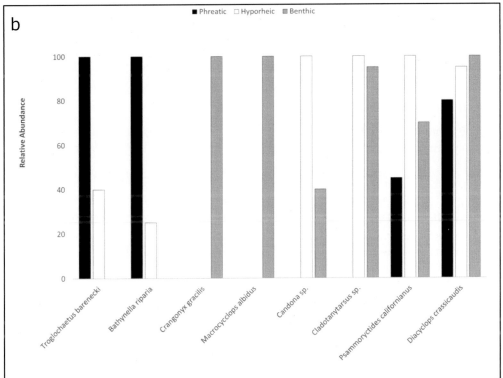

**FIGURE 4.1** (a) Distribution of invertebrates in relation to the hyporheic zone: the case of the Rhône River, France. The *dashed line* indicates the extent of the hyporheic zone. (b) Examples of distribution of representative taxa in the phreatic, hyporheic, and benthic zones of the South Platte River in Colorado, USA. Species include strictly subterranean species such as the polychaete *Troglochaetus beranecki*, surface species such as the amphipod *Crangonyx gracilis*, as well as species with broader and intermediate distributions, such as the copepod *Diacyclops crassicaudis*. (a) From Gibert et al. (1990). Used with permission of Taylor & Francis. (b) Data from Ward and Voelz (1997).

phreatic, hypogean, and benthic zones (Fig. 4.1B), the approximate equivalent of the entrance, twilight, and aphotic zones of caves. Some species were limited to a single zone and others were common in all three. Peralta-Maraver et al. (2018) were able to identify distinct benthic and hyporheic communities in a German stream but species from both communities co-occurred throughout the water column.

A similar zonation pattern at a larger scale might exist when subterranean habitats with different degrees of isolation from the surface and different degrees of oligotrophy are compared, rather than different parts of the same cave. However, Culver and Pipan (2015) showed that the frequency of eyeless and depigmented species was no greater in relatively isolated and deep caves compared to the fauna of epikarst and seepage springs (Table 4.1).

An entirely different hypothesis has been put forward to explain the presence of eyed, reduced eyed, and eyeless species in the same subterranean habitat, one that emphasizes not the equilibrium of distributions, but rather differences in age of the subterranean fauna. Different morphologies are present, according to this hypothesis, because of different lengths of time different species have been isolated in caves. The idea that degree of morphological modification and the age of colonization are correlated was popularized by Poulson (1963) for the amblyopsid fish in North American caves and Wilkens and Strecker (2017) for the Mexican cavefish, *Astyanax mexicanus*. However, the idea dates back at least to neo-Lamarckians such as Eigenmann (1909) and Jeannel (1950). Regardless of its origin, there is little contemporary support for the hypothesis. The molecular phylogeny of amblyopsid fish (Niemiller et al., 2012) does not correlate well with their morphological modification.

## Modern ecological classifications

While Schiødte's classification scheme was the first one, modern classifications are based on the system developed by Schiner (1854) and Racovitza (1907). They recognized three categories:

- Troglobites (troglobionts)—species whose habitat is exclusively in the subterranean domain and who show a preference for deeper parts of the habitat.
- Troglophiles—species permanently inhabiting the subterranean domain, but preferably in superficial regions; they frequently reproduce there but may also be found outside.
- Trogloxenes—occasional or accidental visitors to caves attracted by humidity or food, but that do not live continuously or reproduce in caves.

**TABLE 4.1**  Percent of eyeless aquatic species in different subterranean habitats.

|                                         | Epikarst          | Hypotelminorheic                        | Caves                                            |
| --------------------------------------- | ----------------- | --------------------------------------- | ------------------------------------------------ |
| Number of sites                         | 9                 | 3                                       | 4                                                |
| Geographic location                     | Slovenia (8), USA (1) | Croatia (1), Slovenia (1), USA (1)   | UK (1), Bosnia & Hercegovina (1), USA (2)        |
| Mean percent of eyeless aquatic species | 86                | 57                                      | 47                                               |
| Range (minimum − maximum)               | 40−100            | 47−75                                   | 0−97                                             |

*Data from Culver and Pipan (2015).*

Racovitza's definition of troglophiles included many types of species—those that completed their life cycle in subterranean habitats but could also reproduce outside, and those species that required both surface and subterranean habitats to complete their life cycle. There is, however, ambiguity in his definitions of both troglophile and trogloxene. The trogloxene category is generally interpreted as species that are occasional visitors to caves. Trajano and de Carvalho (2017) define trogloxenes as species that have source populations in surface habitats and sink populations in caves. Pavan (1944), later endorsed by Sket (2008), suggested two kinds of troglophiles—*eutroglophiles*, which spend their entire life cycle in subterranean habitats or in surface habitats, depending on the population, and *subtroglophiles*, which either leave subterranean habitats periodically (e.g., bats foraging at night) or spend part of their life cycle in subterranean habitats and part in surface habitats [e.g., the stonefly *Isocapnia* (Stanford and Gaufin, 1974) and the grotto salamander *Eurycea spelaea* (Gorički et al., 2019)]. A widely used modification is that of Barr (1968), who basically renamed Pavan's subtroglophiles as trogloxenes and eutroglophiles as troglophiles.

Gibert et al. (1994) in the first edition of the enormously influential *Groundwater Ecology* book, proposed a scheme (Fig. 4.2) more like Racovitza's scheme, with stygoxenes being accidental species, species permanently inhabiting the hyporheic (Pavan's troglobites), species using both subterranean and surface habitats to complete their life cycle (amphibites—Pavan's subtroglophiles), and stygophilic species that may either spend their entire life in either surface habitats or the hyporheic zone (Pavan's eutroglophiles). This last category illustrates one of the problems with all of these schemes—there are always cases that fit uncomfortably into any scheme. There does not seem to be an aquatic equivalent of species like many bats and crickets that leave caves most nights to forage. This may be because no one has looked for aquatic subterranean species in surface habitats at night, such as the benthos of streams or spring runs. Bressi et al. (1999) did report that the olm *Proteus anguinus* is frequently found at night in spring runs associated with caves.

Gibert et al. (1994) were the first to use the terms stygobite (biont), stygophile, and stygoxene for aquatic subterranean species and restricting troglo- to terrestrial species. However, this has not been universally accepted (Sket, 2008). Finally, Gibert et al. (1994) distinguished two types of stygobionts—ubiquitous stygobionts and phreatobites (bionts). Phreatobites are species only found deep in alluvial aquifers, and not in other habitats such as caves or seepage springs. The temptation to give names for specialists living in other habitats, e.g., hypotelminorheobionts for hypotelminorheic specialist species, should be firmly resisted.

A more recent elaboration and clarification of the ecological classification of subterranean species is that of Trajano (Trajano, 2012; Trajano and de Carvalho, 2017). She emphasized the importance of thinking about source and sink populations. Source populations are ones whose birth rate exceeds death rate while the opposite is true for sink populations, which are maintained by immigration from source populations (Pulliam, 1988). This is a very useful distinction in cases where a few individuals are observed in the "wrong" habitat, for example, a stygobiont found in a surface stream. As part of the definition of stygophiles (and troglophiles), Trajano and de Carvalho (2017) include the observation that source populations occur in both surface and subterranean habitats. As part of the definition of stygoxenes (and trogloxenes), they posit that they are "source populations in epigean habitats, with individuals using subterranean resources" but the opposite statement could also be made—that they are source populations in subterranean habitats using surface resources. For

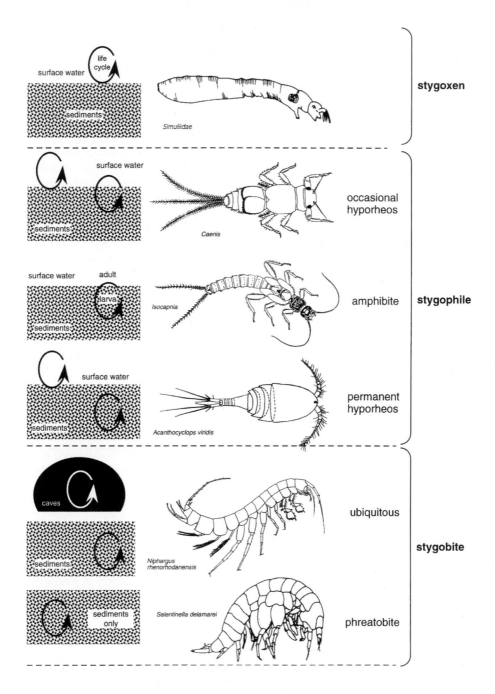

**FIGURE 4.2** A classification of groundwater fauna based on its life cycle and its presence in surface and subterranean habitats. *From Gibert et al. (1994).*

example, *Isocapnia* stoneflies spend nearly their entire life cycle in the hyporheic zone (Stanford and Gaufin, 1974). Other difficulties may arise in subterranean habitats such as the hypotelminorheic zone that are patchy and connected by dispersing individuals, i.e., a metapopulation in the sense of Levins (1969). In this model, which is very widely used for species with multiple, small, semiisolated populations, individual populations often go extinct and are recolonized from other sub-populations. The metapopulation consists of both source and sink populations, even though all of them are of the same type, e.g., hypotelminorheic. Finally, it may be difficult to objectively determine whether a population is a source or a sink, since it depends on the population growth rate, not population size (Pulliam, 1988).

## Critique and alternative classifications

Some of the difficulties with the above classifications become obvious when we hypothetically apply it to a surface habitat, for example lakes. Then we have for example,

- lentobionts, species found only in lakes and specialized for life in lakes,
- lentophiles, species found in lakes and elsewhere, with some modification for life in lakes, and
- lentoxenes, species visiting lakes without modification for life in lakes.

There is another way to categorize aquatic (and terrestrial) subterranean populations, and that is by the standard ecological terms of specialist and generalist. Specialist species in aquatic subterranean habitats include stygobionts and those stygoxenes that have an obligate dependence on subterranean habitats even though they also have an obligate dependence on surface habitats. Stygobionts can be even more specialized, with an obligate dependence on a particular aquatic subterranean habitat rather than subterranean habitats in general. Perhaps most stygobionts fall into this category, including what Gibert et al. (1994) termed phreatobites. Obvious examples of trogloxenic specialists are those bats that always hibernate in caves, such as *Myotis grisescens*. But there are aquatic examples as well. The grotto salamander *Eurycea spelaea* is always found in caves as adults but most larvae develop in surface stream and springs (Gorički et al., 2019). Interestingly, it is almost always listed as a stygobiont, when it is clearly a stygoxene, but just clearly a specialist. There are many generalist species of salamanders, ones that can complete their life cycles either in surface streams or cave streams. Examples include a number of species of *Eurycea* in North America, including *E. lucifuga* and *E. longicauda* (Gorički et al., 2019). While the generalist-specialist dichotomy does not capture the complexities of the association of species with subterranean habitats, it does have the advantage of using a common ecological terminology.

## Colonization and speciation

## Climatic Relict and Adaptive Shift hypotheses and their historical background

Several hypotheses were proposed for the origin of aquatic subterranean animals (reviewed by Coineau and Boutin, 1992; Holsinger, 2000; Stoch, 2004). Among these, the

Climatic Relict Hypothesis (CRH) and Adaptive Shift Hypothesis (ASH) grew the deepest roots in the speleobiological community and stimulated the liveliest debates. The terms CRH and ASH imply that there is something inherently unique and special to the origin of subterranean life that escapes the general ecological and evolutionary vocabulary.

The CRH postulates that subterranean species originated through the following scenario. A surface species exapted (*sensu* Gould and Vrba, 1982) to subterranean life experienced climate change, causing unfavorable environmental conditions in the species' native (surface) habitat. Consequently, the species retreated underground, where conditions were more suitable, and formed subterranean populations. As its movement to the new habitat was forced, the colonization is said to be passive. The ancestral surface populations either died out or migrated to other surface habitats with more favorable conditions. In either case, surface and subterranean populations were geographically isolated and eventually allopatric speciation occurred.

This hypothesis was inspired by the fact that many subterranean animals in temperate latitudes are geographic and/or phyletic relicts. It can be traced back to the early ideas of Jeannel (1950) and their summary in Vandel's (1964) influential book on subterranean life. Originally, it was permeated with the idea that subterranean species are living fossils, primitive taxa, and evolutionary dead-ends. Such Lamarckism was shaken off after the hypothesis was modified and advanced by American speleobiologists in a neo-Darwinian fashion, starting with Barr (1960). He applied it to explain the distributional patterns of the *Pseudoanophthalmus* trechinae beetles in eastern North America. Pleistocene glacial fluctuations were since considered the dominant environmental change driving animals underground, especially terrestrial but also aquatic species (Juberthie, 1984; Barr and Holsinger, 1985). Later, the hypothesis embraced new discoveries of tropical relicts and associated non-Pleistocene climatic perturbations, such as late Tertiary droughts (e.g., Humphreys, 1993). So, in roughly 6 decades the CRH manifested in minor variants but its essentials remained constant, i.e., passive colonization and allopatric speciation.

CRH was challenged with the development of the Adaptive Shift Hypothesis (ASH). This postulates that after a subterranean environment becomes available, either by its formation or expansion of an exapted surface species distribution, the latter colonizes the new habitat and exploits novel resources. It is not forced underground by unfavorable conditions at the surface and thus colonization is said to be active. A dramatic difference in environmental conditions of both habitats subsequently drives the adaptive divergence between surface and subterranean populations and leads to speciation despite geographic contact and gene flow. The essential components of ASH are thus active colonization and parapatric speciation.

ASH was inspired by the discovery of young, nonrelictual subterranean faunas in the tropics, characterized by extant closely related diverging species pairs with parapatric distributions. It was championed by Howarth (1973, 1987) and Chapman (1982) based on their work on subterranean arthropods of Hawaii and Southeast Asia, respectively, among which Hawaiian planthoppers *Oliarus* stood out as the most prominent example. In years that followed, similar examples from Galapagos and Canary Islands (Oromí et al., 1991; Peck and Finston, 1993) but also of animals from temperate latitudes, such as *Astyanax mexicanus* (Wilkens and Hüppop, 1986) and *Gammarus minus* (Culver et al., 1995), provided further support to the ASH. Part of its pervasiveness in the community must have arisen due to the coincident shift in the general evolutionary doctrine from the view that speciation is invariably allopatric (Mayr, 1970) to the recognition of its nonallopatric modes (White, 1978).

In summary, both CRH and ASH are hypotheses of evolutionary transition from surface to subterranean life. They were first proposed for terrestrial animals, later extended to aquatic ones, and the initial motivation seems to have been to explain biogeographical patterns. Differently from ecological classifications and troglomorphy, they did not start as strict verbal definitions by their original authors, but as ideas that were gradually developed as loosely formulated hypotheses. CRH and ASH agree that success of colonization depends on the exaptation of colonizing surface species, but assume different colonization and speciation modes. In the context of CRH and ASH the distinction of passive and active colonization has always been one of unfavorable versus nonunfavorable environmental conditions as drivers of colonization, and not one of involuntary displacement versus self-propelled locomotion. This later contrast predates the CRH and ASH, and was fueled by the once popular perspective that subterranean environment is inhospitable, harsh, and offers no benefits to the colonizers (Romero and Green, 2005). Last but not least, lumping of colonization and speciation, two distinct and largely independent processes, into one hypothesis is the source of many of the issues discussed next.

## Critique of CRH and ASH

Several difficulties accompany CRH and ASH, which might discourage nonspeleobiologists from considering the related literature. First is the confusion regarding what CRH and ASH are hypotheses about. Many authors are either vague or imprecise in their assertions. The phenomena they ascribe to both hypotheses are either too broad, such as evolution or origin of subterranean life, or too narrow, such as solely colonization or speciation. CRH and ASH are actually hypotheses of evolutionary transition from surface to subterranean life, more precisely, a particular combination of colonization and speciation mode. The origin of a subterranean species from a surface ancestor is however not an evolutionary dead-end. CRH and ASH thus ignore much of the complexity of the origin of subterranean life, i.e., the diversification and speciation of subterranean lineages within the subterranean realm (e.g., Leijs et al., 2012; Trontelj et al., 2012), or even the phenomenon of recolonization of surface habitats by subterranean species (e.g., Copilas-Ciocianu et al., 2018).

Second, the CRH and ASH dichotomy implies that these are the sole competing alternatives of the transition from surface to subterranean life. However, several authors realized these hypotheses are neither competing nor sole alternatives. The relictual fauna of temperate regions and nonrelictual tropical fauna that inspired CRH and ASH, respectively, likely reflect two different time points of the same continuum where historical events (e.g., climatic and geological perturbations) obscured earlier patterns of an older fauna (Holsinger, 2000). Desutter-Grancolas and Grandcolas (1996) demonstrated that CRH and ASH are each defined by a unique combination of past environmental conditions (proxy for colonization mode), speciation mode, and present distribution, and that several other scenarios, such as active colonization followed by allopatric speciation, are reasonable alternatives as well.

Third, the debate about passive versus active colonization mode has largely disappeared from the literature and is considered resolved or perhaps uninteresting. This owes to the realization that stochastic events of accidental displacement or the onset of diverse conditions at surface cannot explain the observation that surface-dwelling animals can and do enter underground all the time. Recognition of shallow subterranean habitats revealed that the subterranean environment is continuous with the surface and that the two are much better interconnected than previously thought (Culver and Pipan, 2014). Furthermore, movement of surface-dwelling species to subterranean habitats is now well-documented and apparently a ubiquitous phenomenon, continuously feeding the subterranean realm with a steady influx of colonists (Culver and Pipan, 2019b). Contemporary evidence thus translates the dilemmas of surface-subterranean colonization to general questions of invasion biology (Trontelj, 2018), with "why surface ancestors colonize subterranean habitats" giving way to "why certain colonizations are successful and others are not" (Ribera et al., 2018). Resolving the debate on colonization modes effectively reduces the distinction of CRH and ASH to allopatric versus parapatric speciation, in other words, whether divergence proceeds without or with gene flow (Niemiller et al., 2008). The necessity for related speleobiological jargon disappears at this point.

Finally and most importantly, CRH and ASH do not give mutually exclusive predictions, making them inherently difficult if not impossible to falsify (Desutter-Grancolas, 1997a; Plath and Tobler, 2010; Niemiller and Soares, 2015). This remains true even when colonization is dissected out of both hypotheses and the CRH versus ASH discourse translates to the general question of the geography of speciation. Testing whether past speciation was allopatric or not is extremely difficult with currently available methods (Losos and Glor, 2003). For example, it is improbable to falsify nonallopatric speciation if surface ancestors are extinct or species geographic distributions changed significantly, which is the case in practically all older subterranean species. In such scenario, even when time calibrated lineage splits agree with independently calibrated periods of past climate change, like Leys et al. (2003) demonstrated for diving beetles from Australian calcretes, one still cannot exclude that speciation was parapatric and geographic isolation occurred after the fact. Similarly, allopatric speciation is hard to falsify in young adjacent surface-subterranean species pairs. Their current parapatry could be due to initial allopatry followed by range expansion and secondary contact and available methods cannot differentiate between the two scenarios (Coyne and Orr, 2004). However, promising approaches stemming from early evidence that present-day biodiversity and genomic patterns carry detectable and distinguishable fingerprints of specific speciation histories are emerging (e.g., Feder et al., 2013; Skeels and Cardellio, 2019). Until these are rigorous, CRH and ASH will remain unfalsifiable and with little power to explain how subterranean species originate from surface ancestors.

## Alternative terminology: swinging the pendulum from geography to processes generating reproductive isolation

As demonstrated above, clinging to CRH and ASH will give only limited insight into the origin of subterranean species. We believe time is ripe to reconcile both hypotheses with modern treatments of speciation. In the last 2 decades speciation research and classification underwent a major shift from the traditional process-free geographic modes to evolutionary processes driving divergence and generation of reproductive isolation, such as selection and genetic drift (Schluter, 2001; Via, 2001). Irrespective of why animals colonize underground and the geographic mode of subsequent speciation, subterranean habitats are lightless, stable, and usually food deprived. Transition from surface to subterranean life will thus always be characterized by a dramatic change in environment. It has been convincingly shown that in such cases reproductive isolation typically evolves as a result of divergent selection originating from ecological differences in environments (Nosil, 2012). According to modern classification this process is ecological speciation and can proceed in any geographic context. Speleobiologists have recognized that ASH includes ecological speciation (e.g., Trontelj, 2019), however it has been less clear that CRH does just as well, albeit in its allopatric mode. We propose ecological speciation is considered a null hypothesis of transition from surface to subterranean life. As any hypothesis it should be tested against its alternatives, in this case speciation without selection (Coyne and Orr, 2004; Schluter, 2009; Nosil, 2012; Langerhans and Riesch, 2013). Ecological speciation gives explicit predictions and tests for it are available (Nosil, 2012); the hypothesis thus meets the falsifiability criteria.

Just as recognition of nonallopatric modes of speciation helped to establish ASH, which greatly extended our knowledge on the origin of subterranean animals, we believe it will be rewarding if the modern, process-oriented approach to speciation is more widely integrated into speleobiology. Shifting focus to testing the predictions of ecological speciation by posing questions like "do subterranean-derived traits emerge barriers to gene flow and constitute reproductive isolation and which traits are these?" will move us forward. So far, only few studies have tested reproductive barriers between a subterranean species and its surface ancestor (e.g., Tobler, 2009; Tobler et al., 2009; Riesch et al., 2011). As speciation occurs at the population — species interface, phylogenetically young systems, like *Astyanax mexicanus*, *Poecillia mexicana*, *Asellus aquaticus*, and *Gammarus minus* (see Chapters by Gross et al., Protas et al., and Fong et al., this volume), where transition from surface to subterranean environment is still ongoing or finished recently, will provide most insight into the process.

## Should we retire the CRH and ASH as formal categories?

Yes. This is not because the conundrum on the origin of subterranean species has been resolved, but because the terms raise several inherent issues, some of which are obsolete. Furthermore, we showed that what remains of CRH and ASH today has a suitable term in the general literature, i.e., ecological speciation, making speleobiology-specific terminology redundant. As CRH and ASH are historically important pillars of our field they will surely persist at least in discourses on its historical development. In other cases, we propose abandoning CRH and ASH as formal categories of classifications of origin of subterranean species. We expect this suggestion to be much less controversial than those proposed for other terms discussed in this chapter. In fact, CRH and ASH are already loosing viability as evident from a number of recent papers (e.g., Schilthuizen et al., 2005; Niemiller et al., 2008; Riesch et al., 2011; Borowsky and Cohen, 2013).

## Morphological modification for subterranean life

### Troglomorphy and the nature of selection

The ecological patterns of species occurrence in aquatic subterranean habitats are correlated, although not perfectly so, with morphological features. In particular, stygobionts (and troglobionts) typically have reduced or absent eyes and pigment, and stygophiles (and troglophiles) have eyes and pigment reduced to a lesser degree, if at all. There are other habitats where eyelessness is common, such as deep soil (Peck, 1990), so the connection is imperfect at best. Of course, there is a long-standing controversy about the causes of these reductions and losses. Among the suggested causes are use and disuse (Jeannel, 1950), neutral mutation and genetic drift (Wilkens and Strecker, 2017), natural selection—acting either directly for energetic savings (Moran et al., 2014, 2015) or indirectly via pleiotropy (Jeffery, 2005), and phenotypic plasticity (Blin et al., 2018; Bilandžija et al., 2020). All of these hypothesized causes have a sense of inevitability of the direction of change, given enough time, which is in agreement with observations.

Christiansen (1962) suggested that a separate terminology was needed to describe morphological patterns of subterranean species, and coined the now frequently used term "troglomorphy" to describe the suite of changes, both losses such as eyes, and gains, such as increases in extra-optic sensory structures (Table 4.2). It has come to include not only morphological traits, but also behavioral and physiological traits. Christiansen's purpose was largely to inject neo-Darwinian concepts in subterranean biology, and he wrote in French, both because French biologists dominated subterranean biology, and because neo-Darwinian concepts were not widely held in France at the time.

Christiansen proposed three new terms to classify subterranean species morphologically:

- Epigeomorphs, species showing no modification for subterranean life;
- Ambimorphs, species showing some modification for subterranean life, but still retaining some morphological features of surface-dwelling species; and
- Troglomorphs, species clearly modified for cave life, totally different from their surface-dwelling counterparts.

**TABLE 4.2**    Troglomorphic characters divided into ones that are elaborated and ones that are reduced in subterranean habitats.

| Elaborated | Reduced |
| --- | --- |
| Specialization of sensory organs (e.g., touch, chemoreception) | Reduction of eyes, pigment, wings |
| Pseudophysogastry (Coleoptera) | Scale reduction or loss (teleost fishes) |
| Compressed or depressed body form (Hexapoda) | Loss of pigment cells and deposits |
| Increased egg volume | Cuticle thinning |
| Increased size (Collembola, Arachnida) | Reduction or loss of swim bladder (teleost fishes) |
| Unguis elongation (Collembola) | |
| Foot modification (Collembola, planthoppers) | |
| Elongate body form (teleost fishes, Arachnida) | |
| Depressed, shovel-like heads (teleost fishes, salamanders) | |
| Reduced femur length/cropy-empty live weight ratio (crickets) | |

*From Christiansen (2012).*

Fortunately, only the last term is widely accepted, because one could read the definitions in such a way that all but the most highly modified species in each lineage was an ambimorph. The advantages of the term troglomorph are:

- It is a testable hypothesis about evolution in subterranean habitats. It is however not necessarily a hypothesis about natural selection for reduced characters because other forces, especially neutral mutation, can account these changes, at least for the reduction of eyes and pigment.
- It effectively decouples morphology and habitat, and ensures that morphology is not used in ecological classifications.

Troglomorphy refers to the suite of characters that change with isolation in subterranean habitats as a result of convergent forces of natural selection (Table 4.2), and is akin to the general evolutionary concept of convergence, which is typically used in the context of overall morphology, or facies. Some authors have questioned the decoupling of morphology (troglomorphy) and habitat specialization (e.g., troglobiont) and suggested troglomorphy be included in the definition of troglobiont (Peck, 2002). In addition, Trajano and de Carvalho (2017) discuss conceptual problems, but its usage has become widespread. Although the list in Table 4.2 includes many elaborated features, only eye and pigment loss are often taken as indicating troglomorphy, without considering other characters. This is not in the spirit of Christiansen's definition, since it involved natural selection, and eye and pigment loss does not necessarily involve natural selection (Wilkens and Strecker, 2017).

At about the same time as he coined the phrase troglomorphy, Christiansen (1961) developed the terms cave-dependent and cave-independent for individual morphological characters. Cave-dependent characters were ones that showed convergence among subterranean species, even when the effects of shared ancestry have been removed (see Christman et al., 1997). The presence of cave-dependent characters can also be tested, as we discuss below.

## Case studies of troglomorphy

Desutter-Grandcolas (1997b), on the basis of a cladistic analysis of the cricket clade (Amphiacustae), was perhaps the first to question whether there really was convergence (troglomorphy) among subterranean species. Her conclusion was that there was no evidence for troglomorphy aside from eye and pigment loss, although it should be noted that she analyzed crickets specialized for cave life but foraging for food on the surface, i.e., cave specialists that are trogloxenes.

In their study of spring and cave populations of the amphipod *Gammarus minus*, Christman et al. (1997) found a similar pattern. They used the technique of phylogenetic subtraction to determine (1) the expected relative antennal and eye size for cave and spring populations in two drainage basins using an mtDNA based phylogeny, and (2) the deviation of observed morphology from the predicted morphology. Their results are summarized in Table 4.3. Based on mtDNA phylogeny, eye size of the two cave populations is somewhat smaller than that of the spring populations. This includes the phylogenetic effect, as well as the effect of natural selection or neutral mutation. However, the residual effects, with phylogeny removed, were both large and negative for the cave populations, indicating that they are smaller than expected from phylogenetic analysis, and subject to convergent selection. There

**TABLE 4.3**  Results of analysis of four populations of *Gammarus minus*, partitioning antennal and eye size into phylogenetic components and residuals. If convergence is important, it will be seen in the residuals. Phylogeny based on mtDNA sequences. See Christman et al. (1997) for details. Y is the relative expected size based on phylogeny and $\varepsilon$ is the residual. Patterns of the residuals may indicate homoplasy.

| | Antenna | | Eye | |
|---|---|---|---|---|
| Population | Y | $\varepsilon$ | Y | $\varepsilon$ |
| Davis spring | −0.0125 | −0.1691 | 0.3601 | 0.462 |
| Benedict cave | 0.1122 | 0.6632 | 0.0751 | −1.344 |
| Organ spring | 0.0324 | 0.0132 | 0.4536 | 0.1945 |
| Organ cave | −0.0962 | −0.7159 | 0.0559 | −1.3385 |

was a different pattern with respect to antenna size. Amphipods from one spring (Davis Spring) and one cave (Organ Cave, in another basin) had phylogenetically small antennae. Unlike the case with eye size, residuals showed no evidence of convergence.

Simčič and Sket (2019) found another case, this one involving a physiological trait, where subterranean species did not show convergence. They compared metabolic rates of a surface- and subterranean-living *Niphargus* species pair and *Asellus aquaticus* population pair and found that metabolic rates were not reduced in either subterranean representative, contrary to expectation.

The most extensive study to date of the prevalence of convergence (troglomorphy) is that of Konec et al. (2015) and Protas et al. (this volume). Konec et al. compared the morphology of two surface-subterranean population pairs of the isopod *Asellus aquaticus* from two drainages in Slovenia and Romania for a total of 63 morphological characters. Their results are summarized in Table 4.4. The most common pattern was one where only one population changed, which happened in 30 traits (48%); convergence occurred in 18 traits (29%), and divergence was found in two traits (3%). Protas et al. (this volume) extended the analysis with four additional population pairs (mostly from Slovenia), of which two included only a partially

**TABLE 4.4**  Summary of character changes in Slovenian and Romanian cave populations of *Asellus aquaticus*.

| Type of change | Number of traits | Percentage of traits |
|---|---|---|
| Convergence | 18 | 29 |
| Divergence | 2 | 3 |
| Change in one population only (autapomorphy) | 22 in Slovenia, 8 in Romania | 48 |
| Increase in variance, mean unchanged | 2 in Romania | 3 |
| No change | 11 | 17 |

*Adapted from Konec et al. (2015).*

depigmented cave population. Considering only the four pairs with a completely depigmented cave population, 39 traits showed a significant change from its surface counterpart in at least three pairs. A total of 13 traits showed divergence among cave populations, 15 were convergently reduced, and only 11 were convergently elaborated.

The work of Fišer et al. (2012) and Trontelj et al. (2012) demonstrated that morphological variability among *Niphargus* amphipod species in both interstitial and cave communities was greater, not less, than expected by chance. Thus, divergence rather than convergence was occurring at this scale. They did show that species in different caves tended to share particular morphologies, e.g., lake giants, a form of convergence, but at a different scale.

## Critique and alternative classifications

Given the real data on convergence and troglomorphism, there is evidence that convergence does occur but that it by no means predominates. Convergence due to selection occurs because of a shared selective environment. Pipan and Culver (2012) argue that only the absence [or near absence (Mejía-Ortìz et al., 2018)] of light is shared among all subterranean environments. Low food resources are often invoked as a nearly universal feature of caves and other subterranean environments (Kováč, 2018), but there are many cases of relatively high resource levels in subterranean habitats that also have species with reduced eyes and pigment (Culver and Pipan, 2014). On the other hand, there are factors that promote divergence, including interspecific competition, differing habitat dimensions, flux of organic matter, and proximity to the surface. Additionally, unique adaptations (autapomorphies), which predominated in Konec et al.'s study, may be common in subterranean communities in general (see Bichuette et al., 2015). The newly emerging paradigm of a mixture of convergence and divergence differs profoundly from the old troglomorphy paradigm, but it does share the feature that morphology of subterranean organisms is molded by natural selection, rather than genetic drift or other non-Darwinian forms of evolution. For example, in Konec et al.'s (2015) study, 52 of the 63 morphological characters measured showed evidence of directional selection in the subterranean populations, strong support for the neo-Darwinian school of evolution.

## Overall recommendations

If there were generally agreed upon definitions of speleobiological jargon, and these terms had no equivalent in the general ecological and evolutionary literature, then the use and continued refinement of these terms would be justified. However, neither of these conditions are met. There continue to be discussions and disputes about terminology (see especially Sket, 2008; Giachino and Vailati, 2017; Trajano and de Carvalho, 2017), and while these discussions are interesting, it is not clear that refinement of terminology actually increases understanding of the underlying ecological and evolutionary processes.

Of course, these terms will never entirely disappear but a reduction in their use and an increase in the use of the standard neo-Darwinian vocabulary should speed up the integration of subterranean biology studies into the broader fields of ecology and evolutionary biology, as well as increase readership and citations of cave biology related papers (Martinez and Mammola, 2021). Historically, subterranean biology developed along a separate path due to the apparent importance of Lamarckian processes involved in the loss and reduction of structures. These non-Darwinian ideas persisted well into the mid-20th century at least in France (Jeannel, 1950; Vandel, 1964). The neo-Darwinian school of subterranean biology, developed by Barr, Christiansen, and Poulson in the 1960s in North America, signaled the beginning of the reintegration of subterranean biology into the mainstream (Mammola, 2019; Mammola et al., 2020). This process continues, as witnessed by the ever expanding number of papers on subterranean biology published in mainstream journals, thus enriching the wider community with knowledge gained from the unique subterranean fauna more effectively. Putting the scientific romanticism to a more practical perspective, such practice will lead to increased number of citations and open better funding options, a crucial element fueling new discoveries in every field.

## Glossaries

### Aim and scope of two glossaries

We provide definitions and commentary on specialized ecological and evolutionary terms used in aquatic subterranean biology. We have excluded a number of terms little if ever used. Both Sket (2008) and Trajano and de Carvalho (2017) discussed a number of these. We have also not included the very large number of terms used to describe aquatic subterranean habitats, which are covered by Robertson et al. (this volume). Many of the terms defined below are ones that we recommend be used rarely if ever, and we have included them as a guide to the literature in what we call the Retired Speleobiological Glossary. As a guide to a transition to a more general ecological and evolutionary terminology, we included the most relevant terms in what we call the Eco-Evo Glossary. We provide their common definitions, although we recognize that other definitions exist and should be consulted.

## Eco-Evo Glossary

**Allopatric speciation** Evolution of geographically isolated populations, with little or no possibility of mutual gene flow, into distinct species.

**Convergence** The independent evolution of similar traits in distinct evolutionary lineages, often as a result of similar selective pressures. Troglomorphy is the occurrence of convergence, especially in morphology, due to natural selection acting upon subterranean populations or species.

**Divergence** The independent evolution of different traits in distinct evolutionary lineages, often as a result of different selective pressures. Although widespread among subterranean species, it was not given a separate name in speleobiology.

**Ecological generalist** Species that generally lives in a wide range of habitats and feeds on a wide range of food sources. Troglophiles (both eutroglophiles and subtroglophiles) fall into this category as well as trogloxenes that do not depend on subterranean habitats to complete their life cycle.

**Ecological specialist** Species that generally occupies a very restricted range or even a single habitat and feeds on a small range or even a single food source. Troglobionts and those trogloxenes with an absolute dependence on subterranean habitats to complete their life cycle are in this category.

**Ecological speciation** Evolution of reproductive isolation between populations due to ecologically based divergent selection between environments.

**Exaptation** A trait, feature, or structure of an organism or taxonomic group that takes on a function when none previously existed or that differs from its original function.

**Parapatric speciation** Evolution of geographically separate but adjacent populations, with limited possibility of mutual gene flow, into distinct species.

**Speciation** The process by which populations become distinct species by evolving mutual reproductive isolation.

**Sympatric speciation** Evolution of populations occupying the same geographical area, with unlimited possibility of mutual gene flow, into distinct species.

# Retired Speleobiological Glossary

**Accidental** Individuals that enter subterranean habitats by chance and are unable to survive and reproduce there. They form sink populations, but sometimes apparently accidental species successfully establish reproducing populations, as is the case with the frog *Rana iberica* in the cave Serra da Estrela in Portugal (Rosa & Penado, 2013). It is the fourth term (together with trogloxene, troglophile, and troglobiont) in many classification schemes, such as Barr's (1968). In the original Schiner-Racovitza scheme, they are equivalent to trogloxenes.

**Adaptive Shift Hypothesis** A hypothesis originally championed by Howarth (1973) that surface species actively colonized subterranean habitats to exploit new resources, and that subsequent speciation was parapatric.

**Ambimorph** A term coined by Christiansen (1962) for a species with a morphology intermediate between that of a surface dweller and a highly modified subterranean dweller (troglomorph). Its use is rare and would introduce a lot of confusion since only the most extreme case of modification in a lineage would be a troglomorph. For example, only *Speoplatyrhinus poulsoni* would be a troglomorph in the Amblyopsidae fish lineage. All others, such as *Amblyopsis spelaea*, would be ambimorphs.

**Amphibites (biont)** Introduced by Gibert et al. (1994), these are species (aquatic insects) that complete part of their life cycle in either phreatic or hyporheic waters (both aphotic) and part of their life cycle in the benthos and other surface waters. The classic case is the stonefly *Isocapnia* (Stanford & Gaufin, 1974) from the Flathead River in Montana, USA. These species spend most of their life cycle in the deep hyporheic, dispersing to the benthic in order to emerge as mating adults. Sket (2008) recommends, on a linguistic basis, use of the suffix "-biont" rather than "-bite" for this and other terms.

**Cave-dependent character** Introduced by Christiansen (1961), it denotes a character that is modified during the course of isolation in a subterranean habitat. He used it in the *a priori* sense of being predicted change, such as the elongation of relative appendage length. It can also be used in the *a posteriori* sense of discovering a cave-dependent character after morphological analysis (Konec et al., 2015).

**Cavernicole** Generally taken to mean populations reproducing in caves (or subterranean habitats in general), although some authors include nonreproducing accidentals in the definition (Humphreys, 2000). Its use in the former sense seems more appropriate.

**Climatic Relict Hypothesis** A hypothesis popularized by Barr (1960), but drawing from ideas of R. Jeannel and A. Vandel, that surface species passively colonized subterranean habitats because of unfavorable surface conditions caused by changed climate which also resulted in extinction or migration of the surface population(s), and thus subsequent speciation was allopatric.

**Epigeomorph** A rarely used term introduced by Christiansen (1962) for individuals and species without morphological modifications for subterranean life, in contrast with ambimorphs and troglomorphs.

**Eutroglophiles** Initially used by Pavan (1944) and resurrected by Sket (2008), these are species that can complete their life cycles in subterranean habitats or in surface habitats. This is essentially the troglophile category of many other authors, such as Barr (1968).

**Phreatobite (biont)** Species or populations only found in the phreatic zone and unable to complete their life cycle elsewhere, even in other subterranean habitats. Gibert et al. (1994) state that "phreatobites are stygobionts restricted to the deep groundwater substrate of alluvial aquifers." The term is rarely used. Similar terms could be suggested for other aquatic subterranean habitats such as epikarst, but fortunately no one has done so yet. Culver and Pipan (2014) use the phrase epikarst specialists rather than introducing new terminology such as epikarstobite. Sket (2008) recommends, on a linguistic basis, use of the suffix "-biont" rather than "-bite" for this and other terms.

I. Setting the scene: groundwater as ecosystems

**Stygobite (biont)** Species or populations that are unable to complete their life cycle outside of aquatic subterranean habitats. For the most part, all authors agree with this definition, although some (e.g., Sket, 2008) prefer the use of troglobite (troglobiont) for both aquatic and terrestrial species to avoid an unnecessary proliferation of terms. Sket (2008) recommends, on a linguistic basis, use of the suffix "-biont" rather than "-bite" for this and other terms.

**Stygomorph** A term sometimes used for the suite of morphological characters that are convergent among subterranean aquatic species, although most authors seem to use the term troglomorph (see below). Both terms are hypotheses about the expected outcome of evolution in subterranean environments. Expected convergent morphologies include both reduced features, such as eyes and pigment, and elaborated features, such as elongation of appendages (see Table 4.2).

**Stygophile** Facultative permanent species, resident of aquatic subterranean habitats. It can also occur in surface habitats. Eustygophile according to the Pavan-Sket scheme (Sket, 2008), but this term is almost not used. According to Trajano and de Carvalho (2017), both subterranean and surface populations are source populations.

**Stygoxene** Species using both surface and aquatic subterranean habitats during their life cycle. The transition between the two habitats can occur once during the life cycle (e.g., emergence of *Isocapnia* from the hyporheic) or on a daily basis. The daily transition appears to be rare among aquatic species but is common among terrestrial species, including many bats. Trajano and de Carvalho (2017) suggest the surface populations are source populations, but more accurately, both source and sink populations occur in both habitats. According to the Schiner-Racovitza scheme, stygoxenes (trogloxenes) are species that are accidentally and sporadically found in aquatic subterranean habitats.

**Subtroglophile** In the Pavan-Sket scheme, species that use both surface and aquatic subterranean habitats during their life cycle. They are the equivalent of trogloxenes (stygoxenes) in other classifications.

**Troglobite (biont)** Obligate permanent resident of terrestrial subterranean habitats. Used by some authors (e.g., Sket, 2008) for aquatic subterranean species (stygobionts) as well. Except for this distinction, the term is used in the same way by all authors. Sket (2008) recommends, on a linguistic basis, use of the suffix "-biont" rather than "-bite" for this and other terms.

**Troglodyte** A human cave dweller, occasionally incorrectly applied to cave populations of other species.

**Troglomorph (troglobiomorph)** A term for the suite of morphological characters that are convergent among subterranean species. It is a hypothesis about the expected outcome of evolution in subterranean environments. Expected convergent morphologies include both reduced features, such as eyes and pigment, and elaborated features, such as elongation of appendages (see Table 4.2).

**Troglophile** Facultative permanent species, resident of terrestrial subterranean habitats, but it can be used for aquatic subterranean species as well. They can also occur in surface habitats. They are eutroglophile according to the Pavan-Sket scheme (Sket, 2008). According to Trajano and de Carvalho (2017), both subterranean and surface populations are source populations.

**Trogloxene** Species using both surface and subterranean terrestrial habitats during their life cycle, and some authors (e.g., Sket, 2008) use the term for aquatic species as well. The transition between the two habitats can occur once during the life cycle or on a daily basis. The daily transition appears to be rare among aquatic species but is common among terrestrial species, including many bats. Trajano and de Carvalho (2017) suggest the surface populations are source populations, but more accurately, both source and sink populations occur in both habitats. According to the Schiner-Racovitza scheme, trogloxenes are species that are accidentally and sporadically found in terrestrial subterranean habitats.

## Acknowledgments

The work of T.P. was supported by the EU H2020 projects eLTER PPP and eLTER PLUS and RI—SI—LifeWatch. The work of Ž.F. was supported by the Slovenian Research Agency through Research Core Funding P1-0184 and research project N1-0069. We thank S. Mammola and F. Malard for constructive comments that improved this chapter.

## References

Barr, T.C., 1960. A synopsis of cave beetles of the genus *Pseudanophthalmus* of the Mitchell Plain in southern Indiana (Coleoptera, Carabidae). The American Midland Naturalist 63, 307—320.

Barr, T.C., 1968. Cave ecology and the evolution of troglobites. Evolutionary Biology 2, 35—102.

Barr, T.C., Holsinger, J.R., 1985. Speciation in cave faunas. Annual Review of Ecology and Systematics 16, 313—337.

Bichuette, M.E., Rantin, B., Hingst-Zaher, E., Trajano, E., 2015. Geometric morphometrics throws light on evolution of the subterranean catfish *Rhamdiopsis krugi* (Teleostei: Siluriformes: Heptapteridae) in eastern Brazil. Biological Journal of the Linnean Society 114, 136—151.

Bilandžija, H., Hollifield, B., Steck, M., Meng, G., Ng, M., Koch, A.D., Gračan, R., Ćetković, H., Porter, M.L., Renner, K.J., Jeffery, W., 2020. Phenotypic plasticity as an important mechanism of cave colonization and adaptation in *Astyanax* cavefish. eLife 9, e51830.

Blin, M., Tine, E., Meister, L., Elipot, Y., Bibliowicz, J., Espinasa, L., Rétaux, S., 2018. Developmental evolution and developmental plasticity of the olfactory epithelium and olfactory skills in Mexican cavefish. Developmental Biology 441 (2), 242—251.

Borowsky, R., Cohen, D., 2013. Genomic consequences of ecological speciation in *Astyanax* cavefish. PLoS One 8 (11), e79903.

Bressi, L., Aljančič, M.M., Lapini, L., 1999. Notes on the presence and feeding of *Proteus anguinus* Laurenti 1768, outside caves. Rivue Idrobiologia 38, 431—435.

Chapman, P., 1982. The origin of troglobites. Proceedings of the University of Bristol Speleological Society 16, 133—141.

Christiansen, K.A., 1961. Convergence and parallelism in cave Entomobryinae. Evolution 15, 288—301.

Christiansen, K.A., 1962. Proposition pour la classification des animaux cavernicoles. Spelunca 2, 76—78.

Christiansen, K.A., 2012. Morphological adaptations. In: White, W.B., Culver, D.C. (Eds.), Encyclopedia of Caves, second ed. Academic/Elsevier Press, Amsterdam, pp. 517—528.

Christman, M.C., Jernigan, R.W., Culver, D.C., 1997. A comparison of two models for estimating phylogenetic effect on trait variation. Evolution 51, 262—266.

Coineau, N., Boutin, C., 1992. Biological processes in space and time. Colonization, evolution and speciation in interstitial stygobionts. In: Camacho, A.I. (Ed.), The Natural History of Biospeleology. Monografías 7, Museo Nacional de Ciencias Naturales, Madrid, Spain, pp. 423—452.

Copilas-Ciocianu, D., Fišer, C., Borza, P., Petrusek, A., 2018. Is subterranean lifestyle reversible? Independent and recent large-scale dispersal into surface waters by two species of the groundwater amphipod genus *Niphargus*. Molecular Phylogenetics and Evolution 119, 37—49.

Coyne, J.A., Orr, H.A., 2004. Speciation. Sinauer Associates, Sunderland, MA.

Culver, D.C., Kane, T.C., Fong, D.W., 1995. Adaptation and Natural Selection in Caves: The Evolution of *Gammarus Minus*. Harvard University Press, Cambridge.

Culver, D.C., Pipan, T., 2014. Shallow Subterranean Habitats. Ecology, Evolution, and Conservation. Oxford University Press, Oxford.

Culver, D.C., Pipan, T., 2015. Shifting paradigms of the evolution of cave life. Acta Carsologica 44, 415—425.

Culver, D.C., Pipan, T., 2019a. Ecological and evolutionary classifications of subterranean organisms. In: White, W.B., Culver, D.C., Pipan, T. (Eds.), Encyclopedia of Caves, third ed. Academic/Elsevier Press, Amsterdam, pp. 376—379.

Culver, D.C., Pipan, T., 2019b. The Biology of Caves and Other Subterranean Habitats, second ed. Oxford University Press, Oxford.

Desutter-Grandcolas, L., 1997a. Studies in cave life evolution: a rationale for future theoretical developments using phylogenetic inference. Journal of Zoological Systematics and Evolutionary Research 35, 23—31.

Desutter-Grandcolas, L., 1997b. Are troglobitic taxa troglobiomorphic? A test using phylogenetic inference. International Journal of Speleology 26, 1—19.

Desutter-Grandcolas, L., Grandcolas, P., 1996. The evolution toward troglobitic life: a phylogenetic reappraisal of climatic relict and local habitat shift hypotheses. Mémoires de Biospéologie 23, 57—63.

Eigenmann, C.H., 1909. Cave Vertebrates of America. A Study in Degenerative Evolution. Carnegie Institution of Washington, Washington, DC.

Feder, J.L., Flaxman, S.M., Egan, S.P., Comeault, A.A., Nosil, P., 2013. Geographic mode of speciation and genomic divergence. Annual Review of Ecology, Evolution, and Systematics 44, 73—97.

Fišer, C., Blejec, A., Trontelj, P., 2012. Niche-based mechanisms operating within extreme habitats: a case study of subterranean amphipod communities. Biology Letters 8, 578—581.

Giachino, P.M., Vailati, D., 2017. Considerations on biological and terminological aspects of the subterranean and endogean environments. Diversity, correlations and faunistic interchange. Atti della Accademia Nazionale Italiana di Entomologia 65, 157–166.

Gibert, J., Dole-Olivier, M.J., Marmonier, P., Vervier, P., 1990. Surface water/groundwater ecotones. In: Naiman, R.J., Décamps, H. (Eds.), Ecology and Management of Aquatic-Terrestrial Ecotones. Parthenon Publishing, Carnforth, UK, pp. 199–225.

Gibert, J., Stanford, J.A., Dole-Olivier, M.J., Ward, J.V., 1994. Basic attributes of groundwater ecosystems and prospects for research. In: Gibert, J., Danielopol, D.L., Stanford, J.A. (Eds.), Groundwater Ecology. Academic Press, San Diego, pp. 8–40.

Gorički, Š., Niemiller, M.L., Fenolio, D.B., Gluesenkamp, A.G., 2019. Salamanders. In: White, W.B., Culver, D.C., Pipan, T. (Eds.), Encyclopedia of Caves, third ed. Elsevier/Academic Press, Amsterdam, pp. 871–884.

Gould, S.J., Vrba, E.S., 1982. Exaptation—a missing term in the science of form. Paleobiology 8, 4–15.

Holsinger, J.R., 2000. Ecological derivation, colonization, and speciation. In: Wilkens, H., Culver, D.C., Humphreys, W.F. (Eds.), Ecosystems of the World. Vol. 30. Subterranean Ecosystems. Elsevier, Amsterdam, pp. 399–415.

Howarth, F.G., 1973. The cavernicolous fauna of Hawaiian lava tubes, 1. Introduction. Pacific Insects 15, 139–151.

Howarth, F.G., 1980. The zoogeography of specialized cave animals: a bioclimatic model. Evolution 28, 365–389.

Howarth, F.G., 1987. The evolution of non-relictual tropical troglobites. International Journal of Speleology 16, 1–16.

Howarth, F.G., Moldovan, O.T., 2018. The ecological classification of cave animals. In: Moldovan, O.T., Kováč, L., Halse, S. (Eds.), Cave Ecology. Springer, Cham, Switzerland, pp. 41–67.

Howarth, F.G., Stone, F.D., 1990. Elevated carbon dioxide levels in Bayliss Cave, Australia: implications for the evolution of obligate cave species. Pacific Science 44, 207–218.

Humphreys, W.F., 1993. The significance of the subterranean fauna in biogeographical reconstruction: examples from Cape Range peninsula, Western Australia. In: Humphreys, W.F. (Ed.), The Biogeography of Cape Range, Western Australia. Records of the Western Australian Museum, Perth, pp. 165–192. Supplement 45.

Humphreys, W.F., 2000. Background and glossary. In: Wilkens, H., Culver, D.C., Humphreys, W.F. (Eds.), Subterranean Ecosystems. Elsevier, Amsterdam, pp. 3–14.

Jeannel, R., 1950. La marche de l'Évolution. Presses Universitaires de France, Paris.

Jeffery, W.R., 2005. Evolution of eye degeneration in cavefish: the return of pleiotropy. Subterranean Biology 3, 1–11.

Juberthie, C., 1984. La colonisation du milieu souterrain; théories et modèles, relations avec la spéciation et l'évolution souterraine. Mémoires de Biospéologie 11, 65–102.

Konec, M., Prevorčnik, S., Sarbu, S.M., Verovnik, R., Trontelj, P., 2015. Parallels between two geographically and ecologically disparate cave invasions by the same species, Asellus aquaticus (Isopoda, Crustacea). Journal of Evolutionary Biology 28, 864–875.

Kováč, L., 2018. Caves as oligotrophic ecosystems. In: Moldovan, O.T., Kováč, L., Halse, S. (Eds.), Cave Ecology. Springer, Cham, Switzerland, pp. 297–307.

Kozel, P., Pipan, T., Mammola, S., Culver, D.C., Novak, T., 2019. Distributional dynamics of a specialized subterranean community oppose the classical understanding of the preferred subterranean habitats. Invertebrate Biology 138 (3), e12254.

Langerhans, R.B., Riesch, R., 2013. Speciation by selection: a framework for understanding ecology's role in speciation. Current Zoology 59, 31–52.

Leijs, R., van Nes, E.H., Watts, C.H., Cooper, S.J.B., Humphreys, W.F., Hogendoorn, K., 2012. Evolution of blind beetles in isolated aquifers: a test of alternative modes of speciation. PLoS One 7 (3), e34260.

Levins, R., 1969. Some demographic and genetic consequences of environmental heterogeneity for biological control. Bulletin of the Entomological Society of America 15 (3), 237–240.

Leys, R., Watts, C.H.S., Cooper, S.J.B., Humphreys, W.F., 2003. Evolution of subterranean diving beetles (Coleoptera: Dytiscidae: Hydroporini, Bidessini) in the arid zone of Australia. Evolution 57, 2819–2834.

Losos, J.B., Glor, R.E., 2003. Phylogenetic comparative methods and the geography of speciation. Trends in Ecology & Evolution 18, 220–227.

Mammola, S., 2019. Finding answers in the dark: caves as models in ecology fifty years after Poulson and White. Ecography 42, 1331–1351.

Mammola, S., Amorim, I.R., Bichuette, M.E., Borges, P.A.V., Cheeptham, N., Cooper, S.J.B., Culver, D.C., Deharveng, L., Eme, D., Ferreira, R.L., Fišer, C., Fišer, Ž., Fong, D.W., Griebler, C., Jeffery, W.R., Jugovic, J.,

Kowalko, J.E., Lilley, T.M., Malard, F., Manenti, R., Martínez, A., Meierhofer, M.B., Niemiller, M.L., Northup, D.E., Pellegrini, T.G., Pipan, T., Protas, M., Reboleira, A.S.P.S., Venarsky, M.P., Wynne, J.J., Zagmajster, M., Cardoso, P., 2020. Fundamental research questions in subterranean biology. Biological Reviews 95 (6), 1855–1872.

Martinez, A., Mammola, S., 2021. Specialized terminology reduces the number of citations of scientific papers. Proceedings of the Royal Society B 288 (1948), 20202581.

Mayr, E., 1970. Populations, Species, and Evolution. Harvard University Press, Cambridge.

Mejía-Ortíz, L.M., Pipan, T., Culver, D.C., Sprouse, P., 2018. The blurred line between photic and aphotic environments: a large Mexican cave with almost no dark zone. International Journal of Speleology 37, 69–80.

Mohr, C.E., Poulson, T.L., 1966. The Life of the Cave. McGraw-Hill, New York.

Moran, D., Softley, R., Warrant, E.J., 2014. Eyeless Mexican cavefish save energy by eliminating the circadian rhythm in metabolism. PLoS One 9 (9), e107877.

Moran, D., Softley, R., Warrant, E.J., 2015. The energetic cost of vision and the evolution of eyeless Mexican cavefish. Science Advances 1 (8), e1500363.

Niemiller, M.L., Soares, D., 2015. Cave environments. In: Riesch, R., Tobler, M., Plath, M. (Eds.), Extremophile Fishes: Ecology, Evolution, and Physiology of Teleosts in Extreme Environments. Springer, Cham, Switzerland, pp. 161–191.

Niemiller, M.L., Fitzpatrick, B.M., Miller, B.T., 2008. Recent divergence with gene flow in Tennessee cave salamanders (Plethodontidae: Gyrinophilus) inferred from gene genealogies. Molecular Ecology 17, 2258–2275.

Niemiller, M.L., Fitzpatrick, B.M., Shah, P., Schmitz, L., Near, T.J., 2012. Evidence for repeated loss of selective constraint in rhodopsin of amblyopsid cavefishes (Teleostei: Amblyopsidae). Evolution 67, 732–748.

Nosil, P., 2012. Ecological Speciation. Oxford University Press, Oxford.

Oromí, P., Martin, J.L., Medina, A.L., Izquierdo, I., 1991. The evolution of the hypogean fauna in the Canary Islands. In: Dudley, E.C. (Ed.), The Unity of Evolutionary Biology, Portland, OR: Proceedings of the Fourth International Congress of Systematic and Evolutionary Biology, vol. 1. Dioscorides Press, pp. 380–395.

Pavan, M., 1944. Considerazioni sui concetti di troglobio, troglofilo e troglosseno. Le Grotte d'Italia, ser 2 5, 35–41.

Peck, S.B., 1990. Eyeless arthropods on the Galapagos Islands, Equador: composition and origin of the crtyptozoic fauna of a young, tropical, oceanic archipelago. Biotropica 22, 366–381.

Peck, S.B., 2002. Florida and Caribbean karsts: an overview of diversity and origins of the obligate invertebrate cave faunas. In: Martin, J.B., Wicks, C.M., Sasowsky, I.D. (Eds.), Hydrogeology and Biology of Post-Paleozoic Carbonate. Karst Waters Special Publication 7 Aquifers. Karst Waters Institute, Charles Town, WV, pp. 54–60.

Peck, S.B., Finston, T.L., 1993. Galapagos Islands troglobites: the questions of tropical troglobites, parapatric distributions with the eyed sister-species, and their origin by parapatric speciation. Mémoires de Biospéologie 20, 19–37.

Peralta-Maraver, I., Galloway, J., Posselt, M., Arnon, S., Reiss, J., Lewandowski, J., Robertson, A.L., 2018. Environmental filtering and community delineation in the streambed ecotone. Scientific Reports 8, 15871.

Pipan, T., Culver, D.C., 2012. Convergence and divergence in the subterranean realm: a reassessment. Biological Journal of the Linnean Society 107, 1–14.

Plath, M., Tobler, M., 2010. Subterranean fishes of Mexico (Poecilia mexicana, Poeciliidae). In: Trajano, E., Bichuette, M.E., Kapoor, B.G. (Eds.), Biology of Subterranean Fishes. Science Publishers, Enfield, NH, pp. 281–330.

Poulson, T.L., 1963. Cave adaptation in amblyopsid fishes. The American Midland Naturalist 70, 257–290.

Pulliam, H.R., 1988. Sources, sinks, and population regulation. The American Naturalist 132, 652–661.

Racovitza, E.G., 1907. Essai sur les problèmes biospéologiques. Archives de Zoologie Expérimentale et Générale 6, 371–488.

Ribera, I., Cieslak, A., Faille, A., Fresneda, J., 2018. Historical and ecological factors determining cave diversity. In: Moldovan, O.T., Kovac, L., Halse, S. (Eds.), Cave Ecology. Springer Nature, Cham, Switzerland, pp. 229–254.

Riesch, R., Plath, M., Schlupp, I., 2011. Speciation in caves: experimental evidence that permanent darkness promotes reproductive isolation. Biology Letters 7, 909–912.

Romero, A., 2009. Cave Biology. Life in Darkness. Cambridge University Press, Cambridge.

Romero, A., Green, S.M., 2005. The end of regressive evolution: examining and interpreting the evidence from cave fishes. Journal of Fish Biology 67, 3–32.

Rosa, G., Penado, A., 2013. Rana iberica (Boulenger, 1879) goes underground: subterranean habitat usage and new insights on natural history. Subterranean Biology 11, 15–29.

Schiner, J.R., 1854. Fauna der Adelsberger-, Luegger-, and Magdalenen Grotte. In: Schmidl, A. (Ed.), Die Grotten und Höhlen von Adelsberg, Lueg, Planina und Laas. Braunmü, Wien, pp. 231–272.

Schiødte, J.C., 1849. Bidrag til den underjordisje Fauna. In: Transactions of the Royal Danish Society. 5th ser, Division of Natural History and Mathematics, vol. 2, pp. 1–39.

Schluter, D., 2001. Ecology and the origin of species. Trends in Ecology & Evolution 16, 372–380.

Schluter, D., 2009. Evidence for ecological speciation and its alternative. Science 323, 737–741.

Schilthuizen, M., Cabanban, A.S., Haase, M., 2005. Possible speciation with gene flow in tropical cave snails. Journal of Zoological Systematics and Evolutionary Research 43, 133–138.

Simčič, T., Sket, B., 2019. Comparison of some epigean and troglobitic animals regarding their metabolism intensity. Examination of a classical assertion. International Journal of Speleology 48, 133–144.

Skeels, A., Cardillo, M., 2019. Reconstructing the geography of speciation from contemporary biodiversity data. The American Naturalist 193, 240–255.

Sket, B., 2008. Can we agree on an ecological classification of subterranean animals? Journal of Natural History 42, 1549–1563.

Stanford, J.A., Gaufin, A.R., 1974. Hyporheic communities of two Montana rivers. Science 185, 700–702.

Stoch, F., 2004. Colonization. In: Gunn, J. (Ed.), Encyclopedia of Caves and Karst Science. Taylor & Francis, Inc, New York, USA, pp. 483–488.

Tobin, B.W., Hutchins, B.T., Schwartz, B.F., 2013. Spatial and temporal changes in invertebrate assemblage structure from entrance to deep-cave zone of a temperate marble cave. International Journal of Speleology 42, 203–214.

Tobler, M., 2009. Does a predatory insect contribute to the divergence between cave- and surface-adapted fish populations? Biology Letters 5, 506–509.

Tobler, M., Riesch, R., Tobler, C.M., Schulz-Mirbach, T., Plath, M., 2009. Natural and sexual selection against immigrants maintains differentiation among micro-allopatric populations. Journal of Evolutionary Biology 22, 2298–2304.

Trajano, E., 2012. Ecological classification of subterranean organisms. In: White, W.B., Culver, D.C. (Eds.), Encyclopedia of Caves, second ed. Academic/Elsevier Press, Amsterdam, pp. 275–277.

Trajano, E., de Carvalho, M.R., 2017. Towards a biologically meaningful classification of subterranean organisms: a critical analysis of the Schiner-Racovitza system from a historical perspective, difficulties of its application and implications for conservation. Subterranean Biology 22, 1–26.

Trontelj, P., 2018. Structure and genetics of cave populations. In: Moldovan, O.T., Kováč, L., Halse, S. (Eds.), Cave Ecology. Springer, Cham, pp. 269–295.

Trontelj, P., 2019. Adaptation and natural selection in caves. In: White, W.B., Culver, D.C., Pipan, T. (Eds.), Encyclopedia of Caves. Elsevier/Academic Press, Amsterdam, pp. 40–46.

Trontelj, P., Blejec, A., Fišer, C., 2012. Ecomorphological convergence in cave communities. Evolution 66, 3852–3865.

Vandel, A., 1964. Biospéologie: la biologie des animaux cavernicoles. Gauthier-Villars, Paris.

Via, S., 2001. Sympatric speciation in animals: the ugly duckling grows up. Trends in Ecology & Evolution 16, 381–390.

Ward, J.V., Voelz, N.J., 1997. Interstitial fauna along an epigean-hypogean gradient in a Rocky Mountain river. In: Gibert, J., Mathieu, J., Fournier, F. (Eds.), Groundwater/surface Water Ecotones: Biological and Hydrological Interactions and Management Options. Cambridge University Press, Cambridge, pp. 37–41.

White, M.J.D., 1978. Modes of Speciation. W. H. Freeman and Company, San Francisco.

Wilkens, H., Hüppop, K., 1986. Sympatric speciation in cave fishes? Studies on a mixed population of epi- and hypogean Astyanax (Characidae, Pisces). Journal of Zoological Systematics and Evolutionary Research 24 (3), 223–230.

Wilkens, H., Strecker, U., 2017. Evolution in the Dark. Darwin's Loss Without Selection. Springer, Cham.

# Drivers and patterns of groundwater biodiversity

# Groundwater biodiversity and constraints to biological distribution

*Pierre Marmonier[1], Diana Maria Paola Galassi[2], Kathryn Korbel[3], Murray Close[4], Thibault Datry[5] and Clemens Karwautz[6]*

[1]Univ Lyon, Université Claude Bernard Lyon 1, CNRS, ENTPE, UMR 5023 LEHNA, Villeurbanne, France; [2]Department of Life, Health and Environmental Sciences, University of L'Aquila, L'Aquila, Italy; [3]School of Natural Sciences, Macquarie University, Sydney, Australia; [4]Institute of Environmental Science and Research, Christchurch, Canterbury, New Zealand; [5]INRAE, UR-RiverLY, Lyon, France; [6]University of Vienna, Department of Functional & Evolutionary Ecology, Vienna, Austria

## Introduction

Groundwater biodiversity is invisible to most people outside the field of groundwater ecology. Media typically convey the idea of subterranean life being restricted to a few hyper-specialized creatures. In fact, microbes densely colonize groundwater and even metazoans are many and show varying degrees of specialization. The first mentions of blind and depigmented groundwater metazoans probably date back to the report in 1541 of the Chinese hyaline fish *Sinocyclocheilus hyalinus* (see Romero, 2001), followed by the report of the European cave salamander, *Proteus anguinus*, in 1689 (Aljancic et al., 1993). With microorganisms, already Antonie van Leeuwenhoek (1677) discovered bacteria in the water of large wells and later, Hassall (1850) reported on microbes in a London waterwork. Since the first discoveries of living organisms in groundwater, knowledge of groundwater biodiversity has increased worldwide (Griebler and Lueders, 2009; Zagmajster et al., 2018; Niemiller et al., 2019). Today, we know that obligate groundwater organisms represent an important, yet largely underestimated, component of the Earth's total biological diversity (Culver and Holsinger, 1992; Danielopol et al., 2000; Malard, 2022). The difficulty in assessing the true diversity of groundwater life is primarily due to a low sampling effort. Groundwater represents the largest reservoir of liquid freshwater on Earth, but groundwater ecologists have only explored a minor fraction of it.

Groundwater metazoan communities comprise a mix of species that show different degrees of dependence on the subterranean environment. In the first edition of the Groundwater Ecology textbook, Gibert et al. (1994a) recognized three main ecological categories: stygoxenes, stygophiles, and stygobionts. Stygoxenes have no affinities with groundwater: they occur accidently in aquifers. Stygophiles are represented by a variety of metazoans that all have higher affinities for the groundwater environment than do stygoxenes. They include insect species whose early instar larvae can reside in the subsurface sediments of surface streams as well as species with a life cycle necessitating both a groundwater and surface water stage (for example the stonefly *Isocapnia*) or those that retreat to groundwater in times of stress (e.g., prolonged drought). Stygophiles are also represented by species with no aerial stage that can complete their entire life cycle either in surface water or in groundwater (for example some nematodes, oligochaete worms, crustaceans, and tardigrades). Stygobionts are obligate groundwater species that complete their entire life cycle exclusively in groundwater. Culver and his coauthors (this volume) question the need and justification for such an elaborated ecological classification and terminology of groundwater metazoans. They prefer to use the general terms of specialist and generalist species. Specialist groundwater species have a strict dependence on groundwater because they complete their entire life cycle exclusively in groundwater (stygobionts) or because they necessarily complete a part of their life cycle there. Generalist species exploit a wide range of resources that they can draw from both groundwater and surface water. The three ecological categories described above do not apply well to microorganisms. Microbes are continuously imported into the subsurface by surface water infiltrating into aquifers. Only a minute fraction of the cells in pore water seem well adapted to the environmental conditions and is highly active. The active part of groundwater microbial communities is attached to the rocks and sediment matrix as single cells, small colonies, and in exceptional cases as a surface covering biofilm (Griebler et al., 2022). Only zones that are hydrologically isolated from the surface for hundreds and thousands of years, may harbor strains not found anywhere else. In general, the groups of microorganisms found in groundwater do not contain endemic taxa but are known already from soils and surface waters (Griebler and Lueders, 2009; Fillinger et al., this volume).

The diversity of obligate groundwater organisms is unevenly distributed across taxa. Some metazoan groups, particularly within the hexapods, show very few species in groundwater, whereas other groups, particularly within the crustaceans, are more diversified in groundwater than in surface water. In Europe, the number of obligate groundwater crustaceans exceeds the number of surface-dwelling crustacean species, even though the description of groundwater species lags behind that of surface species (Stoch and Galassi, 2010). As a rule of thumb, the obligate groundwater fauna is taxonomically dominated by crustaceans, followed by molluscs and oligochaete worms. An extensive survey of stygobionts in six European countries - Belgium, France, Italy, Portugal, Slovenia, and Spain— showed that crustaceans, molluscs, and oligochaete worms represented 70%, 16%, and 6.1% of the obligate-groundwater species diversity (Deharveng et al., 2009; Venarsky et al., chapter 10, this volume). However, these proportions vary across continents. In Europe, only two species of stygobiotic insects, belonging to the dytiscid beetles, have been discovered, whereas over 100 species of stygobiotic dytiscid species have been described from calcrete aquifers in Australia (Langille et al., 2021).

Groundwater organisms are likely to be found wherever there are open spaces within the water-filled subterranean rock and sediment matrix and a reconcilable temperature prevails. Some species make use of the food resources available in superficial twilight habitats (Culver and Pipan, 2014), whereas others colonized habitats at considerable depths below the water table: 500 m in lakes, 600 m in wells, and 3600 m in mines (Fišer et al., 2014). In all cases, microbes are found in even deeper zones (Pedersen, 2000), with their abundance and distribution mainly governed by the availability of energy and temperature (Griebler and Lueders, 2009). The diversity and abundance of obligate groundwater metazoan assemblages are spatially heterogeneous at local (Malard et al., 2002; Gutjahr et al., 2014), regional (Johns et al., 2015) and continental scales (Eme et al., 2015; Iannella et al., 2020, 2021). Striking features in the spatial distribution of groundwater metazoans are described in the next chapter of this book (Zagmajster et al., this volume).

In this chapter, we provide a brief account of the taxonomic diversity of living forms in groundwater including viruses, prokaryotes, microeukaryotes, and metazoans. Then, we present the main constraints to biological distributions in groundwater. Emphasis is put on the diversity and constraints to the distribution of metazoans since another chapter of this book is dedicated to microbial diversity and processes in groundwater (Fillinger et al., this volume). For sake of clarity, we separately consider physical and chemical constraints and biotic interactions. However, we acknowledge that these constraints interact across all ranges of spatial and temporal scales to produce complex distribution patterns (see Gibert et al., 1994a for a discussion of scales, hierarchy, and processes in groundwater ecology). Among physical constraints, we focus on the importance of the size and connectedness of voids, hydrological exchanges between groundwater and surface water, groundwater chemistry, and temperature. Then, we provide examples of how competition, predation, and other forms of biotic interactions may affect the distribution of organisms. We end this chapter by considering how the effects of past constraints linked to major paleogeographic events and historical climates have left their imprints on the present-day distribution patterns of groundwater biodiversity.

## An overview of groundwater biodiversity

All groundwater ecosystems are characterized by the absence of light; hence, they lack photosynthetic primary production. In the absence of light-driven production, particulate organic carbon production at the base of groundwater food webs essentially comes from two pathways: chemoautotrophic primary production, where bacteria fix inorganic carbon to produce biomass, and heterotrophic production, where bacteria produce biomass from assimilation of dissolved organic compounds, which are mainly imported from surface environments. Most groundwater food webs are mainly supported by heterotrophic production (Simon, 2019). Shallow subterranean habitats and deep habitats directly connected to the surface by conduit flow may receive appreciable amounts of organic matter (OM) from the surface. However, heterotrophic production is severely limited in many groundwater systems because a substantial fraction of OM produced at the surface is intercepted in the soil layers and the vadose zone. Datry et al. (2005) estimated that the annual flux of dissolved organic carbon (DOC) reaching the water table of an unconsolidated sediment aquifer was one to two orders of magnitude lower than the annual flux of DOC in the soil organic layer. Herrmann et al. (2020) provided evidence that the importance of microbial biomass produced by chemo-lithoautotrophs relative to that of heterotrophs might be seriously underestimated. In any way, quantitative data on organic carbon import to the subsurface as well as rates of chemo-autotrophic inorganic carbon fixation are lacking (Griebler et al., 2022).

Although low food availability ranks most groundwater habitats among the most energy-limited habitats on Earth, they nevertheless support a high diversity of living forms occupying a reduced number of levels in the trophic cascade (Gibert and Deharveng, 2002). Chemoli-thoautotrophic bacteria act as primary producers, heterotrophic bacteria and fungi as primary consumers, whereas protozoans and metazoans, including invertebrates, fish and salaman-ders, occupy the higher trophic levels (Herrmann et al., 2020; Griebler et al., 2022; Venarsky et al., chapter 10, this volume). Below, we provide a brief overview of the current knowledge of the diversity of living forms in groundwater.

## Viruses

While it is widely accepted that microbes occupy almost every corner on this planet, including groundwater and the deep terrestrial subsurface (Griebler and Lueders, 2009), it has been overlooked for a long time that whenever there are microbes there are viruses. The vast majority of viruses in natural ecosystems infect prokaryotes (i.e., Bacteria and Archaea; see below). These viruses we call (bacterio) phages. Phages are the most abundant biological entities in the biosphere (Suttle, 2005). Unfortunately, almost nothing is known about phages in groundwater. On one hand, recent estimates indicate that >80% of all pro-karyotic biomass on Earth is located in the subsurface (e.g., terrestrial aquifers, vadose zones, and below the seafloor; Bar-On et al., 2018). On the other hand, as has been mentioned already above, groundwater ecosystems are typically energy poor exhibiting comparatively low productivity, which results in low concentrations and activity of prokaryotes, the pre-dominant hosts of viruses in groundwater.

The abundance of phages in aquatic environments is generally higher than the number of prokaryotes with a virus-to-prokaryote ratio (VPR) of 10 reported frequently. Considering all available data, the VPR in groundwater ranges between 0.08 and 728, with the majority of studies reporting values < 10 (Schweichhart et al., 2022). While there is a handful of studies that quantified virus abundance in groundwater, viral diversity is untapped. One study in granitic groundwater observed phages down to a depth of 450 m and found a diverse set of viral morphotypes resembling tailed phages, polyhedral phages, filamentous phages, and fusiform archaeal phages (Kyle et al., 2008). To date, only a few phages have been iso-lated from groundwater microbes (Eydal et al., 2009; Hylling et al., 2020). Metagenomics is the tool of choice for characterizing viruses because they do not share a conserved gene that could be used for phylogenetic analysis and taxonomic classification. Besides the pres-ence of families commonly occurring in surface waters, i.e., Myoviridae , Podoviridae , Sipho-viridae , other groups have been particularly detected in groundwater, i.e., Microviridae and Inoviridae (Smith et al., 2013; Roux et al., 2019). The functional role of these phages has not been explored, however, an intimate involvement in microbial food web interactions and car-bon flow is expected (Schweichhart et al., 2022).

## Prokaryotes: Archaea and Bacteria

The majority of the Earth's prokaryotic biomass is located in the subsurface (Bar-On et al., 2018). Prokaryotic communities not only represent the major fraction of total biomass in groundwater, they are also exceptional in their taxonomic and functional diversity. The presence of a diverse and abundant prokaryotic community in groundwater has been known for a long time (van Leeuwenhoek 1677; Hvid-Hansen, 1951). In the past decades, research indicating that microbes are present even in the most extreme subterranean ecosystems (e.g., volcanic groundwater-fed springs, Segawa et al., 2015; Bornemann et al., 2022) and at depths of 1000s of meters below the soil surface have shifted the perspective on the limits of life (Pedersen, 2000; Newman and Banfield, 2002).

Prokaryotic groundwater communities play a central role in the cycling of nitrogen, carbon, sulfur, and iron within aquifers (Fillinger et al., this volume). Compared to surface waters, the densities of prokaryotes are much lower in the subsurface. They typically range from $10^4$ to $10^8$ cells gram$^{-1}$ of sediment and $10^3$ to $10^5$ cells mL$^{-1}$ of groundwater (Griebler and Lueders, 2009). The majority of groundwater bacteria are attached rather than free-living, with suspended bacteria only contributing to $\leq 10\%$ of the total prokaryotic biomass (Smith et al., 2018). Attached bacteria occur predominantly in small, unevenly distributed colonies rather than large biofilms (Smith et al., 2018). Due to the sampling limitations of subsurface sediments, the majority of research data on microbial groundwater communities is attained from pumped water samples (Griebler et al., 2022).

Prokaryotic communities in shallow aquifers with high connectivity to the surface are more dynamic in their composition and more active than communities in deeper aquifers. However, the strong heterogeneity within and among aquifers as well as methodological limitations complicate attempts to summarize the overall diversity of groundwater microbes. Today, the characterization of the prokaryotic diversity in aquifers is largely based on 16S rDNA amplicon sequencing data. Taxa that are commonly identified in groundwater include a mixture of bacteria and archaea that do not seem to be exclusive to the subsurface (Griebler and Lueders, 2009). Deep groundwater aquifers tend to have fewer and less diverse communities, commonly containing substantial proportions of methanogens, sulfate-reducing bacteria, thermophilic prokaryotes, and an increased archaeal abundance (Purkamo et al., 2018). The most abundant bacteria in aquifers are found within the bacterial lineages of Alphaproteobacteria, Gammaproteobacteria, Betaproteobacteria, Flavobacteriia, Actinobacteria, Sphingobacteria, and Bacilli.

Environmental parameters such as temperature, pH, and nutrient supply, influence microbial community composition (Stegen et al., 2018; Fillinger et al., 2019a), but microscale conditions (e.g., redox processes) are highly relevant for the activity of cells (Peiffer et al., 2021). The array of functions within groundwater prokaryotes spans almost all electron acceptors and donors that provide enough energy to survive. Specific processes and model organisms have been studied in great detail such as sulfate reduction (e.g., Desulfobacterales), denitrification (e.g., Pseudomonadales), ammonia oxidation by bacteria (e.g., Nitrosomondales), and archaea (e.g., Nitrosopumilales), iron reduction (Desulfuromondales: *Geobacter*) and methanogenesis (e.g., Methanosarcinales).

Recently, a combination of high-resolution electron microscopy and molecular techniques revealed an even more diverse and previously unknown community of prokaryotic cells

- many of which are smaller than 0.2 µm and lack several biosynthetic pathways (Luef et al., 2015; He et al., 2021). Metagenomic data highlight the importance of species interactions and metabolic handoff (e.g., one microbe's waste is another microbe's food) in the subsurface (Anantharaman et al., 2016; Hug and Co, 2018). Many of these newly discovered prokaryotes belong to the candidate phyla radiation (CPR) group, which are difficult to be cultivated and are even missed by the classic 16S rRNA approach. Many of the CPR bacteria live as obligate symbionts or parasites (Chaudhari et al., 2021; He et al., 2021). The superphyla Patescibacteria and DPANN archaea (Diapherotrites, Parvarchaeota, Aenigmarchaeota, Nanoarchaeota, and Nanohaloarchaeota) are representative of this lifestyle.

## Microeukaryotes

Single-cell eukaryotic microorganisms, including groups commonly referred to as fungi and protozoa, play an important role in soils and surface aquatic systems. Their diversity, functions, and implication in groundwater food webs are increasingly being targeted by the use of novel sequencing technologies (Sohlberg et al., 2015; Korbel et al., 2017; Perkins et al., 2019; Herrmann et al., 2020; Retter and Nawaz, 2022). Although microeukaryotes are less abundant and diverse in groundwater communities than representatives of the Bacteria and Archaea (Griebler and Lueders, 2009), they substantially contribute to the diversity of microbial communities.

Owing to their diverse life strategies as decomposers, symbionts, parasites, and pathogens, fungi, similar to the prokaryotes, play vital ecological roles in aquatic environments. While the functional importance of fungi in surface waters is long appreciated, i.e., the quantitative contribution to leave litter breakdown in streams, their taxonomic diversity has been historically understudied (Grossart et al., 2019). The reasons are manifold. Reproduction life cycles in which fungi produce distinct fruiting bodies are often unnoticed. There are biases in the molecular analysis due to the choice of inappropriate primers and DNA barcode markers. Finally, only a few aquatic ecosystems have been systematically investigated for fungi. Thus, it is not surprising that there are only a few studies conducted on fungi in groundwater (Retter and Nawaz, 2022).

Fungi in groundwater habitats so far are mostly affiliated with pathogenic and/or saprotrophic lineages. Early phylogenetically diverging lineages include Cryptomycota and Microsporidia of which many are endoparasites (Livermore and Mattes, 2013). Members of another ancient branch of fungi comprise chytrids and blastocladiomycetes. Nawaz et al. (2018) reported the Chytridiomycota fraction to make up the third-largest phylum detected in their groundwater study. In groundwater, chytrids are mainly parasites of other fungi or invertebrates (Grabner et al., 2020). Wurzbacher et al. (2020) found that around 8% of the fungal diversity recovered from volcanic springs was made up of early diverging lineages of fungi, with a significant part belonging to yet unexplored yeast lineages. Most sequences recovered from groundwater metabarcoding efforts are affiliated with Dikarya (Ascomycota + Basidiomycota), many of them yeasts or yeast-like taxa. Yeast species are all placed within Dikarya but do not form a monophyletic group. Yeasts are especially remarkable, as they could be recovered from every habitat of the world, including the deep subsurface (Sohlber et al., 2015; Ivarsson et al., 2018; Purkamo et al., 2018). Dikarya grows on all kinds of organic substrates, such as carbohydrates, some being able to act well under oxygen-deprived conditions (facultative anaerobes). Lategan et al. (2012) identified 89 fungal strains in unconsolidated sediment aquifers of New South Wales, Australia. Fungi and yeasts represented <5% of the total abundance of the heterotrophic microbial community. The assessment of fungal diversity in groundwater samples from 16 boreholes around Lakes Stechlin and Grosse Fuchskuhle, northeast Germany, yielded a total of 1,070 operational taxonomic units (OTUs) essentially belonging to the Ascomycota and Basidiomycota (Perkins et al., 2019). Similarly, Sohlberg et al. (2015) revealed a surprisingly high diversity of active fungi in deep oligotrophic groundwater of crystalline rock fractures, at Olkiluoto, Finland. Today, a significant proportion of the sequences retrieved remains unknown/unclassified, termed the "fungal dark matter" by Grossart et al. (2016), the composition and ecology of which are largely unknown. An update on groundwater studies is found in Retter and Nawaz (2022). Fungi are often most abundant and diversified in the shallow recharge zone of aquifers, where they are represented by taxa commonly found in soil and surface aquatic environments (Lategan et al., 2012; Perkins et al., 2019), indicating that they are either directly imported from the surface or that fungal populations feed on soil derived OM.

The Protista are known to occur in both shallow and deep groundwater environments (Novarino et al., 1997). They are represented by flagellates, ciliates, and ameboid organisms, with flagellates being numerically dominant. Systematic studies with respect to their taxonomic diversity in groundwater ecosystems are entirely missing, with only a few attempts that have been made (Euringer and Lueders, 2008; Holmes et al., 2013). Bacterivorous flagellates grazing on bacteria were shown to exert top-down control on bacterial biomass in an organically contaminated aquifer (Kinner et al., 2002). However, studies on microbial food web interactions in groundwater are extremely rare. A recent study targeting a shallow pristine alpine aquifer provided evidence for bottom-up control of prokaryotic biomass and production, with protozoan grazing and viral lysis playing only a minor role in carbon flow (Karwautz et al., 2022). Hardly addressed so far are the interactions between prokaryotes, microeukaryotes, and metazoans (Griebler et al., 2022). An occurrence network analysis of 2,261 bacterial and 480 eukaryotic OTUs (protists and fungi) as identified in a limestone-dominated aquifer of the Upper Muschelkalk in Central Germany, resulted in a network of significant positive and negative interactions between 730 bacterial and 94 microeukaryotic OTUs (Herrmann et al., 2020). Research on direct interactions between microbes and fauna in groundwater ecosystems is exceptional. A few studies in karst environments found a close connection between microbial biofilms and subterranean metazoans (see citations in Griebler et al., 2022). Such relationships have been suggested to exist also in sedimentary environments, but are not yet supported by empirical data. Taking into account that the microbial biomass in natural pristine aquifers is low and dense biofilms are largely absent, a detritus-based food web is more likely than a food web based on microbial production alone (Griebler et al., 2022).

## Metazoans

We lack recent estimates of the number of obligate groundwater metazoan species worldwide. The faunistic, distributional, and ecological synthesis of the World fauna inhabiting subterranean waters published by Botosaneanu (1986) listed more than 5,500 species.

Zagmajster et al. (2018) used per-country species numbers to map the global distribution of obligate groundwater crustaceans: species richness is higher at temperate latitudes, especially in the Palearctic and Nearctic regions. According to more recent estimates of species richness in Europe, 1,047 stygobiotic species are known from Belgium, France, Italy, Portugal, Slovenia, and Spain (Deharveng et al., 2009) and 1,570 species of obligate groundwater crustaceans are known from Europe (Zagmajster et al., 2014). Niemiller et al. (2019) reported a total of 469 stygobiotic species in their account of the subterranean biodiversity in the United States and Canada. Crustaceans represent by far the most species-rich taxon in groundwater but many other taxa are present. We provide below a nonexhaustive description of major groundwater metazoan taxa.

Very few sponges were described from groundwater: the sole true stygobiotic species is *Eunapius subterraneus* from Croatia. In addition, *Racekiela cavernicola* was described from a large cave system in northeastern Brazil as a stygobiont, but its ecological status needs confirmation (Volkmer-Ribeiro et al., 2010). Groundwater cnidarians are also represented by a single species, the hydrozoan *Velkovrhia enigmatica* described from the Dinaric karst but other species are likely to be present.

The Platyhelminthes are represented in groundwater by 150 species, mostly Dendrocoelidae. The genus *Dendrocoelum* is highly diversified in Palearctic groundwater: some species are even present in hypoxic or anoxic habitats, such as *D. obstinatum* described from the sulfidic Movile Cave, Romania, and *D. leporii* from sulfidic lakes of the Frasassi Cave, Italy. Some Planariidae (e.g., *Atrioplanaria* and *Polycelis*) and Catenulidae (e.g., *Dasyhormus stygios*) are also present in Europe. Stygobiotic Platyhelminthes are also known from South America, Dimarcusidae (*Hausera hauseri*) from Brazil, and Dugesiidae (genus *Girardia*) from Brazil and Mexico, and North America (Kenkiidae of the genus *Sphalloplana*). The class Temnocephalida is known in southern Europe and New Guinea.

Among the Annelida, the Polychaeta is predominantly confined to the marine habitat, but a few species occur in groundwater. *Troglochaetus beranecki* is widely distributed in the hyporheic zone of streams in Europe and it also occurs in Colorado, USA. *Marifugia cavatica* (Serpulidae) is known from the Dinaric karst, Europe, and the genus *Namanereis* (Nereididae) from groundwater of Australia, New Zealand, Yemen, Mexico, and Canary Islands. The Clitellata (oligochaete worms and leeches) is the most species-rich class of annelids in groundwater with stygobiotic species belonging to the families Potamodrilidae and Dorydrilidae (two obligate groundwater families), Lumbriculidae (7 stygobiotic genera), Tubificidae (5 genera, Fig. 5.1A), Enchytraeidae and Naididae. Leeches (Hirudinea) are represented by a few stygobitic species belonging to the genera *Dina, Trocheta, Croatobranchus,* and *Haemopis*.

The Mollusca is the second-most diversified group of stygobionts after the Crustacea. Most global diversity is within the Gastropoda but a few species of Bivalvia (e.g., *Congeria* and *Pisidium*) also occur in groundwater (Prié, 2019). Among the Gastropoda, Hydrobiidae is by far the most species-rich family with over 350 species described. However, within-family diversity varies across continents: Hydrobiidae have no styobiotic species in North America, the highest diversity is within the family Cochliopidae. Despite their high diversity, snails remain among the most understudied of all subterranean groups, even though freshwater snails have the highest documented extinction rate of any taxonomic group (Gladstone et al., 2021).

The Crustacea are the most diverse metazoan group in groundwater (Zagmajster et al., 2014), with a vast majority of species belonging to the Copepoda and the Amphipoda. Within

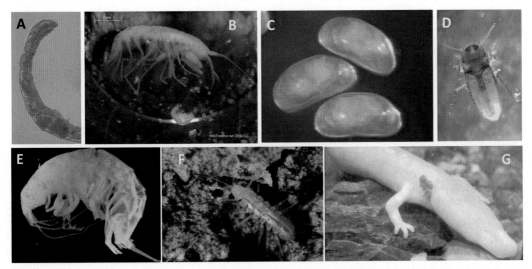

**FIGURE 5.1** An overview of the diversity of obligate groundwater metazoans. (A) *Gianus aquaedulcis* (Clitellata, France, Photo M. Des Châtelliers, body size: 1.7 mm), (B) *Paraleptamphopus subterraneus* (Amphipod, New Zealand, Photo A. Bolton, body size 10 mm), (C) *Fabaeformiscandona breuili* (Ostracoda, France, Photo P. Marmonier, body size: 500 μm), (D) *Paroster sp.* (Ditycidae, Australia, Photo K. Korbel, body size: 2.5 mm), (E) *Niphargus virei* (Amphipod, France, Photo P. Marmonier, body size: 18 mm), (F) *Proasellus valdensis* (Isopod, France, Photo R. Lepennec, body size: 7 mm), (G) *Proteus anguinus* (Urodela, France, Photo CNRS Moulis, body size: 25 cm).

Copepoda, Calanoida (11 species), Cyclopoida (400 species), Harpacticoida (700 species), and Gelyelloida (2 species) have successfully colonized fresh groundwater (Galassi et al., 2009). About 400 stygobiotic species of Ostracoda are described from porous and karstic groundwater habitats (Fig. 5.1C). However, the diversity of that group is probably underestimated. From an extensive survey of groundwater in the arid Pilbara region of northwestern Australia, Reeves et al. (2007) identified 111 species belonging to 21 genera, among which 73 were new to science. In contrast, Cladocera are poorly diversified in groundwater with only seven species of Chydoridae (Brancelj and Dumont, 2007). In the Peracarida, the Spelaeogriphacea are known from nonsaline groundwater of the southern hemisphere, the Thermosbaenacea from brackish and fresh groundwater with a Tethyan distribution, and the Mysida from anchialine and fresh groundwater of Mexico and the Mediterranean basin. The Syncarida contains 256 species worldwide (Camacho and Valdecasas, 2008), but new species are continuously being described both within the Bathynellidae (mostly in the Palearctic region) and the Parabathynellidae (in Europe, Africa, Australia, and Asia). The Isopoda are highly diversified in groundwater with stygobiotic species present in several suborders including the Asellota (Fig. 5.1F), Microcerberidea, Sphaeromatoidea, Cymothoida, and Phreatoicidea. Some species rich-families and genera of isopods are known only from groundwater (for example, the Stenasellidae), whereas others contain a mix of obligate groundwater species and surface aquatic species (for example, the Asellidae, the Lepidocharontidae, and the Microparasellidae). The Amphipoda is the most diversified order of Crustacea in groundwater with more than 740 species described worldwide according to Holsinger (1993). Species-rich families vary across continents: the Niphargidae in Europe

(Fig. 5.1E), the Crangonyctidae in North America, the Paramelitidae in Australia, the Paraleptamphopidae in New Zealand (Fig. 5.1B), the Gammaridae in China, the Bogidiellidae with a circum-global distribution, and the Hadziidae distributed across the Palearctic, Nearctic and Afrotropical regions. The stygobitic Decapoda belongs to several genera (e.g., *Typhlocaris, Orconectes, Troglocaris*) known from North America, West Indies, Australia, and the Mediterranean regions (Stern et al., 2017).

Hardly any Hexapoda group colonized groundwater with the exception of Coleoptera (i.e., Dryopidae, Dytiscidae, Elmidae, Hydrophilidae, and Noteridae). They are recorded from Asia, South America, Europe, Africa, North America, and Australia. The most diversified stygobitic beetle fauna in the world is from Australia with more than 100 species of Dytiscidae known from calcrete aquifers (Langille et al., 2021) (Fig. 5.1D).

Among vertebrates, 288 species of fish are reported from caves and groundwater (Proudlove, 2006). Most stygobiotic fish are tropical and subtropical from Asia, Africa, South and North America, and Australia but one isolated cave population of *Barbatula barbatula* was recently reported in Europe (Behrmann-Godel et al., 2017). Obligate cave salamanders of the family Plethodontidae are represented by 13 species in the USA (Goricki et al., 2019), whereas a single species of cave salamander, *Proteus anguinus*, is known from Europe (Fig. 5.1G).

## Physical constraints to biological distribution

### Size of voids and their interconnectedness

The presence of microorganisms is hardly restricted by the void size with the exception of massive rock matrix (e.g., basalt) and dense clay material. The density of microbial communities is potentially more influenced by the sediment surface' area, hence by grain size, than by the size of voids, although the interconnectedness of voids may also be of particular importance in regulating the fluxes of nutrients available to microbes (Fillinger et al., 2019b). However, void and sediment grain size are strongly related to groundwater flow and replenishment of OM, nutrients, and DO. Moreover, the distribution of microbes between the porewater and sediment matrix, as well as intra- and interspecies interactions are governed by subsurface permeability (Fillinger et al. this volume).

The occurrence and movement of metazoans within aquifers are not only constrained by the presence of habitats of sufficient size but also by the presence of interconnected habitats that can provide corridors for migration. Hence, the size of voids, which can be pores in unconsolidated sediment aquifers or fissures in consolidated sediment aquifers, and their interconnections have long been claimed to explain differences in groundwater metazoan composition among different types of aquifers (Dole-Olivier et al., 2009; Johns et al., 2015; Korbel and Hose, 2015). Clay sediments can only be colonized by burrowers because pore size diameter ($<25\ \mu m$) is less than the body size of most metazoans. The pore space available between sand particles (up to $250\ \mu m$) allows for the occurrence of some meiofaunal organisms (body size: $63-1000\ \mu m$), whereas all meiofaunal organisms and some macroinvertebrates (body size $>1\ mm$) can colonize pore space of gravels. Void size is no longer a physical constraint in karst aquifers where the size of fissures is considerably enlarged by chemical weathering. However, little quantitative information is available to assess the relationship between body and void sizes because most field studies do not measure the size of habitat spaces (Pipan and Culver, 2017). Korbel et al. (2019) experimentally tested the ability of stygobiotic harpacticoids, cyclopoids, syncarids, and amphipods to use clay, sand, and gravel sediments and demonstrated that all taxa showed a clear preference for coarse sediments with large interstitial voids.

The movement of water, nutrients, dissolved oxygen, and organisms through an aquifer depends on the occurrence of connected voids. In hydrogeology, the ability of an aquifer to transmit water is a key hydraulic parameter known as permeability (Aquilina et al., this volume). Alluvial systems are heterogeneous formations containing distinct hydrofacies such as sandy gravel matrices, sand lenses, open framework gravels (OFG), and clay-bound gravels (Dann et al., 2009) (Fig. 5.2A). OFG consist of small to large gravels with no fine sediments that provide interconnected habitats available to stygobionts. The proportion of OFG in a sediment profile is variable, ranging from 3% to 22% (Burbery et al., 2017). Within the OFG of alluvial aquifers near Christchurch, South Island, New Zealand, Dann et al. (2008) measured mean porosity values of 34%. They calculated that between 90% and 95% of the flow occurred through these OFG, even though the proportion of connected OFG in the profile was only about 1%. Large void size in OFG and the interconnectedness of OFG lenses over reasonably large distances provide both living habitats and migration pathways to large body-sized stygobionts in alluvial gravel systems (Hahn and Fuchs, 2009; Johns et al., 2015; Malard et al., 2017). Despite the overriding importance of connected voids on the distribution and composition of groundwater metazoans, measurements of sediment permeability are lacking in most field ecological studies, although fast, effective, and low-cost methods do exist to measure permeability.

The size and amount of pores available to metazoans in the hyporheic zone of streams can be limited by streambed clogging. This is a widespread process that consists of the

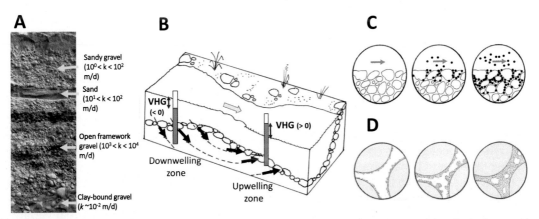

**FIGURE 5.2** (A) Sediment heterogeneity in alluvial systems showing the example of four hydrofacies with contrasted void/grain (pore) sizes and values of permeability (k in m/day; photo by M. Close). (B) Location of downwelling and upwelling zones along a stream riffle showing measurements of vertical hydraulic gradients (VHG) in piezometers. (C) Physical clogging of sediment by deposition and infiltration of fine mineral particles (black dots). (D) Biological clogging of sediment by overgrowth of bacterial biofilm (gray layers around sand grains).

deposition, accumulation, and infiltration of fine sediments into the bed sediment of streams (Fig. 5.2C). This definition underestimates the role of the surface algal mat and interstitial bacterial communities that contribute to the clogging through proliferation and overgrowth (Fig. 5.2D; Marmonier et al., 2004). In its early stages, clogging generates anaerobic microenvironments and increases the surface available to biofilm, thereby potentially increasing the functional diversity of microbes. In its later stages, the complete filling of pores with fine sediments impedes interstitial water flow, generates steep vertical gradients in biogeochemical processes, and restricts the use of hyporheic sediments by surface aquatic taxa, especially fish larvae, large benthic insect larvae (Descloux et al., 2013) and macro-crustaceans (Mathers et al., 2014).

Karst systems also show strong spatial heterogeneities in the size and connectivity of fractures. Beyond cave streams and large phreatic conduits that can harbor large-sized organisms, are small-size fissures through which move organisms, energy, and matter (Simon, 2019). As in alluvial systems, the distribution and dynamics of metazoan communities largely depend on the distribution and connection of groundwater flow between the conductive fractures (i.e., the equivalent of OFG in alluvial aquifers) and adjacent thinly fissured matrix of the saturated zone (Gibert et al., 1994b; Di Lorenzo et al., 2018). In the Lez Karst system, France, Malard et al. (1996) showed that wells separated by distances of only a few meters had contrasted metazoan community composition depending on the size, hydraulic conductivity, and surface hydrological connections of water-bearing fractures intersected by the wells. The fauna of the epikarst aquifer provides an example of species-rich communities dominated by small-sized organisms, including many copepods (Pipan and Culver, 2007).

## Hydrological connection to the surface environment

The degree of hydrological connection of subsurface habitats to the surface environment is a key parameter controlling the distribution of organisms in alluvial and karst aquifers (Hahn and Fuchs, 2009; Hermann et al., 2020). Aquatic subterranean habitats that exhibit a strong hydrological connection to the surface show increased temporal variability, higher fluxes of nutrients, OM, and dissolved oxygen, and are prone to the import of microorganisms with infiltrating surface water, as well as to colonization by aquatic surface metazoans. In river alluvial aquifers, the strength and direction of hydrological exchanges between the river and interstitial water are assessed by measuring the vertical hydraulic gradient (VHG) defined as the difference in height (cm) between the surface water level and the hyporheic water level (Baxter et al., 2003) (Fig. 5.2B). A negative VHG reflects the infiltration of surface water into the riverbed sediments (downwelling zones; Hendricks and White, 1991) that results in sediment patches characterized by elevated concentrations of biodegradable OM and higher densities of benthic organisms. A positive VHG reflects the exfiltration of groundwater into the riverbed (upwelling zones), which results in oligotrophic sediment patches characterized by constant temperature and higher densities of obligate groundwater organisms (Dole-Olivier and Marmonier, 1992; Brunke and Gonser, 1999; Korbel et al., 2022). Spatial variation in the infiltration of surface water also drives community composition in karst aquifers (Gibert et al., 1994b). In the Foussoubie karst system, France, subsurface habitats fed by sinking surface streams were dominated by surface organisms (e.g., Ephemeroptera, Diptera

Chironomidae, and Cladocera), whereas karstic springs fed by diffuse infiltration of rainwater were dominated by stygobionts (e.g., *Niphargus virei, Acanthocyclops venustus;* Vervier and Gibert, 1991).

## Temperature

Subsurface habitats are among the most thermally stable habitats on Earth. Temperature seasonality of groundwater decreases with increasing depth below the soil surface until the temperature becomes constant throughout the year at a depth of $\sim 10-20$ m. Temperate caves also show steep gradients in air temperature seasonality from the outer (annual temperature variation similar to the outside) to the inner passages (annual variation few tenths of degrees) (Mammola et al., 2019a). The climatic variability hypothesis predicts that species living in a thermally buffered environment should have reduced thermal tolerance breadth (Stevens, 1989). In agreement with theory, studies performed using subterranean terrestrial arthropods showed that specialization in deep subterranean life involved a reduction in thermal tolerance (Mammola et al., 2019a; Colado et al., 2022). The lower thermal tolerance among specialized subterranean terrestrial taxa may in turn constrain the size of their geographic and elevation ranges because dispersers are more likely to encounter temperatures outside their thermal tolerance. A formal test of the climatic variability hypothesis is lacking for groundwater taxa but preliminary data suggest that their thermal niche breadth correlates with the range of temperatures they encounter in their geographic range (Mermillod et al., 2013). Nevertheless, there is substantial variation in the thermal niche features among groundwater taxa (Eme et al., 2014; Di Lorenzo and Galassi, 2017; Castaño-Sánchez et al., 2020a). Similar to metazoans, microbial species do show a specific thermal optimum and tolerance. Groundwater ecosystems are typically inhabited by psychrophilic and mesophilic microbes. Groundwater temperature changes were shown to affect biodiversity (Brielmann et al., 2009; Griebler et al., 2016). Understanding groundwater temperature variation and its causes are key to assessing the differential ability of groundwater species to cope with anthropogenic warming of groundwater (e.g., due to geothermal energy use) and global warming (Griebler et al., 2016; Retter et al., 2020; Becher et al., 2022).

## Chemical constraints to biological distribution

## Groundwater chemistry

Groundwater chemical characteristics are determined by the lithology of the aquifer, solution kinetics, and flow patterns, as well as by natural or artificial recharge of groundwater with surface water (Foulquier et al., 2011). While groundwater composition does play an important role in the diversity and distribution of microorganisms, it is often difficult, at regional scales, to disentangle the relative influence of water chemistry from the physical features of habitats on the diversity of groundwater metazoan communities. Indeed, water chemistry and habitat feature both covary with the lithology and hydrodynamics of aquifers. Aquifers in carbonated rocks have both water with higher dissolved solid content and larger void size than aquifers in metamorphic rocks. However, physical habitat features explained a greater proportion of variation in metazoan community than groundwater chemistry in most regional-scale studies (Dole-Olivier et al., 2009; Hahn and Fuchs, 2009; Johns et al., 2015; Korbel and Hose, 2015). Human activities also cause major changes in the chemistry of groundwater by introducing pollutants that may affect the growth and survival of groundwater microorganisms and metazoans (Castaño-Sánchez et al., 2020b; Becher et al., 2022, Fillinger et al., this volume). However, in the present section, we restrict ourselves to biological constraints imposed by natural variation in groundwater chemistry: the tolerance of groundwater organisms to chemical polutants is presented by Di Lorenzo et al. (this volume).

Temporal and spatial changes in dissolved ions represent a stress for metazoans because they imply an osmoregulation activity, which is energetically expensive (Brooks and Mills, 2011). A major ion such as calcium is essential for molluscs and crustaceans because it controls the shell or exoskeleton development and the rate of molting (Rukke, 2002). However, groundwater metazoans appear to tolerate a wide range of total dissolved solids (i.e., the sum of dissolved salt content). Acute toxicity tests performed using a number of freshwater stygobiotic crustaceans including harpacticoids, cyclopoids, syncarids, and asellid isopods reported LC50 values (lethal concentrations for 50% of the population at 96 h) ranging from 2.84 to 12.7 g NaCl/L (Castaño-Sánchez et al., 2020a, 2021). Diversified obligate groundwater communities of metazoans were collected in a broad range of habitats showing highly contrasted concentrations of total dissolved solids. These include the hyporheic zone of Alpine glacial rivers containing water with specific conductance $<50\ \mu S/cm$ (Malard, 2003) and highly saline aquifers in Australia with dissolved solid concentrations ranging from 22.5 to 32.5 g/L (Schulz et al., 2013). Several groundwater habitats that exhibit steep gradients of salinity, including anchialine caves (Iliffe, 2000), coastal aquifers (Shapouri et al., 2016), calcrete aquifers (Humphreys et al., 2009), and sulfidic caves (Galassi et al., 2017), support diverse biological communities of obligate groundwater metazoans, even though the latter is rarely confined to the highest salinity areas. However, in the Frasassi cave system, central Italy, the most abundant copepod species colonized high-salinity sulfidic lakes, whereas they were poorly represented in dripping pools fed by diffuse infiltration of low-salinity meteoric water (Galassi et al., 2017).

## Oxygen content and organic matter

The availability of OM and dissolved oxygen (DO) are key factors shaping the distribution of groundwater organisms (Strayer et al., 1997; Malard and Hervant, 1999; Datry et al., 2005; Shen et al., 2015; Hofmann et al., 2020). The dynamics of OM and DO are intimately linked because there is no substantial primary production of OM (except through chemolithoautotrophy) nor the production of oxygen in groundwater. Thus, the supply of OM and DO to groundwater essentially depends on fluxes from the surface environment. OM and DO are also linked because aerobic degradation of OM consumes oxygen. Thus, the concentration of DO in groundwater is determined by the rate of oxygen transport from the surface and by the rate of oxygen consumption, which itself depends on the input rate of OM to groundwater. This interplay between OM and DO results in heterogeneous patterns of DO concentrations at macro- (km), meso- (m) and micro- (cm) scales (Strayer et al., 1997; Malard and Hervant, 1999) (Fig. 5.3). Dissolved organic carbon (DOC) supply can induce steep decreasing gradients of DO in the recharge zone of shallow aquifers and result in extensive areas of

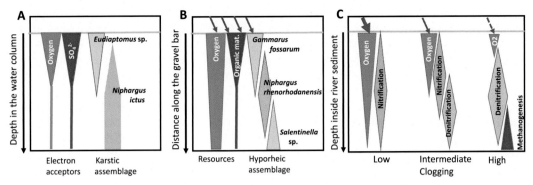

**FIGURE 5.3** (A) Vertical gradients of dissolved oxygen and oxidized sulfur form $SO_4^{2-}$ concentrations in a lake of the Frasassi chemoautotrophic cave, with the distribution of the *Copepod Eudiaptomus* sp. (which avoids deep sulfidic layers) and the amphipod *Niphargus ictus* (which withstands hypoxic conditions, based on Galassi et al., 2017). (B) Distribution of dissolved oxygen and organic matter along a gravel bar (from downwelling to upwelling zone), showing the distribution of the surface amphipod *Gammarus fossarum*, the ubiquitous groundwater amphipod *Niphargus rhenorhodanensis* and the phreatic groundwater amphipod *Salentinella* sp. (based on Dole-Olivier and Marmonier, 1992). (C) Effect of clogging of a riverbed on the vertical distribution of dissolved oxygen and vertical succession of microbial processes (nitrification, denitrification and methanogenesis).

anoxic groundwater, whenever DO replenishment from the influx of atmospheric oxygen is low. Inversely, oxic conditions often prevail over extensive areas in deep water-table aquifers because fluxes of DOC reaching groundwater are not sufficient for microbial respiration to deplete DO. Microbial respiration generates declining gradients of DO along hyporheic paths as surface water moves away, either vertically or subhorizontally, from downwelling zones (Dole-Olivier and Marmonier, 1992; Reeder et al., 2018) (Fig. 5.3B). Gradients become steeper as clogging proceeds (Fig. 5.3C). Steep gradients of DO also occur at the downstream ends of groundwater flow paths when deep anoxic groundwater comes into contact with shallow oxygenated groundwater fed by infiltration of meteoric water, as is the case in a number of sulfidic caves (Galassi et al., 2017) (Fig. 5.3A).

How do these spatial patterns of DO and OM affect the distribution of groundwater organisms? A simple answer is that OM inputs to groundwater support more complex food webs containing higher biomass and higher diversity at least for the metazoans, until a decrease in DO concentration caused mainly by microbial activities restricts the number of metazoan species. However, that relationship is probably dependent on the scale of heterogeneity in OM and DO. An overview of adaptive strategies of groundwater crustaceans to withstand hypoxia and food deprivation (Hervant and Renault, 2002; Hervant and Malard, 2019) suggests that many species are probably able to experience highly variable oxygen concentrations, although they cannot survive severe hypoxia for more than a few days. Indeed, groundwater metazoan communities show extremely low diversity and abundance in large areas of anoxic groundwater (Stein et al., 2012). There is also evidence that the diversity and abundance of metazoans are limited by OM supply in well-oxygenated groundwater of deep water-table aquifers in unconsolidated sediment (Datry et al., 2005). However, DO and OM supply may be less constraining to metazoans in heterogeneous settings made of a mosaic of small-scale patches of anoxic and oxic groundwater, owing to their capacity to exploit the food and oxygen resources occurring in these contrasting patches. A transition from aerobic to anaerobic microbial respiration and fermentation occurs with much greater diversity in redox processes (Fillinger et al., this volume). Hermann et al. (2020) brought evidence that the spatial alternation of oxic and anoxic groundwater in the Hainich oligotrophic aquifer, Germany, sustained chemolithoautotrophic production that supported more complex food webs. In this aquifer, evidence for metazoan top consumers such as Cyclopoida and Platyhelminthes were only found in wells characterized by higher abundances of functional genes associated with chemolithoautotrophy.

The interplay between OM and DO could potentially shape large-scale patterns of groundwater species richness. Eme et al. (2015) showed that the regional richness of subterranean crustaceans in Europe increased monotonically with annual actual evapotranspiration (AET) — a distal surrogate of OM supply to groundwater. In a recent and extensive survey of the distribution of groundwater bogidiellids (Amphipoda) and anthurids (Isopoda) in tropical New Caledonia, Mouron et al. (2022) found that these taxa were confined to the most oxic and oligotrophic subsurface water. This might indicate that the relationship between AET and the regional richness of subterranean crustaceans can be hump-shaped: richness increases with increasing OM supply until DO deficiency becomes a limiting factor. A similar pattern is known from microbial ecology, i.e. the energy-diversity relationship, but it has not yet empirically been proved for groundwater (Griebler et al., 2022).

## Species interactions

Biotic interactions are manifold and their effect can express at multiple scales (Delić and Fišer 2019). Over short-term scales, they exert a major control on the assembly of ecological communities and microdistribution patterns of species. Over longer time scales, they shape functional species' traits through evolutionary processes and may even drive speciation. Here, we successively consider species interactions in surface water—groundwater ecotones and groundwater habitats. The focus is on metazoans because species interaction and food web studies for groundwater microbial communities are extremely scarce (e.g., Karwautz et al., 2022).

## Species interactions in surface water—groundwater ecotones

Species interactions are frequently assumed to explain the distribution of surface water and groundwater species along surface water - groundwater ecotones. Brunke and Gonser (1999) brought evidence that the occurrence of surface aquatic taxa at shallow sediment depths in the hyporheic zone of the Töss River, Switzerland, reflected their fundamental niches, whereas the occurrence of stygobionts at greater depths reflected their realized niches. According to their clinal model of interstitial communities, the penetration deep into the sediment by surface taxa is limited by the availability of food sources, whereas the colonization of shallow sediment layers by stygobionts is restricted by asymmetric competition. Fišer et al. (2007) and Luštrik et al. (2011) explored the mechanisms that allowed the coexistence of the groundwater amphipod *Niphargus timavi* and the epigean amphipod *Gammarus fossarum* in a small surface Slovenian stream. Among different mechanisms, including a differential habitat choice between species, they found that juveniles of *N. timavi* more actively hid in the sand to avoid cannibalism by their conspecifics and predation by *G. fossarum*. Epigean species that actively colonize groundwater, such as the amphipod *Gammarus minus*, a widespread inhabitant of springs and caves in West Virginia, USA, can also predate on groundwater species. In a laboratory stream experiment, Culver et al. (1991) showed that the loss rate of the asellid groundwater isopod *Caecidotea holsingeri* increased nearly 10-fold in the presence of *G. minus* relatively to its loss rate when alone.

## Species interactions in groundwater

Several studies reported the occurrence of ecological patterns that might predictably result from competition for scarce resources and/or predation including morphological specialization to different microhabitats (Trontelj et al., 2012), trophic niche partitioning (Fišer et al., 2019), and character displacement (Delić et al., 2016). Trontelj et al. (2012) revealed four distinct ecomorphs within *Niphargus* species: ecomorphs were geographically and phylogenetically unrelated but they were associated with specific cave microhabitats. Fišer et al. (2019) showed that micro- and macro-feeders of interstitial species of *Niphargus* evolved several times independently and found a positive albeit weak relationship between trophic differentiation and cooccurrence. Delić et al. (2016) provided an example of character displacement where populations of *Niphargus croaticus* living in sympatry with *Niphargus subtypicus* were larger and had longer appendages than populations living in allopatry. However, in all these studies, competitive interactions were inferred from observed patterns rather than from manipulative experiments.

Culver and coworkers combined field observations and experiments conducted in the field and the laboratory to document interactions among stygobiotic amphipods and isopods in several Appalachian cave streams, USA (see review in Culver, 1994). In the streams, individuals reside on the underside of stones, and encounters between them either result in the washout of individuals due to nonlethal antagonistic interactions, loss of individuals due to predation, or a reduction in washout due to commensalism and mutualistic interactions. Culver (1994) evidenced a diversity of interactions including competition, commensalism, amensalism, and predation: their relative importance varied among caves and they could have major effects on microdistribution patterns.

Predation in groundwater remains unstudied with little if no quantitative assessment of the importance of omnivory, cannibalism, predator specialization and body-size dependent predation (Delić and Fišer, 2019). Until the 2000s, subterranean food chains were predicted to be truncated at the top with few or no strict predators (Gibert and Deharveng, 2002). However, quantitative investigation of trophic linkages using stable isotopes showed that food chains could contain four or even five trophic levels and include strict predators in the most species-rich aquifers (Premate et al., 2021). Jugovic et al. (2010) provided a remarkable evolutionary example of a defense mechanism developed by prey in response to its predator.

They found that cave shrimp populations of *Troglocaris* s. str. living with the amphibian predator *Proteus* had a larger carapace bearing a longer rostrum with more teeth than *Troglocaris* populations that were not living with *Proteus*. The rostrum's defense role was confirmed experimentally: frontal attacks of *Proteus* failed on shrimps with long rostra and the amphibian needed more time to swallow long-rostra shrimps.

Other biotic interactions were described between groundwater metazoans and epibiontic ciliate protists, microsporidian parasites, or bacteria, although the nature of these interactions is often not fully established (Delić and Fišer, 2019). Gudmundsdóttir et al. (2018) isolated undescribed ciliate epibionts belonging to the orders Apostomatida and Philasterida that were attached to the ventral surface of the Icelandic endemic groundwater amphipod *Crangonyx islandicus*. The authors hypothesized that the ciliates fed on the exuvial fluid of the groundwater amphipod, which in turn fed on the ciliates. Grabner et al. (2020) found five species of microsporidian parasites from a screening of 69 specimens of groundwater amphipods belonging to 19 species. Microsporidians are known to cause mortality and induce a female-biased sex ratio in their host populations. Weigand et al. (1996) reported four microsporidian species shared by syntopic populations of the surface amphipod *Gammarus fossarum* and the stygobiotic *Niphargus schellenbergi* living in the Tunnel of Huldange, Luxembourg, suggesting horizontal transmission of microsporidian parasites between surface and groundwater amphipods. Sulfur-oxidizing *Thiothrix* bacteria were found attached to the hairs of *Niphargus* in the sulfidic Frasassi cave, Italy, and partly sulfidic aquifers of the Southern Dobrogea region, Romania. However, the potential benefits of this association to either *Thiothrix* or their amphipod hosts are not yet fully clear (Bauermeister et al., 2012; Flot et al., 2014). The biofilm-feeding amphipod may serve as a shuttle between the oxic and anoxic world for the epizoic sulfide oxidizing bacteria (Dattagupta et al., 2009).

## The effect of the past: paleogeographic events and historical climates

The idea that the present-day, large-scale distribution patterns of groundwater biodiversity primarily reflect the biological effects of past environmental constraints associated with large-scale paleogeographic and climatic events has long been a hallmark of groundwater biodiversity studies (Humphreys, 2000). As in any other habitats, historical events undoubtedly influenced two of the three processes responsible for the number of species present in a region, namely speciation and extinction (the last process being dispersal). However, the imprint of historical events is thought to persist for longer in groundwater due to the expectedly low dispersal capacity of obligate groundwater species (Foulquier et al., 2008; Zagmajster et al., 2014). Two nonmutually exclusive mechanisms were proposed to increase the rate of speciation in groundwater. The first one is a paleogeographic or climatic event, which extirpates surface populations, thereby causing the isolation and eventually the speciation of surface populations that had colonized groundwater. Leys et al. (2003) provided phylogenetic evidence that the diversification of diving beetle species in isolated calcrete aquifers of the Yilgarn region, Western Australia, was associated with the Late Miocene and Early Pliocene aridification events. Marine regressions might also have been instrumental in causing the speciation of coastal marine populations that had colonized groundwater but this mechanism still awaits rigorous testing (Notenboom, 1991). The second mechanism is that of a paleogeographic event that provides new subterranean habitats in which lineages diversify. Borko et al. (2021) showed that the speciation rate and ecological and morphological diversification within the obligate groundwater genus *Niphargus* coincided with the uplift of carbonate massifs in South-Eastern Europe.

The effect of historical events on the extinction rate of groundwater species is more difficult to assess because of the absence of fossil records and difficulties in estimating extinction from phylogenetic trees. Extinction may be high because of the narrow ranges of many groundwater species or low because of the environmental stability of deep groundwater systems. However, it is expected to vary strongly among regions: low regional species richness in northern areas of the Paleartic and Nearctic relative to Southern areas has repeatedly been attributed to mass extinction of narrowly-distributed species caused by the southward expansion of Quaternary glaciers and permafrost (Strayer et al., 1995; Hof et al., 2008) (Fig. 5.4A and B). The imprint of cold Quaternary climates on the distribution of groundwater species has not been overwritten by subsequent dispersal phases, even though intense postglacial sediment deposition has provided hyporheic migration corridors for some species to recolonize depopulated areas (Eme et al., 2013; Malard et al., 2017).

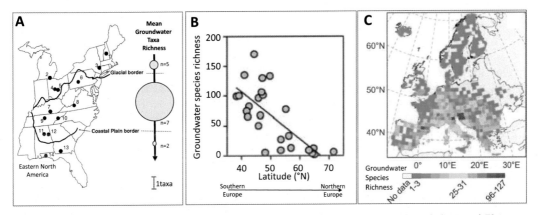

**FIGURE 5.4** (A) Taxonomic richness of stygobionts in the eastern United States with limits of Pleistocene glaciation and coastal plain. Black dots show the sampling sites (n = number of sites). The diameter of blue circles is proportional to the mean taxonomic richness of sites (blue vertical scale for one species) (based on Strayer et al., 1995). (B) Latitudinal gradient of groundwater species richness in Europe. Dots correspond to the number of species in large biogeographical regions (based on Hof et al., 2008). (C) Species richness of obligate groundwater crustaceans across Europe (cell size is 100 × 100 km; richness per cell ranges from 0 to 127 species) (based on Eme et al., 2015). Note that richness pattern depends on spatial resolution. Using a coarse resolution (i.e., biogeographical regions in B), species richness declines with increasing latitude, whereas a ridge of high species richness becomes apparent between 42 and 46 degrees N when using a finer resolution (100 × 100 km cells in B).

The relative importance of historical climates in shaping the range size and species richness of groundwater crustaceans in Europe was revisited in macroecological studies published by Zagmajster et al. (2014) and Eme et al. (2015), respectively (but see also Zagmajster et al., 2018). The findings of these studies provided mixed support for the primary role of historical climates. Zagmajster et al. (2014) showed that the patterns of increasing range size and decreasing species turnover at higher latitudes in Europe were primarily driven by historical temperature oscillations. However, Eme et al. (2015) showed that energy and spatial heterogeneity were important predictors of species richness at mid and Southern latitudes, whereas the role of historical climate stability overlapped with the two other mechanisms in Northern Europe and energy in Southern Europe (Fig. 5.4C). The independent effect of historical climates on species richness was low all over Europe, thereby mitigating the prominent role attributed to the disproportionate extinction of groundwater species in regions of high historical climate instability. Similar studies that tackle the biogeographic distribution of microorganisms in groundwater and its causes are rare. Fillinger et al. (2019a) tested if differences in community composition across different regions are caused by dispersal limitation or selection and if the selection is caused by local environmental conditions alone or by additional broad-scale region-specific factors. The study revealed that differences in microbial community composition were mainly the product of selection imposed by local environmental conditions and to a smaller but still significant extent by dispersal limitation and drift across regions (Fillinger et al., 2019a).

## Conclusion

Our knowledge of biodiversity and species distributions in groundwater is no longer "rudimentary" as stated by Strayer (1994) in the first edition of the groundwater ecology book (Gibert et al., 1994c). However, it is still largely incomplete. Hortal et al. (2015) identified seven shortfalls that beset large-scale knowledge of biodiversity to which Ficetola et al. (2019) added an eighth even more basal shortfall for the case of the subterranean fauna, which they named the "Racovitzan impediment". Typically, the biodiversity of an environment that has not been explored cannot be described or analyzed. Indeed, life in vast areas of the Earth's subsurface has not yet been explored. Ensuring sustained efforts to inventory and map groundwater living forms is crucial to revive, understand and protect a yet invisible component of biodiversity on Earth (Iannella et al., 2020, 2021).

Much of our knowledge on factors constraining biological distribution is from analyses of the distribution of species and biodiversity patterns based on taxonomic diversity measures (e.g., species richness, species composition). We foresee two other important research avenues. First, extending pattern analyses to other facets of diversity including functional and phylogenetic diversity may provide a more robust eco-evolutionary understanding of factors constraining species distribution (Zagmajster et al., 2018; Mammola et al., 2020). Second, many of the supposedly strong constraints to biological distribution would warrant being experimentally tested (Stumpp and Hose, 2013; Korbel et al., 2019).

The size of voids, their interconnectedness, and their hydrological connections to the surface environment are three key physical variables controlling the number of species and relative proportion of obligate groundwater species and surface aquatic species in groundwater metazoan communities. Hydrological connection to the surface may in turn exert a major control on three other key variables, the availability of DO and OM, and thermal instability. These six integrative features probably account for the largest proportion of variance in groundwater metazoan richness and composition among localities within a region (Dole-Olivier et al., 2009; Johns et al., 2015; Korbel and Hose, 2015). Although other variables may be of importance at peculiar sites, systematically measuring these six integrative variables in ecological studies paves the way for developing simple models of groundwater biodiversity distribution within and among habitats of a region. With microbial communities, more factors must be considered with respect to their local, regional biodiversity, and biogeographic distribution, e.g., aquifer geology and groundwater chemistry. More realistic distribution models would require integrating bottom-up and top-down linkages between microbial and metazoan communities as well as the effect of spatio-temporal heterogeneity and environmental disturbances on these linkages (Foulquier et al., 2011; Griebler et al., 2022).

While our understanding of factors driving biological distribution is growing, efforts should be made to predict the fate of groundwater biodiversity in response to global environmental changes. A handful of studies have attempted to predict the fate of subterranean species and communities under future global warming scenarios (Mammola et al., 2019b; Retter et al., 2020). However, groundwater biodiversity alterations in response to anticipated changes in groundwater recharge rate, groundwater organic carbon, or continuity of hyporheic corridors have not yet been evaluated.

## Acknowledgments

We thank C. Griebler and F. Malard for constructive comments that improved this chapter.

## References

Aljancic, M., Bulog, B., Kranjc, A., Habic, P., Josipovic, D., Sket, B., Skoberne, P., 1993. *Proteus:* The Mysterious Ruler of Karst Darkness. Speleo Projects. Caving Publications International, Ljubiana, Slovenia.

Anantharaman, K., Brown, C.T., Hug, L.A., Sharon, I., Castelle, C.J., Probst, A.J., Thomas, B.C., Singh, A., Wilkins, M.J., Karaoz, U., 2016. Thousands of microbial genomes shed light on interconnected biogeochemical processes in an aquifer system. Nature Communications 7 (2), 13219.

Bar-On, Y.M., Phillips, R., Milo, R., 2018. The biomass distribution on earth. Proceedings of the National Academy of Sciences 115 (25), 6506–6511.

Bauermeister, J., Ramette, A., Dattagupta, S., 2012. Repeatedly evolved host-specific ectosymbioses between sulfur-oxidizing bacteria and amphipods living in a cave ecosystem. PLoS One 7, 11.

Baxter, C., Hauer, F.R., Woessner, W.W., 2003. Measuring groundwater–stream water exchange: new techniques for installing minipiezometers and estimating hydraulic conductivity. Transactions of the American Fisheries Society 132 (3), 493–502.

Becher, J., Englisch, C., Griebler, C., Bayer, P., 2022. Groundwater fauna downtown – drivers, impacts and implications for subsurface ecosystems in urban areas. Journal of Contaminant Hydrology 248, 104021.

Behrmann-Godel, J., Nolte, A.W., Kreiselmaier, J., Berka, R., Freyhof, J., 2017. The first European cave fish. Current Biology 27 (7), 257–258.

Borko, Š., Trontelj, P., Seehausen, O., Moškrič, A., Fišer, C., 2021. A subterranean adaptive radiation of amphipods in Europe. Nature Communications 12, 3688.

Bornemann, T.L.V., Adam, P.S., Turzynski, V., Schreiber, U., Figueroa-Gonzalez, P.A., Rahlff, J., Köster, D., Schmidt, T.C., Schunk, R., Krauthausen, B., Probst, A.J., 2022. Genetic diversity in terrestrial subsurface ecosystems impacted by geological degassing. Nature Communications 13 (1), 284.

Botosaneanu, L., 1986. Stygofauna Mundi: A Faunistic, Distributional, and Ecological Synthesis of the World Fauna Inhabiting Subterranean Waters. E.J. and Dr. W. Backhuys, Leiden, The Netherlands.

Brancelj, A., Dumont, H.J., 2007. A review of the diversity, adaptations and groundwater colonization pathways in Cladocera and Calanoida (Crustacea), two rare and contrasting groups of stygobionts. Fundamental and Applied Limnology 168 (1), 3–17.

Brielmann, H., Griebler, C., Schmidt, S.I., Michel, R., Lueders, T., 2009. Effects of thermal energy discharge on shallow groundwater ecosystems. FEMS Microbiology Ecology 68, 273–286.

Brooks, S.J., Mills, C.L., 2011. Osmoregulation in hypogean populations of the freshwater amphipod, *Gammarus pulex* (L.). Journal of Crustacean Biology 31 (2), 332–338.

Brunke, M., Gonser, T., 1999. Hyporheic invertebrates: the clinal nature of interstitial communities structured by hydrological exchange and environmental gradients. Journal of the North American Benthological Society 18 (3), 344–362.

Burbery, L., Moore, C.R., Jones, M.A., Abraham, P.M., Humphries, B., Close, M.E., 2017. Study of connectivity of open framework gravel facies in the Canterbury Plains Aquifer using smoke as a tracer. In: Ventra, D., Clarke, L.E. (Eds.), Geology and Geomorphology of Alluvial and Fluvial Fans: From Terrestrial to Planetary Perspectives. Geological Society of London, London, pp. 327–344.

Camacho, A., Valdecasas, A.G., 2008. Global diversity of syncarids (syncarida; Crustacea) in freshwater. Hydrobiologia 595, 257–266.

Castaño-Sánchez, A., Hose, G.C., Reboleira, A.S.P., 2020a. Salinity and temperature increase impact groundwater crustaceans. Scientific Reports 10 (1), 1–9.

Castaño-Sánchez, A., Hose, G.C., Reboleira, A.S.P., 2020b. Ecotoxicological effects of anthropogenic stressors in subterranean organisms: a review. Chemosphere 244, 125422.

Castaño-Sánchez, A., Malard, F., Kalčíková, G., Reboleira, A.S.P., 2021. Novel protocol for acute in situ ecotoxicity test using native Crustaceans applied to groundwater ecosystems. Water 13 (8), 1132.

Chaudhari, N.M., Overholt, W.A., Figueroa-Gonzalez, Taubert, M., Bornemann, T.L.V., Probst, A.J., Hölzer, M., Marz, M., Küsel, K., 2021. The economical lifestyle of CPR bacteria in groundwater allows little preference for environmental drivers. Environmental Microbiome 16, 24.

Colado, R., Pallarés, S., Fresneda, J., Mammola, S., Rizzo, V., Sánchez-Fernández, D., 2022. Climatic stability, not average habitat temperature, determines thermal tolerance of subterranean beetles. Ecology 103 (4), e3629.

Culver, D.C., 1994. Species interactions. In: Gibert, J., Danielopol, D.L., Stanford, J.A. (Eds.), Groundwater Ecology. Elsevier, Academic Press, London, pp. 271–286.

Culver, D.C., Holsinger, J.R., 1992. How many species of troglobites are there? NSS Bulletin 54, 79–80.

Culver, D.C., Pipan, T., 2014. Shallow Subterranean Habitats. Ecology, Evolution and Conservation. Oxford University Press, Oxford.

Culver, D.C., Fong, D.W., Jernigan, R.W., 1991. Species interactions in cave stream communities: experimental results and microdistribution effects. The American Midland Naturalist 126 (2), 364–379.

Danielopol, D.L., Pospisil, P., Rouch, R., 2000. Biodiversity in groundwater: a large-scale view. Trends in Ecology & Evolution 15 (6), 223–224.

Dann, R.L., Close, M.E., Flintoft, M.J., Hector, R., Barlow, H., Thomas, S., Francis, G., 2009. Characterization and estimation of hydraulic properties in an alluvial gravel vadose zone. Vadose Zone Journal 8 (3), 651–663.

Dann, R.L., Close, M.E., Pang, L., Flintoft, M.J., Hector, R.P., 2008. Complementary use of tracer and pumping tests to characterise a heterogeneous channelized aquifer system. Hydrogeology Journal 16, 1177–1191.

Datry, T., Malard, F., Gibert, J., 2005. Response of invertebrate assemblages to increased groundwater recharge rates in a phreatic aquifer. Journal of the North American Benthological Society 24 (3), 461–477.

Dattagupta, S., Schaperdoth, I., Montanari, A., Mariani, S., Kita, N., Valley, J.W., Macalady, J.L., 2009. A novel symbiosis between chemoautotrophic bacteria and a freshwater cave amphipod. The ISME Journal 3, 935–943.

Deharveng, L., Stoch, F., Gibert, J., Bedos, A., Galassi, D., Zagmajster, M., Brancelj, A., Camacho, A., Fiers, F., Martin, P., Giani, N., Magniez, G., Marmonier, P., 2009. Groundwater biodiversity in Europe. Freshwater Biology 54, 709–726.

Delić, T., Fišer, C., 2019. Species interactions. In: White, W., Culver, D., Pipan, T. (Eds.), Encyclopedia of Caves. Elsevier, Academic Press, London, pp. 967–973.

Delić, T., Trontelj, P., Zakšek, V., Fišer, C., 2016. Biotic and abiotic determinants of appendage length evolution in a cave amphipod. Journal of Zoology 299 (1), 42–50.

Descloux, S., Datry, T., Marmonier, P., 2013. Benthic and hyporheic invertebrate assemblages along a gradient of increasing streambed colmation by fine sediment. Aquatic Sciences 75 (4), 493–507.

Di Lorenzo, T., Galassi, D.M.P., 2017. Effect of temperature rising on the stygobitic crustacean species *Diacyclops belgicus*: does global warming affect groundwater populations? Water 9, 951.

Di Lorenzo, T., Cipriani, D., Fiasca, B., Rusi, S., Galassi, D.M.P., 2018. Groundwater drift monitoring as a tool to assess the spatial distribution of groundwater species into karst aquifers. Hydrobiologia 813 (1), 137–156.

Dole-Olivier, M.J., Marmonier, P., 1992. Patch distribution of interstitial communities: prevailing factors. Freshwater Biology 27 (2), 177–191.

Dole-Olivier, M.J., Malard, F., Martin, D., Lefébure, T., Gibert, J., 2009. Relationships between environmental variables and groundwater biodiversity at the regional scale. Freshwater Biology 54 (4), 797–813.

Eme, D., Malard, F., Colson-Proch, C., Jean, P., Calvignac, S., Konecny-Dupré, L., Hervant, F., Douady, C.J., 2014. Integrating phylogeography, physiology and habitat modelling to explore species range determinants. Journal of Biogeography 41 (4), 687–699.

Eme, D., Malard, F., Konecny-Dupré, L., Lefébure, T., Douady, C.J., 2013. Bayesian phylogeographic inferences reveal contrasting colonization dynamics among European groundwater isopods. Molecular Ecology 22, 5685–5699.

Eme, D., Zagmajster, M., Fišer, C., Galassi, D., Marmonier, P., Stoch, F., Cornu, J.F., Oberdorff, T., Malard, F., 2015. Multi-causality and spatial non-stationarity in the determinants of groundwater crustacean diversity in Europe. Ecography 38 (5), 531–540.

Euringer, K., Lueders, T., 2008. An optimised PCR/T-RFLP fingerprinting approach for the investigation of protistan communities in groundwater environments. Journal of Microbiological Methods 75 (2), 262–268.

Eydal, H.S.C., Jägevall, S., Hermansson, M., Pedersen, K., 2009. Bacteriophage lytic to *Desulfovibrio aespoeensis* isolated from deep groundwater. The ISME Journal 3 (10), 1139–1147.

Ficetola, G.F., Canedoli, C., Stoch, F., 2019. The Racovitzan impediment and the hidden biodiversity of unexplored environments. Conservation Biology 33 (1), 214–216.

Fillinger, L., Hug, K., Griebler, C., 2019a. Selection imposed by local environmental conditions drives differences in microbial community composition across geographically distinct groundwater aquifers. FEMS Microbiology Ecology 95 (11), 1–12.

Fillinger, L., Zhou, Y., Kellermann, C., Griebler, C., 2019b. Non-random processes determine the colonization of groundwater sediments by microbial communities in a pristine porous aquifer. Environmental Microbiology 21 (1), 327–342.

Fišer, C., Delić, T., Luštrik, R., Zagmajster, M., Altermatt, F., 2019. Niches within a niche: ecological differentiation of subterranean amphipods across Europe's interstitial waters. Ecography 42 (6), 1212–1223.

Fišer, C., Keber, R., Kereži, V., Moškrič, A., Palandančić, A., Petkovska, V., Potocnik, H., Sket, B., 2007. Coexistence of species of two amphipod genera: *Niphargus timavi* (Niphargidae) and *Gammarus fossarum* (Gammaridae). Journal of Natural History 41 (41–44), 2641–2651.

Fišer, C., Pipan, T., Culver, D.C., 2014. The vertical extent of groundwater metazoans: an ecological and evolutionary perspective. BioScience 64 (11), 971–979.

Flot, J.F., Bauermeister, J., Brad, T., Hillebrand-Voiculescu, A., Sarbu, S.M., Dattagupta, S., 2014. *Niphargus–Thiothrix* associations may be widespread in sulphidic groundwater ecosystems: evidence from southeastern Romania. Molecular Ecology 23 (6), 1405–1417.

Foulquier, A., Malard, F., Lefébure, T., Douady, C.J., Gibert, J., 2008. The imprint of Quaternary glaciers on the present-day distribution of the obligate groundwater amphipod *Niphargus virei* (Niphargidae). Journal of Biogeography 35, 552–564.

Foulquier, A., Malard, F., Mermillod-Blondin, F., Montuelle, B., Dolédec, S., Volat, B., Gibert, J., 2011. Surface water linkages regulate trophic interactions in a groundwater food web. Ecosystems 14 (8), 1339–1353.

Galassi, D.M.P., Fiasca, B., Di Lorenzo, T., Montanari, A., Porfirio, S., Fattorini, S., 2017. Groundwater biodiversity in a chemoautotrophic cave ecosystem: how geochemistry regulates microcrustacean community structure. Aquatic Ecology 51 (1), 75–90.

Galassi, D.M.P., Huys, R., Reid, J.W., 2009. Diversity, ecology and evolution of groundwater copepods. Freshwater Biology 54, 691−708.

Gibert, J., Vervier, P., Malard, F., Laurent, R., Reygrobellet, J.-L., 1994b. Dynamics of communities and ecology of karst ecosystems: example of three karsts in eastern and southern France. In: Gibert, J., Danielopol, D.L., Stanford, J.A. (Eds.), Groundwater Ecology. Academic Press, San Diego, USA, pp. 425−450.

Gibert, J., Deharveng, L., 2002. Subterranean ecosystems: a truncated functional biodiversity. BioScience 52 (6), 473−481.

Gibert, J., Danielopol, D.L., Stanford, J.A., 1994c. Groundwater Ecology. Academic Press, San Diego, USA.

Gibert, J., Stanford, J.A., Dole-Olivier, M.-J., Ward, J.V., 1994a. Basic attributes of groundwater ecosystems and prospects for research. In: Gibert, J., Danielopol, D.L., Stanford, J.A. (Eds.), Groundwater Ecology. Academic Press, San Diego, USA, pp. 7−40.

Gladstone, N.S., Niemiller, M.L., Hutchins, B., Schwartz, B., Czaja, A., Slay, M.E., Whelan, N.V., 2021. Subterranean freshwater gastropod biodiversity and conservation in the United States and Mexico. Conservation Biology 36 (1), e13722.

Gorički, Š., Niemiller, M.L., Fenolio, D.B., Gluesenkamp, A.G., 2019. Salamanders. In: Culver, D.C., White, W.B., Pipan, T. (Eds.), Encyclopedia of Caves, third ed. Elsevier, London, UK, pp. 871−884.

Grabner, D., Weber, D., Weigand, A.M., 2020. Updates to the sporadic knowledge on microsporidian infections in groundwater amphipods (Crustacea, Amphipoda, Niphargidae). Subterranean Biology 33, 71.

Griebler, C., Lueders, T., 2009. Microbial biodiversity in groundwater ecosystems. Freshwater Biology 54 (4), 649−677.

Griebler, C., Brielmann, H., Haberer, C.M., Kaschuba, S., Kellermann, C., Stumpp, C., Hegler, F., Kuntz, D., Walker-Hertkorn, S., Lueders, T., 2016. Potential impacts of geothermal energy use and storage of heat on groundwater quality, biodiversity and ecosystem processes. Environmental Earth Sciences 75, 1391.

Griebler, C., Fillinger, L., Karwautz, C., Hose, G.C., 2022. Knowledge gaps, obstacles, and research frontiers in groundwater microbial ecology. In: Tockner, K., Mehner, T. (Eds.), Encyclopedia of Inland Waters, 2nd. Elsevier, pp. 611−624.

Grossart, H.-P., Wurzbacher, C., James, T.Y., Kagami, M., 2016. Discovery of dark matter fungi in aquatic ecosystems demands a reappraisal of the phylogeny and ecology of zoosporic fungi. Fungal Ecology 19, 28−38.

Grossart, H.P., Van den Wyngaert, S., Kagami, M., Wurzbacher, C., Cunliffe, M., Rojas-Jimenez, K., 2019. Fungi in aquatic ecosystems. Nature Reviews Microbiology 17, 339−354.

Gudmundsdóttir, R., Kornobis, E., Kristjánsson, B.K., Pálsson, S., 2018. Genetic analysis of ciliates living on the groundwater amphipod *Crangonyx islandicus* (Amphipoda: Crangonyctidae). Acta Zoologica 99 (2), 188−198.

Gutjahr, S., Schmidt, S., Hahn, H., 2014. A proposal for a groundwater habitat classification at local scale. Subterranean Biology 14, 25−49.

Hahn, H.J., Fuchs, A., 2009. Distribution patterns of groundwater communities across aquifer types in south-western Germany. Freshwater Biology 54 (4), 848−860.

Hassall, A.H., 1850. A Microscopic Examination of the Water Supplied to the Inhabitants of London and the Suburban Districts. London: Samuel Highley, 1850.

He, C., Keren, R., Whittaker, M.L., Farag, I.F., Doudna, J.A., Cate, J.H.D., Banfield, J.F., 2021. Genome-resolved metagenomics reveals site-specific diversity of episymbiotic CPR bacteria and DPANN archaea in groundwater ecosystems. Nature Microbiology 6 (3), 354−365.

Hendricks, S.P., White, D.S., 1991. Physicochemical patterns within a hyporheic zone of a northern Michigan river, with comments on surface water patterns. Canadian Journal of Fisheries and Aquatic Sciences 48 (9), 1645−1654.

Herrmann, M., Geesink, P., Yan, L., Lehmann, R., Totsche, K.U., Küsel, K., 2020. Complex food webs coincide with high genetic potential for chemolithoautotrophy in fractured bedrock groundwater. Water Research 170, 115306.

Hervant, F., Malard, F., 2019. Adaptations: low oxygen. In: White, W.B., Culver, D.C., Pipan, T. (Eds.), Encyclopedia of Caves. Academic Press, London, UK, pp. 8−15.

Hervant, F., Renault, D., 2002. Long-term fasting and realimentation in hypogean and epigean isopods: a proposed adaptive strategy for groundwater organisms. Journal of Experimental Biology 205 (14), 2079−2087.

Hof, C., Brändle, M., Brandl, R., 2008. Latitudinal variation of diversity in European freshwater animals is not concordant across habitat types. Global Ecology and Biogeography 17 (4), 539−546.

Hofmann, R., Uhl, J., Hertkorn, N., Griebler, C., 2020. Linkage between DOM transformation, bacterial carbon production and diversity in a shallow oligotrophic aquifer - results from flow-through sediment microcosm experiments. Frontiers in Microbiology 11, 543567.

Holmes, D., Giloteaux, L., Williams, K.H., Wrighton, K.C., Wilkins, M.J., Thompson, C.A., Roper, T.J., Long, P.E., Lovley, D.R., 2013. Enrichment of specific protozoan populations during in situ bioremediation of uranium-contaminated groundwater. The ISME Journal 7, 1286—1298.

Holsinger, J.R., 1993. Biodiversity of subterranean amphipod crustaceans: global patterns and zoogeographic implications. Journal of Natural History 27 (4), 821—835.

Hortal, J., de Bello, F., Diniz, J.A.F., Lewinsohn, T.M., Lobo, J.M., Ladle, R.J., 2015. Seven shortfalls that beset large-scale knowledge of biodiversity. Annual Review of Ecology Evolution and Systematics 46, 523—549.

Hug, L.A., Co, R., 2018. It takes a village: microbial communities thrive through interactions and metabolic handoffs. mSystems 3 (2) e00152-17.

Humphreys, W.F., 2000. Relict faunas and their derivation. In: Wilkens, H., Culver, D.C., Humphreys, W.F. (Eds.), Subterranean Ecosystems. Elsevier, Amsterdam, pp. 417—432.

Humphreys, W.F., Watts, C.H.S., Cooper, S.J.B., Leijs, R., 2009. Groundwater estuaries of salt lakes: buried pools of endemic biodiversity on the western plateau, Australia. Hydrobiologia 626 (1), 79—95.

Hvid-Hansen, N., 1951. Sulphate-reducing and hydrocarbon-producing bacteria in ground-water. Acta Pathologica Microbiologica Scandinavica 29 (3), 314—334.

Hylling, O., Carstens, A.B., Kot, W., Hansen, M., Neve, H., Franz, C.M.A.P., Johansen, A., Ellegaard-Jensen, L., Hansen, L.H., 2020. Two novel bacteriophage genera from a groundwater reservoir highlight subsurface environments as underexplored biotopes in bacteriophage ecology. Scientific Reports 10 (1), 1—9.

Iannella, M., Fiasca, B., Di Lorenzo, T., Biondi, M., Di Cicco, M., Galassi, D.M.P., 2020. Jumping into the grids: mapping biodiversity hotspots in groundwater habitat types across Europe. Ecography 43, 1825—1841.

Iannella, M., Fiasca, B., Di Lorenzo, T., Di Cicco, M., Biondi, M., Mammola, S., Galassi, D.M.P., 2021. Getting the 'most out of the hotspot' for practical conservation of groundwater biodiversity. Global Ecology and Conservation 31, e01844.

Iliffe, T.M., 2000. Anchialine cave ecology. In: Wilkens, H., Culver, D.C., Humphreys, W.F. (Eds.), Subterranean ecosystems. Ecosystems of the World, 30. Elsevier Science, Amsterdam, pp. 59—76.

Ivarsson, M., Bengtson, S., Drake, H., Francis, W., 2018. Fungi in deep subsurface environments. Advances in Applied Microbiology 102, 83—116.

Johns, T., Jones, J.I., Knight, L., Maurice, L., Wood, P., Robertson, A., 2015. Regional-scale drivers of groundwater faunal distributions. Freshwater Science 34 (1), 316—328.

Jugovic, J., Prevorcnik, S., Aljancic, G., Sket, B., 2010. The atyid shrimp (Crustacea: Decapoda: Atyidae) rostrum: phylogeny versus adaptation, taxonomy versus trophic ecology. Journal of Natural History 44 (41—42), 2509—2533.

Karwautz, C., Zhou, Y., Kerros, M.-E., Weinbauer, M.G., Griebler, C., 2022. Bottom-up control of the groundwater microbial food-web in an alpine aquifer. Frontiers in Ecology and Evolution 10, 854228.

Kinner, N.E., Harvey, R.W., Shay, D.M., Metge, D.W., Warren, A., 2002. Field evidence for a protistan role in an organically-contaminated aquifer. Environmental Science & Technology 36, 4312—4318.

Korbel, K.L., Hose, G.C., 2015. Habitat, water quality, seasonality, or site? Identifying environmental correlates of the distribution of groundwater biota. Freshwater Science 34 (1), 329—343.

Korbel, K.L., Rutlidge, H., Hose, G.C., Eberhard, S.M., Andersen, M.S., 2022. Dynamics of microbiotic patterns reveal surface water groundwater interactions in intermittent and perennial streams. Science of the Total Environment 811, 152380.

Korbel, K.L., Stephenson, S., Hose, G.C., 2019. Sediment size influences habitat selection and use by groundwater macrofauna and meiofauna. Aquatic Sciences 81 (2), 39.

Korbel, K., Chariton, A., Stephenson, S., Greenfield, P., Hose, G.C., 2017. Wells provide a distorted view of life in the aquifer: implications for sampling, monitoring and assessment of groundwater ecosystems. Scientific Reports 7, 40702.

Kyle, J.E., Eydal, H.S.C., Grant Ferris, F., Pedersen, K., 2008. Viruses in granitic groundwater from 69 to 450 m depth of the Aspö hard rock laboratory, Sweden. The ISME Journal 2 (5), 571—574.

Langille, B.L., Hyde, J., Saint, K.M., Bradford, T.M., Stringer, D.N., Tierney, S.M., Humphreys, W.F., Austin, A.D., Cooper, S.J.B., 2021. Evidence for speciation underground in diving beetles (Dytiscidae) from a subterranean archipelago. Evolution 75, 166—175.

Lategan, M.J., Torpy, F.R., Newby, S., Stephenson, S., Hose, G.C., 2012. Fungal diversity of shallow aquifers in Southeastern Australia. Geomicrobiology Journal 29 (4), 352—361.

Leys, R., Watts, C.H.S., Cooper, S.J.B., Humphreys, W.F., 2003. Evolution of subterranean diving beetles (Coleoptera: Dytiscidae: Hydroporini, Bidessini) in the arid zone of Australia. Evolution 57, 2819–2834.

Livermore, J.A., Mattes, T.E., 2013. Phylogenetic detection of novel *Cryptomycota* in an Iowa (United States) aquifer and from previously collected marine and freshwater targeted high-throughput sequencing sets. Environmental Microbiology 15, 2333–2341.

Luef, B., Frischkorn, K.R., Wrighton, K.C., Holman, H.Y.N., Birarda, G., Thomas, B.C., Singh, A., Williams, K.H., Siegerist, C.E., Tringe, S.G., Downing, K.H., Comolli, L.R., Banfield, J.F., 2015. Diverse uncultivated ultra-small bacterial cells in groundwater. Nature Communications 6 (1), 6372.

Luštrik, R., Turjak, M., Kralj-Fišer, S., Fišer, C., 2011. Coexistence of surface and cave amphipods in an ecotone environment. Contributions to Zoology 80 (2), 133–141.

Malard, F., 2003. Interstitial fauna. In: Ward, J.V., Uehlinger, U. (Eds.), Ecology of a Glacial Flood Plain. Aquatic Ecology Series. Kluwer Academic Publishers, Dordrecht, pp. 175–198.

Malard, F., 2022. Groundwater metazoans. In: Tockner, K., Mehner, T. (Eds.), Encyclopedia of Inland Waters, 2nd. Elsevier, pp. 1–14.

Malard, F., Hervant, F., 1999. Oxygen supply and the adaptations of animals in groundwater. Freshwater Biology 41 (1), 1–30.

Malard, F., Capderrey, C., Churcheward, B., Eme, D., Kaufmann, B., Konecny-Dupré, L., Léna, J.-P., Liébault, F., Douady, C.J., 2017. Geomorphic influence on intraspecific genetic differentiation and diversity along hyporheic corridors. Freshwater Biology 62, 1955–1970.

Malard, F., Mathieu, J., Reygrobellet, J.-L., Lafont, M., 1996. Biomonitoring groundwater contamination: application to a karst area in Southern France. Aquatic Sciences 58 (2), 158–187.

Malard, F., Tockner, K., Dole-Olivier, M.J., Ward, J.V., 2002. A landscape perspective of surface-subsurface hydrological exchanges in river corridors. Freshwater Biology 47 (4), 621–640.

Mammola, S., Piano, E., Malard, F., Vernon, P., Isaia, M., 2019a. Extending Janzen's hypothesis to temperate regions: a test using subterranean ecosystems. Functional Ecology 33 (9), 1638–1650.

Mammola, S., Arnedo, M.A., Fiser, C., Cardoso, P., Dejanaz, A.J., Isaia, M., 2020. Environmental filtering and convergent evolution determine the ecological specialization of subterranean spiders. Functional Ecology 34 (5), 1064–1077.

Mammola, S., Piano, E., Cardoso, P., Vernon, P., Domínguez-Villar, D., Culver, D.C., Isaia, M., 2019b. Climate change going deep: the effects of global climatic alterations on cave ecosystems. The Anthropocene Review 6 (1–2), 98–116.

Marmonier, P., Delettre, Y., Lefebvre, S., Guyon, J., Boulton, A.J., 2004. A simple technique using wooden stakes to estimate vertical patterns of interstitial oxygenation in the beds of rivers. Archiv für Hydrobiologie 160 (1), 133–143.

Mathers, K.L., Millett, J., Robertson, A.L., Stubbington, R., Wood, P.J., 2014. Faunal response to benthic and hyporheic sedimentation varies with direction of vertical hydrological exchange. Freshwater Biology 59 (11), 2278–2289.

Mermillod-Blondin, F., Lefour, C., Lalouette, L., Renault, D., Malard, F., Simon, L., Douady, C.J., 2013. Thermal tolerance breadths among groundwater crustaceans living in a thermally constant environment. Journal of Experimental Biology 216 (9), 1683–1694.

Mouron, S., Eme, D., Bellec, A., Bertrand, M., Mammola, S., Liébault, F., Douady, C.J., Malard, F., 2022. Unique and shared effects of local and catchment predictors over distribution of hyporheic organisms: does the valley rule the stream? Ecography 5, e06099.

Nawaz, A., Purahong, W., Lehmann, R., Herrmann, M., Totsche, K.U., Küsel, K., Wubet, T., Buscot, F., 2018. First insights into the living groundwater mycobiome of the terrestrial biogeosphere. Water Research 145, 50–61.

Newman, D.K., Banfield, J.F., 2002. Geomicrobiology: how molecular-scale interactions underpin biogeochemical systems. Science 296 (5570), 1071–1077.

Niemiller, M.L., Taylor, S.J., Slay, M.E., Hobbs, H.H., 2019. Biodiversity in the United States and Canada. In: White, W., Culver, D.C., Pipan, T. (Eds.), Encyclopedia of Caves. Academic Press, Amsterdam, The Netherland, pp. 163–176.

Notenboom, J., 1991. Marine regressions and the evolution of groundwater dwelling amphipods (Crustacea). Journal of Biogeography 18, 437–454.

Novarino, G., Warren, A., Butler, H., Lambourne, G., Boxshall, A., Bateman, J., Kinner, N.E., Harvey, R.W., Mosse, R.A., Teltsch, B., 1997. Protistan communities in aquifers: a review. FEMS Microbiology Reviews 20 (3–4), 261–275.

Pedersen, K., 2000. Exploration of deep intraterrestrial microbial life: current perspectives. FEMS Microbiology Letters 185 (1), 9–16.

Peiffer, S., Kappler, A., Haderlein, S.B., Schmidt, C., Byrne, J.M., Kleindienst, S., Vogt, C., Richnow, H.H., Obst, M., Angenent, L.T., Bryce, C., McCammon, C., Planer-Friedrich, B., 2021. A biogeochemical–hydrological framework for the role of redox-active compounds in aquatic systems. Nature Geoscience 14 (5), 264–272.

Perkins, A.K., Ganzert, L., Rojas-Jimenez, K., Fonvielle, J., Hose, G.C., Grossart, H.P., 2019. Highly diverse fungal communities in carbon-rich aquifers of two contrasting lakes in Northeast Germany. Fungal Ecology 41, 116–125.

Pipan, T., Culver, D.C., 2007. Epikarst communities: biodiversity hotspots and potential water tracers. Environmental Geology 53 (2), 265–269.

Pipan, T., Culver, D.C., 2017. The unity and diversity of the subterranean realm with respect to invertebrate body size. Journal of Cave and Karst Studies 79 (1), 1–9.

Premate, E., Borko, Š., Delić, T., Malard, F., Simon, L., Fišer, C., 2021. Cave amphipods reveal co-variation between morphology and trophic niche in a low-productivity environment. Freshwater Biology 66, 1876–1888.

Prié, V., 2019. Molluscs. In: White, W.B., Culver, D.C., Pipan, T. (Eds.), Encyclopedia of Caves. Academic Press, Amsterdam, The Netherland, pp. 725–731.

Proudlove, G.S., 2006. Subterranean fishes of the world. An account of the subterranean (hypogean) fishes described up to 2003 with a bibliography 1541–2004. International Society for Subterranean Biology, Moulis, France.

Purkamo, L., Kietäväinen, R., Miettinen, H., Sohlberg, E., Kukkonen, I., Itävaara, M., Bomberg, M., 2018. Diversity and functionality of archaeal, bacterial and fungal communities in deep Archaean bedrock groundwater. FEMS Microbiology Ecology 94 (8), 116.

Reeder, W.J., Quick, A.M., Farrell, T.B., Benner, S.G., Feris, K.P., Tonina, D., 2018. Spatial and temporal dynamics of dissolved oxygen concentrations and bioactivity in the hyporheic zone. Water Resources Research 54, 2112–2128.

Reeves, J.M., De Deckker, P., Halse, S.A., 2007. Groundwater Ostracods from the arid Pilbara region of northwestern Australia: distribution and water chemistry. Hydrobiologia 585, 99–118.

Retter, A., Nawaz, A., 2022. The Groundwater mycobiome: fungal diversity in terrestrial aquifers. In: Tockner, K., Mehner, T. (Eds.), Encyclopedia of Inland Waters, 2nd. Elsevier, pp. 385–396.

Retter, A., Karwautz, C., Griebler, C., 2020. Groundwater microbial communities in times of climate change. Current Issues in Molecular Biology 41, 509–537.

Romero, A., 2001. Scientists prefer them blind: the history of hypogean fish research. Environmental Biology of Fishes 62, 43–71.

Roux, S., Krupovic, M., Daly, R.A., Borges, A.L., Nayfach, S., Schulz, F., Sharrar, A., Matheus Carnevali, P.B., Cheng, J.F., Ivanova, N.N., Bondy-Denomy, J., Wrighton, K.C., Woyke, T., Visel, A., Kyrpides, N.C., Eloe-Fadrosh, E.A., 2019. Cryptic inoviruses revealed as pervasive in bacteria and archaea across Earth's biomes. Nature Microbiology 4 (11), 1895–1906.

Rukke, N.A., 2002. Effects of low calcium concentration on two common freshwater crustaceans, *Gammarus lacustris* and *Astacus astacus*. Functional Ecology 16, 357–366.

Schulz, C., Steward, A.L., Prior, A., 2013. Stygofauna presence within fresh and highly saline aquifers of the border rivers region in southern Queensland. Proceedings of the Royal Society of Queensland 118, 27–35.

Schweichhart, J.S., Pleyer, D., Winter, C., Retter, A., Griebler, C., 2022. Presence and role of prokaryotic viruses in groundwater environments. In: Tockner, K., Mehner, T. (Eds.), Encyclopedia of Inland Waters, 2nd. Elsevier, pp. 373–384.

Segawa, T., Sugiyama, A., Kinoshita, T., Sohrin, R., Nakano, T., Nagaosa, K., Greenidge, D., Kato, K., 2015. Microbes in groundwater of a volcanic mountain, Mt. Fuji; 16S rDNA phylogenetic analysis as a possible indicator for the transport routes of groundwater. Geomicrobiology Journal 32 (8), 677–688.

Shapouri, M., Cancela da Fonseca, L., Iepure, S., Stigter, T., Ribeiro, L., Silva, A., 2016. The variation of stygofauna along a gradient of salinization in a coastal aquifer. Hydrology 47 (1), 89–103.

Shen, Y., Chapelle, F.H., Strom, E.W., Benner, R., 2015. Origins and bioavailability of dissolved organic matter in groundwater. Biogeochemistry 122, 61–78.

Simon, K.S., 2019. Cave ecosystems. In: White, W.B., Culver, D.C., Pipan, T. (Eds.), Encyclopedia of Caves. Academic Press, London, UK, pp. 223–226.

II. Drivers and patterns of groundwater biodiversity

Smith, H.J., Zelaya, A.J., De León, K.B., Chakraborty, R., Elias, D.A., Hazen, T.C., Arkin, A.P., Cunningham, A.B., Fields, M.W., 2018. Impact of hydrologic boundaries on microbial planktonic and biofilm communities in shallow terrestrial subsurface environments. FEMS Microbiology Ecology 94 (12), fiy191.

Smith, R.J., Jeffries, T.C., Roudnew, B., Seymour, J.R., Fitch, A.J., Simons, K.L., Speck, P.G., Newton, K., Brown, M.H., Mitchell, J.G., 2013. Confined aquifers as viral reservoirs. Environmental Microbiology Reports 5 (5), 725–730.

Sohlberg, E., Bomberg, M., Miettinen, H., Nyyssönen, M., Salavirta, H., Vikman, M., Itävaara, M., 2015. Revealing the unexplored fungal communities in deep groundwater of crystalline bedrock fracture zones in Olkiluoto, Finland. Frontiers in Microbiology 6, 573.

Stegen, J.C., Johnson, T., Fredrickson, J.K., Wilkins, M.J., Konopka, A.E., Nelson, W.C., Arntzen, E.V., Chrisler, W.B., Chu, R.K., Fansler, S.J., 2018. Influences of organic carbon speciation on hyporheic corridor biogeochemistry and microbial ecology. Nature Communications 9 (1), 585.

Stein, H., Griebler, C., Berkhoff, S., Matzke, D., Fuchs, A., Hahn, H.J., 2012. Stygoregions: a promising approach to a bioregional classification of groundwater systems. Scientific Reports 2 (1), 1–9.

Stern, D.B., Breinholt, J., Pedraza-Lara, C., López-Mejía, M., Owen, C.L., Bracken-Grissom, H., Fetzner Jr., J.W., Crandall, K.A., 2017. Phylogenetic evidence from freshwater crayfishes that cave adaptation is not an evolutionary dead-end. Evolution 71, 2522–2532.

Stevens, G.C., 1989. The latitudinal gradient in geographical range: how so many species coexist in the tropics. The American Naturalist 133, 240–256.

Stoch, F., Galassi, D.M.P., 2010. Stygobiotic crustacean species richness: a question of numbers, a matter of scale. Hydrobiologia 653, 217–234.

Strayer, D.L., 1994. Limits to biological distributions. In: Gibert, J., Danielopol, D., Stanford, J.A. (Eds.), Groundwater Ecology. Academic Press, San Diego, USA, pp. 287–310.

Strayer, D.L., May, S.E., Nielsen, P., Wollheim, W., Hausam, S., 1997. Oxygen, organic matter, and sediment granulometry as controls on hyporheic animal communities. Archiv für Hydrobiologie 140 (1), 131–144.

Strayer, D.L., May, S.E., Nielsen, P., Wollheim, W., Hausam, S., 1995. An endemic groundwater fauna in unglaciated eastern North America. Canadian Journal of Zoology 73 (3), 502–508.

Stumpp, C., Hose, G.C., 2013. The impact of water table drawdown and drying on subterranean aquatic fauna in invitro experiments. PLoS One 8 (11), e78502.

Suttle, C.A., 2005. Viruses in the sea. Nature 437 (7057), 356–361.

Trontelj, P., Blejec, A., Fišer, C., 2012. Ecomorphological convergence of cave communities. Evolution 66 (12), 3852–3865.

Van Leeuwenhoek, A., 1677. About little animals observed in rain-well-sea- and snow-water; as also in water wherein pepper had lain infused. Philosophical Transactions of the Royal Society of London 12, 821–831.

Vervier, P., Gibert, J., 1991. Dynamics of surface water/groundwater ecotones in a karstic aquifer. Freshwater Biology 26, 241–250.

Volkmer-Ribeiro, C., Bichuette, M.E., de Sousa Machado, V., 2010. Racekiela cavernicola (Porifera: Demospongiae) new species and the first record of cave freshwater sponge from Brazil. Neotropical Biology and Conservation 5 (1), 53–58.

Weigand, A.M., Kremers, J., Grabner, D.S., 2016. Shared microsporidian profiles between an obligate (Niphargus) and facultative subterranean amphipod population (Gammarus) at sympatry provide indications for underground transmission pathways. Limnologica 58, 7–10.

Wurzbacher, C., Kreiling, A.K., Svantesson, S., Van den Wyngaert, S., Larsson, E., Heeger, F., Nilsson, H.R., Pálsson, S., 2020. Fungal communities in groundwater springs along the volcanic zone of Iceland. Inland Waters 10, 1–10.

Zagmajster, M., Eme, D., Fišer, C., Galassi, D., Marmonier, P., Stoch, F., Cornu, J.F., Malard, F., 2014. Geographic variation in range size and beta diversity of groundwater crustaceans: insights from habitats with low thermal seasonality. Global Ecology and Biogeography 23 (10), 1135–1145.

Zagmajster, M., Malard, F., Eme, E., Culver, D.C., 2018. Subterranean biodiversity patterns from global to regional scales. In: Moldovan, O.T., Kováč, L., Halse, S. (Eds.), Cave Ecology. Springer, Cham, Switzerland, pp. 195–225.

# Patterns and determinants of richness and composition of the groundwater fauna

*Maja Zagmajster[1], Rodrigo Lopes Ferreira[2], William F. Humphreys[3], Matthew L. Niemiller[4] and Florian Malard[5]*

[1]University of Ljubljana, Biotechnical Faculty, Department of Biology, Ljubljana, Slovenia; [2]Universidade Federal de Lavras (UFLA), Centro de Estudos em Biologia Subterrânea, Departamento de Ecologia e Conservação, Lavras, Minas Gerais, Brazil; [3]University of Western Australia, School of Biological Sciences, Crawley, WA, Australia; [4]Department of Biological Sciences, The University of Alabama in Huntsville, Huntsville, AL, United States; [5]Univ Lyon, Université Claude Bernard Lyon 1, CNRS, ENTPE, UMR 5023 LEHNA, Villeurbanne, France

## Introduction

Over the last 20 years, groundwater ecology has shifted from the study of the ecology of local communities to the analysis of large-scale, multi-species ecological patterns and processes, a scientific field referred to as macroecology (Brown, 1995; Lawton, 1999; Beck et al., 2012; Griebler et al., 2014; Zagmajster et al., 2018). This shift from local to global perspectives primarily results from two major conceptual breakthroughs that have reshaped the principles of community ecology. First, ecologists have become increasingly aware of the importance of processes occurring at larger spatial scales in shaping local communities. The composition and richness of local species assemblages necessarily depend on the composition and richness of the regional species pool (Ricklefs, 1987; Gibert and Deharveng, 2002; Malard et al., 2009). Second, the integration of concepts from evolutionary biology has provided community ecology with a robust theoretical framework to analyze biodiversity patterns. Across a range of spatial scales, biodiversity patterns are ultimately shaped by four processes: "species are added to communities via speciation and dispersal, and the relative abundances of these species are then shaped by drift and selection, as well as ongoing dispersal, to drive community dynamics" (Vellend, 2010, p. 202). In this theoretical framework, drift refers to random changes in the relative abundances of species and extinction is considered as the outcome of selection and drift, rather than as a distinct process.

In nature, multiple factors can influence the processes shaping biodiversity patterns, but a common usage is to group them into three broad environmental factors, namely spatial heterogeneity, climate/productivity, and history (Morueta-Holme et al., 2013; Zagmajster et al., 2014). History refers to the long-lasting effect of past climatic and paleogeographic events on present-day pattern of biodiversity. The greatest challenge consists in drawing the links between factors, processes, and patterns. Even the clearest and most consistent patterns, such as the latitudinal gradient of biodiversity (Hillebrand, 2004), the species-area relationship (Rosenzweig, 1995), or the latitudinal increase in species range size in the northern hemisphere (i.e., the Rapoport effect, Stevens, 1989) may have multiple though not mutually exclusive explanations. Hence, the question is not so much about which explanation is correct but about the relative contributions of a plurality of mechanisms, i.e., links between factors and processes (Lawton, 1996). Multi-causality is likely to be the rule, and even the most consistent patterns may be due to multiple mechanisms pulling in the same direction. Cold Pleistocene climates may have caused the extinction of many groundwater species in Northern Europe (Zagmajster et al., 2014), while low productivity and spatial heterogeneity may limit speciation rates (Eme et al., 2015). Both mechanisms are equally plausible and potentially add up to explain lower species richness in groundwater of northern Europe. Furthermore, the relative contribution of different mechanisms in shaping biodiversity patterns vary across continents and regions, a concept that is referred to as spatial nonstationarity (Eme et al., 2015).

Analyses of large-scale groundwater biodiversity patterns have lagged behind analyses of surface biodiversity patterns for many reasons. Large datasets that enable spatial analyses at finer spatial units than countries or bioregions, have only recently been gathered (Deharveng et al., 2009; Zagmajster et al., 2014). Collecting exact spatial data using GPS technology has become much easier, spatial datasets are becoming increasingly available and citable due to open data policies of journals and funding sources. In addition, many groundwater ecologists long considered that biodiversity inventories were too incomplete for undertaking quantitative analysis of patterns in the diversity and composition of groundwater species communities. Culver et al. (2013) outlined a series of potential pitfalls in estimating subterranean biodiversity, but when considered, none is really unsurmountable. Methods exist to account for spatial differences in sampling effort; many of them have been successfully applied also to the analysis of groundwater biodiversity patterns (Zagmajster et al., 2010; Christman and Zagmajster, 2012). Until recently, the lack of large-scale groundwater-oriented environmental data sets has hindered the quantitative analysis of biodiversity patterns. Such data sets are progressively becoming available and used to generate large-scale groundwater environmental predictors. For example, the groundwater habitat map published by Cornu et al. (2013) provides a useful resource for quantifying the availability and heterogeneity of groundwater habitats at the scale of Europe (Iannella et al., 2020a). Moreover, surface-oriented environmental data sets, especially those depicting actual and historical climates, have been used to approximate spatial variation in temperature and productivity in groundwater (see review in Mammola and Leroy, 2018).

There is considerable variation in the description and understanding of groundwater biodiversity patterns among continents, as comparatively more studies have been conducted in Europe and North America (Zagmajster et al., 2018). In this chapter, we provide an overview of the most striking features of patterns in species richness and taxonomic composition of the obligate groundwater fauna at regional to continental scales. Using studies in Europe, America, and Australia, we focus on patterns that are common enough among taxonomic groups and continents to be potentially recognized as "rules." We report on the most probable combination of mechanisms shaping these patterns and emphasize their implications for the assessment and conservation of groundwater biodiversity. We end this chapter by calling for a broader perspective of diversity patterns, which incorporates the taxonomic, phylogenetic, and functional facets of groundwater biodiversity.

## Patterns of species richness

### Defining local and regional species richness

Local species richness (LSR) of groundwater communities is defined differently among studies, but it usually refers to the number of obligate groundwater species present in an aquifer, cave, spring, well, or the hyporheic zone of a stream reach. Although these spatial units have different areas, they all correspond to finite and continuous subterranean hydrological systems within which organisms encounter each other over ecological time. Regional species richness (RSR) is the number of all species in the region that can potentially colonize and inhabit a focal groundwater habitat area (Malard et al., 2009). The spatial boundaries of regions are defined based on environmental variables, on the species found in them, or a combination of both (Abell et al., 2008; Gao and Kupfer, 2018); they are intended to describe broad patterns of species composition and associated ecological and evolutionary processes.

Ideally, large-scale patterns of species richness can be documented using either LSR or RSR as a dependent variable. In practice, however, spatial grain size often increases with the spatial extent of studies due to sampling limitation. Moreover, spatial entities used for documenting large-scale patterns of species richness — most often grid cells — are arbitrarily defined rather than spatially delineated to represent areas with distinct regional species pools (but see Iannella et al., 2020b). In this chapter, the term RSR is also used more widely to denote the species richness of any spatial entity encompassing multiple caves or aquifers.

### Local species richness patterns

LSR of obligate groundwater metazoans exhibits a number of striking features. First, LSR of groundwater habitats is low comparatively to that of surface aquatic habitats. This is likely associated to the harshness of groundwater habitats, being aphotic habitats with very limited food resources. In Europe, karst aquifers and unconsolidated sediment aquifers contained on average 11.7 species (range: 1−32 species, n = 60 aquifers) and 12 species (range: 2−27 species, n = 51 aquifers), respectively (Malard et al., 2009). In southwestern Germany, the median and mean taxonomic richness are 1.0 and 1.6 taxa per sampled well (n = 304 wells; Hahn and Fuchs, 2009). In Tennessee, USA, the mean number of obligate groundwater species in caves of the Interior Plateau, Southwestern Appalachians, and Ridge and Valley (RV) ecoregions is $1.1 \pm 1.1$ (mean $\pm$ standard deviation, $n = 278$ caves), $1.3 \pm 1.2$ ($n = 289$) and $1.1 \pm 1.0$ ($n = 75$) species per cave, respectively (Niemiller and Zigler, 2013). In the Jura Mountains, France, the mean number of species per sampled locality ($n = 269$ localities), including caves, springs, wells, and the hyporheic zone of stream reaches, is four, with a maximum of 15 species collected at a single locality (Dole-Olivier et al., 2009a). In the Dinaric karst of Slovenia, one of the most species-rich and best-studied cave regions in the world, the mean number of groundwater species per cave is $3.9 \pm 6.3$ ($n = 84$ caves; Culver et al., 2004a). At a global scale, the list of local hotspots of groundwater biodiversity (Culver et al., 2021) includes 11 groundwater systems, essentially karst systems, that average 46 obligate groundwater species (range: 28–72 species). For comparison, a surface stream often contains hundreds of species: Schmid-Araya and Schmid (1995) reported 569 and 1000 species of invertebrates from the small mountain streams Oberer Seebach, Austria and Breteinbach, Germany, respectively. The difference in LSR between the two habitats still stands even though facultative groundwater species are taken into consideration for measuring LSR of groundwater habitats.

A second striking feature of LSR patterns is the highly skewed distribution of species richness among localities within a region (Fig. 6.1). Only a few localities harbor a high number of species to the point that differences in species richness among grid cells within a region can largely depend upon the distribution of species-rich localities among cells (Culver et al., 2004a; Pipan et al., 2018). Species-rich localities are due to the overlap of ranges of "widespread" species rather than from the occurrence of many species endemic to these localities (Culver et al., 2004a; Christman et al., 2005; Bregović et al., 2019). Local biodiversity hotspots have been reported from several regions in the World (Culver et al., 2021), having high

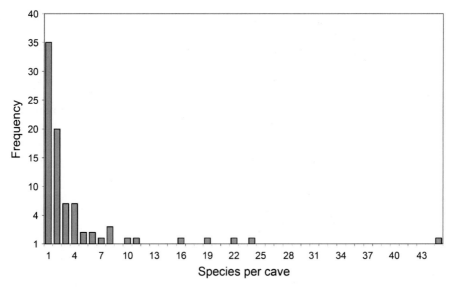

**FIGURE 6.1** Distribution of number of obligate groundwater species per cave in the Slovenian portion of the Dinaric karst. Frequency is the number of caves with a particular number of species ($n = 84$ sampled caves). *After Culver et al. (2004a).*

numbers of obligate groundwater species too: these includes the Postojna Planina Cave System, Slovenia (72 species), the Walsingham Cave, Bermuda (65 species), the San Marcos Artesian Well, USA (55 species), and the Robe Valley wells, Australia (48 species).

We are not aware of any quantitative studies that have examined the causes of variation in LSR of groundwater fauna among a large number of caves and/or aquifers within a region. The most influential factors would probably include the size of caves (Culver et al., 2004b; Kováč et al., 2016), the heterogeneity of microhabitats within caves (Trontelj et al., 2012), and the amount of surface-derived organic matter reaching groundwater (Datry et al., 2005). However, LSR may also be controlled by dispersal opportunities between neighboring localities. Using data from nine regions in Europe, Malard et al. (2009) found that porous aquifers of a region had on average more species and similar species assemblages than karst aquifers of the same region. A higher hydrological connectivity between porous aquifers could promote dispersal, thereby increasing mean LSR and similarity in species composition among aquifers. If subterranean dispersal is important, then species richness in adjoining areas should influence diversity locally. The difficulty is that LSR may be spatially autocorrelated because of some factors that influence species dispersal or because some unmeasured environmental parameters influencing LSR are themselves spatially autocorrelated (Christman et al., 2005). The problem trying to understand spatial patterns in LSR within a region is not deciding among a number of explanations but rather quantifying the relative contributions of different explanations. In the community literature, variation partitioning has become a standard approach for quantifying the variation in species richness and composition accounted for by the unique and shared effects of local environmental factors and dispersal-related factors (Li et al., 2021). This approach has yet to be applied to groundwater species communities.

Explaining spatial patterns of LSR over large areas encompassing multiple regions requires taking into consideration macroevolutionary processes that influenced the size of the regional species pool. In particular, a greater rate of speciation would increase the number of potential species capable to colonize a focal groundwater habitat in a region. One way to test for the importance of macroevolutionary processes in shaping large-scale spatial patterns of LSR is to regress LSR as a function of RSR (Srivastava, 1999). LSR of groundwater species communities appears to be positively related to RSR, but the shape of the relationship varies across studies and habitats (Fig. 6.2) (Malard et al., 2009; Stoch and Galassi, 2010). The best fit is obtained using either a linear model or a second-order polynomial model indicating that LSR may or may not reach a maximum value within the range of observed RSR. In Europe, LSR of karst aquifers is higher in southern regions that probably experienced both an uninterrupted history of speciation and a reduced extinction rate during cold Pleistocene climates (Malard et al., 2009). Hence, macroevolutionary processes causing regional differences in the size of the regional species pool indirectly influence LSR, even though the rate of macroevolutionary processes is negligible relative to other processes at a local scale.

A third striking feature of LSR is that the mean number of species in a single locality (either a cave or an aquifer) is a minor component of the total number of species present in a region. Most localities possess a small fraction of the regional diversity (Culver and Sket, 2000; Niemiller and Zigler, 2013; Pipan et al., 2018). This is largely due to the overrepresentation of narrowly distributed species in groundwater. A direct consequence is that a considerable sampling effort is needed to obtain an unbiased estimate of the regional species pool. Hence, regional species accumulation curves rarely reach saturation (Castellarini et al., 2007; Stoch

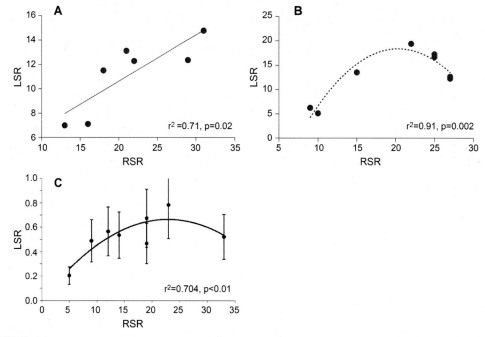

**FIGURE 6.2**   Relationships between local species richness (LSR) and regional species richness (RSR) of obligate groundwater metazoan communities. (A). Localities are karst aquifers belonging to six European regions (B). The same as in A for unconsolidated sediment aquifers (A and B after Malard et al., 2009). (C). Localities are caves belonging to eight karst regions in northeastern Italy. LSR is log transformed and error bars represent standard errors (C: after Stoch and Galassi, 2010).

and Galassi, 2010; Dole-Olivier et al., 2009b, 2015; Niemiller and Zigler, 2013; Iepure et al., 2016, 2017). In Europe, species accumulation curves tend to level off in the most Northern regions containing both lower species richness and lower proportions of narrowly distributed species (for example the Wallonia and Jura regions, Fig. 6.3). Dole-Olivier et al. (2009b) explored how accounting for different sources of within-region environmental heterogeneity including catchment boundaries, aquifer types (i.e., karst vs. unconsolidated sediment), and habitat types within aquifers improved the assessment of RSR. The relative importance of different sources appears to vary across regions.

## Regional species richness patterns

Studies documenting variation in groundwater species richness over large spatial extents have used a coarse spatial grain size corresponding either to biogeographical regions (Hof et al., 2008), geological regions (Culver et al., 2003), or grid cells of $100-10,000$ km$^2$ in size (most often 400 km$^2$; Christman et al., 2005; Niemiller and Zigler, 2013; Zagmajster et al., 2014; Eme et al., 2015, 2018; Bregović et al., 2019). Not surprisingly, patterns of regional species richness and the mechanisms put forward to explain them are sensitive to the spatial grain size used to map species richness (Fig. 6.4). Studies using large biogeographical regions

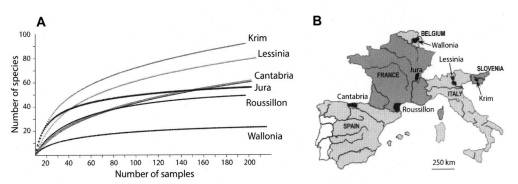

**FIGURE 6.3** Regional species rarefaction curves (A) of observed species richness in groundwater for six regions in Europe: Krim (Slovenia), Lessinia (Italy), Cantabria (Spain), Jura (France), Roussillon (France) and Wallonia (Belgium). Samples are from caves, springs, wells and the hyporheic zone of stream reaches in regions shown in (B). After Dole-Olivier et al. (2009b).

or countries as a spatial grain size revealed a northward monotonic decline in the number of groundwater species in Europe, which has been attributed to a mass extinction of narrowly distributed species during cold Pleistocene climates (Hof et al., 2008; Stoch and Galassi, 2010; Zagmajster et al., 2018) (Fig. 6.4A). Mapping using 10,000 km$^2$ cells revealed a hump-shaped pattern of species richness, which has been best explained by variation in actual evapotranspiration, a surrogate of the amount of energy reaching groundwater (Eme et al., 2015) (Fig. 6.4B).

Subterranean biogeographers have long viewed history as an important factor driving biodiversity patterns. Paleogeographic events, such as marine regressions and aridification, can cause the isolation and eventually speciation of surface populations that had colonized

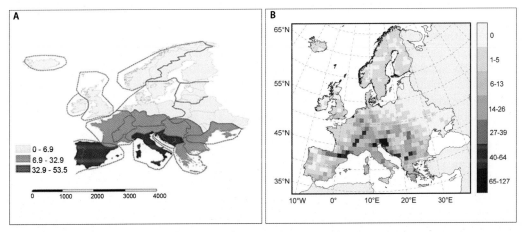

**FIGURE 6.4** Spatial patterns of groundwater crustacean species richness in Europe obtained using different areal units. (A). Latitudinal bands of species richness, joining countries with similar species richness. The legend presents species richness per area (number of species per km$^2$) within each band (after Zagmajster et al., 2018). (B). Observed species richness for 10,000 km$^2$ cells (after Zagmajster et al., 2014).

groundwater (Boutin and Coineau, 2000; Leys et al., 2003). Uplift of carbonate massifs can provide new subterranean habitats for species diversification (Borko et al., 2021), and many groundwater species may have gone extinct during cold Quaternary climates (Zagmajster et al., 2014). However, the spatial imprints of historical events can only persist over time if they are not overwritten by subsequent dispersal phases or by the effects of contemporary factors. Since the 2000s, a broader perspective of groundwater species richness patterns has emerged, one that revives the role of energy and spatial heterogeneity, in addition to history, and attempts to decipher the relative contribution of these three broad factors (Culver et al., 2006; Eme et al., 2015; Zagmajster et al., 2018). Here, we provide an overview of the state-of-the-art in understanding large-scale spatial patterns of groundwater species richness, using representative studies in Europe, North America, and South America.

### Europe

Zagmajster et al. (2014) used 21,700 species occurrence records from 1570 species and subspecies to map species richness of groundwater crustaceans at European scale using 10,000 km$^2$ grid cells. In a follow-up study, Eme et al. (2015) provided a quantitative assessment of the relative importance of three broad factors, history, energy and spatial heterogeneity, in shaping species richness patterns of groundwater crustaceans in Europe. They used regressions models and variance partitioning to quantify the unique and shared effects of the three broad factors and geographically weighted regression models to show variation in their relative importance across Europe. Three important results emerge from these studies. First, species richness peaks at latitudes ranging from 42° to 46° N. This mid-latitude ridge of high species richness appears to be robust to sampling bias (Zagmajster et al., 2014) (Fig. 6.4B). Second, all three broad factors play a role (multi-causality), but their relative importance varies across Europe (spatial nonstationarity) (Eme et al., 2015). In Northern Europe, the effect of cold Pleistocene climates, reduced spatial heterogeneity (topographic heterogeneity and habitat diversity), and lower productivity all contribute to limit species richness. In Southwestern Europe, the increasing aridity since the last glacial maximum and lower productivity limit species richness. Species richness peaks in regions of mid-latitude where the combined beneficial effects of high productivity and high habitat heterogeneity have not been counteracted by cold or arid historical events. However, correlative analyses alone are not sufficient to infer the influence of environmental factors on evolutionary processes that ultimately shape species richness patterns. High topographic heterogeneity can promote speciation by imposing barriers to dispersal and generating steep ecological gradients. It can also increase the probability that species survive changing climate by migrating to nearby refugia. Third, interactions among factors are important. Eme et al. (2015) found that the positive relationship between topographic heterogeneity and species richness is steeper in regions of high productivity as well as in regions of high habitat diversity. This may indicate that the potential for topographic heterogeneity to increase speciation or decrease extinction is higher when productivity allows species to maintain large population size (Evans et al., 2005). Similarly, barriers to dispersal imposed by mountains may lead to greater speciation rate in a habitat-rich landscape that promotes ecological specialization.

Patterns of groundwater crustacean species richness in Europe are robust to the effect of sampling intensity (Zagmajster et al., 2014; Eme et al., 2015). Also, the continental pattern of species richness based on morphologically delimited groundwater crustacean species remains unchanged when molecularly delimited taxonomic units are used (Eme et al., 2018). Similar observations were made at a regional scale, for the Western Balkans (Fig. 6.5). Two species-rich areas of *Niphargus* are apparent, no matter whether species are morphologically or molecularly delimited (Borko et al., 2022).

### North America

Knowledge of groundwater biodiversity in North America is largely restricted to karst areas, especially those rich with caves. Culver et al. (2000) compiled the first comprehensive list of described obligate groundwater species and generated the first county-level map of groundwater species richness for the United States (Fig. 3 in Culver et al., 2000). Their dataset included county-level data for 360 obligate groundwater species. The number of species has greatly increased in the last 20+ years, as Niemiller et al. (2019) reported 469 groundwater taxa from the United States and Canada.

Groundwater species richness is not distributed evenly among major cave biogeographic regions (Christman and Culver, 2001). It is highest in Appalachians (103 species), Interior Low Plateau (82 species), Edwards Plateau & Balcones Escarpment (75 species), Ozarks (53 species), and Florida Lime Sinks (29 species) karst regions (Niemiller et al., 2019). All other karst regions have six or fewer obligate groundwater species. However, 39.7% (186 species) of U.S. and Canadian groundwater biodiversity is known from outside of the 10 major cave biogeographic regions. Several additional but smaller karst regions exist in the United States and Canada, but these regions have not yet received the level of study from speleobiologists as the significant cave biogeographic regions, which is changing in recent years.

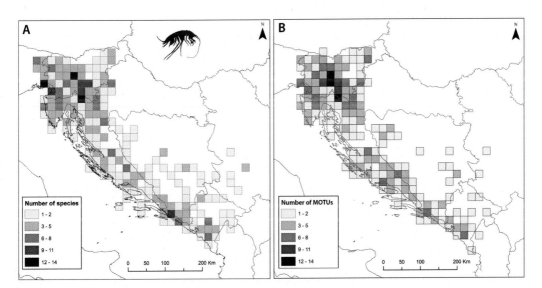

**FIGURE 6.5**  Patterns of groundwater amphipod species richness in the Western Balkans are robust to the effect of species delimitation methods. (A). Species richness pattern, based on morphologically delimited species (after Bregović et al., 2019). (B). Species richness pattern, based on molecularly delimited taxonomic units (after Borko et al. 2022).

Differences in spatial patterns of species richness among groundwater species communities at varying spatial scales suggest that they are influenced by different factors, such as habitat availability, opportunity for dispersal, climate and historical factors, and surface productivity, among others. It is likely that these factors are not mutually exclusive but operate in concert to shape patterns of species richness. Studies have demonstrated that the number of caves is a reliable predictor of regional species richness (Christman and Culver, 2001; Culver et al., 2003, 2006; Christman et al., 2016). For example, species richness in eastern North America is greatest in regions with the greatest cave density, such as the Mammoth Cave region and the southern Cumberland Plateau in the Interior Low Plateau (Fig. 6.6A). Species richness in North America is correlated with number of caves within a cave biogeographic region but not necessarily karst area (Culver et al., 2003; Niemiller and Zigler, 2013), suggesting that degree of karst development (e.g., number and density of cave entrances, habitat diversity) rather than extent of karst is a more important determinant of species richness (Culver et al., 2003). Greater exposures of karst and cave habitat may support higher species richness by facilitating larger populations, lower extinction rates, and greater opportunities for colonization of and dispersal within subterranean habitats (Christman and Culver, 2001; Culver et al., 2006). If we assume that cave density is positively correlated with cave connectivity, areas of greater cave density have higher connectivity among caves, offering more dispersal opportunities. Increased dispersal may decrease local extirpation of populations (i.e., rescue effect) and extinction rates in general, and also can lead to greater homogenization (i.e., similarity) of groundwater species communities among caves, aquifers, and regions (Malard et al., 2009).

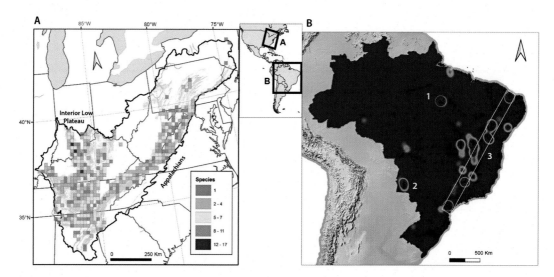

**FIGURE 6.6** Spatial patterns of species richness in two regions in Northern and Southern America (depicted by black rectangles in the middle panel). (A). Groundwater species richness of 20 × 20 km cells, in the Interior Low Plateau and Appalachians karst regions (data from Christman et al., 2016). Gray areas show karst landscapes. (B). Pattern of groundwater species richness in Brazil. Blue, green and red colors indicate areas of low, intermediate and high species richness, respectively. Black dots correspond to caves. (1) Carajás regions with iron ore caves; (2) Bodoquena region; (3) diagonal axis of spots, from Rio Grande do Norte state in northeastern Brazil to southern São Paulo state.

Past and contemporary climatic, geological, and hydrological factors likely also influence regional patterns of species richness. For example, while obligate groundwater species are known north of southern extent of ice sheets during Pleistocene glacial maxima (e.g., Holsinger, 1980; Bousfield and Holsinger, 1981), diversity is low and karst biogeographic regions with the greatest species richness all are located south of maximum glacial extent. Culver et al. (2006) identified a midlatitude ridge between 33° and 35° N in North America where the richness of the cave terrestrial fauna peaks. Because most available food resources in cave systems originate from the surface, high surface productivity has been hypothesized to be an important predictor of subterranean biodiversity (Culver et al., 2006; Christman et al., 2016). This hypothesized ridge may correspond to higher long-term levels of precipitation, warmer temperatures, and subsequently higher surface productivity, particularly over recent geological times in the Pleistocene. Caves in this region are likely to have more energy available to support larger populations, more species, and more diverse communities. Although the mid-latitude ridge hypothesis was framed for terrestrial cave fauna, there is evidence that species richness of the obligate fauna in the Interior Low Plateau also increases within this hypothesized ridge, with a hotspot occurring in Northeastern Alabama and south-central Tennessee (Fig. 6.6A; see also Niemiller and Zigler, 2013). However, additional hotspots exist to the north in the Mammoth Cave region of Kentucky as well as south-central Indiana. In contrast, the diversity gradient declines north to south in the Appalachians karst region with greatest species richness in western Virginia and eastern West Virginia but corresponds with the gradient in cave density in this karst region. Moreover, the Edwards Aquifer & Balcones Escarpment in Texas is especially diverse (Hutchins et al., 2021) and appears to be supported primarily by chemolithoautotrophy (Hutchins et al., 2016). Hutchins et al. (2016) hypothesized that chemolithoautotrophy may facilitate reduced extinction rates during periods of unfavorable climate and might also support high species richness by increasing resource exploitation and reducing competition. Habitat heterogeneity may also be an important predictor of species richness in the hydrogeologically complex Edwards Aquifer of central Texas (Hutchins et al., 2021).

### South America

Knowledge of groundwater biodiversity patterns in South America is rudimentary like most tropical regions of the world. Most species occurrence data are from karst regions in Brazil where research in groundwater ecology has intensified since the 1980s. Brazil has 75 obligate groundwater species described to date. However, the total number of known species exceeds 180, and several species await discovery because most karst areas in the country remain unexplored (Ferreira et al., 2022). Groundwater species-rich regions are located along a diagonal axis running from Northeastern regions (Rio Grande do Norte State) to Southeastern regions (northern São Paulo state) (Fig. 6.6B). All these regions occur in the most significant karst regions in South America, including several important geological formations, such as the Bambui, Una, Caatinga, Canudos, and Apodi carbonate rocks. However, they also correspond to the most sampled regions in the country. Consequently, it is likely that other areas of high diversity will emerge in the future. Two regional hotspots of groundwater species richness lie outside this northeastern to southeastern axis. The Carajás area, in northern Brazil (Amazon region) contains hundreds of iron ore caves, many of which are colonized by widely distributed species. The Bodoquena region, in Mato Grosso do Sul state (Central-western Brazil) is a shallow groundwater-table area where several groundwater species occur in phreatic caves.

II. Drivers and patterns of groundwater biodiversity

## Global pattern of species richness

We lack precise estimates of the number of obligate groundwater species on Earth (see for example Culver and Holsinger, 1992). The few attempts to map global patterns of species richness are plagued with many taxonomic and sampling uncertainties, so that the patterns need to be interpreted with great caution (Culver et al., 2003). Zagmajster et al. (2018) used the most recent per-country species richness data to map global patterns of species richness of groundwater crustaceans. Species richness corrected for by the area effect is higher in temperate regions including Europe and North America as well as in arid regions of Australia than in the tropics. That pattern is in agreement with the worldwide distribution of local hotspots of groundwater species richness. Of the 11 local hotspots containing more than 25 obligate groundwater species listed by Culver et al. (2021), only one, Robe Valley wells, Australia, occurs in the tropics. This suggests that the relationship between species richness of the obligate groundwater fauna and surface productivity may be hump-shaped at a global scale. An excessive supply of organic matter can lead to a deficiency of dissolved oxygen in groundwater (Mouron et al., 2022), limit the acquisition of troglomorphic traits (Zagmajster et al., 2018), and/or may lower the speciation from surface ancestors in the absence of cold or arid historical events. Another simpler and more easily testable explanation is that the comparatively low species richness of tropical regions is due to too low sampling effort. Clearly, more studies in diverse groundwater habitats of tropical regions are needed (Rabelo et al., 2021).

## Patterns of species composition

The taxonomic composition of groundwater species communities is dominated by crustaceans. They represent 71.2%, 65.0%, 78.3%, and 65.7% of obligate groundwater species known in the World, Europe, North America, and Western Australia, respectively (Guzik et al., 2010; Stoch and Galassi, 2010; Niemiller and Zigler, 2013). For comparison, crustaceans represent only 10.8% and 13.5% of the total number of surface aquatic species known in the World and Europe, respectively, whereas insects represent 68.5% and 57.0% (Stoch and Galassi, 2010). Other most-diverse taxonomic groups in groundwater are gastropods, annelids, and mites. However, there are large variations in the numerical importance of non-crustacean taxa among continents. Europe has only two species of groundwater beetles, while Australia has more than 100 taxa (Deharveng et al., 2009; Langille et al., 2021). Groundwater vertebrates are also unevenly distributed across the World. Most of the 288 species of obligate groundwater fishes are from tropical and subtropical regions. The olm (*Proteus anguinus*) is the only exclusively cave-dwelling salamander found in Europe, whereas salamanders of the family Plethodontidae are represented by 13 species in USA (Gorički et al., 2019).

## Variation in species composition among localities within a region

The obligate groundwater fauna shows a high species turnover among localities within a region. Malard et al. (2009) found that the average pairwise dissimilarity in crustacean species between aquifers of six European regions ranged from 40% to 72%. The high spatial turnover is in part due to the high proportion of rare species within communities, a characteristic feature of the obligate groundwater fauna. For the six European regions, the proportion of species occurring in less than 3% of the sampled localities (i.e., cave, spring, well, or hyporheic locality) was 40%, 53%, 57%, 60%, 63%, and 77% in the Jura Mountains (eastern France, $n = 192$ localities), Lessinian Mountains (Italy, $n = 196$), Krim Massif (Slovenia, $n = 187$), Wallonia (Belgium, $n = 206$), Roussillon (southern France, $n = 187$) and Cantabria (Spain, $n = 192$), respectively (Dole-Olivier et al., 2009b). In Tennessee, 19 of the 40 aquatic obligate cave species occur in less than six caves ($n = 661$ sampled caves) (Niemiller and Zigler, 2013). In the Dinaric karst region, 23% of 145 groundwater amphipod species are single-locality endemics (Bregović et al., 2019).

Several studies have explored the environmental drivers of species community changes among localities within a region often using a combination of variables related to aquifer hydrogeology, water physicochemistry, landscape heterogeneity (e.g., elevation and land cover), and paleogeography (e.g., distance to Quaternary glaciers). These include studies in the Jura Mountains, France ($n = 269$ localities; Dole-Oliver et al., 2009a), the Wallon region, Belgium ($n = 202$, Martin et al., 2009), the Lessinian Moutains, Italy ($n = 197$; Galassi et al., 2009), the Baden–Württemberg region, Germany ($n = 304$; Hahn and Fuchs, 2009), the Gwydir Valley aquifer, Australia ($n = 20$; Korbel and Hose, 2015), southwestern England ($n = 221$; Johns et al., 2015) and the Tajo River basin, Central Spain ($n = 24$; Iepure et al., 2017). However, as for species richness, no studies have examined the relative importance of environmental factors versus dispersal-related factors in shaping spatial patterns of species community structure. In most studies, hydrogeological type of aquifers, rather than groundwater physicochemistry or landscape heterogeneity, accounted for the largest proportion of variation in groundwater metazoan composition among localities. In the six regions sampled in Europe as part of the European project PASCALIS (Gibert et al., 2009), the proportion of species exclusive to karst aquifers and unconsolidated sediment aquifers averaged $32.1 \pm 12.2\%$ (range: 12.2%–47.5%) and $27.6 \pm 11.9\%$ (range: 13.1%–46.9%), respectively (Dole-Olivier et al., 2009b). The proportion of species exclusive to a single aquifer type increased with RSR, indicating that habitat specialization was more pronounced in species-rich regions (Malard et al., 2009). Ferreira et al. (2007) found that 47% and 29% of obligate groundwater species collected in France were exclusive to karst and unconsolidated sediment aquifers, respectively, and 20% occurred in both habitats. Aquifer type is a broad proxy of habitat conditions, which encompasses a number of biologically important features such as the size of voids, their interconnectedness, which determines the ability of the rock to transmit water (permeability), and hydrological connection of groundwater to the surface environment. Void size influences the species composition by setting an upper limit on the maximal body size (Pipan and Culver, 2017; Korbel et al., 2019). The interconnectedness of the voids not only influence the foraging activity of organisms but also the fluxes of nutrients, dissolved oxygen, and organic carbon that support assemblages of prokaryotes and metazoans. Aquifers showing strong hydrological connections to the surface show increased temporal variability and higher fluxes of nutrients, organic matter, and dissolved oxygen. A too high hydrological connection to the surface environment promotes the colonization of subterranean habitats by aquatic surface organisms, which can outcompete groundwater obligate taxa (Vervier and Gibert, 1991; Brunke and Gonser, 1999).

Results of regional scale studies of species-environment relationships do not imply that groundwater species do not respond to variation in groundwater physical-chemistry, however. Temporal changes in water physical-chemistry caused by aquifer replenishment or surface-water groundwater exchanges are important drivers of the spatiotemporal dynamics of species assemblages within aquifers and the hyporheic zone of streams (Gibert et al., 1994; Malard et al., 2002). Using a network of 30 boreholes spatially distributed over a 3.5-km$^2$ area, Saccò et al. (2019) showed that changes in water temperature, pH, and salinity induced by different rainfall regimes constrained the spatial distribution of groundwater species over time at the Sturt Meadows calcrete aquifer, Western Australia.

## Variation in species composition among regions

The regional obligate groundwater fauna shows a high turnover in species composition along distances of only a few hundreds of km. Zagmajster et al. (2014) found that groundwater crustacean assemblages were almost entirely replaced within a distance of less than 500 km across Europe. Of the 313 obligate groundwater species collected in six regions of Europe (see location of regions in Fig. 6.3B), only one, three, four and 27 species were shared by six, four, three, and two regions, respectively (Dole-Olivier et al., 2009b). Hence, hierarchical cluster analyses of groundwater invertebrate assemblages systematically group aquifers according to their geographic region of origin, even though dissimilar aquifer types are included in the analysis (Malard et al., 2009; Stein et al., 2012). The dissimilarity in species composition among regions is typically higher than the dissimilarity in species composition among habitats of a region. Biogeographic regions that are spatially delineated to contain distinct assemblages of groundwater species are typically smaller than that of surface biogeographic regions (Ferreira et al., 2005; Stoch and Galassi, 2010; Stein et al., 2012; Niemiller and Zigler, 2013). Ferreira et al. (2007) showed that nearly 70% of the 380 obligate groundwater species collected in France, essentially crustaceans, were restricted to the French landscape whereas the surface aquatic fauna of Ephemeroptera and Odonata (i.e., 252 species) had 80% and 100% species in common with the Italian and Swiss fauna, respectively.

Within biogeographic regions, nearby areas often display high turnover in species composition. In the Northeastern groundwater biogeographic region of Italy, Stoch and Galassi (2010) showed that dissimilarities between crustacean assemblages of regional aquifers were consistently above 0.5 (Fig. 6.7A). At the genus level, each regional aquifer harbors distinct species as exemplified by the distribution of species belonging to the sphaeromatid isopod genus *Monolistra* (Fig. 6.7B). Each regional community apparently evolved in isolation within narrow karst areas delimited in the north by the southward limit of Pleistocene glaciers and in the south by the alluvial sediments of the Padanian Plain. Another striking example of the high species turnover in groundwater is provided by shallow groundwater calcrete aquifers (about 210 major calcretes) in the arid zone of Central and Western Australia (Humphreys et al., 2009). These aquifers support a diverse subterranean water beetle fauna comprising more than 100 species with up to four sympatric species within the genera *Limbodessus*, *Paroster*, *Bidessodes* and *Exocelina* (Leys et al., 2003; Watts and Humphreys, 2009). These are often sympatric with a variety of Amphipoda, Parabathynellidae (Syncarida),

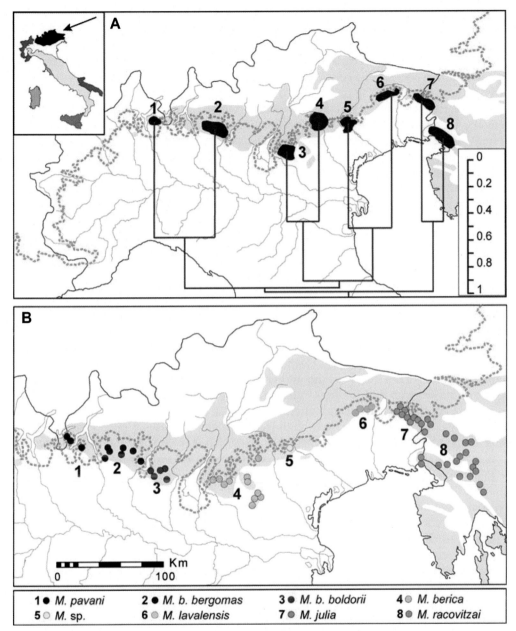

**FIGURE 6.7** Regional patterns of groundwater species composition in northeastern Italy. (A). Dendrogram showing dissimilarity in species composition among regional aquifers (scale on the right). Map in the upper right corner shows the distribution of groundwater biogeographic regions in Italy. Gray areas represent karstic areas; dotted line represents the southern border of Alpine ice sheet during the Last Glacial Maximum. (B). Spatial distribution of isopod species of the genus *Monolistra*. *After Stoch and Galassi (2010).*

Oniscidea (Scyphacidae), and Copepoda (Cyclopoidea) (Humphreys et al., 2009). Each of the more than 100 aquifers sampled has a unique fauna irrespective of the size of the calcrete (ranging between 50 and 1000 km$^2$) or distance from the nearest calcrete, variously between approximately 4–45 km apart. While the area considered is in the order of 660,000 km$^2$, the species turnover needs to be considered only within the modest area of the calcrete aquifers.

We are aware of only a single study that documented patterns of regional species turnover in groundwater at a continental scale. Zagmajster et al. (2014) used 10,000 km$^2$ cells to document the latitudinal patterns in turnover and nestedness components of beta diversity of groundwater crustacean assemblages across Europe. The turnover component of beta diversity represents differences in species composition caused by species replacements whereas its nestedness component represents differences in species composition caused by species losses or gains (Baselga, 2012). Beta diversity was high all over Europe: the Jaccard index of multiple-cell dissimilarity was 0.90 in 500 × 500 km cells and was largely due to species replacement rather than species losses or gains. Species replacement was higher in southern Europe (latitude <47.7 N), as predicted in regions of stable climates where ecological and/or nonecological speciation events can accumulate over time (Baselga et al., 2012). The nestedness component of beta diversity was higher in northern European regions where species extinction (species losses) and recolonization (species gains) might have occurred during Pleistocene glacials and interglacials, respectively. However, we still lack a quantitative understanding of the contribution of environmental filtering and dispersal limitations to continental-scale variation in the turnover and nestedness components of beta diversity in groundwater.

The strong regional turnover in groundwater fauna has several implications. First, although the contribution of a landscape unit (i.e., cave, aquifer, catchment, and region) to overall species richness increases monotonically with its size, hierarchical additive partitioning of groundwater species richness indicates that beta diversity among regions makes by far the highest contribution to overall species richness (Malard et al., 2009; Stoch and Galassi, 2010; Pipan et al., 2018). Second, although local species richness is typically lower in groundwater than in surface water, continental species richness may be higher in groundwater because of the accumulation of species from different aquifers in multiple regions. In Europe, the total number of groundwater crustacean species is higher than the total number of surface aquatic crustacean species (Stoch and Galassi, 2010). Third, the presence of local biodiversity hotspots in regions containing distinct groundwater fauna implies that it is possible to protect a large proportion of species by focusing habitat conservation efforts on a few complementary species-rich aquifers located in multiple regions. Culver et al. (2000) found that 50% of obligate cave fauna occurred in less than 1% of the North American landscape. In Europe, a reserve network arbitrarily limited to 10% of the landscape containing groundwater fauna would allow to protect 80% of groundwater species (Michel et al., 2009). Iannella et al. (2021) showed that reserving only 1.9% of the European landscape would allow to protect 44% of groundwater harpacticoid species (Copepoda, Crustacea) and 93% of endemics.

## Toward a multifaceted approach to groundwater biodiversity patterns

Our present understanding of subterranean biodiversity patterns essentially derives from studies that consider variation in the taxonomic richness and composition of communities.

Here, we emphasize the need to expand our understanding of mechanisms contributing to the assembly of communities by considering the taxonomic, phylogenetic, and functional facets of biodiversity (Cavender-Bares et al., 2009; Cadotte et al., 2011).

Biologists increasingly rely on several criteria—among which morphological and molecular criteria—to delimit species boundaries (Fišer et al., 2018). Eme et al. (2018) explored how incorporating morphological and molecular criteria into the analysis of species richness patterns of groundwater crustaceans in Europe could foster the understanding of the contribution of ecological and nonecological speciation processes. The authors hypothesized that the use of molecular criteria would increase the importance of spatial heterogeneity in shaping biodiversity patterns because genetically divergent but morphologically similar species are more likely along physical barriers separating ecologically similar regions than along resource gradients promoting ecologically based divergent selection. This hypothesis was rejected because the proportion of variance in species richness explained by spatial heterogeneity, productivity, and history was stable whatever the method used to delimit species. However, Eme et al. (2018) found that using molecular instead of morphological criteria decreased by 10-fold the average species range size and reinforced the importance of historical climates in explaining the pattern of increasing range size at higher latitudes (i.e., the Rapoport effect; Stevens, 1989).

A handful of studies have analyzed groundwater biodiversity patterns in ways that account for phenotypic differences among species. Functional traits denote all morphological, physiological, and phenological features, which influence an individual's growth, reproduction, and survival (Violle et al., 2007). They can offer greater explanatory power of niche-based and dispersal-based processes controlling the assembly of communities than measures based only on the presence and abundance of species. Trontelj et al. (2012) showed that morphological specialization to different microhabitats consistently occurred across geographical space and clades in the groundwater amphipod genus *Niphargus* by revealing four distinct geographically and phylogenetically unrelated ecomorphs, each associated to a specific microhabitat. Fišer et al. (2019) and Premate et al. (2021a) provided strong evidence of trophic specialization of morphological structures among *Niphargus* species, which could promote the stable coexistence of species in the same habitat. Delić and Fišer (2019) reported a case of character displacement in which populations of *Niphargus croaticus* living in sympatry with *Niphargus subtypicus* had larger body and longer appendages than populations living in allopatry. Fine-level niche partitioning and character displacement may be driven by competitive interactions between species. Predator—prey relationships can lead to the evolution of specialized defensive morphological structures (Premate et al., 2021b). However, species interactions are rarely accounted for to explain spatial patterns of species composition in groundwater.

Assuming that the size of a species' range is controlled by the magnitude of barriers to dispersal between suitable habitats (but see Lester et al., 2007), geographic patterns of range size may provide useful information about the importance of dispersal in shaping patterns of species composition. Zagmajster et al. (2014) showed that the median range size of European groundwater crustacean communities increased abruptly above a latitudinal threshold of 43° N in regions that had experienced large amplitude Quaternary climate oscillations. Increased habitat connectivity during periods of intense postglacial sediment deposition might have increased the opportunity for dispersal (Eme et al., 2013), thereby resulting in a pattern of decreasing species replacement and increasing nestedness of communities with increasing latitudes (Zagmajster et al., 2014). In general, geographic patterns of species richness is heavily influenced by widely distributed species because their large number of occurrences have a disproportionate contribution to the species richness counts (Šizling et al., 2009). Hence, separately documenting and analyzing geographic patterns of widely and narrowly distributed species can provide deeper insights into the determinants of biodiversity patterns (Jetz and Rahbek, 2002). Bregović et al. (2019) found that spatial patterns of subterranean aquatic and terrestrial species richness in the Dinaric were driven by widely distributed species, even though the subterranean fauna displays an exceptionally high proportion of narrowly distributed species. Christman et al. (2005) found a positive but nonlinear relationship between the number of single-cave endemics and nonendemics in the Eastern North American terrestrial cave fauna; species-rich cave communities had a higher fraction of single-cave endemics. However, no study has yet quantitatively tested for differences in factors shaping species richness of widely and narrowly distributed species in the subterranean fauna.

Evidence is rising that understanding the ecological specialization—dispersal trade-off (Jocque et al., 2010) is key to explain geographic patterns also in subterranean biodiversity. Mammola et al. (2019a) showed that specialization to deep subterranean life involved a reduction in the breadth of the fundamental thermal niche of subterranean spiders (but see also Colado et al., 2022 for terrestrial subterranean beetles) and that this specialization could in turn restrict a species' ability to disperse across elevation gradient. Mermillod-Blondin et al. (2013) brought preliminary evidence that the same mechanism could operate in the groundwater fauna. Elucidating the thermal specialization—dispersal trade-off is key to assess the differential ability of groundwater species to cope with global warming (Mammola et al., 2019b).

Integrating phylogenetic measures of biodiversity has proved a crucial step for understanding processes shaping the assembly of community across space (Cavender-Bares et al., 2009; Mouquet et al., 2012; Tucker et al., 2017). This integration has lagged behind in groundwater ecology due to difficulties in assembling large-scale phylogenies of groundwater species-rich taxa (Fišer et al., 2008; Leijs et al., 2012; Morvan et al., 2013). Similar to studies by Davies and Buckley (2011) and Fritz and Rahbek (2012) on global patterns of mammal and amphibian diversity, respectively, Zagmajster et al. (2018) proposed using measures of phylogenetic diversity to distinguish among groundwater regions that might correspond to cradles of diversity, museums and refugia. By comparing species and phylogenetic richness of *Niphargus* species, Borko et al. (2022) showed that different evolutionary processes had probably shaped the two regional hotspots of species richness occurring in Western Balkans. Data for species-rich clades of groundwater crustaceans such as the amphipod family Niphargidae may soon become available to jointly analyze patterns of taxonomic, functional and phylogenetic diversity across large geographic areas (see for example Bregović et al., 2019; Fišer et al., 2019; Borko et al., 2022).

In this chapter, we highlighted a broad perspective of groundwater biodiversity patterns that explicitly considers the role of multiple mechanisms, their interactions, and changes in their relative contribution across space and scales. In the near future, this perspective will profit from fully integrating multiple facets of biodiversity.

# Acknowledgments

We thank S. Rétaux and C. Griebler for editing an earlier version of this chapter.

# References

Abell, R., Thieme, M.L., Revenga, C., Bryer, M., Kottelat, M., Bogutskaya, N., Coad, B., Mandrak, N., Contreras, B.S., Bussing, W., Stiassny, M.L.J., Skelton, P., Allen, G.R., Unmack, P., Naseka, A., Ng, R., Sindorf, N., Robertson, J., Armijo, E., Higgins, J.V., Heibel, T.J., Wikramanayake, E., Olson, D., López, H.L., Reis, R.E., Lundberg, J.G., Pérez, M.H.S., Petry, P., 2008. Freshwater ecoregions of the world: a new map of biogeographic units for freshwater biodiversity conservation. BioScience 58 (5), 403–414.

Baselga, A., 2012. The relationship between species replacement, dissimilarity derived from nestedness, and nestedness. Global Ecology and Biogeography 21, 1223–1232.

Baselga, A., Gomez-Rodriguez, C., Lobo, J.M., 2012. Historical legacies in world amphibian diversity revealed by the turnover and nestedness components of beta diversity. PLoS One 7 (2), e32341.

Beck, J., Ballesteros-Mejia, L., Buchmann, C.M., Dengler, J., Fritz, S.A., Gruber, B., Hof, C., Jansen, F., Knapp, S., Kreft, H., Schneider, A.-K., Winter, M., Dormann, C.F., 2012. What's on the horizon for macroecology? Ecography 35, 673–683.

Borko, Š., Trontelj, P., Seehausen, O., Moškrič, A., Fišer, C., 2021. A subterranean adaptive radiation of amphipods in Europe. Nature Communications 12, 3688.

Borko, Š., Altermatt, F., Zagmajster, M., Fišer, C., 2022. A hotspot of groundwater amphipod diversity on a crossroad of evolutionary radiations. Diversity and Distributions 28 (12), 2765–2777.

Bousfield, E.L., Holsinger, J.R., 1981. A second new subterranean amphipod crustacean of the genus *Stygobromus* (Crangonyctidae) from Alberta, Canada. Canadian Journal of Zoology 59 (9), 1827–1830.

Boutin, C., Coineau, N., 2000. Evolutionary rates and phylogenetic age in some stygobiontic species. In: Wilkens, H., Culver, D.C., Humphreys, W.F. (Eds.), Ecosystems of the World, 30: Subterranean Ecosystems. Elsevier, Amsterdam, pp. 433–451.

Bregović, P., Fišer, C., Zagmajster, M., 2019. Contribution of rare and common species to subterranean species richness patterns. Ecology and Evolution 9, 11606–11618.

Brown, J.H., 1995. Macroecology. The University of Chicago Press.

Brunke, M., Gonser, T., 1999. Hyporheic invertebrates: the clinal nature of interstitial communities structured by hydrological exchange and environmental gradients. Journal of the North American Benthological Society 18 (3), 344–362.

Cadotte, M.W., Carscadden, K., Mirotchnick, N., 2011. Beyond species: functional diversity and the maintenance of ecological processes and services. Journal of Applied Ecology 48, 1079–1087.

Castellarini, F., Dole-Olivier, M.-J., Malard, F., Gibert, J., 2007. Using habitat heterogeneity to assess stygobiotic species richness in the French Jura region with a conservation perspective. Fundamental and Applied Limnology 169, 69–78.

Cavender-Bares, J., Kozak, K.H., Fine, P.V.A., Kembel, S.W., 2009. The merging of community ecology and phylogenetic biology. Ecology Letters 12, 693–715.

Christman, M.C., Culver, D.C., 2001. The relationship between cave biodiversity and available habitat. Journal of Biogeography 28, 367–380.

Christman, M.C., Zagmajster, M., 2012. Mapping subterranean biodiversity. In: White, W.B., Culver, D.C. (Eds.), Encyclopedia of Caves, second ed. Academic Press, Amsterdam, pp. 474–481.

Christman, M.C., Culver, D.C., Madden, M.K., White, D., 2005. Patterns of endemism of the eastern North American cave fauna. Journal of Biogeography 32, 1441–1452.

Christman, M.C., Doctor, D.H., Niemiller, M.L., Weary, D.J., Young, J.A., Zigler, K.S., Culver, D.C., 2016. Predicting the occurrence of cave-inhabiting fauna based on features of the Earth surface environment. PLoS One 11 (8), e0160408.

Colado, R., Pallarés, S., Fresneda, J., Mammola, S., Rizzo, V., Sánchez-Fernández, D., 2022. Climatic stability, not average habitat temperature, determines thermal tolerance of subterranean beetles. Ecology 103 (4), e3629.

Cornu, J.F., Eme, D., Malard, F., 2013. The distribution of groundwater habitats in Europe. Hydrogeology Journal 21, 949–960.

Culver, D.C., Holsinger, J.R., 1992. How many species of troglobites are there? NSS Bulletin 54, 79–80.

Culver, D.C., Sket, B., 2000. Hotspots of subterranean biodiversity in caves and wells. Journal of Cave and Karst Studies 62, 11–17.

Culver, D.C., Master, L.L., Christman, M.C., Hobbs III, H.H., 2000. Obligate cave fauna of the 48 contiguous United States. Conservation Biology 14, 386–401.

Culver, D.C., Christman, M.C., Elliott, W.R., Hobbs, H.H., Reddell, J.R., 2003. The North American obligate cave fauna: regional patterns. Biodiversity & Conservation 12 (3), 441–468.

Culver, D.C., Christman, M.C., Sket, B., Trontelj, P., 2004a. Sampling adequacy in an extreme environment: species richness patterns in Slovenian caves. Biodiversity and Conservation 13, 1209–1229.

Culver, D.C., Christman, M.C., Šereg, I., Trontelj, P., Sket, B., 2004b. The location of terrestrial species-rich caves in a cave-rich area. Subterranean Biology 2, 27–32.

Culver, D.C., Deharveng, L., Bedos, A., Lewis, J.J., Madden, M., Reddell, J.R., Sket, B., Trontelj, P., White, D., 2006. The mid-latitude biodiversity ridge in terrestrial cave fauna. Ecography 29, 120–128.

Culver, D., Trontelj, P., Zagmajster, M., Pipan, T., 2013. Paving the way for standardized and comparable subterranean biodiversity studies. Subterranean Biology 10, 43–50.

Culver, D.C., Deharveng, L., Pipan, T., Bedos, A., 2021. An overview of subterranean biodiversity hotspots. Diversity 13, 487.

Datry, T., Malard, F., Gibert, J., 2005. Response of invertebrate assemblages to increased groundwater recharge rates in a phreatic aquifer. Journal of the North American Benthological Society 24 (3), 461–477.

Davies, T.J., Buckley, L.B., 2011. Phylogenetic diversity as a window into the evolutionary and biogeographic histories of present-day richness gradients for mammals. Philosophical Transactions of the Royal Society B 366, 2414–2425.

Deharveng, L., Stoch, F., Gibert, J., Bedos, A., Galassi, D., Zagmajster, M., Brancelj, A., Camacho, A., Fiers, F., Martin, P., Giani, N., Magniez, G., Marmonier, P., 2009. Groundwater biodiversity in Europe. Freshwater Biology 54, 709–726.

Delić, T., Fišer, C., 2019. Species interactions. In: White, W., Culver, D., Pipan, T. (Eds.), Encyclopedia of Caves. Elsevier, Academic Press, London, pp. 967–973.

Dole-Olivier, M.-J., Galassi, D.M.P., Fiers, F., Malard, F., Martin, P., Martin, D., Marmonier, P., 2015. Biodiversity in mountain groundwater: the Mercantour National Park (France) as a European hotspot. Zoosystema 37 (4), 529–550.

Dole-Olivier, M.J., Malard, F., Martin, D., Lefébure, T., Gibert, J., 2009a. Relationships between environmental variables and groundwater biodiversity at the regional scale. Freshwater Biology 54 (4), 797–813.

Dole-Olivier, M.-J., Castellarini, F., Coineau, N., Galassi, D.M.P., Martin, P., Mori, N., Valdecasas, A., Gibert, J., 2009b. Towards and optimal sampling strategy to assess groundwater biodiversity: comparison across six European regions. Freshwater Biology 54 (4), 777–796.

Eme, D., Malard, F., Konecny-Dupré, L., Lefébure, T., Douady, C.J., 2013. Bayesian phylogeographic inferences reveal contrasting colonization dynamics among European groundwater isopods. Molecular Ecology 22, 5685–5699.

Eme, D., Zagmajster, M., Fišer, C., Galassi, D., Marmonier, P., Stoch, F., Cornu, J.F., Oberdorff, T., Malard, F., 2015. Multi-causality and spatial non-stationarity in the determinants of groundwater crustacean diversity in Europe. Ecography 38 (5), 531–540.

Eme, D., Zagmajster, M., Delić, T., Fišer, C., Flot, J.-F., Konecny-Dupré, L., Pálsson, S., Stoch, F., Zakšek, V., Douady, C.J., Malard, F., 2018. Do cryptic species matter in macroecology? Sequencing European groundwater crustaceans yields smaller ranges but does not challenge biodiversity determinants. Ecography 41, 424–436.

Evans, K.L., Warren, P.H., Gaston, K.J., 2005. Species-energy relationships at the macroecological scale: a review of the mechanisms. Biological Reviews 80, 1–25.

Ferreira, D., Malard, F., Dole-Olivier, M.-J., Gibert, J., 2005. Hierarchical patterns of obligate groundwater biodiversity in France. In: Gibert, J. (Ed.), Proceedings of the International Symposium on World Subterranean Biodiversity. University Claude Bernard, Lyon, pp. 5–78.

Ferreira, D., Malard, F., Dole-Olivier, M.-J, Gibert, J., 2007. Obligate groundwater fauna of France: diversity patterns and conservation implications. Biodiversity and Conservation 16, 567–596.

Ferreira, R.L., Bernard, E., da Cruz Júnior, F.W., Piló, L.B., Calux, A., Souza-Silva, M., Barlow, J., Pompeu, P.S., Cardoso, P., Mammola, S., García, A.M., Jeffery, W.R., Shear, W., Medellín, R.A., Wynne, J.J., Borges, P.A.V., Kamimura, Y., Pipan, T., Hajna, N.Z., Sendra, A., Peck, S., Onac, B.P., Culver, D.C., Hoch, H., Flot, J.F., Stoch, F., Pavlek, M., Niemiller, M.L., Manchi, S., Deharveng, L., Fenolio, D., Calaforra, J.M., Yager, J., Griebler, C., Nader,

F.H., Humphreys, W.F., Hughes, A.C., Fenton, B., Forti, P., Sauro, F., Veni, G., Frumkin, A., Gavish-Regev, E., Fišer, C., Trontelj, P., Zagmajster, M., Delic, T., Galassi, D.M.P., Vaccarelli, I., Komnenov, M., Gainett, G., da Cunha Tavares, V., Kováč, Ľ., Miller, A.Z., Yoshizawa, K., Di Lorenzo, T., Moldovan, O.T., Sánchez-Fernández, D., Moutaouakil, S., Howarth, F., Bilandžija, H., Dražina, T., Kuharić, N., Butorac, V., Lienhard, C., Cooper, S.J.B., Eme, D., Strauss, A.M., Saccò, M., Zhao, Y., Williams, P., Tian, M., Tanalgo, K., Woo, K.S., Barjakovic, M., McCracken, G.F., Simmons, N.B., Racey, P.A., Ford, D., Labegalini, J.A., Colzato, N., Ramos Pereira, M.J,, Aguiar, L.M.S., Moratelli, R., Du Preez, G., Pérez-González, A., Reboleira, A.S.P.S., Gunn, J., Mc Cartney, A., Bobrowiec, P.E.D., Milko, D., Kinuthia, W., Fischer, E., Meierhofer, M.B., Frick, W.F., 2022. Brazilian cave heritage under siege. Science 375 (6586), 1238-1239.

Fišer, C., Sket, B., Trontelj, P., 2008. A phylogenetic perspective on 160 years of troubled taxonomy of *Niphargus* (Crustacea: Amphipoda). Zoologica Scripta 37, 665−680.

Fišer, C., Robinson, C.T., Malard, F., 2018. Cryptic species as a window into the paradigm shift of the species concept. Molecular Ecology 27, 613−635.

Fišer, C., Delić, T., Luštrik, R., Zagmajster, M., Altermatt, F., 2019. Niches within a niche: ecological differentiation of subterranean amphipods across Europe's interstitial waters. Ecography 42 (6), 1212−1223.

Fritz, S.A., Rahbek, C., 2012. Global patterns of amphibian phylogenetic diversity. Journal of Biogeography 39, 1373−1382.

Galassi, D.M.P., Stoch, F., Fiasca, B., Di Lorenzo, T., Gattone, E., 2009. Groundwater biodiversity patterns in the Lessinian Massif of northern Italy. Freshwater Biology 54, 830−847.

Gao, P., Kupfer, J.A., 2018. Capitalizing on a wealth of spatial information: improving biogeographic regionalization through the use of spatial clustering. Applied Geography 99, 98−108.

Gibert, J., Deharveng, L., 2002. Subterranean ecosystems: a truncated functional biodiversity. BioScience 52 (6), 473−481.

Gibert, J., Culver, D.C., Dole-Olivier, M.-J., Malard, F., Christman, M.C., Deharveng, L., 2009. Assessing and conserving groundwater biodiversity: synthesis and perspectives. Freshwater Biology 54, 930−941.

Gibert, J., Vervier, P., Malard, F., Laurent, R., Reygrobellet, J.-L., 1994. Dynamics of communities and ecology of karst ecosystems: example of two karsts in eastern and southern France. In: Gibert, J., Danielopol, D.L., Stanford, J.A. (Eds.), Groundwater Ecology. Academic Press, San Diego, pp. 425−450.

Gorički, Š., Niemiller, M.L., Fenolio, D.B., Gluesenkamp, A.G., 2019. Salamanders. In: Culver, D.C., White, W.B., Pipan, T. (Eds.), Encyclopedia of Caves, third ed. Elsevier, London, pp. 871−884.

Griebler, C., Malard, F., Lefébure, T., 2014. Current developments in groundwater ecology—from biodiversity to ecosystem function and services. Current Opinion in Biotechnology 27, 159−167.

Guzik, M.T., Austin, A.D., Cooper, S.J.B., Harvey, M.S., Humphreys, W.F., Bradford, T., Eberhard, S.M., King, R.A., Leys, R., Muirhead, K.A., Tomlinson, M., 2010. Is the Australian subterranean fauna uniquely diverse? Invertebrate Systematics 24, 407−418.

Hahn, H.J., Fuchs, A., 2009. Distribution patterns of groundwater communities across aquifer types in south-western Germany. Freshwater Biology 54 (4), 848−860.

Hillebrand, H., 2004. On the generality of the latitudinal diversity gradient. The American Naturalist 163 (2), 192−211.

Hof, C., Brändle, M., Brandl, R., 2008. Latitudinal variation of diversity in European freshwater animals is not concordant across habitat types. Global Ecology and Biogeography 17 (4), 539−546.

Holsinger, J.R., 1980. *Stygobromus canadensis*, a new subterranean amphipod crustacean (Crangonyctidae) from Canada, with remarks on Wisconsin refugia. Canadian Journal of Zoology 58 (2), 290−297.

Humphreys, W.F., Watts, C.H.S., Cooper, S.J.B., Leijs, R., 2009. Groundwater estuaries of salt lakes: buried pools of endemic biodiversity on the western plateau, Australia. Hydrobiologia 626 (1), 79−95.

Hutchins, B.T., Engel, A.S., Nowlin, W.H., Schwartz, B.F., 2016. Chemolithoautotrophy supports macroinvertebrate food webs and affects diversity and stability in groundwater communities. Ecology 97 (6), 1530−1542.

Hutchins, B.T., Gibson, J.R., Diaz, P.H., Schwartz, B.F., 2021. Stygobiont diversity in the san Marcos Artesian well and Edwards aquifer groundwater ecosystem, Texas, USA. Diversity 13 (6), 234.

Iannella, M., Fiasca, B., Di Lorenzo, T., Biondi, M., Di Cicco, M., Galassi, D.M.P., 2020a. Spatial distribution of stygobitic crustacean harpacticoids at the boundaries of groundwater habitat types in Europe. Scientific Reports 10, 19043.

Iannella, M., Fiasca, B., Di Lorenzo, T., Biondi, M., Di Cicco, M., Galassi, D.M.P., 2020b. Jumping into the grids: mapping biodiversity hotspots in groundwater habitat types across Europe. Ecography 43, 1825−1841.

Iannella, M., Fiasca, B., Di Lorenzo, T., Di Cicco, M., Biondi, M., Mammola, S., Galassi, D.M.P., 2021. Getting the 'most out of the hotspot' for practical conservation of groundwater biodiversity. Global Ecology and Conservation 31, e01844.

Iepure, S., Feurdean, A., Bădălută, C., Nagavciuc, V., Persoiu, A., 2016. Pattern of richness and distribution of groundwater Copepoda (Cyclopoida: Harpacticoida) and Ostracoda in Romania: an evolutionary perspective. Biological Journal of the Linnean Society 119 (3), 593−608.

Iepure, S., Rasines-Ladero, R., Meffe, R., Carreño, F., Mostaza, D., Sundberg, A., Di Lorenzo, T., Barroso, J.L., 2017. Exploring the distribution of groundwater Crustacea (Copepoda and Ostracoda) to disentangle aquifer type features—a case study in the upper Tajo basin (Central Spain). Ecohydrology 10, e1876.

Jetz, W., Rahbek, C., 2002. Geographic range size and determinants of avian species richness. Science 297 (5586), 1548−1551.

Jocque, M., Field, R., Brendonck, L., De Meester, L., 2010. Climatic control of dispersal−ecological specialization tradeoffs: a metacommunity process at the heart of the latitudinal diversity gradient? Global Ecology and Biogeography 19, 244−252.

Johns, T., Jones, J.I., Knight, L., Maurice, L., Wood, P., Robertson, A., 2015. Regional-scale drivers of groundwater faunal distributions. Freshwater Science 34 (1), 316−328.

Korbel, K.L., Hose, G.C., 2015. Habitat, water quality, seasonality, or site? Identifying environmental correlates of the distribution of groundwater biota. Freshwater Science 34 (1), 329−343.

Korbel, K.L., Stephenson, S., Hose, G.C., 2019. Sediment size influences habitat selection and use by groundwater macrofauna and meiofauna. Aquatic Sciences 81 (2), 39.

Kováč, L., Parimuchová, A., Miklosová, D., 2016. Distributional patterns of cave Collembola Hexapoda in association with habitat conditions, geography and subterranean refugia in the Western Carpathians. Biological Journal of the Linnean Society 119, 571−592.

Langille, B.L., Hyde, J., Saint, K.M., Bradford, T.M., Stringer, D.N., Tierney, S.M., Humphreys, W.F., Austin, A.D., Cooper, S.J.B., 2021. Evidence for speciation underground in diving beetles (Dytiscidae) from a subterranean archipelago. Evolution 75, 166−175.

Lawton, J.H., 1996. Patterns in ecology. Oikos 75 (2), 145−147.

Lawton, J.H., 1999. Are there general laws in ecology? Oikos 84, 177−192.

Leijs, R., van Nes, E.H., Watts, C.H., Cooper, S.J.B., Humphreys, W.F., Hogendoorn, K., 2012. Evolution of blind beetles in isolated aquifers: a test of alternative modes of speciation. PLoS One 7 (3), e34260.

Lester, S.E., Ruttenberg, B.I., Gaines, S.D., Kinlan, B.P., 2007. The relationship between dispersal ability and geographic range size. Ecology Letters 10 (8), 745−758.

Leys, R., Watts, C.H.S., Cooper, S.J.B., Humphreys, W.F., 2003. Evolution of subterranean diving beetles (Coleoptera: Dytiscidae: Hydroporini, Bidessini) in the arid zone of Australia. Evolution 57, 2819−2834.

Li, Z., Heino, J., Chen, X., Liu, Z., Meng, X., Jiang, X., Ge, Y., Chen, J., Xie, Z., 2021. Understanding macroinvertebrate metacommunity organization using a nested study design across a mountainous river network. Ecological Indicators 121, 107188.

Malard, F., Tockner, K., Dole-Olivier, M.J., Ward, J.V., 2002. A landscape perspective of surface-subsurface hydrological exchanges in river corridors. Freshwater Biology 47 (4), 621−640.

Malard, F., Boutin, C., Camacho, A.I., Ferreira, D., Michel, G., Sket, B., Stoch, F., 2009. Diversity patterns of stygobiotic crustaceans across multiple spatial scales in Europe. Freshwater Biology 54 (4), 756−776.

Mammola, S., Leroy, B., 2018. Applying species distribution models to caves and other subterranean habitats. Ecography 41 (7), 1194−1208.

Mammola, S., Piano, E., Malard, F., Vernon, P., Isaia, M., 2019a. Extending Janzen's hypothesis to temperate regions: a test using subterranean ecosystems. Functional Ecology 33 (9), 1638−1650.

Mammola, S., Piano, E., Cardoso, P., Vernon, P., Domínguez-Villar, D., Culver, D.C., Isaia, M., 2019b. Climate change going deep: the effects of global climatic alterations on cave ecosystems. The Anthropocene Review 6 (1−2), 98−116.

Martin, P., De Broyer, C., Fiers, F., Michel, G., Sablon, R., Wouters, K., 2009. Biodiversity of Belgian groundwater fauna in relation to environmental conditions. Freshwater Biology 54, 814−829.

Mermillod-Blondin, F., Lefour, C., Lalouette, L., Renault, D., Malard, F., Simon, L., Douady, C.J., 2013. Thermal tolerance breadths among groundwater crustaceans living in a thermally constant environment. Journal of Experimental Biology 216 (9), 1683−1694.

Michel, G., Malard, F., Deharveng, L., Di Lorenzo, T., Sket, B., De Broyer, C., 2009. Reserve selection for conserving groundwater biodiversity. Freshwater Biology 54, 861–876.

Morueta-Holme, N., Enquist, B.J., McGill, B.J., Boyle, B., Jørgensen, P.M., Jeffrey, E.O., Peet, R.K., Šímová, I., Sloat, L.L., Thiers, B., Violle, C., Wiser, S.K., Dolins, S., Donoghue, J.C., II, Kraft, N.J.B., Regetz, J., Schildhauer, M., Spencer, N., Svenning, J.-C., 2013. Habitat area and climate stability determine geographical variation in plant species range sizes. Ecology Letters 16, 1446–1454.

Morvan, C., Malard, F., Paradis, E., Lefébure, T., Konecny-Dupré, L., Douady, C.J., 2013. Timetree of Aselloidea reveals species diversification dynamics in groundwater. Systematic Biology 62 (4), 512–522.

Mouquet, N., Devictor, V., Meynard, C.N., Munoz, F., Bersier, L.-F., Chave, J., Couteron, P., Dalecky, A., Fontaine, C., Gravel, D., Hardy, O.J., Jabot, F., Lavergne, S., Leibold, M., Mouillot, D., Münkemüller, T., Pavoine, S., Prinzing, A., Rodrigues, A.S., Rohr, R.P., Thébault, E., Thuiller, W., 2012. Ecophylogenetics: advances and perspectives. Biological Reviews 87, 769–785.

Mouron, S., Eme, D., Bellec, A., Bertrand, M., Mammola, S., Liébault, F., Douady, C.J., Malard, F., 2022. Unique and shared effects of local and catchment predictors over distribution of hyporheic organisms: does the valley rule the stream? Ecography (5), e06099.

Niemiller, M.L., Zigler, K.S., 2013. Patterns of cave biodiversity and endemism in the Appalachians and interior plateau of Tennessee, USA. PLoS One 8, e64177.

Niemiller, M.L., Taylor, S.J., Slay, M.E., Hobbs, H.H., 2019. Biodiversity in the United States and Canada. In: White, W., Culver, D.C., Pipan, T. (Eds.), Encyclopedia of Caves. Academic Press, Amsterdam, pp. 163–176.

Pipan, T., Culver, D.C., 2017. The unity and diversity of the subterranean realm with respect to invertebrate body size. Journal of Cave and Karst Studies 79 (1), 1–9.

Pipan, T., Culver, D.C., Papi, F., Kozel, P., 2018. Partitioning diversity in subterranean invertebrates: the epikarst fauna of Slovenia. PLoS One 13 (5), e0195991.

Premate, E., Borko, Š., Delić, T., Malard, F., Simon, L., Fišer, C., 2021a. Cave amphipods reveal co-variation between morphology and trophic niche in a low-productivity environment. Freshwater Biology 66, 1876–1888.

Premate, E., Zagmajster, M., Fišer, C., 2021b. Inferring predator–prey interaction in the subterranean environment: a case study from Dinaric caves. Scientific Reports 11 (1), 21682.

Rabelo, L.M., Souza-Silva, M., Lopes Ferreira, R., 2021. Epigean and hypogean drivers of Neotropical subterranean communities. Journal of Biogeography 48 (3), 662–675.

Ricklefs, R.E., 1987. Community diversity: relative roles of local and regional processes. Science 235 (4785), 167–171.

Rosenzweig, M.L., 1995. Species Diversity in Space and Time. Cambridge University Press, Cambridge.

Saccò, M., Blyth, A.J., Humphreys, W.F., Karasiewicz, S., Meredith, K.T., Laini, A., Cooper, S.J.B., Bateman, P.W., Grice, K., 2019. Stygofaunal community trends along varied rainfall conditions: deciphering ecological niche dynamics of a shallow calcrete in Western Australia. Ecohydrology 13, e2150.

Schmid-Araya, J.M., Schmid, P.E., 1995. The invertebrate species of a gravel stream. Jber. Biol. Stn Lunz 15, 11–21.

Šizling, A.L., Šizlingová, E., Storch, D., Reif, J., Gaston, K.J., 2009. Rarity, commonness, and the contribution of individual species to species richness patterns. The American Naturalist 174 (1), 82–93.

Srivastava, D.S., 1999. Using local–regional richness plots to test for species saturation: pitfalls and potentials. Journal of Animal Ecology 68, 1–16.

Stein, H., Griebler, C., Berkhoff, S., Matzke, D., Fuchs, A., Hahn, H.J., 2012. Stygoregions: a promising approach to a bioregional classification of groundwater systems. Scientific Reports 2 (1), 1–9.

Stevens, G.C., 1989. The latitudinal gradient in geographical range: how so many species coexist in the tropics. The American Naturalist 133, 240–256.

Stoch, F., Galassi, D.M.P., 2010. Stygobiotic crustacean species richness: a question of numbers, a matter of scale. Hydrobiologia 653, 217–234.

Trontelj, P., Blejec, A., Fišer, C., 2012. Ecomorphological convergence of cave communities. Evolution 66 (12), 3852–3865.

Tucker, C.M., Cadotte, M.W., Carvalho, S.B., Davies, T.J., Ferrier, S., Fritz, S.A., Grenyer, R., Helmus, M.R., Jin, L.S., Mooers, A.O., Pavoine, S., Purschke, O., Redding, D.W., Rosauer, D.F., Winter, M., Mazel, F., 2017. A guide to phylogenetic metrics for conservation, community ecology and macroecology. Biological Reviews 92, 698–715.

Vellend, M., 2010. Conceptual synthesis in community ecology. The Quarterly Review of Biology 85 (2), 183–206.

Vervier, P., Gibert, J., 1991. Dynamics of surface water/groundwater ecotones in a karstic aquifer. Freshwater Biology 26, 241–250.

Violle, C., Navas, M.-L., Vile, D., Kazakou, E., Fortunel, C., Hummel, I., Garnier, E., 2007. Let the concept of trait be functional! Oikos 116 (5), 882–892.

Watts, C.H.S., Humphreys, W.F., 2009. Fourteen new Dytiscidae (Coleoptera) of the genera *Limbodessus* Guignot, *Paroster* Sharp and *Exocelina* Broun, from underground waters in Australia. Transactions of the Royal Society of South Australia 133, 62–107.

Zagmajster, M., Culver, D.C., Christman, M.C., Sket, B., 2010. Evaluating the sampling bias in pattern of subterranean species richness: combining approaches. Biodiversity & Conservation 19, 3035–3048.

Zagmajster, M., Eme, D., Fišer, C., Galassi, D., Marmonier, P., Stoch, F., Cornu, J.F., Malard, F., 2014. Geographic variation in range size and beta diversity of groundwater crustaceans: insights from habitats with low thermal seasonality. Global Ecology and Biogeography 23 (10), 1135–1145.

Zagmajster, M., Malard, F., Eme, E., Culver, D.C., 2018. Subterranean biodiversity patterns from global to regional scales. In: Moldovan, O.T., Kováč, L., Halse, S. (Eds.), Cave Ecology. Springer, Cham, pp. 195–225.

# Phylogenies reveal speciation dynamics: case studies from groundwater

*Steven Cooper[1,2], Cene Fišer[3], Valerija Zakšek[3], Teo Delić[3], Špela Borko[3], Arnaud Faille[4] and William Humphreys[5]*

[1]South Australian Museum, Adelaide, SA, Australia; [2]The University of Adelaide, School of Biological Sciences and Australian Centre for Evolution Biology and Biodiversity, Adelaide, SA, Australia; [3]University of Ljubljana, Biotechnical Faculty, Department of Biology, Ljubljana, Slovenia; [4]Stuttgart State Museum of Natural History, Stuttgart, Germany; [5]Western Australian Museum, Welshpool DC, WA, Australia

## Introduction

Since the publication of the first edition of the "Groundwater Ecology" book by Gibert, Danielopol and Stanford in 1994, significant advances have been made in the field of molecular biology and its applications in systematics and evolution. Techniques such as Polymerase Chain Reaction (PCR) amplification and DNA sequencing, via Sanger sequencing, or more recently using next generation sequencing methods, have allowed the development of robust phylogenies showing the evolutionary relationships of groundwater taxa and their surface (henceforth epigean) relatives. Together with the development of increasingly sophisticated molecular clock analyses using Bayesian inference (Drummond et al., 2006), it is now possible to more rigorously explore fundamental questions in the evolution and biogeographic history of stygofauna, when unambiguous calibration points are available (Juan et al., 2010; Mammola et al., 2020b). In this chapter we highlight the role that phylogenetic analyses have played in elucidating the process of diversification. The number of extant obligate groundwater species within a clade depends on the addition of species through speciation and their subtraction through extinction over time. Thus, diversification reflects the balance between speciation and extinction. However, in the absence of information on the number of species in the past, particularly for organisms like crustacean stygobionts where fossil data are limited, it is difficult to assess and quantify extinction. Therefore, we have focused this chapter on the factors that are likely to generate new stygobiotic species.

Speciation events that can contribute to the addition of stygobiotic species are of two types. The first is that of speciation occurring during the transition of an epigean ancestor to groundwater, referred to as phase-1 speciation by Holsinger (2000). The second is that of speciation caused by the divergence of populations of an obligate groundwater species (phase-2 speciation; Barr and Holsinger, 1985; Holsinger, 2000). In the first case, two main hypotheses have dominated the literature: the Climatic Relict Hypothesis (CRH; Barr, 1968; Barr and Holsinger, 1985) and the Adaptive Shift Hypothesis (ASH; Howarth, 1987) (Fig. 7.1, see also Culver et al., this volume). The CRH and ASH represent different spatial contexts for speciation from surface ancestors; CRH represents an allopatric mode of speciation, where groundwater populations become isolated from disconnected groundwater populations following unfavorable climatic conditions (e.g., aridity or glaciations) and the extinction of surface populations; ASH involves parapatric speciation between surface populations and adjacent subterranean populations, following an adaptive shift of the latter.

We will use the term 'subterranean speciation' (Fig. 7.1) for cases where stygobiotic (or troglobiotic) species evolve from stygobiotic (or troglobiotic) ancestors within underground environments (Trontelj, 2018; Langille et al., 2021). This term does not negate the possibility of dispersal by stygobiotic (or troglobiotic) ancestral species via surface environments to isolated groundwater systems or caves, which is hypothesized for many hypogean lineages of beetles (e.g., Carabidae Trechini and Leiodidae Leptodirini; Faille et al., 2010; Rizzo et al., 2013). Although cases of subterranean speciation are often considered to be rare, particularly for groups of organisms that normally live in surface environments (e.g., vertebrates, insects

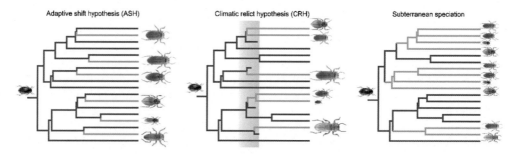

**FIGURE 7.1** Theoretical phylogenetic patterns evident from evolution of stygobiotic beetle species under the Adaptive Shift Hypothesis (ASH), Climatic Relict Hypothesis (CRH) and Subterranean Speciation (SS). Under the ASH, a pattern of epigean (green) and stygobiotic (blue) sister lineages is evident. Under the CRH, epigean species potentially go extinct near the location of stygobiotic species due to climatic changes (e.g., aridity), occurring during a distinct time window across the phylogeny (shown here as a vertical green bar in the middle panel), enhancing allopatric isolation and speciation. Note that, due to extinction of epigean ancestors, divergences among stygobiotic and epigean lineages may predate the period of climate change. Under SS, monophyletic groups of stygobiotic species emerge, with species potentially ecologically differentiated.

and arachnids), we suggest that current studies may have under-estimated the cases of subterranean speciation in these groups (see below). It should further be noted that animals with eyes/vision represent the ultimate ancestral state for each major stygobiotic lineage: i.e., the birth of subterranean species ultimately started from a colonisation event by an epigean ancestor, with subterranean speciation potentially occurring subsequently.

Discriminating the different forms of speciation of stygobionts is challenging, but the analyses of molecular data, using phylogenetic and time tree approaches, may provide valuable insights into the spatial context of speciation and the likely ancestral state (epigean vs. subterranean; Fig. 7.1). For example, under the CRH, it is predicted that there would be synchronous diversification, with multiple groundwater taxa diverging at a common time period that coincided with a hypothesised environmental change (for example, an extended period of aridity; Leys et al., 2003; Faille et al., 2014). Under the ASH, it is predicted that there would be asynchronous diversification, with sister species pairs in the phylogeny showing ecological differentiation (i.e., epigean/groundwater; Fig. 7.1; e.g., Saclier et al., 2018). Subterranean speciation leaves a distinct phylogenetic footprint that is supported by the presence of monophyletic groups of stygobiotic species. The latter hypothesis is further supported when there is a clear phylogeographic pattern (i.e., the geographic distribution of phylogenetic lineages; Avise, 1987) for the relationships among stygobiotic species (e.g., related stygobiotic lineages among adjacent cave habitats via dispersal along groundwater or surface water connections). However, in the real world, a confounding factor in distinguishing the above hypotheses using phylogenetic approaches is the ever-pervasive presence of lineage extinction (Culver et al., 2009) and many studies are further constrained by the limited availability of samples, exacerbated by the nature of the habitat, which is often difficult to access. For example, extinction of epigean ancestors or a failure to collect epigean relatives, may lead to a pattern of monophyly of stygobiotic species that does not reflect their actual evolution under the CRH, or may lead to substantial errors in the estimated time of diversification of the subterranean species (Fig. 7.1). Even in extremely well-studied cases, such as the Mexican cave fish *Astyanax*, distinguishing the ASH and CRH has been problematic with support for a role of both hypotheses during the evolution of the ~30 cave fish populations over the Pleistocene (Dowling et al., 2002; Strecker et al., 2004; Hausdorf et al., 2011; Bradic et al., 2012; Strecker et al., 2012; Fumey et al., 2018; Herman et al., 2018; Wilkens, 2020; Policarpo et al., 2021).

Given the theoretical difficulties in distinguishing the CRH and ASH, which may not be mutually exclusive, we focus this chapter on the question of whether stygobionts were derived from independent colonization from the surface versus from endogenous diversification/subterranean speciation, and emphasize the evidence for the latter hypothesis, which has received less attention in previous reviews of the speciation of stygobionts. We also consider the evidence for allopatric processes (plate tectonics/vicariance) driving subterranean speciation versus ecological speciation with geneflow within groundwater systems. In the second part of the chapter we discuss how historical events influenced speciation patterns.

## Single colonization versus multiple colonizations from surface ancestors

As stated earlier, there is a broad consensus that all subterranean fauna ultimately originated from surface ancestors. Thus, subterranean diversity is not completely independent from diversification processes on the surface. The latter indirectly predetermines the phylogenetic structure and spatial distribution of subterranean diversity.

Using phylogenetic analyses, it is possible to estimate the number of independently-evolved subterranean lineages. This may further suggest which taxa are more likely to colonize subterranean environments and help us to quantify which phyla are overrepresented and which ones are underrepresented in the subterranean realm. A single subterranean lineage in a group of mainly surface taxa may represent an exceptional event of colonization, possibly indicating that members of these groups are poorly preadapted for groundwater colonization. These subterranean species can be considered as rare cases of radiations that predominately unfolded in surface waters. Apart from an obvious deficiency of stygobiotic insects (Sket, 1999), stygobiotic suspension feeders are especially rare. For example, sponges, clams, cnidarians and sedentary polychaetes are globally species-rich groups, represented by only a few stygobiotic species in subterranean environments. In most cases, these species apparently originated through single, apparently exceptional, colonization events (Kupriyanova et al., 2009; Harcet et al., 2010; Bilandžija et al., 2013; Zagmajster et al., 2013).

By contrast, some animal groups are overrepresented in subterranean environments (Stoch and Galassi, 2010). These taxa seem to be predisposed to colonize subterranean environments, and they colonized them multiple times in different geographic regions. Groups with a high potential for colonization of the subterranean environment show an interesting, hierarchically repeated pattern of multiple colonization events. An insightful case are amphipod crustaceans, a large group of approximately 10,000 species (Horton et al., 2016) with roughly 20% of freshwater representatives (Väinölä et al., 2008). A global phylogenetic analysis revealed that species from at least 10 lineages colonized subterranean environments independently (Copila Ciocianu et al., 2020). Among these, at least four represent large and species rich groups comprising mainly or exclusively subterranean species (order Ingolfiellida, superfamily Bogidiellieloidea, family Niphargidae, and genus *Stygobromus*). At a lower hierarchical level, species from some genera, e.g., *Gammarus* or *Hyallela*, independently colonized the subterranean realm in the Palearctic and Neotropics, respectively (Hou et al., 2011; Hou and Sket, 2016; Bastos-Pereira et al., 2018). At the population level, multiple colonizations were reported from the species *Gammarus minus* Say and *Palmorchestia hypogea* Stock & Martin (Villacorta et al., 2008; Carlini et al., 2009; Fong, 2019).

A similar pattern can be found among aquatic isopods. Subterranean clades independently emerged in several clades of higher taxonomic rank, and may count some tens of species, like the sphaeromatid genus *Monolistra* (Delić et al., unpublished data). As in amphipods, phylogenetic analysis provided evidence for multiple colonization events from the surface within single genera, like *Proasellus* (Morvan et al., 2013) and *Haloniscus* (Cooper et al., 2008; Guzik et al., 2019). At the population level, there is also evidence for multiple colonization events in *Asellus aquaticus* (Konec et al., 2015).

An intriguing question is whether multiple colonization events detected at different hierarchical levels possibly indicate the different time points of evolution of subterranean species diversity. For example, in the first stage of evolution a single, phylogeographically structured surface ancestor may have colonized the subterranean environment several times, and several subterranean species emerged. As time passed by, the surface ancestor may have diverged into several species. Phylogenetic reconstruction of these species would show multiple colonization on a species group or genus level. In the final stage, several surface species might have gone extinct. A phylogenetic analysis would recover monophyly of exclusively subterranean species (see Fig 6 in Faille et al. (2013) for a depiction of this process). This scenario raises the question whether these clades evolved through multiple colonization accompanied by massive extinction on the surface, or a single colonization followed by speciation in the subterranean environment. In the next section we argue that speciation from subterranean ancestors is a plausible hypothesis for stygobionts, as already shown for terrestrial groups (e.g., beetles; Faille et al., 2010; Cieslak et al., 2014).

## Speciation from subterranean ancestors

In some groups surface ancestors are not known. These are particularly common among crustaceans, and include higher taxa, such as Remipedia (Neiber et al., 2011), Bathynellacea (Perina et al., 2019; Camacho et al., 2020), Thermosbaenacea (Page et al., 2016), some clades of decapods (Zakšek et al., 2007; von Rintelen et al., 2012), amphipods (Fišer et al., 2008) and isopods (Magniez, 1999; Morvan et al., 2013). In many cases, the origin of these groups is unclear, as their monophyly and relationships with epigean relatives has not been tested. In these cases, it is difficult to discriminate between two competing hypotheses that these species originated through multiple independent colonization events following mass extinction of their surface ancestors, or that these species originated through a radiation that took place entirely in the subterranean realm. The truth may lie somewhere in between. Many of these groups are distributed globally, implying they might be old, potentially dating back to Pangea or the Tethys Ocean (e.g., Thermosbaenacea, Page et al., 2016). If the age estimation of these taxa is accurate, their surface ancestors possibly went extinct in one of the subsequent mass extinction events. Nevertheless, the hypothesis of subterranean speciation has recently gained some interest (Trontelj, 2018). Several studies, each using somewhat different approaches, have provided more or less explicit evidence for speciation from a subterranean ancestor.

Subterranean dytiscid beetles found in Australian groundwater calcretes show an interesting phylogenetic pattern. Although phylogenetic analyses suggest that the majority (approximately 70%) of species evolved independently from several surface dytiscid lineages, many of the subterranean lineages comprise sympatric pairs and triplets of phylogenetic sister species, suggesting that they may have evolved from a subterranean ancestor within the confines of a single calcrete body (Cooper et al., 2002; Leys et al., 2003). The opposing hypothesis of multiple colonization from a nowadays extinct surface ancestor was found less likely in two differently designed studies. Leijs et al. (2012) took a probabilistic approach, and modeled which scenario was likelier. They explored a large range of parameters, considering the probability of colonization and strength of interspecific competition among sympatric subterranean species, and concluded that multiple independent colonizations was a less likely scenario (Leijs et al., 2012). Another study utilized a very different approach involving a comparative evolutionary analysis of the long wavelength opsin (*lwop*) and arrestin (*arr1* and *arr2*) genes that encode proteins involved in the light detection cascade (Langille et al., 2021, 2022). They revealed the presence of shared deleterious mutations (i.e., frameshift mutations or nonsense mutations leading to stop codons in the encoded protein) for a sympatric triplet of subterranean species and, in independent cases, for several phylogenetic sister species (Fig. 7.2). These shared deleterious mutations suggest that they originated from ancestors that were most likely stygobionts living underground within the calcretes or interstitial habitat linking the calcrete bodies along palaeodrainages. Overall, the study provided strong evidence for the role of subterranean speciation in the evolution of at least 11 stygobiotic *Paroster* species, out of a total of 28 stygobiotic *Paroster* species studied to date. This approach, identifying loss of function mutations in genes that are evolving under relaxed selection, offers great potential to disentangle the two types of speciation (i.e., from epigean vs. subterranean ancestors). It also provides an opportunity to pinpoint when transitions from surface water to groundwater occurred along branches of a phylogeny as shown by Lefébure et al. (2017) using opsin gene sequences.

A different approach was taken by Stern et al. (2017), who tested the opposite hypothesis that colonization of subterranean environments and adaptation to ecological conditions of groundwater represented an evolutionary dead-end. Using crayfish as a model, they assumed that subterranean species decreased their ranges and thereby enhanced the chance for extinction, that rates of species diversification declined after colonization of the subterranean environment and that chances to re-colonize surface habitats were zero. Using spatial

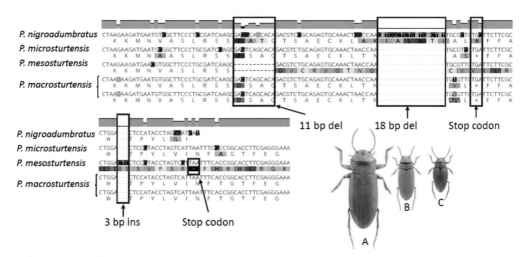

**FIGURE 7.2** Nucleotide and translated sequence of the long wavelength opsin gene (*lwop*) from epigean (*Paroster nigroadumbratus*) and stygobiotic (A. *P. macrosturtensis*; B. *P. mesosturtensis*; C. *P. microsturtensis*) dytiscid beetles showing deletions (del), insertions (ins), and nonsense (stop) mutations in *lwop*. The 18 bp deletion and stop codon is shared among the three subterranean species, with an independent 11 bp deletion in *P. mesosturtensis* resulting in a frameshift mutation, moving the stop codon further downstream. The shared mutations provide evidence that the common ancestor of the three subterranean species was a stygobiont (Langille et al., 2021).

data and simultaneous modeling of speciation rates and species' ecology, they concluded that ranges of cave species indeed are smaller, that dispersal to the surface is unlikely, yet possible, and that there was no evidence for decreased speciation rates after colonization of the subterranean environment (Stern et al., 2017). Although this conclusion should be accepted with care, due to conceptual and methodological issues, such as uncertainty about when transitions from surface to groundwater occurred in the phylogeny and the potential for high speciation rates to increase extinction rates (Greenberg and Mooers, 2017), it implicitly validates the hypothesis of subterranean speciation.

Further evidence for subterranean speciation comes from a third well-studied example: radiation of the amphipod family Niphargidae. The family is sister to another amphipod family comprised of exclusively subterranean members, Pseudoniphargidae. Both families are, with weak support, nested among predominantly marine amphipod families (Copila Ciocianu et al., 2020). Niphargidae diverged from a common ancestor with Pseudoniphargidae between 56 and 39 Mya (Borko et al., 2021), according to reconstructions of ranges in the region that nowadays belongs to Western Europe (McInerney et al., 2014; Borko et al., 2021), most likely as an interstitial amphipod (Fišer et al., 2019). The oldest fossils of age 30 My already showed affinities for subterranean life (Jażdżewski and Kupryjanowicz, 2010). Only a few species live in shallow subterranean habitats on the boundary with surface aquatic habitats, however, their phylogenetic position, eyelessness and depigmented body, as well as a strong photophobic behavior, indicate that this habitat was colonized secondarily (Fišer et al., 2016; Copilas-Ciocianu et al., 2018). As the orogenesis of the Alps, Carpathian arch and Dinaric Karst proceeded, niphargids most likely spread through interstitial and shallow subterranean habitats toward Southern Europe where they massively diversified in at least five radiations (Borko et al., 2021). The first phase of speciation was fast, but later on it slowed down, although speciation events continued until recently. Many studies identified relatively young stygobiotic species nested within different subterranean clades (Delić et al., 2017, 2022; Fišer et al., 2018), therefore, a hypothesis that widespread epigean ancestors lived all across Europe until recently, when the last speciation events occurred, seems unlikely.

Although this chapter has been primarily focused on stygobionts, strong evidence for subterranean speciation has also come from a study of troglobiotic beetles (Leiodidae and Trechinae) in the Pyrénées and the Alps (Faille et al., 2010, 2013; Ribera et al., 2010). Phylogenetic analyses revealed large monophyletic groups of subterranean beetle species with all epigean representatives excluded. Some of these clades even show higher diversification rates, especially when a shift in the standard life cycle occurred in favor of a contraction of the larval life (Cieslak et al., 2014).

## Speciation from subterranean ancestors: likely mechanisms

The evidence that subterranean species continue to speciate in the subterranean environment tells us little about the geographic mode of speciation (e.g., allopatric, parapatric, or sympatric). Detailed mechanisms were rarely explored. Nevertheless, the diverse degrees of morphological and possibly ecological diversification, along with phylogeographic information for descendant species provide a testable framework for studying geographic and ecological modes of subterranean speciation.

One side of a continuum is represented by species that differ little in their morphology, and cannot be delineated without the help of molecular methods. These are often referred to as "cryptic species" or "molecular operational taxonomic units (MOTUs)" and they are common and found virtually in all taxa hitherto examined, including amphibians (Gorički and Trontelj, 2006), fishes (Niemiller et al., 2013), mollusks (Bilandžija et al., 2013), and different groups of crustaceans (Cánovas et al., 2016; Page et al., 2016; Delić et al., 2017; Camacho et al., 2018). A preliminary Europe-wide analysis based on subterranean amphipods and isopods indicated that every nominal name covered on average, two to three cryptic species (Eme et al., 2018). Morphological similarity suggests that descendant species inherited from ancestors their ecological requirements and similar selective regimes that constrained further morphological evolution, i.e., morphological stasis (Struck et al., 2017; Fišer et al., 2018). Most of these species are allopatric or parapatric; sympatry regularly occurs, but seems to be a result of secondary contact (Zakšek et al., 2009; Fišer et al., 2015a,b; Gorički et al., 2017). It seems that many of these species evolved in allopatry, after fragmentation of their initial distributional range. The role of selection in this type of speciation remains unclear. It is possible that parallel selection accelerated speciation through the so-called mutation-order mechanism (Schluter, 2009), where "different, incompatible mutations (alleles) fix in different populations adapting to the same selective pressure" (Nosil and Flaxman, 2011). Nevertheless, explicit tests of these hypotheses are needed on a case-by-case basis.

On the other side of the continuum there is an increasing number of studies showing that descendant species often overlap their ranges and, at the same time, they ecologically differ, which may point to ecological speciation in sympatry or parapatry. The above-mentioned Australian dytiscid sympatric pairs or triplets of sister species showed a striking pattern of body sizes. The ratios of body lengths of coexisting species across 40 studied calcretes are unexpectedly fixed to a ratio of 1:1.61 (Vergnon et al., 2013). This result is in accord with the theory of limiting similarity, which states that species stably co-exist only if their ecological niches are sufficiently differentiated. In three cases in the genus *Paroster*, ecologically divergent species likely evolved from a shared ancestor within the confines of a single calcrete body (Langille et al., 2021). Assuming that the distinct size classes of the beetles reflect differences in their ecological niches and that divergent evolution assisted the evolution of a reproductive barrier, it is possible that these species are potential cases of ecological speciation with gene flow. A similar case of morphological and ecological differentiation within a well-supported lineage was studied in Greek cave Melissotrypa, with freshwater and deep-sulfidic water. Three amphipod species from the genus *Niphargus* show striking differences in body sizes, body shapes and their ecology: one species lives in fissure systems of freshwater, the second one in a freshwater lake and the third one in a sulfidic lake, which might point to ecological speciation (Borko et al., 2019).

Both cases of putative ecological speciation, however, did not test the central tenet of ecological speciation, i.e., that reproductive isolation evolved as a by-product of ecological differentiation, though assortative mating is likely in the stygobiotic dytiscids due to their size variation. The alternative possibilities of microallopatric speciation followed by secondary sympatry also cannot be ruled out. Calcrete bodies are considerably heterogeneous in thickness and connectivity, and previous studies provided evidence for genetic substructuring in beetles (Guzik et al., 2009, 2011a) and amphipods (Bradford et al., 2013), suggesting the potential for allopatric isolation during their speciation, though periods of restricted gene flow were likely to have been short. Indeed, another study of *Niphargus* shows that conclusions about ecological speciation need to be cautious. Zakšek et al. (2019) studied a clade of five morphologically similar species and one morphologically different species. All morphologically similar species live in strict parapatry, whereas morphologically different species overlap their range with sister species suggesting the distributional pattern was driven by intraspecific competition (Delić et al., 2017). This might indicate that intraspecific competition within the ancestor of morphologically divergent sister species prompted ecological divergence and subsequent speciation. However, reconstruction of ancestral ranges imply that speciation began in allopatry, possibly accompanied by morphological differentiation, and that sympatry is only secondary (Zakšek et al., 2019).

# Drivers of subterranean diversity: the role of paleoclimatic and paleogeological events

Many authors acknowledge that diversification processes were under a strong influence of geological history and long-term climate oscillations. Three interlinked phenomena were studied in subterranean biology so far: first, tectonic shifts that grounded cycles of dispersal and vicariance; second, climatic changes that occasionally eradicated surface relatives, and possibly enforced colonization of, or isolation within, groundwater; third, the origin of karstic regions, which represented a new, predator and competitor-free space, the so-called ecological opportunity, that prompted bursts of speciation from a single ancestor.

## Tethys Ocean, plate tectonics and marine regression-transgression cycles

Geographic distribution of numerous taxa suggests that plate tectonics importantly shaped subterranean biodiversity, most likely through opening and closure of dispersal routes. Many biogeographic hypotheses explaining the distribution of subterranean taxa referred to geological history. We highlight three interconnected hypotheses, namely continent-wide vicariance of freshwater organisms, the beginning and end of dispersal along the Tethyan Ocean, and marine regressions-transgressions during the Messinian Salinity Crisis.

Many subterranean taxa exhibit transoceanic distributions, which suggests that these groups may have emerged before the formation of modern continents. If so, some groups, such as ingolfiellid amphipods, Speleogriphacea and several families of Amphipods (Humphreys, 2000; Vonk and Schram, 2003; Fišer et al., 2013), might date back to Pangea (circa 250 Mya) or Gondwana (circa 180 Mya).

The plate tectonics inevitably shaped seaways, and opened dispersal routes through shallow sea. The break-up of Pangea into Laurasia and Gondwana opened up a Tethys Sea approximately 250 Mya. The shallows of the Tethys were presumably connecting regions that nowadays belong to the majority of modern continents. The Tethyan seaway was terminated by the opening of the Atlantic in the Mid-Jurassic (170 Mya onwards) and deepening of the Indian Ocean in the Eocene (53 Mya onward). Globally distributed taxa like thermosbaenaceans, remipedes, amphipods, ostracods and isopods represent the presumed legacy of this period (Hou and Li, 2018).

Besides the global patterns of speciation, tectonics profoundly affected patterns at smaller geographic scales and narrower time frames. A particularly well studied example is the Messinian Salinity Crisis (6—5.33 Myr), when the tectonic uplift of Gibraltar sealed the seaway connection between the Mediterranean Sea and Atlantic Ocean. This led to substantial desiccation of the Mediterranean (Gargani and Rigollet, 2007), i.e., marine regression, establishment of new riverine corridors, and enhanced connectivity for freshwater taxa. Reopening of the Mediterranean — Atlantic seaways refilled the Mediterranean basin and terminated the existing freshwater connections between the coasts of Southern Europe, Middle East and North Africa. These events presumably shaped the distributions of amphipod genera *Pseudoniphargus* and *Niphargus*, shrimps (Atyidae) and the isopod genus *Monolistra* (Boutin and Coineau, 1988; Notenboom, 1991; Jurado-Rivera et al., 2017; Delić et al., 2020; Delić et al., unpublished data).

It can be expected that all these historical changes eventually resulted in vicariant events that should have left footprints on phylogenetic hierarchy. A reasonable expectation is that there should be a congruence between timing of branching events and geographic distribution of terminal taxa or even clades. Molecular phylogenies that explicitly tested these hypotheses are rare and include fish (Chakrabarty et al., 2012), atyid shrimps (Page et al., 2008; Von Rintelen et al., 2012; Botello et al., 2013; Jurado-Rivera et al., 2017) as well as amphipods from families Metacrangonyctidae (Bauzà-Ribot et al., 2012; Pons et al., 2019), Crangonyctidae (Copilas-Ciocianu et al., 2019), Pseudoniphargidae (Stokkan et al., 2018) and Niphargidae (Delić et al., 2020). All these studies show few commonalities and evidence that evolutionary history was most likely more complex (e.g., Phillips et al., 2013) and cannot be explained by vicariance alone. As predicted, there is some evidence that plate tectonics and break up of Tethyan seaways induced vicariant splits among the studied taxa between Mediterranean and Caribbean as well as between Malagasy and Australia (Chakrabarty et al., 2012; Jurado-Rivera et al., 2017; Copilas-Ciocianu et al., 2019; Pons et al., 2019). However, transoceanic vicariances are relatively rare and cannot explain geographic distributions without transoceanic dispersal or the initial existence of a common ancestor. The longest transoceanic dispersal was presumably made by atyid shrimps across the Indian Ocean (Jurado-Rivera et al., 2017). Existence of atyid shrimps on volcanic islands that are younger than Tethys (Botello et al., 2013) still remain unanswered. Logically, the only way these island could be colonized was by dispersal. It has been hypothesized that ancestors lived and dispersed in shallow seas, which provided temporary routes to the emerging islands. This hypothesis was invoked for atyid shrimps and *Pseudoniphargus* amphipods. Interestingly, both of these taxa seem to be halotolerant, which indirectly supports the hypothesis of dispersal through shallow seas (Botello et al., 2013; Stokkan et al., 2018). Likewise, marine regressions-transgressions apparently interchangeably allowed cycles of dispersal and vicariance. *Niphargus* amphipods crossed the Adriatic Sea several times in both directions (Delić et al., 2020), while shrimps of the genus *Typhlocaris* apparently crossed the Mediterranean Sea (Guy-Haim et al., 2018).

## Climatic oscillations

Climatic oscillations, resulting in aridification or glaciations profoundly shaped the biotas on the surface (Hewitt, 2000). Subterranean environments, largely buffered from climatic oscillations on the surface, could have played a role of refugia for taxa that could survive in darkness. This idea is an essential component of the CRH, stating that harsh climates prompted colonization of, or isolation within, subterranean environments and that speciation was enhanced when surface ancestors went extinct owing to increased allopatry. This hypothesis might have spatial and temporal consequences for the diversification patterns of subterranean species (see Fig. 7.1).

The first prediction would assume that speciation rates accelerated in regions influenced by climate perturbations. This is closely linked to a broader question of a general interest, i.e., how speciation rates relate to latitudinal and altitudinal gradients (Schluter and Pennell, 2017). For example, Pleistocene glaciations in the Palearctic and Nearctic could locally accelerate speciation at the glacier-refugia boundaries (Delić et al., 2022), but also extinction in high latitudes (Eme et al., 2015). This spatial nonstationarity of historical climatic oscillations might contribute part of the explanation (e.g., locally high speciation and extinction rates) for a high species richness at mid-latitudes that were more influenced by Pleistocene glaciations than the tropics (Culver et al., 2006; Zagmajster et al., 2014; Eme et al., 2015). To our knowledge, this question has not been addressed, however, it seems that many subterranean lineages often predate Pleistocene glaciations (Leijs et al., 2012; Morvan et al., 2013; Esmaeili-Rineh et al., 2015). Copilas-Ciocianu et al. (2019) found an interesting link between speciation and climates using crangonyctid amphipods, a Holarctic group of amphipods, dominated by subterranean representatives. Amphipods in general prefer cooler climates [although by no means are limited to them (Bradford et al., 2010; King et al., 2012)]. Thus, the authors assumed that speciation was more frequent during the past warmer and more recent cooler periods at higher and lower latitudes, respectively. This premise led to a prediction that in northern latitudes species should be older. Indeed, Copilas-Ciocianu et al. (2019) found a weak but significant correlation between latitude and the age of the species.

Secondly, if unfavorable climates wiped out surface ancestors, we should detect this as an increased proliferation of subterranean species in an exact time window. This hypothesis was explicitly tested using a phylogeny of aselloid isopods (Morvan et al., 2013) and Australian dytiscid beetles (Leijs et al., 2012). The results were opposing. The study on aselloid isopods showed no increase in species richness during Pleistocene glaciations, while diversification of Australian dytiscids remarkably increased during Pliocene aridification, as is the case for many European troglobiotic beetles (Faille et al., 2013). These results need to be considered with care as they represent different time-scales and impacts (glaciation vs. aridification). Furthermore, diversification is an outcome of speciation and extinction, hence an increase in diversification could be a result of higher speciation rates or decreased extinctions. In the absence of fossil records, estimation of extinction rates is almost impossible empirically. Additionally, speciation from subterranean ancestors may have added to increased speciation rates; however, this speciation was not necessarily driven by climate oscillations (Cieslak et al., 2014; Delić et al., 2022).

## New karst areas

Subterranean species evolved only in suitable habitats, such as karstic regions. Karst developed on limestones or dolomites, rocks of marine origin, uplifted by tectonics. New karst regions provide the so-called ecological opportunity, a wealth of unexploited or underutilized ecological resources that permit ecological diversification of species (Schluter, 2000; Stroud and Losos, 2016). Ecological opportunity is a prerequisite for a phenomenon called "adaptive radiation", a rapid species proliferation (i.e., high speciation rate) from a common ancestor, accompanied by ecological diversification (Schluter, 2000). Until recently, subterranean habitats were considered as oligotrophic, deprived of ecological heterogeneity that could not support adaptive radiations. Recent studies have shown that even closely related subterranean species can diversify into ecologically distinct species (Leijs et al., 2012; Trontelj et al., 2012; Fišer et al., 2015a,b; Mammola et al., 2020a). These studies lead to a question: do new karst areas provide ecological opportunities that could prompt adaptive radiations?

Borko et al. (2021) predicted that the Alpine-Dinaric orogenesis in SE Europe 20-15 Mya exposed massive karstic areas of Eastern Alps, Dinarids and Carpathians, which presumably provided an ecological opportunity that could accelerate speciation rates of the studied taxon, subterranean amphipod *Niphargus*. They analyzed the changes of patterns of speciation, ecological diversity and morphological disparity through time, and showed a sharp increase in speciation rates as well as in ecological and morphological diversification rates at approximately 15 Mya. Further analysis of individual clades showed that this massive adaptive radiation comprises several smaller, geographically confined adaptive radiations. Adaptive radiation of *Niphargus* in SE Europe is probably an exceptional example of how ecological opportunity accelerated speciation and diversification in subterranean taxa.

## Synthesis and future prospects

In this chapter, we have explored the use of phylogenetic analyses in addressing major questions on the speciation of stygofauna. We focused on the potential role of subterranean speciation, which we contend has been under-estimated as a generator of stygofaunal diversity. Great advances have been made in the development of techniques for generating DNA sequence data (e.g., via next generation sequencing) and new phylogenetic approaches (e.g., phylogenomic approaches using thousands of genetic markers). However, it is only by combining these techniques with additional approaches including morphology, population genetics, ecology, functional ecology, palaeogeography, or even analyses such as, for example, the molecular evolution of genes specifically involved in regressed traits, such as vision or pigmentation, (see Lefébure et al., 2017; Saclier et al., 2018; Tierney et al., 2018; Langille et al., 2021; Policarpo et al., 2021), that the ancestral states, colonization times, hotspots of origination, speciation tempo, and evolutionary forces that lead to speciation can potentially be resolved.

While phylogenetic analyses and calibration may provide valuable information for distinguishing the ancestral state (epigean or subterranean) and drivers of speciation (spatial and/or ecological) in subterranean animals, it is important to acknowledge the many limitations of these approaches. Phylogenetic patterns and divergence time estimates may be misleading for a variety of reasons, including lineage extinction, particularly the extinction of surface lineages (see Fig. 7.1, CRH tree), limited sampling of taxa (e.g., Botello et al., 2013 vs Jurado-Rivera et al., 2017) and errors in molecular clock estimates of colonization or diversification times for stygobiotic lineages because of inappropriate calibrations. Speciation rates and divergence time estimates may also be significantly distorted by the use of genetic markers that are potentially saturated (e.g., *Cytochrome Oxidase subunit I* (*COI*) third codon positions; Phillips et al., 2013). Saturation may also lead to phylogenies showing short branches and a lack of resolution at internal nodes in the tree, giving the impression of a rapid radiation. These latter technical challenges can potentially be overcome by the use of next generation sequencing approaches (e.g., exon capture, Klopfstein et al., 2019) to obtain sequence data from 100s or 1000s of independent genetic loci, and which have only recently been applied to stygobionts (e.g., *Haloniscus* isopods, Stringer et al., 2021). Problems finding suitable fossil calibrations for many taxonomic groups are not so easily overcome, and palaeogeographic calibrations come with a range of assumptions that may mislead divergence time estimates (e.g., Bauza-Ribot et al., 2012 vs Pons et al., 2019). In many cases, distinguishing the traditional models of speciation, such as the CRH or ASH, is not feasible, given these above-mentioned limitations, but also because it is reasonable to think that both hypotheses of speciation may potentially apply during different stages of the speciation continuum (Butlin et al., 2008). We suggest that it is better to focus on questions that can potentially be resolved, such as whether the ancestor was epigean or stygobiotic or whether there has been a role of hybridization and selection (e.g., through ecological divergence) during speciation.

An intriguing question is whether species that are predisposed to colonize subterranean environments also are more inclined to further speciate within subterranean environments? It is possible that these species bring along more standing genetic variation when colonizing the subterranean realm, enhancing their ability for adaptive radiations in the new groundwater environment. While we have highlighted the potential importance of subterranean speciation it is unknown whether it has contributed more to global subterranean species diversity than speciation from surface ancestors (except in terrestrial groups like subterranean beetles; Faille et al., 2010; Cieslak et al., 2014), or which clades represent the largest share of subterranean biodiversity? These and many other questions remain unanswered, but have the potential to be addressed using phylogenies with comprehensive taxon selection, particularly targeting global hotspots such as the Western Balkans, Caucasus and western half of Australia (Zakšek et al., 2007; Guzik et al., 2011b; Esmaeili-Rineh et al., 2015).

# Acknowledgments

We thank David Eme and Florian Malard for helpful suggestions that improved the earlier version of the manuscript. We thank C. Watts and H. Hamon for providing the beetle photos used in Figs. 7.1 and 7.2. ŠB, TD, VZ and CF were supported by the Slovenian Research Agency (SRA) through core funding programs P1-0184 and projects N1-0069, Z1-9164 and J1-2464. SC and WH were supported by the Australian Biological Resources Study and Australian Research Council, under projects DP120102132 and DP180103851.

# References

Avise, J.C., Arnold, J., Ball Jr., R.M., Bermingham, E., Lamb, T., Neigel, J.E., Reeb, C.A., Saunders, N.C., 1987. Intraspecific phylogeography: the mitochondrial DNA bridge between population genetics and systematics. Annual Review of Ecology and Systematics 18, 489–522.

Barr, T.C., 1968. Cave ecology and the evolution of troglobites. Evolutionary Biology 2, 35–102.

Barr, T.C., Holsinger, J.R., 1985. Speciation in cave faunas. Annual Review of Ecology and Systematics 16, 313–337.

Bastos-Pereira, R., Alves de Oliveira, M.P., Ferreira, R.L., 2018. Anophtalmic and epigean? Description of an intriguing new species of Hyalella (Amphipoda, Hyalellidae) from Brazil. Zootaxa 4407 (2), 254–266.

Bauzà-Ribot, M.M., Juan, C., Nardi, F., Oromí, P., Pons, J., Jaume, D., 2012. Mitogenomic phylogenetic analysis supports continental-scale vicariance in subterranean thalassoid crustaceans. Current Biology 22 (21), 2069–2074.

Bilandžija, H., Morton, B., Podnar, M., Cetković, H., 2013. Evolutionary history of relict Congeria (Bivalvia: Dreissenidae): unearthing the subterranean biodiversity of the Dinaric Karst. Frontiers in Zoology 10 (1), 5.

Borko, Š., Collette, M., Brad, T., Zakšek, V., Flot, J.-F., Vaxevanopoulos, M., Sarbu, S.M., Fišer, C., 2019. Amphipods in a Greek cave with sulphidic and non-sulphidic water: phylogenetically clustered and ecologically divergent. Systematics and Biodiversity 17 (6), 558–572.

Borko, Š., Trontelj, P., Seehausen, O., Moškrič, A., Fišer, C., 2021. A subterranean adaptive radiation of amphipods in Europe. Nature Communications 12, 3688.

Botello, A., Iliffe, T.M., Alvarez, F., Juan, C., Pons, J., Jaume, D., 2013. Historical biogeography and phylogeny of Typhlatya cave shrimps (Decapoda: Atyidae) based on mitochondrial and nuclear data. Journal of Biogeography 40, 594–607.

Boutin, C., Coineau, N., 1988. Pseudoniphargus maroccanus n. sp. (subterranean amphipod), the first representative of the genus in Morocco. Phylogenetic relationships and paleobiogeography. Crustaceana Supplement 13, 1–19.

Bradford, T.M., Adams, M., Humphreys, W.F., Austin, A.D., Cooper, S.J.B., 2010. DNA barcoding of stygofauna uncovers cryptic amphipod diversity in a calcrete aquifer in Western Australia's arid zone. Molecular Ecology Resources 10, 41–50.

Bradford, T.M., Adams, M., Guzik, M.T., Humphreys, W.F., Austin, A.D., Cooper, S.J.B., 2013. Patterns of population genetic variation in sympatric chiltoniid amphipods within a calcrete aquifer reveal a dynamic subterranean environment. Heredity 111 (1), 77–85.

Bradic, M., Beerli, P., García-de León, F.J., Esquivel-Bobadilla, S., Borowsky, R.L., 2012. Gene flow and population structure in the Mexican blind cavefish complex (Astyanax mexicanus). BMC Evolutionary Biology 12 (1), 9.

Butlin, R.K., Galindo, J., Grahame, J.W., 2008. Sympatric, parapatric or allopatric: the most important way to classify speciation? Philosophical Transactions of the Royal Society B 363, 2997–3007.

Camacho, A.I., Mas-Peinado, P., Dorda, B.A., Casado, A., Brancelj, A., Knight, L.R.F.D., Hutchins, B., Bou, C., Perina, G., Rey, I., 2018. Molecular tools unveil an underestimated diversity in a stygofauna family: a preliminary world phylogeny and an updated morphology of Bathynellidae (Crustacea: Bathynellacea). Zoological Journal of the Linnean Society 183, 70–96.

Camacho, A.I., Mas-Peinado, P., Iepure, S., Perina, G., Dorda, B.A., Casado, A., Rey, I., 2020. Novel sexual dimorphism in a new genus of Bathynellidae from Russia, with a revision of phylogenetic relationships. Zoologica Scripta 49, 47–63.

Cánovas, F., Jurado-Rivera, J.A., Cerro-Gálvez, E., Juan, C., Jaume, D., Pons, J., 2016. DNA barcodes, cryptic diversity and phylogeography of a W Mediterranean assemblage of thermosbaenacean crustaceans. Zoologica Scripta 45 (6), 659–670.

Carlini, D.B., Manning, J., Sullivan, P.G., Fong, D.W., 2009. Molecular genetic variation and population structure in morphologically differentiated cave and surface populations of the freshwater amphipod Gammarus minus. Molecular Ecology 18 (9), 1932–1945.

Chakrabarty, P., Davis, M.P., Sparks, J.S., 2012. The first record of a trans-oceanic sister-group relationship between obligate vertebrate troglobites. PLoS One 7 (8).

Cieslak, A., Fresneda, J., Ribera, I., 2014. Life history specialization was not an evolutionary dead-end in Pyrenean cave beetles. Proceedings of the Royal Society B 281 (1781), 20132978.

Cooper, S.J.B., Hinze, S., Leys, R., Watts, C.H.S., Humphreys, W.F., 2002. Islands under the desert: molecular systematics and evolutionary origins of stygobitic water beetles (Coleoptera: dytiscidae) from central Western Australia. Invertebrate Systematics 16, 589–598.

Cooper, S.J.B., Saint, K.M., Taiti, S., Austin, A.D., Humphreys, W.F., 2008. Subterranean archipelago: mitochondrial DNA phylogeography of stygobitic isopods (Oniscidea: Haloniscus) from the Yilgarn region of Western Australia. Invertebrate Systematics 22, 195–203.

Copilas-Ciocianu, D., Fišer, C., Borza, P., Petrusek, A., 2018. Is subterranean lifestyle reversible? Independent and recent large-scale dispersal into surface waters by two species of the groundwater amphipod genus *Niphargus*. Molecular Phylogenetics and Evolution 119, 37–49.

Copilas-Ciocianu, D., Sidorov, D., Gontcharov, A., 2019. A drift across tectonic plates: molecular phylogenetics supports the ancient Laurasian origin of old limnic crangonyctid amphipods. Organisms, Diversity and Evolution 19, 191–207.

Copila-Ciocianu, D., Borko, Š., Fišer, C., 2020. The late blooming amphipods: global change promoted post-Jurassic ecological radiation despite Palaeozoic origin. Molecular Phylogenetics and Evolution 143, 106664.

Culver, D.C., Deharveng, L., Bedos, A., Lewis, J.J., Madden, M., Reddell, J.R., Sket, B., Trontelj, P., White, D., 2006. The mid-latitude biodiversity ridge in terrestrial cave fauna. Ecography 29 (1), 120–128.

Culver, D.C., Pipan, T., Schneider, K., 2009. Vicariance, dispersal and scale in the aquatic subterranean fauna of karst regions. Freshwater Biology 54, 918–929.

Delić, T., Trontelj, P., Rendoš, M., Fišer, C., 2017. The importance of naming cryptic species and the conservation of endemic subterranean amphipods. Scientific Reports 7 (1), 3391.

Delić, T., Stoch, F., Borko, Š., Flot, J.F., Fišer, C., 2020. How did subterranean amphipods cross the Adriatic Sea? Phylogenetic evidence for dispersal–vicariance interplay mediated by marine regression–transgression cycles. Journal of Biogeography 47 (9), 1875–1887.

Delić, T., Trontelj, P., Zakšek, V., Brancelj, A., Simčič, T., Stoch, F., Fišer, C., 2022. Speciation of a subterranean amphipod on the glacier margins in South Eastern Alps, Europe. Journal of Biogeography 49, 38–50.

Dowling, T.E., Martasian, D.P., Jeffery, W.R., 2002. Evidence for multiple genetic forms with similar eyeless phenotypes in the blind cavefish *Astyanax mexicanus*. Molecular Biology and Evolution 19 (4), 446–455.

Drummond, A.J., Ho, S.Y.W., Phillips, M.J., Rambaut, A., 2006. Relaxed phylogenetics and dating with confidence. PLoS Biology 4 (5), e88.

Eme, D., Zagmajster, M., Fišer, C., Galassi, D., Marmonier, P., Stoch, F., Cornu, J.-F., Oberdorff, T., Malard, F., 2015. Multi-causality and spatial non-stationarity in the determinants of groundwater crustacean diversity in Europe. Ecography 38 (5), 531–540.

Eme, D., Zagmajster, M., Delić, T., Fišer, C., Flot, J.-F., Konecny-Dupré, L., Pálsson, S., Stoch, F., Zakšek, V., Douady, C.J., Malard, F., 2018. Do cryptic species matter in macroecology? Sequencing European groundwater crustaceans yields smaller ranges but does not challenge biodiversity determinants. Ecography 41, 424–436.

Esmaeili-Rineh, S., Sari, A., Delić, T., Moškrič, A., Fišer, C., 2015. Molecular phylogeny of the subterranean genus *Niphargus* (Crustacea: Amphipoda) in the Middle East: a comparison with European Niphargids. Zoological Journal of the Linnean Society 174 (4), 812–826.

Faille, A., Ribera, I., Deharveng, L., Bourdeau, C., Garnery, L., Queinnec, E., Deuve, T., 2010. A molecular phylogeny shows the single origin of the Pyrenean subterranean Trechini ground beetles (Coleoptera: Carabidae). Molecular Phylogenetics and Evolution 54, 97–106.

Faille, A., Casale, A., Balke, M., Ribera, I., 2013. A molecular phylogeny of Alpine subterranean Trechini (Coleoptera: Carabidae). BMC Evolutionary Biology 13, 248.

Faille, A., Andújar, C., Fadrique, F., Ribera, I., 2014. Late Miocene origin of a Ibero-Maghrebian clade of ground beetles with multiple colonisations of the subterranean environment. Journal of Biogeography 41, 1979–1990.

Fišer, C., Sket, B., Trontelj, P., 2008. A phylogenetic perspective on 160 years of troubled taxonomy of *Niphargus* (Crustacea: Amphipoda). Zoologica Scripta 37, 665–680.

Fišer, C., Zagmajster, M., Ferreira, R.L., 2013. Two new Amphipod families recorded in South America shed light on an old biogeographical enigma. Systematics and Biodiversity 11 (2), 117–139.

Fišer, C., Luštrik, R., Sarbu, S., Flot, J.-F., Trontelj, P., 2015. Morphological evolution of coexisting amphipod species pairs from sulfidic caves suggests competitive interactions and character displacement, but no environmental filtering and convergence. PLoS One 10 (4), e0123535.

Fišer, C., Robinson, C.T., Malard, F., 2018. Cryptic species as a window into the paradigm shift of the species concept. Molecular Ecology 27 (3), 613–635.

Fišer, C., Delić, T., Luštrik, R., Zagmajster, M., Altermatt, F., 2019. Niches within a niche: ecological differentiation of subterranean amphipods across Europe's interstitial waters. Ecography 42 (6), 1212–1223.

Fišer, Ž., Altermatt, F., Zakšek, V., Knapič, T., Fišer, C., 2015. Morphologically cryptic Amphipod species sre "ecological clones" at regional but not at local scale: a case study of four *Niphargus* species. PLoS One 10, e0134384.

Fišer, Ž., Novak, L., Luštrik, R., Fišer, C., 2016. Light triggers habitat choice of eyeless subterranean but not of eyed surface amphipods. Naturwissenschaften 103 (1–2), 7.

Fong, D.W., 2019. *Gammarus minus*: a model system for the study of adaptation to the cave environment. In: White, W.B., Culver, D.C., Pipan, T. (Eds.), Encyclopedia of Caves, third ed. Elsevier, Amsterdam, pp. 451–458.

Fumey, J., Hinaux, H., Noirot, C., Thermes, C., Rétaux, S., Casane, D., 2018. Evidence for late Pleistocene origin of *Astyanax mexicanus* cavefish. BMC Evol Biol 18 (1), 43.

Gargani, J., Rigollet, C., 2007. Mediterranean Sea level variations during the Messinian salinity crisis. Geophysical Research Letters 34 (10), 1–5.

Gibert, J., Danielopol, D., Stanford, J.A., 1994. Groundwater Ecology. Academic Press, San Diego, CA.

Gorički, Š., Stanković, D., Snoj, A., Kuntner, M., Jeffery, W.R., Trontelj, P., Pavićević, M., Grizelj, Z., Năpăruš-aljančič, M., Aljančič, G., 2017. Environmental DNA in subterranean biology: range extension and taxonomic implications for *Proteus*. Scientific Reports 7, 91–93.

Gorički, Š., Trontelj, P., 2006. Structure and evolution of the mitochondrial control region and flanking sequences in the European cave salamander *Proteus anguinus*. Gene 378 (1–2), 31–41.

Greenberg, D.A., Mooers, A.Ø., 2017. Linking speciation to extinction: diversification raises contemporary extinction risk in amphibians. Evolution Letters 1, 40–48.

Guy-Haim, T., Simon-Blecher, N., Frumkin, A., Naaman, I., Achituv, Y., 2018. Multiple transgressions and slow evolution shape the phylogeographic pattern of the blind cave-dwelling shrimp *Typhlocaris*. PeerJ 6, e5268.

Guzik, M.T., Cooper, S.J.B., Humphreys, W.F., Austin, A.D., 2009. Fine-scale comparative phylogeography of a sympatric sister species triplet of subterranean diving beetles from a single calcrete aquifer in Western Australia. Molecular Ecology 18 (17), 3683–3698.

Guzik, M.T., Cooper, S.J.B., Humphreys, W.F., Ong, S., Kawakami, T., Austin, A.D., 2011a. Evidence for population fragmentation within a subterranean aquatic habitat in the Western Australian desert. Heredity 107 (3), 215–230.

Guzik, M.T., Austin, A.D., Cooper, S.J.B., Harvey, M.S., Humphreys, W.F., Bradford, T., Eberhard, S.M., King, R.A., Leijs, R., Muirhead, K.A., Tomlinson, M., 2011b. Is the Australian subterranean fauna uniquely diverse? Invertebrate Systematics 24, 407–418.

Guzik, M.T., Stringer, D.N., Murphy, N.P., Cooper, S.J.B., Taiti, S., King, R.A., Humphreys, W.F., Austin, A.D., 2019. Molecular phylogenetic analysis of Australian arid zone oniscidean isopods (Crustacea) reveals strong regional endemicity and new putative species. Invertebrate Systematics 33, 556–574.

Harcet, M., Bilandžija, H., Bruvo-Madarić, B., Ćetković, H., 2010. Taxonomic position of *Eunapius subterraneus* (Porifera, Spongillidae) inferred from molecular data- A revised classification needed? Molecular Phylogenetics and Evolution 54 (3), 1021–1027.

Hausdorf, B., Wilkens, H., Strecker, U., 2011. Population genetic patterns revealed by microsatellite data challenge the mitochondrial DNA based taxonomy of *Astyanax* in Mexico (Characidae, Teleostei). Molecular Phylogenetics and Evolution 60, 89–97.

Herman, A., Brandvain, Y., Weagley, J., Jeffery, W.R., Keene, A.C., Kono, T.J.Y., Bilandžija, H., Borowsky, R., Espinasa, L., O'Quin, K., Ornelas-García, C.P., Yoshizawa, M., Carlson, B., Maldonado, E., Gross, J.B., Cartwright, R.A., Rohner, N., Warren, W.C., McGaugh, S.E., 2018. The role of gene flow in rapid and repeated evolution of cave-related traits in Mexican tetra, *Astyanax mexicanus*. Molecular Ecology 27, 4397–4416.

Hewitt, G., 2000. The genetic legacy of the Quaternary ice ages. Nature 405, 907–913.

Holsinger, J.R., 2000. Ecological derivation, colonisation, and speciation. In: Wilkens, H., Culver, D.C., Humphreys, W.F. (Eds.), Ecosystems of the World: Subterranean Ecosystems. Elsevier, Amsterdam, pp. 399–415.

Horton, T., Lowry, J.K., De Broyer, C., 2016. World Amphipoda Database. Retrieved April 19, 2016, from. http://www.marinespecies.org/amphipoda.

Hou, Z., Li, S., 2018. Tethyan changes shaped aquatic diversification. Biological Reviews 93, 874–896.

Hou, Z., Sket, B., Fišer, C., Li, S., 2011. Eocene habitat shift from saline to freshwater promoted Tethyan amphipod diversification. Proceedings of the National Academy of Sciences of the United States of America 108 (35), 14533–14538.

Hou, Z., Sket, B., 2016. A review of Gammaridae (Crustacea: Amphipoda): the family extent, its evolutionary history, and taxonomic redefinition of genera. Zoological Journal of the Linnean Society 176, 323–348.

Howarth, F.G., 1987. The evolution of non-relictual tropical troglobites. International Journal of Speleology 16, 1–16.

Humphreys, W.F., 2000. Relict faunas and their derivation. In: Wilkens, H., Culver, D.C., Humphreys, W.F. (Eds.), Ecosystems of the World: Subterranean Ecosystems. Elsevier, Amsterdam, pp. 417–432.

Jażdżewski, K., Kupryjanowicz, J., 2010. One more fossil niphargid (Malacostraca: Amphipoda) from baltic Amber. Journal of Crustacean Biology 30 (3), 413–416.

Juan, C., Guzik, M.T., Jaume, D., Cooper, S.J.B., 2010. Evolution in caves: Darwin's "wrecks of ancient life" in the molecular era. Molecular Ecology 19, 3865–3880.

Jurado-Rivera, J.A., Pons, J., Alvarez, F., Botello, A., Humphreys, W.F., Page, T.J., Iliffe, T.M., Willassen, E., Meland, K., Juan, C., Jaume, D., 2017. Phylogenetic evidence that both ancient vicariance and dispersal have contributed to the biogeographic patterns of anchialine cave shrimps. Scientific Reports 7, 2852.

King, R.A., Bradford, T., Austin, A.D., Humphreys, W.F., Cooper, S.J.B., 2012. Divergent molecular lineages and not-so-cryptic species: the first descriptions of stygobitic chiltoniid amphipods (Talitroidea: Chiltoniidae) from Western Australia. Journal of Crustacean Biology 32 (3), 465–488.

Konec, M., Prevorčnik, S., Sarbu, S.M., Verovnik, R., Trontelj, P., 2015. Parallels between two geographically and ecologically disparate cave invasions by the same species, Asellus aquaticus (Isopoda, Crustacea). Journal of Evolutionary Biology 28 (4), 864–875.

Klopfstein, S., Langille, B., Spasojevic, T., Broad, G., Cooper, S.J.B., Austin, A.D., Niehuis, O., 2019. Hybrid capture data unravels a rapid radiation in pimpliform parasitoid wasps (Hymenoptera: ichneumonidae: Pimpliformes). Systematic Entomology 44, 361–383.

Kupriyanova, E.K., ten Hove, H.A., Sket, B., Zakšek, V., Trontelj, P., Rouse, G.W., 2009. Evolution of the unique freshwater cave-dwelling tube worm Marifugia cavatica (Annelida: serpulidae). Systematics and Biodiversity 7, 389–401.

Langille, B.L., Hyde, J., Saint, K.M., Bradford, T.M., Stringer, D.N., Tierney, S.M., Humphreys, W.G., Austin, A.D., Cooper, S.J.B., 2021. Evidence for speciation underground in diving beetles (Dytiscidae) from a subterranean archipelago. Evolution 75, 166–175.

Langille, B.L., Tierney, S.M., Bertozzi, T., Beasley-Hall, P.G., Bradford, T.M., Fagan-Jeffries, E.P., Hyde, J., Leijs, R., Richardson, M., Saint, K.M., Stringer, D.N., Villastrigo, A., Humphreys, W.F., Austin, A.D., Cooper, S.J.B., 2022. Parallel decay of vision genes in subterranean water beetles. Molecular Phylogenetics and Evolution 173, 107522.

Lefébure, T., Morvan, C., Malard, F., François, C., Konecny-Dupré, L., Guéguen, L., Weiss-Gayet, M., Seguin-Orlando, A., Ermini, L., Sarkissian, C., Charrier, N.P., Eme, D., Mermillod-Blondin, F., Duret, L., Vieira, C., Orlando, L., Douady, C.J., 2017. Less effective selection leads to larger genomes. Genome Research 27 (6), 1016–1028.

Leijs, R., van Nes, E.H., Watts, C.H., Cooper, S.J.B., Humphreys, W.F., Hogendoorn, K., 2012. Evolution of blind beetles in isolated aquifers: a test of alternative modes of speciation. PLoS One 7 (3), e34260.

Leys, R., Watts, C.H.S., Cooper, S.J.B., Humphreys, W.F., 2003. Evolution of subterranean diving beetles (Coleoptera: dytiscidae: Hydroporini, Bidessini) in the arid zone of Australia. Evolution 57, 2819–2834.

Magniez, G.J., 1999. A review of the family Stenasellidae (Isopoda, Asellota, Aselloidea) of underground waters. Crustaceana 72 (8), 837–848.

Mammola, S., Arnedo, M.A., Fišer, C., Cardoso, P., Dejanaz, A.J., Isaia, M., 2020a. Environmental filtering and convergent evolution determine the ecological specialisation of subterranean spiders. Functional Ecology 34 (5), 1064–1077.

Mammola, S., Amorim, I.R., Bichuette, M.E., Borges, P.A.V., Cheeptham, N., Cooper, S.J.B., Culver, D.C., Deharveng, L., Eme, D., Ferreira, R.L., Fišer, C., Fišer, Ž., Fong, D.W., Griebler, C., Jeffery, W.R., Jogovic, J., Kowalko, J.E., Lilley, T.M., Malard, F., Manenti, R., Martínez, A., Meierhofer, M.B., Niemiller, M.L., Northup, D.E., Pellegrini, T.G., Pipan, T., Protas, M., Reboleira, A.S.P.S., Venarsky, M.P., Wynne, J.J., Zagmajster, M., Cardoso, P., 2020b. Fundamental research questions in subterranean biology. Biological Reviews 95, 1855–1872.

McInerney, C.E., Maurice, L., Robertson, A.L., Knight, L.R.F.D., Arnscheidt, J.J., Venditti, C., Dooley, J.S.G., Mathers, T., Matthijs, S., Eriksson, K., Proudlove, G.S., Hänfling, B., 2014. The ancient Britons: groundwater fauna survived extreme climate change over tens of millions of years across NW Europe. Molecular Ecology 23 (5), 1153–1166.

Morvan, C., Malard, F., Paradis, E., Lefébure, T., Konecny-Dupré, L., Douady, C.J., 2013. Timetree of Aselloidea reveals species diversification dynamics in groundwater. Systematic Biology 62 (4), 512–522.

Neiber, M.T., Hartke, T.R., Stemme, T., Bergmann, A., Rust, J., Iliffe, T.M., Koenemann, S., 2011. Global biodiversity and phylogenetic evaluation of remipedia (crustacea). PLoS One 6, e19627.

Niemiller, M.L., Graening, G.O., Fenolio, D.B., Godwin, J.C., Cooley, J.R., Pearson, W.D., Fitzpatrick, B.M., Near, T.J., 2013. Doomed before they are described? The need for conservation assessments of cryptic species complexes using an amblyopsid cavefish (Amblyopsidae: *typhlichthys*) as a case study. Biodiversity & Conservation 22 (8), 1799–1820.

Nosil, P., Flaxman, S.M., 2011. Conditions for mutation-order speciation. Proceedings of the Royal Society B 278, 399–407.

Notenboom, J., 1991. Marine regressions and the evolution of groundwater dwelling amphipods (Crustacea). Journal of Biogeography 18, 437–454.

Page, T.J., Humphreys, W.F., Hughes, J.M., 2008. Shrimps down under: evolutionary relationships of subterranean crustaceans from Western Australia (Decapoda: Atyidae: *Stygiocaris*). PLoS One 3 (2), e1618.

Page, T.J., Hughes, J.M., Real, K.M., Stevens, M.I., King, R.A., Humphreys, W.F., 2016. Allegory of a cave crustacean: systematic and biogeographic reality of *Halosbaena* (Peracarida: Thermosbaenacea) sought with molecular data at multiple scales. Marine Biodiversity 48, 1185–1202.

Perina, G., Camacho, A.I., Huey, J., Horwitz, P., Koenders, A., 2019. New Bathynellidae (Crustacea) taxa and their relationships in the Fortescue catchment aquifers of the Pilbara region, Western Australia. Systematics and Biodiversity 17 (2), 148–164.

Phillips, M.J., Page, T.J., de Bruyn, M., Huey, J.A., Humphreys, W.F., Hughes, J.M., Santos, S.R., Schmidt, D.J., Waters, J.M., 2013. The linking of plate tectonics and evolutionary divergence. Current Biology 23 (14), R603–R605.

Policarpo, M, Fumey, J, Lafargeas, P, Legendre, L, Naquin, D, Thermes, C, Møller, P.R, Bernatchez, L, García-Machado, E, Rétaux, S, Casane, D, 2021. Contrasted gene decay in subterranean vertebrates: insights from cavefishes and fossorial mammals. Molecular Biology and Evolution 38 (2), 589–605.

Pons, J., Jurado-rivera, J.A., Jaume, D., Vonk, R., Maria, M., Juan, C., 2019. The age and diversification of Metacrangonyctid subterranean amphipod crustaceans revisited. Molecular Phylogenetics and Evolution 140, 106599.

Ribera, I., Fresneda, J., Bucur, R., Izquierdo, A., Vogler, A.P., Salgado, J.M., Cieslak, A., 2010. Ancient origin of a Western Mediterranean radiation of subterranean beetles. BMC Evolutionary Biology 10, 29.

Rizzo, V., Comas, J., Fadrique, F., Fresneda, J., Ribera, I., 2013. Early Pliocene range expansion of a clade of subterranean Pyrenean beetles. Journal of Biogeography 40, 1861–1873.

Saclier, N., François, C.M., Konecny-Dupré, L., Lartillot, N., Guéguen, L., Duret, L., Malard, F., Douady, C.J., Lefébure, T., 2018. Life history traits impact the nuclear rate of substitution but not the mitochondrial rate in isopods. Molecular Biology and Evolution 35, 2900–2912.

Schluter, D., 2000. The Ecology of Adaptive Radiation. Oxford University Press, Oxford.

Schluter, D., 2009. Evidence for ecological speciation and its alternative. Science 323, 737–741.

Schluter, D., Pennell, M.W., 2017. Speciation gradients and the distribution of biodiversity. Nature 546 (7656), 48–55.

Sket, B., 1999. The nature of biodiversity in hypogean waters and how it is endangered. Biodiversity & Conservation 8 (10), 1319–1338.

Stern, D.B., Breinholt, J., Pedraza-Lara, C., López-Mejía, M., Owen, C.L., Bracken-Grissom, H., Fetzner Jr., J.W., Crandall, K.A., 2017. Phylogenetic evidence from freshwater crayfishes that cave adaptation is not an evolutionary dead-end. Evolution 71 (10), 2522–2532.

Stoch, F., Galassi, D.M.P., 2010. Stygobiotic crustacean species richness: a question of numbers, a matter of scale. Hydrobiologia 653 (1), 217–234.

Stokkan, M., Jurado-Rivera, J.A., Oromí, P., Juan, C., Jaume, D., Pons, J., 2018. Species delimitation and mitogenome phylogenetics in the subterranean genus *Pseudoniphargus* (Crustacea: Amphipoda). Molecular Phylogenetics and Evolution 127, 988–999.

Stroud, J.T., Losos, J.B., 2016. Ecological opportunity and adaptive radiation. Annual Review of Ecology and Systematics 47, 507–532.

Strecker, U., Faúndez, V.H., Wilkens, H., 2004. Phylogeography of surface and cave *Astyanax* (Teleostei) from Central and North America based on cytochrome b sequence data. Molecular Phylogenetics and Evolution 33, 469–481.

Strecker, U., Hausdorf, B., Wilkens, H., 2012. Parallel speciation in *Astyanax* cave fish (Teleostei) in Northern Mexico. Molecular Phylogenetics and Evolution 62, 62–70.

Stringer, D.N., Bertozzi, T., Meusemann, K., Delean, S., Guzik, M.T., Tierney, S.M., Mayer, C., Cooper, S.J.B., Javidkar, M., Zwick, A., Austin, A.D., 2021. Development and evaluation of a custom bait design based on 469 single-copy protein-coding genes for exon capture of isopods (Philosciidae: *Haloniscus*). PLoS One 16 (9), e0256861.

Struck, T.H., Feder, J.L., Bendiksby, M., Birkeland, S., Cerca, J., Gusarov, V.I., Kistenich, S., Larsson, K.-H., Liow, L.H., Nowak, M.D., Stedje, B., Bachmann, L., Dimitrov, D., 2017. Finding evolutionary processes hidden in cryptic species. Trends in Ecology & Evolution 33 (3), 153—163.

Tierney, S.M., Langille, B., Humphreys, W.F., Austin, A.D., Cooper, S.J.B., 2018. Massive parallel regression: genetic mechanisms for vision loss amongst subterranean diving beetles. Integrative and Comparative Biology 58, 465—479.

Trontelj, P., 2018. Structure and genetics of cave populations. In: Moldovan, O.T., Kováč, L., Halse, S. (Eds.), Cave Ecology. Springer Nature, Switzerland, pp. 269—296.

Trontelj, P., Blejec, A., Fišer, C., 2012. Ecomorphological convergence of cave communities. Evolution 66 (12), 3852—3865.

Väinölä, R., Witt, J.D.S., Grabowski, M., Bradbury, J.H., Jazdzewski, K., Sket, B., 2008. Global diversity of amphipods (Amphipoda; Crustacea) in freshwater. Hydrobiologia 595, 241—255.

Vergnon, R., Leijs, R., van Nes, E.H., Scheffer, M., 2013. Repeated parallel evolution reveals limiting similarity in subterranean diving beetles. The American Naturalist 182, 67—75.

Villacorta, C., Jaume, D., Oromí, P., Juan, C., 2008. Under the volcano: phylogeography and evolution of the cave-dwelling *Palmorchestia hypogaea* (Amphipoda, Crustacea) at La Palma (Canary Islands). BMC Biology 6 (1), 7.

von Rintelen, K., Page, T.J., Cai, Y., Roe, K., Stelbrink, B., Kuhajda, B.R., Iliffe, T.M., Hughes, J., von Rintelenaet, T., 2012. Drawn to the dark side: a molecular phylogeny of freshwater shrimps (Crustacea: Decapoda: Caridea: Atyidae) reveals frequent cave invasions and challenges current taxonomic hypotheses. Molecular Phylogenetics and Evolution 63 (1), 82—96.

Vonk, R., Schram, F.R., 2003. Ingolfiellidea (Crustacea, Malacostraca, Amphipoda): a phylogenetic and biogeographic analysis. Contributions to Natural Science 72 (1), 39—72.

Wilkens, H., 2020. The role of selection in the evolution of blindness in cave fish. Biological Journal of the Linnean Society 130 (2), 421—432.

Zagmajster, M., Delić, T., Prevorčnik, S., Zakšek, V., 2013. New records and unusual morphology of the cave hydrozoan *Velkovrhia enigmatica* Matjašič & Sket, 1971 (Cnidaria: Hydrozoa: bougainvilliidae). Natura Sloveniae 15 (2), 13—22.

Zagmajster, M., Eme, D., Fišer, C., Galassi, D., Marmonier, P., Stoch, F., Cornu, J.-F., Malard, F., 2014. Geographic variation in range size and beta diversity of groundwater crustaceans: insights from habitats with low thermal seasonality. Global Ecology and Biogeography 23 (10), 1135—1145.

Zakšek, V., Sket, B., Trontelj, P., 2007. Phylogeny of the cave shrimp *Troglocaris*: evidence of a young connection between Balkans and Caucasus. Molecular Phylogenetics and Evolution 42, 223—235.

Zakšek, V., Sket, B., Gottstein, S., Franjević, D., Trontelj, P., 2009. The limits of cryptic diversity in groundwater: phylogeography of the cave shrimp *Troglocaris anophthalmus* (Crustacea: Decapoda: Atyidae). Molecular Ecology 18 (5), 931—946.

Zakšek, V., Delić, T., Fišer, C., Jalžić, B., Trontelj, P., 2019. Emergence of sympatry in a radiation of subterranean amphipods. Journal of Biogeography 46 (3), 657—669.

# Dispersal and geographic range size in groundwater

*Florian Malard[1], Erik Garcia Machado[2], Didier Casane[3,4], Steven Cooper[5,6], Cene Fišer[7] and David Eme[8]*

[1]Univ Lyon, Université Claude Bernard Lyon 1, CNRS, ENTPE, UMR 5023 LEHNA, Villeurbanne, France; [2]Institut de Biologie Intégrative et des Systèmes (IBIS), Pavillon Charles-Eugène-Marchand, Avenue de la Médecine, Université Laval Québec, Québec, Canada; [3]Université Paris-Saclay, CNRS, IRD, UMR Évolution, Génomes, Comportement et Écologie, Gif-sur-Yvette, France; [4]Université Paris Cité, UFR Sciences du Vivant, Paris, France; [5]South Australian Museum, Adelaide, SA, Australia; [6]The University of Adelaide, School of Biological Sciences and Australian Centre for Evolution Biology and Biodiversity, Adelaide, SA, Australia; [7]University of Ljubljana, Biotechnical Faculty, Department of Biology, Ljubljana, Slovenia; [8]INRAE, UR-RiverLY, Lyon, France

## Introduction

Dispersal is defined as "any movement of individuals or propagules with potential consequences for gene flow across space" (Ronce, 2007). Dispersal can be passive, but in groundwater, it is essentially active because water flow velocity is too low to wash out animals, except in fast-flowing galleries of karst aquifers. The terms "natal dispersal" and "breeding dispersal" designate the movement of an individual from its site of birth to its first breeding site and the movement of an individual among successive breeding sites, respectively (Greenwod and Harvey, 1982; Cayuela et al., 2018). Effective dispersal specifically describes the movement of individuals that results in gene flow, that is, when dispersers successfully reproduce in the arrival patches (Broquet and Petit, 2009). In population genetics, it corresponds to the migration rate, that is, the proportion of genes in a local population that originates from immigrants at each generation. Noneffective dispersal describes the movement of individuals regardless of whether dispersers successfully reproduce in the arrival patch. It typically encompasses the seasonal movement of animals from one habitat to another (i.e., migration in the ecological literature) as well as frequent short-distance foraging movements.

Dispersal propensity, that is, the probability to disperse to a new habitat patch, can ideally be conceived as an organism' property, which can be measured at individual to species levels using demographic and genetic methods (Stevens et al., 2010; Cayuela et al., 2018). It can for example be expressed as a distribution of individuals' dispersal distance, that is, the distance between the "start" and "end" points of a dispersal event. A species' dispersal propensity is a function of a number of dispersal-related phenotypic traits, including behavioral, physiological, morphological, and life history traits (Sarremejane et al., 2020). Given that many dispersal-related traits are heritable (Doligez et al., 2009; Saastamoinen et al., 2018), dispersal propensity can evolve in response to several selective pressures stemming from environmental predictability and heterogeneity, population density, and intra- and interspecific interactions (Clobert et al., 2004). Natural selection may act upon the three stages of dispersal by improving emigration decisions (i.e., departure stage), reducing dispersal cost during transfer (transience), or increasing success in selecting and settling in the arrival patch (settlement stage) (Clobert et al., 2004; Matthysen, 2012). Stevens et al. (2014) proposed that lifetime dispersal effort could be defined as a fitness component, thereby highlighting the existence of trade-offs between dispersal and other components of fitness such as survival and fecundity (see also Bonte and Dahirel, 2017). There may be distinct interspecific patterns of covariation between dispersal propensity and phenotypic traits due to shared evolutionary history among species or shared environments. These patterns of covariation have been coined dispersal syndromes (Ronce and Clobert, 2012; Stevens et al., 2014; Beckman et al., 2018).

Dispersal has important consequences at individual, population, species, community, and ecosystem levels. At the individual level, the decision to move may have contrasting outcomes ranging from successful dispersal (i.e., settlement and reproduction) to death during transience. Dispersal exerts a major control on the long-term persistence of spatially structured populations by influencing the growth rate, genetic diversity, adaptation, and extinction risk of local populations (Cayuela et al., 2018). Colonization, the movement of organisms to empty sites, allows for the expansion of a species' range (Kubish et al., 2014). Dispersal together with species sorting are fundamental processes structuring metacommunities (Leibold et al., 2004; Heino et al., 2015). Dispersal has important consequences on ecosystem functioning because dispersers and nondispersers can, for example, differ in their consumption rates of resources (Little et al., 2019). Hence, dispersal creates patterns at individual to ecosystem levels that, through eco-evolutionary feedbacks, set up new selective pressures, which in turn influence the evolution of dispersal (Starrfelt and Kokko, 2012).

Of the big four forces—mutation/speciation, drift, selection, and dispersal—driving microevolutionary and macroevolutionary patterns of groundwater biodiversity, dispersal has received comparatively little attention. The focus has been on determining the contribution of vicariance and dispersal to groundwater species diversification (Culver et al., 2009; Trontelj, 2018; Faille, 2021). The vicariance hypothesis is that after a surface species' population colonizes groundwater and becomes isolated from its surface ancestor, it undergoes little dispersal, thereby occupying a narrow geographic range. The dispersal hypothesis assumes dispersal following colonization and isolation in groundwater, ultimately leading to ecological and/or nonecological speciation among established subterranean species. Until recently, the vicariance hypothesis has prevailed (Culver and Pipan, 2019). However, quantitative estimates of the dispersal propensity of groundwater organisms are scarce because dispersal is challenging to measure in groundwater.

Several methods can be used to track the movement of organisms, including capture—mark—recapture (CMR), transmitters, or stable isotope techniques (Hobson, 2005; Kissling et al., 2014; Cayuela et al., 2018). However, they were rarely applied in groundwater because of animal rareness and the remote nature of their habitats, among other difficulties (but see Trajano (2001), Jugovic et al. (2015), and Balazs et al. (2020) for application examples of CMR). Culver et al. (2009) used an interesting approach to test for the importance of dispersal propensity by assuming that if dispersal propensity is important, then differences among species in occupancy frequencies at small spatial scales should predict occupancy patterns at larger scales. They provided two case-studies where occupancy of drips by crustacean copepods in one cave and occupancy of sites by obligate groundwater invertebrates in a small groundwater basin was a good predictor of occupancy of sites at the scale of a region and county, respectively. However, in neither case was the frequency of occupancy at the finest scale a good predictor of species range size (Culver et al., 2009). In recent decades, estimates of groundwater dispersal have become available from the application of molecular markers and approaches dedicated to quantifying both effective and noneffective dispersal (Asmyhr et al., 2014; Coghill et al., 2014; Malard et al., 2017; Trontelj, 2018; Jordan et al., 2020; van den Berg-Stein et al., 2022).

In part because of the difficulties to measure dispersal in groundwater, the dispersal propensity of groundwater organisms is often inferred from the size of their geographic range (Trontelj et al., 2009; Eme et al., 2013; Zagmajster et al., 2014). However, range size is the outcome of multiple factors including the propensity to disperse, landscape connectivity, availability of suitable settlement patches, and species niche breadth and position (Brown et al., 1996; Alzate and Onstein, 2022). Studies testing for a positive dispersal-range size relationship provided equivocal results with large variation in the relationship among clades, realms, and dispersal proxies (Lester et al., 2007; Sheth et al., 2020; Alzate and Onstein, 2022). Intuitively, a species' range is directly determined by a species' dispersal propensity across a dynamically spatiotemporal mosaic of abiotic conditions and biotic interactions (Kubish et al., 2014). That dynamic mosaic in turn sets up selective pressures driving the evolution of a species' dispersal propensity.

Several studies documented an active drift of groundwater organisms at the outlets (i.e., springs) of karst groundwater systems, indicating they may have a strong propensity to disperse within the boundaries of karst systems (Gibert et al., 1994; Carlini et al., 2009; Trontelj, 2018; Manenti and Piazza, 2021). Many groundwater species are restricted to a single karst system, potentially because they cannot disperse in inhospitable surface habitats between isolated karst systems (Fišer et al., 2016). However, empirical evidence suggests barriers to surface dispersal of groundwater species are labile (Eme et al., 2013; Trontelj, 2018; Delić et al., 2020). Within the boundaries of groundwater systems, higher environmental stochasticity in shallower habitats may have selected for increased dispersal propensities, causing a directional movement of individuals toward deeper and more stable habitats. For example, Pipan et al. (2006) documented an important drift of copepods into Organ Cave, West Virginia, the United States, from the upper layer of the karst (i.e., epikarst). However, the selection differential in dispersal propensity and hydrological connectivity between habitats can both be reset by a long-term change in groundwater recharge regime.

In this chapter, we review the factors that may drive the evolution of dispersal in groundwater. Given the paucity of data available, this review mainly aims to provide an evolutionary framework for the study of dispersal in groundwater. Then, we document spatial and temporal patterns of species range size and highlight major factors controlling the eco-evolutionary dynamics of dispersal, relevant for understanding range dynamics. We identify the main barriers to dispersal of groundwater organisms and provide a case study showing how groundwater landscape resistance controls the movement of organisms. We end this chapter by identifying priority dispersal-related research areas for better understanding and forecasting species distributions.

## Evolution of dispersal

For the sake of clarity, we distinguish between two categories of factors that might have influenced the evolution of dispersal in groundwater. These are abiotic and biotic factors that are extrinsic to the species under consideration, and intrinsic factors including body size, reproductive strategies, inbreeding, and intraspecific interactions. We acknowledge that distinguishing between the two types of factors is not always clear-cut and that other classification schemes are possible (see for example Clobert et al., 2004; Kubish et al., 2014).

### Extrinsic factors

#### Environmental heterogeneity in space and time

Environmental predictability is often considered as an important factor driving the evolution of dispersal (Clobert et al., 2004). Increased dispersal is selected in temporally unpredictable environments (Holt and McPeek, 1996; Ronce, 2007; Hidalgo et al., 2016). On the contrary, spatial heterogeneity in reproductive potential among patches alone does not select for dispersal in a temporally stable environment (Hastings, 1983; Parvinen et al., 2020). There is evidence among plants and animals that environmental unpredictability can jointly select for higher dispersal ability and strategic demography leading to fast population growth (Burton et al., 2010; Stevens et al., 2014; Beckman et al., 2018). For example, in butterflies, high mobility is typically part of the r-strategy (Stevens et al., 2012). By contrast, groundwater environment is temporally stable, and groundwater organisms exhibit a "troglomorphic syndrome" which encompasses delayed maturity, slow growth rate, long life span, and multiple reproductive events per lifetime (Malard, 2022; Vernarsky et al. chapter 19, this volume). Therefore, it is tempting to propose that the predictability of the groundwater environment may have selected for a low intrinsic dispersal propensity. In addition, the occurrence of trade-offs between dispersal and reproduction (Hughes et al., 2003; Guera, 2011) may cause groundwater organisms to avoid energy expenditure and predation risks associated with dispersal for maintaining a high life-span fecundity in a resource poor environment.

The troglomorphic syndrome is a hypothesis that obligate groundwater metazoans should show convergent traits because of a convergent selective environment that includes no light, reduced food availability, and low temporal variability (Christiansen, 2012; Culver and Pipan, 2015). However, many selective pressures, except the absence of light, differ markedly among groundwater habitats, thereby potentially resulting in contrasting life-history traits (Culver and Pipan, 2012, 2015; Simčič and Sket, 2019). Food supply is typically higher and environmental predictability is typically lower in subterranean habitats that have intimate hydrological connection with the surface environment, including the hyporheic zone, springs, sinking streams, and saturated soils, than in deep subterranean habitats (Culver and Pipan, 2014). Groundwater species colonizing these ecotonal habitats may have faster population growth and a higher propensity to disperse along the surface and shallow subsurface corridors than species occupying deep and predictable groundwater habitats. Hence, we propose that groundwater ecologists take advantage of the diversity of selective pressures operating in different groundwater habitats for quantifying how dispersal covaries with life-history traits.

### Environmental cues to disperse

Natural selection may also influence decisions to disperse by shaping responses to certain stimuli (Benton and Bowler, 2012). The study by Fišer et al. (2016) on the behavioral response to light in several species of the groundwater amphipod *Niphargus* suggests light may function as a cue to avoid costly dispersal in a lighted environment. Another potential environmental cue of dispersal is food shortage. Many groundwater organisms likely experience episodes of food shortage, which may act as cues influencing departure. Dispersal costs may be reduced among groundwater species because they accumulate large energy reserve stores that can be fueled during the transience stage of dispersal. The individual responses to environmental cues, however, depend on the individual conditions (e.g., amount of reserve storage) which may trigger or not dispersal (Gyllenberg et al., 2008; Kisdi et al., 2012). Individuals may decide to monopolize their reserve to resist food shortages on place until conditions become favorable again, thereby avoiding any dispersal hazards. Laboratory experiments using several groundwater organisms indicate that they drastically reduce their locomotory activity during periods of food and oxygen deprivation (Hervant and Renault, 2002; Hervant and Malard, 2019). Whether this "sit-and-wait" strategy also occurs in their natural habitats remains to be demonstrated.

### The dispersal-ecological specialization trade-off

The trade-off between dispersal ability and ecological specialization or local adaptation is central to several hypotheses seeking to explain large-scale patterns of biodiversity (Janzen, 1967; Stevens, 1989; Jansson and Dynesius, 2002; Jocque et al., 2010). Ecological specialization increases the cost of dispersing in nonsuitable habitat patches and decreases the probability of finding suitable habitat patches elsewhere, thereby selecting against dispersal. On the contrary, dispersal increases gene flow between populations, thereby acting against local adaptations (Kirkpatrick and Barton, 1997). Some studies indeed suggest that subterranean organisms may have undergone extreme ecological specialization that could act against dispersal (Mammola et al., 2019a). Local adaptation leading to a narrowing thermal niche is expected to proceed in subterranean habitats because seasonal variation in temperature is substantially lower than between-site heterogeneity in mean annual temperature. Mammola et al. (2019a) tested for a relationship between thermal specialization and elevational range among congeneric alpine spiders occurring along a steep gradient of decreasing thermal seasonality with increasing cave depth. They hypothesized that deep-cave spiders would have narrower thermal ranges, and consequently, smaller elevational ranges, because they would be more likely to encounter temperatures outside their thermal tolerance during dispersal. Indeed, the authors found that a spider species' elevational range extent and the variation of temperature encountered across its range decreased with increasing specialization to thermally stable deep subterranean habitats. The study by Eme et al. (2014) indicated that recent gene flow among populations of the widely distributed groundwater isopod *Proasellus valdensis* (Chappuis, 1948) might explain the weak local adaptation of populations to the stable thermal conditions of low elevation habitats (mean annual groundwater temperature >10°C) and high elevation habitats (mean annual groundwater temperature <6°C).

### *Interspecific interactions*

Culver and coworkers' studies in Appalachian cave streams, the United States, illustrate how interspecific interactions may contribute to dispersal (Culver et al., 1991, see review in Culver and Pipan, 2019). In the streams, amphipods and isopods are highly aggregated in riffles where they reside on the underside of rocks to avoid being washed out by the current. Individuals compete for space so that encounters result in one individual moving to another stone or being washed out by the current. Culver and coworkers' show that the washout rate of the amphipod *Stygobromus spinatus* (Holsinger, 1967) in an artificial riffle almost doubles in the presence of the isopod *Caecidotea holsingeri* (Steeves, 1963) indicating that *C. holsingeri* is a superior competitor. Forced dispersal may, however, be extremely costly because washout individuals can suffer appendage damage and predation by salamanders. Hence, increased costs of dispersal due to competitive interactions may lead to spatial segregations and control dispersal rate at the interspecific range boundaries.

## Intrinsic factors

### *Body size*

Dispersal cost may also be reduced in large-bodied organisms because the energy cost of locomotion decreases with increasing body weight (Schmidt-Nielsen, 1972, but see also Bale et al., 2014). Large-bodied species were found to disperse further in several taxa with different locomotion modes (Stevens et al., 2014). Culver and Pipan (2014) suggested that larger groundwater species of the amphipod genus *Stygobromus* could better disperse through the soil litter between seepage springs of the mid-Atlantic region of the United States than small-sized species. They found a positive relationship between the body size of *Stygobromus* species and the number of springs they occupied (Fig. 8.1A). The range size of the amphipods also increased with body size (Fig. 8.1B). Kralj-Fišer et al. (2020) suggested body size and relative appendage lengths of *Niphargus* amphipod species can be used as proxies of a species' dispersal propensity. In a laboratory experiment, they showed that large, stout, and long-legged species living in lakes use a variety of locomotion modes and move comparably faster than small, slender, and short-legged species that move using crawling or tail-flipping in cave streams. In another study conducted in the shallow subterranean layer of limestone rocks in central Slovenia, Culver and Pipan (2014) found no relationship between the body size of groundwater copepods and the number of drips they occupied in caves. In fissured rocks and unconsolidated sediments, the size of voids potentially sets a body size threshold above which locomotor performance decreases with increased body size because large body-sized organisms are less agile (Pipan and Culver, 2017). Moreover, a correlation between body

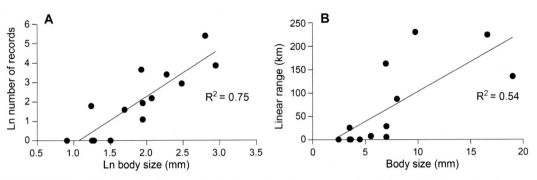

FIGURE 8.1 (A) Relationship between body size and number of seepage springs occupied by species of *Stygobromus* amphipods in the mid-Atlantic region, the United States. (B) Relationship between body size and maximum linear range size for mid-Atlantic *Stygobromus* species found in seepage springs. *After Culver and Pipan (2014)*.

size and dispersal distance does not necessarily imply direct causation because body size influences virtually all components of fitness. Larger body-size crustaceans produce more offspring, which may result in higher kin competition that in turn promotes dispersal (see below).

### Reproductive strategies

Factors related to reproductive strategy have been recognized as a major force driving the evolution of dispersal (Hargreaves and Eckert, 2014). Groundwater populations exhibit a high reproductive asynchrony so that, at any time, receptive females can be extremely rare in a population (Henry, 1976). Asynchrony may result in an increased proportion of females failing to mate, potentially causing an Allee effect, where population growth rate decreases at low population densities (Calabrese et al., 2008). This negative effect of asynchrony in low-density groundwater populations may be compensated by increased mate-searching efforts, which can be accomplished by increasing dispersal ability. If mate searching is sex-specific, then sex-specific dispersal can be expected. Mating system type (e.g., monogamy, polygyny and polyandry) and resource defense were proposed to promote the evolution of between-gender differences in dispersal but the link between mating system and sex-biased dispersal is still controversial (Trochet et al., 2016; Li and Kokko, 2019).

### Inbreeding

Dispersal is often presented as a behavior that can decrease the extinction risk of population due to inbreeding depression (Gandon, 1999; Nonaka et al., 2019). Mating among relatives (i.e., inbreeding) can yield offspring that have reduced fitness due to fixation of recessive deleterious allele variants and loss of genetic diversity, a mechanism known as inbreeding depression. This mechanism was empirically shown to decrease the persistence of local populations and small metapopulations (Saccheri et al., 1998; Nonaka et al., 2019). It may also be an important factor in the evolution of sex-biased dispersal rates (Gandon, 1999). In several groundwater species, local populations are presumably small (Lefébure et al., 2017) and females can lay relatively large clusters of eggs. A brood of the groundwater isopod *Proasellus cavaticus* (Leydig, 1871) may contain up to 66 juveniles (Henry, 1976) and the groundwater amphipod *Niphargus virei* Chevreux, 1896 can release up to 70 juveniles during a single reproduction event (Ginet, 1960). In the absence of dispersal of individuals from their natal patch, there is an elevated probability of encountering siblings while looking for mates. Hence, natal dispersal may have been selected for in groundwater to decrease the susceptibility of local populations to extinction due to inbreeding depression. In the Jura Mountains, France, females of the amphipod *N. virei* releases their offspring in early autumn, which results in a massive drift of juveniles at the springs of karst systems during the autumnal floods (Turquin and Barthélémy, 1985; Malard et al., 1997). Juveniles drifting at the outlet of the karst system represent dispersers that die on route without reproducing (i.e., the cost of dispersal) but an unknown and potentially a majority number of juveniles disperse within the aquifer (Turquin, 1981).

## Intraspecific interactions

Intraspecific competition, more specifically kin competition for space and resources (i.e., between parents and offspring or among offspring), promotes dispersal even when dispersal is costly (Hamilton and May, 1977; Gandon, 1999; Parvinen et al., 2020). Bearing in mind the patchy distribution of many groundwater metazoans, individuals within a patch are likely to be more related than individuals from different patches if the dispersal rate is low. Increased dispersal will cause emigrating individuals to arrive in patches without relatives and nondispersing relatives will benefit from more resources in their natal patch. Dispersal can be perceived as an outcome of kin selection: emigrants may gain no benefit, but they alleviate kin competition in their natal patch. Despite the potentially important role of intraspecific interactions, either between kin or between unrelated individuals, in groundwater, we are aware of no experiments explicitly designed to test its effect on dispersal.

Cannibalism, the killing and consumption of conspecifics, is another form of intraspecific interaction that can select for dispersal (Rudolf et al., 2010). Cannibalism has costs and benefits: filial cannibalism can decrease inclusive fitness, but it provides a rich resource with the correct stoichiometric ratio and can decrease density-dependent mortality in offspring. Thus, cannibalism may be favored in habitats where species experience strong resource limitations such as groundwater (Nishimura and Isoda, 2004). Cannibalism is probably widespread in groundwater, as it has been observed in several taxa including isopods (Magniez, 1975), amphipods (Culver et al., 1995; Lustrik et al., 2011), and fishes (Parzefall and Trajano, 2010). Luštrik et al. (2011) showed that juveniles of the groundwater amphipod *Niphargus timavi* S. Karaman, 1954 altered their habitat use in response to cannibalistic conspecifics by taking, more often, refuge in fine sands. The dispersal of offspring away from their cannibalistic parents is another way to reduce the loss of inclusive fitness while cannibalism of nonrelatives still provides energy gain and removes competitors. In the absence of dispersal, cannibalism may also reinforce inbreeding depression if unrelated individuals are cannibalized more often than related individuals are (Dobler and Kölliker, 2010). Rudolf et al. (2010) modeled the interplays between the spatial structure of the population, the evolution of cannibalism, and the evolution of dispersal. The spatial structure of the population can alter the evolution of cannibalism, but cannibalism can alter dispersal that influences the spatial structure itself. They found that the coevolution of cannibalism and dispersal can result in evolutionary branching leading to the cooccurrence of strategies that have different dispersal and cannibalism regimes. The strategy predominating in different groundwater habitats may potentially be determined by the level of trophic resources influencing the initial cannibalism rate.

# Range size

The frequency distribution of range size of groundwater species is typically right-skewed: a substantial proportion of species are restricted to a single groundwater system and a few species are widely distributed (Culver et al., 2000; Zagmajster el al. 2014; Niemiller et al., 2019) (Fig. 8.2A). This pattern is by no means specific to groundwater fauna as it has been documented for a wide range of taxa including invertebrates, birds, and mammals (Gaston, 1998).

Groundwater species are predicted to have smaller range sizes than surface aquatic species, although we are aware of a single phylogenetically controlled comparison in range size between surface and groundwater species. Using a large phylogeny of crayfishes (i.e., one with 733 species), Stern et al. (2017) showed that cave-adapted species had the lowest median range size when compared to other specialized species (i.e., primary burrowers, or lotic or lentic water species) and generalist species (i.e., species occupying more than one habitat type). Here, we show that groundwater asellids have on average a significantly smaller range size than surface aquatic asellids, although that comparison does not account for phylogenetic relationships among species (Fig. 8.3A). By contrast, Esmaeili-Rineh et al. (2020) found

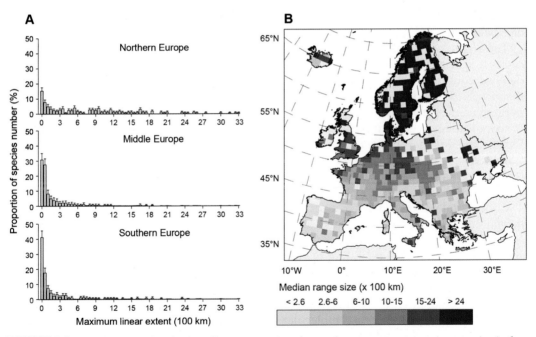

**FIGURE 8.2** (A) Frequency distribution of linear range size of groundwater crustacean species occurring in three latitudinal bands in Europe. Northern Europe: 47.6–54°N, 171 species; Middle Europe: 41.3–47.6°N, 1194 species; Southern Europe: 35–41.3°N, 396 species. One hundred species were resampled 100 times in each band to account for differences in the number of species among bands. Bars and whiskers show the average and standard deviation, respectively. (B) Geographic distribution of median linear range size (maximum linear extent) of groundwater crustacean communities in Europe (n = 1570 species). The cell area is 10,000 km². *Panel B after Zagmajster et al. (2014).*

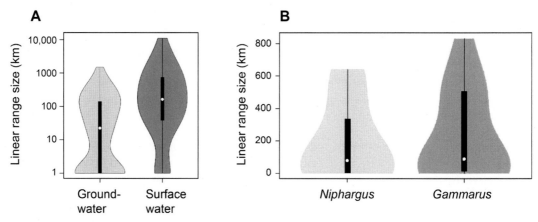

**FIGURE 8.3**  Comparison of linear range sizes (MLE: maximum linear extent in km). (A) between surface aquatic asellid isopods (n = 87) and groundwater asellid isopods (n = 259) (worldwide data from the World Asellidae database, Malard et al., 2022). (B) between surface aquatic amphipod species of *Gammarus* (n = 13) and groundwater species of *Niphargus* (n = 8) in Iran. The white dot, thick black bar, and thin black line show the median value, interquartile range, and 95% of all data, respectively. *After Esmaeili-Rineh et al. (2020).*

no significant differences in the median range size between surface aquatic amphipod species of *Gammarus* (n = 13) and groundwater species of *Niphargus* (n = 8) in Iran (Fig. 8.3B).

The shape of the species-range size distribution is ultimately determined by three processes - speciation, extinction and transformation of range size (see Gaston (1998) for a detailed consideration of the role played by each process). Speciation adds new ranges to the distribution, extinction causes loss of ranges and range transformation refers to the temporal dynamics of species' range sizes over the course of their lifetimes. These processes vary in space leading to geographic variation in the shape of the species-range size distribution. In Europe, the skewness of the range size distribution of groundwater crustacean species significantly decreases northward as the proportion of species with large ranges increases (Fig. 8.2A). Range size increases with increasing latitude (Zagmajster et al., 2014), a pattern known as the Rapoport's effect (Stevens, 1989) (Figs. 8.2B and 8.4).

Continental patterns of range size are sensitive to species delimitation. The occurrence of morphologically similar but genetically divergent species, a phenomenon referred to as cryptic diversity (Bickford et al., 2007), can bias the estimation of the size of groundwater species ranges and understanding of range size patterns (Trontelj et al., 2009). To account for this bias, Eme et al. (2018) assembled 2205 sequences of the partial mitochondrial cytochrome *c* oxidase subunit I (COI) gene collectively representing 263 morphologically distinguishable species of groundwater Aselloidea (Isopoda) and Niphargidae (Amphipoda) from Europe. Then, they used several molecular delimitation methods to infer species boundaries and compared the latitudinal patterns of range size obtained when delimiting species with molecular methods and morphology. Molecular methods delimited up to 2.5 more species (n = 646) than morphology (n = 263). The average maximum linear extent (MLE, the straight-line distance between the two most distant known localities of a species' range) of species delimited with molecular methods was typically smaller than that of species delimited with morphology (molecular method: MLE = 22 ± 96 km, n = 477 species;

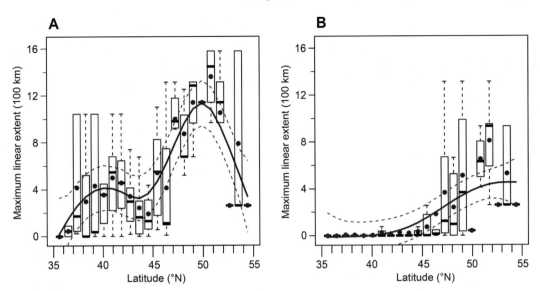

**FIGURE 8.4** Relationship between median range size (maximum linear extent) per 0.9 degrees latitudinal band and latitude for obligate groundwater Aselloidea (Isopoda) and Niphargidae (Amphipoda) in Europe. (A) Species (n = 147) are delimited using morphology. (B) Species (n = 477) are delimited using a molecular method, in particular here the Poisson tree processes (PTP) method applied to the mitochondrial cytochrome c oxydase subunit I (COI) gene. Black horizontal bars, red dots, and boxes show the median, average, and interquartile range, respectively, for latitudinal bands. Dashed red lines, which stand for 95% confidence intervals, and continuous red lines represent the fit of a generalized additive model to the averages of latitudinal bands. *Modified after Eme et al. (2018).*

morphology: MLE = 170 ± 316 km, n = 147 species) (Eme et al., 2018). The Rapoport's effect was more evident with species delimited with molecular methods than with morphology (Fig. 8.4). Below 43°N, the ranges of morphologically distinguishable species were systematically split into a series of narrow ranges by molecular methods, thereby resulting in a flat distribution of narrow range sizes (Fig. 8.4B). Above 43 degrees N, the ranges of morphologically distinguishable species were asymmetrically split in fewer ranges by molecular methods, some of which remained large. This resulted in a higher mean range size and higher variability around the mean in northern than in southern latitudes (Eme et al., 2018) (Fig. 8.4B). A possible, yet untested scenario is that successive range expansions and retractions during Quaternary climatic oscillations at northern latitudes may have left peripheral populations, thereby resulting in a highly asymmetric range size pattern.

Zagmajster et al. (2014) and Eme et al. (2018) tested the relative importance of three broad factors—habitat area/heterogeneity, climate seasonality, and historical climate variability in shaping the pattern of range size of groundwater crustaceans in Europe. Temperature and precipitation anomalies defined, respectively, as the differences in mean annual temperature and annual precipitation between the present and the last glacial maximum (LGM, 21,000 years ago) were used as a measure of historical climate variability. The study by Zagmajster et al. (2014) used range size data for 1570 morphologically distinguishable species of groundwater crustaceans (Fig. 8.2B). The study by Eme et al. (2018) was based on range size data for groundwater Aselloidea (Isopoda) and Niphargidae (Amphipoda), but the authors used morphology and three molecular species delimitation methods to produce four range size data sets (i.e., the two data sets shown in Fig. 8.4 and two additional data sets obtained with two other molecular species delimitation methods). In both studies, historical climate variability, more especially the long-term variability of temperature rather than precipitation, had by far the highest independent contribution (up to 31.3%) to geographic variation in range size, regardless of the method used to delimit species.

Three explanations, two of them emphasizing the effect of higher long-term climatic variability on dispersal, may explain the pattern of increasing range size at higher latitudes (Figs. 8.2 and 8.4). First, cycles of glacials and interglacials may have provided landscape corridors promoting the movement of animals during short-time environmental windows of increased connectivity. Subsurface dispersal potentially occurred along extensive hyporheic corridors generated by intensive postglacial sediment deposition (Eme et al., 2013; Malard et al., 2017). Surface dispersal of groundwater species might have been facilitated during glacial phases or at the onset of glacial melting because competition and predation pressures were presumably low in surface aquatic habitats (Fišer et al., 2010; Eme et al., 2014). Second, the increasing amplitude of Milankovitch climatic oscillations at higher latitudes may have selected for vagility and generalism (Jansson and Dynesius, 2002). Hence, higher-latitude species would have evolved a stronger dispersal propensity. Third, there has been a disproportionate extinction of small-range species in northern regions severely affected by cold Pleistocene climate (Rohde, 1996), because extinction probability is greater for species with small ranges (Gaston, 1998). The three explanations discussed above are nonmutually exclusive; rather, they probably acted in concert. The occurrence of ephemeral corridors is more likely to result in species range expansion if dispersal propensity has been selected for by the environment.

Reconstruction of ancestral ranges using phylogeographic diffusion analysis revealed considerable complexity in the factors shaping the range dynamics of groundwater species during the Pleistocene. Eme et al. (2013) showed considerable spatiotemporal heterogeneity in the speed of range expansion of groundwater asellid isopods, suggesting short-time dispersal corridors could have been instrumental in shaping species ranges. Reconstruction of the range dynamics of the *Niphargus stygius* (Schiödte, 1847) species complex indicated that allopatric speciation was promoted by local adaptation, followed by phases of dispersal leading to secondary contact, and occasional overlap of ranges (Delić et al. 2022). Studies by Zakšek et al. (2019) and Delić et al. (2016) pinpointed the importance of interspecific interactions on dispersal and the evolution of range size. Zakšek et al. (2019) modeled the distribution of a complex of six species of groundwater amphipods *Niphargus*, taking into account phylogenetic relationships and presumed competitive interactions. While ecomorphologically similar species pairs showed allopatry due to putative interspecific competition, the only ecomorphologically differentiated species pair established a broad area of sympatry, presumably due to relaxed interspecific competition. An additional and more detailed analysis further supported the hypothesis of interspecific competition. It showed syntopic populations of this species pair differentiate further morphologically, thereby relaxing interspecific competition presumably through character displacement (Delić et al., 2016).

# Groundwater landscape connectivity modulates dispersal

As already outlined, dispersal depends on intrinsic features of the organism under consideration, that is, its dispersal propensity, and extrinsic factors limiting the movement of organisms, such as geographical distance (i.e., isolation by distance), landscape resistance to movement (i.e., isolation by resistance, Zeller et al., 2012), and barriers to dispersal. Here, we use the term groundwater connectivity to refer collectively to these extrinsic factors.

In recent decades, understanding the landscape features that impede or facilitate the movement of groundwater organisms has extensively relied on molecular techniques and methods in population genetics and phylogeography (Trontelj, 2019). While applying these methods to assess dispersal in groundwater, several difficulties should be kept in mind. First, multiple invasions of groundwater through time by successive and genetically distinct populations of a surface species can produce subterranean genetic patterns that resemble subterranean isolation of subterranean populations following subterranean dispersal. This difficulty is amply illustrated by Trontelj (2018). Second, dispersal between populations may occur with no or little gene flow because of barriers imposed by natural selection. In the Mexican blind cavefish *Astyanax mexicanus* (De Filippi, 1853), large numbers of individuals from surface populations move into the caves and come into contact with individuals of cave populations. However, there is little admixture between the two populations because of the selective advantage of cave individuals in competing for food (Strecker et al., 2012; Coghill et al., 2014). Depending on the objectives of a dispersal study, it may be important to combine methods that can quantify both effective and noneffective dispersal (Cayuela et al., 2018). Third, testing for the potential effect of a landscape connectivity attribute on dispersal ideally requires obtaining independent physical and/or ecological measures of groundwater connectivity. In practice, obtaining such measures has proved difficult so landscape resistance to dispersal is often inferred a posteriori from the results of population genetic structure (Sbordoni et al., 2012).

Many groundwater aquifers may conceptually be perceived as small islands with little if any permanent subsurface hydrological connections between nearby aquifers. Hence, multiple colonization events by surface populations are likely to result in small isolated subterranean populations that have little opportunity to disperse beyond the boundaries of the aquifer or hydrogeologically connected sets of aquifers that they have colonized. Landmark studies in population genetics and phylogeography of the cavefish *Astyanax mexicanus* (Bradic et al., 2012; Strecker et al., 2012), amphipod *Gammarus minus* Say, 1818 (Carlini et al., 2009; Fong and Carlini, this volume), isopod *Asellus aquaticus* (Linnaeus, 1758) (Verovnik et al., 2004), and paramelitid amphipods (Cooper et al., 2007) have largely corroborated this view of groundwater species as being constituted of small populations, that are deeply structured genetically according to geographical proximity and extant hydrological connections. In all these studies, gene flow among cave populations, even when they were geographically close, was lower than gene flow among surface populations or between surface and cave populations. Using presence—absence data and genetic distances computed from amplified fragment length polymorphism, Foulquier et al. (2008) showed that catchment boundaries acted as dispersal barriers restricting the recolonization of formerly glaciated areas by the groundwater amphipod *Niphargus virei*. Konec et al. (2016) showed that the geographic separation between the closely related subterranean aquatic isopods *Asellus aquaticus* on the Danubian catchment and *Asellus kosswigi* Verovnik, Prevorcnik & Jugovic, 2009 on the Adriatic catchment reflected both historical and contemporary groundwater connectivity.

Barriers to dispersal may further constrain gene flow among populations within the boundaries of groundwater systems as demonstrated by a series of studies of intraspecific genetic structuring in calcrete aquifers in Western Australia (Guzik et al., 2009, 2011; Bradford et al., 2013). These studies revealed concordant patterns of genetic structure among groundwater taxa as diverse as amphipods, beetles, and isopods, indicating the same barriers to dispersal may have repeatedly impeded the movement of animals. Candidate barriers included salinity clines (Guzik et al., 2011) and falling groundwater levels leading to short-term isolation of populations on a microgeographic scale (Bradford et al., 2013). However, range expansion during periods of high-water levels may increase gene flow among previously isolated populations, thereby increasing genetic variation within subterranean populations (Buhay and Crandall, 2005; Bradford et al., 2013). In addition to concordant genetic structuring among disparate taxa, Guzik et al. (2009) also reported a case of discordant intraspecific genetic structure that might result from differences in dispersal propensity among species inhabiting the same calcrete. Of three species of sympatric and congeneric diving beetles differing in body size, only the smallest bodied size species, *Paroster microsturtensis* (Watts and Humphreys, 2006), showed no sign of population structuring and isolation by distance. Guzik et al. (2009) suggested that a small body size might have enabled this species to disperse across barriers that were otherwise insurmountable to its congeners.

There is evidence showing that hydrological barriers among aquifers are labile or that they can be overcome, thereby facilitating transient dispersal between distant aquifers. Indications of recent and/or ancient dispersal events across present-day aquifer boundaries come from studies reporting shared haplotypes or derived haplotypes among populations separated by considerable distances, sometimes exceeding several hundred kilometers. These studies are from a variety of groundwater taxa including fish (Hernández et al., 2016), crayfish (Buhay and Crandall, 2005), shrimp (Zakšek et al., 2009), amphipod (Lefébure et al., 2006; Weber et al., 2020), and isopod (Eme et al., 2013). Two dispersal pathways are often invoked to account for the lack of deep genetic structuring in groundwater populations from distant localities; they are surface aquatic habitats and the hyporheic habitats of rivers. Both habitats are typically more continuous than groundwater habitats but their suitability as dispersal corridors might have been extremely variable over geological times.

Specialization to groundwater environments apparently results in a high dispersal cost in the surface environment where groundwater organisms suffer from strong competition with aquatic surface species (see Copilas-Ciocianu et al., 2018). *Niphargus hrabei* S. Karaman, 1932 and *N. valachicus* Dobreanu and Manolache, 1933 provide two actual examples of blind and depigmented amphipods that have secondarily colonized and extensively dispersed in surface waters (Copilas-Ciocianu et al., 2018). Their dispersal in the surface environment is apparently facilitated by their ability to live in hypoxic waters where their competitors from the family Gammaridae do not survive. Buhay and Crandall (2005) suggested dispersal in surface streams during periods of high-water levels might explain extensive gene flow among populations of obligate cave crayfishes in karst systems of the Cumberland Plateau, the United States. Zakšek et al. (2009) also proposed surface dispersal in flooded poljes connecting otherwise isolated caves as the most plausible mechanism for explaining the distribution of the southern Adriatic phylogroup of the cave shrimp *Troglocaris anophthalmus* (Kollar, 1848) across 300 km of the fragmented Dinaric karst landscape. Biogeographical modeling by Delić et al. (2020) shows that ancestors of *Niphargus* amphipod species found on both sides of the Adriatic Sea dispersed three times independently from the Balkan Peninsula to the Apennine Peninsula (Italy) and once back to the Balkans. During narrow time windows associated with marine regressions, groundwater amphipods were able to spread through moderately saline and species-poor habitats of the Adriatic basin and/or they migrated in freshwater subterranean habitats associated with paleo-rivers. Reconstruction of the range dynamics of the isopod *Proasellus cavaticus*, which occurs on both sides of the Channel, indicates dispersal occurred along surface and/or hyporheic habitats of the Pleistocene Channel River (Eme et al., 2013).

Interestingly, whereas surface dispersal may contribute to gene flow among otherwise isolated groundwater populations, subsurface dispersal may contribute to gene flow among otherwise isolated surface populations. Palandacic et al. (2012) provided evidence of recurrent groundwater dispersal maintaining gene flow between geographically isolated populations of the surface aquatic cyprinid fish *Delminichthys adspersus* (Heckel, 1843) that inhabits temporary springs in the Dinaric karst.

The idea that the hyporheic zone (i.e., the interstitial habitats in the bed sediment of rivers, but see Tonina and Buffington, this volume) may function as a stepping stone dispersal corridor between distant aquifers was first introduced by Ward and Palmer (1994). The few studies that examined the dispersal of groundwater species in river alluvial aquifers indicate considerable variation in the functional connectivity of the hyporheic zone. From the results of a microsatellite study in the Macquarie River alluvium, New South Wales, Australia, Asmyhr et al. (2014) found deep intraspecific genetic structuring over short geographical distances (<50 m) for an unspecified groundwater species of Parabathynellidae (Bathynellacea, Crustacea). Individuals were genetically less similar than expected from geographical distance alone for distances higher than about 30 m, indicating strong barriers to gene flow over very small distances in alluvium. In contrast, Jordan et al. (2020) showed no genetic structuring in populations of the groundwater-obligate amphipod *Stygobromus* sp. along a 70-km long hyporheic corridor of the Flathead River basin, Montana, USA. Genetic differentiation measured from single nucleotide polymorphisms between populations of the amphipod was similarly low to that observed for the stonefly *Paraperla frontalis* (Banks, 1902) with aerial (winged) adults, indicating the hyporheic corridor was extremely conducive to the movement of the amphipod.

Malard et al. (2017) showed that the degree to which the hyporheic zone of gravel rivers facilitated the movement of groundwater organisms depended on the importance of sediment deposition, which is itself determined by the balance between the transport and supply of sediment. The authors used microsatellite markers to test for a relationship between genetic differentiation among demes of the minute groundwater isopod *Proasellus walteri* (Chappuis, 1948) and channel morphology along three nearby hyporheic corridors differing widely in their sediment regime (Fig. 8.5A and B). In the sediment-rich corridor of the Drôme River, genetic differentiation among demes was weak and best explained by hydrologic distance. In the two corridors facing a sediment shortage (i.e., the Roubion and Ouvèze Rivers), genetic differentiation was higher and the length of sediment supply limited channels (i.e., geomorphic resistance in Fig. 8.5C) was more important than the hydrologic distance in

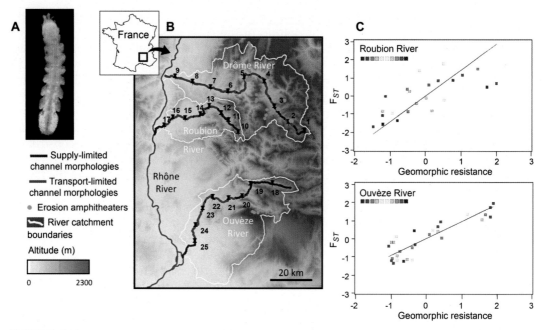

**FIGURE 8.5** Geomorphic influence on intraspecific genetic differentiation along alluvial corridors in the groundwater asellid isopod *Proasellus walteri* (Chappuis, 1948). (A) Photo of *P. walteri*. Body length is 3 mm and appendages are missing. (B) Location of the 25 sampled demes of *P. walteri* and distribution of sediment supply-limited channels and sediment transport-limited channels along three alluvial river corridors. (C) Relationship between $F_{ST}$ and geomorphic resistance for two alluvial river corridors. Values of $F_{ST}$ and geomorphic resistance are normalized per river. Geomorphic resistance increases with the length of sediment supply limited channels between demes. Colors of data points correspond to 10 classes of increasing hydrologic distance from blue to red. *Modified after Malard et al. (2017).*

explaining the longitudinal distribution of genetic differentiation. These findings suggest that the connectivity of the hyporheic zone, and hence its importance as a dispersal corridor, can considerably vary over time in response to climate-mediated changes in the balance between the transport and supply of sediment (Eme et al., 2013; Malard et al., 2017).

## Conclusion

The understanding of factors controlling dispersal and the dynamics of species' range in groundwater has substantially progressed over the last decades, especially thanks to the development of research in population genetics and phylogeography. However, there is a major gap between theoretical developments and the amount of empirical data available to test theoretical predictions. We foresee three important research avenues to fill this gap. First, there is a considerable need for measurements of dispersal propensity and phenotypic traits of species across multiple groundwater habitats with contrasting selective pressures. Searching for patterns of covariation between dispersal and phenotypic traits (i.e., dispersal syndrome) is key to identifying dispersal-related traits among morphological, behavioral, physiological, and life history traits (Stevens et al., 2014; Beckman et al., 2018). This endeavor paves the way for developing a trait database to assess the dispersal propensity of groundwater species (see Sarremejane et al. (2020) for a database of dispersal-related traits of surface aquatic macroinvertebrates). Second, identifying hydrological, geomorphological, and ecological surrogates of groundwater landscape connectivity is prerequisite to determining the importance of groundwater landscape resistance to animal movement (Malard et al., 2017). Third, continued efforts are needed to document the geographic distribution of groundwater species. Combining the sampling of environmental DNA, metabarcoding and DNA taxonomy can both speed up the acquisition of species occurrence data and provide better estimates of species range size (Eme et al., 2018; Saccò et al., 2022). Bringing together data from these three research avenues can tease apart the relative importance of dispersal-related traits, groundwater landscape connectivity, and phylogeny in shaping the range of groundwater species. At last, there is an urgent need to integrate knowledge gained on dispersal and range dynamics into assessing the potential for species range shift in the face of environmental change (Mammola et al., 2019b). A major uncertainty concerns the plasticity of dispersal-related traits and the temporal scale at which they can evolve.

## Acknowledgments

We thank S. Rétaux and C. Griebler for editing an earlier version of this chapter.

## References

Alzate, A., Onstein, R.E., 2022. Understanding the relationship between dispersal and range size. Ecology Letters 25, 2303–2323.

Asmyhr, M.G., Hose, G., Graham, P., Stow, A.J., 2014. Fine-scale genetics of subterranean syncarids. Freshwater Biology 59, 1–11.

Balázs, G., Lewarne, B., Herczeg, G., 2020. Extreme site fidelity of the olm (*Proteus anguinus*) revealed by a long-term capture-mark-recapture study. Journal of Zoology 311, 99–105.

Bale, R., Hao, M., Bhalla, A.P., Patankar, N.A., 2014. Energy efficiency and allometry of movement of swimming and flying animals. Proceedings of the National Academy of Sciences of the United States of America 111 (21), 7517–7521.

Beckman, N.G., Bullock, J.M., Salguero-Gómez, R., 2018. High dispersal ability is related to fast life-history strategies. Journal of Ecology 106, 1349–1362.

Benton, T.G., Bowler, D.E., 2012. Dispersal in invertebrates: influences on individual decisions. In: Clobert, J., Baguette, M., Benton, T.G., Bullock, J.M. (Eds.), Dispersal Ecology and Evolution. Oxford University Press, Oxford, UK, pp. 41–49.

Bickford, D., Lohman, D.J., Sodhi, N.S., Ng, P.K.L., Meier, R., Winker, K., Ingram, K.K., Das, L., 2007. Cryptic species as a window on diversity and conservation. Trends in Ecology & Evolution 22 (3), 148–155.

Bonte, D., Dahirel, M., 2017. Dispersal: a central and independent trait in life history. Oikos 126, 472–479.

Bradford, T.M., Adams, M., Guzik, M.T., Humphreys, W.F., Austin, A.D., Cooper, S.J.B., 2013. Patterns of population genetic variation in sympatric chiltoniid amphipods within a calcrete aquifer reveal a dynamic subterranean environment. Heredity 111, 77–85.

Bradic, M., Beerli, P., García-de León, F.J., Esquivel-Bobadilla, S., Borowsky, R.L., 2012. Gene flow and population structure in the Mexican blind cavefish complex (*Astyanax mexicanus*). BMC Evolutionary Biology 12, 9.

Broquet, T., Petit, E.J., 2009. Molecular estimation of dispersal for ecology and population genetics. Annual Review of Ecology, Evolution, and Systematics 40, 193–216.

Brown, J.H., Stevens, G.C., Kaufman, D.M., 1996. The geographic range: size, shape, boundaries, and internal structure. Annual Review of Ecology and Systematics 27, 597–623.

Buhay, J.E., Crandall, K.A., 2005. Subterranean phylogeography of freshwater crayfishes shows extensive gene flow and surprisingly large population sizes. Molecular Ecology 14, 4259–4273.

Burton, O.J., Phillips, B.L., Travis, J.M.J., 2010. Trade-offs and the evolution of life-histories during range expansion. Ecology Letters 13, 1210–1220.

Calabrese, J.M., Ries, L., Matter, S.F., Debinski, D.M., Auckland, J.N., Roland, J., Fagan, W.F., 2008. Reproductive asynchrony in natural butterfly populations and its consequences for female matelessness. Journal of Animal Ecology 77, 746–756.

Carlini, D.B., Manning, J., Sullivan, P.G., Fong, D.W., 2009. Molecular genetic variation and population structure in morphologically differentiated cave and surface populations of the freshwater amphipod *Gammarus minus*. Molecular Ecology 18, 1932–1945.

Cayuela, H., Rougemont, Q., Prunier, J.G., Moore, J.-S., Clobert, J., Besnard, A., Bernatchez, L., 2018. Demographic and genetic approaches to study dispersal in wild animal populations: a methodological review. Molecular Ecology 27, 3976–4010.

Christiansen, K.A., 2012. Morphological adaptations. In: White, W.B., Culver, D.C. (Eds.), Encyclopedia of Caves, Second edition. Academic/Elsevier Press, Amsterdam, the Netherlands, pp. 517–528.

Clobert, J., Ims, R.A., Rousset, F., 2004. Causes, mechanisms and consequences of dispersal. In: Hanski, I., Gaggiotti, O.E. (Eds.), Ecology, Genetics and Evolution of Metapopulations. Academic Press, pp. 307–335.

Coghill, L.M., Hulsey, C.D., Chaves-Campos, J., García de Leon, F.J., Johnson, S.G., 2014. Next generation phylogeography of cave and surface *Astyanax mexicanus*. Molecular Phylogenetics and Evolution 79, 368–374.

Cooper, S.J.B., Bradbury, J.H., Saint, K.M., Leys, R., Austin, A.D., Humphreys, W.F., 2007. Subterranean archipelago in the Australian arid zone: mitochondrial DNA phylogeography of amphipods from central Western Australia. Molecular Ecology 16, 1533–1544.

Copilas-Ciocianu, D., Fišer, C., Borza, P., Petrusek, A., 2018. Is subterranean lifestyle reversible? Independent and recent large-scale dispersal into surface waters by two species of the groundwater amphipod genus *Niphargus*. Molecular Phylogenetics and Evolution 119, 37–49.

Culver, D.C., Pipan, T., 2019. The Biology of Caves and Other Subterranean Habitats. Oxford University Press, Oxford, UK.

Culver, D.C., Pipan, T., 2012. Convergence and divergence in the subterranean realm: a reassessment. Biological Journal of the Linnean Society 107 (1), 1–14.

Culver, D.C., Pipan, T., 2014. Shallow Subterranean Habitats. Ecology, Evolution and Conservation. Oxford University Press, Oxford, UK.

Culver, D.C., Pipan, T., 2015. Shifting paradigms of the evolution of cave life. Acta Carsologica 44 (3), 415–425.

Culver, D.C., Fong, D.W., Jernigan, R.W., 1991. Species interactions in cave stream communities: experimental results and microdistribution effects. The American Midland Naturalist 126 (2), 364–379.

Culver, D.C., Kane, T.C., Fong, D.W., 1995. Adaptation and Natural Selection in Caves: The Evolution of *Gammarus minus*. Harvard University Press, Cambridge, MA.

Culver, D.C., Master, L.L., Christman, M.C., Hobbs III, H.H., 2000. Obligate cave fauna of the 48 contiguous United States. Conservation Biology 14, 386–401.

Culver, D.C., Pipan, T., Schneider, K., 2009. Vicariance, dispersal and scale in the aquatic subterranean fauna of karst regions. Freshwater Biology 54, 918–929.

Delić, T., Stoch, F., Borko, Š., Flot, J.-F., Fišer, C., 2020. How did subterranean amphipods cross the Adriatic Sea? Phylogenetic evidence for dispersal-vicariance interplay mediated by marine regression-transgression cycles. Journal of Biogeography 47, 1875–1887.

Delić, T., Trontelj, P., Zakšek, V., Brancelj, A., Simčič, T., Stoch, F., Fišer, C., 2022. Speciation of a subterranean amphipod on the glacier margins in South eastern Alps, Europe. Journal of Biogeography 49, 38–50.

Delić, T., Trontelj, P., Zakšek, V., Fišer, C., 2016. Biotic and abiotic determinants of appendage length evolution in a cave amphipod. Journal of Zoology 299 (1), 42–50.

Dobler, R., Kölliker, M., 2010. Kin-selected siblicide and cannibalism in the European earwig. Behavioral Ecology 21 (2), 257–263.

Doligez, B., Gustafsson, L., Pärt, T., 2009. 'Heritability' of dispersal propensity in a patchy population. Proceedings of the Royal Society B 276, 2829–2836.

Eme, D., Malard, F., Colson-Proch, C., Jean, P., Calvignac, S., Konecny-Dupré, L., Hervant, F., Douady, C.J., 2014. Integrating phylogeography, physiology and habitat modelling to explore species range determinants. Journal of Biogeography 41 (4), 687–699.

Eme, D., Malard, F., Konecny-Dupré, L., Lefébure, T., Douady, C.J., 2013. Bayesian phylogeographic inferences reveal contrasting colonization dynamics among European groundwater isopods. Molecular Ecology 22, 5685–5699.

Eme, D., Zagmajster, M., Delić, T., Fišer, C., Flot, J.-F., Konecny-Dupré, L., Pálsson, S., Stoch, F., Zakšek, V., Douady, C.J., Malard, F., 2018. Do cryptic species matter in macroecology? Sequencing European groundwater crustaceans yields smaller ranges but does not challenge biodiversity determinants. Ecography 41, 424–436.

Esmaeili-Rineh, S., Mamaghani-Shishvan, M., Fišer, C., Akmali, V., Najafi, N., 2020. Range sizes of groundwater amphipods (Crustacea) are not smaller than range sizes of surface amphipods: a case study from Iran. Contributions to Zoology 89 (1), 1–13.

Faille, A., 2021. Cave biogeography. In: Guilbert, E. (Ed.), Biogeography: An Integrative Approach of the Evolution of Living. ISTE Ltd and Hoboken, USA: John Wiley & Sons, Inc, London, UK, pp. 143–163.

Fišer, C., Coleman, C.O., Zagmajster, M., Zwittnig, B., Gerecke, R., Sket, B., 2010. Old museum samples and recent taxonomy: a taxonomic, biogeographic and conservation perspective of the *Niphargus tatrensis* species complex (Crustacea: Amphipoda). Organisms, Diversity and Evolution 10, 5–22.

Fišer, Ž., Novak, L., Luštrik, R., Fišer, C., 2016. Light triggers habitat choice of eyeless subterranean but not of eyed surface amphipods. Naturwissenschaften 103 (1–2), 7.

Foulquier, A., Malard, F., Lefébure, T., Douady, C.J., Gibert, J., 2008. The imprint of Quaternary glaciers on the present-day distribution of the obligate groundwater amphipod *Niphargus virei* (Niphargidae). Journal of Biogeography 35, 552–564.

Gandon, S., 1999. Kin competition, the cost of inbreeding and the evolution of dispersal. Journal of Theoretical Biology 200 (4), 345–364.

Gaston, K.J., 1998. Species-range size distributions: products of speciation, extinction and transformation. Philosophical Transactions of the Royal Society of London - B 353, 219–230.

Gibert, J., Vervier, P., Malard, F., Laurent, R., Reygrobellet, J.-L., 1994. Dynamics of communities and ecology of karst ecosystems: example of three karsts in eastern and southern France. In: Gibert, J., Danielopol, D.L., Stanford, J.A. (Eds.), Groundwater Ecology. Academic Press, San Diego, USA, pp. 425–450.

Ginet, R., 1960. Ecologie, éthologie et biologie de *Niphargus* (Amphipodes Gammarides hypogés). Annales de Speleologie 15, 1–254.

Greenwood, P.J., Harvey, P.H., 1982. The natal and breeding dispersal of birds. Annual Review of Ecology and Systematics 13, 1–21.

Guerra, P.A., 2011. Evaluating the life-history trade-off between dispersal capability and reproduction in wing dimorphic insects: a meta-analysis. Biological Reviews 86, 813–835.

Guzik, M.T., Cooper, S.J.B., Humphreys, W.F., Austin, A.D., 2009. Fine-scale comparative phylogeography of a sympatric sister species triplet of subterranean diving beetles from a single calcrete aquifer in Western Australia. Molecular Ecology 18, 3683–3698.

Guzik, M.T., Cooper, S.J.B., Humphreys, W.F., Ong, S., Kawakami, T., Austin, A.D., 2011. Evidence for population fragmentation within a subterranean aquatic habitat in the Western Australian desert. Heredity 107, 215–230.

Gyllenberg, M., Kisdi, É., Utz, M., 2008. Evolution of condition-dependent dispersal under kin competition. Journal of Mathematical Biology 57, 285–307.

Hamilton, W., May, R., 1977. Dispersal in stable habitats. Nature 269, 578–581.

Hargreaves, A.L., Eckert, C.G., 2014. Evolution of dispersal and mating systems along geographic gradients: implications for shifting ranges. Functional Ecology 28, 5–21.

Hastings, A., 1983. Can spatial variation alone lead to selection for dispersal? Theoretical Population Biology 24 (3), 244–251.

Heino, J., Melo, A.S., Siqueira, T., Soininen, J., Valanko, S., Bini, L.M., 2015. Metacommunity organisation, spatial extent and dispersal in aquatic systems: patterns, processes and prospects. Freshwater Biology 60, 845–869.

Henry, J.-P., 1976. Recherches sur les Asellidae hypogés de la lignée cavaticus, Crustacea, Isopoda, Asellota. PhD Thesis. University of Dijon, Dijon, France.

Hernández, D., Casane, D., Chevalier-Monteagudo, P., Bernatchez, L., García-Machado, E., 2016. Go west: a one way stepping-stone dispersion model for the cavefish *Lucifuga dentata* in Western Cuba. PLoS One 11 (4), e0153545.

Hervant, F., Malard, F., 2019. Adaptations: low oxygen. In: White, W.B., Culver, D.C., Pipan, T. (Eds.), Encyclopedia of Caves. Academic Press, London, UK, pp. 8–15.

Hervant, F., Renault, D., 2002. Long-term fasting and realimentation in hypogean and epigean isopods: a proposed adaptive strategy for groundwater organisms. Journal of Experimental Biology 205 (14), 2079–2087.

Hidalgo, J., Casas, R.R., Muñoz, M.Á., 2016. Environmental unpredictability and inbreeding depression select for mixed dispersal syndromes. BMC Evolutionary Biology 16, 71.

Hobson, K.A., 2005. Using stable isotopes to trace long-distance dispersal in birds and other taxa. Diversity and Distributions 11, 157–164.

Holt, R.D., McPeek, M.A., 1996. Chaotic population dynamics favors the evolution of dispersal. The American Naturalist 148 (4), 709–718.

Hughes, C.L., Hill, J.K., Dytham, C., 2003. Evolutionary trade-offs between reproduction and dispersal in populations at expanding range boundaries. Proceedings of the Royal Society B 270, S147–S150.

Jansson, R., Dynesius, M., 2002. The fate of clades in a world of recurrent climatic change: Milankovitch oscillations and evolution. Annual Review of Ecology and Systematics 33, 741–777.

Janzen, D.H., 1967. Why mountain passes are higher in the tropics. The American Naturalist 101, 233–249.

Jocque, M., Field, R., Brendonck, L., De Meester, L., 2010. Climatic control of dispersal-ecological specialization trade-offs: a metacommunity process at the heart of the latitudinal diversity gradient? Global Ecology and Biogeography 19, 244–252.

Jordan, S., Hand, B.K., Hotaling, S., Delvecchia, A.G., Malison, R., Nissley, C., Luikart, G., Stanford, J.A., 2020. Genomic data reveal similar genetic differentiation in aquifer species with different dispersal capabilities and life histories. Biological Journal of the Linnean Society 129 (2), 315–322.

Jugovic, J., Praprotnik, E., Buzan, E.V., Lužnik, M., 2015. Estimating population size of the cave shrimp *Troglocaris anophthalmus* (Crustacea, Decapoda, Caridea) using mark-release-recapture data. Animal Biodiversity and Conservation 38 (1), 77–86.

Kirkpatrick, M., Barton, N.H., 1997. Evolution of a species' range. The American Naturalist 150 (1), 1–23.

Kisdi, E., Utz, M., Gyllenberg, M., 2012. Evolution of condition-dependent dispersal. In: Clobert, J., Baguette, M., Benton, T.G., Bullock, J.M. (Eds.), Dispersal Ecology and Evolution. Oxford University Press, Oxford, UK, pp. 139–151.

Kissling, W.D., Pattemore, D.E., Hagen, M., 2014. Challenges and prospects in the telemetry of insects. Biological Reviews 89, 511–530.

Konec, M., Delić, T., Trontelj, P., 2016. DNA barcoding sheds light on hidden subterranean boundary between Adriatic and Danubian drainage basins. Ecohydrology 9, 1304–1312.

Kralj-Fišer, S., Premate, E., Copilas-Ciocianu, D., Volk, T., Fišer, Ž., Balázs, G., Herczeg, G., Delić, T., Fišer, C., 2020. The interplay between habitat use, morphology and locomotion in subterranean crustaceans of the genus *Niphargus*. Zoology 139, 125742.

Kubisch, A., Holt, R.D., Poethke, H.-J., Fronhofer, E.A., 2014. Where am I and why? Synthesizing range biology and the eco-evolutionary dynamics of dispersal. Oikos 123, 5–22.

Lefébure, T., Douady, C.J., Gouy, M., Trontelj, P., Briolay, J., Gibert, J., 2006. Phylogeography of a subterranean amphipod reveals cryptic diversity and dynamic evolution in extreme environments. Molecular Ecology 15, 1797–1806.

Lefébure, T., Morvan, C., Malard, F., François, C., Konecny-Dupré, L., Guéguen, L., Weiss-Gayet, M., Seguin-Orlando, A., Ermini, L., Der Sarkissian, C., Charrier, N.P., Eme, D., Mermillod-Blondin, F., Duret, L., Vieira, C., Orlando, L., Douady, C., 2017. Less effective selection leads to larger genomes. Genome Research 27, 1016–1028.

Leibold, M.A., Holyoak, M., Mouquet, N., Amarasekare, P., Chase, J.M., Hoopes, M.F., Holt, R.D., Shurin, J.B., Law, R., Tilman, D., Loreau, M., Gonzalez, A., 2004. The metacommunity concept: a framework for multi-scale community ecology. Ecology Letters 7, 601–613.

Lester, S.E., Ruttenberg, B.I., Gaines, S.D., Kinlan, B.P., 2007. The relationship between dispersal ability and geographic range size. Ecology Letters 10 (8), 745–758.

Li, X.-Y., Kokko, H., 2019. Sex-biased dispersal: a review of the theory. Biological Reviews 94, 721–736.

Little, C.J., Fronhofer, E.A., Altermatt, F., 2019. Dispersal syndromes can impact ecosystem functioning in spatially structured freshwater populations. Biology Letters 15, 20180865.

Luštrik, R., Turjak, M., Kralj-Fišer, S., Fišer, C., 2011. Coexistence of surface and cave amphipods in an ecotone environment. Contributions to Zoology 80 (2), 133–141.

Magniez, G., 1975. Observations sur la biologie de Stenasellus virei (Crustacea Isopoda Asellota des eaux souterraines). International Journal of Speleology 7, 79–228.

Malard, F., 2022. Groundwater metazoans. In: Mehner, T., Tockner, K. (Eds.), Encyclopedia of Inland Waters. Elsevier, pp. 474–487.

Malard, F., Capderrey, C., Churcheward, B., Eme, D., Kaufmann, B., Konecny-Dupré, L., Léna, J.-P., Liébault, F., Douady, C.J., 2017. Geomorphic influence on intraspecific genetic differentiation and diversity along hyporheic corridors. Freshwater Biology 62, 1955–1970.

Malard, F., Grison, P., Duchemin, L., Ferrer, M., Lewis, J., Konecny-Dupré, L., Lefébure, T., Douady, C.J., 2022. An introduction to the world Asellidae database. Proceedings of the 18th international congress of speleology, chambéry, fance. Karstologia - Memoires 21 (1), 1–4.

Malard, F., Turquin, M.-J., Magniez, G., 1997. Filter effect of karstic spring ecotones on the population structure of the hypogean amphipod Niphargus virei. In: Gibert, J., Mathieu, J., Fournier, F. (Eds.), Groundwater/Surface Water Ecotones: Biological and Hydrological Interactions and Management Options. Cambridge University Press, Cambridge, UK, pp. 42–50.

Mammola, S., Piano, E., Malard, F., Vernon, P., Isaia, M., 2019a. Extending Janzen's hypothesis to temperate regions: a test using subterranean ecosystems. Functional Ecology 33 (9), 1638–1650.

Mammola, S., Piano, E., Cardoso, P., Vernon, P., Domínguez-Villar, D., Culver, D.C., Isaia, M., 2019b. Climate change going deep: the effects of global climatic alterations on cave ecosystems. The Anthropocene Review 6 (1–2), 98–116.

Manenti, R., Piazza, B., 2021. Between darkness and light: spring habitats provide new perspectives for modern researchers on groundwater biology. PeerJ 9, e11711.

Matthysen, E., 2012. Multicausality of dispersal: a review. In: Clobert, J., Baguette, M., Benton, T.G., Bullock, J.M. (Eds.), Dispersal Ecology and Evolution. Oxford University Press, Oxford, UK, pp. 3–18.

Niemiller, M.L., Taylor, S.J., Slay, M.E., Hobbs, H.H., 2019. Biodiversity in the United States and Canada. In: White, W., Culver, D.C., Pipan, T. (Eds.), Encyclopedia of Caves. Academic Press, Amsterdam, the Netherland, pp. 163–176.

Nishimura, K., Isoda, Y., 2004. Evolution of cannibalism: referring to costs of cannibalism. Journal of Theoretical Biology 226 (3), 293–302.

Nonaka, E., Sirén, J., Somervuo, P., Ruokolainen, L., Ovaskainen, O., Hanski, I., 2019. Scaling up the effects of inbreeding depression from individuals to metapopulations. Journal of Animal Ecology 88, 1202–1214.

Palandačić, A., Matschiner, M., Zupančič, P., Snoj, A., 2012. Fish migrate underground: the example of Delminichthys adspersus (Cyprinidae). Molecular Ecology 21, 1658–1671.

Parvinen, K., Ohtsuki, H., Wakano, J.Y., 2020. Evolution of dispersal in a spatially heterogeneous population with finite patch sizes. Proceedings of the National Academy of Sciences of the United States of America 117 (13), 7290–7295.

Parzefall, J., Trajano, E., 2010. Behavioral patterns in subterranean fishes. In: Trajano, E., Bichuette, M.E., Kapoor, B.G. (Eds.), Biology of Subterranean Fishes. Science Publishers, Enfield, New Hampshire, pp. 81–114.

Pipan, T., Culver, D.C., 2017. The unity and diversity of the subterranean realm with respect to invertebrate body size. Journal of Cave and Karst Studies 79 (1), 1–9.

Pipan, T., Christman, M.C., Culver, D.C., 2006. Dynamics of epikarst communities: microgeographic pattern and environmental determinants of epikarst copepods in Organ Cave, West Virginia. The American Midland Naturalist 156 (1), 75–87.

Rohde, K., 1996. Rapoport's rule is a local phenomenon and cannot explain latitudinal gradients in species diversity. Biodiversity Letters 3, 10–13.

Ronce, O., Clobert, J., 2012. Dispersal syndromes. In: Clobert, J., Baguette, M., Benton, T.G., Bullock, J.M. (Eds.), Dispersal Ecology and Evolution. Oxford University Press, Oxford, UK, pp. 119–138.

Ronce, O., 2007. How does it feel to be like a rolling stone? Ten questions about dispersal evolution. Annual Review of Ecology, Evolution, and Systematics 38 (1), 231–253.

Rudolf, V.H., Kamo, M., Boots, M., 2010. Cannibals in space: the coevolution of cannibalism and dispersal in spatially structured populations. The American Naturalist 175 (5), 513–524.

Saastamoinen, M., Bocedi, G., Cote, J., Legrand, D., Guillaume, F., Wheat, C.W., Fronhofer, E.A., Garcia, C., Henry, R., Husby, A., Baguette, M., Bonte, D., Coulon, A., Kokko, H., Matthysen, E., Niitepõld, K., Nonaka, E., Stevens, V.M., Travis, J.M.J., Donohue, K., Bullock, J.M., Del Mar Delgado, M., 2018. Genetics of dispersal. Biological Reviews 93 (1), 574–599.

Saccheri, I., Kuussaari, M., Kankare, M., Vikman, P., Fortelius, W., Hanski, I., 1998. Inbreeding and extinction in a butterfly metapopulation. Nature 392, 491–494.

Saccò, M., Guzik, M.T., van der Heyde, M., Nevill, P., Cooper, S.J.B., Austin, A.D., Coates, P.J., Allentoft, M.E., White, N.E., 2022. eDNA in subterranean ecosystems: Applications, technical aspects, and future prospects. Science of the Total Environment 820, 153223.

Sarremejane, R., Cid, N., Stubbington, R., Datry, T., Alp, M., Cañedo-Argüelles, M., Cordero-Rivera, A., Csabai, Z., Gutiérrez-Cánovas, C., Heino, J., Forcellini, M., Millán, A., Paillex, A., Pařil, P., Polášek, M., de Figueroa, J.M.T., Usseglio-Polatera, P., Zamora-Muñoz, C., Bonada, N., 2020. DISPERSE, a trait database to assess the dispersal potential of European aquatic macroinvertebrates. Scientific Data 7, 386.

Sbordoni, V., Allegrucci, G., Cesaroni, D., 2012. Population structure. In: White, W.B., Culver, D.C. (Eds.), Encyclopedia of Caves, Second edition. Academic/Elsevier Press, Amsterdam, the Netherlands, pp. 608–618.

Schmidt-Nielsen, K., 1972. Locomotion: energy cost of swimming, flying, and running. Science 177 (4045), 222–228.

Sheth, S.N., Morueta-Holme, N., Angert, A.L., 2020. Determinants of geographic range size in plants. New Phytologist 226, 650–665.

Simčič, T., Sket, B., 2019. Comparison of some epigean and troglobitic animals regarding their metabolism intensity. Examination of a classical assertion. International Journal of Speleology 48, 133–144.

Starrfelt, J., Kokko, H., 2012. The theory of dispersal under multiple influences. In: Clobert, J., Baguette, M., Benton, T.G., Bullock, J.M. (Eds.), Dispersal Ecology and Evolution. Oxford University Press, Oxford, UK, pp. 19–28.

Stern, D.B., Breinholt, J., Pedraza-Lara, C., López-Mejía, M., Owen, C.L., Bracken-Grissom, H., Fetzner, J.W., Crandall, K.A., 2017. Phylogenetic evidence from freshwater crayfishes that cave adaptation is not an evolutionary dead-end. Evolution 71, 2522–2532.

Stevens, G.C., 1989. The latitudinal gradient in geographical range: how so many species coexist in the tropics. The American Naturalist 133, 240–256.

Stevens, V.M., Pavoine, S., Baguette, M., 2010. Variation within and between closely related species uncovers high intra-specific variability in dispersal. PLoS One 5 (6), e11123.

Stevens, V.M., Trochet, A., Van Dyck, H., Clobert, J., Baguette, M., 2012. How is dispersal integrated in life histories: a quantitative analysis using butterflies. Ecology Letters 15, 74–86.

Stevens, V.M., Whitmee, S., Le Galliard, J.-F., Clobert, J., Böhning-Gaese, K., Bonte, D., Brändle, M., Dehling, D.M., Hof, C., Trochet, A., Baguette, M., 2014. A comparative analysis of dispersal syndromes in terrestrial and semi-terrestrial animals. Ecology Letters 17, 1039–1052.

Strecker, U., Hausdorf, B., Wilkens, H., 2012. Parallel speciation in Astyanax cave fish (Teleostei) in Northern Mexico. Molecular Phylogenetics and Evolution 62 (1), 62–70.

Trajano, E., 2001. Ecology of subterranean fishes: an overview. Environmental Biology of Fishes 62, 133–160.

Trochet, A., Courtois, E.A., Stevens, V.M., Baguette, M., Chaine, A., Schmeller, D.S., Clobert, J., 2016. Evolution of sex-biased dispersal. The Quarterly Review of Biology 91 (3), 297–320.

Trontelj, P., 2018. Structure and genetics of cave populations. In: Moldovan, O.T., Kováč, L., Halse, S. (Eds.), Cave Ecology. Springer, Cham, Switzerland, pp. 269–295.

Trontelj, P., 2019. Vicariance and dispersal in caves. In: White, W., Culver, D., Pipan, T. (Eds.), Encyclopedia of Caves. Elsevier, Academic Press, London, pp. 1103–1109.

Trontelj, P., Douady, C.J., Fišer, C., Gibert, J., Goricki, S., Lefébure, T., Sket, B., Zakšek, V., 2009. A molecular test for cryptic diversity in ground water: how large are the ranges of macro-stygobionts? Freshwater Biology 54, 727–744.

Turquin, M.-J., Barthélémy, D., 1985. The dynamics of a population of the troglobitic amphipod Niphargus virei Chevreux. Stygologia 1, 109–117.

Turquin, M.-J., 1981. Profil démographique et environnement chez une population de Niphargus virei (Amphipocle troglobie). Bulletin de la Societe Zoologique de France 106, 457–466.

Van den Berg-Stein, S., Hahn, H.J., Thielsch, A., Schwenk, K., 2022. Diversity and dispersal of aquatic invertebrate species from surface and groundwater: development and application of microsatellite markers for the detection of hydrological exchange processes. Water Research 210, 117956.

Verovnik, R., Sket, B., Trontelj, P., 2004. Phylogeography of subterranean and surface populations of water lice *Asellus aquaticus* (Crustacea: isopoda). Molecular Ecology 13, 1519—1532.

Ward, J.V., Palmer, M.A., 1994. Distribution patterns of interstitial meiofauna aver a range of spatial scales, with emphasis on alluvial river-aquifer systems. Hydrobiologia 287, 147—156.

Weber, D., Flot, J.-F., Weigand, H., Weigand, A.M., 2020. Demographic history, range size and habitat preferences of the groundwater amphipod *Niphargus puteanus* (C.L. Koch in Panzer, 1836). Limnologica 82, 125765.

Zagmajster, M., Eme, D., Fišer, C., Galassi, D., Marmonier, P., Stoch, F., Cornu, J.F., Malard, F., 2014. Geographic variation in range size and beta diversity of groundwater crustaceans: insights from habitats with low thermal seasonality. Global Ecology and Biogeography 23 (10), 1135—1145.

Zakšek, V., Delić, T., Fišer, C., Jalžić, B., Trontelj, P., 2019. Emergence of sympatry in a radiation of subterranean amphipods. Journal of Biogeography 46, 657—669.

Zakšek, V., Sket, B., Gottstein, S., Franjević, D., Trontelj, P., 2009. The limits of cryptic diversity in groundwater: phylogeography of the cave shrimp *Troglocaris anophthalmus* (Crustacea: Decapoda: Atyidae). Molecular Ecology 18, 931—946.

Zeller, K.A., McGarigal, K., Whiteley, A.R., 2012. Estimating landscape resistance to movement: a review. Landscape Ecology 27, 777—797.

# Roles of organisms in groundwater

# Microbial diversity and processes in groundwater

*Lucas Fillinger[1], Christian Griebler[1], Jennifer Hellal[2],
Catherine Joulian[2] and Louise Weaver[3]*

[1]University of Vienna, Department of Functional & Evolutionary Ecology, Vienna, Austria;
[2]BRGM, DEPA, Geomicrobiology and Environmental Monitoring Unit, Orléans, France;
[3]Institute of Environmental Science and Research (ESR) Christchurch, Canterbury, New Zealand

## Introduction

Aquifers are one of the largest habitats for microorganisms on Earth, second probably only to marine environments, accommodating about 25% (i.e., $\sim 3 \times 10^{29}$ cells) of the total global prokaryotic biomass (Griebler and Lueders, 2009; McMahon and Parnell, 2014; Magnabosco et al., 2018; Flemming and Wuertz, 2019). The activity, physiology, and diversity of these microorganisms is largely determined by environmental conditions characteristic for groundwater environments. These conditions are hallmarked by the lack of light, low concentrations of dissolved organic matter (DOM) and other essential nutrients like phosphorus, as well as relatively constant temperatures (Griebler and Lueders, 2009). As the lack of light excludes primary production via photosynthesis, inputs from soils and surface waters constitute a major source of DOM in groundwater. A large fraction of this DOM is being degraded as it travels through overlaying zones before reaching the groundwater table, causing depletion of mainly labile substrates (Hofmann and Griebler, 2018). Hence, the DOM pool in groundwater consists for a large part of stable, recalcitrant compounds of which often only a small fraction ($\leq 1\%$) in the range of a few micrograms per liter is readily bioavailable to microorganisms (Goldscheider et al., 2006; Gooddy and Hinsby, 2009; Shen et al., 2015). As a result, microbial cell densities in groundwater are lower compared to those found in surface waters and in soils, typically ranging between $10^6$ and $10^9$ cells $L^{-1}$ groundwater, and $10^7$ to $10^{11}$ cells $dm^{-3}$ sediment (Griebler and Lueders, 2009).

Microbial communities in groundwater are dominated by heterotrophic and chemolithoautotrophic bacteria and archaea that are well adapted to living under such nutrient-poor conditions. On a physiological level, important features are low maintenance energy

*Groundwater Ecology and Evolution, Second Edition*
https://doi.org/10.1016/B978-0-12-819119-4.00009-3

requirements, high substrate use efficiencies, mixed-substrate utilization, and the ability to take up substrates as soon as they become available by maintaining a constant basal expression of multiple metabolic pathways at once to sustain cell maintenance (Lin et al., 2009; Egli, 2010; Castelle et al., 2013; Marozava et al., 2014). Engaging in metabolic interactions between different species and functional groups on a community level is another key factor that allows microorganisms to exploit scarce nutrients more fully and efficiently by sharing metabolic workloads. Thus, metabolic interactions play an essential role for driving biogeochemical cycles, the attenuation of contaminants, and for sustaining diverse microbial communities in groundwater despite the low nutrient levels (Kantor et al., 2013; Morris et al., 2013; Wrighton et al., 2014; Luef et al., 2015; Anantharaman et al., 2016; Lau et al., 2016; Pacheco et al., 2019). Therefore, mapping and understanding these interactions is an important step toward explaining microbial diversity and biogeochemical fluxes in groundwater ecosystems (Thullner and Regnier, 2019).

Aquifers are open systems that are connected to the surface. In general, groundwater microbial communities are composed of bacteria that are typical for freshwater environments (except for photoautotrophs), including *Proteobacteria, Acidobacteria, Actinobacteria, Bacteroidetes, Chloroflexi, Cytophagales, Firmicutes, Gemmatimonadetes, Nitrospira, Planctomycetes, Verrucomicrobia,* and to a lesser extent archaea like *Thaumarchaeota, Nanoarchaeota,* and *Methanosarcina* or *Methanosaeta* gaining importance in anoxic groundwater (Griebler and Lueders, 2009; Akob and Küsel, 2011; Fillinger et al., 2019a). Overall, there appears to be no evidence for the existence of genuine stygobiontic or endemic groundwater species among prokaryotes to date. However, the advent of modern metagenomic approaches has recently enabled the discovery of several entirely novel bacterial and archaeal phyla not just in the deep subsurface (Probst et al., 2018), but even in shallow groundwater at a depth of less than 10 meters below the land surface (Wrighton et al., 2012; Castelle et al., 2013; Wrighton et al., 2014; Castelle et al., 2015; Anantharaman et al., 2016), suggesting that we have just started to explore the vast diversity of groundwater microbiomes. These phyla were members of the bacterial Candidate Phyla Radiation (CPR) and the archaeal DPANN radiation, which represent major branches in the tree of life that so far have largely eluded cultivation. A widely observed feature of these organisms are exceptionally small cell and genome sizes (Castelle et al., 2018). These features have been proposed to enhance the uptake of scarce substrates due to an increased cell surface-to-volume ratio, streamline gene expression, and reduce energy requirements for genome replication and maintenance, which could be part of the explanation for their considerable abundances in oligotrophic groundwater environments (Kantor et al., 2013; Luef et al., 2015; Herrmann et al., 2019).

While similar microbial lineages may be found across habitats, the transition in microbial community composition between groundwater and overlaying soil zones or surface waters is determined not only by passive transport of cells, but to a strong extent by environmental conditions that select for distinct suites of microorganisms that are able to thrive and compete in groundwater environments (Graham et al., 2017; Herrmann et al., 2019). Consequently, microbial community composition in groundwater can differ strongly compared to other, non-subsurface habitats (Griebler and Lueders, 2009; Smith et al., 2012). Moreover, Taubert et al. (2019) showed that ecosystem functions like the transformation of certain DOM compounds are carried out by distinct lineages in groundwater compared to overlaying surface zones. In the deep terrestrial subsurface at depths of hundreds of meters, there can additionally be considerable time lags in the exchange with surface ecosystems over thousands and even millions of years. This may be considered as de facto isolation of the groundwater through which speciation and evolutionary drift can become significant contributors to differences in microbial community composition across habitats (van Waasbergen et al., 2000; Griebler et al., 2014; Lau et al., 2014).

Spatial heterogeneity is another characteristic feature of aquifers that has a determining effect on microbial community composition. Large variations in hydrochemical conditions, mineralogy, and hydraulic conductivity can exist across different vertical and horizontal zones of an aquifer. Species sorting along these environmental gradients and dispersal limitation can result in significant differences in microbial community composition between these zones (Grösbacher et al., 2016; Fillinger et al., 2019a; Yan et al., 2020), which in turn can bring about a large species richness on the aquifer scale. The interplay between ecological processes like species sorting, dispersal, and drift not only determines microbial community composition, but can also affect community functioning as well as resistance to changing environmental conditions and perturbations (Graham and Stegen, 2017; Stegen et al., 2018a).

Perturbations resulting from anthropogenic activities are among the most serious threats to groundwater ecosystems. There is a long record of studies devoted to microbial community composition in groundwater contaminated with widespread pollutants like petroleum hydrocarbons or chlorinated organic solvents, and mechanisms related to the natural attenuation of these compounds (Lueders, 2017; Nijenhuis et al., 2018; Weatherill et al., 2018). However, fewer studies have investigated how communities in pristine systems respond to sudden contamination events and their ability to recover from such perturbations (Zhou et al., 2014; Herzyk et al., 2017). In addition to these so-called legacy pollutants, groundwater contamination with emerging pollutants like pharmaceuticals, pesticides, and other industrial compounds has become an increasing concern (Lapworth et al., 2012; Fenner et al., 2013; Careghini et al., 2015). Even though these compounds typically occur only at low concentrations in the range of micro-to nanograms per liter, they are widespread and often highly persistent in groundwater, and the interaction of microbial communities with these micropollutants is not yet fully understood (Helbling, 2015; Mauffret et al., 2017).

In this chapter we will discuss the factors that determine the diversity and composition of microbial communities in groundwater, their role in key biogeochemical processes, including the natural attenuation of contaminants and rate-limiting steps therein, as well as their resistance and resilience to perturbations.

## Ecological processes determining microbial community diversity and composition

### Importance of environmental conditions, dispersal, and species interactions

Historically, microbial community composition was assumed to be virtually entirely determined by local environmental conditions, which were supposed to select for specific microorganisms with traits that allow them to colonize a given environment while excluding others that lack the necessary traits (also referred to as species sorting or environmental filtering). Dispersal limitation and drift due to random birth-death events were considered unimportant, because the small body size, large population size, and fast growth of microorganisms were assumed to facilitate ubiquitous dispersal and prevent local extinction. Today we know that this is not the case (Nemergut et al., 2013; Zhou and Ning, 2017; Langenheder and Lindström, 2019). Although selection by environmental conditions certainly is important (Hanson et al., 2012; Lindström and Langenheder, 2012), it is now widely accepted that microorganisms in the environment can indeed be subject to dispersal limitation, and that the importance of species sorting relative to dispersal, dispersal limitation, and drift largely depends on the spatial scale at which communities are being investigated, environmental heterogeneity, and habitat properties (Langenheder and Lindström, 2019).

In general, groundwater hydrochemistry has been shown to be one of the main factors that determine microbial community composition. For instance, comparisons of microbial communities from shallow porous aquifers located in distinct catchment areas indicated that differences in community composition were mainly caused by species sorting, primarily driven by differences in groundwater pH, and concentrations of dissolved oxygen and organic carbon. Dispersal, dispersal limitation, and drift within as well as across aquifers only played a secondary role (Fillinger et al., 2019a). Similarly, Ben Maamar et al. (2015) showed that differences in microbial community composition between three unconnected fractured aquifers were correlated with differences in concentrations of oxygen, nitrate, sulfate, and total iron in the groundwater. Moreover, groundwater temperature, nitrogen availability, and carbon pool composition have been reported important factors that shape microbial community composition (Shabarova et al., 2014; Graham et al., 2017; Stegen et al., 2018b; Wu et al., 2018; Benk et al., 2019; Yan et al., 2020).

However, microbial community composition can be uncoupled from physicochemical conditions by aquifer features that determine species dispersal. Coarse sediments in porous aquifers in combination with high groundwater flow velocities can allow for high dispersal rates, causing homogenization of local communities (Stegen et al., 2013). This may enable taxa to occur even under non-optimal conditions and thus can reduce the importance of species sorting. On the other hand, fine-grained sediments may cause dispersal limitation, which no longer permits microorganisms to be transported along environmental gradients and reach locations with their preferred conditions. Together with random drift, this can cause communities in different aquifer zones to diverge over time (Stegen et al., 2013). Likewise, complex heterogeneous groundwater flow paths can prevent dispersal between spatially adjacent communities (Stegen et al., 2013), which may be a critical factor especially in fractured aquifers (Beaton et al., 2016; Yan et al., 2020). In karst systems, the interplay between species sorting and dispersal can be particularly dynamic. In times of high discharge, communities may be homogenized across groundwater pools, which otherwise are stagnant and hydrologically separated for extensive time periods. During these stagnant periods, dispersal of taxa is limited and differences in community composition between groundwater pools may then develop due to drift, priority effects determined by the order and time of arrival of taxa, or species sorting for example due to changes in DOM composition within pools (Shabarova et al., 2013, 2014; Fukami, 2015; Svoboda et al., 2018; Wu et al., 2018).

Whereas hydrochemistry and the physical structure of the aquifer together determine the distribution of microorganisms across aquifer zones at the scale of several meters to kilometers, biotic interactions between species play an important role for the co-occurrence of taxa at smaller scales. Essentially, these interactions can be negative, such that the presence of some species excludes the presence of others, for example through resource competition, or positive when one or more species facilitate the presence of others for instance by exchanging metabolites (Faust and Raes, 2012). In the light of the nutrient-poor conditions that typically prevail in groundwater, sharing scarce resources may seem like an ill-advised strategy at first glance. However, it has been shown that microorganisms can secrete and share compounds including lipids, protein building blocks, and metabolic intermediates in oligotrophic environments without creating fitness disadvantages for themselves (Pacheco et al., 2019). Especially the characteristically small genomes of CPR and DPANN organisms often lack synthesis pathways for essential metabolites including certain nucleotides, lipids, amino acids, and cofactors, making these organisms depend on interactions with others to grow (Kantor et al., 2013; Luef et al., 2015; Castelle et al., 2017, 2018; Geesink et al., 2020). Furthermore, different species can cooperate syntrophically via metabolic handoffs to consume substrates that neither species could completely metabolize on its own, either because of energetic constraints of the reaction, or because a given species lacks the enzymes for the complete metabolic pathway (Morris et al., 2013; Anantharaman et al., 2016; Lau et al., 2016). Thus, whereas competition certainly affects microbial community composition, cooperation between species could be equally important in nutrient-poor groundwater systems to sustain diverse microbial communities (Zelezniak et al., 2015; Geesink et al., 2018; Pacheco et al., 2019).

Protozoan grazing and viral infections are additional important biotic factors that can affect microbial community composition, although those food web interactions in groundwater ecosystems are still little understood (Karwautz et al., 2022; Schweichhart et al., 2022). Grazing and viral lysis can exert top-down control by directly causing changes in microbial abundances, as well as bottom-up effects by releasing labile organic substrates from cells, which in turn can stimulate the growth of other members of the microbial community. The presence and activity of protozoans and viruses, as well as the extent to which they are affected by hydrochemical conditions, dispersal, and dispersal limitation can further add to the variation in microbial community composition across aquifer zones (Nagaosa et al., 2008; Eydal et al., 2009; Longnecker et al., 2009; Holmes et al., 2013; Pan et al., 2014; Daly et al., 2019).

## Differences between planktonic and surface-attached microbial communities

The majority of microbial cells in groundwater environments live attached to rock surfaces and sediment particles (Alfreider et al., 1997; Lehman et al., 2001; Griebler et al., 2002; Zhou et al., 2012; Flemming and Wuertz, 2019). This attached mode of life can have several advantages over a planktonic lifestyle, for example by providing access to limiting nutrients adsorbed to surfaces or contained in minerals (Griebler and Lueders, 2009; Smith et al., 2018; Flemming and Wuertz, 2019). Accordingly, microbial cell density and bulk activity of attached communities can exceed those of planktonic communities by one to four orders of magnitude (Alfreider et al., 1997; Lehman et al., 2001; Griebler et al., 2002; Zhou et al., 2012; Flemming and Wuertz, 2019). Unlike biofilms in other aquatic environments, which form dense, spatially coherent and heterogeneous structures with a thickness of several hundred micrometers (Battin et al., 2016), surface-attached communities in aquifers typically occur as patchily distributed microcolonies that consist of only a few cells (Wanger et al., 2006; Griebler and Lueders, 2009; Anneser et al., 2010; Schmidt et al., 2017; Smith et al., 2018; Flemming and Wuertz, 2019).

The composition of attached communities can differ substantially from planktonic communities in the surrounding groundwater (Zhou et al., 2012; Flynn et al., 2013; Hug et al., 2015). While planktonic communities are the seed bank for the colonization of surfaces (Griebler et al., 2014), the assembly of attached communities is not governed purely by random colonization events, where cells stochastically attach to surfaces, and subsequently form communities that diverge from planktonic communities over time through drift. Instead, there is evidence suggesting that those attached communities assemble deterministically via species sorting (Graham et al., 2016, 2017; Fillinger et al., 2019b). In contrast to planktonic communities, however, this selection appears to be less determined by hydrochemistry, but more by the mineral composition and surface properties of sediments (Graham et al., 2016; Grösbacher et al., 2016; Stegen et al., 2016; Jones and Bennett, 2017). Additionally, biotic interactions between taxa already established on a surface and newly-arriving propagules from the planktonic species pool can be expected to influence community assembly on sediment particles (Fillinger et al., 2019b), similar to biofilms in nonsubsurface aquatic habitats (Battin et al., 2007, 2016).

In terms of community functions, the differences between surface-attached and planktonic communities are less well understood (Griebler et al., 2022). Indications for functional redundancy come from metagenome-based reconstructions of more than 2000 individual draft genomes from a shallow porous aquifer. Here, abundances of genomes encoding metabolic pathways related to the cycling of carbon, sulfur, nitrogen, hydrogen, and iron only displayed relatively small differences between the two types of communities (Anantharaman et al., 2016). Therefore, sediments might act as a vault for key ecosystem functions that could rescue functions in aquifer zones following disturbances via detachment and subsequent dispersal of cells (Griebler et al., 2014). On the other hand, functional differences have been found for communities in a hydrocarbon-contaminated aquifer at a former coal-gasification site. Here, the degradation of monoaromatic hydrocarbons and sulfate reduction could be inferred from the analysis of groundwater samples, whereas degradation of polyaromatics, iron reduction, and sulfide oxidation were mainly associated with sediments (Anneser et al., 2010). Hence, we may assume that differences in the functional potential between sediment-attached and planktonic communities are to some extent determined by the overall hydrochemical conditions of the aquifer, and the tendency of electron donors and acceptors to preferentially either adsorb to sediments or dissolve in the groundwater. Moreover, Flynn et al. (2013) reported noticeably higher abundances of methanogenic archaea related to Methanosarcinaceae and Methanosaetaceae in sediment-attached communities, while nonmethanogenic *Thaumarchaeota* were mainly suspended in the groundwater. These findings suggest that attached and planktonic communities are not generally functionally redundant, and that the selection of taxa during the assembly of surface-attached communities can indeed have consequences also on a functional level (Griebler et al., 2022).

## Implications of assembly processes for community functioning and knowledge gaps

The mechanisms underlying microbial community assembly can have important implications for community functioning. Graham and Stegen (2017) showed using in silico simulations that strong dispersal can reduce biogeochemical process rates by increasing the number of maladapted taxa in a local community. It was proposed that these maladapted organisms have to invest more energy into maintenance to cope with the nonoptimal conditions, which in turn leaves less energy for the synthesis of enzymes that catalyze biogeochemical reactions. On the other hand, a certain amount of dispersal is required to maintain functional diversity and buffer impacts of environmental changes (Shade et al., 2012; Zha et al., 2016; Graham and Stegen, 2017). However, discerning the exact influence of species sorting and dispersal on groundwater microbial communities is still a challenge. Although studies have found that species sorting imposed by hydrochemistry generally explains more variation in microbial community composition than dispersal represented by the spatial distance between communities, a large fraction of the variation ranging between 25% and 50% often remained unexplained (Beaton et al., 2016; Fillinger et al., 2019a; Yan et al., 2020). While a portion of this unexplained variation may be simply due to random drift, it may also contain undetected effects of species sorting by unmeasured factors like species interactions (Konopka et al., 2015), as well as variation caused by grazing and viral lysis (Nagaosa et al., 2008; Eydal et al., 2009; Longnecker et al., 2009; Holmes et al., 2013; Pan et al., 2014). Moreover, actual dispersal routes for microorganisms are often unknown due to heterogeneous and dynamic groundwater flow paths, which may further add to the unexplained variation (Fillinger et al., 2019a). Thus, combining microbial community analyses with transport models would be important to better understand the influence of dispersal on microbial community assembly in aquifer systems.

Additional critical complications are inherent to the methods that are commonly used to study microbial communities, which are usually based on the analysis of bulk groundwater samples pumped from monitoring wells. These bulk samples homogenize groundwater from different zones surrounding a well, thereby making it impossible to discern the original spatial distribution of organisms found in those samples. Consequently, bulk samples obscure potentially important differences in physicochemical conditions and hydraulic connectivity on the pore-scale where microorganisms actually disperse, interact, and ultimately assemble into communities (Schmidt et al., 2017; Thullner and Regnier, 2019). Especially at these very small scales in the range of micro-to millimeters, factors like species interactions may be even more relevant to community assembly than for example the physicochemical parameters discussed above (Cordero and Datta, 2016; Langenheder and Lindström, 2019). Therefore, designing new sampling strategies (e.g., see Anneser et al., 2008) and experiments to investigate the importance of pore-scale processes on microbial community assembly will be a key task for future research (Griebler et al., 2022).

## Microbial communities and biogeochemical cycles

## The role of microbial networks in the cycling of DOM, nitrogen, and other elements

Microbial communities are the main catalysts of biogeochemical processes that drive the cycling of carbon, nitrogen, and other nutrients in groundwater ecosystems (Griebler and Lueders, 2009; Thullner and Regnier, 2019). Instead of working in isolation, these cycles are intertwined in a network of different functional groups of respiring heterotrophs, chemolithoautotrophs, and fermenting microorganisms (Wrighton et al., 2014; Anantharaman et al., 2016) (Fig. 9.1). However, the exact function of individual members of a microbial community

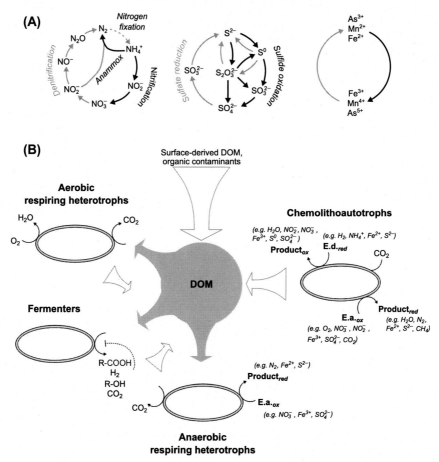

**FIGURE 9.1** (A) Key redox cycles in groundwater for nitrogen, sulfur, and metals (representatively shown for iron, manganese, and arsenic). Reduction of electron acceptors is represented by gray arrows; black arrows represent oxidation of electron donors. (B) Functional groups involved in the cycling of DOM linked to the reduction of oxidized electron acceptors ($E.a._{ox}$) and reduced electron donors ($E.d._{red}$) shown in A. Filled arrows represent consumption of DOM; open arrows represent replenishment of DOM from surface inputs as well as microbially produced compounds. Product inhibition of fermentation reactions is symbolized by the dashed line.

is often unknown since these communities are complex and several microorganisms are able to switch between metabolic modes or between different electron acceptors to adjust to changing redox conditions. Nevertheless, much progress has been made over the past decade, largely thanks to advances in culture-independent molecular tools and meta-omics techniques in particular. These methods have helped to unravel metabolic functions and activities of individual microorganisms, and how they integrate into microbial networks that ultimately drive biogeochemical cycles (Franzosa et al., 2015; Muller et al., 2018).

Members of the *Bacteroidetes*, *Actinobacteria*, *Acidobacteria*, *Chloroflexi*, and CPR have been found to play important roles in the hydrolysis of complex polymeric compounds like proteins, lipids, chitin, or DOM from surface-derived plant material such as cellulose into smaller monomeric molecules that are accessible to a broader group of other heterotrophic organisms (Wrighton et al., 2014; Taubert et al., 2019; Wegner et al., 2019). In the presence of oxygen, these monomeric compounds may then be readily oxidized to $CO_2$ via aerobic respiration. Under hypoxic and anoxic conditions, respiring organisms may depend on fermenters like those mentioned above, but also others like *Firmicutes, Spirochetes*, or members of the *Deltaproteobacteria* that further convert these compounds to small organic acids, alcohols, and hydrogen, which in turn can serve as electron donors for anaerobic respiration (Morris et al., 2013; Dong et al., 2018).

Based on individually reconstructed draft genomes from metagenomic data combined with metaproteomics, Wrighton et al. (2014) provided a comprehensive overview of a metabolic interaction network of fermenting and respiring microorganisms in an unconfined porous aquifer with low oxygen concentrations ($<0.1$ mg $L^{-1}$) that can be regarded as a representative example also for other aquifer systems. Here, organisms related to *Bacteroidetes, Chloroflexi*, as well as CPR organisms including *Cand. Parcubacteria* (formerly OD1) and *Cand. Microgenomates* (formerly OP11) were key players in the transformation of DOM, acting as hydrolyzers of polymeric substrates and fermenters with the ability to produce acetate, lactate, butyrate, formate, ethanol, and hydrogen. Respiring bacteria identified in the system responsible for nitrate reduction (Rhodocyclaceae and Comamonadaceae), iron reduction (*Geobacteraceae*), and sulfate reduction (*Desulfobulbaceae*) could only use a narrow range of carbon substrates, mainly restricted to small organic acids and ethanol that could be supplied by the fermenting organisms. Some of these respiring bacteria further showed signs of autotrophic growth with hydrogen or sulfide as electron donor.

The example above nicely illustrates how respiratory processes involved in the cycling of nitrogen, sulfate, and iron can be fueled by and even depend on the degradation of DOM by fermenting organisms. However, this dependency is not unidirectional. The fermentation of several organic compounds to hydrogen and small organic acids like acetate can be thermodynamically unfeasible, or may not yield sufficient energy to support growth, unless the concentrations of fermentation products are constantly kept low (Morris et al., 2013). Hence, fermenters often depend on respiring microorganisms as syntrophic partners to pull the fermentation forward by consuming fermentation products and thus keeping them below inhibitory concentrations (Morris et al., 2013). Therefore, the flux of DOM through a microbial community in anoxic groundwater is not necessarily controlled exclusively by the fermenting organisms that primarily catalyze the degradation. In fact, Röling et al. (2007) could demonstrate that while the control of organic matter degradation in anoxic environments mainly resides with fermenting organisms under nitrate- or iron-reducing conditions, it can shift toward hydrogen- and acetate-consuming organisms under less favorable redox conditions like sulfate-reducing or methanogenic conditions. Consequently, even autotrophic hydrogen-consuming sulfate reducers or methanogens can determine rates of DOM turnover without directly performing the degradation of organic compounds themselves.

## Chemolithoautotrophy in groundwater

Even though groundwater ecosystems are dominated by heterotrophic processes fueled by the degradation of DOM that ultimately derived from the surface (Foulquier et al., 2009; Baker et al., 2000), the contribution of carbon-fixing chemolithoautotrophs to biogeochemical cycles has started to be increasingly recognized. Studies on a fractured limestone aquifer suggested that up to 17% of the microorganisms relied on autotrophic growth via denitrification coupled to the oxidation of reduced sulfur compounds or hydrogen by bacteria related to *Sulfuricella*, *Thiobacillus*, *Dechloromonas*, and *Hydrogenophaga* under hypoxic and anoxic conditions (Herrmann et al., 2015; Kumar et al., 2018). On the other hand, aerobic ammonium oxidation (nitrification) by *Nitrosomonas*, *Nitrospirae*, and *Thaumarchaeota* was observed in zones of the aquifer where oxygen was present (Wegner et al., 2019). Moreover, sediment-attached microbial communities in a different porous aquifer were dominated by an autotrophic, sulfur- or hydrogen-oxidizing, denitrifying bacterium (*Cand. Sulfuricurvum*) that constituted almost half of the organisms in these communities (Handley et al., 2014). Apart from denitrification and nitrification processes, anaerobic ammonium oxidation (anammox) by *Planctomycetes* is a key process in the cycling of nitrogen, especially in highly oligotrophic groundwater where denitrification is limited by low levels of DOM. In such systems, anammox has been estimated to be responsible for up to 80% of the total nitrogen loss (Kumar et al., 2017; Wang et al., 2020).

Although abundances of chemolithoautotrophs in groundwater microbial communities detected on the DNA level can be lower compared to heterotrophs, they can contribute significantly to the total gene expression in a community (Jewell et al., 2016; Wegner et al., 2019). Therefore, the contribution of chemolithoautotrophs to biogeochemical cycles may be disproportionately larger than their mere abundance in a microbial community may suggest (Griebler et al., 2022). Importantly, this does not only apply to groundwater with low DOM concentrations, where autotrophic growth would be an attractive means for microorganisms to evade carbon limitation, if inorganic electron donors and acceptors were sufficiently available. Chemolithoautotrophs have also been detected and were highly active in groundwater contaminated with high loads of organic contaminants, or during field experiments involving the injection of acetate into the groundwater (Alfreider et al., 2009; Alfreider and Vogt, 2012; Kellermann et al., 2012; Jewell et al., 2016; Müller et al., 2016, 2020).

The quantitative relevance of $CO_2$ fixation as a source of organic carbon in groundwater is still not fully understood. Findings by Shen et al. (2015) suggested that up to 34% of the DOM in groundwater could be traced back to bacterial production. Unfortunately, they were unable to distinguish between compounds produced in situ and those possibly derived from the surface, and to differentiate between heterotrophically and autotrophically produced molecules. However, the considerable abundances of chemolithoautotrophs in the studies discussed above suggest that they could provide a substantial share of that fraction. Furthermore, an often overlooked process is the fixation of $CO_2$ by heterotrophs via carboxylation reactions involved in the central metabolism (Kellermann et al., 2012; Spona-Friedl et al., 2020; Braun et al., 2021). For instance, anaerobic hydrocarbon-degrading Peptococcaceae enriched from contaminated aquifers obtained as much as 50% of their biomass carbon from $CO_2$ during growth on stable isotope-labelled, carbon-rich substrates like benzene and toluene (Winderl et al., 2010; Taubert et al., 2012). Thus, although $CO_2$ fixation might quantitatively be less important for the total DOM pool in groundwater compared to surface-derived inputs, it could still make a considerable contribution to the overall carbon flow in these ecosystems. Therefore, a more detailed understanding of these processes would be critical in order to better understand the role of groundwater as source and sink of carbon.

# Toward predictive models of microbial communities and biogeochemical fluxes

The ability to reconstruct individual high-quality genomes from metagenomic data as well as from sorted single cells has yielded unprecedented insights into the metabolic potential of uncultured microorganisms in groundwater environments (Wrighton et al., 2012; Castelle et al., 2013; Kantor et al., 2013; Rinke et al., 2013; Handley et al., 2014; Wrighton et al., 2014; Castelle et al., 2015; Castelle et al., 2018; Probst et al., 2018). Reconstructing the metabolism of individual community members can give indications of the functions of these organisms in the community, and how they may interact with each other (Wrighton et al., 2014; Anantharaman et al., 2016; Castelle et al., 2018). Apart from enabling qualitative statements about exchanged metabolites based on the potential for substrate use and product formation encoded in individual genomes, this information could further provide the basis for genome-scale models to make quantitative predictions of fluxes through those metabolic networks (Magnúsdóttir et al., 2017).

Flux balance analysis (FBA) is a widely used method for constrained-based genome-scale metabolic modeling. These models are mathematical representations of all known metabolic reactions encoded in a genome in the form of a stoichiometry matrix that links substrates and products of the individual reactions. Given a defined biological objective, for example biomass production, FBA predicts flux distributions through those reactions that optimize the chosen objective. Fluxes through the reactions are constrained by stoichiometries such that each reaction must be balanced, and input-output rates obtained from empirical data. Ultimately, these models can predict growth rates of an organism, or rates of metabolite production, under defined environmental conditions (Orth et al., 2010; Gottstein et al., 2016).

While FBA was originally developed in medical and biotechnological contexts, it has proven to be useful also for studying microbial processes in groundwater. Particular attention in this regard has been paid to *Geobacter* species (Mahadevan et al., 2011). FBA was combined with a hydrological transport model to predict in situ flux distributions, substrate uptake, and growth yields of *Geobacter metallireducens* during growth on acetate with iron as electron acceptor in an anoxic, uranium-contaminated aquifer. These predictions were in agreement with proteome profiles, acetate concentrations, and abundances, respectively, that were observed in situ during a field experiment with acetate amendment of the aquifer (Fang et al., 2012). The same approach was also able to accurately predict rates of uranium reduction by *Geobacter* in relation to the availability of ammonium as nitrogen source in the groundwater (Scheibe et al., 2009). Next to these single-species models, FBA models for *Geobacter sulfurreducens* and *Rhodoferax ferrireducens* were combined into a multispecies model to study the competition between these two iron-reducing organisms for acetate and ammonium as carbon and nitrogen source, respectively. The model was able to predict relative proportions of the two organisms that matched abundance ratios in situ, and provided explanations why one organism could outcompete the other in different zones of the aquifer (Zhuang et al., 2011). Another layer of complexity was later added to the model by including *Shewanella oneidensis* as an additional competitor for iron, while at the same time acting also as potential synergistic partner that could produce acetate from lactate (Zomorrodi et al., 2014). In addition to modeling the outcome of species competition, FBA has furthermore been applied to model synergistic processes relevant to groundwater like syntrophic interactions between fermenters and nitrate reducers or methanogens (Stolyar et al., 2007; Hanemaaijer et al., 2017).

These examples demonstrate the potential of genome-scale constrained-based models to simulate microbial processes in groundwater that are mainly catalyzed by a single organism, or simple consortia of two or three species. However, modeling fluxes and microbial growth during complex processes like DOM turnover in a community with hundreds of different species is currently still out of reach (for a detailed discussion of current limitations see Gottstein et al., 2016).

## Microbial attenuation of groundwater contaminants and bottlenecks

### Petroleum hydrocarbons

The release of petroleum hydrocarbons due to industrial activities and accidental spills is among the leading causes of anthropogenic groundwater contamination. These compounds occur as mixtures of which aromatic hydrocarbons constitute the major, risk-determining fraction due to their toxicity and relatively high water solubility. Aromatic compound-degrading bacteria are widespread in aquifers and are the main drivers of the natural attenuation of these contaminants (Lueders, 2017). Aromatic hydrocarbons serve as carbon source and electron donor for aerobic respiration, anaerobic respiration with various electron acceptors, or fermentation in syntrophic interactions between fermenting and respiring microorganisms (for reviews see Weelink et al., 2010; Kleinsteuber et al., 2012; Lueders, 2017). The release of hydrocarbons from point sources leads to the formation of contaminant plumes that contain high loads of carbon, resulting in an excess of electron donors over electron acceptors. Because the degradation of hydrocarbons with oxygen as electron acceptor is highly favorable and enables fast growth of aerobic bacteria, oxygen is rapidly depleted inside the plume. Therefore, the microbial attenuation of hydrocarbons in groundwater mainly relies on anaerobic biodegradation (Meckenstock et al., 2015; Lueders, 2017). However, compared to aerobic degradation, the degradation of aromatic hydrocarbons under anoxic conditions is considerably slower as it is enzymatically more challenging and offers less energy for microbial growth (Weelink et al., 2010; Fuchs et al., 2011; Lueders, 2017).

Different redox processes are arranged along steep vertical concentration gradients of electron donors and acceptors across the hydrocarbon plume (Fig. 9.2). The reduction of soluble electron acceptors including oxygen, nitrate, and sulfate is limited to the plume fringes, since

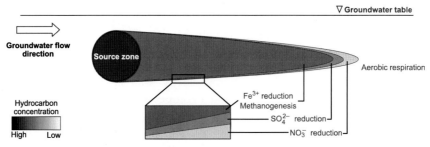

**FIGURE 9.2**  Schematic representation of a hydrocarbon plume in groundwater and the vertical arrangement of redox processes. *(Modified from Meckenstock et al., 2015).*

their depletion in the center of the plume usually cannot be compensated by the limited transverse dispersion of new electron acceptors from the outside of the plume. Hence, biodegradation in the plume core relies on the reduction of either insoluble electron acceptors like iron, or methanogenesis (Thornton et al., 2001; van Breukelen et al., 2003; Meckenstock et al., 2015). Thus, plume fringes are the hot-spots of hydrocarbon degradation, stimulated by the mixing of electron donors and acceptors (Anneser et al., 2008). In these hot-spots, contaminant-degrading microorganisms may actually operate faster than the mixing can occur. Hence, transverse dispersion is a critical rate-limiting step in this process (Thornton et al., 2001; Chu et al., 2005; Meckenstock et al., 2015). Pore water flow velocity can be an additional bottleneck, as it determines the mass transfer of electron donors and acceptors to the cell within pores and attached to sediments (Grösbacher et al., 2018). Moreover, flow velocities might interfere with the exchange of hydrogen and acetate between syntrophic partners that controls biodegradation especially in methanogenic plumes cores (Röling et al., 2007; Meckenstock et al., 2015).

Microbial community composition across contaminant plumes is characterized by distinct degrader populations within the different redox zones. In addition to differences across redox zones, these communities also differ strongly from communities outside the plume and often display a lower species diversity and evenness (Anneser et al., 2010; Yagi et al., 2010; Staats et al., 2011; Larentis et al., 2013; Herzyk et al., 2017). However, contaminant plumes are not static over time, but can expand, shrink, and change position with rising or declining groundwater levels. This ultimately brings about changes in the hydrochemical regime to which microbial communities need to adapt, which can transiently interrupt biodegradation (Eckert et al., 2015; Pilloni et al., 2019).

Microbial activities can at least in part counteract certain biodegradation bottlenecks. For example, chemolithoautotrophic sulfide oxidizers can recycle sulfate from reduced sulfur compounds in zones where sulfate cannot be replenished fast enough by transverse mixing (Kellermann et al., 2012; Einsiedl et al., 2015). A particularly intriguing process involves long-distance electron transfer by sulfide-oxidizing cable bacteria within the *Desulfobulbaceae*. These bacteria form long filaments up to several millimeters in length that could act as short-cuts between different redox zones of a contaminant plume, coupling sulfide oxidation in reduced zones to oxygen respiration or nitrate reduction in oxidized zones (Müller et al., 2016, 2020). In addition, certain bacteria are able to overcome low degradation efficiencies under anoxic conditions by generating oxygen from the dismutation of nitrite, which could enable them to conduct aerobic hydrocarbon degradation in the absence of oxygen (Zedelius et al., 2011; Ettwig et al., 2012; Atashgahi et al., 2018). However, the actual relevance of these organisms and processes for the degradation of hydrocarbons in the subsurface environments still needs to be further explored.

## Chlorinated organic compounds

Chlorinated organic compounds are widely used in various industrial applications, for instance as cleaning, cooling, and degreasing agents, flame retardants, as well as herbicides and pesticides, and thus have become widespread groundwater contaminants. Whereas chlorinated herbicides and pesticides are generally leached into groundwater through soils at low concentrations, others, especially chlorinated ethenes (perchloroethylene (PCE), trichloroethene (TCE); dichloroethene (DCE), and vinyl chloride (VC)) as well as polychlorinated biphenyls (PCBs) typically occur locally at high concentrations resulting from inappropriate waste disposal and spills. Due to their high toxicity, these compounds are of particular concern. Chlorinated ethenes and PCBs typically occur in mixtures that are leached into groundwater as dense nonaqueous phase liquids (DNAPL), which then form plumes of dissolved-phase contaminants (Weatherill et al., 2018). Although historically believed to be insusceptible to biodegradation, various bacteria have meanwhile been described that are able to degrade synthetic chlorinated organic compounds via oxidative or reductive metabolic pathways (Holliger et al., 1998; Chapelle, 2001).

The parent chlorinated ethenes PCE and TCE can be reductively transformed via organohalide respiration to the less chlorinated molecules *cis*-1,2-dichlorethene (*cis*-DCE), VC, and ultimately to nontoxic ethene under anoxic conditions (Adrian and Löffler, 2016). The reduction of PCE and TCE to *cis*-DCE can be catalyzed by various bacteria like *Desulfitobacterium*, *Geobacter*, *Dehalobacter*, *Sulfurospirillum*, and *Dehalococcoides*. However, *Dehalococcoides* is currently the only known genus to harbor organisms capable of reducing PCE completely to nontoxic ethene under anoxic conditions (for a reviews see Adrian and Löffer, 2016; Dolinova et al., 2016; Weatherill et al., 2018). Electron donors that can be used by these bacteria for organohalide respiration are typically limited to fermentation products like small organic acids and hydrogen. Hence, fermenting microorganisms play a crucial role for the microbial attenuation of chlorinated organic compounds without directly catalyzing the degradation themselves (Holliger et al., 1998; Röling et al., 2007; Weatherill et al., 2018). Apart from reductive dehalogenation, which is strictly anaerobic, chlorinated ethenes can also be degraded aerobically, either by serving as electron donors (DCE and VC), or via nonspecific co-metabolic enzyme reactions (Hartmans et al., 1985). Similarly, PCBs can be degraded anaerobically via organohalide respiration to less halogenated aromatic compounds, which then can serve as carbon source for other microorganisms, or aerobically via co-metabolic reactions (Urbaniak, 2007, 2013).

Similar to the degradation of petroleum hydrocarbons, the fate of chlorinated organic compounds in DNAPL plumes in groundwater is determined by mass transfer and mixing of contaminants and suitable electron donors that can fuel organohalide respiration (Weatherill et al., 2018). In addition, metabolic interactions between organohalide-respiring bacteria and other functional groups, apart from fermenters that provide electron donors, are important. For example, *Dehalococcoides* lack synthesis pathways for cobalamin (vitamin $B_{12}$), which is an essential cofactor for the complete dehalogenation of chlorinated ethenes to nontoxic ethene, and thus depend on other organisms in a microbial community for the supply of this compound (Yan et al., 2013). Moreover, organohalide-respiring bacteria have to compete with other respiratory processes (e.g., nitrate or sulfate reduction) for electron donors, posing an additional degradation bottleneck (Weatherill et al., 2018; Hermon et al., 2019). On top of limitations imposed by mass transfer and metabolic interactions, also species dispersal can be an important rate-determining factor. Although organohalide-respiring bacteria, including *Dehalococcoides*, are generally widespread in aquifers and other freshwater environments, cases have been reported for the absence of these bacteria and concomitant lack of complete mineralization of chlorinated organic compounds at these sites (Hendrickson et al., 2002; Rodenburg et al., 2010; Tas et al., 2010; Rossi et al., 2012; Hermon et al., 2019). Thus, although complete absence of certain microorganisms within a given area is difficult to prove (Meckenstock et al., 2015), these examples show that dispersal limitation of sufficient numbers of degrader organisms can limit the biodegradation of chlorinated organic compounds in certain zones of an aquifer.

## Toxic metals and metalloids

All metals and metalloids occur naturally in the environment at various concentrations depending on geological contexts and natural phenomena such as soil alteration or volcanic eruptions. Anthropogenic activities can also cause their dissemination in the environment and the contamination of soils, surface waters, and aquifers through industrial processes, mining, waste disposal, or fossil fuel consumption. While several metals are of concern in groundwater, here we will focus on mercury (Hg) and arsenic (As) as prominent examples of highly toxic metals and metalloids whose geochemical cycling is largely determined by microbial processes.

Mercury occurs mainly in three forms: metallic $Hg°$, which is liquid at ambient temperatures and volatile, ionic mercury ($Hg^{2+}$), and organic methylated forms like methylmercury (MeHg), which can bioaccumulate along the food chain and is considered as the most toxic naturally formed neurotoxin. As a mineral, mercury occurs as cinnabar (HgS). Although abiotic pathways are involved in the cycling of mercury, the reduction of $Hg^{2+}$ or MeHg to volatile $Hg°$, and the methylation of $Hg^{2+}$ to MeHg are mainly microbially driven processes. The reduction of $Hg^{2+}$, genetically coded by the well described *mer* operon, is used as a detoxification mechanism to remove mercury from the cell in the form of volatile $Hg°$ (Barkay et al., 2003). The *mer* operon is located on a plasmid and is found in many bacterial species. In contrast, mercury methylation is a less widespread mechanism restricted to anoxic conditions, mainly occurring in sulfate and iron-reducing bacteria, for which the genetic basis was only recently uncovered (Parks et al., 2013). There are only few studies on the fate and speciation of mercury in groundwater (Hellal et al., 2015). The main focus of the existing research in groundwater focuses on the impact of environmental conditions like redox conditions or carrying phases as these will influence both mercury speciation and bioavailability.

Arsenic (As) contamination is of major concern in groundwater worldwide (Nordstrom, 2002; Smedley and Kinniburgh, 2002), especially in Southern Asia where arsenic occurs naturally in sediments and is mobilized under anoxic conditions (Charlet and Polya, 2006). Although organic methylated forms of arsenic exist, inorganic forms are predominant in aqueous environments. Under oxidizing conditions arsenate (As(V)) dominates whereas under reducing conditions arsenite (As(III)) is dominant. The occurrence of these inorganic forms are mainly determined by microbial processes (Oremland and Stolz, 2003). A variety of heterotrophic and chemolithoautotrophic prokaryotes can oxidize As(III) to As(V) catalyzed by arsenite oxidases (Quéméneur et al., 2008). Chemolithoautotrophs can also use As(III) as an energy source (Santini et al., 2000; Battaglia-Brunet et al., 2002). The reduction of As(V) to As(III) is either performed via the detoxification *Ars* system to excrete arsenic from the cell, or by dissimilatory respiration reactions in which As(V) is the terminal electron acceptor that is reduced with hydrogen or organic electron donors (Mukhopadhyay et al., 2002). Thus, DOM and its transformation play key roles in arsenic cycling (Cui and Jing, 2019). Microbial transformations of inorganic arsenic affect its transport and mobility in groundwater, as the oxidized form As(V) tends to precipitate on minerals such as iron oxy-hydroxides and thus can be more easily trapped than the reduced form As(III). In anoxic groundwater, arsenic can be mobilized as As(III) directly by dissimilatory reduction of As(V) sorbed on minerals, but also by microbial processes like the reduction of iron contained in arsenic-bearing minerals (Ma et al., 2015a; Osborne et al., 2015; Hassan et al., 2016; Chen et al., 2017). Therefore, stimulating the microbial oxidation of As(III) to As(V) is an effective means to limit the dissemination of arsenic in groundwater (Battaglia-Brunet et al., 2006; Wan et al., 2010; Crognale et al., 2017). Although methylation of inorganic arsenic is a common detoxification mechanism in microorganisms, similar to mercury methylation, this process has not been widely studied in groundwater environments so far (Cui and Jing, 2019).

## Emerging organic contaminants and micropollutants

The term emerging organic contaminants—also known as micropollutants due to their low concentrations in the environment—covers a broad range of novel anthropogenic compounds, as well as synthetic compounds that have only recently been detected and classified as environmental contaminants (Lapworth et al., 2012). In addition to pesticides, which have a relatively long history of agricultural use, these contaminants further include pharmaceuticals, food additives, constituents of personal care products, and diverse chemical compounds widely used in industry, for example in the production of plastics, as lubricants, flame retardants, or adhesives (Lapworth et al., 2012; Fenner et al., 2013; Careghini et al., 2015). Whereas spills of hydrocarbons or chlorinated ethenes result in high concentrations of contaminants that spread from the point source as plumes, micropollutants in groundwater mainly derive from diffuse inputs. Wastewater effluents are among the main sources, since micropollutants are often only insufficiently removed during wastewater treatment, and consequently are released back into the water cycle. Additional important sources are untreated urban and agricultural runoffs. Concentrations of individual contaminants in groundwater are usually low, ranging between 0.01 and 10 $\mu g\ L^{-1}$ (Lapworth et al., 2012). However, many of these compounds are highly persistent in groundwater and thus can accumulate over time, which ultimately poses threats also to the health of groundwater-dependent ecosystems and humans (Lapworth et al., 2012; Fenner et al., 2013; Careghini et al., 2015).

The persistence of micropollutants in groundwater reveals a limited potential for their natural attenuation. The contribution of microbial biodegradation relative to abiotic chemical and physical processes depends on the chemical properties of a compound, and physico-chemical conditions in the groundwater like redox conditions (Fenner et al., 2013; Im and Löffler, 2016). Under conditions where biodegradation does occur, the interaction between degraders and other members of the microbial community as well as the availability and quality of ambient DOM influence degradation efficiencies. Due to the low concentrations of individual contaminants, they typically do not serve as main substrate for growth, but are consumed along with other compounds present in the DOM pool (Helbling, 2015). Sufficient concentrations of easily bioavailable DOM can thus fuel the biodegradation of micropollutants by stimulating growth of degrader populations (Horemans et al., 2013; Liu et al., 2014). If DOM concentrations are too low, degraders may even switch into maintenance mode and stop their catabolic activity (Kundu et al., 2019). However, whether stimulation of micropollutant degradation by DOM does occur is determined by the competition of degraders and nondegraders in the community. Thus, if the ability to degrade a specific micropollutant is restricted to a certain group of microorganisms, the degradation will depend on the ability of these degraders to compete with other organisms for available carbon for growth (Liu et al., 2014). Additionally, carbon catabolite repression may occur, that is repression of metabolic pathways for the degradation of micropollutants to favor the consumption of other more easily degradable DOM compounds (Horemans et al., 2013, 2014). However, mixed substrate use is a common strategy among microorganisms to cope with low carbon concentrations as they are typically found in groundwater (Egli, 2010; Castelle et al., 2013; Marozava et al., 2014). Therefore, significant limitations due to carbon catabolite repression likely only occur in the case of additional high inputs of external easily degradable organic carbon (Horemans et al., 2013). Under normal conditions, any reduction in cell-specific degradation activity due to the preferential degradation of alternative compounds will likely be compensated by the enhanced growth of degrader cells (Horemans et al., 2014; Liu et al., 2014).

The reason why individual micropollutants persist in groundwater and are not degraded further below certain concentrations, even if degrading organisms are present, is not fully understood. However, limited pore-scale mixing as well as reduced mass transfer to the cells determined by pore water flow velocities likely present key bottlenecks (Meckenstock et al., 2015; Grösbacher et al., 2018; Kang et al., 2019). Moreover, mass transfer limitation across the cell membrane has been demonstrated to limit biodegradation of the pesticide atrazine at concentrations lower than $10 \ \mu g \ L^{-1}$ when supplied as sole carbon and nitrogen source in continuous culture experiments with *Arthrobacter aurescens* (Ehrl et al., 2019; Kundu et al., 2019).

# Resistance and resilience of groundwater microbial communities to perturbations

Aquifers are considered to be relatively stable environments with less variation in nutrient concentrations and physicochemical conditions over time compared to surface habitats (Goldscheider et al., 2006; Griebler and Lueders, 2009). Given that these conditions exert a strong influence on microbial community composition through species sorting as discussed earlier, we may assume that microorganisms in groundwater are well adapted to their environment and react sensitively to environmental changes (Graham and Stegen, 2017; Stegen et al., 2018a). In addition, mechanisms to cope with stress require energy and can lower the metabolic potential and competitiveness of microorganisms (Ferenci and Spira, 2007). Therefore, microorganisms in typically energy-poor groundwater environments probably only have a narrow stress tolerance.

Perturbations—that is strong or sudden environmental changes that exceed the range of fluctuation to which native organisms are adapted—can change community composition and functioning by inhibiting or inactivating part of a community (Shade et al., 2012). The inactivation of resident taxa opens new niches and reduces competition for other organisms that are dispersed into the community. This in turn can amplify community changes, and increases the influence of stochastic colonization events on community assembly, which makes it difficult to predict the trajectory of a community in response to perturbations (Shade et al., 2012; Ferrenberg et al., 2013; Zhou et al., 2014). Resistance is defined as the ability of a community to withstand significant changes beyond the natural stochastic variation following a perturbation; resilience is the rate of recovery of a perturbed community back to its initial state. If a disturbance is sufficiently strong or persistent, communities may not be able to recover and thus remain changed in an alternative stable state (Pimm, 1984; Beisner et al., 2003). These properties are determined by the physiological responses of microorganisms, dominating mechanisms of community assembly, and the type of perturbation, which can occur as pulses (i.e., short and discrete events) and press disturbances (i.e., continuous changes over longer time periods) (Shade et al., 2012) (Fig. 9.3). Meta-analyses of studies on microbial communities from various habitats have indicated that microbial communities generally display low resistance and resilience to perturbations, especially to press disturbances, both in terms of taxonomic as well as functional composition (Allison and Martiny, 2008; Shade et al., 2012).

Low resistance and resilience of groundwater microbial communities to press disturbances were observed in controlled experiments with aquifer model systems that were exposed to a continuous, high supply of aromatic hydrocarbons over several months (Ma et al., 2015b, 2015c; Herzyk et al., 2017). Hydrocarbon contamination constitutes a severe perturbation as it dramatically increases nutrient concentrations, which subsequently changes redox conditions, and additionally exposes microorganisms to solvent stress (Duldhardt et al., 2007; Trautwein et al., 2008; Meckenstock et al., 2015). Upon the release of hydrocarbons into the initially pristine systems, microbial communities showed low resistance, as taxa from the original communities were replaced by degrader populations in the contaminant plume zones within days, leading to significant changes in community composition. Low resilience became evident as the communities remained significantly different from their initial state

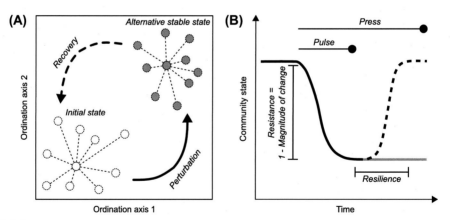

FIGURE 9.3   (A) Changes in microbial community composition in response to a perturbation depicted in an ordination plot (e.g., NMDS of Bray–Curtis dissimilarities). White symbols show the initial state; gray symbols show the changed community composition after the perturbation, which may manifest into an alternative stable state if a community cannot recover to its initial composition. Dashed-lined symbols around the centroids represent differences in community composition due to natural stochastic drift. (B) Community trajectory in response to a perturbation over time and quantitative definitions of resistance and resilience. Resistance can be quantified as the inverse of the magnitude of change after a perturbation (e.g., measured as ordination axis scores); resilience is the time needed until a community recovers back to its initial state after a perturbation (lack of resilience results in an alternative stable state shown in gray). Pulse disturbances are shown as short discrete events; press disturbances are shown as continuous long-term events. *(Modified from Shade et al., 2012).*

even after about 2 years of recovery after the contaminant supply had been shut off and contaminants had been completely removed (Ma et al., 2015b; Herzyk et al., 2017). Ma et al. (2015b) further showed that the low resistance and resilience observed on the taxonomic level were also reflected in terms of functional gene composition. Similar observations were made during and after the perturbation of a natural porous aquifer with high concentrations of emulsified vegetable oil lasting several months (Zhou et al., 2014).

In general, community functioning may have recovered in those studies given enough time despite persistent taxonomic changes in the communities after the perturbation. This would be the case if taxa in the original community are replaced by other functionally redundant organisms that can perform the same functions at comparable rates (Konopka et al., 2015). However, this critically depends on the ability of microorganisms to disperse into perturbed aquifer zones (Zhou et al., 2014). In the case of strong dispersal limitation, certain functions could be erased completely or only recover slowly after extended periods of time (Graham and Stegen, 2017).

Apart from severe impacts like high inputs of hydrocarbons, the effect of perturbations caused by micropollutants have not been widely addressed so far. The antibiotic sulfamethoxazole is among the most widespread pharmaceutical contaminants of groundwater worldwide (Lapworth et al., 2012). Exposure of aquifer microbial communities to a pulse of sulfamethoxazole in situ for 30 days caused changes in microbial community composition and reduced cell viability at concentrations of ~100 $\mu g\ L^{-1}$ (Haack et al., 2012). This concentration is about 100 times higher than average concentrations detected in groundwater (Lapworth et al., 2012), and thus the study likely overestimated the effect of the antibiotic on microbial communities under more realistic conditions. However, community changes as well as reduced denitrification rates were observed in batch cultures from the same aquifer exposed to environmentally relevant concentrations (Underwood et al., 2011). Similarly, significant changes in taxonomic as well as functional composition of groundwater microbial communities have been reported in response to triazine herbicides at concentrations as low as 1 $\mu g\ L^{-1}$ (Mauffret et al., 2017). Perturbations of important microbial community functions can also be expected for other organic micropollutants, which can adsorb to biomolecules and thus could cause enzyme inhibition (Im and Löffler, 2016). However, the effect of these compounds supplied as pulses might not accurately reflect impacts of perturbations caused by micropollutants in the environment. Since micropollutants continuously enter groundwater from diffuse sources and are highly persistent (Lapworth et al., 2012), they are likely more perceived as press disturbances to which microbial communities might be less resistant than to pulses (Shade et al., 2012). These considerations need to be taken into account in future studies to yield better insights into the long-term effect of micropollutants on groundwater microbial communities and biogeochemical processes.

The effect of other types of perturbations still need to be systematically assessed. For instance, the use of shallow aquifers as a place for heat storage and discharge has become an increasingly popular source of renewable energy. These activities constitute broad-scale press disturbances not only by directly altering the temperature regime, but also by causing various changes in hydrochemistry, which can affect microbial community composition as well as functioning (Brielmann et al., 2009; Bonte et al., 2013a, 2013b; Jesußek et al., 2013a, 2013b; Griebler et al., 2016). Similar effects can be expected for gradually increasing groundwater temperatures due to rising mean annual air temperatures resulting from climate change (Kløve et al., 2014). In addition, the increasing frequency and intensity of extreme hydrological events as a result of a changing climate will increasingly expose microbial communities in groundwater to pulse disturbances, including strong fluctuations of the groundwater table during intense droughts and floods, or pulses of contaminants from the surface during flood events (Okkonen et al., 2010; Taylor et al., 2012; Zhou et al., 2012; Retter et al., 2020). Weaver et al. (2015) showed that desiccation due to declining groundwater levels observed in situ and in lab experiments led to reduced enzymatic activities in sediment-attached microbial communities, although activities were able to recover back to original levels upon rewetting. However, the resistance and resilience of groundwater microbial communities to repeated desiccation-rewetting cycles still remains to be investigated. Therefore, the generation and analysis of time-resolved data—both from field observations as well as experiments—within defined concepts of microbial community resistance and resilience will be key to understanding the impacts of those events on groundwater ecosystems.

## Outlook

Research over the past two decades has vastly expanded our understanding of the microbial diversity and functioning in groundwater ecosystems. Nevertheless, groundwater is still largely underrepresented in microbial ecology research compared to other environments (Griebler et al., 2022). A large fraction of the total microbial biomass lives in the terrestrial subsurface (Griebler and Lueders, 2009; McMahon and Parnell, 2014; Magnabosco et al., 2018; Flemming and Wuertz, 2019). Yet, as pointed out by Smith et al. (2018), microorganisms from these habitats are represented by far less than 10% in reference sequence databases, which creates a large risk of misclassification and underestimation of the microbial diversity and metabolic potential in these important ecosystems. Hence, continued efforts to sample, sequence, and—just as importantly—enrich, isolate, and characterize microorganisms from groundwater will be needed to fill this knowledge gap, and ultimately gain a better understanding of biogeochemical processes and fluxes in these environments. Sediment-attached communities should receive special attention in this regard, as they account for the bulk of the microbial biomass and activity in aquifers, but have been understudied so far compared to planktonic communities (Griebler et al., 2022). In addition, we have discussed how the exchange of metabolites and cofactors between species, species competition, dispersal, and mass transfer can affect and even pose critical rate-limiting steps in the cycling of DOM and the attenuation of groundwater contaminants. As these processes operate on the pore-scale, an important task for future research will be to shed more light on these processes and how they affect the functioning of groundwater ecosystems on larger scales (Schmidt et al., 2017). Finally, more systematic studies are needed to investigate the resistance and resilience of groundwater microbial communities to perturbations, especially with respect to the chronic exposure to micropollutants as well as impacts resulting from climate change. To this end, generating long-term, temporally resolved data will be imperative.

## Acknowledgments

We thank A. Probst and M. Herrmann for constructive comments that improved this chapter.

# References

Adrian, L., Löffler, F.E., 2016. Organohalide-respiring bacteria—an introduction. In: Adrian, L., Löffler, F.E. (Eds.), Organohalide-Respiring Bacteria. Springer, Berlin, Heidelberg, pp. 3—6.

Akob, D.M., Küsel, K., 2011. Where microorganisms meet rocks in the Earth's Critical Zone. Biogeosciences 8 (12), 3531—3543.

Alfreider, A., Vogt, C., 2012. Genetic evidence for bacterial chemolithoautotrophy based on the reductive tricarboxylic acid cycle in groundwater systems. Microbes and Environments 27 (2), 209—214.

Alfreider, A., Krössbacher, M., Psenner, R., 1997. Groundwater samples do not reflect bacterial densities and activity in subsurface systems. Water Research 31 (4), 832—840.

Alfreider, A., Vogt, C., Geiger-Kaiser, M., Psenner, R., 2009. Distribution and diversity of autotrophic bacteria in groundwater systems based on the analysis of RubisCO genotypes. Systematic and Applied Microbiology 32 (2), 140—150.

Allison, S.D., Martiny, J.B.H., 2008. Resistance, resilience, and redundancy in microbial communities. Proceedings of the National Academy of Sciences 105, 11512—11519.

Anantharaman, K., Brown, C.T., Hug, L.A., Sharon, I., Castelle, C.J., Probst, A.J., Thomas, B.C., Singh, A., Wilkins, M.J., Karaoz, U., Brodie, E.L., Williams, K.H., Hubbard, S.S., Banfield, J.F., 2016. Thousands of microbial genomes shed light on interconnected biogeochemical processes in an aquifer system. Nature Communications 7, 13219.

Anneser, B., Einsiedl, F., Meckenstock, R.U., Richters, L., Wisotzky, F., Griebler, C., 2008. High-resolution monitoring of biogeochemical gradients in a tar oil-contaminated aquifer. Applied Geochemistry 23 (6), 1715—1730.

Anneser, B., Pilloni, G., Bayer, A., Lueders, T., Griebler, C., Einsiedl, F., Richters, L., 2010. High resolution analysis of contaminated aquifer sediments and groundwater—what can be learned in terms of natural attenuation? Geomicrobiology Journal 27 (2), 130—142.

Atashgahi, S., Hornung, B., van der Waals, M.J., Nunes da Rocha, U., Hugenholtz, F., Nijsse, B., Molenaar, D., van Spanning, R., Stams, A.J.M., Gerritse, J., Smidt, H., 2018. A benzene-degrading nitrate-reducing microbial consortium displays aerobic and anaerobic benzene degradation pathways. Scientific Reports 8, 4490.

Baker, M.A., Maurice, V.H., Dahm, C.N., 2000. Organic carbon supply and metabolism in a shallow groundwater ecosystem. Ecology 81 (11), 3133—3148.

Barkay, T., Miller, S.M., Summers, A.O., 2003. Bacterial mercury resistance from atoms to ecosystems. FEMS Microbiology Reviews 27 (2—3), 355—384.

Battaglia-Brunet, F., Dictor, M.-C., Garrido, F., Crouzet, C., Morin, D., Dekeyser, K., Clarens, M., Baranger, P., 2002. An arsenic(III)-oxidizing bacterial population: selection, characterization, and performance in reactors. Journal of Applied Microbiology 93 (4), 656—667.

Battaglia-Brunet, F., Joulian, C., Garrido, F., Dictor, M.-C., Morin, D., Coupland, K., Barrie Johnson, D., Hallberg, K.B., Baranger, P., 2006. Oxidation of arsenite by *Thiomonas* strains and characterization of *Thiomonas arsenivorans* sp. nov. Antonie van Leeuwenhoek 89 (1), 99—108.

Battin, T.J., Sloan, W.T., Kjelleberg, S., Daims, H., Head, I.M., Curtis, T.P., Eberl, L., 2007. Microbial landscapes: new paths to biofilm research. Nature Reviews Microbiology 5 (1), 76.

Battin, T.J., Besemer, K., Bengtsson, M.M., Romani, A.M., Packmann, A.I., 2016. The ecology and biogeochemistry of stream biofilms. Nature Reviews Microbiology 14 (4), 251—263.

Beaton, E.D., Stevenson, B.S., King-Sharp, K.J., Stamps, B.W., Nunn, H.S., Stuart, M., 2016. Local and regional diversity reveals dispersal limitation and drift as drivers for groundwater bacterial communities from a fractured granite formation. Frontiers in Microbiology 7, 1933.

Beisner, B.E., Haydon, D.T., Cuddington, K., 2003. Alternative stable states in ecology. Frontiers in Ecology and the Environment 1 (7), 376—382.

Benk, S.A., Yan, L., Lehmann, R., Roth, V.-N., Schwab, V.F., Totsche, K.U., Küsel, K., Gleixner, G., 2019. Fueling diversity in the subsurface: composition and age of dissolved organic matter in the critical zone. Frontiers of Earth Science 7, 296.

Ben Maamar, S., Aquilina, L., Quaiser, A., Pauwels, H., Michon-Coudouel, S., Vergnaud-Ayraud, V., Labasque, T., Roques, C., Abbott, B.W., Dufresne, A., 2015. Groundwater isolation governs chemistry and microbial community structure along hydrologic flowpaths. Frontiers in Microbiology 6, 1457.

Bonte, M., van Breukelen, B.M., Stuyfzand, P.J., 2013a. Temperature-induced impacts on groundwater quality and arsenic mobility in anoxic aquifer sediments used for both drinking water and shallow geothermal energy production. Water Research 47 (14), 5088—5100.

Bonte, M., Röling, W.F.M., Zaura, E., van der Wielen, P.W.J.J., Stuyfzand, P.J., van Breukelen, B.M., 2013b. Impacts of shallow geothermal energy production on redox processes and microbial communities. Environmental Science & Technology 47 (24), 14476—14484.

van Breukelen, B.M., Röling, W.F.M., Groen, J., Griffioen, J., van Verseveld, H.W., 2003. Biogeochemistry and isotope geochemistry of a landfill leachate plume. Journal of Contaminant Hydrology 65 (3—4), 245—268.

Braun, A., Spona-Friedl, M., Avramov, M., Elsner, M., Baltar, F., Reinthaler, T., Herndl, G.J., Griebler, C., 2021. Reviews and syntheses: Heterotrophic fixation of inorganic carbon — significant but invisible flux in environmental carbon cycling. Biogeosciences 18, 3689—3700.

Brielmann, H., Griebler, C., Schmidt, S.I., Michel, R., Lueders, T., 2009. Effects of thermal energy discharge on shallow groundwater ecosystems. FEMS Microbiology Ecology 68 (3), 273—286.

Careghini, A., Mastorgio, A.F., Saponaro, S., Sezenna, E., 2015. Bisphenol A, nonylphenols, benzophenones, and benzotriazoles in soils, groundwater, surface water, sediments, and food: a review. Environmental Science and Pollution Research International 22 (8), 5711—5741.

Castelle, C.J., Hug, L.A., Wrighton, K.C., Thomas, B.C., Williams, K.H., Wu, D., Tringe, S.G., Singer, S.W., Eisen, J.A., Banfield, J.F., 2013. Extraordinary phylogenetic diversity and metabolic versatility in aquifer sediment. Nature Communications 4, 2120.

Castelle, C.J., Wrighton, K.C., Thomas, B.C., Hug, L.A., Brown, C.T., Wilkins, M.J., Frischkorn, K.R., Tringe, S.G., Singh, A., Markillie, L.M., Taylor, R.C., Williams, K.H., Banfield, J.F., 2015. Genomic expansion of domain archaea highlights roles for organisms from new phyla in anaerobic carbon cycling. Current Biology 25 (6), 690—701.

Castelle, C.J., Brown, C.T., Thomas, B.C., Williams, K.H., Banfield, J.F., 2017. Unusual respiratory capacity and nitrogen metabolism in a *Parcubacterium* (OD1) of the candidate phyla radiation. Scientific Reports 7, 40101.

Castelle, C.J., Brown, C.T., Anantharaman, K., Probst, A.J., Huang, R.H., Banfield, J.F., 2018. Biosynthetic capacity, metabolic variety and unusual biology in the CPR and DPANN radiations. Nature Reviews Microbiology 16 (10), 629—645.

Chapelle, F.H., 2001. Ground-Water Microbiology and Geochemistry. John Wiley & Sons, New York, NY.

Charlet, L., Polya, D.A., 2006. Arsenic in shallow, reducing groundwaters in southern Asia: an environmental health disaster. Elements 2 (2), 91—96.

Chen, X., Zeng, X.-C., Wang, J., Deng, Y., Ma, T., Guoji, E., Mu, Y., Yang, Y., Li, H., Wang, Y., 2017. Microbial communities involved in arsenic mobilization and release from the deep sediments into groundwater in Jianghan plain, Central China. Science of the Total Environment 579, 989—999.

Chu, M., Kitanidis, P.K., McCarty, P.L., 2005. Modeling microbial reactions at the plume fringe subject to transverse mixing in porous media: when can the rates of microbial reaction be assumed to be instantaneous? Water Resources Research 41, W06002.

Cordero, O.X., Datta, M.S., 2016. Microbial interactions and community assembly at microscales. Current Opinion in Microbiology 31, 227—234.

Crognale, S., Amalfitano, S., Casentini, B., Fazi, S., Petruccioli, M., Rossetti, S., 2017. Arsenic-related microorganisms in groundwater: a review on distribution, metabolic activities and potential use in arsenic removal processes. Reviews in Environmental Science and Biotechnology 16 (4), 647—665.

Cui, J., Jing, C., 2019. A review of arsenic interfacial geochemistry in groundwater and the role of organic matter. Ecotoxicology and Environmental Safety 183, 109550.

Daly, R.A., Roux, S., Borton, M.A., Morgan, D.M., Johnston, M.D., Booker, A.E., Hoyt, D.W., Meulia, T., Wolfe, R.A., Hanson, A.J., Mouser, P.J., Moore, J.D., Wunch, K., Sullivan, M.B., Wrighton, K.C., Wilkins, M.J., 2019. Viruses control dominant bacteria colonizing the terrestrial deep biosphere after hydraulic fracturing. Nature Microbiology 4 (2), 352—361.

Dolinová, I., Czinnerová, M., Dvořák, L., Stejskal, V., Ševců, A., Černík, M., 2016. Dynamics of organohalide-respiring bacteria and their genes following in-situ chemical oxidation of chlorinated ethenes and biostimulation. Chemosphere 157, 276—285.

Dong, X., Greening, C., Brüls, T., Conrad, R., Guo, K., Blaskowski, S., Kaschani, F., Kaiser, M., Laban, N.A., Meckenstock, R.U., 2018. Fermentative Spirochaetes mediate necromass recycling in anoxic hydrocarbon-contaminated habitats. The ISME Journal 12 (8), 2039—2050.

Duldhardt, I., Nijenhuis, I., Schauer, F., Heipieper, H.J., 2007. Anaerobically grown *Thauera aromatica, Desulfococcus multivorans, Geobacter sulfurreducens* are more sensitive towards organic solvents than aerobic bacteria. Applied Microbiology and Biotechnology 77 (3), 705—711.

Eckert, D., Kürzinger, P., Bauer, R., Griebler, C., Cirpka, O.A., 2015. Fringe-controlled biodegradation under dynamic conditions: quasi 2-D flow-through experiments and reactive-transport modeling. Journal of Contaminant Hydrology 172, 100—111.

Egli, T., 2010. How to live at very low substrate concentration. Water Research 44 (17), 4826—4837.

Ehrl, B.N., Kundu, K., Gharasoo, M., Marozava, S., Elsner, M., 2019. Rate-limiting mass transfer in micropollutant degradation revealed by isotope fractionation in chemostat. Environmental Science & Technology 53 (3), 1197—1205.

Einsiedl, F., Pilloni, G., Ruth-Anneser, B., Lueders, T., Griebler, C., 2015. Spatial distributions of sulphur species and sulphate-reducing bacteria provide insights into sulphur redox cycling and biodegradation hot-spots in a hydrocarbon-contaminated aquifer. Geochimica et Cosmochimica Acta 156, 207—221.

Ettwig, K.F., Speth, D.R., Reimann, J., Wu, M.L., Jetten, M.S.M., Keltjens, J.T., 2012. Bacterial oxygen production in the dark. Frontiers in Microbiology 3, 273.

Eydal, H.S.C., Jägevall, S., Hermansson, M., Pedersen, K., 2009. Bacteriophage lytic to *Desulfovibrio aespoeensis* isolated from deep groundwater. The ISME Journal 3 (10), 1139—1147.

Fang, Y., Wilkins, M.J., Yabusaki, S.B., Lipton, M.S., Long, P.E., 2012. Evaluation of a genome-scale in silico metabolic model for *Geobacter metallireducens* by using proteomic data from a field biostimulation experiment. Applied and Environmental Microbiology 78 (24), 8735—8742.

Faust, K., Raes, J., 2012. Microbial interactions: from networks to models. Nature Reviews Microbiology 10 (8), 538—550.

Fenner, K., Canonica, S., Wackett, L.P., Elsner, M., 2013. Evaluating pesticide degradation in the environment: blind spots and emerging opportunities. Science 341 (6147), 752—758.

Ferenci, T., Spira, B., 2007. Variation in stress responses within a bacterial species and the indirect costs of stress resistance. Annals of the New York Academy of Sciences 1113, 105—113.

Ferrenberg, S., O'Neill, S.P., Knelman, J.E., Todd, B., Duggan, S., Bradley, D., Robinson, T., Schmidt, S.K., Townsend, A.R., Williams, M.W., Cleveland, C.C., Melbourne, B.A., Jiang, L., Nemergut, D.R., 2013. Changes in assembly processes in soil bacterial communities following a wildfire disturbance. The ISME Journal 7 (6), 1102—1111.

Fillinger, L., Hug, K., Griebler, C., 2019a. Selection imposed by local environmental conditions drives differences in microbial community composition across geographically distinct groundwater aquifers. FEMS Microbiology Ecology 95 (11), fiz160.

Fillinger, L., Zhou, Y., Kellermann, C., Griebler, C., 2019b. Non-random processes determine the colonization of groundwater sediments by microbial communities in a pristine porous aquifer. Environmental Microbiology 21 (1), 327—342.

Flemming, H.-C., Wuertz, S., 2019. Bacteria and archaea on Earth and their abundance in biofilms. Nature Reviews Microbiology 17 (4), 247—260.

Flynn, T.M., Sanford, R.A., Ryu, H., Bethke, C.M., Levine, A.D., Ashbolt, N.J., Santo Domingo, J.W., 2013. Functional microbial diversity explains groundwater chemistry in a pristine aquifer. BMC Microbiology 13, 146.

Foulquier, A., Simon, L., Gilbert, F., Fourel, F., Malard, F., Mermillod-Blondin, F., 2009. Relative influences of DOC flux and subterranean fauna on microbial abundance and activity in aquifer sediments: new insights from $^{13}$C-tracer experiments. Freshwater Biology 55 (7), 1560—1576.

Franzosa, E.A., Hsu, T., Sirota-Madi, A., Shafquat, A., Abu-Ali, G., Morgan, X.C., Huttenhower, C., 2015. Sequencing and beyond: integrating molecular "omics" for microbial community profiling. Nature Reviews Microbiology 13 (6), 360—372.

Fuchs, G., Boll, M., Heider, J., 2011. Microbial degradation of aromatic compounds — from one strategy to four. Nature Reviews Microbiology 9 (11), 803—816.

Fukami, T., 2015. Historical contingency in community assembly: integrating niches, species pools, and priority effects. Annual Review of Ecology, Evolution, and Systematics 46 (1), 1—23.

Geesink, P., Tyc, O., Küsel, K., Taubert, M., van de Velde, C., Kumar, S., Garbeva, P., 2018. Growth promotion and inhibition induced by interactions of groundwater bacteria. FEMS Microbiology Ecology 94 (11), fiy164.

Geesink, P., Wegner, C.-E., Probst, A.J., Herrmann, M., Dam, H.T., Kaster, A.-K., Küsel, K., 2020. Genome-inferred spatio-temporal resolution of an uncultivated *Roizmanbacterium* reveals its ecological preferences in groundwater. Environmental Microbiology 22 (2), 726—737.

Goldscheider, N., Hunkeler, D., Rossi, P., 2006. Review: microbial biocenoses in pristine aquifers and an assessment of investigative methods. Hydrogeology Journal 14 (6), 926—941.

III. Roles of organisms in groundwater

Gooddy, D.C., Hinsby, K., 2009. Organic quality of groundwaters. In: Edmunds, W.M., Shand, P. (Eds.), Natural Groundwater Quality. Blackwell Publishing Ltd., pp. 59–70

Gottstein, W., Olivier, B.G., Bruggeman, F.J., Teusink, B., 2016. Constraint-based stoichiometric modelling from single organisms to microbial communities. Journal of the Royal Society Interface 13 (124), 20160627.

Graham, E.B., Stegen, J.C., 2017. Dispersal-based microbial community assembly decreases biogeochemical function. Processes 5 (4), 65.

Graham, E.B., Crump, A.R., Resch, C.T., Fansler, S., Arntzen, E., Kennedy, D.W., Fredrickson, J.K., Stegen, J.C., 2016. Coupling spatiotemporal community assembly processes to changes in microbial metabolism. Frontiers in Microbiology 7, 1949.

Graham, E.B., Crump, A.R., Resch, C.T., Fansler, S., Arntzen, E., Kennedy, D.W., Fredrickson, J.K., Stegen, J.C., 2017. Deterministic influences exceed dispersal effects on hydrologically-connected microbiomes. Environmental Microbiology 19 (4), 1552–1567.

Griebler, C., Fillinger, L., Karwautz, C., Hose, G.C., 2022. Knowledge gaps, obstacles, and research frontiers in groundwater microbial ecology. In: Mehner, T., Tockner, K. (Eds.), Encyclopedia of Inland Waters, 2nd Edition. Elsevier, pp. 611–624.

Griebler, C., Lueders, T., 2009. Microbial biodiversity in groundwater ecosystems. Freshwater Biology 54 (4), 649–677.

Griebler, C., Mindl, B., Slezak, D., Geiger-Kaiser, M., 2002. Distribution patterns of attached and suspended bacteria in pristine and contaminated shallow aquifers studied with an in situ sediment exposure microcosm. Aquatic Microbial Ecology 28, 117–129.

Griebler, C., Malard, F., Lefébure, T., 2014. Current developments in groundwater ecology — from biodiversity to ecosystem function and services. Current Opinion in Biotechnology 27, 159–167.

Griebler, C., Brielmann, H., Haberer, C.M., Kaschuba, S., Kellermann, C., Stumpp, C., Hegler, F., Kuntz, D., Walker-Hertkorn, S., Lueders, T., 2016. Potential impacts of geothermal energy use and storage of heat on groundwater quality, biodiversity, and ecosystem processes. Environmental Earth Sciences 75 (20), 1391.

Grösbacher, M., Spicher, C., Bayer, A., Obst, M., Karwautz, C., Pilloni, G., Wachsmann, M., Scherb, H., Griebler, C., 2016. Organic contamination versus mineral properties: competing selective forces shaping bacterial community assembly in aquifer sediments. Aquatic Microbial Ecology 76 (3), 243–255.

Grösbacher, M., Eckert, D., Cirpka, O.A., Griebler, C., 2018. Contaminant concentration versus flow velocity: drivers of biodegradation and microbial growth in groundwater model systems. Biodegradation 29 (3), 211–232.

Haack, S.K., Metge, D.W., Fogarty, L.R., Meyer, M.T., Barber, L.B., Harvey, R.W., Leblanc, D.R., Kolpin, D.W., 2012. Effects on groundwater microbial communities of an engineered 30-day in situ exposure to the antibiotic sulfamethoxazole. Environmental Science & Technology 46 (14), 7478–7486.

Handley, K.M., Bartels, D., O'Loughlin, E.J., Williams, K.H., Trimble, W.L., Skinner, K., Gilbert, J.A., Desai, N., Glass, E.M., Paczian, T., Wilke, A., Antonopoulos, D., Kemner, K.M., Meyer, F., 2014. The complete genome sequence for putative $H_2$- and S-oxidizer Candidatus Sulfuricurvum sp., assembled de novo from an aquifer-derived metagenome. Environmental Microbiology 16 (11), 3443–3462.

Hanemaaijer, M., Olivier, B.G., Röling, W.F.M., Bruggeman, F.J., Teusink, B., 2017. Model-based quantification of metabolic interactions from dynamic microbial-community data. PLoS One 12 (3), e0173183.

Hanson, C.A., Fuhrman, J.A., Horner-Devine, M.C., Martiny, J.B.H., 2012. Beyond biogeographic patterns: processes shaping the microbial landscape. Nature Reviews Microbiology 10 (7), 497–506.

Hartmans, S., de Bont, J.A.M., Tramper, J., Luyben, K.C.A.M., 1985. Bacterial degradation of vinyl chloride. Biotechnology Letters 7 (6), 383–388.

Hassan, Z., Sultana, M., Westerhoff, H.V., Khan, S.I., Röling, W.F.M., 2016. Iron cycling potentials of arsenic contaminated groundwater in Bangladesh as revealed by enrichment cultivation. Geomicrobiology Journal 33 (9), 779–792.

Helbling, D.E., 2015. Bioremediation of pesticide-contaminated water resources: the challenge of low concentrations. Current Opinion in Biotechnology 33, 142–148.

Hellal, J., Guédron, S., Huguet, L., Schäfer, J., Laperche, V., Joulian, C., Lanceleur, L., Burnol, A., Ghestem, J.-P., Garrido, F., Battaglia-Brunet, F., 2015. Mercury mobilization and speciation linked to bacterial iron oxide and sulfate reduction: a column study to mimic reactive transfer in an anoxic aquifer. Journal of Contaminant Hydrology 180, 56–68.

Hendrickson, E.R., Payne, J.A., Young, R.M., Starr, M.G., Perry, M.P., Fahnestock, S., Ellis, D.E., Ebersole, R.C., 2002. Molecular analysis of Dehalococcoides 16S ribosomal DNA from chloroethene-contaminated sites throughout North America and Europe. Applied and Environmental Microbiology 68 (2), 485–495.

Hermon, L., Hellal, J., Denonfoux, J., Vuilleumier, S., Imfeld, G., Urien, C., Ferreira, S., Joulian, C., 2019. Functional genes and bacterial communities during organohalide respiration of chloroethenes in microcosms of multi-contaminated groundwater. Frontiers in Microbiology 10, 89.

Herrmann, M., Rusznyák, A., Akob, D.M., Schulze, I., Opitz, S., Totsche, K.U., Küsel, K., 2015. Large fractions of $CO_2$-fixing microorganisms in pristine limestone aquifers appear to be involved in the oxidation of reduced sulfur and nitrogen compounds. Applied and Environmental Microbiology 81 (7), 2384–2394.

Herrmann, M., Wegner, C.-E., Taubert, M., Geesink, P., Lehmann, K., Yan, L., Lehmann, R., Totsche, K.U., Küsel, K., 2019. Predominance of Cand. Patescibacteria in groundwater is caused by their preferential mobilization from soils and flourishing under oligotrophic conditions. Frontiers in Microbiology 10, 1407.

Herzyk, A., Fillinger, L., Larentis, M., Qiu, S., Maloszewski, P., Hünniger, M., Schmidt, S.I., Stumpp, C., Marozava, S., Knappett, P.S.K., Elsner, M., Meckenstock, R., Lueders, T., Griebler, C., 2017. Response and recovery of a pristine groundwater ecosystem impacted by toluene contamination – a meso-scale indoor aquifer experiment. Journal of Contaminant Hydrology 207, 17–30.

Hofmann, R., Griebler, C., 2018. DOM and bacterial growth efficiency in oligotrophic groundwater: absence of priming and co-limitation by organic carbon and phosphorus. Aquatic Microbial Ecology 81 (1), 55–71.

Holliger, C., Wohlfarth, G., Diekert, G., 1998. Reductive dechlorination in the energy metabolism of anaerobic bacteria. FEMS Microbiology Reviews 22 (5), 383–398.

Holmes, D.E., Giloteaux, L., Williams, K.H., Wrighton, K.C., Wilkins, M.J., Thompson, C.A., Roper, T.J., Long, P.E., Lovley, D.R., 2013. Enrichment of specific protozoan populations during in situ bioremediation of uranium-contaminated groundwater. The ISME Journal 7 (7), 1286–1298.

Horemans, B., Vandermaesen, J., Breugelmans, P., Hofkens, J., Smolders, E., Springael, D., 2013. The quantity and quality of dissolved organic matter as supplementary carbon source impacts the pesticide-degrading activity of a triple-species bacterial biofilm. Applied Microbiology and Biotechnology 98 (2), 931–943.

Horemans, B., Hofkens, J., Smolders, E., Springael, D., 2014. Biofilm formation of a bacterial consortium on linuron at micropollutant concentrations in continuous flow chambers and the impact of dissolved organic matter. FEMS Microbiology Ecology 88 (1), 184–194.

Hug, L.A., Thomas, B.C., Brown, C.T., Frischkorn, K.R., Williams, K.H., Tringe, S.G., Banfield, J.F., 2015. Aquifer environment selects for microbial species cohorts in sediment and groundwater. The ISME Journal 9 (8), 1846–1856.

Im, J., Löffler, F.E., 2016. Fate of Bisphenol A in terrestrial and aquatic environments. Environmental Science & Technology 50 (16), 8403–8416.

Jesußek, A., Köber, R., Grandel, S., Dahmke, A., 2013a. Aquifer heat storage: sulphate reduction with acetate at increased temperatures. Environmental Earth Sciences 69 (5), 1763–1771.

Jesußek, A., Grandel, S., Dahmke, A., 2013b. Impacts of subsurface heat storage on aquifer hydrogeochemistry. Environmental Earth Sciences 69 (6), 1999–2012.

Jewell, T.N.M., Karaoz, U., Brodie, E.L., Williams, K.H., Beller, H.R., 2016. Metatranscriptomic evidence of pervasive and diverse chemolithoautotrophy relevant to C, S, N and Fe cycling in a shallow alluvial aquifer. The ISME Journal 10 (9), 2106–2117.

Jones, A.A., Bennett, P.C., 2017. Mineral ecology: surface specific colonization and geochemical drivers of biofilm accumulation, composition, and phylogeny. Frontiers in Microbiology 8, 491.

Kang, P.K., Bresciani, E., An, S., Lee, S., 2019. Potential impact of pore-scale incomplete mixing on biodegradation in aquifers: from batch experiment to field-scale modeling. Advances in Water Resources 123, 1–11.

Kantor, R.S., Wrighton, K.C., Handley, K.M., Sharon, I., Hug, L.A., Castelle, C.J., Thomas, B.C., Banfield, J.F., 2013. Small genomes and sparse metabolisms of sediment-associated bacteria from four candidate phyla. mBio 4 (5), e00708–e00713.

Karwautz, C., Zhou, Y., Kerros, M.-E., Weinbauer, M.G., Griebler, C., 2022. Bottom-up control of the groundwater microbial food-web in an alpine aquifer. Frontiers in Ecology and Evolution 10, 854228.

Kellermann, C., Selesi, D., Lee, N., Hügler, M., Esperschütz, J., Hartmann, A., Griebler, C., 2012. Microbial $CO_2$ fixation potential in a tar-oil-contaminated porous aquifer. FEMS Microbiology Ecology 81 (1), 172–187.

Kleinsteuber, S., Schleinitz, K.M., Vogt, C., 2012. Key players and team play: anaerobic microbial communities in hydrocarbon-contaminated aquifers. Applied Microbiology and Biotechnology 94 (4), 851–873.

Kløve, B., Ala-Aho, P., Bertrand, G., Gurdak, J.J., Kupfersberger, H., Kværner, J., Muotka, T., Mykrä, H., Preda, E., Rossi, P., Bertacchi Uvo, C., Velasco, E., Pulido-Velazquez, M., 2014. Climate change impacts on groundwater and dependent ecosystems. Journal of Hydrology 518, 250–266.

Konopka, A., Lindemann, S., Fredrickson, J., 2015. Dynamics in microbial communities: unraveling mechanisms to identify principles. The ISME Journal 9 (7), 1488–1495.

Kumar, S., Herrmann, M., Thamdrup, B., Schwab, V.F., Geesink, P., Trumbore, S.E., Totsche, K.-U., Küsel, K., 2017. Nitrogen loss from pristine carbonate-rock aquifers of the Hainich Critical Zone Exploratory (Germany) is primarily driven by chemolithoautotrophic anammox processes. Frontiers in Microbiology 8, 1951.

Kumar, S., Herrmann, M., Blohm, A., Hilke, I., Frosch, T., Trumbore, S.E., Küsel, K., 2018. Thiosulfate- and hydrogen-driven autotrophic denitrification by a microbial consortium enriched from groundwater of an oligotrophic limestone aquifer. FEMS Microbiology Ecology 94 (10), fiy141.

Kundu, K., Marozava, S., Ehrl, B., Merl-Pham, J., Griebler, C., Elsner, M., 2019. Defining lower limits of biodegradation: atrazine degradation regulated by mass transfer and maintenance demand in Arthrobacter aurescens TC1. The ISME Journal 13 (9), 2236–2251.

Langenheder, S., Lindström, E.S., 2019. Factors influencing aquatic and terrestrial bacterial community assembly. Environmental Microbiology Reports 11 (3), 306–315.

Lapworth, D.J., Baran, N., Stuart, M.E., Ward, R.S., 2012. Emerging organic contaminants in groundwater: a review of sources, fate and occurrence. Environmental Pollution 163, 287–303.

Larentis, M., Hoermann, K., Lueders, T., 2013. Fine-scale degrader community profiling over an aerobic/anaerobic redox gradient in a toluene-contaminated aquifer. Environmental Microbiology Reports 5 (2), 225–234.

Lau, M.C.Y., Cameron, C., Magnabosco, C., Brown, C.T., Schilkey, F., Grim, S., Hendrickson, S., Pullin, M., Sherwood Lollar, B., van Heerden, E., Kieft, T.L., Onstott, T.C., 2014. Phylogeny and phylogeography of functional genes shared among seven terrestrial subsurface metagenomes reveal N-cycling and microbial evolutionary relationships. Frontiers in Microbiology 5, 531.

Lau, M.C.Y., Kieft, T.L., Kuloyo, O., Linage-Alvarez, B., van Heerden, E., Lindsay, M.R., Magnabosco, C., Wang, W., Wiggins, J.B., Guo, L., Perlman, D.H., Kyin, S., Shwe, H.H., Harris, R.L., Oh, Y., Yi, M.J., Purtschert, R., Slater, G.F., Ono, S., Wei, S., Li, L., Sherwood Lollar, B., Onstott, T.C., 2016. An oligotrophic deep-subsurface community dependent on syntrophy is dominated by sulfur-driven autotrophic denitrifiers. Proceedings of the National Academy of Sciences 113 (49), E7927–E7936.

Lehman, R.M., Colwell, F.S., Bala, G.A., 2001. Attached and unattached microbial communities in a simulated basalt aquifer under fracture- and porous-flow conditions. Applied and Environmental Microbiology 67 (6), 2799–2809.

Lin, B., Westerhoff, H.V., Röling, W.F.M., 2009. How Geobacteraceae may dominate subsurface biodegradation: physiology of *Geobacter metallireducens* in slow-growth habitat-simulating retentostats. Environmental Microbiology 11 (9), 2425–2433.

Lindström, E.S., Langenheder, S., 2012. Local and regional factors influencing bacterial community assembly. Environmental Microbiology Reports 4 (1), 1–9.

Liu, L., Helbling, D.E., Kohler, H.-P.E., Smets, B.F., 2014. A model framework to describe growth-linked biodegradation of trace-level pollutants in the presence of coincidental carbon substrates and microbes. Environmental Science & Technology 48 (22), 13358–13366.

Longnecker, K., Da Costa, A., Bhatia, M., Kujawinski, E.B., 2009. Effect of carbon addition and predation on acetate-assimilating bacterial cells in groundwater. FEMS Microbiology Ecology 70 (3), 456–470.

Lueders, T., 2017. The ecology of anaerobic degraders of BTEX hydrocarbons in aquifers. FEMS Microbiology Ecology 93 (1), fiw220.

Luef, B., Frischkorn, K.R., Wrighton, K.C., Holman, H.-Y.N., Birarda, G., Thomas, B.C., Singh, A., Williams, K.H., Siegerist, C.E., Tringe, S.G., Downing, K.H., Comolli, L.R., Banfield, J.F., 2015. Diverse uncultivated ultra-small bacterial cells in groundwater. Nature Communications 6, 6372.

Ma, J., Guo, H., Lei, M., Zhou, X., Li, F., Yu, T., Wei, R., Zhang, H., Zhang, X., Wu, Y., 2015a. Arsenic adsorption and its fractions on aquifer sediment: effect of pH, arsenic species, and iron/manganese minerals. Water, Air, & Soil Pollution 226, 260.

Ma, J., Deng, Y., Yuan, T., Zhou, J., Alvarez, P.J.J., 2015b. Succession of microbial functional communities in response to a pilot-scale ethanol-blended fuel release throughout the plume life cycle. Environmental Pollution 198, 154–160.

Ma, J., Nossa, C.W., Alvarez, P.J.J., 2015c. Groundwater ecosystem resilience to organic contaminations: microbial and geochemical dynamics throughout the 5-year life cycle of a surrogate ethanol blend fuel plume. Water Research 80, 119–129.

Magnabosco, C., Lin, L.-H., Dong, H., Bomberg, M., Ghiorse, W., Stan-Lotter, H., Pedersen, K., Kieft, T.L., van Heerden, E., Onstott, T.C., 2018. The biomass and biodiversity of the continental subsurface. Nature Geoscience 11 (10), 707–717.

Magnúsdóttir, S., Heinken, A., Kutt, L., Ravcheev, D.A., Bauer, E., Noronha, A., Greenhalgh, K., Jäger, C., Baginska, J., Wilmes, P., Fleming, R.M.T., Thiele, I., 2017. Generation of genome-scale metabolic reconstructions for 773 members of the human gut microbiota. Nature Biotechnology 35 (1), 81–89.

Mahadevan, R., Palsson, B.Ø., Lovley, D.R., 2011. In situ to in silico and back: elucidating the physiology and ecology of *Geobacter* spp. using genome-scale modelling. Nature Reviews Microbiology 9 (1), 39–50.

Marozava, S., Röling, W.F.M., Seifert, J., Küffner, R., von Bergen, M., Meckenstock, R.U., 2014. Physiology of *Geobacter metallireducens* under excess and limitation of electron donors. Part II. Mimicking environmental conditions during cultivation in retentostats. Systematic and Applied Microbiology 37 (4), 287–295.

Mauffret, A., Baran, N., Joulian, C., 2017. Effect of pesticides and metabolites on groundwater bacterial community. Science of the Total Environment 576, 879–887.

McMahon, S., Parnell, J., 2014. Weighing the deep continental biosphere. FEMS Microbiology Ecology 87 (1), 113–120.

Meckenstock, R.U., Elsner, M., Griebler, C., Lueders, T., Stumpp, C., Aamand, J., Agathos, S.N., Albrechtsen, H.-J., Bastiaens, L., Bjerg, P.L., Boon, N., Dejonghe, W., Huang, W.E., Schmidt, S.I., Smolders, E., Sørensen, S.R., Springael, D., van Breukelen, B.M., 2015. Biodegradation: updating the concepts of control for microbial cleanup in contaminated aquifers. Environmental Science & Technology 49 (12), 7073–7081.

Morris, B.E.L., Henneberger, R., Huber, H., Moissl-Eichinger, C., 2013. Microbial syntrophy: interaction for the common good. FEMS Microbiology Reviews 37 (3), 384–406.

Mukhopadhyay, R., Rosen, B.P., Phung, L.T., Silver, S., 2002. Microbial arsenic: from geocycles to genes and enzymes. FEMS Microbiology Reviews 26 (3), 311–325.

Muller, E.E.L., Faust, K., Widder, S., Herold, M., Martínez Arbas, S., Wilmes, P., 2018. Using metabolic networks to resolve ecological properties of microbiomes. Current Opinion in Systems Biology 8, 73–80.

Müller, H., Bosch, J., Griebler, C., Damgaard, L.R., Nielsen, L.P., Lueders, T., Meckenstock, R.U., 2016. Long-distance electron transfer by cable bacteria in aquifer sediments. The ISME Journal 10 (8), 2010–2019.

Müller, H., Marozava, S., Probst, A.J., Meckenstock, R.U., 2020. Groundwater cable bacteria conserve energy by sulfur disproportionation. The ISME Journal 14, 623–634.

Nagaosa, K., Maruyama, T., Welikala, N., Yamashita, Y., Saito, Y., Kato, K., Fortin, D., Nanba, K., Miyasaka, I., Fukunaga, S., 2008. Active bacterial populations and grazing impact revealed by an in situ experiment in a shallow aquifer. Geomicrobiology Journal 25 (3–4), 131–141.

Nemergut, D.R., Schmidt, S.K., Fukami, T., O'Neill, S.P., Bilinski, T.M., Stanish, L.F., Knelman, J.E., Darcy, J.L., Lynch, R.C., Wickey, P., Ferrenberg, S., 2013. Patterns and processes of microbial community assembly. Microbiology and Molecular Biology Reviews 77 (3), 342–356.

Nijenhuis, I., Stollberg, R., Lechner, U., 2018. Anaerobic microbial dehalogenation and its key players in the contaminated Bitterfeld-Wolfen megasite. FEMS Microbiology Ecology 94 (4), fiy012.

Nordstrom, D.K., 2002. Public health. Worldwide occurrences of arsenic in ground water. Science 296 (5576), 2143–2145.

Okkonen, J., Jyrkama, M., Kløve, B., 2010. A conceptual approach for assessing the impact of climate change on groundwater and related surface waters in cold regions (Finland). Hydrogeology Journal 18 (2), 429–439.

Oremland, R.S., Stolz, J.F., 2003. The ecology of arsenic. Science 300 (5621), 939–944.

Orth, J.D., Thiele, I., Palsson, B.Ø., 2010. What is flux balance analysis? Nature Biotechnology 28 (3), 245–248.

Osborne, T.H., McArthur, J.M., Sikdar, P.K., Santini, J.M., 2015. Isolation of an arsenate-respiring bacterium from a redox front in an arsenic-polluted aquifer in West Bengal, Bengal Basin. Environmental Science & Technology 49 (7), 4193–4199.

Pacheco, A.R., Moel, M., Segrè, D., 2019. Costless metabolic secretions as drivers of interspecies interactions in microbial ecosystems. Nature Communications 10 (1), 103.

Pan, D., Watson, R., Wang, D., Tan, Z.H., Snow, D.D., Weber, K.A., 2014. Correlation between viral production and carbon mineralization under nitrate-reducing conditions in aquifer sediment. The ISME Journal 8 (8), 1691–1703.

Parks, J.M., Johs, A., Podar, M., Bridou, R., Hurt Jr., R.A., Smith, S.D., Tomanicek, S.J., Qian, Y., Brown, S.D., Brandt, C.C., Palumbo, A.V., Smith, J.C., Wall, J.D., Elias, D.A., Liang, L., 2013. The genetic basis for bacterial mercury methylation. Science 339 (6125), 1332–1335.

Pilloni, G., Bayer, A., Ruth-Anneser, B., Fillinger, L., Engel, M., Griebler, C., Lueders, T., 2019. Dynamics of hydrology and anaerobic hydrocarbon degrader communities in A tar-oil contaminated aquifer. Microorganisms 7 (2), 46.

Pimm, S.L., 1984. The complexity and stability of ecosystems. Nature 307, 321–326.

Probst, A.J., Ladd, B., Jarett, J.K., Geller-McGrath, D.E., Sieber, C.M.K., Emerson, J.B., Anantharaman, K., Thomas, B.C., Malmstrom, R.R., Stieglmeier, M., Klingl, A., Woyke, T., Cathryn, R.M., Banfield, J.F., 2018. Differential depth distribution of microbial function and putative symbionts through sediment-hosted aquifers in the deep terrestrial subsurface. Nature Microbiology 3 (3), 328–336.

Quéméneur, M., Heinrich-Salmeron, A., Muller, D., Lièvremont, D., Jauzein, M., Bertin, P.N., Garrido, F., Joulian, C., 2008. Diversity surveys and evolutionary relationships of *aoxB* genes in aerobic arsenite-oxidizing bacteria. Applied and Environmental Microbiology 74 (14), 4567–4573.

Retter, A., Karwautz, C., Griebler, C., 2020. Groundwater microbial communities in times of climate change. Current Issues in Molecular Biology 41, 509–537.

Rinke, C., Schwientek, P., Sczyrba, A., Ivanova, N.N., Anderson, I.J., Cheng, J.-F., Darling, A., Malfatti, S., Swan, B.K., Gies, E.A., Dodsworth, J.A., Hedlund, B.P., Tsiamis, G., Sievert, S.M., Liu, W.-T., Eisen, J.A., Hallam, S.J., Kyrpides, N.C., Stepanauskas, R., Rubin, E.M., Hugenholtz, P., Woyke, T., 2013. Insights into the phylogeny and coding potential of microbial dark matter. Nature 499 (7459), 431–437.

Rodenburg, L.A., Du, S., Fennell, D.E., Cavallo, G.J., 2010. Evidence for widespread dechlorination of polychlorinated biphenyls in groundwater, landfills, and wastewater collection systems. Environmental Science & Technology 44 (19), 7534–7540.

Röling, W.F.M., van Breukelen, B.M., Bruggeman, F.J., Westerhoff, H.V., 2007. Ecological control analysis: being(s) in control of mass flux and metabolite concentrations in anaerobic degradation processes. Environmental Microbiology 9 (2), 500–511.

Rossi, P., Shani, N., Kohler, F., Imfeld, G., Holliger, C., 2012. Ecology and biogeography of bacterial communities associated with chloroethene-contaminated aquifers. Frontiers in Microbiology 3, 260.

Santini, J.M., Sly, L.I., Schnagl, R.D., Macy, J.M., 2000. A new chemolithoautotrophic arsenite-oxidizing bacterium isolated from a gold mine: phylogenetic, physiological, and preliminary biochemical studies. Applied and Environmental Microbiology 66 (1), 92–97.

Scheibe, T.D., Mahadevan, R., Fang, Y., Garg, S., Long, P.E., Lovley, D.R., 2009. Coupling a genome-scale metabolic model with a reactive transport model to describe in situ uranium bioremediation. Microbial Biotechnology 2 (2), 274–286.

Schmidt, S.I., Cuthbert, M.O., Schwientek, M., 2017. Towards an integrated understanding of how micro scale processes shape groundwater ecosystem functions. Science of the Total Environment 592, 215–227.

Shabarova, T., Widmer, F., Pernthaler, J., 2013. Mass effects meet species sorting: transformations of microbial assemblages in epiphreatic subsurface karst water pools. Environmental Microbiology 15 (9), 2476–2488.

Schweichhart, J.S., Pleyer, D., Winter, C., Retter, A., Griebler, C., 2022. Presence and role of prokaryotic viruses in groundwater environments. In: Mehner, T., Tockner, K. (Eds.), Encyclopedia of Inland Waters, 2nd Edition. Elsevier, pp. 373–384.

Shabarova, T., Villiger, J., Morenkov, O., Niggemann, J., Dittmar, T., Pernthaler, J., 2014. Bacterial community structure and dissolved organic matter in repeatedly flooded subsurface karst water pools. FEMS Microbiology Ecology 89 (1), 111–126.

Shade, A., Peter, H., Allison, S.D., Baho, D.L., Berga, M., Bürgmann, H., Huber, D.H., Langenheder, S., Lennon, J.T., Martiny, J.B.H., Matulich, K.L., Schmidt, T.M., Handelsman, J., 2012. Fundamentals of microbial community resistance and resilience. Frontiers in Microbiology 3, 417.

Shen, Y., Chapelle, F.H., Strom, E.W., Benner, R., 2015. Origins and bioavailability of dissolved organic matter in groundwater. Biogeochemistry 122 (1), 61–78.

Smedley, P.L., Kinniburgh, D.G., 2002. A review of the source, behaviour and distribution of arsenic in natural waters. Applied Geochemistry 17 (5), 517–568.

Smith, H.J., Zelaya, A.J., De León, K.B., Chakraborty, R., Elias, D.A., Hazen, T.C., Arkin, A.P., Cunningham, A.B., Fields, M.W., 2018. Impact of hydrologic boundaries on microbial planktonic and biofilm communities in shallow terrestrial subsurface environments. FEMS Microbiology Ecology 94 (12), fiy191.

Smith, R.J., Jeffries, T.C., Roudnew, B., Fitch, A.J., Seymour, J.R., Delpin, M.W., Newton, K., Brown, M.H., Mitchell, J.G., 2012. Metagenomic comparison of microbial communities inhabiting confined and unconfined aquifer ecosystems. Environmental Microbiology 14 (1), 240–253.

Spona-Friedl, M., Braun, A., Huber, C., Eisenreich, W., Griebler, C., Kappler, A., Elsner, M., 2020. Substrate-dependent $CO_2$-fixation in heterotrophic bacteria revealed by stable isotope labelling. FEMS Microbiology Ecology 96, fiaa080.

Staats, M., Braster, M., Röling, W.F.M., 2011. Molecular diversity and distribution of aromatic hydrocarbon-degrading anaerobes across a landfill leachate plume. Environmental Microbiology 13 (5), 1216−1227.

Stegen, J.C., Lin, X., Fredrickson, J.K., Chen, X., Kennedy, D.W., Murray, C.J., Rockhold, M.L., Konopka, A., 2013. Quantifying community assembly processes and identifying features that impose them. The ISME Journal 7 (11), 2069−2079.

Stegen, J.C., Konopka, A., McKinley, J.P., Murray, C., Lin, X., Miller, M.D., Kennedy, D.W., Miller, E.A., Resch, C.T., Fredrickson, J.K., 2016. Coupling among microbial communities, biogeochemistry, and mineralogy across biogeochemical facies. Scientific Reports 6, 30553.

Stegen, J.C., Bottos, E.M., Jansson, J.K., 2018a. A unified conceptual framework for prediction and control of microbiomes. Current Opinion in Microbiology 44, 20−27.

Stegen, J.C., Johnson, T., Fredrickson, J.K., Wilkins, M.J., Konopka, A.E., Nelson, W.C., Arntzen, E.V., Chrisler, W.B., Chu, R.K., Fansler, S.J., Graham, E.B., Kennedy, D.W., Resch, C.T., Tfaily, M., Zachara, J., 2018b. Influences of organic carbon speciation on hyporheic corridor biogeochemistry and microbial ecology. Nature Communications 9 (1), 585.

Stolyar, S., Van Dien, S., Hillesland, K.L., Pinel, N., Lie, T.J., Leigh, J.A., Stahl, D.A., 2007. Metabolic modeling of a mutualistic microbial community. Molecular Systems Biology 3 (1), 92.

Svoboda, P., Lindström, E.S., Ahmed Osman, O., Langenheder, S., 2018. Dispersal timing determines the importance of priority effects in bacterial communities. The ISME Journal 12 (2), 644−646.

Tas, N., van Eekert, M.H.A., de Vos, W.M., Smidt, H., 2010. The little bacteria that can - diversity, genomics and ecophysiology of *Dehalococcoides* spp. in contaminated environments. Microbial Biotechnology 3 (4), 389−402.

Taubert, M., Vogt, C., Wubet, T., Kleinsteuber, S., Tarkka, M.T., Harms, H., Buscot, F., Richnow, H.-H., von Bergen, M., Seifert, J., 2012. Protein-SIP enables time-resolved analysis of the carbon flux in a sulfate-reducing, benzene-degrading microbial consortium. The ISME Journal 6 (12), 2291−2301.

Taubert, M., Stähly, J., Kolb, S., Küsel, K., 2019. Divergent microbial communities in groundwater and overlying soils exhibit functional redundancy for plant-polysaccharide degradation. PLoS One 14 (3), e0212937.

Taylor, R.G., Scanlon, B., Döll, P., Rodell, M., van Beek, R., Wada, Y., Longuevergne, L., Leblanc, M., Famiglietti, J.S., Edmunds, M., Konikow, L., Green, T.R., Chen, J., Taniguchi, M., Bierkens, M.F.P., MacDonald, A., Fan, Y., Maxwell, R.M., Yechieli, Y., Gurdak, J.J., Allen, D.M., Shamsudduha, M., Hiscock, K., Yeh, P.J.-F., Holman, I., Treidel, H., 2012. Ground water and climate change. Nature Climate Change 3, 322−329.

Thornton, S.F., Quigley, S., Spence, M.J., Banwart, S.A., Bottrell, S., Lerner, D.N., 2001. Processes controlling the distribution and natural attenuation of dissolved phenolic compounds in a deep sandstone aquifer. Journal of Contaminant Hydrology 53 (3−4), 233−267.

Thullner, M., Regnier, P., 2019. Microbial controls on the biogeochemical dynamics in the subsurface. Reviews in Mineralogy and Geochemistry 85 (1), 265−302.

Trautwein, K., Kühner, S., Wöhlbrand, L., Halder, T., Kuchta, K., Steinbüchel, A., Rabus, R., 2008. Solvent stress response of the denitrifying bacterium *Aromatoleum aromaticum* strain EbN1. Applied and Environmental Microbiology 74 (8), 2267−2274.

Underwood, J.C., Harvey, R.W., Metge, D.W., Repert, D.A., Baumgartner, L.K., Smith, R.L., Roane, T.M., Barber, L.B., 2011. Effects of the antimicrobial sulfamethoxazole on groundwater bacterial enrichment. Environmental Science & Technology 45 (7), 3096−3101.

Urbaniak, M., 2007. Polychlorinated biphenyls: sources, distribution and transformation in the environment - a literature review. Acta Toxicologica 15 (2), 83−93.

Urbaniak, M., 2013. Biodegradation of PCDDs/PCDFs and PCBs. In: Chamy, R., Rosenkranz, F. (Eds.), Biodegradation − Engineering and Technology. InTech Publishing, Rijeka, pp. 73−100.

van Waasbergen, L.G., Balkwill, D.L., Crocker, F.H., Bjornstad, B.N., Miller, R.V., 2000. Genetic diversity among *Arthrobacter* species collected across a heterogeneous series of terrestrial deep-subsurface sediments as determined on the basis of 16S rRNA and recA gene sequences. Applied and Environmental Microbiology 66 (8), 3454−3463.

Wan, J., Klein, J., Simon, S., Joulian, C., Dictor, M.-C., Deluchat, V., Dagot, C., 2010. AsIII oxidation by *Thiomonas arsenivorans* in up-flow fixed-bed reactors coupled to as sequestration onto zero-valent iron-coated sand. Water Research 44 (17), 5098−5108.

Wang, S., Zhu, G., Zhuang, L., Li, Y., Liu, L., Lavik, G., Berg, M., Liu, S., Long, X.-E., Guo, J., Jetten, M.S.M., Kuypers, M.M.M., Li, F., Schwark, L., Yin, C., 2020. Anaerobic ammonium oxidation is a major N-sink in aquifer systems around the world. The ISME Journal 14, 151−163.

Wanger, G., Southam, G., Onstott, T.C., 2006. Structural and chemical characterization of a natural fracture surface from 2.8 kilometers below land surface: biofilms in the deep subsurface. Geomicrobiology Journal 23 (6), 443−452.

Weatherill, J.J., Atashgahi, S., Schneidewind, U., Krause, S., Ullah, S., Cassidy, N., Rivett, M.O., 2018. Natural attenuation of chlorinated ethenes in hyporheic zones: a review of key biogeochemical processes and in-situ transformation potential. Water Research 128, 362–382.

Weaver, L., Webber, J.B., Hickson, A.C., Abraham, P.M., Close, M.E., 2015. Biofilm resilience to desiccation in groundwater aquifers: a laboratory and field study. Science of the Total Environment 514, 281–289.

Weelink, S.A.B., van Eekert, M.H.A., Stams, A.J.M., 2010. Degradation of BTEX by anaerobic bacteria: physiology and application. Reviews in Environmental Science and Biotechnology 9 (4), 359–385.

Wegner, C.-E., Gaspar, M., Geesink, P., Herrmann, M., Marz, M., Küsel, K., 2019. Biogeochemical regimes in shallow aquifers reflect the metabolic coupling of the elements nitrogen, sulfur, and carbon. Applied and Environmental Microbiology 85 (5), e02346-18.

Winderl, C., Penning, H., von Netzer, F., Meckenstock, R.U., Lueders, T., 2010. DNA-SIP identifies sulfate-reducing *Clostridia* as important toluene degraders in tar-oil-contaminated aquifer sediment. The ISME Journal 4 (10), 1314–1325.

Wrighton, K.C., Thomas, B.C., Sharon, I., Miller, C.S., Castelle, C.J., VerBerkmoes, N.C., Wilkins, M.J., Hettich, R.L., Lipton, M.S., Williams, K.H., Long, P.E., Banfield, J.F., 2012. Fermentation, hydrogen, and sulfur metabolism in multiple uncultivated bacterial phyla. Science 337 (6102), 1661–1665.

Wrighton, K.C., Castelle, C.J., Wilkins, M.J., Hug, L.A., Sharon, I., Thomas, B.C., Handley, K.M., Mullin, S.W., Nicora, C.D., Singh, A., Lipton, M.S., Long, P.E., Williams, K.H., Banfield, J.F., 2014. Metabolic interdependencies between phylogenetically novel fermenters and respiratory organisms in an unconfined aquifer. The ISME Journal 8 (7), 1452–1463.

Wu, X., Wu, L., Liu, Y., Zhang, P., Li, Q., Zhou, J., Hess, N.J., Hazen, T.C., Yang, W., Chakraborty, R., 2018. Microbial interactions with dissolved organic matter drive carbon dynamics and community succession. Frontiers in Microbiology 9, 1234.

Yagi, J.M., Neuhauser, E.F., Ripp, J.A., Mauro, D.M., Madsen, E.L., 2010. Subsurface ecosystem resilience: long-term attenuation of subsurface contaminants supports a dynamic microbial community. The ISME Journal 4 (1), 131–143.

Yan, J., Im, J., Yang, Y., Löffler, F.E., 2013. Guided cobalamin biosynthesis supports *Dehalococcoides mccartyi* reductive dechlorination activity. Philosophical Transactions of the Royal Society B 368 (1616), 20120320.

Yan, L., Herrmann, M., Kampe, B., Lehmann, R., Totsche, K.U., Küsel, K., 2020. Environmental selection shapes the formation of near-surface groundwater microbiomes. Water Research 170, 115341.

Zedelius, J., Rabus, R., Grundmann, O., Werner, I., Brodkorb, D., Schreiber, F., Ehrenreich, P., Behrends, A., Wilkes, H., Kube, M., Reinhardt, R., Widdel, F., 2011. Alkane degradation under anoxic conditions by a nitrate-reducing bacterium with possible involvement of the electron acceptor in substrate activation. Environmental Microbiology Reports 3 (1), 125–135.

Zelezniak, A., Andrejev, S., Ponomarova, O., Mende, D.R., Bork, P., Patil, K.R., 2015. Metabolic dependencies drive species co-occurrence in diverse microbial communities. Proceedings of the National Academy of Sciences 112 (20), 6449–6454.

Zha, Y., Berga, M., Comte, J., Langenheder, S., 2016. Effects of dispersal and initial diversity on the composition and functional performance of bacterial communities. PLoS One 11 (5), e0155239.

Zhou, J., Ning, D., 2017. Stochastic community assembly: does it matter in microbial ecology? Microbiology and Molecular Biology Reviews 81 (4), e00002–e00017.

Zhou, J., Deng, Y., Zhang, P., Xue, K., Liang, Y., Van Nostrand, J.D., Yang, Y., He, Z., Wu, L., Stahl, D.A., Hazen, T.C., Tiedje, J.M., Arkin, A.P., 2014. Stochasticity, succession, and environmental perturbations in a fluidic ecosystem. Proceedings of the National Academy of Sciences 111 (9), E836–E845.

Zhou, Y., Kellermann, C., Griebler, C., 2012. Spatio-temporal patterns of microbial communities in a hydrologically dynamic pristine aquifer. FEMS Microbiology Ecology 81 (1), 230–242.

Zhuang, K., Izallalen, M., Mouser, P., Richter, H., Risso, C., Mahadevan, R., Lovley, D.R., 2011. Genome-scale dynamic modeling of the competition between *Rhodoferax* and *Geobacter* in anoxic subsurface environments. The ISME Journal 5 (2), 305–316.

Zomorrodi, A.R., Islam, M.M., Maranas, C.D., 2014. d-OptCom: dynamic multi-level and multi-objective metabolic modeling of microbial communities. ACS Synthetic Biology 3 (4), 247–257.

# Groundwater food webs

*Michael Venarsky[1,2], Kevin S. Simon[3], Mattia Saccò[4], Clémentine François[5], Laurent Simon[5] and Christian Griebler[6]*

[1]Department of Biodiversity Conservation and Attractions, Kensington, WA, Australia; [2]Australian Rivers Institute, Griffith University, Nathan, QLD, Australia; [3]School of Environment, University of Auckland, Auckland, New Zealand; [4]Subterranean Research and Groundwater Ecology (SuRGE) Group, Trace and Environmental DNA (TrEnD) Laboratory, School of Molecular and Life Sciences, Curtin University, Perth, WA, Australia; [5]Univ Lyon, Université Claude Bernard Lyon 1, CNRS, ENTPE, UMR 5023 LEHNA, Villeurbanne, France; [6]University of Vienna, Department of Functional & Evolutionary Ecology, Vienna, Austria

## Introduction

Ecological studies are built around understanding the relationships between organisms and their environment. Food webs are "the scaffolding around which ecological communities assemble" (Paine, 1996) and thereby provide a flexible framework from which to explore the interactions between organisms and their environment (Power and Dietrich, 2002). The flexibility of this framework can be seen in the different contexts within which food web processes are used to explain ecological patterns. Spatial and temporal patterns in population demographics (recruitment-, survival-, and migration-rates) as well as community structure (biodiversity, relative abundance of species) have been attributed to various food web mechanisms, including bottom-up limitation, top-down pressure from consumption, and resource subsidies (Power, 1992; Polis and Strong, 1996; Shurin et al., 2006). More detailed examinations of consumer—resource interactions have also revealed that community productivity is often supported by multiple sources of energy (autotrophic and detrital) and that the importance of these pathways are a function of spatial factors (headwater stream vs. lowland river), temporal patterns (summer vs. winter), and measures of resource quality (stoichiometric elemental or fatty acid ratios) (Sterner and Elser, 2002; Moore et al., 2004; Marcarelli et al., 2011; Junker et al., 2014). Thus, food web studies have been central in the efforts to understand the processes that support and maintain ecosystem structure and function.

In this chapter, we explore groundwater ecosystem dynamics in the context of food web processes. Every food web is ultimately limited by either the quantity or quality of basal energy production. Thus, we begin by outlining the two basal sources of energy that support groundwater community productivity: particulate and dissolved organic matter (collectively referred to as detritus) from the surface and reduced inorganic compounds from geological sources. Next, we describe how food web processes influence groundwater community dynamics and discuss how the physical habitat template of groundwater ecosystems influences consumer–resource interactions. Groundwater food webs are generally perceived as "less complex" than their surface counterparts and we explore how this perspective is likely an artifact of the cryptic nature of consumer–resource interactions in groundwater food webs. Lastly, we highlight several areas of research that will likely advance our understanding of how food web processes influence the structure and function of groundwater communities.

## Basal energy dynamics in groundwater food webs

### Detrital food webs

Most food webs in aquatic surface ecosystems are supported by two sources of energy: photosynthetic primary production and detritus. In contrast, most groundwater communities have access to only detrital sources of energy due to the lack of photosynthetic primary production (but see below for chemolithoautotrophic primary production) (Simon et al., 2007). In this section we describe the origins and transport of detritus into groundwater ecosystems as well as how detrital energy is transferred to higher trophic levels in groundwater food webs. For the purposes of this chapter, we define detritus as all nonliving particulate and dissolved organic matter originating as either exudates from living organisms or from the decomposition of plants and animals.

#### *Transport*

Detritus enters groundwater ecosystems in two forms: (i) particulate organic matter (POM; leaves, dead wood, guano, and animal carrion) and (ii) dissolved organic matter (DOM), which is derived from the breakdown and leaching of plant organic matter, exudates from plant roots, and the by-products of microbial metabolism (Schiff et al., 1997; Birdwell and Engel, 2010; Simon et al., 2010). Note that POM can be classified into size fractions, including coarse particulate organic matter (CPOM > 1 mm) and fine particulate organic matter (1 mm > FPOM > 0.45 µm). The type of groundwater ecosystem influences the type, quantity, and quality of organic matter inputs. For example, groundwater in unconsolidated sediments receive mostly DOM and FPOM, while karst aquifers can receive relatively high inputs of CPOM due to larger conduits (cave openings, sinking streams) that connect surface and groundwater habitats (Gibert, 1986; Simon and Benfield, 2001; Graening and Brown, 2003; Simon et al., 2003; Huntsman et al., 2011; Venarsky et al., 2012a,b, 2014; Simon, 2013).

Organic matter is also created, processed, and transformed along transport pathways. Transport distances of CPOM (leaves and wood) are generally short, with the majority of CPOM being retained and processed near points of entry into subsurface environments, including soil horizons, interstitial spaces in sand and gravel aquifers, and fractures and large passages in bedrock aquifers (Simon et al., 2001). We know little about FPOM transport in groundwater ecosystems, but FPOM is smaller than CPOM and is thus more likely to be transported throughout groundwater ecosystems. A substantial fraction (15%–34%) of DOM can be derived from the processing of FPOM and CPOM along transport pathways (Shen et al., 2015), whereas a significant proportion of DOM is also retained due to microbial consumption, adsorption to soils, and flocculation (Baker et al., 2000; Pabich et al., 2001; Simon et al., 2010; Mermillod-Blondin et al., 2015; Hofmann and Griebler, 2018; Hofmann et al., 2020). For example, deeper groundwater systems will generally receive less DOM than shallower systems due to differences in water transport times (Pabich et al., 2001; Simon et al., 2010). Additionally, DOM is a complex mixture of organic compounds differing in quality (e.g., carbon: nitrogen: phosphorus ratios) (Hofmann et al., 2020) and the high quality fractions of DOM are generally retained during transport, resulting in a relatively low proportion (7%–42% depending on the study) of DOM that is bioavailable in groundwater ecosystems (Mindl et al., 2000; van Beynen et al., 2000, 2002; Foulquier et al., 2011; Mermillod-Blondin et al., 2015; Shen et al., 2015).

### Transfer to higher trophic levels

Some detrital inputs (e.g., animal carrion and guano, leaf fragments) can be directly assimilated by invertebrate and vertebrate consumers, but most POM and DOM must first be converted to microbial biomass before being transferred to higher trophic levels (Simon et al., 2003; Hancock et al., 2005, Fig. 10.1). Indeed, microbial biofilms can represent up to 80% of the total organic carbon available to consumers in groundwater habitats and are also of higher quality than POM (Francois et al., 2020). While POM often represents a small fraction of the diet of groundwater consumers (Venarsky et al., 2014; Francois et al., 2016b), it can supply key compounds that support the productivity of groundwater species (Saccò et al., 2022a). For example, Mondy et al. (2014) showed that phytosterols present in POM acted as precursors for the synthesis of steroids in the asellid isopod *Proasellus merdianus*.

## Chemolithoautotrophic food webs

Despite the lack of light, primary production does occur in some groundwater ecosystems via chemolithoautotrophic prokaryotes (Engel, 2007; Venarsky and Huntsman, 2018; Fillinger et al., this volume; Fig. 10.1). These bacteria use reduced inorganic compounds (rather than sunlight) as an electron donor ($H_2$, $NH_4^+$, $NO_2^-$, $HS^-$, $S^0$, $S_2O_3^{2-}$, CO, $Fe^{2+}$) coupled with various types of electron acceptors ($O_2$, $NO_3^-$, $SO_4^{2-}$, $S^0$, $S_2O_3^{2-}$, $Fe^{3+}$, $CO_2$) in the appropriate combination (commonly referred to as a redox gradient) to gain the energy necessary to fix inorganic carbon (carbon dioxide, hydrogen carbonate, bicarbonate) into bacteria biomass (Berg, 2011). The types of chemolithoautotrophic prokaryotes found in groundwater ecosystems include nitrifying bacteria (oxidizing ammonium), iron bacteria (oxidizing $Fe^{2+}$), sulfur bacteria (oxidizing $HS^-$, $S^0$, $S_2O_3^{2-}$), and hydrogen bacteria and archaea (oxidizing $H_2$) including hydrogenotrophic methanogens and acetogens (Chapelle et al., 2002; Nealson et al., 2005; Herrmann et al., 2015; Kumar et al., 2017; Engel, 2019; Wang et al., 2020; Elbourne et al., 2022). While bacteria can use these metabolic pathways to grow and reproduce, chemolithoautotrophic $CO_2$ fixation is energy intensive with the main limitation being the availability of favorable redox gradients (Kellermann et al., 2012).

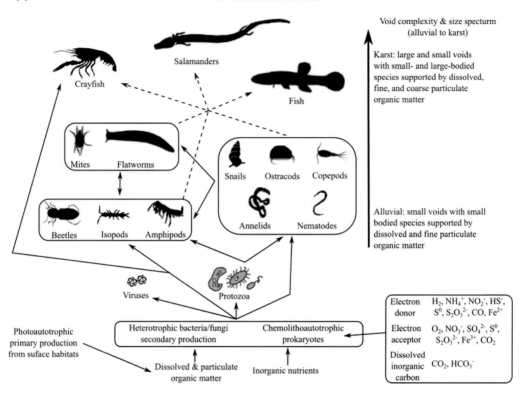

FIGURE 10.1    General groundwater food web illustrating: (i) different sources of basal energy production (heterotrophic bacteria/fungi secondary production vs. chemolithoautotrophic prokaryotes), (ii) general energy flow pathways, and (iii) how void size can influence community composition and food web complexity.

   The environmental conditions necessary to support chemolithoautotrophic primary production (e.g., low oxygen, high hydrogen sulfide concentrations) produces a toxic environment for most eukaryotes (Tobler et al., 2006; Engel, 2007; Tobler, 2008). Despite these extreme conditions, chemolithoautotrophic primary production can support complex food webs that include both vertebrate and invertebrate taxa (Sarbu et al., 1996; Pohlman et al., 1997; Sarbu, 2001; Opsahl and Chanton, 2006; Plath et al., 2007; Tobler, 2008; Dattagupta et al., 2009; Roach et al., 2011; Pohlman, 2011; Neisch et al., 2012; Hutchins, 2013; Tobler et al., 2013; Herrmann et al., 2020). However, the environmental conditions required to facilitate chemolithoautotrophic primary production appear to be found in an increasing number of groundwater systems (Alfreider et al., 2003; Alfreider et al., 2009; Kellermann et al., 2012; Hutchins et al., 2014; Herrmann et al., 2015; Kumar et al., 2017). Moreover, while eukaryotic photoautotrophs all use the Calvin cycle for $CO_2$ fixation, prokaryotes have several pathways to fix inorganic carbon under various environmental conditions (Kinkle and Kane, 2001; Thauer, 2007; Berg, 2011). Thus, although $CO_2$ fixation might quantitatively be less important for the total organic matter pool in groundwater compared to surface-derived organic matter inputs, chemolithoautotrophic production could be more prevalent in groundwater ecosystems than currently appreciated.

Interestingly, $CO_2$-fixation in nature is not restricted to autotrophic organisms. Virtually all heterotrophic organisms (microbes to humans) incorporate $CO_2$ via a variety of central and peripheral metabolic pathways, which involve many kinds of carboxylases (Spona-Friedl et al., 2020). For example, anaplerotic carboxylases incorporate $CO_2$ into biomass and replenish intermediates of the tricarboxylic acid (TCA) cycle, which are constantly withdrawn for the biosynthesis of amino acids and other metabolic products (Erb, 2011). These reactions appear to be so prevalent that they can contribute up to 50% of the cell carbon in methano-trophic bacteria and up to 10% of the carbon locked into global prokaryotic biomass (Braun et al., 2021). Thus, while photolitho- and chemolitho-autotrophy are the most recognized pathways of inorganic carbon fixation, the importance of heterotrophic $CO_2$ fixation remains enigmatic.

## The role of habitat in groundwater food web dynamics

In aquatic surface ecosystems, habitat can be viewed as a set of physical (substrate type and distribution, flow regime) and chemical (conductivity, nutrients) factors that are influenced by a hierarchical set of climatic (precipitation, temperature), terrestrial (vegetation, soil type, geology), and geomorphic (discharge, channel gradient and geometry) elements (Thorp et al., 2006; Winemiller et al., 2010). The same logic can be used to describe the habitat present in groundwater ecosystems, albeit with system-specific factors, such as sediment permeability, void size, distance from the surface, soil depth, and groundwater recharge area. In this section we explore how the structure of the groundwater habitat influences food web processes.

### Network structure

River catchments comprise dendritic networks of channels with predictable characteristics (gradient, width, geometry; Church, 2002), which are known drivers of food web structure (Vannote et al., 1980; Thorp et al., 2006). Similarly, groundwater systems are composed of a network of voids that transport water and materials, but the structure of these voids are specific to the type of groundwater system. In alluvial systems voids are generally of more similar size (i.e., small amount variation in cross-sectional area) compared to karst systems, which contain voids ranging from small fractures to large cave passages (Cornu et al., 2013). These patterns in void size result in alluvial systems having a relatively homogenous distribution of flow paths (i.e., low variation in sediment permeability) and the karst systems having preferential fast and slow flow paths due to the large variation in void sizes (Larned, 2012). These patterns of void sizes can influence food web dynamics via several mechanisms.

In river networks, channel width and depth influences food web dynamics in many ways. Wider channels are less likely to be shaded by riparian vegetation and thereby allow higher rates of primary production, while deeper channels generally have large-bodied predatory species that are more likely to exert top-down control on prey populations (Vannote et al., 1980; Thorp et al., 2006). Similarly, the differences in void structure between alluvial and karst groundwater systems strongly influences energy dynamics and community composition. The small voids in alluvial systems have a high surface area to volume ratio, which results in relatively slow water transport times compared to the larger conduits in karst systems (Larned, 2012). The slow travel time in alluvial systems enhances contact with biological active surfaces (biofilms), resulting in increased processing of DOM and the depletion of bioavailable organic matter. DOM processing can create localized areas of low oxygen concentration, which is a known driver of community composition in groundwater ecosystems(Grimm and Fisher, 1984; Dole-Olivier et al., 2009; Cornu et al., 2013; Culver and Pipan, 2019). While the small conduits in karst systems function in a similar way to those found in alluvial systems, the large conduits allow for more rapid transport of DOM and POM of various sizes (leaf fragments to logs; Gibert and Deharveng, 2002; Stoch et al., 2009; Dole-Olivier et al., 2009; Hahn, 2009). Thus, the differences in void size and permeability within and among groundwater ecosystems can influence food web dynamics by impacting the transport and retention of organic matter (Gnaspini and Trajano, 2001; Huntsman et al., 2011; Venarsky et al., 2012a,b; Pellegrini and Ferreira, 2013;Venarsky et al., 2014; Venarsky and Huntsman, 2018).

Void size also presents a physical limit to the body size of groundwater species (Pipan and Culver, 2017), which may influence food web structure by limiting the types of taxa present. In alluvial systems, the smaller range of void sizes results in relatively small-bodied species contributing to community composition, while karst systems have small-to large-bodied species (microcrustaceans to vertebrates) due to the presence of both small and large void sizes (Fig. 10.1). A larger range in void size potentially facilitates niche differentiation, which can influence food web structure (Hose et al., 2017; Ercoli et al., 2019; Kozel et al., 2019). For example, amphipod species of the genus *Niphargus* in cave streams are larger than species in aquifers with small void sizes and larger-bodied species typically have larger gnathopods and are more likely to be predatory (Fišer et al., 2017, 2019; Premate et al., 2021). Thus, void size can influence microhabitat diversity and thereby trophic niche partitioning, but this hypothesis requires further testing.

## Surface-subsurface connectivity

Hynes (1975) stated that "… in every respect the valley rules the stream", meaning that the terrestrial environment has a strong influence on surface streams, and groundwater ecosystems are no different. Land-cover and precipitation patterns play roles in shaping groundwater communities (Baker et al., 2000; Boulton et al., 2008; Español et al., 2017). Rainfall can increase dissolved oxygen levels and transport organic matter to aquifers, which can trigger changes in community composition (Datry et al., 2005; Dole-Olivier et al., 2009; Brankovits et al., 2017; Hofmann et al., 2020; Saccò et al., 2020a,b). Saccò et al. (2021) illustrated how regional climate patterns can influence consumer—resource interactions by showing that rainfall triggered shifts in the source of energy supporting secondary consumers, which resulted in increased top-down pressure from predatory beetles. Additionally, local- and regional-scale patterns in obligate cave species biodiversity have been qualitatively linked to patterns in surface ecosystem productivity (Culver and Sket, 2000; Christman et al., 2005; Eme et al., 2015).

# The role of food web processes in groundwater community dynamics

Groundwater communities are generally composed of fewer species with lower abundances than surface aquatic ecosystems (Hüppop, 2001; Zagmajster et al., this volume). Bottom-up limitation is considered the primary mechanism structuring groundwater communities and the available data strongly support this hypothesis. Traditional population and community ecology studies using various approaches (natural gradients, pre- and post-organic pollution episodes, experimental manipulations) generally found positive correlations between measures of community composition (biodiversity, abundance, biomass, secondary production) and energy availability (Sinton, 1984; Smith et al., 1986; Madsen et al., 1991; Notenboom et al., 1994; Simon and Buikema, 1997; Sket, 1999, 2005; Wood et al., 2002; Culver and Pipan, 2019; Huntsman et al., 2011; Venarsky et al., 2018). Furthermore, food web analyses that compare energy input versus consumption rates have found that some groundwater communities need to consume nearly all available energy to support community productivity, which indicates that some systems are operating at carrying capacity and that surplus energy is not available to support additional productivity (Venarsky et al., 2014). Importantly though, the influence of bottom-up limitation on groundwater community dynamics is a function of community composition.

Many groundwater communities can be divided into two broad groups, each with a distinct evolutionary history. Obligate groundwater taxa exclusively complete their life cycles in groundwater ecosystems and have evolved a suite of traits in the energy-limited groundwater environment, including low metabolic rates, increased starvation resistance, slow growth rates, delayed maturity, and low fecundity (Hervant et al., 1997; Hüppop, 2001; Lefébure et al., 2017; Venarsky et al., this volume, Chapter 19). In comparison, facultative surface taxa generally have a suite of traits that appear to be adapted to energy richer habitats, including higher metabolic rates, fast growth, early maturity, and high fecundity (Hervant et al., 1997; Hüppop, 2001; Venarsky et al., this volume, Chapter 19). While some facultative taxa are capable of completing their life cycles in groundwater habitats, many of these appear to be sink-populations that are sustained from continuous immigration from surface habitats (Huntsman et al., 2020).

Interestingly, facultative taxa can dominate measures of biodiversity, abundance, and biomass in groundwater systems with strong connections to surface habitats and relaxation of bottom-up limitation appears to facilitate the coexistence of obligate and facultative taxa. For example, minimal energetic surpluses allow obligate cave species to compete in energy-limited groundwater environments with facultative taxa, but when surpluses of energy become available the later are more capable of monopolizing and assimilating the additional resources for growth and reproduction than obligate groundwater taxa (Venarsky et al., 2018). However, under low-energy conditions that favor obligate groundwater taxa, facultative taxa are likely a source of energy rather than a source of competition as a substantial portion of obligate groundwater predator production was attributed to the consumption of facultative surface taxa (Huntsman et al., 2011). Thus, bottom-up limitation shapes groundwater community dynamics on both evolutionary and contemporary ecological timescales.

Importantly though, bottom-up limitation is not necessarily the exclusive mechanism through which food web processes influence groundwater communities as several lines of evidence suggest that top-down forces may also play a role in groundwater community dynamics. In surface ecosystems, bacterial populations can be influenced by the feeding activities of consumers (Pace and Cole, 1994) and groundwater microbial communities appear to be no different. In both laboratory and organically enriched field settings the feeding activities of heterotrophic nanoflagellates and larger consumers (e.g., amphipods) are known to impact microbial production (Kinner et al., 1998; Sintes et al., 2004; Kinsey et al., 2007; Cooney and Simon, 2009). Interestingly, Cooney and Simon (2009) found that amphipods influence bacterial production on rock surfaces but not in fine sediments, suggesting that substrate characteristics (fine sediments vs. rock surfaces) dictate whether epilithic microbial biofilms are top-down controlled (Foulquier et al., 2010a,b). Using a framework to explore consumer–resource interactions in heterotrophic nanoflagellates (Gasol et al., 1994), Sintes et al. (2004) collected evidence for top-down regulation of heterotrophic nanoflagellates in coastal cave systems, but this mechanism may not drive population dynamics in heterotrophic nanoflagellates in all oligotrophic subsurface habitats (see Karwautz et al., 2022). Venarsky et al. (2014) also found that nearly all detrital and prey resources needed to be consumed to support community productivity, which implies cave communities could simultaneously be under both bottom-up and top-down control.

While many groundwater ecosystems are supported via detrital inputs, a select few systems have both detrital and chemolithoautotrophic sources of energy (Pohlman et al., 1997; Sarbu, 2001; Opsahl and Chanton, 2006; Roach et al., 2011; Pohlman, 2011; Neisch et al., 2012; Hutchins, 2013; Tobler et al., 2013; Herrmann et al., 2015) and the presence of both pathways can strongly influence groundwater community composition. The spatially heterogeneous redox conditions allow multiple microbial metabolic pathways to exist in sympatry, which increases microbial biodiversity at the habitat scale (Porter et al., 2009; Pohlman, 2011). Macroconsumer biodiversity is also highest in areas where both detritus inputs and chemolithoautotrophic primary production support in situ secondary productivity and lowest in areas where detrital inputs are the sole energy source (Hutchins, 2013; Herrmann et al., 2020). The higher biodiversity in areas with both energy sources appears to be the result of trophic niche differentiation (Hutchins et al., 2014), where the increased productivity at the base of the food web creates more trophic niches for macroconsumers to exploit (Pianka, 1966; Abrams, 1995). High biodiversity has also been reported from several Mexican sulfidic caves whose communities are supported by both detrital and chemolithoautotrophic energy pathways (Engel, 2007; Roach et al., 2011; Tobler et al., 2013).

## Trophic niche diversification in groundwater ecosystems

Groundwater foods webs are generally viewed as less complex when compared to food webs in aquatic surface ecosystems and several lines of evidence support this perspective (Culver, 1994; Tobler, 2008, Fig. 10.1). For example, a lack of photosynthetic primary production in most groundwater systems coupled with lower biodiversity can reduce the overall complexity of groundwater food webs by reducing the number of potential energy flow pathways (Hancock et al., 2005; Boulton et al., 2008). Additionally, the dominance of omnivory and scarcity of obligate predators in groundwater communities results in redundant feeding strategies and frequent niche overlap among consumers (Mammola et al., 2016; Saccò et al., 2020b). However, a more nuanced view of groundwater food web complexity appears to be emerging due to the availability of more advanced methods (isotopic ecology, metabarcoding) to delineate consumer—resource interactions in groundwater ecosystems (Hervant et al., 1999; Korbel et al., 2017; Saccò et al. 2022b, 2022d).

Despite the dominance of omnivory in groundwater food webs, specialist feeding strategies (diet dominated by a specific resource) as well as trophic specialization (increased consumers performance on a specific resource) do occur among groundwater species. Quintessential examples can be found in the groundwater systems with both detrital and chemolithoautotrophic sources of energy described above. However, trophic specialization does occur within groundwater systems supported by only detritus. For example, Francois et al. (2016b) found that different species of groundwater isopods were specialized on a given trophic resource (sedimentary biofilm in this case) and that this specialization is conserved among individuals within the same population. Ercoli et al. (2019) also showed that two species of sympatric groundwater isopods consumed the same resources in different proportions, which could be viewed as a relatively weak form of specialist feeding strategies. However, the feeding strategies of groundwater species appears to be influenced by the availability of resources: Francois et al. (2020) found: that groundwater isopods fed less selectively in resource limited environments (i.e., consume resources in proportions that is similar to the availability of resources in the habitat). Saccò et al. (2019a) also found that feeding strategies in groundwater beetles can change in relation to local precipitation patterns. Beetles were more predatory during periods of high rainfall, which is presumably a period of high inputs of energy from the surface, while during periods of low rainfall the beetles exhibited more cannibalistic behaviors. Collectively these results suggest that the availability of resources can influence the selection of trophic strategies (specialist to generalist) in groundwater food webs. While further studies in other groundwater systems is needed to determine if this is a widespread pattern, adjusting feeding strategies based on the availability of resources may represent an efficient strategy to cope with resource scarcity in groundwater ecosystems (Francois et al., 2016b, 2020; Saccò et al., 2022c).

# Future directions

Our understanding of food web dynamics has evolved over time, beginning with simple linear food chains like those described by Lindeman (1942). Today we understand that food webs in surface ecosystems often contain thousands of species that are connected via multiple links over various spatial (local to regional) and temporal (days to years) scales (Polis and Strong, 1996). While our understanding of groundwater food webs is expanding by the day, our perspective of groundwater food webs certainly lags behind that of aquatic surface ecosystems (Mammola et al., 2020). For example, highly resolved food webs can be found for aquatic surface ecosystems that describe the number of links among species as well as the transfer of energy through these links (Cross et al., 2013). For various reasons, similar efforts have not yet been applied in groundwater ecosystems. A theoretical groundwater habitat food web with putative interactions is provided in Fig. 10.1. In this section, we highlight several areas of research that can help to provide a more nuanced and comprehensive understanding of the food webs processes that influence the structure and function of groundwater communities.

## Unpacking the microbial compartment of groundwater food webs

One of the least understood aspects of groundwater food webs is the microbial compartment (Griebler et al., 2022), which is surprising given that microbes fundamentally contribute to fluxes of materials and energy through virtually all aquatic ecosystems (Fischer et al., 2002; Sabater et al., 2002; Weitere et al., 2018). It is well-established that bacteria and fungi are limited by both the quantity and quality of organic matter in groundwater ecosystems (Hofmann et al., 2020; Retter and Nawaz, 2022) and that their consumption by macroconsumers (insect larvae, amphipods, isopods) can influence the structure and biomass of the bacterial components of the microbial community (Kinsey et al., 2007; Cooney and Simon, 2009; Weitowitz et al., 2019). However, the microbial compartment is composed of more taxa than just bacteria and fungi, including protozoans such as flagellates, ciliates, ameba, and viruses (Novarino et al., 1997). All of these taxa can be viewed as a smaller food web with its own consumer–resource interactions that influence carbon and nutrient dynamics (*sensu* microbial loop; Meyer, 1994; Pomeroy et al., 2007; Herrmann et al., 2020) (Fig. 10.1).

In many ways, the microbial compartments of surface and groundwater food webs are similar. Microorganisms transform DOM into both microbial biomass and extrapolymeric substances (EPS; polysaccharides) that are the structural backbone for biofilm development on substratum (Hall and Meyer, 1998; Simon et al., 2003; Madigan et al., 2010). In aquatic surface ecosystems it is well-established that bacteria (and potentially fungi) receive feeding pressure from micrograzers, such as ciliates, heterotrophic nanoflagellates, and microcrustaceans (copepods, ostracods) (Wey et al., 2008; Johnke et al., 2014). Furthermore, bacterial lysis via viral infection is also very common in aquatic surface habitats (Jacquet et al., 2010). To date, however, there is little support for these mechanisms significantly influencing microbial dynamics in groundwater ecosystems. Well-developed thick biofilms are generally absent in oligotrophic groundwater habitats (with the exception of highly productive chemolithoautotrophic caves). Thus, the prevailing view is that groundwater prokaryotic biomass is low with protozoan grazing believed inefficient (Griebler et al., 2022). However, thin biofilms can represent the main source of energy in some groundwater habitats due to their ubiquitous distribution (Francois et al., 2020). Karwautz et al. (2022) also found that groundwater bacterial communities and associated microconsumers were bottom-up limited by the availability of organic matter, but did not find evidence that grazing by microeukaryotes (heterotrophic nanoflagellates) or lysis via viral infection caused top-down control on bacteria in an alpine shallow porous aquifer. Additionally, groundwater systems are diluted environments when compared to aquatic surface systems, which results in an extremely low probability that a virus finds a suitable host. In consequence, bacterial lysis via viral infection is unlikely to be of great importance in oligotrophic groundwater ecosystems (Schweichhart et al., 2022).

# Clarifying and quantifying food web structure

Food web studies use various approaches to explore the underlying processes influencing population and community dynamics. Studies have correlated abundance, biomass, or secondary production of individual species or microbial or macroconsumers with energy availability to explore bottom-up limitation (Huntsman et al., 2011; Venarsky et al., 2018). Additionally, bioenergetic studies incorporate turnover rates of basal resources with consumption rates of consumers in groundwater communities to determine the degree to which groundwater macroconsumers are limited by carrying capacity (Venarsky et al., 2014). Using these approaches, however, requires access to the groundwater habitat and many groundwater habitats are not easily sampled, such as aquifers (Ficetola et al., 2019; Mammola et al., 2021). Furthermore, while these approaches provide a broad view of the food web processes that structure groundwater communities, they provide minimal insight into the structural aspects of groundwater food webs, such as number and length of feeding links or a description of the feeding niche of a groundwater organism (Fig. 10.1). Quantifying these aspects of food web structure is difficult because the basal resources in groundwater ecosystems consist of many compounds and particles of varying qualities, including humic-fluvic compounds, amino acids, and fragments of animals and plants. However, these details are often lost as basal resources are often placed into broad groups, such as DOM and POM. Additionally, gut content analysis of groundwater species is problematic as gut contents are often dominated by amorphous detritus and the consumption of high-quality preys can be infrequent, making it difficult to delineate the trophic niche of an animal (Griebler et al., 2022; Saccò et al., 2022d). Moreover, gut content analyses have the potential to describe the consumption of specific resources, but these analyses do not estimate energy flow as they do not account for assimilation. However, a more nuanced view of groundwater food webs appears to be emerging due to the use of various environmental tracers (Hervant et al., 1999; Korbel et al., 2017; Sacco et al., 2019a).

## *Natural abundances of stable isotopes and mixing models*

Stable isotope analyses are the most widely used and accessible environmental tracer method to delineate feeding linkages and energy flow pathways in food webs (Boecklen et al., 2011). To date, the majority of stable isotope analyses in groundwater ecosystems used bulk carbon ($\delta^{13}C$) and nitrogen ($\delta^{15}N$). Stable isotopes of other elements, specifically sulfur ($^{34}S$) and hydrogen ($^{2}H$), have also been used in studies of aquatic surface food webs to differentiate food sources and quantify terrestrial subsidies (Doucett et al., 2007; Deines et al., 2009; Croisetière et al., 2009; Vander Zanden et al., 2016; Carr et al., 2017). Indeed, the integration of these underused isotope proxies could help elucidate groundwater food web relationships and estimate trophic positions. However, the ability of bulk stable isotope approaches to discriminate among the resources supporting consumer growth can be complicated by several issues, including differences in tissue specific (muscle, fat, gonad) fractionation, temporal variability in resource stable isotope composition, and lack of discrimination among resources (Caut et al., 2009; Bunn et al., 2013). Several approaches can overcome the limitations of bulk stable isotope methods, some of which have been applied to groundwater food webs, albeit in a limited number of systems. Below we discuss some of these approaches.

Mixing models are a commonly used tool to reconstruct the diet of an organism. Mixing models assume that there are predictable shifts (commonly referred to as trophic discrimination factors) in the isotopic composition between consumers and their food (Fry, 2006). Trophic discrimination factors are dependent on several variables, including taxonomy (fish vs. insects) and food type and quality (organic matter vs. prey) (Caut et al., 2009). Bayesian Mixing Models (BMMs) are one approach that can account for variability in trophic discrimination factors as well as incorporating other sources of information that could influence the results, such as metabarcoding analyses or gut content measurements (Hette-Tronquart, 2019; Saccò et al., 2020c). In other words, BMMs are more flexible than conventional mixing models as they can better represent and account for the biological variability inherent to stable isotope data (e.g., heterogeneity in the isotopic signature of a bulk resource). Furthermore, recent advances allow integration of multiple tracers in multifactorial analyses, including $\delta^{13}C$, $\delta^{15}N$, $\delta^{34}S$, and $\Delta^{14}C$ (Larsen et al., 2013; Majdi et al., 2018). As a result, BMM has the potential to significantly enhance the accuracy of subterranean diet reconstructions and open interdisciplinary perspectives into groundwater food web analysis (Saccò et al., 2019b). At the time of writing this book chapter, specific analytical packages that use BMMs include Mixsiar (Stock and Semmens, 2016), FRUITS (Fernandes et al., 2014) and tRophicPosition (Quezada-Romegialli et al., 2018), and SIBER (Jackson et al., 2011).

### Radiocarbon ($^{14}C$) natural abundance

This is a radioactive isotope of carbon that is continuously produced by natural processes in the upper atmosphere and is also produced artificially via aboveground nuclear testing, which peaked in the mid-1960s (also known as "Bomb-pulse radiocarbon") (Gäggeler, 1995). This approach is frequently applied to aquatic surface food webs because $\Delta^{14}C$ values do not change between the consumer and its food (Ishikawa et al., 2013). This characteristic of $\Delta^{14}C$ has made it a useful tool to delineate several aspects of organic matter dynamics in surface habitats, including estimation of the age of carbon supporting food webs as well as differentiating the importance of dissolved (organic and inorganic) and particulate carbon in supporting consumer productivity (Briones et al., 2005; Fernandes et al., 2013; Larsen et al., 2013; Keaveney et al., 2015). While this approach has rarely been applied to groundwater food webs, recent applications have illustrated the usefulness of this approach by tracking carbon inputs to groundwater food webs as well as delineating the diet preferences of groundwater consumers (McMahon et al., 2019; Saccò et al., 2020a,c; Saccò et al., 2021).

### Artificially enriched tracer compounds

This approach consists in adding an artificially enriched tracer compound (e.g., acetate, ammonium) to an ecosystem and then tracks the movement of the tracer through the food web. This approach has widely been used in both surface (Carpenter et al., 2005; Norman et al., 2017) and subterranean aquatic ecosystems. $^{13}C$ or/and $^{15}N$ tracer additions have been used for elucidating trophic transfer in cave stream (Simon et al., 2003) and to characterize feeding activities of groundwater organisms in laboratory experiments (Foulquier et al., 2010a,b; Francois et al., 2016a). Dattagupta et al. (2009) used $^{13}C$ labeling in combination with other methods (fluorescence in situ hybridization, secondary ion mass spectrometry) to show a symbiotic relationship between chemolithoautotrophic bacteria and stygobiotic amphipod specimens in a cave ecosystem. Isotope ratio imaging also combines isotopic labeling with other methods (NanoSIMS or Raman spectroscopy) to track the uptake of compounds by the microbial compartment (Wagner, 2009; Musat et al., 2016; Weitere et al., 2018). To date, however, the use of this approach in groundwater studies has been limited to the microbial assimilation of contaminants (Vogt et al., 2016; Wilhelm et al., 2018).

### Compound specific methods

While bulk stable isotope analyses use either the whole animal or specific tissues (muscle, gonad, bone), compound specific stable isotope analyses focus on individual molecular groups, including amino and fatty acids (Chikaraishi et al., 2007; Ohkouchi et al., 2017; Ishikaws, 2018). This approach is advantageous over bulk stable isotopes because the amino and fatty acids from a producer can be directly incorporated into a consumer's tissues, meaning that the isotopic signatures in the producer and consumer are nearly identical (i.e., no trophic fractionation or bioconversion), which implies a direct feeding link. However, consumers regularly synthesize amino and fatty acids from either dietary macromolecules (carbohydrates and lipids) or precursor compounds, such as the conversion of short-chain polyunsaturated fatty acids to long-chain polyunsaturated fatty acids (Bell and Tocher, 2009; Whiteman et al., 2019). Importantly though, the biosynthesis of these compounds as inferred from compounds specific stable isotope analyses can be used to understand both the physiology, diets, and trophic position of consumers (Chikaraishi et al., 2014; Whiteman et al., 2019). Some of these methods were recently used to delineate food webs in calcrete aquifers (Saccò et al., 2019a), illustrating how compound specific stable isotope analyses are a useful tool to delineate carbon flow and trophic position through groundwater food webs.

## Acknowledgments

We thank S. Mammola and F. Malard for constructive comments that improved this chapter.

## References

Abrams, P.A., 1995. Monotonic or unimodal diversity-productivity gradients: what does competition theory predict? Ecology 76, 2019—2027.

Alfreider, A., Vogt, C., Geiger-Kaiser, M., Psenner, R., 2009. Distribution and diversity of autotrophic bacteria in groundwater systems based on the analysis of RubisCO genotypes. Systematic and Applied Microbiology 32, 140—150.

Alfreider, A., Vogt, C., Hoffmann, D., Babel, W., 2003. Diversity of ribulose-1, 5-bisphosphate carboxylase/oxygenase large-subunit genes from groundwater and aquifer microorganisms. Microbial ecology 45, 317—328.

Baker, M.A., Valett, H.M., Dahm, C.N., 2000. Organic carbon supply and metabolism in a shallow groundwater ecosystem. Ecology 81, 3133—3148.

Bell, M.V., Tocher, D.R., 2009. Biosynthesis of polyunsaturated fatty acids in aquatic ecosystems: general pathways and new directions. In: Kainz, M., Brett, M., Arts, M. (eds). Lipids in Aquatic Ecosystems. Springer, New York, NY, pp. 211—236.

Berg, I.A., 2011. Ecological aspects of the distribution of different autotrophic $CO_2$ fixation pathways. Applied and Environmental Microbiology 77, 1925—1936.

Birdwell, J.E., Engel, A.S., 2010. Characterization of dissolved organic matter in cave and spring waters using UV—Vis absorbance and fluorescence spectroscopy. Organic Geochemistry 41, 270—280.

Boecklen, W.J., Yarnes, C.T., Cook, B.A., James, A.C., 2011. On the use of stable isotopes in trophic ecology. Annual Review of Ecology, Evolution, and Systematics 42, 411—440.

Boulton, A.J., Fenwick, G.D., Hancock, P.J., Harvey, M.S., 2008. Biodiversity, functional roles and ecosystem services of groundwater invertebrates. Invertebrate Systematics 22, 103−116.

Brankovits, D., Pohlman, J., Niemann, H., Leigh, M., Leewis, M., Becker, K., Iliffe, T., Alvarez, F., Lehmann, M., Phillips, B., 2017. Methane-and dissolved organic carbon-fueled microbial loop supports a tropical subterranean estuary ecosystem. Nature Communications 8, 1−12.

Braun, A., Spona-Friedl, M., Avramov, M., Elsner, M., Baltar, F., Reinthaler, T., Herndl, G.J., Griebler, C., 2021. Reviews and syntheses: heterotrophic fixation of inorganic carbon−significant but invisible flux in environmental carbon cycling. Biogeosciences 18, 3689−3700.

Briones, M.J.I., Garnett, M., Piearce, T.G., 2005. Earthworm ecological groupings based on $^{14}$C analysis. Soil Biology and Biochemistry 37, 2145−2149.

Bunn, S., Leigh, S., Jardine, T., 2013. Diet-tissue fractionation of $^{15}$N by freshwater consumers and the importance of algae in stream and river food webs. Limnology and Oceanography 58, 765−773.

Carpenter, S.R., Cole, J.J., Pace, M.L., Van de Bogert, M., Bade, D.L., Bastviken, D., Gille, C.M., Hodgson, J.R., Kitchell, J.F., Kritzberg, E.S., 2005. Ecosystem subsidies: terrestrial support of aquatic food webs from $^{13}$C addition to contrasting lakes. Ecology 86, 2737−2750.

Carr, M.K., Jardine, T.D., Doig, L.E., Jones, P.D., Bharadwaj, L., Tendler, B., Chételat, J., Cott, P., Lindenschmidt, K.-E., 2017. Stable sulfur isotopes identify habitat-specific foraging and mercury exposure in a highly mobile fish community. Science of the Total Environment 586, 338−346.

Caut, S., Angulo, E., Courchamp, F., 2009. Variation in discrimination factors ($\Delta^{15}$N and $\Delta^{13}$C): the effect of diet isotopic values and applications for diet reconstruction. Journal of Applied Ecology 46, 443−453.

Chapelle, F.H., O'Neill, K., Bradley, P.M., Methé, B.A., Ciufo, S.A., Knobel, L.L., Lovley, D.R., 2002. A hydrogen-based subsurface microbial community dominated by methanogens. Nature 415, 312−315.

Chikaraishi, Y., Steffan, S.A., Ogawa, N.O., Ishikawa, N.F., Sasaki, Y., Tsuchiya, M., Ohkouchi, N., 2014. High-resolution food webs based on nitrogen isotopic composition of amino acids. Ecology and Evolution 4, 2423−2449.

Chikaraishi, Y., Kashiyama, Y., Ogawa, N.O., Kitazato, H., Ohkouchi, N., 2007. Metabolic control of nitrogen isotope composition of amino acids in macroalgae and gastropods: implications for aquatic food web studies. Marine Ecology Progress Series 342, 85−90.

Christman, M.C., Culver, D.C., Madden, M.K., White, D., 2005. Patterns of endemism of the eastern North American cave fauna. Journal of Biogeography 32, 1441−1452.

Church, M., 2002. Geomorphic thresholds in riverine landscapes. Freshwater Biology 47, 541−557.

Cooney, T.J., Simon, K.S., 2009. Influence of dissolved organic matter and invertebrates on the function of microbial films in groundwater. Microbial Ecology 58, 599−610.

Cornu, J.-F., Eme, D., Malard, F., 2013. The distribution of groundwater habitats in Europe. Hydrogeology Journal 21, 949−960.

Croisetiere, L., Hare, L., Tessier, A., Cabana, G., 2009. Sulphur stable isotopes can distinguish trophic dependence on sediments and plankton in boreal lakes. Freshwater Biology 54, 1006−1015.

Cross, W.F., Baxter, C.V., Rosi-Marshall, E.J., Hall, R.O., Kennedy, T.A., Donner, K.C., Kelly, H.A.W., Seegert, S.E.Z., Behn, K.E., Yard, M.D., 2013. Food-web dynamics in a large river discontinuum. Ecological Monographs 83, 311−337.

Culver, D.C., 1994. Species interactions. In: Gibert, J., Danielopol, D.L., Stanford, J.A. (Eds.), Groundwater Ecology. Academic Press, Waltham, Massachusetts, pp. 271−286.

Culver, D.C., Pipan, T., 2019. The Biology of Caves and Other Subterranean Habitats. Oxford University Press, USA.

Culver, D.C., Sket, B., 2000. Hotspots of subterranean biodiversity in caves and wells. Journal of Cave and Karst Studies 62, 11−17.

Datry, T., Malard, F., Gibert, J., 2005. Response of invertebrate assemblages to increased groundwater recharge rates in a phreatic aquifer. Journal of the North American Benthological Society 24, 461−477.

Dattagupta, S., Schaperdoth, I., Montanari, A., Mariani, S., Kita, N., Valley, J.W., Macalady, J.L., 2009. A novel symbiosis between chemoautotrophic bacteria and a freshwater cave amphipod. The ISME Journal 3, 935−943.

Deines, P., Wooller, M.J., Grey, J., 2009. Unravelling complexities in benthic food webs using a dual stable isotope (hydrogen and carbon) approach. Freshwater Biology 54, 2243−2251.

Dole-Olivier, M.J., Malard, F., Martin, D., Lefébure, T., Gibert, J., 2009. Relationships between environmental variables and groundwater biodiversity at the regional scale. Freshwater Biology 54, 797−813.

Doucett, R.R., Marks, J.C., Blinn, D.W., Caron, M., Hungate, B.A., 2007. Measuring terrestrial subsidies to aquatic food webs using stable isotopes of hydrogen. Ecology 88, 1587−1592.

Elbourne, L.D., Sutcliffe, B., Humphreys, W., Focardi, A., Saccò, M., Campbell, M.A., Paulsen, I.T., Tetu, S.G., 2022. Unravelling stratified microbial assemblages in Australia's only deep anchialine system, The Bundera Sinkhole. Frontiers in Marine Science 9, 872082.

Eme, D., Zagmajster, M., Fišer, C., Galassi, D., Marmonier, P., Stoch, F., Cornu, J.F., Oberdorff, T., Malard, F., 2015. Multi-causality and spatial non-stationarity in the determinants of groundwater crustacean diversity in Europe. Ecography 38, 531−540.

Engel, A.S., 2007. Observations on the biodiversity of sulfidic karst habitats. Journal of Cave and Karst Studies 69, 187−206.

Engel, A.S., 2019. Chemolithoautotrophy. In: White, W.B., Culver, D.C., Pipan, T. (Eds.), Encyclopedia of Caves. Elsevier, pp. 267−276.

Erb, T.J., 2011. Carboxylases in natural and synthetic microbial pathways. Applied and Environmental Microbiology 77, 8466−8477.

Ercoli, F., Lefebvre, F., Delangle, M., Godé, N., Caillon, M., Raimond, R., Souty-Grosset, C., 2019. Differing trophic niches of three French stygobionts and their implications for conservation of endemic stygofauna. Aquatic Conservation: Marine and Freshwater Ecosystems 29, 2193−2203.

Espanol, C., Comín, F.A., Gallardo, B., Yao, J., Yela, J.L., Carranza, F., Zabaleta, A., Ladera, J., Martínez-Santos, M., Gerino, M., 2017. Does land use impact on groundwater invertebrate diversity and functionality in floodplains? Ecological Engineering 103, 394−403.

Fernandes, R., Dreves, A., Nadeau, M.-J., Grootes, P.M., 2013. A freshwater lake saga: carbon routing within the aquatic food web of Lake Schwerin. Radiocarbon 55, 1102−1113.

Fernandes, R., Millard, A.R., Brabec, M., Nadeau, M.-J., Grootes, P., 2014. Food reconstruction using isotopic transferred signals (FRUITS): a Bayesian model for diet reconstruction. PLoS One 9, e87436.

Ficetola, G.F., Canedoli, C., Stoch, F., 2019. The Racovitzan impediment and the hidden biodiversity of unexplored environments. Conservation Biology 33, 214−216.

Fischer, H., Sachse, A., Steinberg, C.E., Pusch, M., 2002. Differential retention and utilization of dissolved organic carbon by bacteria in river sediments. Limnology and Oceanography 47, 1702−1711.

Fišer, C., Delić, T., Luštrik, R., Zagmajster, M., Altermatt, F., 2019. Niches within a niche: ecological differentiation of subterranean amphipods across Europe's interstitial waters. Ecography 42, 1212−1223.

Fišer, C., Konec, M., Alther, R., Švara, V., Altermatt, F., 2017. Taxonomic, phylogenetic and ecological diversity of Niphargus (Amphipoda: Crustacea) in the Hölloch cave system (Switzerland). Systematics and Biodiversity 15, 218−237.

Foulquier, A., Malard, F., Mermillod-Blondin, F., Datry, T., Simon, L., Montuelle, B., Gibert, J., 2010a. Vertical change in dissolved organic carbon and oxygen at the water table region of an aquifer recharged with stormwater: biological uptake or mixing? Biogeochemistry 99, 31−47.

Foulquier, A., Malard, F., Mermillod-Blondin, F., Montuelle, B., Dolédec, S., Volat, B., Gibert, J., 2011. Surface water linkages regulate trophic interactions in a groundwater food web. Ecosystems 14, 1339−1353.

Foulquier, A., Simon, L., Gilbert, F., Fourel, F., Malard, F., Mermillod-Blondin, F., 2010b. Relative influences of DOC flux and subterranean fauna on microbial abundance and activity in aquifer sediments: new insights from $^{13}$C-tracer experiments. Freshwater Biology 55, 1560−1576.

Francois, C.M., Duret, L., Simon, L., Mermillod-Blondin, F., Malard, F., Konecny-Dupré, L., Planel, R., Penel, S., Douady, C.J., Lefébure, T., 2016a. No evidence that nitrogen limitation influences the elemental composition of isopod transcriptomes and proteomes. Molecular Biology and Evolution 33, 2605−2620.

Francois, C.M., Mermillod-Blondin, F., Malard, F., Fourel, F., Lécuyer, C., Douady, C.J., Simon, L., 2016b. Trophic ecology of groundwater species reveals specialization in a low-productivity environment. Functional Ecology 30, 262−273.

Francois, C.M., Simon, L., Malard, F., Lefébure, T., Douady, C.J., Mermillod-Blondin, F., 2020. Trophic selectivity in aquatic isopods increases with the availability of resources. Functional Ecology 34, 1078−1090.

Fry, B., 2006. Stable Isotope Ecology. Springer.

Gasol, J.M., 1994. A framework for the assessment of top-down vs bottom-up control of heterotrophic nanoflagellate abundance. Marine Ecology Progress Series. Oldendorf 113, 291−300.

Gäggeler, H., 1995. Radioactivity in the atmosphere. Radiochimica Acta 70, 345−354.

Gibert, J., 1986. Ecologie d'un système karstique jurassien. Hydrogéologie, dérive animale, transit de matières, dynamique de la population de *Niphargus* (Crustacé Amphipode). Memoires de Biospeologie 13, 1–379.

Gibert, J., Deharveng, L., 2002. Subterranean ecosystems: a truncated functional biodiversity. BioScience 52, 473–481.

Gnaspini, P., Trajano, E., 2001. Guano communities in tropical caves. In: Wilkens, H., Culver, D.C., Humphreys, W.F. (Eds.), Ecosystems of the World: Subterranean Ecosystems. Elsevier Science, New York, pp. 251–268.

Graening, G.O., Brown, A.V., 2003. Ecosystem dynamics and pollution effects in an Ozark cave stream. Journal of the Amercian Water Resources Association 39, 1497–1507.

Griebler, C., Fillinger, L., Karwautz, C., Hose, G., 2022. Knowledge gaps, obstacles and research frontiers in microbial ecology. In: Mehner, T., Tockner, K., (Eds.). Encyclopedia of Inland Waters, 2$^{nd}$ Edition. Elsevier, pp. 611–624.

Grimm, N.B., Fisher, S.G., 1984. Exchange between interstitial and surface water: implications for stream metabolism and nutrient cycling. Hydrobiologia 111, 219–228.

Hahn, H.J., Fuchs, A., 2009. Distribution patterns of groundwater communities across aquifer types in south-western Germany. Freshwater Biology 54, 848–860.

Hall Jr., R.O., Meyer, J.L., 1998. The trophic significance of bacteria in a detritus-based stream food web. Ecology 79, 1995–2012.

Hancock, P.J., Boulton, A.J., Humphreys, W.F., 2005. Aquifers and hyporheic zones: towards an ecological understanding of groundwater. Hydrogeology Journal 13, 98–111.

Herrmann, M., Rusznyák, A., Akob, D.M., Schulze, I., Opitz, S., Totsche, K.U., Küsel, K., 2015. Large fractions of $CO_2$-fixing microorganisms in pristine limestone aquifers appear to be involved in the oxidation of reduced sulfur and nitrogen compounds. Applied and Environmental Microbiology 81, 2384–2394.

Herrmann, M., Geesink, P., Yan, L., Lehmann, R., Totsche, K., Küsel, K., 2020. Complex food webs coincide with high genetic potential for chemolithoautotrophy in fractured bedrock groundwater. Water Research 170, 115306.

Hervant, F., Mathieu, J., Barré, H., 1999. Comparative study on the metabolic responses of subterranean and surface-dwelling amphipods to long-term starvation and subsequent refeeding. Journal of Experimental Biology 202, 3587–3595.

Hervant, F., Mathieu, J., Barré, H., Simon, K., Pinon, C., 1997. Comparative study on the behavioral, ventilatory, and respiratory responses of hypogean and epigean crustaceans to long-term starvation and subsequent feeding. Comparative Biochemistry and Physiology Part A: Physiology 118, 1277–1283.

Hette-Tronquart, N., 2019. Isotopic niche is not equal to trophic niche. Ecology Letters 22, 1987–1989.

Hofmann, R., Griebler, C., 2018. DOM and bacterial growth efficiency in oligotrophic groundwater: absence of priming and co-limitation by organic carbon and phosphorus. Aquatic Microbial Ecology 81, 55–71.

Hofmann, R., Uhl, J., Hertkorn, N., Griebler, C., 2020. Linkage between dissolved organic matter transformation, bacterial carbon production, and diversity in a shallow oligotrophic aquifer: results from flow-through sediment microcosm experiments. Fronters in Microbiololgy 11, 2425.

Hose, G.C., Fryirs, K.A., Bailey, J., Ashby, N., White, T., Stumpp, C., 2017. Different depths, different fauna: habitat influences on the distribution of groundwater invertebrates. Hydrobiologia 797, 145–157.

Huntsman, B.M., Venarsky, M.P., Abadi, F., Huryn, A.D., Kuhajda, B.R., Cox, C.L., Benstead, J.P., 2020. Evolutionary history and sex are significant drivers of crayfish demography in resource-limited cave ecosystems. Evolutionary Ecology 34, 235–255.

Huntsman, B.M., Venarsky, M.P., Benstead, J.P., Huryn, A.D., 2011. Effects of organic matter availability on the life history and production of a top vertebrate predator (Plethodontidae: *Gyrinophilus palleucus*) in two cave streams. Freshwater Biology 56, 1746–1760.

Hüppop, K., 2001. How do cave animals cope with the food scarcity in caves? In: Wilkens, H., Culver, D.C., Humphreys, W.F. (Eds.), Ecosystems of the World: Subterranean Ecosystems. Elsevier Science, New York, pp. 159–188.

Hutchins, B.T., 2013. The Trophic Ecology of Phreatic Karst Aquifers. Texas State University, p. 143.

Hutchins, B.T., Schwartz, B.F., Nowlin, W.H., 2014. Morphological and trophic specialization in a subterranean amphipod assemblage. Freshwater Biology 59, 2447–2461.

Hynes, H.B.N., 1975. Edgardo Baldi memorial lecture. The stream and its valley. Verhandlungen der Internationalen Vereinigung fur theoretische und angewandte Limnologie 19, 1–15.

Ishikawa, N.F., 2018. Use of compound-specific nitrogen isotope analysis of amino acids in trophic ecology: assumptions, applications, and implications. Ecological Research 33, 825–837.

Ishikawa, N.F., Hyodo, F., Tayasu, I., 2013. Use of carbon-13 and carbon-14 natural abundances for stream food web studies. Ecological Research 28, 759–769.

Jackson, A.L., Inger, R., Parnell, A.C., Bearhop, S., 2011. Comparing isotopic niche widths among and within communities: SIBER—Stable Isotope Bayesian Ellipses in R. Journal of Animal Ecology 80, 595–602.

Jacquet, S., Miki, T., Noble, R., Peduzzi, P., Wilhelm, S., 2010. Viruses in aquatic ecosystems: important advancements of the last 20 years and prospects for the future in the field of microbial oceanography and limnology. Advances in Oceanography and Limnology 1, 97–141.

Johnke, J., Cohen, Y., de Leeuw, M., Kushmaro, A., Jurkevitch, E., Chatzinotas, A., 2014. Multiple micro-predators controlling bacterial communities in the environment. Current Opinion in Biotechnology 27, 185–190.

Junker, J.R., Cross, W.F., 2014. Seasonality in the trophic basis of a temperate stream invertebrate assemblage: importance of temperature and food quality. Limnology and Oceanography 59, 507–518.

Karwautz, C., Zhou, Y., Kerros, M.E., Weinbauer, M.G., Griebler, C., 2022. Bottom-up control of the groundwater microbial food-web in an Alpine aquifer. Frontiers in Ecology and Evolution, 10, 854228.

Keaveney, E.M., Reimer, P.J., Foy, R.H., 2015. Young, old, and weathered carbon—Part 2: using radiocarbon and stable isotopes to identify terrestrial carbon support of the food web in an alkaline, humic lake. Radiocarbon 57, 425–438.

Kellermann, C., Selesi, D., Lee, N., Hügler, M., Esperschütz, J., Hartmann, A., Griebler, C., 2012. Microbial $CO_2$ fixation potential in a tar-oil-contaminated porous aquifer. FEMS Microbiology Ecology 81, 172–187.

Kinkle, B.K., Kane, T.C., 2001. Chemolithoautotrophic micro-organisms and their potential role in subsurface environments. In: Wilkens, H., Culver, D.C., Humphreys, W.F. (Eds.), Ecosystems of the World: Subterranean Ecosystems. Elsevier Science, New York, pp. 309–318.

Kinner, N.E., Harvey, R.W., Blakeslee, K., Novarino, G., Meeker, L.D., 1998. Size-selective predation on groundwater bacteria by nanoflagellates in an organic-contaminated aquifer. Applied and Environmental Microbiology 64, 618–625.

Kinsey, J., Cooney, T.J., Simon, K.S., 2007. A comparison of the leaf shredding ability and influence on microbial films of surface and cave forms of Gammarus minus Say. Hydrobiologia 589, 199–205.

Korbel, K., Chariton, A., Stephenson, S., Greenfield, P., Hose, G.C., 2017. Wells provide a distorted view of life in the aquifer: implications for sampling, monitoring and assessment of groundwater ecosystems. Scientific Reports 7, 1–13.

Kozel, P., Pipan, T., Mammola, S., Culver, D.C., Novak, T., 2019. Distributional dynamics of a specialized subterranean community oppose the classical understanding of the preferred subterranean habitats. Invertebrate Biology 138, e12254.

Kumar, S., Herrmann, M., Thamdrup, B., Schwab, V.F., Geesink, P., Trumbore, S.E., Totsche, K.-U., Küsel, K., 2017. Nitrogen loss from pristine carbonate-rock aquifers of the Hainich Critical Zone Exploratory (Germany) is primarily driven by chemolithoautotrophic anammox processes. Frontiers in Microbiology 8, 1951.

Larned, S.T., 2012. Phreatic groundwater ecosystems: research frontiers for freshwater ecology. Freshwater Biology 57, 885–906.

Larsen, T., Ventura, M., Andersen, N., O'Brien, D.M., Piatkowski, U., McCarthy, M.D., 2013. Tracing carbon sources through aquatic and terrestrial food webs using amino acid stable isotope fingerprinting. PLoS One 8, e73441.

Lefébure, T., Morvan, C., Malard, F., François, C., Konecny-Dupré, L., Guéguen, L., Weiss-Gayet, M., Seguin-Orlando, A., Ermini, L., Der Sarkissian, C., 2017. Less effective selection leads to larger genomes. Genome Research 27, 1016–1028.

Lindeman, R.L., 1942. The trophic-dynamic aspect of ecology. Ecology 23, 399–417.

Madigan, M.T., Martinko, J.M., Stahl, D.A., Clark, D.P., 2010. Brock Biology of Microorganisms. Pearson Benjamin Cummings, San Francisco, CA.

Madsen, E.L., Sinclair, J.L., Ghiorse, W.C., 1991. In situ biodegradation: microbiological patterns in a contaminated aquifer. Science 252, 830–833.

Majdi, N., Hette-Tronquart, N., Auclair, E., Bec, A., Chouvelon, T., Cognie, B., Danger, M., Decottignies, P., Dessier, A., Desvilettes, C., 2018. There's no harm in having too much: a comprehensive toolbox of methods in trophic ecology. Food Webs 17, e00100.

Mammola, S., Amorim, I.R., Bichuette, M.E., Borges, P.A., Cheeptham, N., Cooper, S.J., Culver, D.C., Deharveng, L., Eme, D., Ferreira, R.L., Fišer, C., Fišer, Ž., Fong, D.W., Griebler, C., Jeffery, W.R., Jugovic, J., Kowalko, J.E., Lilley, T.M., Malard, F., Manenti, R., Martínez, A., Meierhofer, M.B., Niemiller, M.L., Northup, D.E.,

Pellegrini, T.G., Pipan, T., Protas, M., Reboleira, A.S.P.S., Venarsky, M.P., Wynne, J.J., Zagmajster, M., Cardoso, P., 2020. Fundamental research questions in subterranean biology. Biological Reviews 95, 1855–1872.

Mammola, S., Lunghi, E., Bilandžija, H., Cardoso, P., Grimm, V., Schmidt, S.I., Hesselberg, T., Martínez, A., 2021. Collecting eco-evolutionary data in the dark: impediments to subterranean research and how to overcome them. Ecology and Evolution 11, 5911–5926.

Mammola, S., Piano, E., Isaia, M., 2016. Step back! Niche dynamics in cave-dwelling predators. Acta Oecologica 75, 35–42.

Marcarelli, A.M., Baxter, C.V., Mineau, M.M., Hall Jr., R.O., 2011. Quantity and quality: unifying food web and ecosystem perspectives on the role of resource subsidies in freshwaters. Ecology 92, 1215–1225.

McMahon, K.W., Newsome, S.D., 2019. Amino acid isotope analysis: a new frontier in studies of animal migration and foraging ecology. In: Hobson, K.A., Wassenaar, L.I. (Eds.), Tracking Animal Migration with Stable Isotopes. Academic Press, pp. 173–190.

Mermillod-Blondin, F., Simon, L., Maazouzi, C., Foulquier, A., Delolme, C., Marmonier, P., 2015. Dynamics of dissolved organic carbon (DOC) through stormwater basins designed for groundwater recharge in urban area: Assessment of retention efficiency. Water Research 81, 27–37.

Meyer, J., 1994. The microbial loop in flowing waters. Microbial Ecology 28, 195–199.

Mindl, B., Griebler, C., Wirth, N., Starry, O., 2000. Biodegradability of DOC and metabolic response of heterotrophic bacteria in groundwater. Internationale Vereinigung für theoretische und angewandte Limnologie: Verhandlungen 27, 453–459.

Mondy, N., Grossi, V., Cathalan, E., Delbecque, J.P., Mermillod-Blondin, F., Douady, C.J., 2014. Sterols and steroids in a freshwater crustacean (Proasellus meridianus): hormonal response to nutritional input. Invertebrate Biology 133, 99–107.

Moore, J.C., Berlow, E.L., Coleman, D.C., de Ruiter, P.C., Dong, Q., Hastings, A., Johnson, N.C., McCann, K.S., Melville, K., Morin, P.J., Nadelhoffer, K., Rosemond, A.D., Post, D.M., Sabo, J.L., Scow, K.M., Vanni, M.J., Wall, D.H., 2004. Detritus, trophic dynamics and biodiversity. Ecology Letters 7, 584–600.

Musat, N., Musat, F., Weber, P.K., Pett-Ridge, J., 2016. Tracking microbial interactions with NanoSIMS. Current Opinion in Biotechnology 41, 114–121.

Nealson, K.H., Inagaki, F., Takai, K., 2005. Hydrogen-driven subsurface lithoautotrophic microbial ecosystems (SLiMEs): do they exist and why should we care? Trends in Microbiology 13, 405–410.

Neisch, J., Pohlman, J., Iliffe, T., 2012. The use of stable and radiocarbon isotopes as a method for delineating sources of organic material in anchialine systems. Natura Croatica 21, 83–85.

Norman, B.C., Whiles, M.R., Collins, S.M., Flecker, A.S., Hamilton, S.K., Johnson, S.L., Rosi, E.J., Ashkenas, L.R., Bowden, W.B., Crenshaw, C.L., 2017. Drivers of nitrogen transfer in stream food webs across continents. Ecology 98, 3044–3055.

Notenboom, J., Plénet, S., Turquin, M.J., 1994. Groundwater contamination and its impact on groundwater animals and ecosystems. In: Gibert, J., Danielopol, D.L. (Eds.), Groundwater Ecology. Academic Press, San Diego, USA, pp. 477–504.

Novarino, G., Warren, A., Butler, H., Lambourne, G., Boxshall, A., Bateman, J., Kinner, N.E., Harvey, R.W., Mosse, R.A., Teltsch, B., 1997. Protistan communities in aquifers: a review. FEMS Microbiology Reviews 20, 261–275.

Ohkouchi, N., Chikaraishi, Y., Close, H.G., Fry, B., Larsen, T., Madigan, D.J., McCarthy, M.D., McMahon, K.W., Nagata, T., Naito, Y.I., Ogawa, N.O., Popp, B.N., Steffan, S., Takano, Y., Tayasu, I., Wyatt, A.S.J., Yamaguchi, Y.T., Yokoyama, Y., 2017. Advances in the application of amino acid nitrogen isotopic analysis in ecological and biogeochemical studies. Organic Geochemistry 113, 150–174.

Opsahl, S.P., Chanton, J.P., 2006. Isotopic evidence for methane-based chemosynthesis in the Upper Floridan aquifer food web. Oecologia 150, 89–96.

Pabich, W.J., Valiela, I., Hemond, H.F., 2001. Relationship between DOC concentration and vadose zone thickness and depth below water table in groundwater of Cape Cod, USA. Biogeochemistry 55, 247–268.

Pace, M.L., Cole, J.J., 1994. Comparative and experimental approaches to top-down and bottom-up regulation of bacteria. Microbial Ecology 28, 181–193.

Paine, R.T., 1996. Preface. In: Polis, G.A., Winemiller, K.O. (Eds.), Food Webs: Integration of Patterns & Dynamics. Springer, Boston, MA, pp. ix–x.

Pellegrini, T.G., Ferreira, L.R., 2013. Structure and interactions in a cave guano - soil continuum community. European Journal of Soil Biology 57, 19—26.

Pianka, E.R., 1966. Latitudinal gradients in species diversity: a review of concepts. The American Naturalist 100, 33—46.

Pipan, T., Culver, D.C., 2017. The unity and diversity of the subterranean realm with respect to invertebrate body size. Journal of Cave and Karst Studies 79, 1—9.

Plath, M., Tobler, M., Riesch, R., de Leon, F.J.G., Giere, O., Schlupp, I., 2007. Survival in an extreme habitat: the roles of behaviour and energy limitation. Naturwissenschaften 94, 991—996.

Pohlman, J.W., 2011. The biogeochemistry of anchialine caves: progress and possibilities. Hydrobiologia 677, 33—51.

Pohlman, J.W., Iliffe, T.M., Cifuentes, L.A., 1997. A stable isotope study of organic cycling and the ecology of an anchialine cave ecosystem. Marine Ecology Progress Series 155, 17—27.

Polis, G.A., Strong, D.R., 1996. Food web complexity and community dynamics. The American Naturalist 147, 813—846.

Pomeroy, L.R., Williams, P.J. leB., Azam, F., Hobbie, J.E., 2007. The microbial loop. Oceanography 20, 28—33.

Porter, M.L., Engel, A.S., Kane, T.C., Kinkle, B.K., 2009. Productivity-diversity relationships from chemolithoautotrophically based sulfidic karst systems. International Journal of Speleology 38, 27—40.

Power, M.E., 1992. Top-down and bottom-up forces in food webs: do plants have primacy. Ecology 73, 733—746.

Power, M.E., Dietrich, W.E., 2002. Food webs in river networks. Ecological Research 17, 451—471.

Premate, E., Borko, Š., Delić, T., Malard, F., Simon, L., Fišer, C., 2021. Cave amphipods reveal co-variation between morphology and trophic niche in a low-productivity environment. Freshwater Biology 66, 1876—1888.

Quezada-Romegialli, C., Jackson, A.L., Harrod, C., 2018. tRophicPosition: Bayesian trophic position calculation with stable isotopes. R Package Version 0.7, 5.

Retter, A., Nawaz, A., 2022. The groundwater mycobiome: fungal diversity in terrestrial aquifers. In: Mehner, T., Tockner, K. (Eds.), Encyclopedia of Inland Waters, 2nd. Elsevier, pp. 385—396.

Roach, K.A., Tobler, M., Winemiller, K.O., 2011. Hydrogen sulfide, bacteria, and fish: a unique, subterranean food chain. Ecology 92, 2056—2062.

Sabater, S., Guasch, H., Romaní, A., Muñoz, I., 2002. The effect of biological factors on the efficiency of river biofilms in improving water quality. Hydrobiologia 469, 149—156.

Saccò, M., Blyth, A.J., Humphreys, W.F., Kuhl, A., Mazumder, D., Smith, C., Grice, K., 2019a. Elucidating stygofaunal trophic web interactions via isotopic ecology. PLoS One 14, e0223982.

Saccò, M., Blyth, A., Bateman, P.W., Hua, Q., Mazumder, D., White, N., Humphreys, W.F., Laini, A., Griebler, C., Grice, K., 2019b. New light in the dark-a proposed multidisciplinary framework for studying functional ecology of groundwater fauna. Science of the Total Environment 662, 963—977.

Saccò, M., Blyth, A., Humphreys, W.F., Middleton, J.A., Campbell, M., Mousavi-Derazmahalleh, M., Laini, A., Hua, Q., Meredith, K., Cooper, S.J.B., Christian Griebler, C., Allard, S., Grierson, P., Grice, K., 2020a. Tracking down carbon inputs underground from an arid zone Australian calcrete. PLoS One 15, e0237730.

Saccò, M., Blyth, A.J., Humphreys, W.F., Karasiewicz, S., Meredith, K.T., Laini, A., Cooper, S.J., Bateman, P.W., Grice, K., 2020b. Stygofaunal community trends along varied rainfall conditions: deciphering ecological niche dynamics of a shallow calcrete in Western Australia. Ecohydrology 13, e2150.

Saccò, M., Blyth, A.J., Humphreys, W.F., Cooper, S.J., Austin, A.D., Hyde, J., Mazumder, D., Hua, Q., White, N.E., Grice, K., 2020c. Refining trophic dynamics through multi-factor Bayesian mixing models: a case study of subterranean beetles. Ecology and Evolution 10, 8815—8826.

Saccò, M., Blyth, A.J., Humphreys, W.F., Cooper, S.J., White, N.E., Campbell, M., Mousavi-Derazmahalleh, M., Hua, Q., Mazumder, D., Smith, C., Griebler, C., Grice, K., 2021a. Rainfall as a trigger of ecological cascade effects in an Australian groundwater ecosystem. Scientific Reports 11, 1—15.

Saccò, M., Blyth, A.J., Venarsky, M., Humphreys, W.F., 2022d. Trophic interactions in subterranean environments. In: Mehner, T., Tockner, K. (Eds.), Encyclopedia of Inland Waters, 2nd. Elsevier, pp. 537—547.

Saccò, M., Campbell, M.A., Nevill, P., Humphreys, W.F., Blyth, A.J., Grierson, P.F., White, N.E., 2022a. Getting to the root of organic inputs in groundwaters: stygofaunal plant consumption in a calcrete aquifer. Frontiers in Ecology and Evolution 10, 854591.

Saccò, M., Guzik, M.T., van der Heyde, M., Nevill, P., Cooper, S.J., Austin, A.D., Coates, P.J., Allentoft, M.E., White, N.E., 2022b. eDNA in subterranean ecosystems: applications, technical aspects, and future prospects. Science of The Total Environment 820, 153223.

Saccò, M., Humphreys, W.F., Stevens, N., Jones, M.R., Taukulis, F., Thomas, E., Blyth, A.J., 2022c. Subterranean carbon flows from source to stygofauna: a case study on the atyidshrimp *Stygiocaris stylifera* (Holthuis, 1960) from Barrow Island (WA). Isotopes in Environmental and Health Studies 58 (3), 247–257.

Sarbu, S.M., 2001. Movile Cave: a chemoautotrophically based groundwater ecosystem. In: Wilkens, H., Culver, D.C., Humphreys, W.F. (Eds.), Ecosystems of the World: Subterranean Ecosystems. Elsevier Science, New York, pp. 319–343.

Sarbu, S.M., Kane, T.C., Kinkle, B.K., 1996. A chemoautotrophically based cave ecosystem. Science 272, 1953–1955.

Schiff, S.L., Aravena, R., Trumbore, S.E., Hinton, M.J., Elgood, R., Dillon, P.J., 1997. Export of DOC from forested catchments on the precambrian shield of central Ontario: clues from $^{13}$C and $^{14}$C. Biogeochemistry 36, 43–65.

Schweichhart, J.S., Pleyer, D., Retter, A., Winter, C., Griebler, C., 2022. Presence and role of prokaryotic viruses in groundwater environments. In: Mehner, T., Tockner, K. (Eds.), Encyclopedia of Inland Waters, second ed. Elsevier, pp. 373–384.

Shen, Y., Chapelle, F.H., Strom, E.W., Benner, R., 2015. Origins and bioavailability of dissolved organic matter in groundwater. Biogeochemistry 122, 61–78.

Shurin, J.B., Gruner, D.S., Hillebrand, H., 2006. All wet or dried up? Real differences between aquatic and terrestrial food webs. Proceedings of the Royal Society of London B Biological Sciences 273, 1–9.

Simon, K.S., 2013. Organic matter flux in the epikarst of the Dorvan karst, France. Acta Carsologica 42, 237–244.

Simon, K.S., Benfield, E.F., 2001. Leaf and wood breakdown in cave streams. Journal of the North American Benthological Society 20, 550–563.

Simon, K.S., Benfield, E.F., Macko, S.A., 2003. Food web structure and the role of epilithic biofilms in cave streams. Ecology 84, 2395–2406.

Simon, K.S., Buikema Jr., A.L., 1997. Effects of organic pollution on an Appalachian cave: changes in macroinvertebrate populations and food supplies. The American Midland Naturalist 138, 387–401.

Simon, K.S., Gibert, J., Petitot, P., Laurent, R., 2001. Spatial and temporal patterns of bacterial density and metabolic activity in a karst aquifer. Archiv für Hydrobiologie 151, 67–82.

Simon, K.S., Pipan, T., Culver, D.C., 2007. A conceptual model of the flow and distribution of organic carbon in caves. Journal of Cave and Karst Studies 69, 279–284.

Simon, K.S., Pipan, T., Ohno, T., Culver, D.C., 2010. Spatial and temporal patterns in abundance and character of dissolved organic matter in two karst aquifers. Fundamental and Applied Limnology/Archiv für Hydrobiologie 177, 81–92.

Sintes, E., Martinez-Taberner, A., Moya, G., Ramon, G., 2004. Dissecting the microbial food web: structure and function in the absence of autotrophs. Aquatic Microbial Ecology 37, 283–293.

Sinton, L.W., 1984. The macroinvertebrates in a sewage-polluted aquifer. Hydrobiologia 119, 161–169.

Sket, B., 1999. The nature of biodiversity in hypogean waters and how it is endangered. Biodiversity and Conservation 8, 1319–1338.

Sket, B., 2005. Dinaric karst, diversity in. In: Culver, D.C., White, W.B. (Eds.), Encyclopedia of Caves. Elsevier, New York, pp. 158–165.

Smith, G.A., Nickels, J.S., Kerger, B.D., Davis, J.D., Collins, S.P., Wilson, J.T., McNabb, J.F., White, D.C., 1986. Quantitative characterization of microbial biomass and community structure in subsurface material: a prokaryotic consortium responsive to organic contamination. Canadian Journal of Microbiology 32, 104–111.

Spona-Friedl, M., Braun, A., Huber, C., Eisenreich, W., Griebler, C., Kappler, A., Elsner, M., 2020. Substrate-dependent $CO_2$ fixation in heterotrophic bacteria revealed by stable isotope labelling. FEMS Microbiology Ecology 96, fiaa080.

Sterner, R.W., Elser, J.J., 2002. Ecological Stoichiometry: The Biology of Elements from Molecules to the Biosphere. Princeton University Press.

Stoch, F., Artheau, M., Brancelj, A., Galassi, D.M., Malard, F., 2009. Biodiversity indicators in European ground waters: towards a predictive model of stygobiotic species richness. Freshwater Biology 54, 745–755.

Stock, B., Semmens, B., 2016. MixSIAR GUI User Manual V3. 1. Scripps Institution of Oceanography, UC San Diego, San Diego, California, USA.

Thauer, R.K., 2007. A fifth pathway of carbon fixation. Science 318, 1732–1733.

Thorp, J.H., Thoms, M.C., Delong, M.D., 2006. The riverine ecosystem synthesis: biocomplexity in river networks across space and time. River Research and Applications 22, 123–147.

Tobler, M., 2008. Divergence in trophic ecology characterizes colonization of extreme habitats. Biological Journal of the Linnean Society 95, 517–528.

Tobler, M., Roach, K., Winemiller, K.O., Morehouse, R.L., Plath, M., 2013. Population structure, habitat use, and diet of giant waterbugs in a sulfidic cave. The Southwestern Naturalist 58, 420–426.

Tobler, M., Schlupp, I., Heubel, K.U., Riesch, R., García De León, F.J., Giere, O., Plath, M., 2006. Life on the edge: hydrogen sulfide and the fish communities of a Mexican cave and surrounding waters. Extremophiles 10, 577–585.

van Beynen, P., Ford, D., Schwarcz, H., 2000. Seasonal variability in organic substances in surface and cave waters at Marengo Cave, Indiana. Hydrological Processes 14, 1177–1197.

van Beynen, P.E., Schwarcz, H.P., Ford, D.C., Timmins, G.T., 2002. Organic substances in cave drip waters: studies from Marengo Cave, Indiana. Canadian Journal of Earth Sciences 39, 279–284.

Vander Zanden, H.B., Soto, D.X., Bowen, G.J., Hobson, K.A., 2016. Expanding the isotopic toolbox: applications of hydrogen and oxygen stable isotope ratios to food web studies. Frontiers in Ecology and Evolution 4, 20.

Vannote, R.L., Minshall, G.W., Cummins, K.W., Sedell, J.R., Cushing, C.E., 1980. The river continuum concept. Canadian Journal of Fisheries and Aquatic Sciences 37, 130–137.

Venarsky, M.P., Benstead, J.P., Huryn, A.D., 2012a. Effects of organic matter and season on leaf litter colonisation and breakdown in cave streams. Freshwater Biology 57, 773–786.

Venarsky, M.P., Benstead, J.P., Huryn, A.D., Huntsman, B.M., Edmonds, J.W., Findlay, R.H., Wallace, J.B., 2018. Experimental detritus manipulations unite surface and cave stream ecosystems along a common energy gradient. Ecosystems 21, 629–642.

Venarsky, M.P., Huntsman, B.M., 2018. Food webs in caves. In: Moldovan, O., Kováč, Ľ., Halse, S. (eds) Cave Ecology. Ecological Studies, vol 235. Springer, Cham, pp. 309–328.

Venarsky, M.P., Huntsman, B.M., Huryn, A.D., Benstead, J.P., Kuhajda, B.R., 2014. Quantitative food web analysis supports the energy-limitation hypothesis in cave stream ecosystems. Oecologia 176, 859–869.

Venarsky, M.P., Huryn, A.D., Benstead, J.P., 2012b. Re-examining extreme longevity of the cave crayfish *Orconectes australis* using new mark-recapture data: a lesson on the limitations of iterative size-at-age models. Freshwater Biology 57, 1471–1481.

Vogt, C., Lueders, T., Richnow, H.H., Krüger, M., Von Bergen, M., Seifert, J., 2016. Stable isotope probing approaches to study anaerobic hydrocarbon degradation and degraders. Journal of Molecular Microbiology and Biotechnology 26, 195–210.

Wagner, M., 2009. Single-cell ecophysiology of microbes as revealed by Raman microspectroscopy or secondary ion mass spectrometry imaging. Annual Review of Microbiology 63, 411–429.

Wang, S., Zhu, G., Zhuang, L., Li, Y., Liu, L., Lavik, G., Berg, M., Liu, S., Long, X.-E., Guo, J., 2020. Anaerobic ammonium oxidation is a major N-sink in aquifer systems around the world. The ISME Journal 14, 151–163.

Weitere, M., Erken, M., Majdi, N., Arndt, H., Norf, H., Reinshagen, M., Traunspurger, W., Walterscheid, A., Wey, J.K., 2018. The food web perspective on aquatic biofilms. Ecological Monographs 88, 543–559.

Weitowitz, D.C., Robertson, A.L., Bloomfield, J.P., Maurice, L., Reiss, J., 2019. Obligate groundwater crustaceans mediate biofilm interactions in a subsurface food web. Freshwater Science 38 (3), 491–502.

Wey, J.K., Scherwass, A., Norf, H., Arndt, H., Weitere, M., 2008. Effects of protozoan grazing within river biofilms under semi-natural conditions. Aquatic Microbial Ecology 52, 283–296.

Whiteman, J.P., Elliott Smith, E.A., Besser, A.C., Newsome, S.D., 2019. A guide to using compound-specific stable isotope analysis to study the fates of molecules in organisms and ecosystems. Diversity 11, 8.

Wilhelm, R.C., Hanson, B.T., Chandra, S., Madsen, E., 2018. Community dynamics and functional characteristics of naphthalene-degrading populations in contaminated surface sediments and hypoxic/anoxic groundwater. Environmental Microbiology 20, 3543–3559.

Winemiller, K.O., Flecker, A.S., Hoeinghaus, D.J., 2010. Patch dynamics and environmental heterogeneity in lotic ecosystems. Journal of the North American Benthological Society 29, 84–99.

Wood, P.J., Gunn, J., Perkins, J., 2002. The impact of pollution on aquatic invertebrates within a subterranean ecosystem - out of sight out of mind. Archiv für Hydrobiologie 155, 223–237.

III. Roles of organisms in groundwater

# Role of invertebrates in groundwater ecosystem processes and services

*Florian Mermillod-Blondin[1], Grant C. Hose[2], Kevin S. Simon[3], Kathryn Korbel[2], Maria Avramov[4] and Ross Vander Vorste[5]*

[1]Univ Lyon, Université Claude Bernard Lyon 1, CNRS, ENTPE, UMR 5023 LEHNA, Villeurbanne, France; [2]School of Natural Sciences, Macquarie University, Sydney, Australia; [3]School of Environment, University of Auckland, Auckland, New Zealand; [4]Helmholtz Zentrum München, German Research Center for Environmental Health, Institute of Groundwater Ecology, Neuherberg, Germany; [5]Department of Biology, University of Wisconsin - La Crosse, La Crosse, WI, United States

## Introduction

Since the 1990s, there has been an enormous growth in research aiming to understand the role of biodiversity in sustaining the functioning of ecosystems (i.e., BEF: Biodiversity—Ecosystem Functioning). Early work performed in grasslands (see Tilman et al., 1996) demonstrated a positive effect of biodiversity on plant productivity and ecosystem stability. This positive relationship was linked to (1) more biodiverse areas having a higher probability of containing productive functional groups (i.e., diverse communities are more likely to include representatives of a highly productive functional group) and (2) more biodiverse areas having stronger complementarity among the functional groups (i.e., diverse communities are more likely to have critical functions filled) (Loreau et al., 2001; Bulling et al., 2010). These mechanisms leading to a positive effect of biodiversity on ecosystem functioning have also been observed in freshwater (e.g., Cardinale et al., 2006) and marine (e.g., Worm et al., 2006) ecosystems.

In contrast, relatively little attention has been paid to BEF in groundwater ecosystems despite their importance for global freshwater supplies and the presence of diverse communities thriving in them (e.g., Marmonier et al., this volume). As an exception, the ecotones between surface and groundwater (e.g., the hyporheic zone, sinking streams, and the recharge areas of alluvial aquifers; Fig. 11.1) have been well studied because of their importance as zones of matter and energy exchanges in aquatic ecosystems (e.g., Vervier et al., 1992; Hancock et al., 2005). Particularly, the hyporheic zone of streams, where surface water and groundwater mix, has a key role in organic matter processing, nutrient cycling and water purification processes (e.g., Boulton et al., 1998; Gandy et al., 2007; Tonina and Buffington, this volume). For example, Battin et al. (2003) reported that the hyporheic zone contributed around 40% to the whole stream respiration in a third-order piedmont stream. In addition, apart from stimulating nutrient turnover in the stream ecosystem, the hydrological exchange at the surface water–groundwater interfaces modulates the supply of organic matter to deeper subterranean areas which often is a scarce and limiting resource in heterotrophic groundwater ecosystems.

While microorganisms are recognized as being the main actors in biogeochemical processes such as organic matter processing and nitrogen cycling in all environments (Griebler and Lueders, 2009; Peralta-Maraver et al., 2018, Fillinger et al., this volume), this chapter will explicitly focus on invertebrates. Groundwaters harbor a wide diversity of invertebrates (Marmonier et al., this volume) which can provide ecosystem services like organic matter processing and nutrient cycling depending on their feeding and burrowing traits (e.g., Boulton et al. 2008). Overall, groundwater invertebrates play significant roles in ecosystem functioning via two main types of action: (1) their feeding activities that influence organic matter processing and food web structure and (2) their nontrophic, "engineering" activities (such as sediment reworking, burrowing and bio-irrigation–burrow ventilation) that modify the physical and chemical environment for other organisms.

The relative contributions of the invertebrates' trophic and ecosystem engineering activities are largely modulated by the hydrological and biological connectivity between surface water

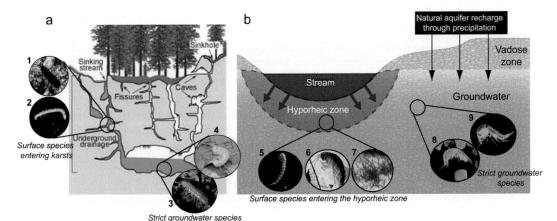

FIGURE 11.1    Examples of downwelling surface-groundwater interfaces associated with (A) karstic and (B) alluvial aquifers with photos of invertebrates living at these interfaces. Presented taxa (name and size): 1- *Proasellus meridianus* (length 0.8 cm), 2- Chironomid larvae (length 1.0 cm), 3- *Proasellus valdensis* (length 0.6 cm), 4- *Niphargus virei* (length 2.5 cm), 5- Coleoptera larvae (length 1.0 cm), 6- *Gammarus fossarum* (length 2.1 cm), 7- Tubificid worms (length 1.0–1.8 cm), 8- *Niphargus kochianus* (length 0.4 cm), and 9- *Salentinella juberthieae* (length 0.2 cm). *(B) Credits: P. Marmonier, LEHNA, University Lyon 1.*

and groundwater. The surface water—groundwater interfaces are highly diverse and can vary substantially among different systems in terms of geological and hydrological properties (as exemplified in Fig. 11.1 with karstic and alluvial interfaces). Thus, the objective of this chapter is to review the present knowledge on the role of invertebrates in groundwater ecosystem functioning within different ecosystem types. To this end, we comparatively explore the provision of ecosystem services for human society in (i) karstic and (ii) alluvial systems, while building on the knowledge from surface water-groundwater interfaces. We also present a conceptual framework that illustrates how the interactions between invertebrate activities and hydrological exchange and connectivity influence ecosystem processes and services.

## Trophic actions of invertebrates

## Karst systems

Compared to aquatic ecosystems on the surface, caves have simpler food webs (Gibert and Deharveng, 2002) but comparable community trophic structures (detritus, biofilms, collectors, shredders, biofilm scrapers, predators, parasites) (Simon et al., 2003, Venarsky et al., chapter 10, this volume). The structure of the food webs in caves is largely dependent on the organic matter supply from surface ecosystems (e.g., Simon and Benfield, 2001; Culver and Pipan, 2014) and the rate of organic matter supply can be a limiting factor for communities (Venarsky et al., 2018). When caves are highly connected to surface ecosystems, large inputs of coarse particulate organic matter (CPOM) modify invertebrate communities by favoring the occurrence of shredder taxa (Venarsky et al., 2018). In the absence of CPOM input, heterotrophic microbial biofilms on inorganic substrates that are fueled by dissolved organic matter (DOM) (Simon and Benfield, 2001) or chemolithoautotrophic biofilms (Hutchins et al., 2016) serve as an energy basis for higher trophic levels. In fact, even when CPOM is abundant, biofilms that use DOM still represent an important energy source for aquatic communities in caves. Using $^{13}$C-actetate as a carbon source for tracing the organic carbon in food webs, Simon et al. (2003) showed that snails (*Fontigens tartarea*, *Gyraulus parvus*, and *Physa* sp.) fed directly on biofilms. Similarly, other primary consumers such as the amphipod *Gammarus minus* and the isopod *Caecidotea holsingeri*, fed on a combination of biofilms and fine particulate organic matter (FPOM). Moreover, the $^{13}$C within the DOM was assimilated by biofilms and recovered from predators (the amphipods *Stygobromus emarginatus*, *Stygobromus spinatus*, and the flatworm *Macrocotyla hoffmasteri*).

Connectivity of karstic ecosystems to surface waters not only supplies CPOM, but can also enhance breakdown rates by facilitating the immigration of shredders. This immigration can be in the form of surface taxa (e.g., larvae and nymphs of stoneflies and midges) that enter with CPOM. For example, Venarsky et al. (2012) found that leaf litter breakdown rates were higher in caves highly connected to surface systems than in caves poorly connected to surface because of the dominance of leaf-shredding surface taxa in connected caves. In some cases, however, the input of CPOM and the presence of effective shredding taxa are disconnected. For example, genetic evidence suggests that the stygophilic amphipod *G. minus*, an effective leaf shredder in caves (Simon and Benfield, 2001; Kinsey et al., 2007), invaded cave systems from downstream springs so that its arrival and evolution had been disparate from the source of CPOM input at the entry of cave systems (Culver et al., 1995). Considering that the leaf breakdown rates measured in cave systems rival the range of values measured in surface streams (Simon and Benfield, 2001; Venarsky et al., 2018), the functional capacity of shredders in these groundwaters can be assumed to be similar to the one observed on the surface. However, stygobitic taxa appear to be poorly adapted to being effective shredders compared to facultative and stygophilic taxa since the measured decomposition rates of leaves were not related to the abundances of stygobiotic taxa (Venarsky et al., 2018). Nevertheless, at present, we are unaware of any comparative approach between stygobitic and stygophilic taxa to evaluate the ability of stygobitic taxa to effectively process CPOM.

The feeding of invertebrates on POM and inorganic particles has also been shown to influence the microbial activity on those substrates (Danielopol, 1989; van de Bund et al., 1994). For example, Kinsey et al. (2007) found that grazing by *G. minus* increased the rate of microbial respiration on leaves by 50%. Likewise, Edler & Dodds (1996) reported that sediment feeding by the subterranean isopod *C. tridentata* stimulated bacterial abundance and activity by 300%–400%. In both cases, the positive effect of feeding on microorganisms may have been due to several mechanisms including (1) the renewal of the biofilm developed on leaf and sediment surfaces and (2) the increased availability of DOM and inorganic nutrients from fragmentation of leaf material and animal excretion. However, this positive microbe-invertebrate interaction was not observed by Cooney and Simon (2009) who showed that biofilm grazing by *G. minus* reduced bacterial production on rocks. It is possible that the discrepancy in the feeding role of *G. minus* on biofilms developed on leaves (Kinsey et al., 2007) and on biofilms developed on rocks (Cooney and Simon, 2009) was associated with differences in biofilm characteristics between the two colonized substrata. Indeed, Sinsabaugh et al. (1991) demonstrated that biofilms developed on organic substratums like leaves were denser and more active than biofilms developed on rocks (inorganic/mineral). In these conditions, biofilms developed on leaves would have a higher resistance (and resilience) to biomass loss induced by invertebrate grazing than biofilms developed on rocks.

Based on a limited number of studies, it thus remains difficult to upscale laboratory results to the field, especially since biofilm characteristics, nutrient availability, and grazer densities vary widely with the different environmental conditions in caves (e.g., Brannen-Donnelly and Engel, 2015). Several studies on gut contents (Weitowitz et al., 2019) and on stable isotopes in bulk tissues (François et al., 2016) and in amino acids (Sacco et al., 2019a) established that a wide range of karstic invertebrates (amphipods, isopods, worms) ingest and can assimilate organic matter from sedimentary biofilms. It has been also demonstrated that some amphipods could affect trophic webs by increasing protozoan abundances in laboratory mesocosms (Weitowitz et al., 2019). Nevertheless, we do not know the consequences of these grazing activities on ecosystem process rates, particularly with respect to nutrient cycling (C, N, P, S cycles) in the environment.

## Alluvial systems

In alluvial systems, the exchange of matter and energy at the surface water-groundwater interfaces varies widely from hyporheic zones highly connected to streams, to deep alluvial aquifers naturally recharged with rain water (Fig. 11.1, Boulton et al., 1998). These hydrological and biochemical exchanges are key factors shaping the community structure in the hyporheic zone (e.g., Dole-Olivier, 1998; Olsen and Townsend, 2003) and in alluvial aquifers (e.g., Datry et al., 2005; Hartland et al., 2011). Consequently, the relative abundance of invertebrate groups in alluvial systems was found to be mainly determined by the availability of organic matter, dissolved oxygen, and interstitial space (e.g., Strayer et al., 1997 for the hyporheic zone and Korbel and Hose, 2015 for alluvial aquifers). Under these conditions, CPOM processing by invertebrates has been exclusively considered in the hyporheic zone where large amounts of leaf detritus can enter during spates (e.g., Boulton and Foster, 1998). Experimental studies performed in the field using leaf litter bags demonstrated that the contribution of hyporheic invertebrates to CPOM processing mainly depends on the ability of the shredders to access the buried leaf litter (Marmonier et al., 2010). Navel et al. (2010) confirmed these conclusions in laboratory experiments that evaluated the influence of sediment porosity (interstitial space) on shredder feeding activity. By manipulating porosity, they showed that porosities of 25% and 35% allowed a consumption of buried leaf litter by surface shredders similar to those observed in the benthic zone of streams (Navel et al., 2010). In contrast, a reduction of porosity from 25% to 12% inhibited the action of surface shredders by limiting their access to buried leaf litter (Fig. 11.2).

With respect to groundwater invertebrates, it can be expected that access to buried leaf litter is less problematic, since they are specifically adapted to interstitial environments (Marmonier et al., this volume) and hence, the results obtained for surface invertebrates will not be applicable to the subsurface. Nevertheless, there is no experimental evidence that strict groundwater species living in the hyporheic zone shred CPOM. For example, the subterranean amphipod *Niphargus rhenorhodanensis*, which has shredder representatives in surface environments, was classified as a collector-gatherer, feeding preferentially on

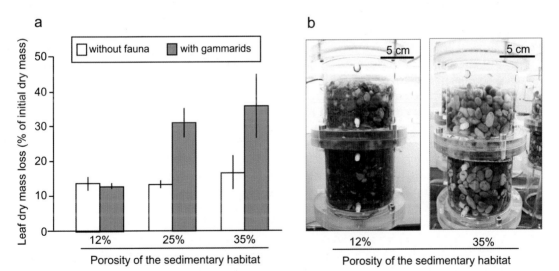

**FIGURE 11.2** (A) Influence of surface shredders (gammarids) on mass loss of leaf litter buried in sediments depending on the porosity of the sedimentary habitat (pore volume). Leaf dry mass loss were obtained after 4 weeks of experiment for three porosity treatments and for two gammarid treatments (mean ± SD, n = 3). (B) Examples of mesocosms of the lowest and highest porosity treatments used. *(A) Modified from Navel et al. (2010). (B) credit S. Navel, LEHNA, University Lyon 1.*

FPOM (Navel et al., 2011a). As observed in cave systems, the contribution of invertebrate feeding to CPOM processing in the hyporheic zone was more likely associated with shredder immigration from surface water. This is not surprising given that CPOM is held back by the top sediment/soil layers, before it can reach the groundwater. Hence, the main contribution of stygofauna to nutrient turnover processes can be expected from their feeding activities on FPOM (Navel et al., 2011a), hetero- and chemoautotrophic biofilms (Foulquier et al., 2010a; Weitowitz et al., 2019), as well as from predation on subordinate trophic levels, scavenging (eating eggs and corpses), and presumably (in places with locally restricted, higher population densities) also cannibalism.

It has been suggested that the feeding activities of invertebrate grazers and deposit-feeders may have direct influences on microorganisms and their activities (Traunspurger et al., 1997). Microbial biofilms coating the large interstitial surface areas of sediment particles provide food for protozoa and grazing invertebrates (Bärlocher and Murdoch, 1989). Microbial activity may be enhanced by this feeding activity (Danielopol, 1989; Traunspurger et al., 1997) as well as fueled by nutrients excreted by the invertebrates (Boulton, 2000; Marshall and Hall, 2004). Moreover, the processing of sediment organic matter into feces by deposit-feeders may stimulate microbial production as fecal pellets can act as new and favorable colonization substrates for microorganisms (Wavre and Brinkhurst, 1971; Mermillod-Blondin et al., 2002).

Based on the studies above, feeding on microorganisms developed on hard substrates, fine sediment and organic materials appears to be an important factor in alluvial ecosystem functioning. However, the outcomes of invertebrate grazing will be highly dependent on grazer population density. Related to this, microbial standing stock cannot be significantly modified by invertebrate grazing when the biofilm renewal rate exceeds the consumption rate by invertebrates (e.g., McManus and Fuhrman, 1988). Indeed, Foulquier et al. (2010a) reported that the subterranean amphipod N. rhenorhodanensis did not influence microbial biomass because bacterial ingestion by this species represented less than 5% of the carbon assimilated by bacteria over 4 days. The authors explained these results by the low metabolism of subterranean species (e.g., Hervant and Renault, 2002), limiting their ability to regulate the microbial compartment. However, this interpretation must be taken with caution because (1) N. rhenorhodanensis is not a strict bacterivore and (2) the use of a highly biodegradable source of dissolved organic carbon (sodium acetate) during experiments does not reproduce the low availability of DOC encountered in an underground environment (Foulquier et al., 2010a). Indeed, the quantity of biodegradable DOC can be more than fourfold lower in groundwater environments than in surface waters (e.g., Mermillod-Blondin et al., 2015a for alluvial aquifers). Thus, it appears necessary to develop field-relevant laboratory experiments on strict bacterivorous groundwater species to evaluate the role of invertebrate grazers on microbial processes in alluvial systems.

While the low metabolic rates and low population densities of invertebrates may be limiting their direct influence on microbial standing stock, their grazing/deposit-feeding activities may still be acting indirectly, through more subtle mechanisms. An example for such a mechanism could be the proposed ability of subterranean invertebrates to selectively feed on microorganisms, thereby modifying the bacterial assemblages in alluvial groundwater ecosystems. For instance, crustaceans (Phreatoicus typicus, Phreatogammarus fragilis, and Paracrangonyx compactus) of a sewage-polluted aquifer ingested large amounts of coliform bacteria, potentially reducing the populations of these harmful biological agents (Sinton, 1984). Positive influences of subterranean invertebrates on protozoan densities, as observed by Weitowitz et al. (2019), could also have an indirect influence on pathogen control (e.g., when protozoa control bacterial and viral densities in aquifers, Deng et al., 2013). Similarly, an increase in protozoan population density may have a positive effect on microbial contaminant biodegradation since protozoan biofilm grazing facilitates oxygen, nutrient and contaminant diffusion into the biofilm and improves the availability of these compounds to microorganisms of the biofilm (Mattison et al., 2005; Peralta-Maraver et al., 2018; Mattison et al., 2005). Notably, microbe—invertebrate interactions in groundwater need not necessarily be only positive for groundwater hygienic quality: Smith et al. (2016) reported that stygofauna may transport pathogenic organisms throughout aquifers, potentially increasing the spatial extent of contamination. However, stygofauna may also act as a vector for the distribution of specialized microbes, thus providing the key to specific biochemical processes (Griebler and Avramov, 2015).

## Ecosystem engineering activities by invertebrates

### Karst systems

Ecosystem engineers are "organisms that directly or indirectly control the availability of resources to other organisms by causing physical state changes in biotic or abiotic materials" (Jones et al., 1997). Surface streams provide the most well-known examples of ecosystem engineers (e.g., beavers, crayfish; Moore, 2006). Cave streams are physically similar to surface streams, characterized by distinct habitats for invertebrates, such as mud pools, bedrock pools, rocks and small gravels in riffles (Culver, 1973). Some taxa (e.g., crayfish) that are well-known engineers on the surface are present in caves but many categories of engineers (e.g., large vertebrates such as benthic fishes) are not. A key feature that determines the degree to which engineering occurs, is population density (Moore, 2006). It may limit engineering significance on whole karstic groundwater ecosystems as densities in karsts, particularly of large taxa, are low. It is possible that engineering may be important on small scales, but not much work has focused on this issue in karst systems. As shown in marine, estuarine, and freshwater soft-bottom benthic habitats, bioturbation by invertebrates (sediment reworking, production of biogenic structure, ventilation of burrows, Kristensen et al., 2012) that redistributes particles and modifies water fluxes at the water—sediment interface, affects the availability of electron acceptors (e.g., dissolved $O_2$), organic matter, and nutrients to sedimentary microorganisms (Mermillod-Blondin and Rosenberg, 2006). Several cave species, such as burrowing worms (Creuzé des Châtelliers et al., 2009) and amphipods (*Niphargus virei*, Ginet, 1960), are able to produce biogenic structures in muddy habitats of caves. Consequently, we cannot exclude a significant role of bioturbation activities in karstic environments but research is needed to evaluate the contribution of ecosystem engineer species on the microbial processes involved in ecosystem functioning.

### Alluvial systems

The influence of ecosystem engineering by invertebrates in alluvial systems, particularly in hyporheic zones, has received more attention than in karstic environments. Several studies (e.g., Griebler, 1996; Mermillod-Blondin et al., 2002) suggest that burrowing activities of tubificid worms increase the microbial activity in hyporheic sediments. By stimulating microorganisms, tubificid worms enhanced the organic matter degradation and water purification processes such as denitrification (Mermillod-Blondin et al., 2004) in river streambeds. However, the effects of burrowing invertebrates on the biogeochemical processes in hyporheic sediments varied depending on the bioturbation modes of the ecosystem engineers and on the water flow rates in sediments (Mermillod-Blondin and Rosenberg, 2006). For these reasons, the influence of invertebrates on aerobic microbial processes (measured via changes in $O_2$ consumption) in hyporheic zones can range from less than 5% (Pusch and Schwoerbel, 1994) to more than 50% (Marshall and Hall, 2004). This variability in invertebrate effects can be due to different assemblages of species with distinct modes of activity (sediment reworking, biogenic structure building, bio-irrigation). For example, Mermillod-Blondin et al. (2002) showed that deep burrow galleries of worms stimulated microbial processes (aerobic respiration, denitrification), whereas fine sediment reworking by isopods did not.

The importance of invertebrates as engineers likely varies spatially as a function of hydrological conditions in hyporheic zones. The conceptual framework of Boulton et al. (2002) suggests that the role of ecosystem engineering in sedimentary environments is negatively correlated with the advection rate of water in sediments (Mermillod-Blondin and Rosenberg, 2006; Mermillod-Blondin, 2011). Since ecosystem engineering mainly influences microbial communities by modifying the transport of nutrients and electron acceptors to sedimentary biofilms (Hölker et al., 2015), bioturbation processes may be a "vector of fluxes" in systems characterized by low advection fluxes and a "modulator of water fluxes" in advection-dominated systems (Boulton et al., 2002). This means that the same ecosystem engineer species can stimulate hydrological exchanges in sedimentary systems with a low permeability but have little or no effect in highly permeable sediments (as demonstrated with tubificid worms by Navel et al., 2012). Such a framework could apply in both the hyporheic zone and deep alluvial aquifers, since amphipods (*Niphargus inopinatus*) living in alluvial aquifers alter the hydraulic properties of sedimentary environments by creating burrow networks (Stumpp and Hose, 2017; Hose and Stumpp, 2019). Ecosystem engineers are likely most influential in downwelling zones (i.e., where water exchange is occurring from surface to groundwater) where they can modulate material flux into groundwater. In the opposite situation (upwelling), hydraulic exchanges are less controlled by sediment characteristics (as upwelling can prevent clogging) and hot spots of biological activity are not observed in the groundwater environment but in the surface environment (e.g., increased algal growth, Boulton et al., 1998). Consequently, we focus on the downwelling zones in the following parts of this chapter.

## Conceptual model of the role of invertebrates on ecosystem processes and consequences for ecosystem services

Based on the previous sections, we propose a conceptual framework (Fig. 11.3) that - illustrates how hydrological exchange and connectivity at surface water-groundwater interfaces may influence the functional traits of invertebrate communities. Invertebrate activities are largely dependent on the functional traits of organisms: their feeding mode, feeding rate, sediment mixing rate, burrowing depth, type of biogenic structure produced, and bio-

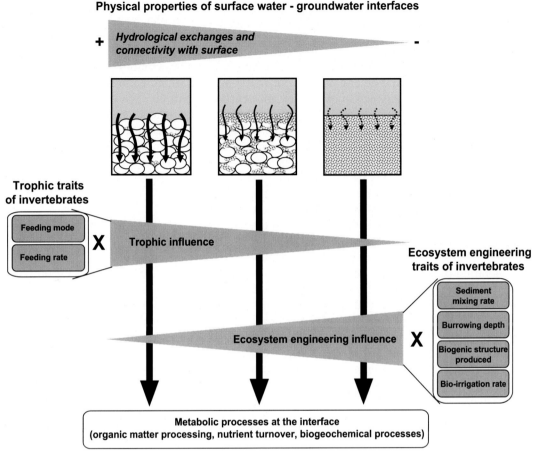

**FIGURE 11.3** Conceptual framework hypothesizing the relative contributions of trophic and ecosystem engineering activities to ecosystem processes at surface water-groundwater interfaces depending on their physical properties (hydrological exchanges and connectivity) and the functional traits of the invertebrates. In absolute terms, the role of groundwater invertebrates in the provision of ecosystem services is also modulated by local invertebrate population densities. *Modified from Mermillod-Blondin (2011).*

irrigation rates. Thus the collective traits of a community influence the significance of invertebrates on ecosystem processes and services.

High hydrological exchange and connectivity at the interface leading to large inputs of organic matter, nutrients and electron acceptors favor the trophic contribution of invertebrates on ecosystem functioning by supporting relatively high invertebrate population densities (Strayer et al., 1997; Datry et al., 2005; Humphreys, 2009), modifying the trophic assemblage of species (Culver and Pipan, 2014), and increasing the potential colonization of groundwaters by surface invertebrates (Venarsky et al., 2018). In deeper groundwater and at interfaces characterized by low hydrological exchange and connectivity, top-down control of microbial biomass through trophic activities of invertebrates is expected to be less significant as both surface-dwelling and groundwater invertebrates would have low abundances and groundwater taxa with a low metabolism would dominate the invertebrate communities. Thus, we expect a decreasing significance of trophic activities of invertebrates from systems having a high hydrological exchange to systems having a low hydrological exchange with surface water (Fig. 11.3).

The opposite trend is expected for ecosystem engineering activities. For example, under high hydrological exchange and connectivity with surface water, ecosystem engineering activities of invertebrates would have a low influence on microbial processes as microorganisms are predominantly controlled by physical fluxes of nutrients and electron acceptors (Mermillod-Blondin, 2011). At the other end of the spectrum, engineering activities are expected to have the highest role in interfaces characterized by low hydrological exchange as they can stimulate hydrological exchange and hence increase the availability of oxygen and nutrients to microorganisms. Thus, the functional significance of ecosystem engineering activities would increase from systems having a high hydrological exchange to systems having a low hydrological exchange with surface water (Fig. 11.3).

As the relative contributions of trophic and ecosystem engineering activities to ecosystem processes are inversely linked to the physical properties of the environment in our conceptual framework (Fig. 11.3), invertebrates are expected to provide essential ecosystem processes by accelerating organic matter processing and nutrient cycling in all surface water—groundwater interfaces. By stimulating nutrient recycling and biofilm activities, invertebrates contribute also to ecosystem services such as water purification (e.g., stimulation of denitrification in hyporheic sediments, Mermillod-Blondin et al., 2004). In chemically polluted groundwater systems, microorganisms are key actors in biodegradation of organic contaminants such as hydrocarbons (Chapelle, 1999; Meckenstock et al., 2015; Fillinger et al., this volume). Consequently, invertebrate activities that stimulate microbial activities could also indirectly promote bioremediation. In addition, invertebrates consuming and digesting viruses and other pathogenic organisms (e.g., coliform bacteria, Sinton, 1984; Boulton et al., 2008; Deng et al., 2013) directly contribute to the biological elimination of pathogens and the purification of groundwater.

The positive influence of invertebrates on the hydraulic conductivity of sediments (Nogaro et al., 2006; Hose and Stumpp, 2019) could help maintain aquifer properties and hence the performance of managed aquifer recharge systems (Gette-Bouvarot et al., 2015). In hyporheic sediments, this same hydrological function of invertebrates like tubificid worms (Navel et al., 2012) is expected to enhance the exchange of water, dissolved oxygen and nutrients at the water-sediment interface, supporting the functioning of the hyporheic zone in its role as an efficient biogeochemical reactor for the stream (Krause et al., 2011) and as a favorable zone for the provision of other valuable ecosystem services, such as salmon egg development (Malcolm et al., 2004).

The use of groundwater invertebrates for bio-indication, biomonitoring, and ecotoxicological purposes (Danielopol et al., 2003; Di Lorenzo et al., 2019) might also be considered as ecosystem services as they inform us about groundwater quality and ecosystem status, as well as on hydrology. As indicated in the previous sections, the occurrence of surface-dwelling species in groundwater ecosystems can infer environmental conditions, including hydrological exchanges and connectivity between surface and groundwater ecosystems (Dole-Olivier, 1998; Graillot et al., 2014). In association with microorganisms, invertebrates can also serve as indicators of chemical contamination in the groundwater (e.g., Hose,

2005; Hahn, 2006; Korbel and Hose, 2011). Although not yet fully integrated in policies like the European Union Groundwater Directive, the application of ecological criteria partly based on invertebrate communities is promising for groundwater quality assessment in the future (Hose et al., this volume).

## Environmental impacts on surface water–groundwater interfaces and consequences for the provision of ecosystem services by invertebrates

Based on our conceptual framework, we hypothesize that the contributions of invertebrates to ecosystem services will be largely influenced by the environmental impacts that affect the hydrological exchanges and connectivity between surface and groundwaters (Fig. 11.4). Therefore, while there are also other environmental issues that pose serious threats to invertebrate communities and their activities like chemical contaminations (see Di Lorenzo et al., Hose et al., this volume) and that the impact of multiple stressors is likely (Strayer et al., 1997; Korbel et al., 2013), in the discussion below we focus on three environmental stressors: (1) drying of streams and reduction of groundwater levels as a result of droughts and groundwater overabstraction, (2) increased artificial recharge of aquifers, and (3) excessive sedimentation and colmation of stream beds.

Global climate change and anthropogenic pressures (e.g., water overabstraction) are increasing the frequency and duration of drying events in aquatic ecosystems (e.g., Prudhomme et al., 2014). This can lead to the complete drying of perennial rivers (Jaeger et al., 2014), and—through lowering the groundwater table—the loss of habitat for groundwater

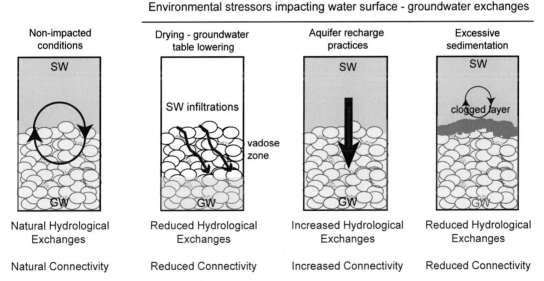

FIGURE 11.4 Simplified diagram of the expected impacts of global change on hydrological exchanges and connectivity at the surface water - groundwater interface of streams. Abbreviations: *GW*, groundwater; *SW*, surface water.

species dwelling in the otherwise saturated zones, as well as for surface species, occupying the hyporheic zone (Vander Vorste et al., 2016). The drying process is also associated with the formation of a vadose zone beneath streams (Fig. 11.4). In aquifers, it has been demonstrated that small quantities of water retained in this zone following a lowering of the water table can sustain invertebrates, at least in the short term (Stumpp and Hose, 2013). Moreover, the vadose zone plays a key role in the retention of particulate and DOM during water infiltration from surface to groundwater (Datry et al., 2005; Voisin et al., 2018). The occurrence of a vadose zone in drying streams would thus reduce the supply of nutrients and organic matter for the biota (microorganisms and invertebrates), consequently reducing the amount of C and nutrients processed by invertebrates living at the vadose zone—groundwater interface.

Groundwater exploitation has grown rapidly, and has challenged human capability to sustain the resource (Famiglietti, 2014). To address this challenge, artificial aquifer recharge is being increasingly applied in Australia, USA, and Europe as a means of recycling stormwater or treated sewage effluent for nonpotable and indirect potable reuse in urban and rural areas (Dillon et al., 2019). Such practices strongly increase the import of surface water into groundwater ecosystems (Fig. 11.4). As waters used to artificially recharge aquifers are usually richer in organic matter than natural groundwater, enrichments in organic carbon are commonly observed in the recharge zones of aquifers (Foulquier et al., 2010b). Consequently, more energy is available for groundwater biota, and both biofilm growth and invertebrate abundances are increased in the recharge zones (Datry et al., 2005; Hartland et al., 2011). Under such conditions, we can expect that invertebrate activities would stimulate the mineralization of organic matter and reduce the clogging processes associated with biofilm growth up to a certain extent. However, this is contingent on the artificial recharge water not being toxic to biota, and not leading to oxygen depletion in the aquifer, as this would reduce invertebrate population densities and hence, diminish their contribution (as observed in aquifers recharged with urban runoffs rich in dissolved organic carbon, Foulquier et al., 2010b; Marmonier et al., 2013).

Excessive sedimentation in streams and rivers is a global problem (Owens et al., 2005). Disturbance of the land surface due to agriculture, forestry, construction, mining activities or urbanization increases the import of fine inorganic particles and sedimentation and affects riverbed physical characteristics by clogging the top sediment layers (Wood and Armitage, 1997; Rehg et al., 2005). Sedimentation reduces the availability of interstitial spaces in the hyporheic zone, displaces the invertebrates that use interstitial spaces (Descloux et al., 2013), and hampers the migration of surface invertebrates into the hyporheic zone (Fig. 11.3). Consequently, surface shredders like gammarids do not have access to CPOM buried in the hyporheic zone and CPOM processing is expected to be lower compared to non-impacted conditions (Navel et al., 2010, Fig. 11.2). Moreover, if surface invertebrates cannot enter deeper zones due to clogging of the stream sediments, they lose an important refuge that would otherwise offer survival under unfavorable conditions such as droughts and floods, or temperature extremes due to climate change (Vander Vorste et al., 2017). In addition, a reduction of the hydrological exchanges between surface water and groundwater limits the supply of organic matter and dissolved oxygen from the surface to the deeper groundwater, impacting the stygofaunal communities in alluvial aquifers. Moreover, a reduction in the supply of oxygen from surface water into groundwater ecosystems could lead to anoxic conditions for microorganisms and invertebrates, thus reducing the mineralization rates of organic matter (Navel et al., 2011b) and invertebrate population densities (Sarriquet et al., 2007), the latter resulting in a negative feedback-loop for invertebrate-driven controls of ecosystem processes. These negative impacts could be partially mitigated by ecosystem engineering activities (e.g., gallery building) of invertebrate taxa, such as tubificid worms, which can reestablish hydrological exchanges in systems clogged by fine sediments (Nogaro et al., 2006; Navel et al., 2012). Nevertheless, depending on the severity of clogging, this biological lever may not always be strong enough to effectively reverse the damage caused by colmation to the surface water-groundwater continuum and the provision of ecosystem services.

## Suggestions for future research directions

The conceptual framework presented in this chapter predicts the relative contribution of invertebrates to ecosystem processes depending on the hydrological exchanges and connectivity at surface water–groundwater interfaces. The framework is grounded in empirical evidence, but awaits further evaluation under laboratory and field conditions. To do so, we identified key knowledge gaps and research opportunities that will improve our understanding of the roles of invertebrates, and the characterization of the groundwater environment and exchange zones.

A key step to predicting the functional capacity of groundwater communities is to characterize the environment, particularly the degree of hydrological exchange. Geophysical approaches like ground-penetrating radar could be promising to map groundwater sedimentary habitats and zones with different hydrological conditions, which may harbor distinct communities of hyporheic and groundwater invertebrates (e.g., Mermillod-Blondin et al., 2015b). Studies using isotopic or biochemical (fatty acids) tracers to quantify the fluxes of C, N and contaminants in groundwater ecosystems remain scarce but have potential in identifying areas of exchange (see Pombo et al., 2002; Simon et al., 2003) as has the use of microbiota (Zhu et al., 2020; Korbel et al., 2022).

Much of our current knowledge is derived from laboratory studies, but there is often a disconnect between field and laboratory approaches, which limits the ability to scale up the quantities of functions measured in laboratory experiments to the field. To date, quantifying invertebrate density (particularly in aquifers) and species distributions in groundwater is a difficult task, but it is necessary to develop field-relevant laboratory experiments and to make predictions of invertebrate contributions at the aquifer scale. The use of mark—recapture methods comparable to those used for cave isopods (Simon and Buikema, 1997) and crayfishes (Huntsman et al., 2020) could be a time-consuming but appropriate approach for evaluating population sizes of aquifer invertebrates. However, in the era of molecular approches, we can also expect that environmental DNA [eDNA] approaches will permit the quantification of population sizes in groundwater ecosystems (see Baldigo et al., 2017 for an example of size estimations of trout populations in streams using eDNA).

Most obviously, a deeper understanding of the functional traits of stygofauna, as well as of the trophic interactions in groundwater food webs, is needed. Classifications of groundwater invertebrates (e.g., Claret et al., 1999; Hose et al., 2022) based on their functional traits are still in their infancy and unavailable for most poorly known strict groundwater species (Boulton et al., 2008) whereas approaches based on functional groups and traits have been successfully developed to link biodiversity and ecosystem functioning in many environments (e.g., Flynn et al.,

2011). To date, it has been often assumed that obligate groundwater invertebrates should affect ecosystem processes in a similar way to that observed for surface-dwelling aquatic invertebrates (Boulton et al., 2008). However, this assumption is not confirmed by experimental data (e.g., the amphipod *Niphargus rhenorhodanensis* is not a shredder like surface-dwelling gammarids, Navel et al., 2011a). Only a few studies (e.g., Foulquier et al., 2010a; Hose and Stumpp 2019) have explored the potential for these to be classified in functional groups based on their roles in facilitating groundwater ecosystem processes and services such as water purification, toxin and waste material breakdown, maintenance of hydraulic conductivity and connectivity, organic matter decomposition, and nutrient recycling. Therefore, laboratory experiments aimed at the functional classification of invertebrates in groundwater ecosystems are needed. In addition, detailed knowledge of trophic webs in aquifers is required in order to assess trophic traits and quantify the impact of trophic activities of stygobionts in their environments (see Venarsky et al., chapter 10, this volume). Following the recommendations of Sacco et al. (2019b), carefully designed multidisciplinary field approaches based on stable isotope analyses, nutrient flux measurements, assessments of invertebrate and microbial diversity, activity rates, as well as population densities will be crucial in order to quantify the role of stygobitic invertebrates and biodiversity in groundwater ecosystem functioning.

## Acknowledgments

We thank C. Griebler and an anonymous reviewer for constructive comments that improved this chapter.

## References

Baldigo, B.P., Sporn, L.A., George, S.D., Ball, J.A., 2017. Efficacy of environmental DNA to detect and quantify brook trout populations in headwater streams of the Adirondack Mountains, New York. Transactions of the American Fisheries Society 146, 99–111.

Barlocher, F., Murdoch, J.H., 1989. Hyporheic biofilms—a potential food source for interstitial animals. Hydrobiologia 184, 61–67.

Battin, T.J., Kaplan, L.A., Newbold, J.D., Hendricks, S.P., 2003. A mixing model analysis of stream solute dynamics and the contribution of a hyporheic zone to ecosystem function. Freshwater Biology 48, 995–1014.

Boulton, A.J., 2000. The functional role of the hyporheos. Internationale Vereinigung für theoretische und angewandte Limnologie: Verhandlungen 27, 51–63.

Boulton, A.J., Foster, J.G., 1998. Effects of buried leaf litter and vertical hydrologic exchange on hyporheic water chemistry and fauna in a gravel-bed river in northern New South Wales, Australia. Freshwater Biology 40, 229–243.

Boulton, A.J., Hakenkamp, C.C., Palmer, M.A., Strayer, D.L., 2002. Freshwater meiofauna and surface water-sediment linkages: a conceptual framework for cross-system comparisons. In: Rundle, S.D., Robertson, A.L., Schmid-Araya, J.M. (Eds.), Freshwater Meiofauna: Biology and Ecology. Backhuys Publishers, The Netherlands, pp. 241–259.

Boulton, A.J., Fenwick, G.D., Hancock, P.J., Harvey, M.S., 2008. Biodiversity, functional roles and ecosystem services of groundwater invertebrates. Invertebrate Systematics 22, 103–116.

Boulton, A.J., Findlay, S., Marmonier, P., Stanley, E.H., Valett, H.M., 1998. The functional significance of the hyporheic zone in streams and rivers. Annual Review of Ecology and Systematics 29, 59–81.

Brannen-Donnelly, K., Engel, A.S., 2015. Bacterial diversity differences along an epigenic cave stream reveal evidence of community dynamics, succession, and stability. Frontiers in Microbiology 6, 729.

Bulling, M.T., Hicks, N., Murray, L., Paterson, D.M., Raffaelli, D., White, P.C.L., Solan, M., 2010. Marine biodiversity-ecosystem functions under uncertain environmental futures. Philosophical Transactions of the Royal Society 365, 2107–2116.

Cardinale, B.J., Srivastava, D.S., Duffy, J.E., Wright, J.P., Downing, A.L., Sankaran, M., Jouseau, C., 2006. Effects of biodiversity on the functioning of trophic groups and ecosystems. Nature 443, 989–992.

Chapelle, F.H., 1999. Bioremediation of petroleum hydrocarbon-contaminated ground water: the perspectives of history and hydrology. Groundwater 37, 122–132.

Claret, C., Marmonier, P., Dole-Olivier, M.J., Creuzé Des Châtelliers, M., Boulton, A.J., Castella, E., 1999. A functional classification of interstitial invertebrates: supplementing measures of biodiversity using species traits and habitat affinities. Archiv für Hydrobiologie 145, 385–403.

Cooney, T.J., Simon, K.S., 2009. Influence of dissolved organic matter and invertebrates on the function of microbial films in groundwater. Microbial Ecology 58, 599–610.

Creuzé des Châtelliers, M., Juget, J., Lafont, M., Martin, P., 2009. Subterranean aquatic oligochaeta. Freshwater Biology 54, 678–690.

Culver, D.C., 1973. Competition in spatially heterogeneous systems: an analysis of simple cave communities. Ecology 54, 102–110.

Culver, D.C., Pipan, T., 2014. Shallow Subterranean Habitats: Ecology, Evolution, and Conservation. Oxford University Press, USA.

Culver, D.C., Kane, T.C., Fong, D.W., 1995. Adaptation and Natural Selection in Caves: The Evolution of *Gammarus minus*. Harvard University Press, USA.

Danielopol, D.L., 1989. Groundwater fauna associated with riverine aquifers. Journal of the North American Benthological Society 8, 18–35.

Danielopol, D.L., Griebler, C., Gunatilaka, A., Notenboom, J., 2003. Present state and future prospects for groundwater ecosystems. Environmental Conservation 30, 104–130.

Datry, T., Malard, F., Gibert, J., 2005. Response of invertebrate assemblages to increased groundwater recharge rates in a phreatic aquifer. Journal of the North American Benthological Society 24, 461–477.

Deng, L., Krauss, S., Feichtmayer, J., Hofmann, R., Arndt, H., Griebler, C., 2013. Grazing of heterotrophic flagellates on viruses is driven by feeding behaviour. Environmental Microbiology Reports 6, 325–330.

Descloux, S., Datry, T., Marmonier, P., 2013. Benthic and hyporheic invertebrate assemblages along a gradient of increasing streambed colmation by fine sediment. Aquatic Sciences 75, 493–507.

Di Lorenzo, T., Di Marzio, W.D., Fiasca, B., Galassi, D.M.P., Korbel, K., Iepure, S., Pereira, J., Reboleira, A.S.P.S., Schmidt, S., Hose, G.C., 2019. Recommendations for ecotoxicity testing with stygobiotic species in the framework of groundwater environmental risk assessment. Science of the Total Environment 681, 292–304.

Dillon, P., Stuyfzand, P., Grischek, T., Lluria, M., Pyne, R.D.G., Jain, R.C., Bear, J., Schwarz, J., Wang, W., Fernandez, E., Stefan, C., Pettenati, M., van der Gun, J., Sprenger, C., Massmann, G., Scanlon, B.R., Xanke, J., Jokela, P., Zheng, Y., Rossetto, R., Shamrukh, M., Pavelic, P., Murray, E., Ross, A., Bonilla Valverde, J.P., Palma Nava, A., Ansems, N., Posavec, K., Ha, K., Martin, R., Sapiano, M., 2019. Sixty years of global progress in managed aquifer recharge. Hydrogeology Journal 27, 1–30.

Dole-Olivier, M.J., 1998. Surface water–groundwater exchanges in three dimensions on a backwater of the Rhône River. Freshwater Biology 40, 93–109.

Edler, C., Dodds, W.K., 1996. The ecology of a subterranean isopod, *Caecidotea tridentata*. Freshwater Biology 35, 249–259.

Famiglietti, J.S., 2014. The global groundwater crisis. Nature Climate Change 4, 945–948.

Flynn, D.F., Mirotchnick, N., Jain, M., Palmer, M.I., Naeem, S., 2011. Functional and phylogenetic diversity as predictors of biodiversity–ecosystem-function relationships. Ecology 92, 1573–1581.

Foulquier, A., Malard, F., Mermillod-Blondin, F., Datry, T., Simon, L., Montuelle, B., Gibert, J., 2010b. Vertical change in dissolved organic carbon and oxygen at the water table region of an aquifer recharged with stormwater: biological uptake or mixing? Biogeochemistry 99, 31–47.

Foulquier, A., Simon, L., Gilbert, F., Fourel, F., Malard, F., Mermillod-Blondin, F., 2010a. Relative influences of DOC flux and subterranean fauna on microbial abundance and activity in aquifer sediments: new insights from [13]C-tracer experiments. Freshwater Biology 55, 1560–1576.

Francois, C.M., Mermillod-Blondin, F., Malard, F., Fourel, F., Lécuyer, C., Douady, C.J., Simon, L., 2016. Trophic ecology of groundwater species reveals specialization in a low-productivity environment. Functional Ecology 30, 262–273.

Gandy, C.J., Smith, J.W.N., Jarvis, A.P., 2007. Attenuation of mining-derived pollutants in the hyporheic zone: a review. Science of the Total Environment 373, 435–446.

III. Roles of organisms in groundwater

Gette-Bouvarot, M., Volatier, L., Lassabatere, L., Lemoine, D., Simon, L., Delolme, C., Mermillod-Blondin, F., 2015. Ecological engineering approaches to improve hydraulic properties of infiltration basins designed for groundwater recharge. Environmental Science and Technology 49, 9936−9944.

Gibert, J., Deharveng, L., 2002. Subterranean ecosystems: a truncated functional biodiversity. BioScience 52, 473−481.

Ginet, R., 1960. Ecologie, éthologie et biologie de «*Niphargus*» (Amphipodes Gammaridés hypogés). Annales de Spéléologie 15, 127−382.

Graillot, D., Paran, F., Bornette, G., Marmonier, P., Piscart, C., Cadilhac, L., 2014. Coupling groundwater modeling and biological indicators for identifying river/aquifer exchanges. SpringerPlus 3, 68.

Griebler, C., 1996. Some applications for the DMSO-reduction method as a new tool to determine the microbial activity in water saturated sediments. Archiv für Hydrobiologie Suppl. 113, 405−410.

Griebler, C., Avramov, M., 2015. Groundwater ecosystem services: a review. Freshwater Science 34, 355−367.

Griebler, C., Lueders, T., 2009. Microbial biodiversity in groundwater ecosystems. Freshwater Biology 54, 649−677.

Hahn, H.J., 2006. The GW-Fauna-Index: a first approach to a quantitative ecological assessment of groundwater habitats. Limnologica 36, 119−137.

Hancock, P.J., Boulton, A.J., Humphreys, W.F., 2005. Aquifers and hyporheic zones: towards an ecological understanding of groundwater. Hydrogeology Journal 13, 98−111.

Hartland, A., Fenwick, G.D., Bury, S.J., 2011. Tracing sewage-derived organic matter into a shallow groundwater food web using stable isotope and fluorescence signatures. Marine and Freshwater Research 62, 119−129.

Hervant, F., Renault, D., 2002. Long-term fasting and realimentation in hypogean and epigean isopods: a proposed adaptive strategy for groundwater organisms. Journal of Experimental Biology 205, 2079−2087.

Hölker, F., Vanni, M.J., Kuiper, J.J., Meile, C., Grossart, H.P., Stief, P., Adrian, R., Lorke, A., Dellwig, O., Brand, A., Hupfer, M., Mooij, W.M., Nützmann, G., Lewandowski, J., 2015. Tube-dwelling invertebrates: tiny ecosystem engineers have large effects in lake ecosystems. Ecological Monographs 85, 333−351.

Hose, G.C., 2005. Assessing the need for groundwater quality guidelines for pesticides using the species sensitivity distribution approach. Human and Ecological Risk Assessment 11, 951−966.

Hose, G.C., Chariton, A.A., Daam, M., Di Lorenzo, T., Galassi, D.M.P., Halse, S.A., Reboleira, A.S.P.S., Robertson, A.L., Schmidt, S.I., Korbel, K., 2022. Invertebrate traits, diversity and the vulnerability of groundwater ecosystems. Functional Ecology 36, 2200−2214.

Hose, G.C., Stumpp, C., 2019. Architects of the underworld: bioturbation by groundwater invertebrates influences aquifer hydraulic properties. Aquatic Sciences 81, 20.

Humphreys, W.F., 2009. Hydrogeology and groundwater ecology: does each inform the other? Hydrogeology Journal 17, 5−21.

Huntsman, B.M., Venarsky, M.P., Abadi, F., Huryn, A.D., Kuhajda, B.R., Cox, C.L., Benstead, J.P., 2020. Evolutionary history and sex are significant drivers of crayfish demography in resource-limited cave ecosystems. Evolutionary Ecology 34, 235−255.

Hutchins, B.T., Engel, A.S., Nowlin, W.H., Schwartz, B.F., 2016. Chemolithoautotrophy supports macroinvertebrate food webs and affects diversity and stability in groundwater communities. Ecology 97, 1530−1542.

Jaeger, K.L., Olden, J.D., Pelland, N.A., 2014. Climate change poised to threaten hydrologic connectivity and endemic fishes in dryland streams. Proceedings of the National Academy of Sciences 111, 13894−13899.

Jones, C.G., Lawton, J.H., Shachak, M., 1997. Positive and negative effects of organisms as physical ecosystem engineers. Ecology 78, 1946−1957.

Kinsey, J., Cooney, T.J., Simon, K.S., 2007. A comparison of the leaf shredding ability and influence on microbial films of surface and cave forms of *Gammarus minus* Say. Hydrobiologia 589, 199−205.

Korbel, K.L., Hose, G.C., 2011. A tiered framework for assessing groundwater ecosystem health. Hydrobiologia 661, 329−349.

Korbel, K.L., Hancock, P.J., Serov, P., Lim, R.P., Hose, G.C., 2013. Groundwater ecosystems vary with land use across a mixed agricultural landscape. Journal of Environmental Quality 42, 380−390.

Korbel, K.L., Hose, G.C., 2015. Habitat, water quality, seasonality or site? Identifying environmental correlates of the distribution of groundwater biota. Freshwater Sciences 34, 329−343.

Korbel, K.L., Rutlidge, H., Hose, G.C., Eberhard, S., Andersen, M.A., 2022. Dynamics of microbiotic patterns reveal surface water groundwater interactions in intermittent and perennial streams. Science of the Total Environment 811, 152380.

Krause, S., Hannah, D.M., Fleckenstein, J.H., Heppell, C.M., Kaeser, D., Pickup, R., Pinay, G., Robertson, A.L., Wood, P.J., 2011. Inter-disciplinary perspectives on processes in the hyporheic zone. Ecohydrology 4, 481–499.

Kristensen, E., Penha-Lopes, G., Delefosse, M., Valdemarsen, T., Quintana, C.O., Banta, G.T., 2012. What is bioturbation? The need for a precise definition for fauna in aquatic sciences. Marine Ecology Progress Series 446, 285–302.

Loreau, M., Naeem, S., Inchausti, P., Bengtsson, J., Grime, J.P., Hector, A., Hooper, D.U., Huston, M.A., Raffaelli, D., Schmid, B., Tilman, D., Wardle, D.A., 2001. Biodiversity and ecosystem functioning: current knowledge and future challenges. Science 294, 804–808.

Malcolm, I.A., Soulsby, C., Youngson, A.F., Hannah, D.M., McLaren, I.S., Thorne, A., 2004. Hydrological influences on hyporheic water quality: implications for salmon egg survival. Hydrological Processes 18, 1543–1560.

Marmonier, P., Maazouzi, C., Foulquier, A., Navel, S., François, C., Hervant, F., Mermillod-Blondin, F., Vienney, A., Barraud, S., Togola, A., Piscart, C., 2013. The use of crustaceans as sentinel organisms to evaluate groundwater ecological quality. Ecological Engineering 57, 118–132.

Marmonier, P., Piscart, C., Sarriquet, P.E., Azam, D., Chauvet, E., 2010. Relevance of large litter bag burial for the study of leaf breakdown in the hyporheic zone. Hydrobiologia 641, 203–214.

Marshall, M.C., Hall, R.O.J., 2004. Hyporheic invertebrates affect N cycling and respiration in stream sediment microcosms. Journal of the North American Benthological Society 23, 416–428.

Mattison, R.G., Taki, H., Harayama, S., 2005. The soil flagellate Heteromita globosa accelerates bacterial degradation of alkylbenzenes through grazing and acetate excretion in batch culture. Microbial Ecology 49, 142–150.

McManus, G.B., Fuhrman, J.A., 1988. Control of marine bacterioplankton populations: measurement and significance of grazing. Hydrobiologia 159, 51–62.

Meckenstock, R.U., Elsner, M., Griebler, C., Lueders, T., Stumpp, C., Dejonghe, W., Bastiaens, L., Sprigael, D., Smolders, E., Boon, N., Agathos, S., Sorensen, S.R., Aamand, J., Albrechtsen, H.-J., Bjerg, P., Schmidt, S.I., Huang, W., van Breukelen, B., 2015. Biodegradation: updating the concepts of control for microbial clean-up in contaminated aquifers. Environmental Science and Technology 49, 7073–7081.

Mermillod-Blondin, F., Gérino, M., Creuzé des Châtelliers, M., Degrange, V., 2002. Functional diversity among 3 detritivorous hyporheic invertebrates: an experimental study in microcosms. Journal of the North American Benthological Society 21, 132–149.

Mermillod-Blondin, F., Rosenberg, R., 2006. Ecosystem engineering: the impact of bioturbation on biogeochemical processes in marine and freshwater benthic habitats. Aquatic Sciences 68, 434–442.

Mermillod-Blondin, F., 2011. The functional significance of bioturbation and biodeposition on biogeochemical processes at the water–sediment interface in freshwater and marine ecosystems. Journal of the North American Benthological Society 30, 770–778.

Mermillod-Blondin, F., Gaudet, J.P., Gerino, M., Desrosiers, G., Jose, J., Creuzé des Châtelliers, M., 2004. Relative influence of bioturbation and predation on organic matter processing in river sediments: a microcosm experiment. Freshwater Biology 49, 895–912.

Mermillod-Blondin, F., Simon, L., Maazouzi, C., Foulquier, A., Delolme, C., Marmonier, P., 2015a. Dynamics of dissolved organic carbon (DOC) through stormwater basins designed for groundwater recharge in urban area: assessment of retention efficiency. Water Research 81, 27–37.

Mermillod-Blondin, F., Winiarski, T., Foulquier, A., Perrissin, A., Marmonier, P., 2015b. Links between sediment structures and ecological processes in the hyporheic zone: ground-penetrating radar as a non-invasive tool to detect subsurface biologically active zones. Ecohydrology 8, 626–641.

Moore, J.W., 2006. Animal ecosystem engineers in streams. BioScience 56, 237–246.

Navel, S., Mermillod-Blondin, F., Montuelle, B., Chauvet, E., Simon, L., Piscart, C., Marmonier, P., 2010. Interactions between fauna and sediment control the breakdown of plant matter in river sediments. Freshwater Biology 55, 753–766.

Navel, S., Mermillod-Blondin, F., Montuelle, B., Chauvet, E., Marmonier, P., 2012. Sedimentary context controls the influence of ecosystem engineering by bioturbators on microbial processes in river sediments. Oikos 121, 1134–1144.

Navel, S., Mermillod-Blondin, F., Montuelle, B., Chauvet, E., Simon, L., Marmonier, P., 2011b. Water–sediment exchanges control microbial processes associated with leaf litter degradation in the hyporheic zone: a microcosm study. Microbial Ecology 61, 968–979.

Navel, S., Simon, L., Lécuyer, C., Fourel, F., Mermillod-Blondin, F., 2011a. The shredding activity of gammarids facilitates the processing of organic matter by the subterranean amphipod *Niphargus rhenorhodanensis*. Freshwater Biology 56, 481–490.

Nogaro, G., Mermillod-Blondin, F., François-Carcaillet, F., Gaudet, J.P., Lafont, M., Gibert, J., 2006. Invertebrate bioturbation can reduce the clogging of sediment: an experimental study using infiltration sediment columns. Freshwater Biology 51, 1458–1473.

Olsen, D.A., Townsend, C.R., 2003. Hyporheic community composition in a gravel-bed stream: influence of vertical hydrological exchange, sediment structure and physicochemistry. Freshwater Biology 48, 1363–1378.

Owens, P.N., Batalla, R.J., Collins, A.J., Gomez, B., Hicks, D.M., Horowitz, A.J., Kondolf, G.M., Marden, M., Page, M.J., Peacock, D.H., Petticrew, E.L., Salomons, W., Trustrum, N.A., 2005. Fine-grained sediment in river systems: environmental significance and management issues. River Research and Applications 21, 693–717.

Peralta-Maraver, I., Reiss, J., Robertson, A.L., 2018. Interplay of hydrology, community ecology and pollutant attenuation in the hyporheic zone. Science of the Total Environment 610, 267–275.

Pombo, S.A., Pelz, O., Schroth, M.H., Zeyer, J., 2002. Field-scale $^{13}$C-labeling of phospholipid fatty acids (PLFA) and dissolved inorganic carbon: tracing acetate assimilation and mineralization in a petroleum hydrocarbon-contaminated aquifer. FEMS Microbiology Ecology 41, 259–267.

Prudhomme, C., Giuntoli, I., Robinson, E.L., Clark, D.B., Arnell, N.W., Dankers, R., Fekete, B.M., Franssen, W., Gerten, D., Gosling, S.N., Hagemann, S., Hannah, D.M., Kim, H., Masaki, Y., Satoh, Y., Stacke, T., Wada, Y., Wisser, D., 2014. Hydrological droughts in the 21$^{st}$ century, hotspots and uncertainties from a global multimodel ensemble experiment. Proceedings of the National Academy of Sciences of the United States of America 111, 3262–3267.

Pusch, M., Schwoerbel, J., 1994. Community respiration in hyporheic sediments of a mountain stream (Steina, Black Forest). Archiv für Hydrobiologie 130, 35–52.

Rehg, K.J., Packman, A.I., Ren, J.H., 2005. Effects of suspended sediment characteristics and bed sediment transport on streambed clogging. Hydrological Processes 19, 413–427.

Saccò, M., Blyth, A.J., Humphreys, W.F., Kuhl, A., Mazumder, D., Smith, C., Grice, K., 2019a. Elucidating stygofaunal trophic web interactions via isotopic ecology. PLoS One 14 (10), e0223982.

Saccò, M., Blyth, A., Bateman, P.W., Hua, Q., Mazumder, D., White, N., Humphreys, W.F., Laini, A., Grieber, C., Grice, K., 2019b. New light in the dark-a proposed multidisciplinary framework for studying functional ecology of groundwater fauna. Science of the Total Environment 662, 963–977.

Sarriquet, P.E., Bordenave, P., Marmonier, P., 2007. Effects of bottom sediment restoration on interstitial habitat characteristics and benthic macroinvertebrate assemblages in a headwater stream. River Research and Applications 23, 815–828.

Simon, K.S., Benfield, E.F., 2001. Leaf and wood breakdown in cave streams. Journal of the North American Benthological Society 20, 550–563.

Simon, K.S., Buikema Jr., A.L., 1997. Effects of organic pollution on an Appalachian cave: changes in macroinvertebrate populations and food supplies. The American Midland Naturalist 138, 387–401.

Simon, K.S., Benfield, E.F., Macko, S.A., 2003. Food web structure and the role of epilithic biofilms in cave streams. Ecology 84, 2395–2406.

Sinsabaugh, R.L., Golladay, S.W., Linkins, A.E., 1991. Comparison of epilithic and epixylic biofilm development in a boreal river. Freshwater Biology 25, 179–187.

Sinton, L.W., 1984. The macroinvertebrates in a sewage-polluted aquifer. Hydrobiologia 119, 161–169.

Smith, R.J., Paterson, J.S., Launer, E., Tobe, S.S., Morello, E., Leijs, R., Marri, S., Mitchell, J.G., 2016. Stygofauna enhance prokaryotic transport in groundwater ecosystems. Scientific Reports 6, 32738.

Stein, H., Kellermann, C., Schmidt, S.I., Brielmann, H., Steube, C., Berkhoff, S.E., Fuchs, A., Hahn, H.J., Thulind, B., Griebler, C., 2010. The potential use of fauna and bacteria as ecological indicators for the assessment of groundwater quality. Journal of Environmental Monitoring 12, 242–254.

Strayer, D.L., May, S.E., Nielsen, P., Wollheim, W., Hausam, S., 1997. Oxygen, organic matter, and sediment granulometry as controls on hyporheic animal communities. Archiv für Hydrobiologie 140, 131–144.

Stumpp, C., Hose, G.C., 2013. Impact of water table drawdown and drying on subterranean aquatic fauna in in-vitro experiments. PLoS One 8 (11), e78502.

Stumpp, C., Hose, G.C., 2017. Groundwater amphipods alter aquifer sediment structure. Hydrological Processes 31, 3452–3454.

Tilman, D., Wedin, D., Knops, J., 1996. Productivity and sustainability influenced by biodiversity in grassland ecosystems. Nature 379, 718–720.

Traunspurger, W., Bergtold, M., Goedkoop, W., 1997. The effects of nematodes on bacterial activity and abundance in a freshwater sediment. Oecologia 112, 118–122.

van de Bund, W.J., Goedkoop, W., Johnson, R.K., 1994. Effects of deposit-feeder activity on bacterial production and abundance in profundal lake sediment. Journal of the North American Benthological Society 13, 532–539.

Vander Vorste, R., Mermillod-Blondin, F., Hervant, F., Mons, R., Datry, T., 2017. *Gammarus pulex* (Crustacea: Amphipoda) avoids increasing water temperature and intraspecific competition through vertical migration into the hyporheic zone: a mesocosm experiment. Aquatic Sciences 79, 45–55.

Vander Vorste, R., Mermillod-Blondin, F., Hervant, F., Mons, R., Forcellini, M., Datry, T., 2016. Increased depth to the water table during river drying decreases the resilience of *Gammarus pulex* and alters ecosystem function. Ecohydrology 9, 1177–1186.

Venarsky, M.P., Benstead, J.P., Huryn, A.D., 2012. Effects of organic matter and season on leaf litter colonisation and breakdown in cave streams. Freshwater Biology 57, 773–786.

Venarsky, M.P., Benstead, J.P., Huryn, A.D., Huntsman, B.M., Edmonds, J.W., Findlay, R.H., Wallace, J.B., 2018. Experimental detritus manipulations unite surface and cave stream ecosystems along a common energy gradient. Ecosystems 21, 629–642.

Vervier, P., Gibert, J., Marmonier, P., Dole-Olivier, M.J., 1992. A perspective on the permeability of the surface freshwater-groundwater ecotone. Journal of the North American Benthological Society 11, 93–102.

Voisin, J., Cournoyer, B., Vienney, A., Mermillod-Blondin, F., 2018. Aquifer recharge with stormwater runoff in urban areas: influence of vadose zone thickness on nutrient and bacterial transfers from the surface of infiltration basins to groundwater. Science of the Total Environment 637, 1496–1507.

Wavre, M., Brinkhurst, R.O., 1971. Interactions between some tubificid oligochaetes and bacteria found in the sediments of Toronto Harbour, Ontario. Journal of the Fisheries Board of Canada 28, 335–341.

Weitowitz, D.C., Robertson, A.L., Bloomfield, J.P., Maurice, L., Reiss, J., 2019. Obligate groundwater crustaceans mediate biofilm interactions in a subsurface food web. Freshwater Science 38, 491–502.

Wood, P.J., Armitage, P.D., 1997. Biological effects of fine sediment in the lotic environment. Environmental Management 21, 203–217.

Worm, B., Barbier, E.B., Beaumont, N., Duffy, J.E., Folke, C., Halpern, B.S., Jackson, J.B.C., Lotze, H.K., Micheli, F., Palumbi, S.R., Sala, E., Selkoe, K.A., Stachowicz, J.J., Watson, R., 2006. Impacts of biodiversity loss on ocean ecosystem services. Science 314, 787–790.

Zhu, A., Yang, Z., Liang, Z., Gao, L., Li, R., Hou, L., Li, S., Xie, Z., Wu, Y., Chen, J., 2020. Integrating hydrochemical and biological approaches to investigate the surface water and groundwater interactions in the hyporheic zone of the Liuxi River basin, southern China. Journal of Hydrology 583, 124622.

# Principles of evolution in groundwater

# Voices from the underground: animal models for the study of trait evolution during groundwater colonization and adaptation

*Sylvie Rétaux[1] and William R. Jeffery[2]*

[1]Paris-Saclay Institute of Neuroscience, Université Paris-Saclay and CNRS, Saclay, France;
[2]Department of Biology, University of Maryland, College Park, MD, United States

## Introduction

A diverse group of animals inhabit groundwater in subterranean voids and caves (Culver and Pipan, 2019), and how they colonized and adapted to these unique and challenging habitats is largely unknown. Subterranean species are considered to be the direct descendants of surface-dwelling ancestors and some of the best examples of character polarity in evolutionary biology. Adaptation to perpetual darkness, limited resources, and extreme physical conditions resulted in the evolution of a suite of specialized traits, highlighted by elongated appendages, depigmentation, and the loss of eyes. The specialized traits of subterranean species are called troglomorphic traits, and the species having them are known as troglomorphic species. When the close relatives of troglomorphic species are present in surface habitats, they often serve as proxies for the ancestral underground colonizers. However, it must be kept in mind that the surface forms may not represent the strict ancestral state as they have evolved for the same amount of time as the cave species and could have been exposed to their own specialized ecological conditions. Similarly, the coexistence of relatively nontroglomorphic (e.g., with rudimentary eyes and pigmentation) and troglomorphic taxa in groundwater habitats is useful for interrogating trait evolution in groundwater models. The systems of nontroglomorphic and troglomorphic species are "voices" from the underground of similar value for understanding the mechanisms of trait evolution as wild-type and mutant forms in traditional genetic models, such as *Drosophila* and *Caenorhabditis*. Here, we evaluate four of the

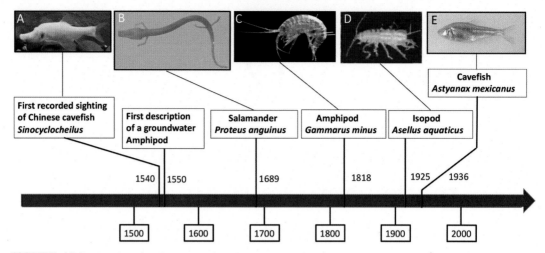

**FIGURE 12.1** Timeline for discovery of groundwater models in this chapter and volume. (A) The Chinese cavefish *Sinocyclocheilus*, (B) The cave salamander *Proteus anguinus*, (C). The cave amphipod *Gammarus minus*, (D) The cave isopod *Asellus aquaticus*, and (E) The cavefish *Astyanax mexicanus*. *Adapted from Balázs et al. (2021). (A) Photo courtesy of Jiahu Lan, (B) Photo courtesy of Špela Gorički, and (C) Photo courtesy of Daniel Fong.*

unique animal systems comprised of troglomorphic and nontroglomorphic forms used in modern groundwater research, the urodele amphibian *Proteus* (Fig. 12.1B), the amphipod *Gammarus* (Fig. 12.1C), the isopod *Asellus* (Fig. 12.1D), and the teleost *Astyanax* (Fig. 12.1E), describe the mechanisms responsible for the evolution of their troglomorphic traits and discuss future goals using these model systems to advance our understanding of the evolutionary and developmental biology of subterranean fauna.

## Brief historical timeline

The fascination of humans with subterranean animals probably emerged when their ghostly pale and eyeless forms were first encountered by our cave-dwelling (or cave-exploring) ancestors and has continued to the present day. Figure 12.1 shows a historical timeline for the discovery and scientific description of the model groundwater species covered in the following Chapters.

Groundwater animals were probably encountered throughout human history, as indicated by the carving of the blind cave salamander *Proteus anguinus* on a medieval tombstone in Stolac, Bosnia, and Herzegovina. However, the first written record of a groundwater animal was a *Sinocycloceilus* cavefish (Fig. 12.1A) observed in Yunnan, China in 1540 (Ma et al., 2019). Later, a troglomorphic amphipod (probably *Niphargus*) from alpine northern Italy was the first groundwater species mentioned in a scientific publication (Alberti, 1550). This was followed by scientific descriptions of *Proteus anguinus* (Valvasor, 1689; Laurenti, 1768) from the Balkan Dinaric Karst, an extraordinary animal that became the flagship subterranean species (Fig. 12.1B; Kostanjšek et al., this volume). During the 1800s, after cave science was initiated in the United States, the troglomorphic amphipod *Gammarus minus* was described from the Appalachian Karst (Fig. 12.1C; Say, 1818), and the blind cavefish *Amblyopsis spelea* was discovered in Mammoth Cave, Kentucky (DeKay, 1842). This cavefish species was the subject of Charles Darwin's statement on troglomorphic evolution in *The Origin of Species*: "As it is difficult to imagine that eyes, though useless, could be in any way injurious to animals living in darkness, I attribute their loss wholly to disuse" (Darwin, 1859). Groundwater species discoveries continued in Europe and North America, and troglomorphic populations of the European isopod *Asellus aquaticus* (Fig. 12.1D; Racovitza, 1925) and the Mexican cavefish *Astyanax mexicanus* (originally *Anoptichthys* sp., (Jordan, 1936)) (Fig. 12.1E; Gross et al., this volume) were described. By the middle of the 20th century, subterranean animals were featured in mainstream books on genetics, speciation, and evolutionary biology (e.g., Dobzhansky, 1941; White, 1978).

## Groundwater model systems

What are the key attributes of a model system for studying troglomorphic trait evolution, and why are *Proteus*, *Gammarus*, *Asellus*, and *Astyanax* good candidates? The two most critical attributes of a groundwater model system are (1) the existence of closely related nontroglomorphic taxa that can be used for comparison with the troglomorphic forms and (2) the ability to be cultivated and bred from generation to generation in the laboratory. *Gammarus*, *Asellus* and *Astyanax* have troglomorphic cave-dwelling and nontroglomorphic surface-dwelling populations, and troglomorphic *Proteus*, which itself consists of multiple cryptic species, can be compared to a nontroglomorphic subterranean form, the black *Proteus*, or its closest surface-dwelling relative, the North American mudpuppy *Necturus* (See Kostanjšek et al. this volume). *Astyanax* surface fish and cavefish have been raised in the laboratory continuously almost since their discovery, first by Charles Breeder at The American Museum of Natural History, then by Horst Wilkens at the University of Hamburg, Germany, and later by Richard Borowsky at New York University and William Jeffery at the University of Maryland. They are now successfully cultured in laboratories around the world. Laboratory propagation has been developed to varying degrees in the other models: *Proteus* have been raised in captivity for more than 50 years at the CNRS Subterranean Laboratory in Moulis, France, and the Tular Cave Laboratory in Kranj, Slovenia, and more recently in an underground laboratory within Postonja Cave, Slovenia (See Kostanjšek et al., this volume). Laboratory culture of cave *Asellus* was recently developed by Meredith Protas at Dominican University, California (Protas et al., 2011), and *Gammarus* laboratory cultivation is under development by Daniel Fong and David Carlini at American University.

A groundwater model should consist of species that are fertile for most of the year, have a tractable generation time, produce sufficient offspring for deep experimental analysis, and its troglomorphic and nontroglomorphic taxa should be able to hybridize, which is useful for genetic analyses and gene discovery. *Astyanax* surface and cave forms routinely spawn hundreds of eggs in the laboratory throughout most of the year, have a generation time of about 6–8 months, and cavefish can be hybridized with the nontroglomorphic surface form or the troglomorphic forms from different caves. *Asellus* has a generation time of less than a year, also conducive to genetic analysis, but a genetic approach with *Proteus* is problematic due to a generation time of 36.5 years (Voituron et al., 2011). Genetic analysis applying the hybridization approach has resulted in the identification of numerous quantitative trait loci (QTL) underpinning troglomorphic traits in *Astyanax* and *Asellus* (Protas et al., 2008; Protas et al., 2011; O'Quin et al., 2013, and others). Further investigation of the *Astyanax* QTL resulted in discovery of the *oculocutaneous albinism type 2* (*oca2*) (Protas et al., 2006) and *cystathionine ß-synthase a* (*cbsa*) (Ma et al., 2020) genes, which are responsible for the loss of melanin pigment and contribute to the loss of eyes, respectively. Sufficient offspring for various laboratory studies can also be obtained in *Gammarus* and *Proteus*, but hybridization between their troglomorphic and nontroglomorphic forms has not been accomplished, although this might be possible between the *Proteus* white and black varieties, once the latter is established in the laboratory.

Another desirable quality of a groundwater model system is the availability of multiple troglomorphic populations for studying the repeated evolution of troglomorphic traits. About 30 *Astyanax* cavefish populations have been identified from separate Mexican caves, and the *Asellus*, *Gammarus*, and *Proteus* systems also feature multiple cave-adapted populations (see Fong and Carlini, this volume; Gross et al., this volume; Kostanjšek et al., this volume; Protas et al., this volume). Laboratory crosses between different cave populations can provide important information about the number of genetic factors responsible for troglomorphic evolution. For example, crosses between different cavefish populations have shown that the same gene (*oca2*) is responsible for albinism (Protas et al., 2006). Moreover, intercavefish crosses show that eye loss is controlled by multiple genes, and that some of these genes are the same and others are unique in different cavefish populations (Borowsky, 2008). In the case of *Asellus*, crosses between two cave populations indicate that the same genes are responsible for eye and pigment degeneration in both populations (Re et al., 2018). Thus, variations in genetic architecture are likely to underlie trait evolution in different groundwater animals. Additionally, all stages of development and the life cycle should be available for study in a groundwater model, and at least the early developmental stages should be of sufficient cytological clarity to conduct molecular expression studies using in situ mRNA and protein detection procedures. Protein and mRNA localization studies are routine in *Astyanax* embryos but have yet to be developed in the other model systems.

The subterranean habitats of the model species should be reasonably accessible for repeated sampling, ecological studies, and behavioral observations. Access by humans to subterranean systems is complicated by small spaces, vertical pitches requiring specialized techniques, and periodic floods, as is typical in the *Astyanax* cavefish locations, where cave entry can be hazardous at certain times of the year. The cave habitats of *Astyanax*, *Gammarus* and *Asellus* are reasonably accessible for sampling and collection, at least by expert cave scientists, but access to *Proteus* locations sometimes requires SCUBA diving to enter groundwater resurgences.

The use of groundwater models to study trait evolution at the molecular level requires an extensive collection of genomic tools and resources. Sequenced genomes and transcriptomes are in place for *Astyanax* (Gross et al., 2013; Hinaux et al., 2013; Warren et al., 2021; McGaugh et al., 2014) and *Asellus* (Gross et al., 2020; Bakovic et al., 2021). In the *Astyanax* system, gene manipulations such as gene knockdown with morpholinos (Alié et al., 2018; Ma et al., 2020), TALENs or CRISPR-Cas9 knock-out (Ma et al., 2015; Klaassen et al., 2018), gene overexpression by injection of synthetic mRNAs or DNA expression constructs (Yamamoto et al., 2009; Hinaux et al., 2015; Alié et al., 2018), transgenesis methods (Elipot et al., 2014; Stahl et al., 2019), and gene knock-in (Devos et al., 2021) have been performed. Transcriptomes of troglomorphic and nontroglomorphic *Gammarus* populations have been published (Carlini and

Fong, 2017), and the massive *Proteus* genome (about 50 gigabases) has recently been sequenced (Kostanjšek et al., 2022), but the development of gene manipulation techniques needs further attention in these models.

In the future, it will be important to develop additional groundwater models. A reasonable candidate is the Chinese cavefish *Sinocycloceilus* (Fig. 12.1A), which consists of a large number of different surface and cave dwelling species (Ma et al., 2019). Genomic and transcriptomic resources are available for Chinese cavefish (Meng et al., 2013; Yang et al., 2016) but laboratory breeding procedures will need to be established for further progress. The blind cichlid *Lamprologus lethops* and its closely related eyed species *L. tigripictilis,* or the several blind species of *Mastacembelus* spiny eels are also promising because they combine the challenges of living in deep waters with highly turbulent conditions in the Congo river rapids, and genomic resources are available (Alter et al., 2015; Aardema et al., 2020). Groundwater model systems outside of the crustaceans and vertebrates should also be developed. The bivalve *Congeria* sp. (Bilandžija et al., 2013) and the tubeworm *Marifugia cavatica* (Kupriyanova et al., 2009), each with extant surface dwelling relatives, are promising candidates. Ultimately, for a complete understanding of subterranean trait evolution, terrestrial cave model species will also need to be developed. The Hawaiian planthopper *Oliarus*, which has closely related surface and lava cave-dwelling (*O. polyphemus*) species (Bilandžija et al., 2012), is a potential candidate.

## Troglomorphic traits

Troglomorphic traits were originally defined as morphological adaptations of animals living in the constant darkness of caves (Christiansen, 1962). More recent studies of cave-dwelling animals at the molecular, physiological, and behavioral levels justify the expansion of this definition. In this chapter, we define troglomorphic traits as changes in the morphology, physiology, or behavior of a subterranean species relative to a closely related surface counterpart (Fig. 12.2). Troglomorphic traits are subdivided into (1) reductive (or regressive) traits, which are directed downward in subterranean species (Fig. 12.2A), (2) constructive traits, which are directed upward (Fig. 12.2B), and (3) other traits that cannot be easily classified as reductive or constructive (Fig. 12.2C) (Jeffery, 2001).

### Reductive traits

The loss of vision is a reductive trait that has converged in many different groundwater species. The parts of the eyes, the brain regions receiving inputs from visual sensory organs, and the connections between photoreceptors and brain processing regions can be affected to various degrees in different groundwater species. For example, the lens of the eye is more affected than the retina in *Astyanax* cavefish (Jeffery, 2009), whereas the retina is more affected than the lens in the Somalian cavefish *Phreatichthys andruzzii* (Stemmer et al., 2015). The loss of eyes is a polygenic trait in *Astyanax* cavefish (Borowsky and Wilkens, 2002), and more than one gene is also responsible for missing eyes in *Asellus* cave populations (Protas et al., 2011). Parts of the visual system outside the eyes are also subject to reductive changes. The optic tecta of *Astyanax* cavefish and cave adapted *Gammarus* (Culver et al., 1995)

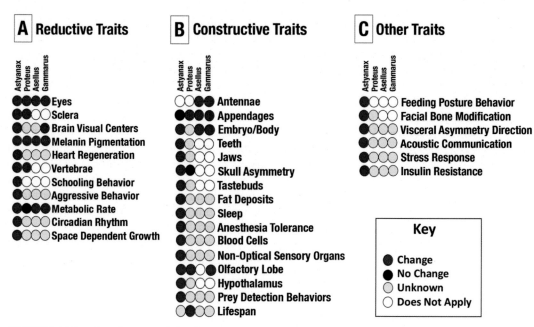

**FIGURE 12.2**    Classification and distribution of (A) reductive, (B) constructive, and (C) other troglomorphic traits in the groundwater models in this chapter and volume.

are reduced in size, and in both *Astyanax* and *Proteus,* eye degeneration is accompanied by loss or modification of the sclera, the fibrous, and skeletal coat surrounding vertebrate eyes (O'Quin et al., 2015).

Pigmentation is a reductive trait that is expressed at levels ranging from reduction of pigment cell numbers to complete loss of melanin (albinism) in groundwater models. Pigmentation occurs in the eyes, brain, and body in surface-dwelling animals. Vertebrates have several different types of pigment cells, including black melanophores, yellow xanthophores, and silvery iridophores. *Astyanax* cave populations have reduced or completely lost melanophores but xanthophores (and perhaps iridophores) do not seem to be modified (Jeffery et al., 2016). Nontroglomorphic *Asellus* has red and orange pigmented eyes and a brown pigmented body, and all three types of pigment are lost in the troglomorphic form (Protas et al., 2011). Surface dwelling *Gammarus* and nontroglomorphic *Proteus* have dark eye and body pigmentation, probably melanin, which is reduced in troglomorphic forms. In some groundwater species, such as *Astyanax*, the intensity of pigmentation is not changed when animals are exposed to light, consistent with loss of function of the *oca2* gene. In other groundwater species, such as troglomorphic *Proteus*, black pigmentation is increased by exposure to light, indicating that the responsible pathway is interrupted rather than suppressed. The *oca2* gene functions at the first step of the melanin biosynthesis pathway, the conversion of L-tyrosine to L-DOPA. This step may be a control point for the evolution of albinism in *Astyanax* cavefish, as suggested by the restoration of melanin pigmentation after provision of L-DOPA, the missing downstream factor in albino cavefish (McCauley et al.,

2004). Melanin biosynthesis is interrupted at the same step in the melanin biosynthesis pathway of independently evolved underground fish, such as the blind cichlid *L. lethops* (*oca2*; Aardema et al., 2020) or the cave-dwelling planthopper *O. polyphemus* (Bilandžija et al., 2012), implying convergence in the mechanisms underlying loss of pigmentation in different albino subterranean species. Similarly, another gene involved in regression of pigmentation in a different metabolic pathway, the melanocortin 1 receptor (*mc1R*), is also mutated in some *A. mexicanus* populations (Gross et al., 2009) and the cave-dwelling catfish *Astroblepus pholeter* (Espinasa et al., 2018), again showing repeated targeting of the same pigmentation gene during evolution underground.

Metabolism is decreased as a part of an energy saving program in food-deficient cave environments (Poulson, 1963), and lower metabolic rates have been measured in cave populations of *Astyanax*, *Asellus*, and *Gammarus* relative to their nontroglomorphic counterparts (Culver et al., 1995; Simčič & Sket, 2019; Hüppop, 2000). The basal metabolic rate of *Proteus* matches the low levels of metabolism typical of nontroglomorphic salamander species (Voituron et al., 2011), which could be interpreted as a preadaptation to subterranean life. Aggression, schooling behavior, and circadian rhythms are also reduced or lost in *Astyanax* cavefish (Beale et al., 2013; Elipot et al., 2013; Kowalko et al., 2013a; Moran et al., 2014). Aggressive behavior is shifted to constant food searching based on an embryonic sonic hedgehog (shh)-dependent increase of serotonergic neurons in the hypothalamus. The loss of light-dark rhythms in cavefish has a genetic basis: cyclic transcriptional activity in the clock gene per1 is absent in dark caves (Beale et al., 2013), although cavefish can be entrained to undergo rhythmic per1 expression by light cycles in the laboratory. The loss of heart regenerative capacity (Stockdale et al., 2018), the reduction in the number of spinal vertebrae, which converges between *Astyanax* and *Proteus* (Dowling et al., 2002, Kostanjšek et al., this volume; Gross et al., this volume), and the absence of space dependent (confined) growth (Gallo and Jeffery, 2012) are intriguing reductive traits, but their relationship to groundwater adaptation needs further investigation.

## Constructive traits

Constructive troglomorphic traits are enhancements of preexisting characters that are related to the survival and adaptation of subterranean species in challenging groundwater environments. Increased appendage size (including antennae in arthropods) is a constructive trait that is expressed convergently in many subterranean animals. Longer antennae and legs in *Asellus* and *Gammarus* (Fong and Carlini, this volume; Protas et al., this volume) are mediators of enhanced nonvisual sensory processes and probably also used to stabilize locomotion in the dark. Furthermore, *Proteus* troglomorphic forms have longer limbs than the nontroglomorphic form. Although appendages (fins) are not different between *Astyanax* surface fish and cavefish, other constructive changes have appeared, particularly in the head and brain (Fig. 12.2B; Gross et al., this volume). The size of the olfactory organs and lobes is increased, which probably improves sensitivity to odors and prey detection (Hinaux et al., 2016; Blin et al., 2018). An increase in the size of olfactory centers also occurs in cave populations of *Gammarus* compared to surface counterparts (Culver et al., 1995). Moreover, large olfactory epithelia have been reported in the remarkably elongated and flattened *Proteus* head (Tesařova et al., 2022). Thus, enhancement of the olfactory system may be widespread to partially compensate for the loss of vision in groundwater species. Cave populations of

*Asellus* and *Gammarus* have larger embryos and adult bodies relative to their surface counterparts (Fong and Carlini, this volume; Protas et al., this volume), and large eggs are a constructive trait in the cave forms of *Astyanax* (Hüppop and Wilkens, 2009). Larger eggs with more yolk may be adaptive for survival when food is in short supply.

Additional constructive traits have been discovered in *Astyanax* cavefish. The preoptic region and hypothalamus of the cavefish brain are larger compared to the surface form (Menuet et al., 2007). The cavefish hypothalamic center controlling sleep contains more hypocretin (a sleep-controlling neuropeptide) secreting neurons (Gross et al., this volume). Thus, cavefish are hyperactive and sleep less than surface fish (Duboué et al., 2011), which may be an adaptation to increase foraging behavior. Sleep and a related constructive trait, anesthesia tolerance, may be coupled to albinism, a reductive trait, via the pleiotropic *oca2* gene (Bilandžija et al., 2018; O'Gorman et al., 2021). The adult cavefish skull has evolved an unusual lateral bend (Gross and Powers, 2016), an asymmetry that may promote navigation in darkness. Jaw span, the number of oral taste buds and teeth (Varatharasan et al., 2009; Yamamoto et al., 2009; Atukorala et al., 2013), also possible adaptions to feeding style, and the size and number of cranial neuromasts, including the specialized neuromasts involved in prey detection by vibration (vibration attraction behavior; Yoshizawa et al., 2010), are increased in cavefish. Cavefish have more fat reserves (Xiong et al., 2018, 2022), which adapt them for survival during periods of low food input. They also develop more and larger erythrocytes with higher hemoglobin concentrations as an adaptation to oxygen depletion in their underground habitats (Boggs et al., 2022; van der Weele and Jeffery, 2022). The other groundwater models also probably experience periodic decreases in food and oxygen, but their survival "tactics" remain to be investigated.

Longevity is a key constructive trait of groundwater species because it can provide a model for aging research. The average lifespan of *Proteus* is increased by threefold (to a maximum of about 102 years!) compared to its most closely related salamander species (Voituron et al., 2011). Increased lifespan is also documented in Amblyopsid cavefish (Poulson, 1963) and fossorial mammals (Fang et al., 2014), suggesting significant convergence in this trait across subterranean species, but longevity still needs to be studied in the *Astyanax* system and the crustacean models.

## Other traits

The *Astyanax* system exhibits some troglomorphic traits that cannot be categorized as reductive or constructive (Fig. 12.1C). Feeding posture behavior, a change in feeding angle in darkness from 90° to the substrate, as occurs in the surface form, to 45° in the cave form, is probably conducive to cavefish bottom feeding using their wide shovel-like jaws. Feeding posture behavior may be advantageous for feeding in the dark (Hüppop, 1987). According to QTL analysis, the genes underpinning feeding posture behavior are different in at least two cavefish populations and therefore this trait may have evolved by convergence (Schemmel;, 1980; Kowalko et al., 2013b). Changes in acoustic communication and associated behaviors (Hyacinthe et al., 2019), stress response (Pierre et al., 2020), insulin resistance (Riddle et al., 2018), the direction of visceral asymmetry (Ma et al., 2021), and facial bone fragmentation (Powers et al., 2018) are mysterious traits that also fall into this category. They could be adaptive but further study is required to determine if they have special roles in subterranean life.

## Timeline of troglomorphic trait evolution

How long did it take for troglomorphic traits to evolve? It is impossible to directly determine the timelines of each individual trait but morphological, molecular, and geological "clocks" can be used to estimate the ages of different troglomorphic species, which are indicative of the maximal times for troglomorphic trait evolution. Troglomorphic *Proteus* and *Asellus* are estimated to have evolved between 2 and 10 million years ago (Trontelj et al., 2007), and cave adapted *Gammarus* from 100,000 to 500,000 years ago (Culver et al., 1995). Remarkedly, the timeline for establishment of *Astyanax* cavefish populations is estimated to be only 10,000–20,000 years ago, during the late Pleistocene (Mitchell et al., 1977; Fumey et al., 2018; Policarpo et al., 2021). Therefore, at least in *Astyanax*, troglomorphic traits can evolve in a relatively short time. This brings up the question of what evolutionary forces can act quickly enough to be responsible for the evolution of troglomorphic traits? For answers, we can turn to EvoDevo and genomics.

## Evolutionary developmental biology of groundwater organisms

Developmental evolution, or EvoDevo, is a discipline at the crossroads of evolutionary and developmental biology, which seeks to identify variations in the developmental processes responsible for the extraordinary morphological diversity of life, as well as its major evolutionary transitions and innovations. Making a new trait appear, losing a structure, changing its organization, size, shape, color, or functional modalities are some of the key topics in the realm of EvoDevo. To study developmental evolution, researchers use a variety of organisms, rather than the traditional genetic models (*Drosophila*, *Caenorhabditis*, zebrafish, and the mouse), which nevertheless fulfill a certain number of technical criteria as good laboratory models. As described in the previous section, groundwater creatures are particularly well suited for EvoDevo questions. Here we focus on the EvoDevo of eye loss, a classic troglomorphic trait in groundwater animals.

### Loosing eyes

The Chapter by Gross and coauthors in this book discusses current knowledge on the detailed developmental mechanisms by which *Astyanax* cavefish, the best-studied cave-dwelling organism, lost their eyes. In adult fish, eye cysts sunken inside the orbits are the remnants of initial development of the visual organs. By rewinding cavefish eye development from juveniles back into the early embryonic stages, three main processes governing eye loss emerge (Fig. 12.3). The last step is *degeneration*: the eyes that initially form regress during larval stages and this degeneration is triggered by the defective lens (Yamamoto and Jeffery, 2000; Alunni et al., 2007; Ma et al., 2014; Hinaux et al., 2015, 2017) and the defective optic vasculature (Ma et al., 2020). Before that, the embryonic eyes suffer *malformation*: the cavefish eyes are small, misshapen, and truncated because of abnormal morphogenetic movements during the formation of the optic vesicles, which are themselves affected by changes in gene expression in and around the presumptive optic field at the end of gastrulation (Pottin

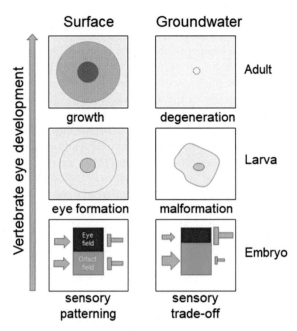

**FIGURE 12.3**  Developmental processes involved in eye loss in groundwater vertebrates. The sensory trade-off due to changes in signaling in early embryos (represented by *arrows* and *broken arrows*) has only been demonstrated in *Astyanax*. Olfact: olfactory.

et al., 2011; Devos et al., 2021; Agnès et al., 2022). Third and precociously, *trade-offs*, due to changes in pleiotropic signaling in early embryos (some of maternal origin), the presumptive optic territories are reduced in size but other territories are in turn expanded, as is the case for the olfactory placodes or the preoptic and hypothalamic regions in the brain (Yamamoto et al., 2004; Pottin et al., 2011; Hinaux et al., 2016; Ren et al., 2018; Torres-Paz et al., 2019). Thus, development and degeneration of the *Astyanax* cavefish eye provides a near complete sampling of the "EvoDevo toolbox." It illustrates the morphological outcomes of subtle heterotopies and heterochronies of gene expression and subtle variations in tissue movements and interactions. The maternal genetic effects involved in cavefish eye loss also provide one of the first examples of extremely precocious developmental evolution that starts in the mother's gonad (Ma et al., 2018; Torres-Paz et al., 2019).

## The many ways to lose eyes during development

In all vertebrates examined so far, including *Astyanax* as well as the blind salamander *Proteus anguinus*, the Somalian cavefish *Phreatichthys andruzzi*, the blind catfish *Rhamdia zongolicensis*, and the sulfur cave molly *Poecilia mexicana*, eyes begin to develop during embryogenesis but subsequently degenerate during later development (Durand, 1976; Wilkens, 2001; Riesch et al., 2011; Stemmer et al., 2015). Not only did this occur independently in all these different species, but also the mechanisms that drive secondary eye degeneration are distinct (reviewed in Rétaux and Casane, 2013). As making an eye is an energetically costly process for embryos, this repeated pattern suggests a strong constraint for the maintenance of eye development. In fact, cave-dwelling vertebrates provide one of the most compelling examples of a developmental constraint in morphological evolution. This is because during embryogenesis the formation of the optic vesicles is tightly linked to the formation of other forebrain structures, whose malformation would be embryonically lethal if eyes did not initially develop (Devos et al., 2021) (see also Gross et al., this volume).

The case of invertebrates is less clear. The two best-studied models are cave-dwelling crustaceans, the isopod *Asellus aquaticus,* and the amphipod *Gammarus minus* (Fong and Carlini, this volume; Protas et al., this volume). Both are direct developers and do not undergo metamorphosis, meaning that their eyes and antennae do not develop from a larval imaginal disc as they do in metamorphosing insects such as *Drosophila* (Aldaz and Escudero, 2010). In contrast to vertebrates, no ommatidia develop in the cave isopod, even in the youngest embryos (Mojaddidi et al., 2018). In the amphipod *Niphargus virei,* eye defects are similarly established in embryogenesis (Turquin, 1969), which tends to suggest that eye development in these animals may show less constraints than in vertebrates. However, early studies by Kosswig and Kosswig reported the presence of degenerate eye regions described as "eye nuclei" in adult *A. aquaticus,* which opens the possibility that photoreceptors could develop internally and then degenerate (Kosswig and Kosswig, 1940). In *Gammarus* as well, eyes are greatly reduced but can show variability from a few to no discernible ommatidia (Culver et al., 1995). Detailed, cell- and molecular-level analyses of compound eye development in these species are needed to reach a definitive conclusion on the developmental mechanisms of eye loss and associated developmental constraints.

## Exchanging eyes for other senses during development

In *Astyanax* cavefish embryos, the territories that are fated to become eye tissues (retina, lens) are reduced at the expense of the expansion of other sensory organs. The increase in size of the olfactory epithelia in the nose can be traced to larger olfactory placodes at the end of gastrulation and is based on an embryonic shh- and bmp (bone morphogenetic protein)-signaling dependent trade-off with the adjacent lens placode, which becomes smaller (Hinaux et al., 2016). Eye degeneration in cavefish is also linked to taste bud number, a constructive trait, through increased expression of the pleiotropic *shh* gene along the embryonic midline (Yamamoto et al., 2009). The supernumerary neuromasts of the lateral line, which drive the emergence of the adaptive vibration attraction behavior, develop postembryonically in the space of the eye orbit left after eye degeneration (Yoshizawa et al., 2010). Although not a strict sensory trade-off, shh and fgf (fibroblast growth factor) signaling changes in cavefish also affect the development of peptidergic neurons in the basal forebrain, which control energy homeostasis and sleep (Alié et al., 2018). Together these EvoDevo discoveries support the idea that eyes could be lost as a sort of collateral damage, by indirect selection for mutations that affect other, constructive traits that are advantageous for life underground (Jeffery, 2005; Rétaux and Casane, 2013). The pleiotropy of developmental genes is instrumental in this scenario.

Much less is known in this regard for other groundwater organisms. Comparison of the developmental transcriptome in *Asellus* surface and cave populations did not yield obvious candidate genes (Gross et al., 2020). In *Gammarus minus,* however, the *hedgehog* gene is downregulated in adult heads of several cave populations as compared to surface populations (Aspiras et al., 2012). *Shh* expression is regulated in the opposite direction (upregulated) in *Astyanax* cavefish embryos (Yamamoto et al., 2004), and the embryonic mechanisms driving arthropod and vertebrate eye development are very different, but the important point is that this signaling pathway is dysregulated in both cases. Therefore, developmental pleiotropy might be a shared feature underlying eye loss and the appearance of constructive trait(s) in groundwater animals. Developmental studies on holometabolous metamorphosing insects inhabiting ground waters would be enlightening, because bona fide trade-offs, such as the optic/olfactory trade-off, could be explored easily. In these species, the antenna and the eyes develop from a single imaginal disc (Aldaz and Escudero, 2010). Variations in the proportions of the disc territory devoted to the two sensory tissues, visual or olfactory, exist in *Drosophila* species (Keesey et al., 2019), and some of the possible underlying genetic mechanisms are known (Ramaekers et al., 2019).

## Evolutionary genomics of groundwater organisms

Once developmental mechanisms underlying troglomorphic evolution are identified, researchers can seek for their genomic origins. With the advent of next-generation sequencing and genome assembling methods, genomes and developmental transcriptomes of all groundwater model organisms should be available soon. Casane and co-authors (this volume) discuss evolutionary genomics and population genomics of groundwater animals in depth. Here, we cover the respective contributions of coding mutations and gene losses versus regulatory mutations, focusing on developmental evolution.

### Coding mutations

Comparative transcriptomics and genomics can relatively easily reveal coding mutations. In the *Astyanax* cavefish early embryonic transcriptome, the contribution of coding mutations seems minimal (Leclercq et al., 2022), which is relatively expected as the establishment of embryonic axes and patterning is highly constrained and deleterious mutations should be strongly counterselected. Later, in the cavefish larval transcriptome, some genes carry potential deleterious mutations (Hinaux et al., 2013). Most of these genes are known to be expressed in the visual system, suggesting relaxed selection for "vision genes" during the evolution in the absence of light. The list of 11 genes provided by Hinaux et al. (2013) that are potential candidates for having a role in cavefish visual system degeneration has yet to be functionally tested. In *Gammarus*, adult transcriptomes identified relaxed selection on a photolyase gene (Carlini and Fong, 2017). The *Asellus* developmental transcriptome was recently available but has not been examined in this respect (Gross et al., 2020).

Some general principles on the extent and tempo of the decay of genes involved in vision, circadian clock, and pigmentation after relaxation of functional constraints came from pangenomic analyses in three genera of cavefishes from Mexico, Cuba, and China (Policarpo et al., 2021). While two Cuban *Lucifuga* species present massive loss of eye-specific genes and nonvisual opsin genes (19 pseudogenes), the recently evolved *Astyanax mexicanus* has only one vision pseudogene (*pde6b*). The Chinese tetraploid species of *Sinocyclocheilus* on the other hand can reveal the combined effects of the level of eye regression, time, and genome ploidy on eye-specific gene pseudogenization. As confirmed by simulations, the very small number of loss-of-function mutations per pseudogene in all these cavefishes suggests that their eye degeneration may be quite recent, from early to late Pleistocene (Policarpo et al., 2021). The same holds true for the blind cichlid *L. lethops* (Aardema et al., 2020). This is in sharp contrast with more ancient fossorial mammals, in which many vision pseudogenes carry multiple loss-of-function mutations that have accumulated over time. A corollary of these analyses is that blind fishes probably cannot thrive more than a few million years in cave ecosystems.

There are about 12–15 loci involved in developmental evolution of eye regression in *Astyanax* cavefish (Protas et al., 2007). Yet, only one of them has been identified, and by a candidate gene approach rather than by evolutionary genomics. The *cbsa* gene, which encodes the key enzyme of the trans-sulfuration pathway, is mutated in cavefish. Its inactivation induces defects in optic vasculature, which results in aneurysms and eye hemorrhages and contributes to eye degeneration (Ma et al., 2020).

## Regulatory changes

Changes in the regulation of developmental gene expression can be inferred from quantitative transcriptomics. In *Asellus aquaticus* or *Astyanax mexicanus* embryos at various stages, differential gene expression between cave and surface morphs is extensive (Gross et al., 2013; Torres-Paz et al., 2019; Gross et al., 2020). Using F1 hybrid embryos resulting from crosses between cave and surface morphs, together with the assessment of allelic expression ratios in their transcriptomes, the *cis*- or *trans*-regulatory origins of these substantial evolutionary developmental variations in gene expression can be explored (Fig. 12.4). In both cave-dwelling *Asellus* and *Astyanax*, the major contribution comes from changes in *cis*-acting mutations that occurred in enhancers, promoters or other noncoding regulatory elements of the corresponding genes (Gross et al., 2020; Leclercq et al., 2022). Indeed, *trans*-acting regulatory changes affecting the expression or function of transcription factors or other molecules are expected to cause drastic pleiotropic effects and to be less frequent. The *long-wavelength sensitive opsin* gene of *Asellus* (Gross et al., 2020) and the eyefield transcription factor *rx3* of *Astyanax* (Leclercq et al., 2022) are flagship examples of eye development and specification genes that have evolved due to transcriptional *cis*-regulation in cave morphs, respectively.

Reciprocal F1 hybrids, obtained from crosses of a female cave morph and a male surface morph and vice versa, are also used to explore maternal genetic effects (Fig. 12.4). If a maternal effect is present, gene expression in reciprocal F1 embryos should be different and should be close to that of the maternal morph, respectively. In *Astyanax*, a very significant maternal signature exists in the transcriptome of eggs spawned by river-dwelling versus cave-dwelling females (Torres-Paz et al., 2019). As the egg transcriptome is exclusively of maternal origin (the zygotic genome will be activated much later), this pushes back the regulatory origins of differential development control in cave and surface embryos to oogenesis. The causal mechanisms are unknown and should be explored, as well as the possibility that these maternal effects could be a shared feature during groundwater evolution.

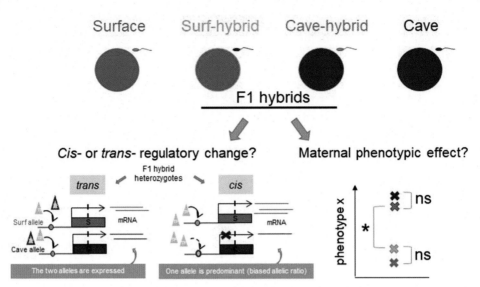

**FIGURE 12.4** Principles of F1 hybrid analysis to decipher the origins of developmental regulatory changes and developmental evolution (Surf: surface). *Left*: after a change in a *trans*-regulator such as a transcription factor (TF, triangle), the two parental alleles will be expressed in F1 hybrids because the proper wildtype transcription factor will act in *trans* to drive expression of both alleles. However, after a change in a *cis*-regulatory element such as an enhancer (circle), the mutated parental allele will not be expressed, leading to a biased allelic ratio in F1 heterozygotes. Of note, the two parental alleles need to carry sequence polymorphisms to be sorted from one another and to calculate allelic ratios. *Right*: Maternal effects can be inferred from reciprocal hybrid comparisons.

Finally, very little is known about the role of epigenetics in developmental evolution in groundwater. In *Astyanax* larval eyes, excess DNA methylation-based epigenetic silencing seems to promote eye degeneration (Gore et al., 2018). Interestingly, the most significantly hyper-methylated and down-regulated genes in the cave morph are also linked to human eye disorders. A future challenge is to identify the cause and origin of the dysregulation of the DNA methyl transferase that is responsible for the abnormal epigenetic marks and silencing observed in the cavefish eye genes.

In sum, functional genomics and transcriptomics are still at their beginnings in groundwater species development and evolution. Tools and resources have been developed and must be implemented as technologies progress. For example, the advent of single-cell multiomics for profiling of the genome, transcriptome, and epigenome has taken the field to the next level (Tritschler et al., 2019), opening the possibility to compare developmental trajectories of specific cell types of major interest in cave-dwelling species and their epigean counterparts and to decipher the fine variations of their developmental gene regulatory networks.

## Conclusions

The present chapter aimed to introduce Groundwater Ecology and Evolution at the organismal level. Groundwater organisms have always fascinated humans. Ironically, they may also provide some interesting insights into human health. Traits such as longevity, metabolic diseases, photoreceptor degeneration, cataract, albinism, or sensory hypersensitivity should be relevant to explore at molecular level.

We believe that the observation and study of life in subterranean waters should be multi-disciplinary and integrative in nature, combine laboratory and field studies, and benefit from experimental and model-centered approaches (Torres-Paz et al., 2018). The following chapters of this book will illustrate the efforts in these directions that have been pursued in four species, including the "flagship species" *Proteus anguinus*. To understand better the constraints and convergences of the adaptation to groundwater, the development of additional models, in all phyla, will also be necessary. To promote and stimulate research on cave organisms, scientific communities should continue to organize, collaborate, meet, and exchange in a multidisciplinary spirit. The dynamism of the field is exemplified by a number of special issues on cave organisms published recently in multiple journals (EvoDevo 2013; Developmental Biology 2018; The Journal of Experimental Zoology Part B Molecular and Developmental Evolution 2020; Diversity 2020; Zoological Research 2022).

Finally, the concept of ecological developmental biology, introduced by Gilbert et al. (2015) and now coined as Eco-Evo-Devo, could not be more appropriate for groundwater organisms. Ecologists, developmental, and evolutionary biologists will get together to understand how the environment shapes genomes, influences development, and changes phenotypes in the realm of dark waters.

## Acknowledgments

We thank C. Griebler and F. Malard for editing an earlier version of this chapter.

## References

Aardema, M.L., Stiassny, M.L.J., Alter, S.E., 2020. Genomic analysis of the only blind cichlid reveals extensive inactivation in eye and pigment formation genes. Genome Biology and Evolution 12, 1392–1406.

Agnès, F., Torres-Paz, J., Michel, P., Rétaux, S., 2022. A 3D molecular map of the cavefish neural plate illuminates eye-field organization and its borders in vertebrates. Development 149 (7), dev199966.

Alberti, L. (1550). Descrittione di tutta Italia. Bologna, Anselmo Giacarrelli.

Aldaz, S., Escudero, L.M., 2010. Imaginal discs. Current Biology 20, R429–R431.

Alié, A., Devos, L., Torres-Paz, J., Prunier, L., Boulet, F., Blin, M., Elipot, Y., Retaux, S., 2018. Developmental evolution of the forebrain in cavefish, from natural variations in neuropeptides to behavior. Elife 7, e32808

Alter, S.E., Brown, B., Stiassny, M.L., 2015. Molecular phylogenetics reveals convergent evolution in lower Congo River spiny eels. BMC Evolutionary Biology 15, 224.

Alunni, A., Menuet, A., Candal, E., Penigault, J.B., Jeffery, W.R., Rétaux, S., 2007. Developmental mechanisms for retinal degeneration in the blind cavefish *Astyanax mexicanus*. Journal of Comparative Neurology 505, 221–233.

Aspiras, A.C., Prasad, R., Fong, D.W., Carlini, D.B., Angelini, D.R., 2012. Parallel reduction in expression of the eye development gene hedgehog in separately derived cave populations of the amphipod *Gammarus minus*. Journal of Evolutionary Biology 25, 995–1001.

Atukorala, A.D.S., Hammer, C., Dufton, M., Franz-Odendaal, T.A., 2013. Adaptive evolution of the lower jaw dentition in Mexican tetra *(Astyanax mexicanus)*. EvoDevo 4, 28.

Bakovic, V., Cerezo, M.L.M., Höglund, A., Fogelhom, G., Hendricksen, R., Harbegby, A., Wright, D., 2021. The genomics of phenotypically differentiated *Asellus aquaticus* cave, surface stream and lake ecotypes. Molecular Ecology 30, 3530–3547.

Balázs, G., Biró, A., Fišer, Ž., Fišer, C., Herczeg, G., 2021. Parallel morphological evolution and habitat-dependent sexual dimorphism in cave- vs. surface populations of the *Asellus aquaticus* (Crustacea: isopoda: Asellidae) species complex. Ecology and Evolution 11(21), 15389—15403.

Beale, A.D., Guibal, C., Tamai, T.K., Klotz, L., Cowen, S., Peyric, E., Reynoso, V.H., Yamamoto, Y., Whitmore, D., 2013. Circadian rhythms in Mexican blind cavefish *Astyanax mexicanus*: in the lab and in the field. Nature Communications 4, 2769.

Bilandžija, H., Abraham, L., Ma, L., Renner, K.J., Jeffery, W.R., 2018. Behavioural changes controlled by catecholaminergic systems explain recurrent loss of pigmentation in cavefish. Proceedings of the Royal Society B: Biological Sciences 285, 20180243.

Bilandžija, H., Cetković, H., Jeffery, W.R., 2012. Evolution of albinism in cave planthoppers by a convergent defect in the first step of the melanin biosynthesis pathway. Evolution and Development 14, 196—203.

Bilandžija, H., Morton, B., Podnar, M., Ćetković, H., 2013. Evolutionary history of relict *Congeria* (Bivalvia: Dreissenidae): unearthing the subterranean biodiversity of the Dinaric Karst. Frontiers in Zoology 10, 5.

Blin, M., Tine, E., Meister, L., Elipot, Y., Bibliowicz, J., Espinasa, L., Rétaux, S., 2018. Developmental evolution and developmental plasticity of the olfactory epithelium and olfactory skills in Mexican cavefish. Developmental Biology 441, 242—251.

Boggs, T.E., Friedman, J.S., Gross, J.B., 2022. Alterations to cavefish red blood cells provide evidence of adaptation to reduced subterranean oxygen. Scientific Reports 12, 3735.

Borowsky, R., 2008. Restoring sight in blind cavefish. Current Biology 18, R23—R24.

Borowsky, R., Wilkens, H., 2002. Mapping a cave fish genome. Polygenic systems and regressive evolution. Journal of Heredity 93, 19—21.

Carlini, D.B., Fong, D.W., 2017. The transcriptomes of cave and surface populations of *Gammarus minus* (Crustacea: Amphipoda) provide evidence for positive selection on cave downregulated transcripts. PLoS One 12, e0186173.

Christiansen, K.M., 1962. Proposition pour la classification des animaux cavernicoles. Spelunca 2, 75—78.

Culver, D.C., Pipan, T., 2019. The Biology of Cave and Other Subterranean Habitats. Oxford University Press, Oxford, UK.

Culver, D.C., Kane, T.C., Fong, D.W., 1995. Adaptation and Natural Selection in Caves. The Evolution of *Gammarus Minus*. Harvard University Press, Cambridge, MA.

Darwin, C., 1859. On the Origin of Species by Means of Natural Selection, or the Preservation of Favoured Races in the Struggle for Life. John Murray, London, UK.

DeKay, J.E., 1842. Geology of New York, or the New York Fauna, Part IV, Fishes. Albany NY: W. & A. White & J. Visscher.

Devos, L., Agnès, F., Edouard, J., Simon, V., Legendre, L., Elkhallouki, N., Barbachou, S., Sohm, F., Rétaux, S., 2021. Eye morphogenesis in the blind Mexican cavefish. Biology Open 15 (10), bio059031.

Dobzhansky, T., 1941. Genetics and the Origin of Species. Columbia. University Press, New York.

Dowling, T.E., Martasian, D.P., Jeffery, W.R., 2002. Evidence for multiple genetic lineages with similar eyeless phenotypes in the blind cavefish, *Astyanax mexicanus*. Molecular Biology and Evolution 19, 446—455.

Duboué, E.R., Keene, A.C., Borowsky, R.L., 2011. Evolutionary convergence on sleep loss in cavefish populations. Current Biology 21, 671—676.

Durand, J.P., 1976. Ocular development and involution in the European cave salamander, *Proteus anguinus* laurenti. The Biological Bulletin 151, 450—466.

Elipot, Y., Hinaux, H., Callebert, J., Rétaux, S., 2013. Evolutionary shift from flighting to foraging in blind cavefish through changes in the serotonin network. Current Biology 23, 1—10.

Elipot, Y., Legendre, L., Père, S., Sohm, F., Rétaux, S., 2014. *Astyanax* transgenesis and husbandry: how cavefish enters the laboratory. Zebrafish 11 (4), 291—299.

Espinasa, L., Robinson, J., Espinasa, M., 2018. Mc1r gene in *Astroblepus pholeter* and *Astyanax mexicanus*: convergent regressive evolution of pigmentation across cavefish species. Developmental Biology 441, 305—310.

Fang, X., Seim, I., Huang, Z., Gerashchenko, M.V., Xiong, Z., Turanov, A.A., Zhu, Y., Lobanov, A.V., Fan, D., Yim, S.H., Yao, X., Ma, S., Yang, L., Lee, S.G., Kim, E.B., Bronson, R.T., Šumbera, R., Buffenstein, R., Zhou, X., Krogh, A., Park, T.J., Zhang, G., Wang, J., Gladyshev, V.N., 2014. Adaptations to a subterranean environment and longevity revealed by the analysis of mole rat genomes. Cell Reports 8 (5), 1354—1364.

Fumey, J., Hinaux, H., Noirot, C., Thermes, C., Rétaux, S., Casane, D., 2018. Evidence for late Pleistocene origin of *Astyanax mexicanus* cavefish. Biomedical Central Evolutionary Biology 18, 43.

Gallo, N.D., Jeffery, W.R., 2012. Evolution of space dependent growth in the teleost *Astyanax mexicanus*. PLoS One 7 (8), e41443.

Gilbert, S.F., Bosch, T.C., Ledón-Rettig, C., 2015. Eco-Evo-Devo: developmental symbiosis and developmental plasticity as evolutionary agents. Nature Review Genetics 16 (10), 611–622.

Gore, A.V., Tomins, K.A., Iben, J., Ma, L., Castranova, D., Davis, A.E., Parkhurst, A., Jeffery, W.R., Weinstein, B.M., 2018. An epigenetic mechanism for cavefish eye degeneration. Nature Ecology & Evolution 2, 1155–1160.

Gross, J.B., Powers, A.K., 2016. The evolution of the cavefish craniofacial complex. In: Keene, A., Yoshizawa, M., McGaugh, S. (Eds.), Biology and Evolution of the Mexican Cavefish. Elsevier, New York, pp. 193–207.

Gross, J.B., Borowsky, R., Tabin, C.J., 2009. A novel role for *Mc1r* in the parallel evolution of depigmentation in independent populations of the cavefish *Astyanax mexicanus*. PLoS Genetics 5, e1000326.

Gross, J.B., Furterer, A., Carlson, B.M., Stahl, B.A., 2013. An integrated transcriptome-wide analysis of cave and surface dwelling *Astyanax mexicanus*. PLoS One 8, e55659.

Gross, J.B., Sun, D.A., Carlson, B.M., Brodo-Avo, S., Protas, M.E., 2020. Developmental transcriptome analysis of the cave dwelling Crustacean, *Asellus aquaticus*. Genes 11 (1), 42.

Hinaux, H., Blin, M., Fumey, J., Legendre, L., Heuze, A., Casane, D., Rétaux, S., 2015. Lens defects in *Astyanax mexicanus* cavefish: evolution of crystallins and a role for alphaA-crystallin. Developmental Neurobiology 75 (5), 505–521.

Hinaux, H., Devos, L., Bibliowicz, J., Elipot, Y., Alié, A., Blin, M., Rétaux, S., 2016. Sensory evolution in blind cavefish is driven by early events during gastrulation and neurulation. Development 143, 4521–4532.

Hinaux, H., Poulain, J., Da Silva, C., Noirot, C., Jeffery, W.R., Casane, D., Rétaux, S., 2013. De Novo sequencing of *Astyanax mexicanus* surface fish and Pachón cavefish transcriptomes reveals enrichment of mutations in cavefish putative eye genes. PLoS One 8, e53553.

Hinaux, H., Recher, G., Alié, A., Legendre, L., Blin, M., Rétaux, S., 2017. Lens apoptosis in the *Astyanax* blind cavefish is not triggered by its small size or defects in morphogenesis. PLoS One 12, e0172302.

Hüppop, K., 1987. Food-finding ability in cave fish (*Astyanax fasciatus*). International Journal of Speleology 16, 59–66.

Hüppop, K., 2000. How do cave animals cope with the food scarcity in caves? In: Wilkens, H., Culver, D.C., Humphreys, W.F. (Eds.), Subterranean Ecosystems. Elsevier, Amsterdam, pp. 159–188.

Hüppop, K., Wilkens, H., 2009. Bigger eggs in subterranean *Astyanax fasciatus* (Characidae, Pisces). Journal of Zoological Systematics and Evolutionary Research 29, 280–288.

Hyacinthe, C., Attia, J., Rétaux, S., 2019. Evolution of acoustic communication in blind cavefish. Nature Communications 10, 4231.

Jeffery, W.R., 2001. Cavefish as a model system in evolutionary developmental biology. Developmental Biology 231, 1–12.

Jeffery, W.R., 2005. Adaptive evolution of eye degeneration in the Mexican blind cavefish. Journal of Heredity 96, 185–196.

Jeffery, W.R., 2009. Regressive evolution in *Astyanax* cavefish. Annual Review of Genetics 43, 25–47.

Jeffery, W.R., Ma, L., Parkhurst, A., Bilandžija, H., 2016. Pigment regression and albinism in *Astyanax* cavefish. In: Keene, A., Yoshizawa, M., McGaugh, S. (Eds.), Biology and Evolution of the Mexican Cavefish. Elsevier, New York, pp. 155–173.

Jordan, C.B., 1937. Bringing in the new cave fish *Anoptichthys jordani* Hubbs and Innes. Aquarium, Philadelphia 5 (10), 203–204.

Keesey, I.W., Grabe, V., Gruber, L., Koerte, S., Obiero, G.F., Bolton, G., Khallaf, M.A., Kunert, G., Lavista-Llanos, S., Valenzano, D.R., Rybak, J., Barrett, B.A., Knaden, M., Hansson, B.S., 2019. Inverse resource allocation between vision and olfaction across the genus *Drosophila*. Nature Communications 10, 1162.

Klaassen, H., Wang, Y., Adamski, K., Rohner, N., Kowalko, J.E., 2018. CRISPR mutagenesis confirms the role of *oca2* in melanin pigmentation in *Astyanax mexicanus*. Developmental Biology 441, 313–318.

Kosswig, C., Kosswig, L., 1940. Die Variabilität bei *Asellus aquaticus*, unter besonderer Berücksichtigung der Variabilität in isolierten unter- und oberirdischen Populationen. Revue de Faculté des Sciences (Istanbul) B5 1–55.

Kostanjšek, R., Diderichsen, B., Recknagel, H., Gunde-Cimerman, N., Gostinčar, C., Fan, G., Guangyi, F., Kordiš, D., Trontelj, P., Jiang, H., Bolund, L., Luo, Y., 2022. Toward the massive genome of *Proteus anguinus* —illuminating longevity, regeneration, convergent evolution, and metabolic disorders. Annals of the New York Academy of Sciences 1507 (1), 5–11.

Kowalko, J.E., Rohner, N., Rompani, S.B., Peterson, B.K., Linden, T., Yoshizawa, M., Kay, E.H., Hoekstra, H.E., Jeffery, W.R., Borowsky, R., Tabin, C.J., 2013a. Genetic analysis of the loss of schooling behavior in cavefish reveals both sight-dependent and independent mechanisms. Current Biology 23, 1874–1883.

IV. Principles of evolution in groundwater

Kowalko, J.E., Rohner, N., Linden, T.A., Rompani, S.B., Warren, W.C., Borowsky, R., Tabin, C.J., Jeffery, W.R., Yoshizawa, M., 2013b. Convergence in feeding posture occurs through different genetic loci in independently evolved cave populations of *Astyanax mexicanus*. Proceedings of the National Academy of Sciences of the United States of America 110 (42), 16933−16938.

Kupriyanova, E.K., ten Hove, H.A., Sket, B., Zakšek, V., Trontelj, P., Rouse, G.W., 2009. Evolution of the unique freshwater cave-dwelling tubeworm *Marifugia cavatica*. Systematics and Biodiversity 7, 389−401.

Laurenti, J.N., 1768. Specimen medicum, exhibens synopsin repitilian emendatam cum experimentis circa venena et antidota reptilian Austriacorum. Joan Thomae, Wien.

Leclercq, J., Torres-Paz, J., Policarpo, M., Agnès, F., Rétaux, S., 2022. Evolution of the regulation of developmental gene expression in blind Mexican cavefish bioRxiv/10.1101/2022.07.12.499770.

Ma, L., Gore, A.V., Castranova, D., Shi, J., Ng, M., Tomins, K.A., van der Weele, C.M., Weinstein, B.M., Jeffery, W.R., 2020. A hypomorphic cystathionine β-synthase gene contributes to cavefish eye loss by disrupting optic vasculature. Nature Communications 11, 2772.

Ma, L., Jeffery, W.R., Essner, J.J., Kowalko, J.E., 2015. Genome editing using TALENs in blind Mexican cavefish. PLoS One 10 (3), e0119370.

Ma, L., Ng, M., Shi, J., Gore, A.V., Castranova, D., Weinstein, B.M., Jeffery, W.R., 2021. Maternal control of visceral asymmetry evolution in *Astyanax* cavefish. Scientific Reports 11, 10312.

Ma, L., Parkhurst, A., Jeffery, W.R., 2014. The role of a lens survival pathway including sox2 and alphaA-crystallin in the evolution of cavefish eye degeneration. EvoDevo 5, 28.

Ma, L., Strickler, A.G., Parkhurst, A., Yoshizawa, M., Shi, J., Jeffery, W.R., 2018. Maternal genetic effects in *Astyanax* cavefish development. Developmental Biology 441, 209−220.

Ma, L., Zhao, Y.H., Yang, J.X., 2019. Cavefish of China. In: White, W.B., Culver, D.C., Pipan, T. (Eds.), Encyclopedia of Caves, third ed. Elsevier, New York, pp. 237−254.

McCauley, D.W., Hixon, E., Jeffery, W.R., 2004. Evolution of pigment cell regression in the cavefish *Astyanax*: a late step in melanogenesis. Evolution and Development 6, 209−218.

McGaugh, S.E., Gross, J.B., Aken, B., Blin, M., Borowsky, R., Chalopin, D., Hinaux, H., Jeffery, W.R., Keene, A., Ma, L., Minx, P., Murphy, D., O'Quin, K.E., Rétaux, S., Rohner, N., Searle, S.M.J., Stahl, B., Tabin, C., Volff, J.N., Yoshizawa, M., Warren, W., 2014. The cavefish genome reveals candidate genes for eye loss. Nature Communications 5, 5307.

Meng, F., Braasch, I., Phillips, J.B., Lin, X., Titus, T., Zhang, C., Postlethwait, J.H., 2013. Evolution of the eye transcriptome under constant darkness in *Sinocyclocheilus* cavefish. Molecular Biology and Evolution 30 (7), 1527−1543.

Menuet, A., Alunni, A., Joly, J.S., Jeffery, W.R., Rétaux, S., 2007. Expanded expression of sonic hedgehog in *Astyanax* cavefish: multiple consequences on forebrain development and evolution. Development 134, 845−855.

Mitchell, R.B., Russell, W.H., Elliott, W.R., 1977. Mexican Eyeless Characin Fishes: Genus *Astyanax*. Environment, Distribution, and Evolution, vol 12. Special Publications of the Museum of Texas Tech University, pp. 1−89.

Mojaddidi, H., Fernandez, F.E., Erickson, P.A., Protas, M.E., 2018. Embryonic origin and genetic basis of cave associated phenotypes in the isopod crustacean *Asellus aquaticus*. Scientific Reports 8, 16589.

Moran, D., Softley, R., Warrant, E.J., 2014. Eyeless Mexican cavefish save energy by eliminating the circadian rhythm in metabolism. PLoS One 9, e107877.

O'Gorman, M., Thakur, S., Imrie, G., Moran, R.L., Choy, S., SifuentesRomero, I., Bilandžija, H., Renner, K.J., Duboué, E., Rohner, McGaugh, S.E., Keene, A.C., Kowalko, J.E., 2021. Pleiotropic function of the *oca2* gene underlies the evolution of sleep loss and albinism in cavefish. Current Biology 31, 3694−3701.

O'Quin, K.E., Yoshizawa, M., Doshi, P., Jeffery, W.R., 2013. Quantitative genetic analysis of retinal degeneration in the blind cavefish. PLoS One 8 (2), e57281.

O'Quin, K.E., Doshi, P., Lyon, A., Hoenemeyer, E., Yoshizawa, M., Jeffery, W.R., 2015. Complex evolutionary and genetic patterns characterize the loss of scleral ossification in the blind cavefish *Astyanax mexicanus*. PLoS One 10 (12), e0142208.

Pierre, C., Pradère, N., Froc, C., Ornelas-Garcia, P., Callebert, J., Rétaux, S., 2020. A mutation in *monoamine oxidase* (MAO) affects the evolution of stress behavior in the blind cavefish *Astyanax mexicanus*. Journal of Experimental Biology 223 (Pt 18), jeb226092.

Policarpo, M., Fumey, J., Lafargeas, P., Naquin, D., Thermes, C., Naville, M., Dechaud, C., Volff, J.N., Cabau, C., Klopp, C., Møller, P.R., Bernatchez, L., García-Machado, E., Rétaux, S., Casane, D., 2021. Contrasting gene decay in subterranean vertebrates: insights from cavefishes and fossorial mammals. Molecular Biology and Evolution 38, 589−605.

Pottin, K., Hinaux, H., Rétaux, S., 2011. Restoring eye size in *Astyanax mexicanus* blind cavefish embryos through modulation of the Shh and Fgf8 forebrain organising centres. Development 138, 2467–2476.

Poulson, T.L., 1963. Cave adaptation in amblyopsid fishes. The American Midland Naturalist 70, 257–290.

Powers, A.K., Kaplan, S.A., Boggs, T.E., Gross, J.B., 2018. Facial bone fragmentation in blind cavefish arises through two unusual ossification processes. Scientific Reports 8, 7015.

Protas, M.E., Hersey, C., Kochanek, D., Zhou, Y., Wilkens, H., Jeffery, W.R., Zon, L.I., Borowsky, R., Tabin, C.J., 2006. Genetic analysis of cavefish reveals molecular convergence in the evolution of albinism. Nature Genetics 38, 107–111.

Protas, M.E., Trontelj, P., Patel, N.H., 2011. Genetic basis of eye and pigment loss in the cave crustacean, *Asellus aquaticus*. Proceedings of the National Academy of Sciences of the United States of America 108, 5702–5707.

Protas, M., Conrad, M., Gross, J.B., Tabin, C., Borowsky, R., 2007. Regressive evolution in the Mexican cave tetra, *Astyanax mexicanus*. Current Biology 17, 452–454.

Protas, M., Tabansky, I., Conrad, M., Gross, J.B., Vidal, O., Tabin, C.J., Borowsky, R., 2008. Multi-trait evolution in a cave fish, *Astyanax mexicanus*. Evolution and Development 10, 196–209.

Racovitza, E.G., 1925. Notes sur les isopodes. 13. Morphologie et phylogénie des antennes II. Archives de Zoologie Experimentale et Generale 63, 533–622.

Ramaekers, A., Claeys, A., Kapun, M., Mouchel-Vielh, E., Potier, D., Weinberger, S., Grillenzoni, N., Dardalhon-Cuménal, D., Yan, J., Wolf, R., Flatt, T., Buchner, E., Hassan, B.A., 2019. Altering the temporal regulation of one transcription factor drives evolutionary trade-offs between head sensory organs. Developmental Cell 23, 780–792.

Re, C., Fišer, Ž., Perez, J., Tacdol, A., Trontelj, P., Protas, M.E., 2018. Common genetic basis of eye and pigment loss in two distinct cave populations of the isopod crustacean *Asellus aquaticus*. Integrative and Comparative Biology 58 (3), 421–430.

Ren, X., Hamilton, N., Muller, F., Yamamoto, Y., 2018. Cellular rearrangement of the prechordal plate contributes to eye degeneration in the cavefish. Developmental Biology 441, 221–234.

Rétaux, S., Casane, D., 2013. Evolution of eye development in the darkness of caves: adaptation, drift, or both? Evo-Devo 4, 26.

Riddle, M.R., Aspiras, A.C., Gaudenz, K., Peuß, R., Sung, J.Y., Martineau, B., Peavey, M., Box, A.C., Tabin, J.A., McGaugh, S., Borowsky, R., Tabin, C.J., Rohner, N., 2018. Insulin resistance in cavefish as an adaptation to a nutrient-limited environment. Nature 555, 647–651.

Riesch, R., Schlupp, I., Langerhans, R.B., Plath, M., 2011. Shared and unique patterns of embryo development in extremophile poeciliids. PLoS One 6, e27377.

Say, T., 1818. An account of the crustacea of the United States. Journal of Academy of Natural Science Philadelphia 1, 374–401.

Schemmel, C., 1980. Studies on the genetics of feeding behaviour in the cave fish *Astyanax mexicanus* f. *Anoptichthys*. An example of apparent monofactorial inheritance by polygenes. Zeitschrift für Tierpsychologie 53, 9–22.

Simčič, T., Sket, B., 2019. Comparison of epigean and some troglobiotic animals regarding their metabolism intensity. Examination of a classical assertation. International Journal of Speleology 48 (2), 133–144.

Stahl, B.A., Peuß, R., McDole, B., Kenzior, A., Jaggard, J.B., Gaudenz, K., Krishnan, J., McGaugh, S.E., Duboue, E.R., Keene, A.C., Rohner, N., 2019. Stable transgenesis in *Astyanax mexicanus* using the Tol2 transposase system. Developmental Dynamics 248, 679–687.

Stemmer, M., Schuhmacher, L.N., Foulkes, N.S., Bertolucci, C., Wittbrodt, J., 2015. Cavefish eye loss in response to an early block in retinal differentiation progression. Development 142, 743–752.

Stockdale, W.T., Lemieux, M.E., Killen, A.C., Zhao, J., Hu, Z., Riepsaame, J., Hamilton, N., Kudoh, T., Riley, P.R., van Aerle, R., Yamamoto, Y., Mommersteeg, M.T.M., 2018. Heart regeneration in the Mexican cavefish. Cell Reports 25, 1997–2007.

Tesařová, M., Mancini, L., Mauri, E., Aljančič, G., Năpărus-Aljančič, M., Kostanjšek, R., Bizjak Mali, L., Zikmund, T., Kaucká, M., Papi, F., Goyens, J., Bouchnita, A., Hellander, A., Adameyko, I., Kaiser, J., 2022. Living in darkness: exploring adaptation of *Proteus anguinus* in 3 dimensions by X-ray imaging. GigaScience 11 giac030.

Torres-Paz, J., Hyacinthe, C., Pierre, C., Rétaux, S., 2018. Towards an integrated approach to understand Mexican cavefish evolution. Biology Letters 14, 20180101.

Torres-Paz, J., Leclercq, J., Rétaux, S., 2019. Maternally regulated gastrulation as a source of variation contributing to cavefish forebrain evolution. Elife 8, e50160.

Tritschler, S., Buttner, M., Fischer, D.S., Lange, M., Bergen, V., Lickert, H., Theis, F.J., 2019. Concepts and limitations for learning developmental trajectories from single cell genomics. Development 146 (12), dev170506.

Trontelj, P., Gorički, Š., Polak, S., Verovnik, R., Zakšek, V., Sket, B., 2007. Age estimates for some subterranean taxa and lineages in the Dinaric Karst. Acta Carsologica 36, 183—189.

Turquin, M.J., 1969. Le développement du système nerveux de *Niphargus virei* (Crustacé; Amphipode Hypogé). Bulletin de la Société Zoologique de France 94 (4), 649—656.

Valvasor, J.W., 1689. Die Ehre des Herzogthums Krain. Endter, Ljubljana, Slovenia.

van der Weele, C.M., Jeffery, W.R., 2022. Cavefish cope with environmental hypoxia by developing more erythrocytes and overexpression of hypoxia-inducible genes. Elife 11, e69109.

Varatharasan, N., Croll, R.P., Franz-Odendaal, T., 2009. Taste bud development and patterning in sighted and blind morphs of *Astyanax mexicanus*. Developmental Dynamics 238, 3045—3065.

Voituron, Y., de Fraipont, M., Issartel, J., Guillaume, O., 2011. Extreme lifespan of the human fish (*Proteus anguinus*): a challenge for ageing mechanisms. Biology Letters 7, 105—107.

Warren, W.C., Boggs, T.E., Borowsky, R., Carlson, B.M., Ferrufino, E., Gross, J.B., Hillier, L., Hu, Z., Keene, A.C., Kenzior, A., Kowalko, J.E., Tomlinson, C., Kremitzki, M., Lemieux, M.E., Graves-Lindsay, T., McGaugh, S.E., Miller, J.T., Mommersteeg, M.T.M., Moran, R.L., Peuß, R., Rice, E.S., Riddle, M.R., Sifuentes-Romero, I., Stanhope, B.A., Tabin, C.J., Thakur, S., Yamamoto, Y., Rohner, N., 2021. A chromosome-level genome of *Astyanax mexicanus* surface fish for comparing population-specific genetic differences contributing to trait evolution. Nature Communications 12 (1), 1447.

White, M.J.D., 1978. Modes of Speciation. W.H. Freeman and Co., San Francisco.

Wilkens, H., 2001. Convergent adaptations to cave life in the *Rhamdia laticauda* catfish group (Pimelodidae, Teleostei). Environmental Biology of Fishes 62, 251—261.

Xiong, S., Krishnan, J., Peuß, R., Rohner, N., 2018. Early adipogenesis contributes to excess fat accumulation in cave populations of *Astyanax mexicanus*. Developmental Biology 441, 297—304.

Xiong, S., Wang, W., Kenzior, A., Olsen, L., Krishnan, J., Persons, J., Medley, K., Peuß, R., Wang, Y., Chen, S., Zhang, N., Thomas, N., Miles, J.M., Alvarado, A.S., Rohner, N., 2022. Enhanced lipogenesis through Ppargamma helps cavefish adapt to food scarcity. Current Biology 32 (10), 2272—2280 e6.

Yamamoto, Y., Jeffery, W.R., 2000. Central role for the lens in cave fish eye degeneration. Science 289, 631—633.

Yamamoto, Y., Byerly, M.S., Jackman, W.R., Jeffery, W.R., 2009. Pleiotropic functions of embryonic sonic hedgehog expression link jaw and taste bud amplification with eye loss during cavefish evolution. Developmental Biology 330, 200—211.

Yamamoto, Y., Stock, D.W., Jeffery, W.R., 2004. Hedgehog signalling controls eye degeneration in blind cavefish. Nature 431, 844—847.

Yang, J., Chen, X., Bai, J., Fang, D., Qiu, Y., Jiang, W., Yuan, H., Bian, C., Lu, J., He, S., Pan, X., Zhang, Y., Wang, X., You, X., Wang, Y., Sun, Y., Mao, D., Liu, Y., Fan, G., Zhang, H., Chen, X., Zhang, X., Zheng, L., Wang, J., Cheng, L., Chen, J., Ruan, Z., Li, J., Yu, H., Peng, C., Ma, X., Xu, J., He, Y., Xu, Z., Xu, P., Wang, J., Yang, H., Wang, J., Whitten, T., Xu, X., Shi, Q., 2016. The *Sinocyclocheilus* cavefish genome provides insights into cave adaptation. Biomedical Central Biology 14, 1.

Yoshizawa, M., Gorički, Š., Soares, D., Jeffery, W.R., 2010. Evolution of a behavioral shift mediated by superficial neuromasts helps cavefish find food in darkness. Current Biology 20, 1631—1636.

# 13

# The olm (*Proteus anguinus*), a flagship groundwater species

*Rok Kostanjšek, Valerija Zakšek, Lilijana Bizjak-Mali and Peter Trontelj*

University of Ljubljana, Biotechnical Faculty, Department of Biology, Ljubljana, Slovenia

## Introduction

Nature conservation relies strongly upon attractive and lovable organisms to raise sympathies and funds among the wide human society. These are termed flagship species and their nature is described as charismatic, both terms being used metaphorically (Barua, 2011; Macdonald et al., 2017). This approach, while subjective and not scientifically grounded, can still be used to maximize resources and conciliate them with objectively determined conservation goals (McGowan et al., 2020).

Caves and subterranean habitats are notorious for their lack of charismatic flagship species (Whitten, 2009; Hutchins, 2018). Caves are considered attractive and worthy of protection for the sake of themselves, not because of the biodiversity they harbor. Most groundwater habitats and their inhabitants are even less charismatic than dry caves. The main conservation impetus for groundwater comes from our concern for safe drinking water, which is reflected also in several national and international legislative acts (Niemiller et al., 2018). This is the reverse of what has been found for surface wetlands where flagship species are preferred over ecosystem services (Senzaki et al., 2017). According to anecdotal reports, the awareness of fauna living in drinking water can be disturbing for people who expect their tap water to be "germ free". This brings up the question whether promotion of groundwater biodiversity can and should be tackled via charismatic flagship species in a way that is customary in conservation of other habitats.

The present chapter will try to exemplify that we do have potential flagship species even in groundwater and that their potential has yet to be realized. In Europe, the central role of the groundwater flagship species has been taken by *Proteus anguinus* also known as the olm, or proteus. We show how this has happened gradually and unplanned, and went hand in hand with the development of subterranean biology. Therefore, we can regard the olm as an ambassador not only for the conservation of cave life but also for subterranean biology and the scientific study of groundwater. Further, we review the most important and newest research on olms with the aim to encourage the use of this and other subterranean salamanders as much needed groundwater flagships.

## The historical rise to fame

Unlike the two groundwater model organisms, *Astyanax mexicanus* and *Asellus aquaticus* (see Gross et al., this volume; Protas et al., this volume), the olm is anything but a good biological model. It is rare and hard to collect in the wild, it is exceedingly difficult to breed in captivity, it has long generation times, requires about as much as humans to mature sexually, and reproduces once every six or so years. With a genome size of over 40 giga base pairs (the human genome measures 3.1 giga base pairs) that is laden with repetitive sequences, it also remains a challenge from the genomic perspective (Kostanjšek et al., 2021b).

Despite these obvious obstacles, the olm is arguably the best-known cave animal worldwide, and, especially in Europe, popular with tourists, cavers and scientists. Every year, close to a million visitors from around the world get to see live olms in the subterranean aquaria of Postojna Cave in Slovenia. Here it is one of the country's national symbols and children learn about it in school. Along with the sympathies for the olm, people receive important take-home messages about groundwater biology and its protection. There are several reasons for this unprecedented popularity among subterranean fauna. It is one of very few tetrapods among obligate subterranean species, it is of considerable size, and is often in motion displaying various behaviors. This unique combination of characteristics explains the admiration by the general public, but one has to dig into history to fully appreciate its scientific value.

The olm is not only the first scientifically documented obligate cave animal, but probably also the first that humans came into contact with when individuals sporadically got washed out of its habitat. Its earliest representation may be a Venetian stone carving from the 10th or 11th century (Shaw, 1999). Johann Weichard von Valvasor wrote about it in 1689 in his work The Glory of the Duchy of Carniola (Slava Vojvodine Kranjske), almost 100 years before the first attempt was made to describe it scientifically. Joannes A. Scopoli examined it in detail, provided a description and drawings, and sent his report to the eminence of biological systematics Carl Linné in 1762. Linné, unaware of the pedomorphic nature (retention of juvenile characters in adults) of olms, believed that Scopoli described a newt larva and asked him to find adults before including it in Systema Naturae. Although Scopoli greatly respected Linné, he disagreed with his judgment and described the animal precisely in his work Zoological Observations (Observationes zoologicae) in 1772. Unfortunately too late, because a few years earlier, in 1768, Nicolas J. Laurenti published a valid description along with the name *Proteus anguinus* Laurenti, 1768 (Aljančič, 2019).

The olm gained the interest of the scientific community later in the 19th century when it became the best-known subterranean organism. It played a formative role in the emerging theories of evolution. Being blind and depigmented, it exemplified Lamarck's theory of use and disuse; the latter explanation came handy also to Darwin. In the early 20th century, the Viennese zoologist Paul Kammerer claimed to have found experimental support for the Lamarckian inheritance of acquired characteristics. In his notorious experiments on the midwife toad he allegedly induced a switch from terrestrial to aquatic reproduction by altering environmental factors. An even greater challenge he saw in the reproduction of olms. Although there were already some observations of egg laying in captivity by the end of the 19th century, Kammerer claimed that his experimental animals were viviparous, reproduced annually, and newborns were 10−12 cm long (Kammerer, 1907, 1912). He had also reported that the olms, when kept in daylight, developed pigment and fully functional eyes. Whether Kammerer, including his experiments on olms, should be celebrated as the founding father of epigenetics or scolded for one of the greatest scientific frauds of all times has remained in dispute to the present time (Pennisi, 2009). Notwithstanding, scrutiny of Kammerer's work seems to point toward some fabrication (van Alphen and Arntzen, 2016; Blackburn, 2019).

Research intensified again in 1948, when the French National Center for Scientific Research (CNRS) set up a modern cave laboratory at Moulis in the French Pyrénées, where olms were bred and systematically studied. Establishing a breeding program in seminatural conditions required great effort: after 6 years, scientists led by Alfred Vandel succeeded and for the first time recorded olms laying eggs and hatching in captivity. The mystery of reproduction was solved. The first systematic ecological studies in the natural habitat were conducted by Wolfgang Briegleb (1962) in the Postojna-Planina Cave System (PPCS) and supplemented by experiments in captivity.

Since the late 1960s, University of Ljubljana, Slovenia, has emerged as a new research center on *Proteus anguinus*, addressing various topics, from development and functional morphology, histology, sensory system, environmental research on bioaccumulation in tissues, all the way to evolution, phylogeography, morphology and ecology. A captive breeding program has been established in the private cave laboratory Tular in Kranj, Slovenia (Aljančič, 2008).

The latest milestone in olm research was reached 330 years after Valvasor, in 2019, when sequencing of the huge genome was completed as an international collaborative project, the Proteus Genome Project (Kostanjšek et al., 2021b).

## Systematics and evolution

Tetrapods are sparsely represented in groundwater and caves. The 14 currently accepted obligate subterranean species belong to the salamander families Plethodontidae and Proteidae. Proteids are distributed in North America and Europe and are exclusively pedomorphic with a permanently aquatic life. The family includes surface-dwelling mudpuppies or waterdogs (genus *Necturus*) from North America, and olms living in southeastern Europe. The split between *Necturus* and *Proteus* has been dated to the Jurassic or Cretaceous, roughly 160−120 million years ago (Zhang and Wake, 2009). Surface relatives were widespread in Europe, leaving numerous proteid fossils across the continent in a fairly continuous sequence dating from the lower Oligocene to the Pleistocene (Márton and Codrea, 2018). Today, *Proteus* is distributed in karstic groundwaters of the Dinaric Karst (Fig. 13.1), reaching from the extreme northeast of Italy via Slovenia and Croatia to southern Herzegovina. Recent investigations of environmental DNA (eDNA) revealed that the southern edge of the range might reach all the way to Montenegro (Gorički et al., 2017).

The currently accepted taxonomy recognizes two subspecies, *Proteus anguinus anguinus*, corresponding to the whitish, troglomorphic morph, and *P. a. parkelj* Sket and Arntzen (1994), which is a dark-colored, nontroglomorphic morph (Fig. 13.1). Despite living a completely subterranean life, the latter has small but fully functional eyes covered with transparent skin. While still formally valid, the two-subspecies taxonomy is not supported by molecular phylogenies (Gorički and Trontelj, 2006; Trontelj et al., 2009). Molecular systematic studies showed that the olm comprises more than one species. Using allozymes, Sket and Arntzen (1994) already reported high genetic distances between some populations of the "white morph", while they found the "black morph" to be closely related to white morph populations from the same region. Further analyses of mitochondrial DNA (Gorički, 2006;

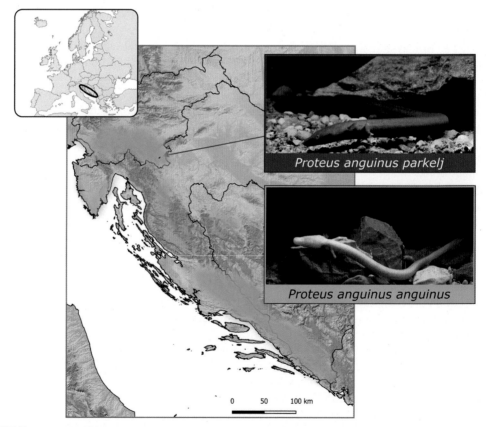

**FIGURE 13.1** Distribution of *Proteus anguinus* in the Dinaric Karst in Southeastern Europe. The map is based on published data and data collected in the database of SubBio Lab (SubBio Database) at the University of Ljubljana. *Photo Credit: Domin Dalessi.*

Goricki and Trontelj, 2006; Trontelj et al., 2009) and nuclear markers (Trontelj et al., 2017; Zakšek et al., 2018; Vörös et al., 2019) brought insights into evolutionary history and populations structure. The entire taxon is phylogenetically deeply subdivided into three main groups: (1) the Istra lineage, (2) the Slovenian and Italian group and (3) the central and southern Dinaric group. The major split between the Istra lineage, which is restricted to the geologically isolated Istra Peninsula, and the remaining groups is from the Miocene, 20—10 million years ago (Fig. 13.2). The morphologically most contrasting and extremely rare black morph, distributed in a very small area of southeastern Slovenia (Bela Krajina) is phylogenetically affined with the Slovenian and Italian group. Since the black morph has retained the ancestral, nontroglomorphic characteristics, it presumably resembles the common ancestor of all extant olm lineages. Fig. 13.2A demonstrates how deeply the black morph is nested among white-morph lineages. Under the phylogenetic principle of parsimony, a single evolutionary reversal of the troglomorphic to the nontroglomorphic phenotype is needed to explain this phylogeny (Ivanovič et al., 2013; Sessions et al., 2015). However, this scenario requires multiple breaches of Dollo's law stating that complex traits lost during the course of evolution do

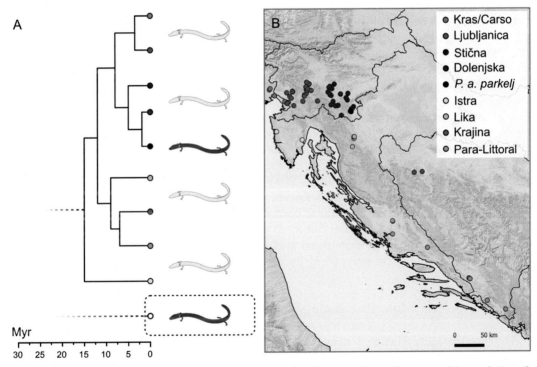

**FIGURE 13.2** Phylogenetic relationships and geographic distribution of *Proteus* lineages and/or evolutionarily significant units (ESUs). (A): Simplified chronogram, based on mtDNA gene sequences with mapped troglomorphic and nontroglomorphic morphs. Although the outgroup is merely hypothetical, the surface ancestor probably resembled the nontroglomorphic, black morph. (B): Map of sampling sites with *Proteus* populations with known lineage membership. *Credit: Data for the chronogram and map are from Goricki and Trontelj (2006), Goricki (2006), Goricki et al. (2017), and Trontelj et al. (2017). The clock rate was estimated from the timing of the* Proteus—Necturus *split in Zhang and Wake (2009).*

not evolve again to the same condition in a given lineage. Therefore, a plausible, even though less parsimonious alternative explanation is that multiple lineages independently invaded caves and evolved troglomorphic characteristics, followed by extinctions of the surface ancestor. Convergent evolution producing multiple similar troglomorphic phenotypes is the rule, not the exception in subterranean biology.

Our current understanding of the deeper phylogeographic structure is based chiefly on mitochondrial phylogenetic analyses (Fig. 13.2). According to these, the Slovenian and Italian group comprises five lineages: (1) the Kras/Carso lineage from the border region between Italy and Slovenia, (2) the Ljubljanica lineage from southwestern Slovenia, (3) the Dolenjska lineage from southeastern Slovenia, (4) the Stična lineage from a small area at the town of Stična, and (5) the black morph, taxonomically assigned to *Proteus anguinus parkelj*.

The central and southern Dinaric group consists of three lineages: (1) the Lika lineage from the Lika region in central Croatia, (2) the Para-Littoral lineage from Dalmatia and southeastern Herzegovina, and (3) the Krajina lineage from northwestern Bosnia and Herzegovina.

The ranges of these lineages or phylogroups are mostly disjunct or parapatric with small areas of overlap and sporadic syntopy. The first area of sympatry between lineages is in southeastern Slovenia where the tiny range of the black morph lineage partly overlaps with the range of the Dolenjska lineage. Examination of eDNA in this area suggested that the two lineages live in contact with each other in the same groundwater body (Gorički et al., 2017). The second known area of overlapping ranges and sympatric co-occurrence is in the northwest of the Dinarides where some populations of the Ljubljanica lineage have dispersed beyond the boundaries of the Danubian drainage basin into the Mediterranean drainage basin. Here, they are coming into contact with the Kras/Carso lineage (Fig. 13.2, blue and orange dots, respectively). An analysis of population genetic structure using microsatellites revealed a narrow hybrid zone between the two lineages in the subterranean Reka (Timavo) River. The effect of hybridization is limited to a small number of F1-hybrids and some F2-hybrids and backcrosses, while no introgression of foreign alleles has been detected deeper in the genepool of either lineage. The behavior of lineages in sympatry suggests that they might be reproductively isolated against each other, both by postmating or premating barriers. The exact mechanisms of these barriers are not yet understood. Differences in ecology and behavior are possibly contributing to the isolating barrier between the black and the white morph. The formal taxonomic status of lineages remains to be clarified, but the status of evolutionarily significant units (ESUs) seems undisputable.

## Molecular ecology and conservation genetics

The nine lineages, whether we regard them as hypothetical species or ESUs, differ considerably in range size, population size and internal population structure, although our knowledge is still insufficient. In order to assess the status and to propose appropriate conservation strategies, information about the geographic subdivision, demography, genetic diversity, effective population size as well as census size and population density is mandatory. These data cannot be obtained as easily as for surface dwelling amphibians. Major portions of the habitat are essentially unknown and remain inaccessible even to cave divers. Where access is possible, individuals are difficult to catch and to sample. A recently developed fieldwork technique combines catching of live animals by snorkeling and scuba diving using hand nets and taking DNA samples using skin swabs (Trontelj and Zakšek, 2017). At sites where karstic springs are the only window to subterranean habitat, trapping and accidental catches are the only way of obtaining individuals for genotyping. A wide set of lineage-specific microsatellite markers already serves as conservation genetic tool (Zakšek et al., 2018; Vörös et al., 2019).

The Postojna-Planina Cave System (PPCS) is one of the best sites to study population parameters due to the high population density and the large stretch of accessible subterranean river. Nowhere else so many animals from a single population can be sampled. The olm population of the PPCS is accessible via the subterranean river Pivka in Postojna Cave and Planina Cave. Both sites are hydrologically connected via explored and unexplored subterranean channels and about 10 km apart (Fig. 13.3). The population genetic structure inferred from 23 polymorphic microsatellites resembled almost complete panmixia (Zakšek et al., 2018). This and a few other genetic studies suggest that olms are able to disperse over tens of kilometers of subterranean rivers within short enough timespans to maintain a nearly panmictic genetic pattern. This could be facilitated by the good mobility of the animals as well as by their longevity. In view of recent efforts to remove sources of pollution of karstic aquifers (Manenti et al., 2021) these results are encouraging as they indicate that, if suitable hydrological connections are available, olms have the ability to recolonize remote habitats. On the other hand, cases of extreme philopatry of individual olms over several years have been reported (Balázs et al., 2020). These are not a consequence of external factors restricting dispersal but can probably be explained by strong territorial attachment.

Population size or some proxy thereof is a crucial parameter in conservation. Standardized and unbiased census methods are required to monitor populations through time. Standard techniques such as counts by sight from fixed points or transect counts along cave rivers and water-filled channels have been applied to estimate population decline at some caves (Sket, 1997; Hudoklin, 2016) and to obtain comparative population density data (Vörös et al., 2019). In the latter study, animals were counted on line transects by cave divers four times per year in three caves in Croatia, and estimated densities varied from 0.45 to 11.45 individuals per 10 m$^2$ of channel ground. Other observations showed that local numbers may vary from zero to several hundred individuals depending on water levels, and that standard census methods do not yield reliable estimates for monitoring (Trontelj et al., 2017).

One way to obtain unbiased population estimates is by applying capture-mark-recapture techniques and models. These work well as long as the caught and marked animals are a random sample of a larger but closed population. Also, they require relatively high numbers of sampled and resampled individuals, which makes individual tagging (e.g., Balázs et al., 2020) impracticable. In the alternative genetic approach, individuals are not actually marked but their DNA is taken using skin swabs. Each individual can be identified by a specific combination of alleles across a set of microsatellite loci, and consequently recaptured individuals can readily be pinpointed. This approach was developed and tested on selected populations in Slovenia during two consecutive years (Trontelj et al., 2017). With a recapture rate of about 15% it was possible to obtain population estimates with reasonable error intervals. The estimated population size exceeded the maximum number of actually observed and counted individuals at any time by a factor of four to five. This suggests that a large proportion of the

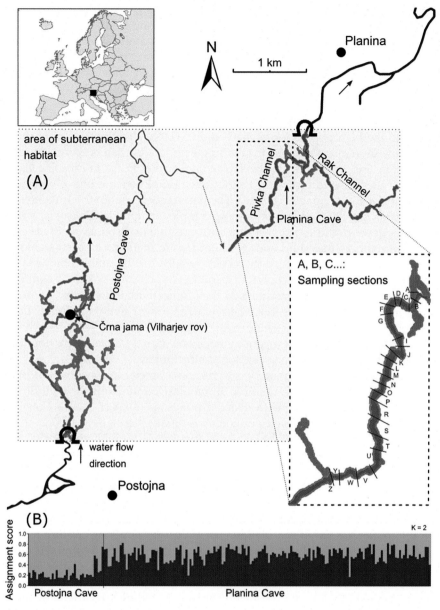

**FIGURE 13.3**   Map of the Postojna-Planina Cave System (PPCS) with genetic structure of the local olm population. (A): Position of sampling sites Črna jama and Pivka Channel. The latter is divided into sampling sections in order to detect movements of recaptured individuals. (B): Genetic clustering of samples ($K = 2$) based on 23 microsatellite loci using Bayesian structure analysis without prior information on the sampling location of individuals. Each column represents an individual. *Credit: Map and structure plot are adapted from Zakšek et al. (2018).*

population remains hidden to the researchers regardless of the particular counting technique they employ. Genetic mark-recapture may be the only reliable method for long-term monitoring of census population size. As with other mark-recapture methods, the downside is that it requires catching and genotyping of hundreds of individuals. This is only possible in a few caves where either large sections of subterranean channels are accessible or population densities are exceptionally high. The method is time consuming, expensive and potentially hazardous as it often requires technical cave diving. The catching with hand nets and handling during genotyping probably causes stress to the animals. On the positive side, because of the long reproductive cycle and longevity of olms (described later in the text), the survey frequency can be kept lower than in surface-dwelling urodelans, with monitoring intervals of about 10 years.

## Morphology and sensory systems of a groundwater top predator

Comparative studies of the white and black morphs have improved our understanding of the biology and evolution of groundwater-related traits in a similar way as have comparisons between cave- and surface-dwelling *Astyanax* or *Asellus* (Gross et al., this volume and Protas et al., this volume, respectively). The upper jaws of both subspecies lack maxillary bones, a pedomorphic trait shared among other members of the family Proteidae (i.e., *Necturus* species). Common to both olm morphs is a reduced number of toes, three on the forelimb and two on the hindlimb, unique among troglobiotic salamanders. Compared with the white morph, the black morph has a wider and shorter skull (Fig. 13.4A and B), larger premaxillae, shorter and more widely spaced vomers, a smaller number of teeth, shorter legs and tail, and a longer trunk with a higher number of vertebrae (Sket and Arntzen, 1994; Ivanovic et al., 2013; Bizjak-Mali and Sket, 2019).

Although reduced eyes is one of the most prominent troglomorphisms of the white morph, eye development begins as in other amphibians. The regression of the almost normally formed larval eyes starts soon after hatching and gradually leads to a considerable reduction or even absence of the lens and the sclera and to a reduced stratification of retinal cells in the adult eye. The degree of reduction varies between and within populations of the white morph, but the process always results in barely visible, eyes about half a millimeter in diameter, positioned deep below the dermis (Fig. 13.4C). As in *Astyanax mexicanus*, eye degeneration in the olm seems to be controlled by intrinsic genetic factors and cannot be altered by cultivating individuals under light or by grafting the olm's developing eye onto another species of salamanders with normal eyes (Durand, 1976). The eye of the adult black morph is small (0.7 mm in diameter), and pedomorphic compared to *Necturus*, with normal larval morphology with a distinct stratification of cells in the retina (Kos et al., 2001; Bizjak-Mali and Sket, 2019) (Fig. 13.4D). Despite anatomical regression and a decreased number of photoreceptor cells in the retina, the eyes of both morphs are sensitive to light. A single subtype of rod cells (i.e., principal rods) and red-sensitive cones were identified in the retina of both morphs, while blue- or UV-sensitive cones were identified only in the black morph (Kos et al., 2001). In addition, red-sensitive visual pigments were detected in the pineal photoreceptors and in the basal layer of the skin epithelium of both morphs, suggesting the presence of active alternative light sensitivity mechanisms.

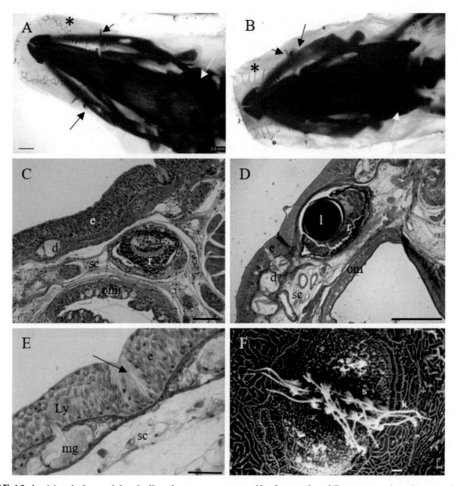

**FIGURE 13.4**  Morphology of the skull and sensory organs of both morphs of *Proteus anguinus* (images A, C, E, F showing the white morph, and images B, D showing the black morph). (**A and B**) Skull morphology; clearing and staining technique for bone (red) and cartilage (blue). Asterisk − cartilages of nasal cavity, long arrow - eye, short arrow - antorbital cartilage, white arrow − otic capsule. (**C and D**) Cross section through the head with regressed eye in the white morph (**C**) and normal eye in the black morph (**D**). e − epidermis, d − dermis with multicelullar mucous gland, l − lens, r − retina with pigment epithelium, om − oral mucosa, sc − subcutis. (**E**) Skin histology of the white morph with visible mechanoreceptive neuromast (arrow) in the epidermis (e). Ly - Leydig cells in epidermis, mg - dermal multicellular mucous glands, sc − subcutis. (**F**) Electron-microscopy micrograph of a neuromast with visible stereocilia (s) and kinocilia (k) of sensory hair cells. Stainings: H & E (**C**), Silver nitrate - Pollak (**D**), Masson trichrome (**E**). Scale bars: 500 μm (**A, B, and D**), 200 μm (**C**), 100 μm (**E**), 1 μm (**F**).

As in several other subterranean animals, the sensitivity of extra-optical sensory organs is increased. These include olfactory receptors, chemoreceptive taste buds in the mouth, and unidentified receptors enabling the olm to orientate in Earth's magnetic field. However, the primary sensory organs of the olm are electro- and mechanoreceptors including the auditory system (Schlegel et al., 2009). With a hearing ability in the frequency range of 10−15,000 Hz and the highest sensitivity at 1500 Hz for the white and 2000 Hz for the black morph, the olm exceeds the hearing capacity of other amphibians and most fish. Nevertheless, since acoustic communication has not been confirmed, underwater orientation and localization of prey remain the most likely function of its hearing ability (Schlegel et al., 2009). This exceptional hearing ability is associated with the air-filled lungs, which in olm serves for retaining buoyancy and oral cavity as resonators and the close anatomical proximity of the oral cavity with the oval window of the inner ear. The general anatomy of the latter is similar in both morphs, the only differences being a slightly shorter and less flattened membranous labyrinth in the black morph (Konec and Bulog, 2010). Detection of frequencies below 50 Hz proceeds probably via mechanosensitive neuromasts of the lateral line sensory system (Fig. 13.4E and F) that are primarily responsible for the detection of water currents and the movement of prey animals. Weak electric fields (up to 0.1 mV/cm) are detected by the ampullary organs of the lateral line sensory system in the epidermis (Schlegel et al., 2009).

Behavioral studies of captive-bred olm individuals also show the importance of chemical signals alongside prey detection, in homing, social behavior, and reproduction, with conspecific attraction leading to group behavior of nonsexually active adults by occupying common shelters (see in Guillaume, 2000). Interestingly, despite having multiple kinds of receptors, evidence has been presented that the ability of olms to detect predators is lost as part of a long history of adaptation to subterranean habitats (Manenti et al., 2020).

## Reproductive peculiarities

The reproductive biology of the olm has been the focus of research for many years, but mostly from observations of animals in captivity. Compared to other vertebrates, including the only other cave-adapted salamanders (all in the family Plethodontidae), olms are characterized by several reproductive peculiarities. Some of these, such as prolonged cycles, decreased growth rate, and iteroparity, are also found in certain other cave-adapted organisms (Poulson, 1963; Kováč, 2018). In olms, sexual maturity is reached in females at 15 years and in males at 11 years of age and reproduction does not occur before the age of 15 (Juberthie et al., 1996), which is far longer than in any other amphibian. Females lay eggs at very long intervals, presumably of 6–12.5 years, and their reproductive period extends until late adulthood of at least 80 years (Voituron et al., 2011; Ipsen and Knolle, 2017). In contrast to most other salamanders, male and female olms cannot be distinguished by external morphology, with the exception of a short period during sexual maturity when maturing oocytes are visible through the abdominal wall in females, while males can be distinguished by a swollen cloaca. After courtship, the female picks up the spermatophore deposited by the male with her cloaca for internal fertilization of her oocytes (Briegleb, 1962). Egg deposition in captivity occurs throughout the year with a slight preference for the winter period (Juberthie et al., 1996). The eggs are laid in clutches of 35–70 eggs and are guarded by the female until hatching, which takes about 4 months at 11–12°C. Despite high embryonic and posthatching mortality (>50%) (Juberthie et al., 1996), captive survival can be improved by providing optimal artificial conditions, as recently demonstrated in the tourist cave of Postojna (Bizjak-Mali et al., 2017).

Detailed gonad morphology shows a positive correlation between body length and egg maturation, and the absence of seasonality of the process in females (Bizjak-Mali and Bulog, 2010; Bizjak-Mali, 2017). On the other hand, body size seems to be a poor predictor of the reproductive status of olm males, while spermatogenesis, which takes place in 2-year cycles, seems to be seasonal and occurs more frequently from late summer to mid-winter. Considering the stable conditions of subterranean groundwater habitats, the seasonality of spermatogenesis is surprising. The loose synchronicity in gametogenesis between the sexes clearly underlines our incomplete understanding of reproduction in this amphibian.

Olms have a female-biased sex ratio of about two to one, and gonadal anomalies, such as a high proportion of degenerated (atretic) ovarian follicles, egg cells in testes (testis-ova) in ~30% of the males examined, and hermaphroditism with apparently functional chimeric gonads with synchronously developing male and female gametocytes (Bizjak-Mali and Sessions, 2016; Bizjak-Mali, 2017). To our knowledge, this suite of reproductive peculiarities has not been reported for any other cave-dwelling vertebrate. The proximate mechanism of both the sex ratio and gonad anomalies is not known, but might be related to recently reported cytogenetic evidence of an X/Y sex chromosome translocation (Sessions et al., 2016), which is responsible for homomorphic male and female karyotypes (Fig. 13.5). Both genera of the family Proteidae, *Proteus* and *Necturus*, are characterized by 19 pairs of chromosomes, a chromosome number that is unique among salamander families (Sessions, 2008). Unlike *Proteus*, all species of the genus *Necturus* have strongly differentiated heteromorphic X/Y sex chromosomes (Fig. 13.5) (Sessions and Wiley, 1985). The translocation in the olm renders sex identification impossible even at the level of chromosomes. Such a translocation can disrupt the function of genes involved in sex determination through random "position effects" with reproductive consequences such as the observed intersexes and hermaphrodites. The apparently negative reproductive consequences of such a chromosomal rearrangement suggests that it may have some other biological significance, perhaps related to cave adaptations, a possibility that is currently under investigation.

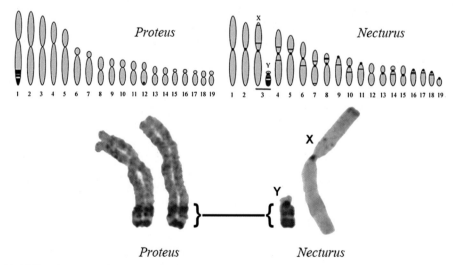

**FIGURE 13.5**  Idiograms of both genera of the family Proteidae, *Proteus* and *Necturus*, with a distinctive C-banding pattern of heterochromatin at the end of the long arms of *Proteus* chromosome one corresponding to the pattern on Y chromosome of *Necturus*. *Modified from Sessions et al. (2016).*

# The overlooked part of groundwater ecology: symbioses, pathogens and parasites

Apart from the recognized role of the olm as the top predator of subterranean waters of the Dinaric karst, several aspects of its ecology, including interactions with microorganisms and parasites, remain understudied. The inaccessibility of a considerable part of the underground environment, the relative scarcity of biological material and the limited taxonomic knowledge of researchers in the past are reflected in sporadic and often generalized descriptions of olm's parasites and symbionts. The only two attempts of comprehensive coverage of interactions with other organisms include the work on the parasites of subterranean organisms dating back more than half a century (Vandel, 1965), and a brief overview of the parasites of the olm (Kostanjšek et al., 2017).

Among the specialized parasites described in the olm are the Myxozoan *Chloromyxum protei*, which inhabits its renal ducts (Joseph, 1905), the trematode *Plagioporus protei* (Prudhoe, 1945), which colonizes the small intestine, and a more recent description of the thorny-headed worm *Acantocephalus anguillae balcanicus*, which infests the intestinal wall and various other organs in the body cavity of the olm (Amin et al., 2019) (Fig. 13.6A). The digestive tract and peritoneal cavity of the olm are frequently colonized by nematodes (Bizjak-Mali and Bulog, 2004). Although the first attempts of scientific classification of nematodes in the olm (e.g., *Nematoideum protei anguinii*) were made as early as the beginning of the 19th century (Rudolphi, 1819), their taxonomic position remains unclear. In addition to free-living nematodes, which occasionally colonize the intestine and even the body cavity (Fig. 13.6E), encysted nematodes entrapped in spherical formations in the wall of the pancreas and duodenum (Fig. 13.6C and D), occur as common parasites of olm's digestive system (Schreiber, 1933). Other less specific records of parasites include unidentified adult trematodes in the gut (Matjašič, 1956) and monogenean flatworms on the skin and gills of the black olms, and recently also white morph. Besides the parasites, ubiquitous protozoa, including amebae and ciliates from the genera *Trichodina* and *Vorticella*, occasionally colonize the skin as epibionts.

The accessibility of high-throughput DNA sequencing technologies over the last decade has significantly improved our knowledge of the importance of skin-associated microbial communities as a source of metabolic capabilities and protection against pathogens in amphibians. Recent threats to amphibian diversity from emerging pathogens, the general lack of information on skin microbiota in neotenic amphibians and conservation potentials initiated the study of the composition and potential role of skin bacteria in olm. Next generation sequencing analysis of skin bacterial communities in individuals from five different genetic lineages of olms and their aquatic environment revealed the environmental origin of skin bacteria. Rather than simply reflecting the bacterial community in the surrounding water, the bacteria colonizing the skin of analyzed individuals consisted of resident microbial community dominated by five bacterial taxa, indicating the selective pressure of the skin environment on colonizing bacteria (Kostanjšek et al., 2019). The high similarity of the resident skin bacteriome between individuals from different unpolluted sampling sites and the replacement of a significant proportion of the resident microbiota by Enterobacteria detected in individuals exposed to pollution from wastewater and agriculture (Fig. 13.7.)

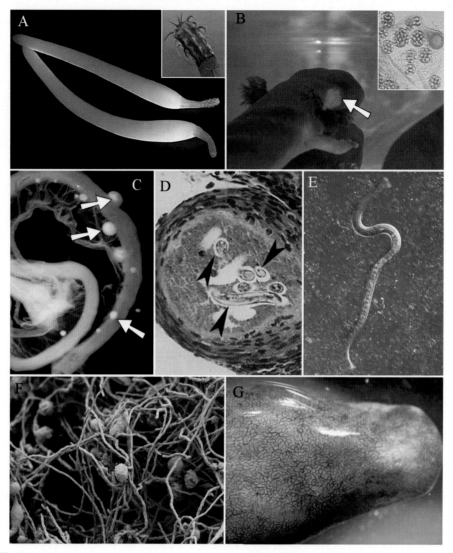

**FIGURE 13.6**  Panel summarizing main known parasites and pathogens of olm. (A) Two adult individuals of *Acanthocephalus anguillae balcanicus* from the olm's gut, and a detail of its hooked proboscis (inset); (B) *Saprolegnia* on the gills of the black morph (arrow) and *Saprolegnia* sp. isolate (inset); C) Nodular formations in the wall of small intestine (arrows) and (D) a section of individual nodule with nematodes (arrowheads); (E) Free-living nematode from the gut contents; (F) Electronic micrograph of *Exophiala salmonis* hyphae in a skin ulcer; (G) Skin hemorrhages on the head caused by *Aeromonas* infection.

demonstrates the potential of the resident skin microbiota to provide protection against invading microbes by occupying the niche. At the same time, the composition of the skin microbiota of olms could serve as an indicator of the status of environmental pressure and the well-being of the host.

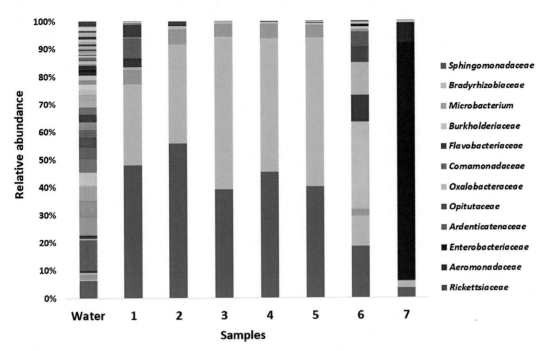

**FIGURE 13.7** Relative abundance of prevailing bacterial families in the skin of *Proteus anguinus* individuals from different populations (Samples 1–7) and in cave water. *Credit: Figure adapted from Kostanjšek et al. (2019).*

Due to the inaccessibility of its natural habitat, individuals of *P. anguinus* are often kept in artificial or seminatural *ex-situ* environments for scientific purposes or recovery, breeding and for exhibitions. Despite the best efforts, the controlled conditions of the ex-situ environment can still cause stress to captive olms, often resulting in increased susceptibility to opportunistic microbial infections. Recorded opportunistic infections caused by fungi include oomycete water molds from the genus *Saprolegnia* sp., which infect the outer surface (Kogej, 1999; Lukač et al., 2019) (Fig. 13.6B), as the most common opportunistic infection of the olm, and skin lesions caused by the black yeast *Exophiala salmonis* (Bizjak-Mali et al., 2018) (Fig. 13.6F). Various fungi and oomycetes have also been identified in the outer coat of olm eggs laid in captivity (Zalar et al., 2016). The most common bacterial pathogen associated with opportunistic infections in the olm is *Aeromonas hydrophila*, which causes a systemic infection associated with cutaneous erythema and haemorrhagic changes in other tissues, and is commonly known as "red leg syndrome" (Densmore and Green, 2007; Lukač et al., 2019) (Fig. 13.6G).

Well-being of captive animals and the establishment of a much-needed sustainable breeding program for olms depends heavily on providing *ex-situ* conditions closely mimicking the natural environment and on the detection and treatment of potential medical conditions of captive animals. To address the latter two issues, several diagnostic protocols have been developed in recent years, including monitoring of hematological parameters of olms (Gredar et al., 2018) and treatment of symptomatic individuals (Bizjak-Mali et al.,

2018; Lukač et al., 2019). At the same time, a study is underway on possible monitoring of stress in captive animals by analysing changes in the composition of their cutaneous microbiota.

Besides opportunistic infections and frequent amphibian diseases with bacterial etiology such as flavobacteriosis, mycobacteriosis and chlamydiosis (Densmore and Green, 2007), newly emerging pathogens pose further microbial threat to the olm. In particular, chytrid fungi of the genus *Batrachochytrium* and ranaviruses from the Iridioviridae family (Densmore and Green, 2007; Latney and Klaphake, 2013), which are responsible for the extinction of several hundred species of amphibians, causing a dramatic global decline in their diversity in recent decades (Scheele et al., 2019). Due to the high mortality and the rapid spread of these pathogens, their occurrence is monitored in amphibian populations throughout Central and Western Europe. Despite recently shown tolerance of the olm to *Batrachochytrium salamandrivorans* infection under laboratory conditions (Li et al., 2020) and the absence of chytrid fungi or ranaviruses in so far examined olm populations in Slovenia (Kostanjšek et al., 2021a) and Croatia (Lukač et al., 2019), continuous monitoring of these pathogens in surface and subterranean amphibians should be established, as their introduction to underground water systems of the Dinaric Karst still poses a considerable threat to olm populations.

Despite the constant presence of microorganisms in their environment and the frequent occurrence of parasites in their tissues, the olm individuals in the natural environment tend to remain in good physical condition, which indicates the effectiveness of their defense mechanisms and their ability to defend themselves against naturally occurring parasites and pathogens. At the same time, the increased sensitivity of olms to pathogens and parasites under controlled conditions indicates a suboptimal response of their immune system, even under relatively moderate captive stress. In this respect, the increased susceptibility of olms to pathogens under the constant pressure of pollutants and other stressors in their natural environment (Bulog, 2007) represents a reasonable threat. In order to assess the severity of these indirect threats and the infestations rate of olm as an indicator of environmental condition, a survey on the occurrence of parasites in archived olm specimens was recently initiated. In addition to providing information on the occurrence of parasites in olms over the last five decades, the results of the survey may contribute to an objective assessment of the threat to natural populations of the olm and to the preparation of appropriate conservation measures.

## Conservation

The nominal species *Proteus anguinus* is protected in all range states including Montenegro, where its presence has only recently been indicated by eDNA. Internationally, *P. anguinus* is listed as vulnerable with a decreasing population trend on the IUCN Red List of Threatened Species (Arntzen et al., 2009), and it is protected as a priority species by the European Union Habitat Directive (Annex II, IV; Council of the European Communities, 1992). The Habitats Directive is a key instrument for conservation of biodiversity adopted in 1992 to protect habitats and species in freshwater, terrestrial and marine habitats in Europe. Although subterranean biodiversity is particularly diverse in Europe (Gibert and Culver 2009), only two more obligate subterranean species are included in the Habitats Directive: the terrestrial coleopteran *Leptodirus hochenwartii* and the bivalve *Congeria kusceri*. The cave bivalve is tiny and difficult to observe, so the olm's role as a flagship species for the protection of groundwater ecosystems is underscored by its position in the EU legislation.

The most important known olm sites within the EU member states Italy, Slovenia and Croatia are part of the Natura 2000 network of protected areas. However, the inclusion within Natura 2000 sites has contributed little to mitigate and eliminate threats and even to halt population decline. Examples of major threats include groundwater overpumping in Italy (Bressi, 2004), heavy organic pollution by farming and sewage in the subterranean Pivka river and the Kočevje area in Slovenia (Hudoklin, 2011), poisoning from heavy metals (Bulog et al., 2002) and illegally disposed polychlorinated biphenyl (PCB) waste at the Krupa karst spring in Slovenia (Pezdirc et al., 2011), habitat destruction through damming and artificial channeling of sinking rivers in the Trebišnjica and Neretva basin in Bosnia and Herzegovina (Aljančič et al., 2014). Most of these threats have severely affected olm populations causing local extinctions and drastic population decline. Besides these obvious major factors, many olm populations are facing a number of local threats with less clear effects. Examples are urbanization in catchment areas, poor wastewater treatment, landfills, small-scale farming with high usage of fertilizer and pesticides, altering of karst springs for various forms of human use, invasive changes to the landscape above karstic aquifers such as construction of roads and infrastructure or mining and limestone quarrying (Bressi, 2004; Hudoklin, 2016; Trontelj et al., 2017). Finally, direct pressure by collecting and disturbance has been known from the past and is emerging in new forms, e.g., as cave tourism and increased caving activity, but their impact has yet to be assessed.

A common denominator of all but the most obvious and direct threats is that their impacts are difficult to assess and impossible to predict. There are three main reasons: (1) The actual position and extent of the olm's subterranean habitat are unknown; what is mapped and formally protected are merely cave entrances or springs that might be kilometers away from the true habitat. (2) The hydrological connectivity between potential sources of pollution from the surface and the subterranean habitat are not well known, as are connections among habitat patches itself. (3) Because of the long lifecycles, low metabolism and relative robustness of adults, population size is believed to respond to worsened conditions with a delay of several years or decades. Early developmental stages that are most sensitive to pollution are usually hidden and cannot be monitored.

One last point that importantly affects conservation strategies and priorities is the taxonomic subdivision of the genus *Proteus* For example, a small marginal population attracts little conservation attention as long as it is part of a species that has several strong and healthy populations. But if it turns out that this marginal population is the last survivor of a relict lineage or represents a distinct species, it may become top priority. Therefore, formalization of the taxonomic status of the nine known lineages of *Proteus* is under way. Already it is clear that these lineages differ substantially with respect to range size, population size and trends as well as exposure to threats (Table 13.1).

It can be concluded that the lack of vision of how the species can be helped in places where it is believed to be declining and safeguard where it is still stable represents a major obstacle to effective conservation. In this sense, too, the olm is representative of groundwater biodiversity in general.

IV. Principles of evolution in groundwater

**TABLE 13.1**    *Proteus* lineages and/or evolutionarily significant units (ESU) as inferred from analyses of mtDNA sequences and nuclear DNA markers, along with estimated gross area, trend and threats.

| Lineage/ ESU[a] | Area (km²)[b] | Cumulative long-term trend[c] | Main conservation issues |
|---|---|---|---|
| Kras/ Carso | 200 | Unknown; local decline likely | Groundwater overpumping in the most western part of range; pollution from sinking Reka river; urbanization pressure, intensive agriculture |
| Ljubljanica | 1100 | Moderate decline | Sinking river pollution by agriculture and sewage; pollution from industry; cave tourism in the Postojna-Planina Cave System; excessive collecting in the past |
| Stična | <1 | Strong decline | Urbanization of the entire catchment area; destruction of habitat by construction and filling; groundwater pollution by diffuse infiltration of sewage and fertilizer; overcollecting in the past; inappropriate use of single remaining site by landowners |
| Dolenjska | 1300 | Moderate decline with local extinctions | Sinking river pollution by agriculture, sewage and industrial waste; illegal landfills and toxic waste dumps; groundwater pollution by diffuse infiltration of sewage and fertilizer |
| *P. a. parkelj* | 3 | Moderate to strong decline | Groundwater pollution by diffuse infiltration of sewage and fertilizer from local biogas plant, intensive agriculture and industrial waste dumps; urbanization and disturbance at karst springs |
| Istra | 300 | Strong decline | Groundwater pollution caused by intensive agriculture and urbanization; filling of caves; alternation of hydrological regime; population probably strongly fragmented |
| Lika | 900 | Unknown | Unknown |
| Krajina | 700 | Unknown | Unknown |
| Para-Littoral | 9000 | Moderate to strong decline | Habitat destruction by damming and channeling of sinking rivers in the Trebišnjica basin; construction of hydropower plants; groundwater overexploitation |

[a]*Listed by geographic order from North-West to South-East.*
[b]*Rough estimate of geographic area covered by known sites; the actual habitat is unknown and probably much smaller.*
[c]*Nonmethodical estimate based on scattered scientific studies, anecdotal reports, comparison of current records with historical data and own observations.*

## Conclusive remarks on flagship species in groundwater

In conclusion, it can be said that the olm fulfills the role of flagship species of groundwater ecosystems in many ways. This can be demonstrated by a check against commonly applied selection criteria (e.g., Kalinkat et al., 2017). Olms meet these criteria more than most other groundwater species, but cannot compete with traditional flagships like larger mammal or bird species (Table 13.2). In addition to the standard criteria, a historical aura of mysterious discoveries and scientific research underscores the flagship value (Aljančič, 2019). Over the centuries, the olm has gained the status of an unofficial national symbol in Slovenia, and it is arguably considered an iconic species in Croatia, Bosnia and Herzegovina and parts

**TABLE 13.2** Evaluation of the suitability of *Proteus anguinus* as conservation flagship species. Criteria follow Kalinkat et al., 2017.

| Criterion | Meets criterion | Does not meet criterion |
| --- | --- | --- |
| Ease of observation | The only groundwater animal typically observed by cave visitors | Caves and groundwater are inaccessible to most people |
| Anthropomorphic features | Skin color, longevity, heart rate | Unattractive, snake-like, slimy |
| Body size | One of the largest stygobionts | Insignificant in comparison to surface tetrapods |
| Taxonomic proximity to humans | Closest to humans among obligate subterranean species | Mammals would be ideal |
| Publicly perceived danger of extinction | Perceived as extremely rare and narrowly endemic | Low understanding of actual threats |

of Italy. Its popularity is perhaps best exemplified by how the news of successful reproduction in the Postojna Cave aquarium in 2016 made the cover story in national and international media (Lučić, 2021, Fig. 13.8). International conservation of the olm rests on its popularity and scientifically founded role as top predator and indicator species. In the course of the 2004 enlargement of the European Union, it was assigned priority status in the Habitats Directive. The relatively detailed data on the distribution of olms has guided the designation of large parts of the Natura 2000 network in Slovenia and Croatia. Olms have been in

**FIGURE 13.8** The Slovenian magazine Delo observed the hatching of 22 olm larvae in the Postojna Cave show aquarium in 2016 with a cover-story cartoon. Illustration by courtesy of Marko Kočevar.

the center of several conservation projects aiming at safeguarding karstic groundwater eco-systems as a whole (e.g., Prelovšek, 2016; Lewarne, 2018).

Nevertheless, groundwater conservation based on flagship species is still at its beginnings, and so is conservation of groundwater as a wildlife habitat in general. The full potential of this approach has therefore yet to be realized. Other species will have to take over the role of umbrella flagships (Kalinkat et al., 2017) in different karstic areas and on other continents. In North America, a dozen of troglobiotic salamanders are the most obvious candidates. Subterranean fishes, as the only remaining groundwater vertebrates could be ambassadors of groundwater biodiversity on other continents where their diversity is much higher than in Europe and North America. In the absence of suitable vertebrate species, large, free ranging subterranean decapod crustaceans lend themselves as invertebrate flagships of choice. The olm, occurring in white and black morphs, could become – in a sense – the "groundwater panda" and be utilized to raise awareness of the threatened life in this habitat worldwide, following the example of Sir David Attenborough, who included the olm among the 10 species of his personal Ark.

## Acknowledgments

The work was supported by the Slovenian Research Agency through Research Core Funding P1-0184 and research projects N1-0096 and J1-2469. We thank E. Lunghi, S. Mammola and S. Goricki for constructive comments that improved this chapter.

## References

Aljančič, G., 2008. Jamski laboratorij Tular in človeška ribica. Proteus 70, 242–258.

Aljančič, G., 2019. History of research on *Proteus anguinus* Laurenti 1768 in Slovenia. Folia Biologica et Geologica 60, 39–69.

Aljančič, G., Goricki, Š., Năpărus, M., Stanković, D., Kuntner, M., 2014. Endangered *Proteus*: combining DNA and GIS analyses for its conservation. In: Sackl, P., Durst, R., Kotrošan, D., Stumberger, B. (Eds.), Dinaric Karst Poljes – Floods for Life. EuroNatur, Radolfzell, pp. 70–75.

Amin, O.M., Heckmann, R.A., Fiser, Z., Zakšek, V., Herlyn, H., Kostanjšek, R., 2019. Description of *Acanthocephalus anguillae balkanicus* subsp. n. (Acanthocephala: Echinorhynchidae) from *Proteus anguinus* Laurenti (Amphibia: Proteidae) and the cave ecomorph of *Asellus aquaticus* (Crustacea: Asellidae) in Slovenia. Folia Parasitologica 66, 015.

Arntzen, J.W., Denoël, M., Miaud, C., Andreone, F., Vogrin, M., Edgar, P., Crnobrnja Isailovic, J., Ajtic, R., Corti, C., 2009. *Proteus anguinus*. The IUCN Red List of Threatened Species 2009.

Balázs, G., Lewarne, B., Herczeg, G., 2020. Extreme site fidelity of the olm (*Proteus anguinus*) revealed by a long-term capture–mark–recapture study. Journal of Zoology 311, 99–105.

Barua, M., 2011. Mobilizing metaphors: the popular use of keystone, flagship and umbrella species concepts. Biodiversity and Conservation 20 (7), 1427–1440.

Bizjak-Mali, L., 2017. Variability of testes morphology and the presence of testis-ova in the European blind cave salamander (*Proteus anguinus*). Acta Biologica Slovenica 60 (1), 53–74.

Bizjak-Mali, L., Bulog, B., 2004. Histology and ultrastructure of the gut epithelium of the neotenic cave salamander, *Proteus anguinus* (Amphibia, Caudata). Journal of Morphology 259, 82–89.

Bizjak-Mali, L., Bulog, B., 2010. Ultrastructure of previtellogene oocytes in the neotenic cave salamander *Proteus anguinus anguinus* (Amphibia, Urodela, Proteidae). Protoplasma 246, 33–39.

Bizjak-Mali, L., Dolenc Batagelj, K., Gnezda, P., Weldt, S., Sessions, S.K., 2017. Moving away from the "cave lab paradigm": successful captive breeding and raising of the European blind cave salamander *Proteus anguinus* using "Optimal artificial" conditions. In: Programme & Abstracts, 19th European Congress of Herpetology. Facultas Verlags- und Buchhanddels AG Wien, Salzburg, p. 145.

Bizjak-Mali, L., Sessions, S., 2016. Testis-ova and male gonad variability in the European blind cave salamander, *Proteus anguinus* (Amphibia: Urodela): consequence of sex-chromosome turnover? The Anatomical Record 299 (1), 109–110.

Bizjak-Mali, L., Sket, B., 2019. History and biology of the «black proteus« (*Proteus anguinus parkelj* Sket & Arntzen 1994; Amphibia: Proteidae): a review. Folia Biologica et Geologica 60 (1), 5–37.

Bizjak-Mali, L., Zalar, P., Turk, M., Babič, M.N., Kostanjšek, R., Gunde-Cimerman, N., 2018. Opportunistic fungal pathogens isolated from a captive individual of the European blind cave salamander *Proteus anguinus*. Diseases of Aquatic Organisms 129 (1), 15–30.

Blackburn, D.G., 2019. The oviparous olm: analysis & refutation of claims for viviparity in the cave salamander *Proteus anguinus* (Amphibia: Proteidae). Zoologischer Anzeiger 281, 16–23.

Bressi, N., 2004. Underground and unknown: updated distribution, ecological notes and conservation guidelines on the Olm *Proteus anguinus anguinus* in Italy (Amphibia, Proteidae). Italian Journal of Zoology 71 (1), 55–59.

Briegleb, W., 1962. Zur Biologie und Ökologie des Grottenolms (*Proteus anguinus* Laur. 1768). Zeitschrift für Morphologie und Ökologie der Tiere 51, 271–334.

Bulog, B., 2007. Okoljske in funkcionalno-morfološke raziskave močerila (*Proteus anguinus*). Proteus 70 (3), 102–114.

Bulog, B., Mihajl, K., Jeran, Z., Toman, M.J., 2002. Trace elements concentrations in the tissues of *Proteus anguinus* (Amphibia, Caudata) and the surrounding environment. Water, Air, and Soil Pollution 136, 147–163.

Council of the European Communities, 1992. Council Directive 92/43/EEC of 21 May 1992 on the conservation of natural habitats and of wild fauna and flora. Official Journal of the European Communities L206, 7–50.

Densmore, C.L., Green, D.E., 2007. Diseases of amphibians. ILAR Journal 48 (3), 235–254.

Durand, J.P., 1976. Ocular development and involution in european cave salamander, *Proteus anguinus* Laurenti. Biological Bulletin 151 (3), 450–466.

Gibert, J., Culver, D., 2009. Assessing and conserving groundwater biodiversity: an introduction. Freshwater Biology 54, 639–648.

Gorički, Š., 2006. Filogeografska in morfološka analiza populacij močerila (*Proteus anguinus*). (Phylogeographic and Morphological Analysis of European Cave Salamander (*Proteus anguinus*) Population). Doctoral Dissertation. University of Ljubljana, Ljubljana, Slovenia.

Gorički, Š., Trontelj, P., 2006. Structure and evolution of the mitochondrial control region and flanking sequences in the European cave salamander *Proteus anguinus*. Gene 378, 31–41.

Gorički, Š., Stanković, D., Snoj, A., Kuntner, M., Jeffery, W.R., Trontelj, P., Pavićević, M., Grizelj, Z., Năpărus-Aljančič, M., Aljančič, G., 2017. Environmental DNA in subterranean biology: range extension and taxonomic implications for *Proteus*. Scientific Reports 27 (7), 45054.

Gredar, T., Prša, P., Bizjak Mali, L., 2018. Comparative analysis of hematological parameters in wild and captive *Proteus anguinus*. Natura Sloveniae 20 (2), 57–59.

Guillaume, O., 2000. Role of chemical communication and behavioural interactions among conspecifics in the choice of shelters by the cave-dwelling salamander *Proteus anguinus* (Caudata, Proteidae). Canadian Journal of Zoology 78, 167–173.

Hudoklin, A., 2011. Are we guaranteeing the favourable status of the *Proteus anguinus* in the Natura 2000 network in Slovenia? In: Prelovšek, M., Hajna, N.Z. (Eds.), Pressures and Protection of the Underground Karst — Cases from Slovenia and Croatia. Inštitut za raziskovanje krasa ZRC SAZU, Postojna, pp. 169–181.

Hudoklin, A., 2016. Stanje človeške ribice v omrežju NATURA 2000 v Sloveniji. (The status of *Proteus* in the NATURA 2000 network in Slovenia). Natura Sloveniae 18, 43–44.

Hutchins, B.T., 2018. The conservation status of Texas groundwater invertebrates. Biodiversity and Conservation 27, 475–501.

Ipsen, A., Knolle, F., 2017. The olm of Hermann's Cave, Harz Mountains, Germany — eggs laid after more than 80 years. Natura Sloveniae 19 (1), 51–52.

Ivanović, A., Aljančič, G., Arntzen, J.W., 2013. Skull shape differentiation of black and white olms (*Proteus anguinus anguinus* and *Proteus a. parkelj*): an exploratory analysis with micro-CT scanning. Contributions to Zoology 82, 107–114.

Joseph, H., 1905. *Chloromyxum protei* n. sp. Zoologischer Anzeiger 29 (14), 450–451.

Juberthie, C., Durand, J., Dupuy, M., 1996. La reproduction des Protées (*Proteus anguinus*): Bilan de 35 ans d'élevage dans les grottes laboratoires de Moulis et d'Aulignac. Mémoires de Biospéologie: Tome XXIII, 53–56.

Kalinkat, G., Cabral, J.S., Darwall, W., Ficetola, G.F., Fisher, J.L., Giling, D.P., Gosselin, M.P., Grossart, H.P., Jähnig, S.C., Jeschke, J.M., Knopf, K., Larsen, S., Onandia, G., Pätzig, M., Saul, W.C., Singer, G., Sperfeld, E., Jarić, I., 2017. Flagship umbrella species needed for the conservation of overlooked aquatic biodiversity. Conservation Biology 31, 481–485.

Kammerer, P., 1907. Die Fortpflanzung des Grottenolmes (*Proteus anguinus* Laurenti). Verhandlungen der Zoologisch-Botanischen Gesellschaft in Wien 57, 277–292.

IV. Principles of evolution in groundwater

Kammerer, P., 1912. Experimente über Fortpflanzung, Farbe, Augen und Körperreduction bei *Proteus anguinus* Laur. (zu-gleich: Vererbung erzwungener Farbveränderungen, III. Mitteilung). Archiv für Entwicklungsmechanik der Organismen 33, 349–461.

Kogej, T., 1999. Infekcija človeške ribice (*Proteus anguinus*) z glivami rodu *Saprolegnia*. (MSc Thesis). University of Ljubljana, Ljubljana.

Konec, M., Bulog, B., 2010. Three-dimensional reconstruction of the inner ear of *Proteus anguinus* (Amphibia: Urodela). In: Moškrič, A., Trontelj, P. (Eds.), Abstract Book of 20th International Conference on Subterranean Biology. Organizing Committee, 20th International Conference on Subterranean Biology, Postojna, pp. 119–120.

Kos, M., Bulog, B., Röhlich, A.S.P., 2001. Immunocytochemical demonstration of visual pigments in the degenerate retinal and pineal photoreceptors of the blind cave salamander (*Proteus anguinus*). Cell and Tissue Research 303, 15–25.

Kostanjšek, R., Bizjak Mali, L., Gunde Cimerman, N., 2017. Microbial and parasitic threats to proteus. Natura Sloveniae 19 (1), 31–32.

Kostanjšek, R., Prodan, Y., Stres, B., Trontelj, P., 2019. Composition of the cutaneous bacterial community of a cave amphibian, *Proteus anguinus*. FEMS Microbiology Ecology 95 (3) fiz007.

Kostanjšek, R., Turk, M., Vek, M., Gutierrez-Aguirre, I., Gunde Cimerman, N., 2021a. First screening for *Batrachochytrium dendrobatidis*, *B. salamandrivorans* and *Ranavirus* infections in wild and captive amphibians in Slovenia. Salamandra 57 (1), 162–166.

Kostanjšek, R., Diderichsen, B., Recknagel, H., Gunde-Cimeerman, N., Gostinčar, C., Fan, G., Kordiš, D., Trontelj, P., Jiang, H., Bolund, L., Luo, Y., 2021b. Toward the massive genome of *Proteus anguinus* - illuminating longevity, regeneration, convergent evolution, and metabolic disorders. Annals of the New York Academy of Sciences 1507, 5–11.

Kováč, Ľ., 2018. Caves as oligotrophic ecosystems. In: Moldovan, O., Kováč, Ľ., Halse, S. (Eds.), Cave Ecology. Springer, Cham, pp. 297–307.

Latney, L.V., Klaphake, E., 2013. Selected emerging diseases of amphibia. Veterinary Clinics of North America: Exotic Animal Practice 16 (2), 283–301.

Lewarne, B., 2018. The "Trebinje *Proteus* observatorium and *Proteus* rescue and care facility", Bosnia and Herzegovina. Natura Sloveniae 20, 73–75.

Li, Z., Verbrugghe, E., Kostanjšek, R., Lukač, M., Pasmans, F., Cizelj, I., Martel, A., 2020. Dampened virulence and limited proliferation of *Batrachochytrium salamandrivorans* during subclinical infection of the troglobiont olm (*Proteus anguinus*). Scientific Reports 10, 16480.

Lučić, I., 2021. An underworld tailored to tourists: a dragon, a photo-model, and a bioindicator. Journal of Cave and Karst Studies 83, 57–65.

Lukač, M., Cizelj, I., Mutschmann, F., 2019. Health research and *ex situ* keeping of *Proteus anguinus*. In: Koller Šarič, K., Jelić, D., Konrad, P., Jalžić, B. (Eds.), Proteus. Udruga Hyla, Zagreb, pp. 219–232.

Macdonald, E.A., Hinks, A., Weiss, D.J., Dickman, A., Burnham, D., Sandom, C.J., Malhi, Y., Macdonald, D.W., 2017. Identifying ambassador species for conservation marketing. Global Ecology and Conservation 12, 204–214.

Manenti, R., Melotto, A., Guillaume, O., Ficetola, G.F., Lunghi, E., 2020. Switching from mesopredator to apex predator: how do responses vary in amphibians adapted to cave living? Behavioral Ecology and Sociobiology 74, 126.

Manenti, R., Piazza, B., Zhao, Y., Padoa Schioppa, E., Lunghi, E., 2021. Conservation studies on groundwaters' pollution: challenges and perspectives for stygofauna communities. Sustainability 13, 7030.

Márton, V., Codrea, V.A., 2018. A new proteid salamander from the early Oligocene of Romania with notes on the paleobiogeography of Eurasian proteids. Journal of Vertebrate Paleontology 38 (5), e1508027.

Matjašič, J., 1956. Trematod iz jamske kozice *Troglocaris*. Bioloski Vestnik 5, 74–75.

McGowan, J., Beaumont, L.J., Smith, R.J., Chauvenet, A.L., Harcourt, R., Atkinson, S.C., Mittermeier, J.C., Esperon-Rodriguez, M., Baumgartner, J.B., Beattie, A., Dudaniec, R.Y., 2020. Conservation prioritization can resolve the flagship species conundrum. Nature Communications 11 (1), 1–7.

Niemiller, M.L., Taylor, S.J., Bichuette, M.E., 2018. Conservation of cave fauna, with an emphasis on Europe and the Americas. In: Moldovan, O.T., Kováč, L., Halse, S. (Eds.), Cave Ecology. Springer, Cham, pp. 451–478.

Pennisi, E., 2009. History of science. The case of the midwife toad: fraud or epigenetics? Science 325, 1194–1195.

Pezdirc, M., Heath, E., Bizjak Mali, L., Bulog, B., 2011. PCB accumulation and tissue distribution in cave salamander (*Proteus anguinus anguinus*, Amphibia, Urodela) in the polluted karstic hinterland of the Krupa River, Slovenia. Chemosphere 84, 987–993.

Poulson, T.L., 1963. Cave adaptation in amblyopsid fishes. The American Midland Naturalist 70, 257–290.

Prelovšek, M., 2016. *Proteus* survival close to an industrial, agricultural and urbanized basin-the case of Kočevsko polje. Natura Sloveniae 18, 47–49.

Prudhoe, S., 1945. Two notes on trematodes. Annals and Magazine of Natural History 11, 378–383.

Rudolphi, C.A., 1819. Entozoorum Synopsis, p. 189. Berlin.

Scheele, B., Pasmans, F., Skerratt, L.F., Berger, L., Martel, A., Beukema, W.,., Canessa, S., 2019. Amphibian fungal panzootic causes catastrophic and ongoing loss of biodiversity. Science 363 (6434), 1459–1463.

Schlegel, P.A., Steinfartz, S., Bulog, B., 2009. Non-visual sensory physiology and magnetic orientation in the Blind Cave Salamander, *Proteus anguinus* (and some other cave-dwelling urodele species). Review and new results on light-sensitivity and non-visual orientation in subterranean urodeles (Amphibia). Animal Biology 59, 351–384.

Schreiber, G., 1933. Sui Nematodi parassiti nel pancreas del Proteo. In: I. Congresso Speleobilogico Nazionale, pp. 3–8. Trieste.

Senzaki, M., Yamaura, Y., Shoji, Y., Kubo, T., Nakamura, F., 2017. Citizens promote the conservation of flagship species more than ecosystem services in wetland restoration. Biological Conservation 214, 1–5.

Sessions, S.K., 2008. Evolutionary cytogenetics in salamanders. Chromosome Research 16, 1 83–201.

Sessions, S.K., Wiley, J.E., 1985. Chromosome evolution in salamanders of the genus *Necturus*. Brimleyana 10, 37–52.

Sessions, S.K., Bulog, B., Bizjak-Mali, L., 2015. The Phoenix rises: reversal of cave adaptations in the blind cave salamander, *Proteus anguinus*? The FASEB Journal 29, LB36.

Sessions, S.K., Bizjak Mali, L., Green, D.M., Trifonov, V., Ferguson-Smith, M., 2016. Evidence for sex chromosome turnover in proteid salamanders. Cytogenetic and Genomic Research 148, 305–313.

Shaw, T., 1999. *Proteus* for sale and for science in the 19th century. Acta Carsologica 28, 229–304.

Sket, B., 1997. Distribution of *Proteus* (Amphibia: Urodela: Proteidae) and its possible explanation. Journal of Biogeography 24, 263–280.

Sket, B., Arntzen, J.W., 1994. A black, non-troglomorphic amphibian from the karst of Slovenia: *Proteus anguinus parkelj* n. ssp. (Urodela: Proteidae). Bijdragen tot de Dierkunde 64 (1), 33–53.

Trontelj, P., Zakšek, V., 2017. Genetic monitoring of *Proteus* populations. Natura Sloveniae 18, 53–54.

Trontelj, P., Douady, C.J., Fišer, C., Gibert, J., Gorički, Š., Lefébure, T., Sket, B., Zakšek, V., 2009. A molecular test for cryptic diversity in ground water: how large are the ranges of macro-stygobionts? Freshwater Biology 54, 727–744.

Trontelj, P., Zakšek, V., Skrbinšek, T., Gabrovšek, F., Kostanjšek, R., 2017. Toward the conservation of the European cave salamander (*Proteus anguinus*): monitoring guidelines, current status estimation and identification of evolutionary significant units. Final project report. http://mop.arhiv-spletisc.gov.si/fileadmin/mop.gov.si/pageuploads/podrocja/narava/cloveska_ribica_zakljucno_porocilo_2017.pdf. Slovenian Government.

Van Alphen, J.J.M., Arntzen, J.W., 2016. Paul Kammerer and the inheritance of aquired characteristics. Contributions to Zoology 85 (4), 457–470.

Vandel, A., 1965. In: Kerkut, G.A. (Ed.), Biospeleology, the Biology of Cavernicolous Animals, first ed. Pergamon Press, Oxford.

Voituron, Y., de Fraipont, M., Issartel, J., Guillaume, O., Clobert, J., 2011. Extreme lifespan of the human fish (*Proteus anguinus*): a challenge for ageing mechanisms. Biology Letters 7 (1), 105–107.

Vörös, J., Ursenbacher, S., Jelić, D., 2019. Population genetic analyses using 10 new polymorphic microsatellite loci confirms genetic subdivision within the olm, *Proteus anguinus*. Journal of Heredity 110, 211–218.

Whitten, T., 2009. Applying ecology for cave management in China and neighbouring countries. Journal of Applied Ecology 46, 520–523.

Zakšek, V., Konec, M., Trontelj, P., 2018. First microsatellite data on *Proteus anguinus* reveal weak genetic structure between the caves of Postojna and Planina. Aquatic Conservation: Marine and Freshwater Ecosystems 28, 241–246.

Zalar, P., Turk, M., Novak, M., Kostanjšek, R., Bizjak Mali, L., Gunde Cimerman, N., 2016. Opportunistic pathogens of the cave salamander *Proteus anguinus*. In: Otoničar, B., Gostinčar, P. (Eds.), Paleokarst: Abstracts & Guide Book, 24th International Karstological School Classical Karst. ZRC SAZU Press, Postojna, p. 35.

Zhang, P., Wake, D.B., 2009. Higher-level salamander relationships and divergence dates inferred from complete mitochondrial genomes. Molecular Phylogenetics and Evolution 53 (2), 492–508.

# 14

# The *Asellus aquaticus* species complex: an invertebrate model in subterranean evolution

Meredith Protas[1], Peter Trontelj[2], Simona Prevorčnik[2] and
Žiga Fišer[2]

[1]Dominican University of California, San Rafael, CA, United States; [2]University of Ljubljana,
Biotechnical Faculty, Department of Biology, SubBio Lab, Ljubljana, Slovenia

## Introduction

For a successful model system, several features are desired such as the ability to raise the animals in the lab, having both subterranean and surface populations that can interbreed and produce fertile offspring, and having multiple subterranean populations in which to study independent evolution of subterranean characteristics. These qualities are rare to find in a species, which is why it has been difficult to establish model systems in subterranean biology. The species that has been most extensively developed as a model is the teleost *Astyanax mexicanus* (see Gross et al., this volume). Despite the advances that have been made using it as a model, studying *Astyanax mexicanus* tells us only how subterranean evolution occurs in one vertebrate species. If we are interested in subterranean evolution as a whole, additional species must be examined. As an invertebrate counterpart, *Asellus aquaticus* Linnaeus (1758) (Fig. 14.1), has previously been recognized as a model system (reviewed in Protas and Jeffery, 2012; Lafuente et al., 2021).

The freshwater isopod crustacean *A. aquaticus*, along with a few closely related species, is distributed over almost the entire European continent excluding the extreme western and northern parts. It is eurytopic and abundant in most types of freshwater habitats apart from oligotrophic fast-flowing rivers and creeks. Its populations can be found in a wide range of temperature regimes and adverse water quality conditions such as organic and inorganic pollutants, moderate salinity and low oxygen concentration. Individual animals, especially

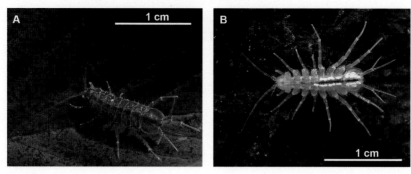

**FIGURE 14.1**    Surface (A) and subterranean (B) ecomorphs of the *Asellus aquaticus* species complex. Both animals are adult males from Pivka Polje and Postojna-Planina Cave System (PPCS), respectively.

juveniles, tend to hide in narrow spaces between and beneath rocks and temporarily enter the interstitial habitat.

Throughout most of its range, the taxon is represented by pigmented surface phenotypes with fully developed eyes, resembling the type population from Sweden (Verovnik et al., 2009). However, several surface populations can be distinguished readily from this classical form; a light-colored stonewort ecomorph inhabiting *Chara* beds in southern Swedish lakes evolved in less than 50 generations (Eroukhmanoff et al., 2011). Older and more stable surface morphological variants known from the western Balkan Peninsula have been formally named as subspecies. Their diagnosability and/or monophyly were refuted by multivariate morphological and molecular scrutiny (Prevorčnik et al., 2004; Verovnik et al., 2004, 2005), leading to abandonment of traditional subspecific taxonomy in *A. aquaticus*. Nevertheless, we refer to animals of such morphology as the **surface ecomorph** (Fig. 14.1A). In addition to the differentiation of the surface ecomorph, several local populations have successfully invaded groundwater and, to a greater or lesser extent, evolved characteristics of the **subterranean ecomorph** (Fig. 14.1B), such as depigmentation and other traits described in this chapter.

This chapter discusses advances that have been made in developing this invertebrate model, the *Asellus aquaticus* species complex. Comparisons between this model, *Astyanax mexicanus* and other emerging subterranean model organisms will allow for a broader understanding of evolution and ecology in this mysterious environment.

## Phylogeography and population structure

### A widespread surface species with local subterranean populations budding off

The above-mentioned natural model system for the study of the evolutionary transitions to life in groundwater essentially comprises eight subterranean populations — along with their extant surface relatives — that invaded groundwater at different times and places and in different ecological settings. This allows researchers to test hypotheses about common principles of subterranean evolution and separate them from contingencies and the influence of local factors. To that end, it is crucial to understand the phylogeny and genetic affinities of subterranean populations. Published molecular phylogenies are based on nucleotide sequences of the mitochondrial cytochrome oxidase subunit I gene (COI) and of a variable fragment of the nuclear 28S rRNA gene (28S rDNA) (Verovnik et al., 2004, 2005; Konec et al., 2015; Sworobowicz et al., 2015).

The evolutionary history of *A. aquaticus* as deduced from the mitochondrial phylogeny (Fig. 14.2) begins about eight million years ago (MYA) with a split between two major geographical clades or phylogroups: (1) the **Trans-Alpine clade** that is restricted to areas north and south of the Alps reaching south to the Apennine Peninsula, and a large clade containing all other lineages including all known subterranean populations. This second clade is further structured into (2) a paraphyletic series of highly divergent lineages from the western and southern Balkan Peninsula ("**Balkan Clades**"), (3) the **Northern Dinaric Karst Clade** and its sister clade, *A. kosswigi*, and (4) the widely distributed **Central Europe Clade**. The clades are of late Miocene and Pliocene age and possibly evolved in response to the extensive hydrological changes of the vanishing Western Paratethys. These clades are further structured into numerous lineages that evolved in Pleistocene times.

Although age and genetic differences between major mitochondrial clades correspond to species-level divergence, the surface ecomorph of *A. aquaticus* remains taxonomically uniform when one takes into consideration nuclear 28S rDNA (Verovnik et al., 2005) and ribosomal internal transcribed spacer (ITS) sequences (Sworobowicz et al., 2020) as well as microsatellite markers (Konec et al., 2015). This suggests that mitochondrial DNA has retained the signal of an ancient phylogeographic structure that in part broke down through dispersal and range expansion accompanied by admixture.

Some subterranean populations depart from this pattern. In Mangalia, Postojna-Planina Cave System (PPCS), Molnár János Cave, and the subterranean Reka River (Fig. 14.2), the subterranean ecomorph lives in close contact with the surface ecomorph but they do not interbreed. Subterranean populations seem to be reproductively isolated from surface populations in a manner that matches the biological species concept. On this grounds, *A. kosswigi* from the subterranean Reka River has been described as a separate species (Verovnik et al., 2009). Based on similar evidence presented by Konec et al. (2015), we propose here to raise the taxonomic status of *A. aquaticus infernus*, the Mangalia subterranean population, to full species. Other new species are likely to follow. Although good species, they are phylogenetically still part of a single large clade dominated by surface *A. aquaticus*. We therefore refer to this model system as the ***Asellus aquaticus* species complex.** We use asellus, hereafter, as a common name to denote any member of the *Asellus aquaticus* species complex.

## Subterranean populations

Distinct subterranean populations evolved several times independently and can be grouped according to their affinity with one of the major clades. The naming used in Table 14.1 is used throughout the rest of the text.

The three **subterranean populations of the Central Europe Clade** are scattered over the profusely ramified mitochondrial DNA genealogy and a vast geographic area (Konec et al., 2015; Sworobowicz et al., 2020). Their large mutual distances in geographic and phylogenetic space (Fig. 14.2) imply independent evolutionary origins from different local surface

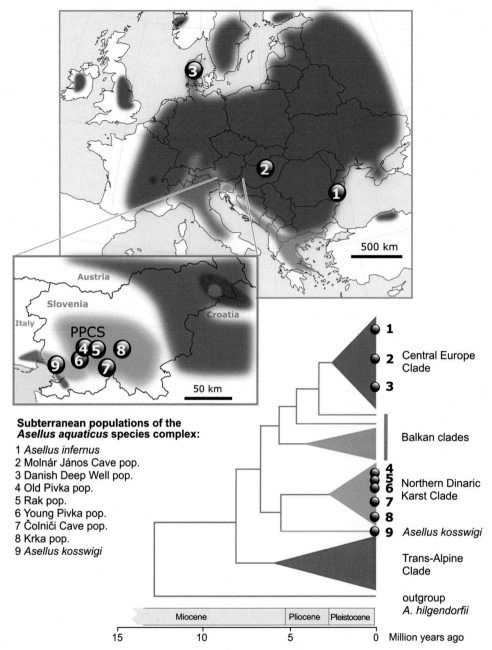

**FIGURE 14.2**  Mitochondrial phylogeography and taxonomic structure of the *Asellus aquaticus* species complex. Subterranean populations and taxa are numbered and marked by dots. Populations 4, 5, and 6 inhabit different sections of the Postojna-Planina Cave System (PPCS) and partially overlap. It should be noted that speciation of subterranean species *A. kosswigi* and *A. infernus* by budding off from *A. aquaticus* renders the mitochondrial phylogeny of the ancestral species paraphyletic. The phylogenetic position of the Danish Deep Well population is hypothetical; only Danish surface individuals were sequenced. *Credit: Maps and tree are simplified, based on data from Konec et al. (2015), Sworobowicz et al. (2015, 2020), Verovnik et al. (2004, 2005) as well as personal information by A. S. Reboleira. The dating of branching events is from Sworobowicz et al. (2015) and complements the timing given in Table 14.1.*

**TABLE 14.1** Subterranean populations of the *Asellus aquaticus* species complex.

| Population/ taxon | Distribution/habitat | Estimated age (million years)[a] | Degree of evolutionary change[b] | Remarks | Source |
|---|---|---|---|---|---|
| **Central Europe Clade** | | | | | |
| *A. infernus* (Turk-Prevorčnik and Blejec, 1998) | Sulfidic thermal aquifer with chemoautotrophic production at Mangalia, Romania | 0.2–3.8 | III | Raised here to full species based on evidence of reproductive isolation in sympatry | Turk-Prevorčnik and Blejec (1998); Konec et al. (2015) |
| Molnár János Cave | Thermal aquifer in Budapest, Hungary | <0.1 | II | | Pérez-Moreno et al. (2017) |
| Danish Deep Well | Drinking water wells in Denmark | Unknown | II (preliminary estimate) | Clade affinity predicted based on geographic position | A. S. Reboleira, pers. comm. |
| **Northern Dinaric Karst Clade** | | | | | |
| Old Subterranean Pivka | Lower section of the Pivka part of PPCS, Slovenia | 0.2–1.3 | III | Described as *A. a. cavernicolus* Racovitza (1925) | Verovnik et al. (2004); Konec et al. (2015) |
| Subterranean Rak | Rak part of PPCS and associated groundwater bodies, Slovenia | Unknown, but probably same range as Old Subterranean Pivka | III | Described as *A. a. cavernicolus* Racovitza (1925) | Verovnik et al. (2004); Konec (2015) |
| Young Subterranean Pivka | Upper section of the Pivka part of PPCS, Slovenia | <0.01 | I | *In statu nascendi*, genetically inseparable from adjacent surface population | Verovnik et al. (2004); Konec (2015) |
| Čolniči Cave | Upper subterranean section of Rak River (local name Obrh), Slovenia | Unknown | III (preliminary estimate) | Small and isolated population | Konec (2015) |
| Subterranean Krka | Subterranean Krka River System, Slovenia | <0.8 | II | Described as *A. a. cyclobranchialis* Sket (1965) | Verovnik et al. (2004) |
| ***A. kosswigi* Clade** | | | | | |
| *A. kosswigi* Verovnik et al. (2009) | Subterranean Reka/ Timavo River, Slovenia and Italy | 0.8–4.3 | III | No extant surface sister lineage | Verovnik et al. (2009); Konec et al. (2016) |

[a]*Estimates refer to possible time since cave colonization calculated as interval between separation from closest surface relative and coalescence time of exclusively subterranean haplotypes.*

[b]*Approximate classification according to the degree of evolutionary change. Level I: body depigmentation partial or incomplete and polymorphic, no eye depigmentation; level II: substantial body depigmentation throughout population, eye pigmentation still present or polymorphic, detectable changes in several other traits; level III: fully depigmented body and eye region, significant changes in numerous other traits.*

*PPCS, the Postojna-Planina Cave System.*

populations. They are relatively young, i.e., of Pleistocene age. *A. infernus* is part of a thoroughly explored chemoautotrophic subterranean community (Sarbu et al., 1996) and has been used in comparative studies of subterranean evolution. The Molnár János Cave population is a new discovery that is already taking a visible role in the model system (Pérez-Moreno et al., 2018). In 2019, a third subterranean population was discovered in drinking water wells in Denmark (A. S. Reboleira, pers. comm). This find came as a surprise because the area is not otherwise known for a specialized subterranean fauna and has been under direct impact of the last glacial maximum.

In contrast to the previous group, all subterranean populations of the **Northern Dinaric Karst Clade** are concentrated in an area measuring about 50 km in diameter, and even the surface range of this clade is less than 100 km. The entire clade has probably evolved in isolation from the rest of the species since the Pliocene, while subterranean populations are of Pleistocene age. There are at least five distinct subterranean populations that inhabit different karstic hydrological systems or different sections of sinking rivers (Table 14.1, Fig. 14.2). All of these populations are characterized by private COI haplotypes derived from different parts of the surface haplotype network, but they have an unresolved, multifurcated phylogeny. They also form distinct microsatellite clusters indicating that they have evolved independently and in isolation from each other (Konec, 2015). In Planina Cave of the PPCS, two distinct populations — the Old Subterranean Pivka and the Subterranean Rak populations — came into secondary contact after a major redirection of subterranean flows about 50 KYA (Šušteršič et al., 2003). Despite being in physical contact for thousands of generations and despite some occasionally observed hybridization, both populations have retained genetic and ecological individuality thus underscoring their evolutionary independence. The Young Subterranean Pivka population is probably no older than 100 years and co-inhabits the Pivka part of the PPCS with the Old Subterranean Pivka population but does not interbreed with it. It should be noted that all these subterranean populations have evolved from a relatively small clade and are therefore sharing essentially the same ancestral genetic stock still present in and around the Poljes of the northern Dinaric Karst.

The population from the **subterranean Reka River**, *A. kosswigi*, has no extant surface sister lineage. It forms an independent clade, with the entire Northern Dinaric Karst Clade as its closest relative. Molecular dating with COI sequences suggests it is up to four million years old and hence the oldest of all subterranean asellus populations. It is in secondary contact with different surface populations at the resurgence and possibly at the sink side, but does not interbreed with them (Verovnik et al., 2009; Konec et al., 2016). This evidence led to the realization that several local subterranean populations have become stable, reproductively isolated genetic entities following independent evolutionary trajectories.

## Phenotypic evolution of subterranean populations

### Morphology

The earliest mention of troglomorphism (i.e., the morphological adaptation of subterranean animals *sensu* Christiansen, 1962) in asellus can be traced back to the first descriptions of more or less depigmented populations, mostly treated as 'varieties', forms, or subspecies.

Their reduced or absent pigmentation and eyes were constantly emphasized, but detailed studies of pigmentation variation have been made only in populations along the subterranean Pivka River and in *A. kosswigi* (De Lattin, 1939; Kosswig, 1939; Kosswig and Kosswig, 1936, 1940). High polymorphism of body and eye pigmentation as well as eye structures were demonstrated in the Young Subterranean Pivka population, while in the Old Subterranean Pivka population and *A. kosswigi* only the degenerated eye structure varies. The early experimenters' expectation that such prominent variation is mediated by multiple genes was recently confirmed for the Old Subterranean Pivka population (Protas et al., 2011). The subterranean Krka System is inhabited by less variable individuals with body pigmentation reaching from homogenously pale-greyish to totally depigmented, and mostly pigmented eyes with a few depigmented exceptions (Sket, 1965, 1985, 1994).

Other troglomorphic traits were first described without any comparative information of their actual extent based on a small number or even a single individual (e.g., Racovitza, 1925; Stammer, 1932; Birštejn, 1951; Karaman, 1952; Schneider, 1887). Sket (1965) was the first to provide numerical data on 20 traits for simple cross-trait comparisons of subterranean and surface ecomorphs. Additional populations and traits have been addressed in extensive morphometric studies with multivariate statistical analyses that began at the turn of the millennium (Turk et al., 1996; Turk-Prevorčnik and Blejec, 1998; Prevorčnik et al., 2004): 59 traits covered the body and appendage proportions, and cuticular and sensory structures elaborations (listed in Prevorčnik et al., 2004) of adult males from multiple surface and subterranean localities. These comparative studies revealed: (1) a great complexity of troglomorphies based on the unique combination of supposedly constructive, regressive and paedomorphic traits specific to each subterranean population, and (2) a considerable variation in these traits within and between these populations, supporting observations of Kosswig and Kosswig (1940) and Ludwig (1942). Based on the ability to separate four totally depigmented subterranean populations from all others, 24 traits were recognized as "troglomorphic." Additionally, a decrease in number, length, and sharpness of spines on the male gonopod was observed in depigmented males, but not quantified. The analysis of quantitative trait change in the Old Subterranean Pivka population and *A. infernus* relative to their reference surface populations showed a dramatic change of overall morphology at both sites (Konec et al., 2015). Of 59 traits 18 changed in a parallel direction and represent candidates for narrow-sense troglomorphies, waiting to be examined for their adaptive function(s) in subterranean environment. Two traits changed in opposing directions, possibly pointing to specifics of the local environment. In two traits mean values remained unchanged while their variance was significantly higher in subterranean populations, possibly indicating relaxed selection. No conclusive interpretation was feasible for the 30 traits that changed in only one of the subterranean populations.

We provide an updated pattern of trait change in four totally (Old Subterranean Pivka and Rak populations, *A. kosswigi*, *A. infernus*) and two partially depigmented subterranean populations (Subterranean Krka and Young Subterranean Pivka), as opposed to their reference surface populations (Table 14.2; based on data from Prevorčnik, 2002 reanalyzed as in Konec et al., 2015). The supposed reduced traits and paedomorphies both outnumber the supposed constructive traits. The former two are mainly represented by the shortening of certain setae on walking legs, pleotelson, and uropod, less numerous robust spines on the claws of pereopods, and less numerous plumose and simple setae on pleopods. At the same time,

**TABLE 14.2**  Changes in quantitative traits of subterranean asellus populations in relation to their reference surface populations.

| Trait | Trait tag | TN | OPIV | RAK | KOSS | INF | KRK | YPIV |
|---|---|---|---|---|---|---|---|---|
| | | | totally depigmented | | | | partially depigmented | |
| Body **pigmentation** | | | - | - | - | - | +,- | +,- |
| Eye **pigmentation** | | | - | - | - | - | +,- | +,- |
| Pereopod IV art. 5, superior robust **setae** N | PE45SU | c,t | ↑ | ↑ | ↑ | ↑ | ↑ | ↑ |
| Pereopod VII art. 5, superior robust **setae** N | PE75SU | c | ↑ | ↑ | ↑ | ↑ | ↑ | ns |
| Pereopod IV art. 5, distal **setae** N | PE45D | c,t | ↑ | ↑ | ↑ | ↑ | ↑ | ns |
| Pereopod IV art. 5, inferior **setae** N | PE45BN | c,t | ↑ | ↑ | ↑ | ↑ | ns | ns |
| Aesthetascs **rL**(mean aesthetasc L vs antenna I rL) | *A1A* | c | ↑ | ↑ | ↑ | ↑ | ns | ↓ |
| Antenna II **rL** (vs body L) | *A2* | c | ↑ | ↑ | ns | ↑ | ↑ | ↑ |
| Antenna II, flagellum **articles** N | A2N | c | ↑ | ↑ | ns | ↑ | ↑ | ns |
| Pereopod VII art. 5, inferior **setae** N | PE75BN | c,t | ↑ | ↑ | ns | ↑ | ↑ | ns |
| Pereopod IV art. 4, distal **setae** N | PE44D | c,t | ↑ | ↑ | ns | ↑ | ↑ | ↓ |
| Pereopod VII art. 4, distal **setae** N | PE75D | c,t | ↑ | ↑ | ↑ | ns | ns | ns |
| Uropod endopodit, plumose **setae** N | UBN | c,t | ↑ | ↑ | ↑ | ns | ns | ↓ |
| Pereopod VII, art. 5 **rL** (vs art. 3 L) | *PE753* | c | ↑ | ↑ | ↑ | ↓ | ↑ | ↑ |
| Pereopod VII art. 5, superior plumose **setae** N | PE75BU | c,t | ↑ | ↑ | ↑ | ↓ | ↑ | ns |
| Pleopod V, exopodit **rW** (vs exopodit L) | *PL5* | | ↑ | ↑ | ↑ | ↓ | ns | ↓ |
| Pereopod I, art. 5 **rW** (vs art. 5 L) | *PE11* | p,t | ↓ | ↓ | ↓ | ↓ | ↓ | ↓ |
| Pereopod VII art. 5, longest robust **seta rL** (vs art. 5 L) | *PE75S* | r,t | ↓ | ↓ | ↓ | ↓ | ↓ | ↓ |
| Uropod endopodit, longest robust **seta rL** (vs endopodit L) | *US* | r,t | ↓ | ↓ | ↓ | ↓ | ↓ | ↓ |
| Uropod, exopodit **rL** (vs endopodit L) | *U23* | p,t | ↓ | ↓ | ↓ | ↓ | ↓ | ↓ |
| Pereopod IV art. 4, longest distal **seta rL** (vs art. 4 L) | *PE44S* | r,t | ↓ | ↓ | ↓ | ↓ | ↓ | ns |
| Pereopod VII art. 3, longest distal robust **seta rL** (vs art. 3 L) | *PE73S* | r,t | ↓ | ↓ | ↓ | ↓ | ↓ | ns |
| Pleopod I exopodit, concavity **rD** (vs exopodit W) | Z | p | ↓ | ↓ | ↓ | ↓ | ns | ↑ |
| Pleopod I exopodit, plumose **setae** N | PL1F | r,t | ↓ | ↓ | ↓ | ↓ | ↑ | ↓ |
| Antenna I **rL** (vs body L) | *A1* | r,t | ↓ | ↓ | ↓ | ↓ | ns | ns |
| Pereopod I, art.3 **rW** (vs art. 3 L) | *PE13* | | ↓ | ↓ | ns | ↓ | ns | ↓ |
| Pereopod VII art. 6, inferior robust **setae** N | PE76N | r,t | ↓ | ↓ | ↓ | ns | ns | ↓ |
| Pleopod II exopodit, **setae** N | PL2B | r,t | ↓ | ↓ | ↓ | ns | ↑ | ↓ |
| Pereopod I art. 6, inferior robust **setae** N | PE1N | r,t | ↓ | ↓ | ↓ | ns | ↑ | ↓ |
| Pereopod IV art. 5, longest distal seta **rL** (vs art. 5 L) | *PE45S* | r | ↓ | ↓ | ↓ | ns | ns | ns |
| Antenna I, flagellum **articles** N | A1N | r,t | ↓ | ↓ | ns | ↓ | ↑ | ↓ |
| Pleotelson, longest marginal **seta rL** (vs shortest marginal seta L) | *PTT* | r | ↓ | ↓ | ↑ | ↓ | ns | ↓ |
| Pereopod I art. 6, inferior robust **setae rL** (vs art. 5 L) | *PE1S* | | ↓ | ↑ | ↑ | ns | ns | ns |
| Pleopod II, protopodit **rW** (vs protopodit L) | *PL21* | | ↓ | ns | ↑ | ↑ | ns | ns |
| Pereopod IV art. 5, inferior robust **setae** N | PE45SN | t | ↑ | ns | ↓ | ↑ | ↑ | ns |
| Pleopod I protopodit, retinaculum **denticles** N | PL1R | | ns | ↑ | ↓ | ↑ | ns | ns |
| Pereopod VII art. 6, longest inferior robust **seta rL** (vs art. 6 L) | *PE76S* | | ↓ | ↑ | ↓ | ns | ↓ | ↓ |
| Pleotelson **rW** (vs pleotelson L) | *PT* | | ns | ↑ | ↓ | ↓ | ↓ | ns |
| Pleotelson marginal **setae** N | PTS | | ↓ | ↑ | ↓ | ↑ | ↓ | ↓ |
| Uropod **rL** (vs body L) | *U* | t | ↑ | ns | ↓ | ↓ | ↑ | ↑ |
| Pereopod VII **rL** (vs body L) | *PE7* | | ↑ | ↓ | ↓ | ns | ↑ | ↑ |

Only traits differing significantly in at least three totally depigmented subterranean populations are listed, along with pigmentation data. Trait tags are adopted from Prevorčnik et al. (2004). Trait presence and absence are marked with »+« and »−«, respectively. The direction and significance of change in trait mean value are shown by arrows (increase: ↑ − $P \leq .01$, ↑ − $P \leq .05$; decrease: ↓ − $P \leq .01$, ↓ − $P \leq .05$; ns denotes nonsignificant change). Abbreviations used in traits are: art − article, r − relative (written in italics), N − number, L − length, W − width, D − depth. The supposed nature of trait (TN) *sensu* Prevorčnik et al. (2004) is: constructive (c), reductive (r), paedomorphic (p), troglomorphic (t). Subterranean populations and their reference surface populations in parentheses: OPIV − Old Subterranean Pivka from Planina Cave (Planina Polje); RAK − Subterranean Rak from Planina Cave (Rakov Škocjan Polje); KOSS − *A. kosswigi* from Grotta di Trebiciano (lake Doberdò); INF − *A. infernus* from well F4 and F8 in Mangalia (Cismigiu spring near Bucuresti and Limanu Bridge spring in Mangalia); KRK − Subterranean Krka, individuals with pigmented eyes from cave Viršnica (water ditch Curnovec near Ljubljana); YPIV − Young Subterranean Pivka from caves Črna jama and Pivka jama (Postojna Cave entrance).

simultaneous multiplication of some setae on walking legs and uropod occur as the main constructive feature. Unfortunately, the adaptive value, if any, of the described changes in the subterranean asellus, remains enigmatic. The elongation of appendages, generally the most common constructive troglomorphic trait (Christiansen, 1962), is quite variable. Only second antennae, which are elongated in four out of five analyzed subterranean populations, show the expected pattern. The elongation corresponds to an article number increase, except in the Young Subterranean Pivka population where elongation is due to longer antennal articles. Walking legs in all subterranean populations have elongated distal articles, which might be an indication of faster locomotion (Hildebrand, 1985). In crustaceans, the appendage elongation is usually attributed to natural selection acting to enhance nonvisual sensory perception (Culver et al., 1995). Although aesthetascs (i.e., extra-optic sensory structures of crustaceans) are longer in all totally depigmented asellus populations, the first antenna that bears them is always relatively shorter than in the surface ecomorph, mainly due to the article number decrease. Neither energy economy nor pleiotropy has been verified as a possible evolutionary mechanism. The same applies to K-selection as a possible basis for the supposed paedomorphies: i.e., slenderer prehensile first leg, shallower concavity on first pleopod and homonomous uropod. It seems that morphological evolution in subterranean asellus represents a mosaic of convergent and divergent traits, at least some of which probably reflect the heterogeneity within and between subterranean environments these isopods inhabit. Finally, a recent morphological study by Balázs et al. (2021) analysed not just males but also females in six cave and nine surface populations. Their most salient finding was that eight functional traits showed divergent sexual dimorphism between the cave and surface ecomorphs. Thus, if we want to fully understand the evolution of subterranean-related morphology in asellus, we should ideally consider both sexes.

## Behavior

Behavior is probably the least known aspect of the biology of asellus. Many of its behaviors are likely exaptations that facilitate evolutionary transitions to subterranean life, e.g., nocturnal activity, photophobia, successful mating in darkness, and opportunistic feeding. Further behavioral changes are expected to evolve in subterranean populations, similarly to the cavefish *Astyanax mexicanus* (Kowalko, 2020). So far, behavioral traits of subterranean asellus are poorly understood.

It is not surprising that feeding-related behaviors were among the first behaviors investigated in a comparative way, as change in food availability is considered one of the main drivers of subterranean evolution (Hüppop, 2000). Mösslacher and Creuzé des Châtelliers (1996) found that *A. infernus* fed roughly 10 times less and moved roughly 10 times more than the surface ecomorph from Austria. The authors interpreted the high movement activity of the former as extensive food-searching behavior, presumably an adaptation to low and patchy food supply in the wells they inhabit. On the other hand, Herczeg et al. (2020) observed that individuals from the Molnár János Cave population were more likely to feed than the surface ones from nearby locations and ascribed this to increased boldness due to the negligible predation pressure in this specific cave. Additionally, Herczeg et al. (2022) found no differences between ecomorphs regarding plasticity of foraging behavior.

Negligible predation pressure was also put forth by Berisha et al. (2022) to explain the higher movement activity of isopods from the same subterranean population compared to their surface counterparts, as well as the male-biased sexual dimorphism of this behavior in the subterranean population.

Asellus are frequent prey items. The surface ecomorph typically avoids predators and other adverse environmental factors by sheltering within decaying organic material and gravel, in crevice-like tight and dark spaces to which it is guided by strong positive thigmotaxis (Janzer and Ludwig, 1952). Because unfavorable factors like light and predation pressure are absent or diminished in groundwater, natural selection should favor a reduction of shelter-seeking behavior. This was indeed demonstrated by Fišer et al. (2019) for the Old Subterranean Pivka population. Compared to surface individuals from Planina Polje subterranean ones spent less time beneath the shelter providing thigmotactic stimulation. In both ecomorphs, males sheltered less than females, possibly revealing their higher tendency for risk-taking behaviors. However, individuals from the Subterranean Rak population sheltered similarly as surface individuals from Cerknica Polje. Using a slightly different set-up, Horváth et al. (2021) demonstrated reduced sheltering behavior also in the Molnár János Cave population. The reasons for unchanged shelter-seeking behavior in the Subterranean Rak population remain elusive, but thigmotactic shelter-seeking is apparently another trait that changes in some but not in all subterranean populations and defies the paradigm of strictly parallel phenotypic evolution in groundwater. The surface ecomorph also shows strong aggregation behavior, another antipredator strategy expected to be reduced upon diminished predation pressure in groundwater. However, Horváth et al. (2021) found no such pattern in the Molnár János Cave population.

Benthic walkers like asellus require a grippy substrate for normal locomotion. When placed in a dish with a half-smooth half-rough bottom, surface individuals from Planina and Cerknica Poljes as well as individuals from the Old Subterranean Pivka and Subterranean Rak populations clearly preferred the rough substrate (Fišer et al., 2019). Yet, the preference of the latter populations was significantly stronger. Unlike their surface counterparts, subterranean individuals would often grip even the tiniest scratch in a smooth-bottomed dish with solely the distal claw of one pereopod. Such manic preference for grip-providing substrate is obviously an important behavior in groundwater.

Reproductive behaviors like mate-searching and copulation can be executed in total darkness in both asellus ecomorphs. As females are receptive only for a short time after molting, the male secures its chances for copulation by firmly holding the female already several days prior her molt; a behavior called precopulary mate-guarding. In several subterranean crustaceans, like the isopod *Stenasellus virei* or amphipods *Niphargus*, this behavior is severely shortened or absent (Magniez, 1978; Marin and Palatov, 2019). However, in subterranean asellus mate-guarding is present and it seems not to be shortened either. Lattinger-Penko (1979) observed Old Subterranean Pivka individuals and found that mate-guarding lasted about 2 weeks in most cases, similar to the timeframe reported for the surface ecomorph. Apart from that, it has long been recognized that surface and subterranean ecomorphs can mate with each other (Baldwin and Beatty, 1941) and the same was more recently demonstrated for individuals of distinct subterranean populations (Re et al., 2018). Future experiments might reveal if ecomorph assortative mating nevertheless exists between surface and subterranean asellus. This phenomenon often evolves during ecological and parallel speciation

(Schluter, 2001; Nosil, 2012) and has been shown also for the recently evolved surface stonewort ecomorph in relation to the classical surface ecomorph from Swedish lakes (Eroukhmanoff et al., 2011).

## Physiology

Many subterranean animals live on limited resources and their metabolic rates have been shown to be reduced (Hüppop, 2000). In *A. infernus*, however, higher metabolic rates were demonstrated, supposedly due to its higher motility (Mösslacher and Creuzé des Châtelliers, 1996). Although it consumed less oxygen per individual per hour, the authors stressed that the mass specific oxygen consumption (respiration rate) should be considered instead; subterranean individuals had a higher respiration rate than the surface ones. The Subterranean Rak individuals from cave Zelške jame did not differ significantly from surface individuals from Cerknica Polje in the metabolic activity and oxygen consumption rate (Simčič and Sket, 2019). The only case of lower metabolic but also locomotor activity was indicated by the reduced activity of enzymes acetylcholinesterase and glutathione S-transferase in the Old Subterranean Pivka population as opposed to the surface ecomorph from Planina Polje (Jemec et al., 2017). Lower seasonal fluctuation in both enzyme activities in subterranean individuals was consistent with that observed for their lipid, carbohydrate, and protein energy reserves (Zidar et al., 2018). The reserves, however, did not differ significantly between the ecomorphs. Currently, there is insufficient data to address questions of evolutionary parallelism and convergence of physiological traits.

## Life history

Life history characteristics of the subterranean asellus are mostly unknown. Lattinger-Penko (1979) reported similarities between the Old Subterranean Pivka population and the surface ecomorph regarding the duration of molt, intermolt, marsupial development, and number of descendants in one brood. In a preliminary analysis, Prevorčnik (2002) also found approximately the same number of eggs in marsupia of females from Old Subterranean Pivka and Subterranean Rak populations and surface females from Planina and Cerknica Poljes (from about 60–80 eggs in 6 mm to about 130–150 eggs in 9 mm females). Hatchling body size is the only trait that has been observed to differ significantly between the subterranean and surface ecomorphs so far (Mojaddidi et al., 2018); hatchlings from the Subterranean Rak population are bigger than surface hatchlings of the same age. Further investigations are needed for more rigorous comparisons of life history traits.

## Raising and breeding in the laboratory

Raising and breeding organisms in the laboratory is a necessary prerequisite for many types of studies including those on traits' development, genetic background and plasticity. Fortunately, the surface ecomorph of asellus is relatively easy to breed in the lab, making it a commonly used model in fields other than groundwater biology, such as ecotoxicology

(O'Callaghan et al., 2019). It can be raised in diverse conditions, but commonly 5—20 individuals are housed in medium-sized (0.5 L) plastic or glass containers at about 12°C in filtered tap water or standardized artificial freshwater medium (such as M4). Aeration is not necessary but regular water changes, e.g., once every 2 weeks, are a good practice. They can be raised under a diurnal light cycle or in constant darkness. Decaying leaves (especially those of black alder, *Alnus glutinosa*) are used as a food source. Standard Petri dishes (90 × 15 mm) are practical to keep single individuals, pairs used for crosses, juvenile siblings, and even whole families until the offspring gets too big. The subterranean ecomorph can be kept in similar conditions but generally is not as robust as the surface ecomorph. So far, the most success in rearing and breeding the subterranean ecomorph has been with the Subterranean Rak population (Mojaddidi et al., 2018).

## Genetic basis of subterranean-related traits

A genetic mapping approach (Fig. 14.3A) and/or phenotype-genotype association tests are ways of identifying genomic regions responsible for subterranean traits. This approach has been used extensively in the cavefish *Astyanax mexicanus*, for which morphological, behavioral, and physiological traits have been mapped and in some cases even the genes responsible have been identified (reviewed in Casane and Rétaux, 2016; Kowalko, 2020). Application of genetic mapping is, however, a fairly recent addition to the studies of subterranean evolution in asellus. Eye, pigment, and antennal traits have thus far been examined in either one or two subterranean populations, while additional traits wait to be examined across multiple other populations.

Mapping studies in asellus first started with the goal of identifying genomic regions responsible for eye and pigment loss in the Old Subterranean Pivka population (Protas et al., 2011). Eight linkage groups were found, matching the eight pairs of homologous chromosomes of *A. aquaticus* (Salemaa, 1979) (Fig. 14.3B). Backcrosses displayed diverse eye and body pigmentation and ommatidia phenotypes (Fig. 14.3C—F). For each of the four qualitative pigmentation traits examined, i.e., presence versus absence of eye pigment, red versus orange/brown eye pigment, orange versus red/brown eye pigmentation, and stellate versus diffuse body pigmentation pattern, a single genomic region was identified responsible for a majority of the variation. Eyes of backcrosses had either normal ommatidia, fragmented ommatidia of various sizes, or completely lacked even the ommatidial fragments. A single genomic region was mapped for each examined eye structure trait, i.e., presence versus absence of ommatidia and eye size. Genetic basis of another common trait related to subterranean life, i.e., antenna elongation, was examined more recently in the Subterranean Rak population by Mojaddidi et al. (2018) who found it was associated with two genomic regions.

One striking result was that several of the four regions mapped so far coincided for multiple traits (Fig. 14.3B). For example, stellate versus diffuse body pigmentation pattern and eye presence versus absence were located in the same general region. Additionally, the region mapped for eye size was at the same general location as the region responsible for eye pigment presence versus absence (Protas et al., 2011). Likewise, the region responsible for presence versus absence of eye pigment and the region responsible for red versus orange/

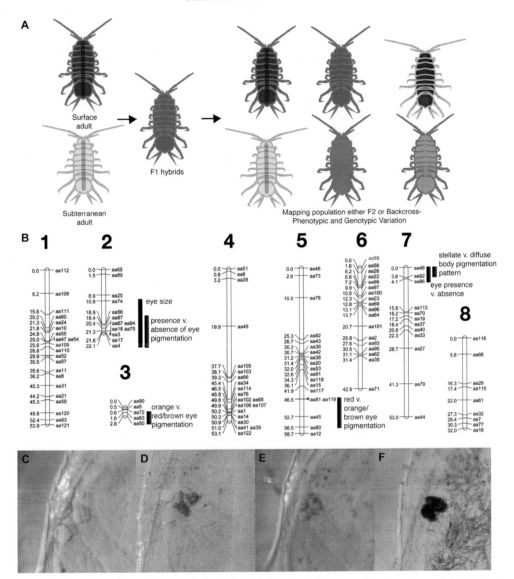

**FIGURE 14.3** Linkage mapping approach in asellus populations. (A) Schematic of generating an F2 or backcross population between a surface and subterranean ecomorph. (B) Linkage map generated using a backcross population from the Old Subterranean Pivka population. Vertical black bars mark regions mapped for eye and pigment traits. Vertical red bars indicate that a genetic marker within a region was associated to the same traits in the Subterranean Rak population. Genetic markers with a red square were associated with number of antennal articles in the Subterranean Rak population. (C–F) Eyes of F2 hybrids from the Subterranean Rak population. *Credit: (A) Schematics courtesy of Dennis Sun. (B) From Protas et al. (2011), adding data from Re et al., 2018 and Mojaddidi et al. (2018). (C–F) From Re et al., 2018.*

brown eye pigment were significantly associated with antennal article number (Mojaddidi et al., 2018). Whether this pattern is due to different but linked genes responsible for each trait, or same genes responsible for multiple traits, remains unknown. However, clustering of traits in the same genomic regions is a common theme in the evolution of natural populations (Protas et al., 2008; Miller et al., 2014; Morris et al., 2019). In *Astyanax mexicanus*, genetic association of eye and pigment traits has been documented (Protas et al., 2008; Gross et al., 2016), similarly to asellus. Nevertheless, additional subterranean species must be tested to see if this specific association is widespread.

Another interesting outcome of mapping experiments in asellus was that there were multiple ways to reduce both pigment and eyes within a single subterranean population (Protas et al., 2011). For example, one locus was mapped for presence versus absence of eye pigment. However, when loci for red and orange eye pigment both showed homozygous subterranean genotypes, individuals also had unpigmented eyes. In addition, one locus was found for presence versus absence of ommatidia, and another locus for eye size. One explanation for these apparent redundancies in mechanisms is that pigment and eyes are not needed in subterranean environments and over time degenerate due to accumulation of selectively neutral mutations. And, in fact, certain subterranean populations are variable in the degree of reduction of the eye phenotype. Another explanation is that the alleles causing orange and red pigment may be advantageous for subterranean life through pleiotropy. Support for the latter explanation is that these alleles are found in both subterranean populations examined (see below). Multiple ways of pigment reduction were also identified in *Astyanax mexicanus* (Protas et al., 2006; Gross et al., 2009; Gross and Wilkens, 2013; Stahl and Gross, 2015) in which mutations in one of the genes (*oca2*) causing pigment reduction may have beneficial pleiotropic consequences on behavior (Bilandžija et al., 2013, 2018).

More recently, phenotype-genotype association tests were performed in the Subterranean Rak population. Markers within regions responsible for pigmentation phenotypes and presence versus absence of ommatidia in the Old Subterranean Pivka population were significantly associated for the same or similar phenotypes in the Subterranean Rak population, while eye size was not investigated (Re et al., 2018). A complementation test further revealed that at least one of the genes responsible for pigment loss is the same in both subterranean populations. These results defy the accumulation of neutral mutations as an explanation for regressive evolution of eyes and pigment as one would not expect these traits to degenerate the same way and with redundant mechanisms in independently evolved populations. Thus, gene pleiotropy seems a likelier explanation. Another reason for the same regions being responsible for the same traits is that the observed variation might have been present in the ancestral surface population and was repeatedly fixed in multiple subterranean populations. This seems a likely scenario as we have observed red and orange eyes in surface individuals raised over multiple generations in the lab and think that alleles responsible for red and orange eyes exist at low frequencies in the current surface populations. Such involvement of standing genetic variation in evolutionary change has been documented in multiple species (reviewed in Seehausen, 2015; Peichel and Marques, 2017).

## Evolutionary development (evo-devo)

Asellus is a convenient evo-devo model as females can have over 100 eggs in a single brood allowing collection of a large number of embryos. Furthermore, females can be anesthetized, embryos removed from an external brood pouch and raised in vitro until hatching. So far, embryological studies have mostly focused on the surface ecomorph (Dohrn, 1867; Vick et al., 2009; Vick and Blum, 2010) and little is known about rearing and embryology of the subterranean ecomorph. Sket (1965) was the first to compare wild caught surface and subterranean embryos. Those from the Old Subterranean Pivka and Subterranean Rak populations had already unpigmented body and eyes, just like the adults from the same subterranean populations.

More recently, comparisons between lab reared surface and subterranean embryos from the Subterranean Rak population have been performed by Mojaddidi et al. (2018) who investigated whether the phenotypic differences of eye and pigment loss as well as second antenna elongation were already established by the end of embryonic development. Both pigmentation and ommatidia were not observed in subterranean embryos at any developmental stage (Fig. 14.4). Similarly, in *Astyanax mexicanus*, albino subterranean individuals do not show pigmentation at any developmental stage. However, eye development in *Astyanax mexicanus* is similar in subterranean and surface forms early on in development and then degenerates in the subterranean form (reviewed in Protas and Jeffery, 2012). In asellus, we do not yet know if eyes begin to develop and then degenerate but future experiments including sectioning of the eye region and/or using photoreceptor markers will investigate the process and timeline of eye loss. In addition, the subterranean asellus hatchlings examined were bigger and had longer antennae. However, the relative length of the antennae to body size was not increased and therefore the relatively longer antennae of adult subterranean individuals must be established post embryonically. Multiple other features have not yet been compared between surface and subterranean embryos. Interesting characters to study would be weight, protein and fat content, and the timing of development of different appendages and bristles. A resource that would be helpful in the establishment of asellus as a developmental model would be a staging table similar to those of the crustacean models *Porcellio scaber*, an isopod, and *Parhyale hawaiensis*, an amphipod (Browne et al., 2005; Wolff, 2009). In addition to morphological examination of embryos, embryonic transcriptomes have been sequenced and analyzed; these are discussed in the next section.

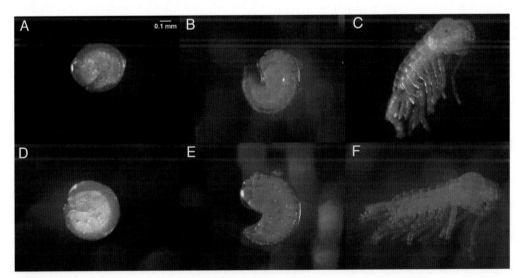

**FIGURE 14.4**  Surface (A−C) and subterranean (D−F) asellus embryos in progressively advancing stages of development. *Credit: From Mojaddidi et al. (2018).*

Establishment of tools used in asellus evo-devo research is still in very early stages (Table 14.3). One of the traditionally most utilized techniques, i.e., in situ hybridization, has been performed in the surface ecomorph and focused on Hox genes (Vick et al., 2009; Vick and Blum, 2010). Theoretically, comparative in situ hybridization should be possible in surface and subterranean embryos and could be used to test candidate genes involved in subterranean phenotypes as has been done on many occasions for *Astyanax mexicanus* (reviewed in Jeffery, 2009). Also important are functional techniques, such as RNAi, RNA overexpression, and genome editing, which allow for testing if particular candidate genes are in fact responsible or rather, affect a particular trait by knocking out a gene or altering the expression level of a gene (Mallarino and Abzhanov, 2012). All of these techniques have been established in *Astyanax mexicanus* (Yamamoto et al., 2009; Ma et al., 2015; Klaassen et al., 2018) though none have been established yet for asellus (Table 14.3). Once this is achieved, it will be possible to investigate the role of specific genes and pathways in the evolution of subterranean characteristics in this species as well.

## Comparative transcriptomics

Transcriptome sequencing is becoming a commonly used method for studying the evolution of subterranean animals (Pérez-Moreno et al., 2016) and thus obtained resources are important for further development of asellus as a model system. First, transcriptomes are the main source of sequence information. Second, transcriptomic studies on differential expression and allele specific expression can pave the way for an understanding of the genes and pathways responsible for the differences between subterranean and surface ecomorphs. In asellus, transcriptomes were first sequenced from the head of an adult individual from the Old Subterranean Pivka population, Planina Polje surface population, an F1 hybrid between these two populations, and from a pooled sample of embryos and hatchlings from the Planina Polje surface population. The presence of light interacting genes was examined in all of the transcriptomes using phylogenetic-informed annotation (PIA) (Speiser et al., 2014). In the subterranean sample, several members of the light interacting toolkit were found, including *purple, vermillion, Alas1/2, omega crystallin,* and *slowpoke.* Stahl et al. (2015) further used these transcriptome sequences to identify SNPs between the two populations and place additional candidate genes on the linkage map generated by Protas et al. (2011). In addition, in the single hybrid sequenced, allele specific expression was observed in genes

**TABLE 14.3**  Tools commonly needed for evo-devo research. Existence of tools, yes (Y) or no (N), is shown for *Astyanax mexicanus* and asellus (reviewed in: Vick et al., 2009; Vick and Blum, 2010; Protas et al., 2011; Mojaddidi et al., 2018; Jeffery, 2020; Bakovic et al., 2021).

| Model | Set up crosses | Obtain embryos | In situ hybridization | Transcriptome | Genome | RNAi | Overexpression | Genome editing |
|---|---|---|---|---|---|---|---|---|
| *Astyanax mexicanus* | Y | Y | Y | Y | Y | Y | Y | Y |
| asellus | Y | Y | Y | Y | Y* | N | N | N |

* = *draft genome.*

*peroxiredoxin 1 like protein* and *tnf receptor associated protein*, biased toward the subterranean allele, and in *chymotrypsin like protein* biased toward the surface allele.

Whole-body transcriptomes of two adult specimens were also sequenced from both Molnár János Cave and a nearby surface population (Pérez-Moreno et al., 2018). The PIA method was again used to identify phototransduction pathway genes and most of them were expressed though the expression level was not quantified. Two opsins were found, a short and long wavelength sensitive one, and no obvious coding differences were seen that would make the opsins nonfunctional. Expression of functional opsins has been observed in other subterranean animals such as *Astyanax mexicanus*, *Sinocyclocheilus*, and *Gammarus minus*, though in some cases reduced expression was detected (Carlini et al., 2013; Meng et al., 2013; Simon et al., 2019).

Recently, Gross et al. (2020) focused on embryonic transcriptomes of the Subterranean Rak population at the time point when eye and pigment differences appear between surface and subterranean individuals. They compared pooled samples for three broods of surface, subterranean, and their hybrid embryos each and investigated differential gene expression. Among the top 100 differentially expressed genes (50 overexpressed and 50 underexpressed in the subterranean ecomorph), were several genes, e.g., *long wavelength sensitive opsin*, *scarlet*, and *atonal*, that are involved in pigmentation and eye traits. The hybrid samples were used to examine whether any of these top differentially expressed genes also had allele specific expression, which might indicate mutations in *cis*-regulatory regions. Allele specific expression was seen in 45 of these genes including: *pygopus*, *long wavelength sensitive opsin*, *lines*, and *annulin*, which are candidate genes for subterannean phenotypes like eyelessness and appendage elongation.

Although multiple transcriptomic resources are already present for asellus, much remains to be investigated including additional embryonic and postembryonic timepoints as well as additional subterranean populations. Furthermore, genomic resources, which have recently been generated (Bakovic et al., 2021), are a necessary next step as they can be used with mapping techniques to narrow regions responsible for different phenotypes, potentially to the gene responsible. Also, genomic resources could allow for the identification of mutations responsible for different phenotypes, even *cis*-regulatory mutations, which have been shown to be responsible for evolutionary change in multiple species (reviewed in Wray, 2007).

## Conclusions and prospect

Significant progress has been made in the last 3 decades in developing *A. aquaticus* species complex as a model for subterranean evolution and ecology. Much is known about the morphology of adult individuals and the phylogenetic relationships of the populations. Inroads have been made regarding the genetic architecture of certain subterranean traits, the establishment of sequencing data, and the basic framework and tools to study genetic and developmental questions have been established.

However, the finer scale relationships of subterranean and surface populations remain elusive, as do recent and historical population size and the actual timescale of subterranean evolution. New genomic and analytical approaches are being applied to fill this knowledge gap. But also the classical fields of morphology, experimental biology, and ethology hold great potential. We do not yet understand the actual functionality of the numerous dramatic phenotypic changes associated with the evolutionary transition to groundwater. Without this knowledge, it is impossible to understand the process of ecological speciation and the evolving reproductive barriers that are continuously fueling the budding of new and new subterranean species.

The genes and mutations responsible for the differences between subterranean and surface ecomorphs still elude us. To achieve this goal, much progress is needed regarding reliable breeding and culture of subterranean populations, establishment of more sophisticated behavioral analyses, generation of additional genome sequences, and development of functional techniques such as CRISPR genome editing. In addition, one of the most powerful features of this model system is the existence of multiple subterranean populations. Once all of the above experiments have been performed in a single population, they can be translated to other populations to examine similarities or differences in the evolution of subterranean characteristics between populations. And finally, much of what has been so far examined in this species complex is with lab bred animals. To truly understand the ecology of these isopods and to help with its conservation, field experiments must be performed. In sum, asellus has all of the necessary features to be a model in understanding subterranean biology. Much work still needs to be done to establish the tools and techniques to allow for an in depth dissection of the evolution of subterranean characteristics; then, in concert with information from other subterranean models, a better understanding of the evolution of subterranean characteristics can be achieved.

## Acknowledgments

Research reported in this publication was supported by the National Eye Institute of the National Institutes of Health under Award Number NEI R15EY029499 to M.E.P. The content is solely the responsibility of the authors and does not necessarily represent the official views of the National Institutes of Health. The work of Ž.F., S.P., and P.T. was supported by the Slovenian Research Agency through Research Core Funding P1-0184 and research projects N1-0069 and N1-0096. We thank H. Bilandzija and J. Gross for constructive comments that improved this chapter.

## References

Bakovic, V., Cerezo, M.L., Höglund, A., Fogelholm, J., Henriksen, R., Hargeby, A., Wright, D., 2021. The genomics of phenotypically differentiated *Asellus aquaticus* cave, surface stream and lake ecotypes. Molecular Ecology 30 (14), 3530–3547.

Balázs, G., Biró, A., Fišer, Ž., Fišer, C., Herczeg, G., 2021. Parallel morphological evolution and habitat-dependent sexual dimorphism in cave- vs. surface populations of the *Asellus aquaticus* (Crustacea: Isopoda: Asellidae) species complex. Ecology and Evolution 11, 15389–15403.

Baldwin, E., Beatty, R.A., 1941. The pigmentation of cavernicolous animals. Journal of Experimental Biology 18, 136–143.

Berisha, H., Horváth, G., Fišer, Ž, Balázs, G., Fišer, C., Herczeg, G., 2022. Sex-dependent increase of movement activity in the freshwater isopod *Asellus aquaticus* following adaptation to a predator-free cave habitat. Current Zoology, zoac063.

Bilandžija, H., Abraham, L., Ma, L., Renner, K.J., Jeffery, W.R., 2018. Behavioural changes controlled by catecholaminergic systems explain recurrent loss of pigmentation in cavefish. Proceedings of the Royal Society B: Biological Sciences 285, 20180243.

Bilandžija, H., Ma, L., Parkhurst, A., Jeffery, W.R., 2013. A potential benefit of albinism in *Astyanax* cavefish: down-regulation of the *oca2* gene increases tyrosine and catecholamine levels as an alternative to melanin synthesis. PLoS One 8 (11), e80823.

Birštejn, J.A., 1951. Presovodnye osliki (Asellota). In: Pavlovskii, E.N., Sakelberg, A.A. (Eds.), Fauna SSSR, Rakoobraznye, vol. 7. Akademii Nauk SSSR, Moscou, pp. 58−69.

Browne, W.E., Price, A.L., Gerberding, M., Patel, N.H., 2005. Stages of embryonic development in the amphipod crustacean, *Parhyale hawaiensis*. Genesis 42, 124−149.

Carlini, D.B., Satish, S., Fong, D.W., 2013. Parallel reduction in expression, but no loss of functional constraint, in two opsin paralogs within cave populations of *Gammarus minus* (Crustacea: Amphipoda). BMC Evolutionary Biology 13, 89.

Casane, D., Rétaux, S., 2016. Evolutionary genetics of the cave fish *Astyanax mexicanus*. Advances in Genetics 95, 117−159.

Christiansen, K.A., 1962. Proposition pour la classification des animaux cavernicoles. Spelunca 2, 76−78.

Culver, D.C., Kane, T.C., Fong, D.W., 1995. *Adaptation and Natural Selection in Caves: The Evolution of* Gammarus Minus. Harvard University Press, Cambridge.

De Lattin, G., 1939. Untersuchungen an Isopodenaugen. Zoologisches Jahrbuch Der Anatomie 65, 418−468.

Dohrn, A., 1867. Die embryonale Entwicklung des *Asellus aquaticus*. Zeitschrift Für Wissenschaftliche Zoologie 17 (17), 221−278.

Eroukhmanoff, F., Hargeby, A., Svensson, E.I., 2011. The role of different reproductive barriers during phenotypic divergence of isopod ecotypes. Evolution 65 (9), 2631−2640.

Fišer, Ž., Prevorčnik, S., Lozej, N., Trontelj, P., 2019. No need to hide in caves: shelter-seeking behavior of surface and cave ecomorphs of *Asellus aquaticus* (Isopoda: Crustacea). Zoology 134, 58−65.

Gross, J.B., Borowsky, R., Tabin, C.J., 2009. A novel role for *Mc1r* in the parallel evolution of depigmentation in independent populations of the cavefish *Astyanax mexicanus*. PLoS Genetics 5 (1), e1000326.

Gross, J.B., Powers, A.K., Davis, E.M., Kaplan, S.A., 2016. A pleiotropic interaction between vision loss and hypermelanism in *Astyanax mexicanus* cave x surface hybrids. BMC Evolutionary Biology 16 (1), 145.

Gross, J.B., Sun, D.A., Carlson, B.M., Brodo-Abo, S., Protas, M.E., 2020. Developmental transcriptomic analysis of the cave-dwelling crustacean, *Asellus aquaticus*. Genes 11, 42.

Gross, J.B., Wilkens, H., 2013. Albinism in phylogenetically and geographically distinct populations of *Astyanax* cavefish arises through the same loss-of-function *Oca2* allele. Heredity 111, 122−130.

Herczeg, G., Hafenscher, V.P., Balázs, G., Fišer, Ž., Kralj-Fišer, S., Horváth, G., 2020. Is foraging innovation lost following colonization of a less variable environment ? A case study in surface- vs. cave- dwelling *Asellus aquaticus*. Ecology and Evolution 10 (12), 5323−5331.

Herczeg, G., Nyitrai, V., Balázs, G., Horváth, G., 2022. Food preference and food type innovation of surface- *vs.* cave-dwelling waterlouse (*Asellus aquaticus*) after 60,000 years of isolation. Behavioral Ecology and Sociobiology 76 (1).

Hildebrand, M., 1985. Walking and running. In: Hildebrand, M., Bramble, D.M., Liem, K.F., Wake, D.B. (Eds.), Functional Vertebrate Morphology. Harvard University Press, Cambridge, pp. 38−57.

Horváth, G., Sztruhala, S.S., Balázs, G., Herczeg, G., 2021. Population divergence in aggregation and sheltering behaviour in surface- versus cave-adapted *Asellus aquaticus* (Crustacea: Isopoda). Biological Journal of the Linnean Society 134 (3), 667−678.

Hüppop, K., 2000. How do cave animals cope with the food scarcity in caves? In: Wilkens, H., Culver, D.C., Humphreys, F.W. (Eds.), Ecosystems of the World: Subterranean Ecosystems. Elsevier, Amsterdam, pp. 159−188.

Janzer, W., Ludwig, W., 1952. Versuche zur evolutorischen Entstehung der Höhlentiermerkmale. Zeitschrift für induktive Abstammungs- und Vererbungslehre 84, 462−479.

Jeffery, W.R., 2009. Evolution and development in the cavefish *Astyanax*. Current Topics in Developmental Biology 86, 191−221.

Jeffery, W.R., 2020. *Astyanax* surface and cave fish morphs. EvoDevo 11, 14.

Jemec, A., Škufca, D., Prevorčnik, S., Fišer, Ž., Zidar, P., 2017. Comparative study of acetylcholinesterase and glutathione S-transferase activities of closely related cave and surface *Asellus aquaticus* (Isopoda: Crustacea). PLoS One 12 (5), e0176746.

Karaman, S.L., 1952. *Asellus aquaticus* i njegove podvrste na Balkanu. Prirodoslovni Istraživački Institut, Jugoslovenska Akademija Znanosti i Umjetnosti 25, 80−85.

Klaassen, H., Wang, Y., Adamski, K., Rohner, N., Kowalko, J.E., 2018. CRISPR mutagenesis confirms the role of *oca2* in melanin pigmentation in *Astyanax mexicanus*. Developmental Biology 441, 313−318.

IV. Principles of evolution in groundwater

Konec, M., 2015. Genetic Differentiation and Speciation in Subterranean and Surface Populations of *Asellus aquaticus* (Crustacea: Isopoda). Doctoral Dissertation. University of Ljubljana.

Konec, M., Delić, T., Trontelj, P., 2016. DNA barcoding sheds light on hidden subterranean boundary between Adriatic and Danubian drainage basins. Ecohydrology 9 (7), 1304—1312.

Konec, M., Prevorčnik, S., Sarbu, S.M., Verovnik, R., Trontelj, P., 2015. Parallels between two geographically and ecologically disparate cave invasions by the same species, *Asellus aquaticus* (Isopoda, Crustacea). Journal of Evolutionary Biology 28 (4), 864—875.

Kosswig, C., 1939. Zur Farbvariabilität bei unterirdisch Wasserasseln, *Asellus aquaticus*, sensu Racovitzai. Mitteilungen Über Höhlen- Und Karstforschung 2 (4), 94—102.

Kosswig, C., Kosswig, L., 1936. Über Augenrück- und -missbildung bei *Asellus aquaticus cavernicolus*. Verhandlungen Der Deutschen Zoologischen Gesellschaft, 1936, Zoologischer Anzeiger Supplement 9, 274—281.

Kosswig, C., Kosswig, L., 1940. Die Variabilität bei *Asellus aquaticus*, unter besonderer Berücksichtigung der Variabilität in isolierten unter- und oberirdischen Populationen. Revue de La Faculté Des Sciences Forestières de l'Université d'Istanbul. Série B 5, 1—56.

Kowalko, J., 2020. Utilizing the blind cavefish *Astyanax mexicanus* to understand the genetic basis of behavioral evolution. Journal of Experimental Biology 223, jeb208835.

Lafuente, E., Moritz, L., Moritz, R, Matthews, B., Buser, C., Vorburger, C., Räsänen, K., 2021. Building on 150 Years of knowledge: the freshwater isopod *Asellus aquaticus* as an integrative eco-evolutionary model system. Frontiers in Ecology and Evolution 9, 748212.

Lattinger-Penko, R., 1979. Data on the biology of an underground crustacean, *Asellus aquaticus cavernicolus* Racovitza (Crustacea, Isopoda). Ekologija 14, 83—95.

Ludwig, W., 1942. Zur evolutorischen Erklärung der Höhlentiermerkmale durch Allelelimination. Biologisches Zentralblatt 62, 447—482.

Ma, L., Jeffery, W.R., Essner, J.J., Kowalko, J.E., 2015. Genome editing using TALENs in blind mexican cavefish, *Astyanax mexicanus*. PLoS One 10 (3), e0119370.

Magniez, G., 1978. Précopulation et vie souterraine chez quelques Péracarides (Crustacea Malacostraca). Archives de Zoologie Expérimentale et Générale 119, 471—478.

Mallarino, R., Abzhanov, A., 2012. Paths less traveled: evo-devo approaches to investigating animal morphological evolution. Annual Review of Cell and Developmental Biology 28, 743—763.

Marin, I., Palatov, D., 2019. An occasional record of the amplexus in epigean *Niphargus* (Amphipoda: Niphargidae) from the Russian Western Caucasus. Zootaxa 4701 (1), 97—100.

Meng, F., Zhao, Y., Postlethwait, J.H., Zhang, C., 2013. Differentially-expressed opsin genes identified in *Sinocyclocheilus* cavefish endemic to China. Current Zoology 59 (2), 170—174.

Miller, C.T., Glazer, A.M., Summers, B.R., Blackman, B.K., Norman, A.R., Shapiro, M.D., Cole, B.L., Peichel, C.L., Schluter, D., Kingsley, D.M., 2014. Modular skeletal evolution in sticklebacks is controlled by additive and clustered quantitative trait Loci. Genetics 197 (1), 405—420.

Mojaddidi, H., Fernandez, F.E., Erickson, P.A., Protas, M.E., 2018. Embryonic origin and genetic basis of cave associated phenotypes in the isopod crustacean *Asellus aquaticus*. Scientific Reports 8 (1), 16589.

Morris, J., Navarro, N., Rastas, P., Rawlins, L.D., Sammy, J., Mallet, J., Dasmahapatra, K.K., 2019. The genetic architecture of adaptation: convergence and pleiotropy in *Heliconius* wing pattern evolution. Heredity 123, 138—152.

Mösslacher, F., Creuzé des Châtelliers, M., 1996. Physiological and behavioural adaptations of an epigean and a hypogean dwelling population of *Asellus aquaticus* (L.) (Crustacea, Isopoda). Archiv für Hydrobiologie 138 (2), 187—198.

Nosil, P., 2012. Ecological Speciation. Oxford University Press, New York, NY.

O'Callaghan, I., Harrison, S., Fitzpatrick, D., Sullivan, T., 2019. The freshwater isopod *Asellus aquaticus* as a model biomonitor of environmental pollution: a review. Chemosphere 235, 498—509.

Peichel, C.L., Marques, D.A., 2017. The genetic and molecular architecture of phenotypic diversity in sticklebacks. Philosophical Transactions of the Royal Society of London Series B Biological Sciences 372, 20150486.

Pérez-Moreno, J.L., Balázs, G., Bracken-Grissom, H.D., 2018. Transcriptomic insights into the loss of vision in Molnár János Cave's crustaceans. Integrative and Comparative Biology 58 (3), 452—464.

Pérez-Moreno, J.L., Balázs, G., Wilkins, B., Herczeg, G., Bracken-Grissom, H.D., 2017. The role of isolation on contrasting phylogeographic patterns in two cave crustaceans. BMC Evolutionary Biology 17, 247.

Pérez-Moreno, J.L., Iliffe, T.M., Bracken-Grissom, H.D., 2016. Life in the underworld: anchialine cave biology in the era of speleogenomics. International Journal of Speleology 45 (2), 149—170.

Prevorčnik, S., 2002. Racial Differentation in Water Louse, *Asellus Aquaticus* (Crustacea: Isopoda: Asellidae). Doctoral Dissertation. University of Ljubljana, Ljubljana, Slovenia.

Prevorčnik, S., Blejec, A., Sket, B., 2004. Racial differentiation in *Asellus aquaticus* (L.) (Crustacea: Isopoda: Asellidae). Archiv für Hydrobiologie 160 (2), 193–214.

Protas, M., Hersey, C., Kochanek, D., Zhou, Y., Wilkens, H., Jeffery, W.R., Zon, L.I., Borowsky, R., Tabin, C.J., 2006. Genetic analysis of cavefish reveals molecular convergence in the evolution of albinism. Nature Genetics 38, 107–111.

Protas, M.E., Jeffery, W.R., 2012. Evolution and development in cave animals: from fish to crustaceans. Wiley Interdisciplinary Reviews: Developmental Biology 1 (6), 823–845.

Protas, M.E., Trontelj, P., Patel, N.H., 2011. Genetic basis of eye and pigment loss in the cave crustacean, *Asellus aquaticus*. Proceedings of the National Academy of Sciences of the United States of America 108 (14), 5702–5707.

Protas, M., Tabansky, I., Conrad, M., Gross, J.B., Vidal, O., Tabin, C.J., Borowsky, R., 2008. Multi-trait evolution in a cave fish, *Astyanax mexicanus*. Evolution and Development 10 (2), 196–209.

Racovitza, E.G., 1925. Notes sur les Isopodes 1. Archives de Zoologie Expérimentale et Générale 68, 533–622.

Re, C., Fišer, Ž., Perez, J., Tacdol, A., Trontelj, P., Protas, M.E., 2018. Common genetic basis of eye and pigment loss in two distinct cave populations of the isopod crustacean *Asellus aquaticus*. Integrative and Comparative Biology 58 (3), 421–430.

Salemaa, H., 1979. The chromosomes of *Asellus aquaticus* (L.): a technique for isopod karyology. Crustaceana 36 (3), 316–318.

Sarbu, S.M., Kane, T.C., Kinkle, B.K., 1996. A chemoautotrophically based cave ecosystem. Science 272 (5270), 1953–1955.

Schluter, D., 2001. Ecology and the origin of species. Trends in Ecology & Evolution 16 (7), 372–380.

Schneider, R., 1887. Ein bleicher *Asellus* in den Gruben von Freiberg im Erzgebirge (*Asellus aquaticus*, var. Fribergensis). Sitzungsberichte Der Königlichen Preussische Akademie Des Wissenschaften Zu Berlin 2, 723–742.

Seehausen, O., 2015. Process and pattern in cichlid radiations – inferences for understanding unusually high rates of evolutionary diversification. New Phytologist 207, 304–312.

Simčič, T., Sket, B., 2019. Comparison of some epigean and troglobiotic animals regarding their metabolism intensity. Examination of a classical assertion. International Journal of Speleology 48 (2), 133–144.

Simon, N., Fujita, S., Porter, M., Yoshizawa, M., 2019. Expression of extraocular opsin genes and light-dependent basal activity of blind cavefish. PeerJ 7, e8148.

Sket, B., 1965. Taxonomische Problematik der Art *Asellus aquaticus* (L.) Rac. mit besonderer rücksicht auf die populationen Sloweniens. Slovenska Akademija Znanosti in Umetnosti, Razred za prirodoslovne in medicinske vede, IV, Razprave 8, 177–221.

Sket, B., 1985. Why all cave animals do not look alike – a discussion on adaptive value of reduction processes. Bulletin of the National Speleological Society 47 (2), 78–85.

Sket, B., 1994. Distribution of *Asellus aquaticus* (Crustacea: Isopoda: Asellidae) and its hypogean populations at different geographic scales, with a note on *Proasellus istrianus*. Hydrobiologia 287 (1), 39–47.

Speiser, D.I., Pankey, M.S., Zaharoff, A.K., Battelle, B.A., Bracken-Grissom, H.D., Breinholt, J.W., Bybee, S.M., Cronin, T.W., Garm, A., Lindgren, A.R., Patel, N.H., Porter, M.L., Protas, M.E., Rivera, A.S., Serb, J.M., Zigler, K.S., Crandall, K.A., Oakley, T.H., 2014. Using phylogenetically-informed annotation (PIA) to search for light-interacting genes in transcriptomes from non-model organisms. BMC Bioinformatics 15 (1), 350.

Stahl, B.A., Gross, J.B., 2015. Alterations in *Mc1r* gene expression are associated with regressive pigmentation in *Astyanax* cavefish. Development Genes and Evolution 225 (6), 367–375.

Stahl, B.A., Gross, J.B., Speiser, D.I., Oakley, T.H., Patel, N.H., Gould, D.B., Protas, M.E., 2015. A transcriptomic analysis of cave, surface, and hybrid isopod crustaceans of the species *Asellus aquaticus*. PLoS One 10 (10), e0140484.

Stammer, H.J., 1932. Die Fauna des Timavo. Ein Beitragzur Kenntnis der Höhlengewässer, des Süss- und Brackwasser im Karst. Zoologische Jahrbucher, Abteilung für Systematik, Geographie und Biologie der Tiere 63, 521–656.

Šušteršič, F., Šušteršič, S., Stepišnik, U., 2003. The Late Quaternary dynamics of Planinska jama, south-central Slovenia. Cave and Karst Science 30 (2), 89–96.

Sworobowicz, L., Grabowski, M., Mamos, T., Burzyński, A., Kilikowska, A., Sell, J., Wysocka, A., 2015. Revisiting the phylogeography of *Asellus aquaticus* in Europe: insights into cryptic diversity and spatiotemporal diversification. Freshwater Biology 60 (9), 1824–1840.

IV. Principles of evolution in groundwater

Sworobowicz, L., Mamos, T., Grabowski, M., Wysocka, A., 2020. Lasting through the ice age: the role of the proglacial refugia in the maintenance of genetic diversity, population growth, and high dispersal rate in a widespread freshwater crustacean. Freshwater Biology 65 (6), 1028–1046.

Turk-Prevorčnik, S., Blejec, A., 1998. *Asellus aquaticus infernus*, new subspecies (Isopoda: Asellota: Asellidae), from Romanian hypogean waters. Journal of Crustacean Biology 18 (4), 763–773.

Turk, S., Sket, B., Sarbu, S., 1996. Comparison between some epigean and hypogean populations of *Asellus aquaticus* (Crustacea: Isopoda: Asellidae). Hydrobiologia 337 (1), 161–170.

Verovnik, R., Sket, B., Trontelj, P., 2005. The colonization of Europe by the freshwater crustacean *Asellus aquaticus* (Crustacea: Isopoda) proceeded from ancient refugia and was directed by habitat connectivity. Molecular Ecology 14 (14), 4355–4369.

Verovnik, R., Prevorčnik, S., Jugovic, J., 2009. Description of a neotype for *Asellus aquaticus* Linné, 1758 (Crustacea: Isopoda: Asellidae), with description of a new subterranean *asellus* species from Europe. Zoologischer Anzeiger 248 (2), 101–118.

Verovnik, R., Sket, B., Trontelj, P., 2004. Phylogeography of subterranean and surface populations of water lice *Asellus aquaticus* (Crustacea: Isopoda). Molecular Ecology 13 (6), 1519–1532.

Vick, P., Blum, M., 2010. The isopod *Asellus aquaticus*: a novel arthropod model organism to study evolution of segment identity and patterning. Palaeodiversity 3, 89–97.

Vick, P., Schweickert, A., Blum, M., 2009. Cloning and expression analysis of the homeobox gene abdominal-A in the isopod *Asellus aquaticus*. In: Kurzfield, N.C. (Ed.):, Development Gene Expression Regulation, Hauppauge, Nova Science Publishers, pp. 285–295.

Wolff, C., 2009. The embryonic development of the malacostracan crustacean *Porcellio scaber* (Isopoda, Oniscidea). Development Genes and Evolution 219, 545–564.

Wray, G.A., 2007. The evolutionary significance of *cis*-regulatory mutations. Nature Reviews Genetics 8, 206–216.

Yamamoto, Y., Byerly, M.S., Jackman, W.R., Jeffery, W.R., 2009. Pleiotropic functions of embryonic sonic hedgehog expression link jaw and taste bud amplification with eye loss during cavefish evolution. Developmental Biology 330, 200–211.

Zidar, P., Škufca, D., Prevorčnik, S., Kalčikova, G., Jemec Kokalj, A., 2018. Energy reserves in the water louse *Asellus aquaticus* (Isopoda, Crustacea) from surface and cave populations: seasonal and spatial dynamics. Fundamental and Applied Limnology 191 (3), 253–265.

# Developmental and genetic basis of troglomorphic traits in the teleost fish *Astyanax mexicanus*

*Joshua B. Gross*[1], *Tyler E. Boggs*[1], *Sylvie Rétaux*[2] *and Jorge Torres-Paz*[2]

[1]Department of Biological Sciences, University of Cincinnati, Cincinnati, OH, United States;
[2]Paris-Saclay Institute of Neuroscience, Université Paris-Saclay and CNRS, Saclay, France

Since their discovery in 1936, blind morphs of the genus *Astyanax* have served as an important model for genetic and developmental studies of trait evolution. An indispensable component of this model system is the presence of extant surface-dwelling morphs, which enable comparative studies with cave-dwelling morphs. This comparative paradigm has provided insight to the genetic and developmental basis of a number of constructive and regressive traits evolving in this system. An entire book has recently been devoted to the "Biology and Evolution of the Mexican Cavefish" (Keene, Yoshizawa, McGaugh Eds, 2016). Here, we will review major discoveries in the developmental regime of cave-dwelling animals, as well as the historic pattern of genetic discoveries. Collectively, this body of work informs how examinations of the developmental and genetic basis for trait evolution has provided deep insights to the origin of morphological differences in cave animals, and animal groups more broadly.

## The history of genetic and genomic studies of troglomorphy in *Astyanax*

### Classical genetic studies

Following a rather modest discovery, the blind Mexican cavefish has developed over the past several decades into a robust model system for cave adaptation and evolutionary biology at large. This emergence can be traced to early experiments in which cave- and surface-dwelling forms were intercrossed, resulting in production of viable hybrid individuals

(Sadoğlu, 1957). This finding enabled a "classical genetics" era that led to the identification of Mendelian phenotypes in cavefish, as well as insight to the basic genetic architecture of more complex phenotypes (e.g., eye degeneration). By the 1980s, the same classical approach was adapted for pioneering studies of cave-associated behaviors, such as feeding (Schemmel, 1980). Around the same time, the rise of recombinant DNA technology ushered in candidate gene studies that aimed to test the role of key regulators in conspicuous cave-associated traits, such as vision loss and sensory expansion (Jeffery et al., 2000). At the start of the 21st century, hybrid pedigrees were first used to assess the segregation of genotypes (using large panels of genomic markers), which were polymorphic between cave and surface forms (Borowsky and Wilkens, 2002). This approach yielded the first direct identification of quantitative trait loci (QTL) and ushered in a new era, wherein the genetic bases for cave evolution were pursued through identification of regions of the genome associated with variation in a trait of interest (Protas et al., 2008; Casane and Rétaux, 2016). Most recently, this field has moved into the genomic era, which has been supported by genome-level resources available for both cave- and surface-dwelling forms.

The earliest era of *Astyanax* genetic studies commenced shortly after the discovery by Basil C Jordan of cavefish from the Chica cave locality, near Cuidad Valles in San Luis Potosi, Mexico. Owing to their stark phenotypic differences, cavefish were originally classified into a distinct genus from local surface-dwelling forms. Indeed, the new genus, *Anoptichthys* (lit, "bony fish with no eyes"), was coined by the influential Prof. Carl Hubbs, an eminent ichthyologist and curator (Hubbs and Innes, 1936). This nomenclature persisted in the literature for several decades and was not revised until cave and surface forms were found to be interfertile. Indeed, the degree of enzymatic similarity between surface and cave forms did not support the assignment of these morphs to different genera or species (Avise and Selander, 1972). An early experimental cross between cave and surface forms yielded the first described Mendelian trait of albinism (Sadoğlu, 1957). The initial discovery of the Chica cave was followed by many others as there are now over 30 known cave localities hosting cavefish populations (Mitchell et al., 1977; Elliott, 2018). In the scientific community, cavefish are named according to the cave where they are found.

These classical genetics studies continued from the late 1960s through the mid-1980s following the same approach of cave x surface matings. Analyses of trait distributions in resultant pedigrees could be interrogated for a variety of different phenotypes. Among these included the characterization of degenerative morphological structures in cavefish, such as the eye. Wilkens (1970a) found that F1 hybrids, generated between surface and Pachón or Sabinos cavefish, developed an eye that was roughly intermediate in size between both morphs. Further, expanded pedigrees (F$_2$ generations or backcrosses) yielded normal distribution of eye sizes, suggesting a complex, multigenic contribution to eye loss in cave-adapted forms. Using the method of Lande (1981), the minimum number of freely segregating genetic "factors" ($n_E$) implicated in trait evolution could be calculated. Wilkens (1970a), who utilized ~17,000 F$_2$ individuals, from a Pachón cave x surface fish cross, to estimate the number of loci involved in eye loss, reported ~six to seven genes implicated in eye regression. Interestingly, despite having evolved in distinct caves, similar estimates were obtained for two additional localities: Sabinos (~7.47 genetic loci) and Pachón (~6.88 genetic loci; Wilkens, 1970a, 1988).

Similar results were obtained when examining other visual structures, such as the pupillary opening, lens muscle, lens size, the anterior chamber of the eye, and the retinal pigmented epithelium (Wilkens, 1970a, 1972, 1976). Around the same period, a number of genetic studies of regressive pigmentation traits were performed. Alongside simple Mendelian pigmentation traits first discovered in the late 1950s and 1960s (Sadoğlu, 1957; Sadoğlu and McKee, 1969) came additional studies evaluating complex pigmentation changes (e.g., melanophore density, Wilkens, 1970b). Interestingly, unequal estimates were obtained when estimating the number of loci implicated in (the complex phenotype of) numerical reduction of melanophores. For this regressive trait, fewer loci were estimated for Sabinos cavefish ($\sim$1.47 loci) compared to Pachón cavefish ($\sim$3.35 loci) suggesting that, despite the fact that each locality is geographically near one another, the genetic architecture of trait regression differs between populations.

This classical approach was not limited to regressive morphological phenotypes. A number of studies began exploring different behaviors exhibited by each morphotype. For instance, Burchards et al. (1985) analyzed the genetic basis for aggressive behavior by quantifying the frequency of "ramming," a behavior found mostly in surface dwelling forms. While this behavior was reduced in fish drawn from the Pachón and Piedras localities (Burchards et al., 1985; Wilkens, 1988), the Micos population appeared to retain aggressive behavior that was scored intermediately between surface fish and other cavefish populations. Shortly thereafter, it was found that cavefish are far less likely to school together compared to surface-dwelling forms (Parzefall and Senkel, 1986). In an experimental $F_2$ pedigree, this trait appeared to segregate according to a complex, polygenic pattern, implying a heritable contribution to schooling behavior. Similarly, the "fright" reaction (i.e., a behavioral response to alarm) was evaluated using a classical genetic approach and found to segregate in experimental hybrid individuals, although this behavior showed age-dependence (Pfeiffer, 1966).

Classical analyses also shifted to the examination of constructive traits. This included the distribution of taste buds in surface fish and the Pachón cavefish population (Schemmel, 1967, 1974); which identified a genetic contribution to the roughly three-to four-fold increase in these peripheral sensory organs in cave morphs. The numerical increase of taste buds in cavefish compared to surface fish appears to be mediated by $\sim$12.27 loci (Wilkens, 1970a). Schemmel (1980) hypothesized that more taste buds may influence the "angle of attack," a behavioral feeding difference between cave and surface fish. Accordingly, cavefish may have a reduced angle of contact with substrate perhaps to allow better contact between food items and ventrally distributed taste buds (Wilkens, 1988). The estimated number of loci involved in feeding behavior was calculated as relatively simple, with an estimate of $\sim$2.1 loci. Hüppop (1989) performed a classical analysis of oxygen consumption (a metabolic adaptation to the oxygen-poor cave environment) using surface, Pachón, $F_1$ hybrids, and an $F_2$ pedigree. However, owing to the relative minor differences in phenotype between parental forms, an estimate of the number of involved genes could not be calculated (Hüppop, 1989).

## Candidate gene studies

Following this productive era of classical genetic analyses, technical advancements gave way to a series of candidate gene cloning studies, beginning in the 1990s. An early report focused on the structure of *opsin* visual pigment genes. Yokoyama and Yokoyama (1990a) reported successful isolation and DNA sequence for *Astyanax* genes that are homologous to the human forms of red and green visual *opsin*. Interestingly, based on these comparisons, the authors concluded that red *opsins* evolved from green *opsins* via convergent substitution at key amino acid residues (Yokoyama and Yokoyama, 1990a, 1990b). Cloning and sequence analyses continued for other candidate vision genes, including *rhodopsin*, which revealed a novel substitution in cavefish at position 261. This position is normally occupied by a highly conserved phenylalanine residue, but is replaced with a tyrosine in *Astyanax* cavefish, leading to functional shift in absorbance.

Candidate studies then shifted to some of the first gene expression analyses in *Astyanax*. Using in situ hybridization, Langecker et al. (1995) demonstrated that (in contrast to surface-dwelling counterparts) an integral *crystallin* gene is not expressed in the lens of cave morphs. Studies in this era continued to focus on candidate genes associated with vision loss, including additional studies of *opsin* and *crystallin* family members. However, attention also turned toward regulation, with the first studies of sequence and expression of the transcription factor, *pax6*. This indispensable vision gene appears to be conserved in structure and (to some degree) function across distantly related animal taxa. Interestingly, no changes were reported for *pax6* sequence or expression when comparing surface fish and cavefish embryos (Behrens et al., 1997). However, later in development, extensive cell death and down-regulation of *pax6* expression were observed in the lens, suggesting a role for this structure in eye degeneration (Strickler et al., 2001).

Candidate gene studies then shifted toward constructive traits. An early example was expression analyses of the gene, *prox1* (Jeffery et al., 2000), a transcription factor that plays a key role in lens fiber elongation in other animal systems. Interestingly, this study revealed unaltered expression in the visual system of cave and surface fish. However, this gene was up-regulated in developing taste buds and neuromasts of cavefish, suggesting it may facilitate expansion of nonvisual senses to compensate for the loss of vision (Jeffery et al., 2000). Similarly, the gene *hedgehog* is implicated in eye loss in cavefish, through expanded midline signaling, which pleiotropically impacts jaws and taste buds (Yamamoto et al., 2004). These features are expanded in cavefish, essentially tying together the evolution of constructive and regressive features through a single genetic pathway (Yamamoto et al., 2009).

## QTL mapping

The recent era of genetic analysis in *Astyanax* cavefish has been marked by genome-wide analyses of trait evolution between cave and surface morphs. Borowsky and Wilkens (2002) provided the first direct evidence of a complex genetic basis for regressive loss, identifying multiple QTL associated with vision loss, melanophore reductions, and condition factor. This early map, based on a backcross pedigree of Pachón cavefish and surface fish, represented roughly half of the size of the genome and relied on anonymous RAPD (random amplification of polymorphic DNA) markers (Borowsky and Wilkens, 2002). Subsequent linkage studies based on $F_2$ pedigrees developed between more cavefish populations (i.e., the Molino locality) and surface fish, and utilized hundreds of microsatellite markers (Protas et al., 2006). This productive period in *Astyanax* quantitative genetics confirmed the genetic architecture of several phenotypes first explored using a classical genetics approach, but extending beyond to many additional traits (reviewed in O'Quin and McGaugh, 2016).

Phenotypic evolution in cave-dwelling organisms can be categorically divided in regressive and constructive traits (Jeffery, 2001). While constructive traits are generally believed to evolve through selection, the evolutionary mechanism(s) mediating trait loss remains contested (Gross, 2012; Rétaux and Casane, 2013). The most salient regressive features are vision and pigmentation loss, which have arguably received the most attention in the literature. The first formal analysis using a backcross pedigree only yielded three QTL, perhaps owing to the incomplete nature of this early genomic map (Borowsky and Wilkens, 2002). Additional markers, and a deeper genotypic survey yielded estimates of eight loci (Protas et al., 2007), or five loci (Yoshizawa et al., 2012), mediating eye size in an $F_2$ pedigree. Interestingly, the number of loci predicted by Protas et al. (2007), closely approximates the estimated six to seven genes involved in eye loss from several decades earlier using a classical approach (Wilkens, 1970a).

The genetic basis for pigmentation, a highly complex phenotype, includes some features demonstrating a Mendelian pattern of inheritance, such as albinism and *brown* melanophores. When subjected to QTL mapping, these qualitative traits (first examined several decades earlier) yielded a single QTL peak consistent with their presumed monogenic basis. The gene underlying albinism, *Oca2*, was functionally validated using both in vitro (Protas et al., 2006) and in vivo approaches (Klaassen et al., 2018). The gene mediating *brown* melanophores, *Mc1r*, harbored diverse genetic lesions throughout the El Abra cavefish localities, including a single nucleotide substitution that impacts an orthologous arginine residue in humans associated with red hair color (Gross et al., 2009).

The impacted genes in more complex pigmentation phenotypes have proved more elusive to identify. For instance, a study examining numerical variation in melanophore numbers revealed diverse polarities for loci associated with this highly complex phenotype (Protas et al., 2007). Candidate genes residing near many of these QTL remain anonymous, however a recent study provided evidence that the gene *Pmela*, which encodes a pigment cell-specific transmembrane glycoprotein, maps near complex pigmentation loci and is both downregulated, and structurally altered, in cavefish. Similarly, the melanosomal enzyme encoded by *Tyrp1b*, which also maps near pigmentation loci, may influence coloration in *Astyanax* cavefish through alterations to the melanin biosynthetic pathway (Stahl et al., 2018).

Among the constructive phenotypes assessed using a QTL approach have been expansions of nonvisual sensory systems, such as mechanoreceptive neuromasts and taste buds. Constructive trait evolution also extends to behaviors, and recent work has clarified the genomic architecture of feeding posture, a behavioral phenotype that likely enhances foraging success (Kowalko et al., 2013). Through the diverse examination of regressive and constructive features, this collection of work has provided deeper insight to the regions of the genome influencing trait evolution. With the advent of draft genomic resources, and the ability to fine-scale map the coarser genomic intervals identified from earlier maps using microsatellites (Fig. 15.1), comes the exciting prospect of identifying candidate genes, and the precise lesions, mediating these traits.

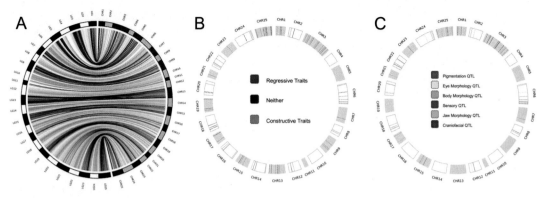

**FIGURE 15.1  A visual incorporation of genetic information derived by multiple techniques in *Astyanax mexicanus*.** Technological advances in science have allowed for nearly continuous improvement of genetic resources. It is important however, to be able to utilize previous work within new resources. As is such in this model. A great deal of synteny exists in the transition from linkage maps (A left (Carlson et al., 2015)) built by anonymous genetic markers, to the finer annotation of the genome (A right (Asty2.0, NCBI). It is now possible to derive new interpretations of unnamed QTL by locating these markers in the genome. For example, QTL may be categorized and visualized in the genome to determine potential patterns or regions of particular interest (B&C).

## Transcriptomics

The first transcriptome surveys in *Astyanax* included de novo assemblies of exhaustive Sanger-sequenced clones, as well as next-generation sequencing technology. The work of Hinaux et al. (2013) revealed many mutated genes derived from cave morphs are expressed in the zebrafish visual system. This broad scale resource of sequence information and clones revealed 11 vision-related genes with sequence alterations, serving as candidate loci for the genetic basis of vision loss in this species (Hinaux et al., 2013). Around the same time, a next-generation sequencing approach yielded a transcriptome based on the integration of reads from cave and surface morphs. This resource further enhanced the existing gene sequence information for the *Astyanax* community, providing coding information for over 20,000 expressed genes (Gross et al., 2013). The combined resources of both transcriptomes substantially elevated the quality of available genetic information in this model system from ∼40 accessioned sequences at public databases to thousands. This, in turn, enabled the first genome-wide analyses of vision-related genes, as well as the first assessment of enriched gene ontology terms between the contrasting morphotypes.

Shortly thereafter, a developmental transcriptome assessed polarity and expression differences between morphs across ontogeny (Stahl and Gross, 2017). This study revealed broader features of the transcriptomic architecture, such as the abundance of down-regulated genes in cavefish compared to surface fish. Further, this work addressed the question of convergence on gene expression patterns by comparing two independent cave populations to extant surface-dwelling forms. An examination of ontology terms revealed that similar functional terms are evolving in two populations, yet the gene expression patterns that underlie these terms may differ. Further, a constellation of gene expression changes, although slightly different between cave populations, appear to stem from two key regulators, *mitf1* and *otx2* (Stahl and Gross, 2017). The value of this approach has certainly accelerated in recent years, reflected in more studies utilizing next-generation sequencing and bioinformatics analyses for a variety of topics, including sleep (McGaugh et al., 2020), basal activity (Simon et al., 2019), and phenotypic plasticity (Bilandžija et al., 2020).

## Genomics

Recently, the contemporary "genomic" era was ushered in with the first draft cavefish genome, published in 2014 (McGaugh et al., 2014). The first genome was created using Pachón cavefish, providing the first insight to changes in genomic architecture that mediate troglomorphic evolution. A second *Astyanax* genome project, this time focused on the surface-dwelling form (Warren et al., 2021), promises to identify differences between morphs that would have been unimaginable, such as a catalog of every nucleotide-through-chromosomal alteration between morphs. These advancements present a powerful new dimension through which the genetic basis of phenotypes, the composition of the genome, and the evolutionary history of this remarkable system could be more rigorously explored. When used in the context of prior QTL studies, RNA-sequencing, and candidate gene selection, this resource has dramatically shifted and accelerated genetic insights to cave biology. This resource has illuminated many features of *Astyanax* biology, including the identification of candidate genes underlying vision loss (e.g., *pitx3*, *rx3* and *olfm2a*), and the discovery of repeat elements present in this species (~9.5% of the entire genome). Subsequent analyses have provided new insight to the amount of gene flow present between cave and surface forms through the recent geologic past (Herman et al., 2018). A forthcoming draft of the first surface fish genome will provide even further insight to the nature of cave and surface fish genomic evolution, enabling finer-scale identification of candidate genes underlying trait loci, and informing the chromosomal evolution between surface and cave morphs. The value of this resource could therefore shed new light on topics ranging from the origins of multiple independent cave populations, to the question of whether convergent losses and gains evolved through the same loci. Additionally, this resource enables a finer collation of previously known genetic information, providing a new opportunity to study genome-wide patterning and association of traits (Fig. 15.1). In sum, the genetic basis for cave-associated trait evolution in cavefish began with the first demonstrations of hybridization between cave and surface forms of *Astyanax*, and the advancement from early classical genetics studies to candidate gene studies, to linkage mapping and contemporary genomic studies.

## Developmental basis of troglomorphy in *Astyanax*

Evolutionary developmental biology (Evo-devo) searches for the origin of phenotypic changes in the mechanisms by which organisms develop. Decades of Evo-devo research have led to the conclusion that animal diversity has arisen mostly through modifications in embryogenesis (Gould, 1977; Carroll, 2008; De Robertis et al., 2017). Great phenotypic changes can occur just by subtle alterations in developmental programs regulating cell division, apoptosis, fate specification or migration. Variations can occur on either time or space, that is, heterochrony and heterotopy respectively, but also quantitatively in the levels of gene expression or in the strength in which a particular signaling pathway is activated or repressed.

The extreme phenotypic change observed in *Astyanax mexicanus* eco-morphotypes has attracted the curiosity of many developmental biologists interested in the evolution of the brain, sensory systems and craniofacial structures. The external fertilization gives access to the full process of embryogenesis, from fertilization onwards, allowing a comprehensive understanding of the embryology of the species. In addition, embryos from both morphotypes develop at similar rates and thus have enabled comparative studies, revealing the critical stages when eye regression takes place (Hinaux et al., 2011).

## Developmental basis of eye regression

The adult cave-dwelling morph is a completely eyeless fish, yet the early larvae develop a relatively normal eye primordium that starts undergoing degeneration the second day after fertilization, through apoptotic mechanisms. The eye in the cavefish larvae is smaller compared to the surface fish, the ventral optic fissure is abnormal, and the lens is also reduced in size. These optic phenotypes are a consequence of developmental defects occurring during the morphogenesis of the anterior nervous system, bringing along changes in other forebrain structures including the hypothalamus and the olfactory system.

The forebrain takes shape through complex cellular rearrangements at the anterior end of the neural plate (ANP) (Fig. 15.2A). This territory becomes regionalized by the expression of several developmental genes in restricted domains, delimitating the future forebrain compartments and driving particular cell behaviors during morphogenesis. The positional information and identity of cells within each morphogenetic field is given by the integration of several gradients of signaling molecules (morphogens), produced and secreted locally by organizer centers. Changes in the exposure of the responding cells within the ANP to morphogens (timing or levels) can affect their identities, leading to changes in the relative sizes of the different prospective forebrain domains. In *A. mexicanus* cavefish, modifications in the properties of midline signaling centers during embryogenesis are known to play a major role in eye regression, as well as in the emergence of some constructive traits.

Eye development is one of the most complex morphogenetic processes that takes place within the ANP (Fig. 15.2A). It begins with the evagination of the optic vesicles from the lateral walls of the forming anterior neural tube. The distal part of the vesicles then forms a cup-like shape, and remain connected to the brain by the optic stalk. The optic cup, that is, the prospective retina, induces the formation of the lens placode in the overlying nonneural ectoderm, an epithelial structure that will differentiate into the lens of the eye.

Globally, these steps occur "relatively properly" in cavefish embryos. However, the size of eye primordium at the end of the process is importantly reduced compared to surface fish. In addition, there are evident morphological defects in the cavefish developing eye: the optic cup lacks its ventral/temporal domain and the lens placode is small (Hinaux et al., 2016; Devos et al., 2019) (Fig. 15.2B). During the second day of development apoptotic events start in the optic tissue, initially restricted to the lens vesicle, and then extend to the adjacent retina (Yamamoto and Jeffery, 2000; Alunni et al., 2007). This massive and specific apoptotic wave is

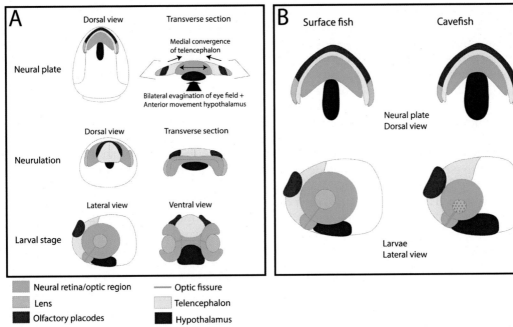

**FIGURE 15.2   Brain developmental evolution in *Astyanax mexicanus*. A.** At neural plate stage (top) the forebrain morphogenetic domains are organized in discrete regions at the anterior end of the neural plate (top left). The morphogenesis of the forebrain is shown at the same stage in a transverse section (top right). Double arrows indicate evagination of optic cups laterally (green). Arrows pointing to the center indicate convergence of telencephalic cells in the midline (yellow). The big arrow inside the drawing points to the anterior movement of the prospective hypothalamus (blue). During neurulation (center), the optic vesicles contact the lens ectoderm (light blue), the hypothalamus locates ventral to the optic region and the telencephalon fuses in the midline contacting laterally with the future olfactory placodes (red). At larval stage (bottom), different forebrain regions are compartmentalized, as well as olfactory and lens placodes. In the ventral region of the neural retina/optic region, indicated by the optic fissure (dark green), is the location where the epithelium fuses during optic morphogenesis. **B.** In *A. mexicanus* morphotypes, the prospective forebrain domains show size differences already at the neural plate stage (top), with trade-offs between neighboring domains also occurring at this stage. The hypothalamic field is larger, and the eye field is smaller in cavefish embryos. At the placodal field, there are also differences—the cavefish olfactory placodal domain is larger and the lens placode is reduced. Changes in the relative size of morphogenetic domains have an effect on the size of forebrain compartments in the larval stage (bottom). In addition to the small optic cup and lens in the cavefish, the optic fissure fuses incorrectly and the lens is apoptotic. Olfactory placodes and the hypothalamus are larger in cavefish compared to surface fish.

a hallmark of the eye degeneration process observed in *Astyanax* cavefish. Another optic phenotype observed during cavefish development is retarded growth of the retinal tissue, with disorganized cellular layers and little photoreceptor differentiation. The degenerated eyes finally sink under the facial skin and remain as vestigial cysts inside the orbits of the adult fish.

During eye development, reciprocal interactions between lens and retina are necessary in order to maintain tissue homeostasis and normal growth. Lens replacement experiments between the two morphs demonstrated that inductive or maintenance signals from the cavefish lens are missing, triggering the complete regression of the eye. In fact, transplantation of a healthy lens from a surface fish embryo into a cavefish optic cup is sufficient to rescue eye growth, photoreceptor differentiation, and the development of the cornea and iris. These experiments demonstrate that the cavefish optic cup retained the responsive properties to lens signals (Yamamoto and Jeffery, 2000). In addition, signals from the retinal pigmented epithelium (RPE, located at the back of the retina) are defective in cavefish, enhancing the phenotypes dependent on lens defects. The RPE is the epithelium that surrounds the neural retina and is important for the differentiation of photoreceptor neurons (Strickler et al., 2007; Devos et al., 2019; Ma et al., 2020).

Further analyses have aimed at identifying the molecular mechanisms implicated in the degenerative process. Comparative de novo sequencing analyses of transcriptomes revealed mutations producing radical amino acid substitutions in 11 cavefish genes expressed in the visual system (Hinaux et al., 2013) as well as one gene loss (*Pde6b*; Policarpo et al., 2021). Due to relaxed selection on these genes in caves, these mutations might also contribute to retinal degeneration.

Defective lenses that undergo abnormal apoptosis trigger eye degeneration. The origin of these events has thus been investigated. The lens is composed of fiber cells surrounded by an epithelium. Differentiation of fiber cells involves the synthesis and massive accumulation of proteins (crystallins) that give transparency to the lens. Among the crystallin genes, four fixed mutations were found in cavefish, two leading to radical substitutions (*crybb1a* and *crybg3*). In addition, comparative analyses of gene expression during development demonstrated that four crystallin genes (*crygm5, crybgx, cryaa,* and *crybb1c*) were downregulated/not expressed in cavefish compared to the surface morph. Functional analyses showed that knock-down of *cryaa* in surface fish triggers lens apoptosis and that induced expression of *cryaa* in the cavefish lens has a protective effect against lens cell death (Ma et al., 2014; Hinaux et al., 2015).

Together, these findings on cavefish eye developmental evolution have medical relevance as mutations in the human orthologue of *cryaa* can cause congenital cataract (Litt et al., 1998) and *Pde6b* is associated with night-blindness and *retinitis pigmentosa* in humans (McLaughlin et al., 1993; Gal et al., 1994).

## Developmental basis of constructive traits

Eye loss in cavefish is accompanied by major morphological changes in head structures such as nonvisual sensory organs and craniofacial structures. The cranial skeleton forms in close intimacy with the brain and sensory organs, thus eye regression influences the shape of facial bones in cavefish, as shown by experimental manipulations in embryos. For example, eye reduction produces an expansion in size of the bones around the eye orbits (circumorbital bones) (Dufton et al., 2012). The extent of the expansion is limited by the degree of eye reduction. Another direct consequence of eye regression is the loss of the scleral ossicles ring, which in cavefish forms as a reduced cartilaginous ring. Ossification within the sclera can be restored in cavefish by lens replacement (Yamamoto et al., 2003).

Other craniofacial changes are not linked to the eye phenotype. For example, there is an increase in the number of maxillary teeth in the cave morph, independent of eye loss (Yamamoto et al., 2003). The expansion of teeth number could be a trophic adaptation to darkness and new food regimes. Left-right cranial asymmetry and bone fragmentation is another feature found in several *Astyanax* cavefish populations that is independent of eye regression. In cavefish, some facial bones exhibit particular fragmentation or fusion patterns, never observed in surface fish. One of the most remarkable examples is the unilateral fragmentation of the third suborbital bone (SO3). Although the development of the chondrocranium occurs symmetrically in cavefish juveniles, bilateral asymmetries become evident during the ossification process (Powers et al., 2017). Fragmentation of SO3 occurs through the formation of ectopic ossification centers as well as through postossification bone remodeling (Powers et al., 2018).

The potential adaptive value of asymmetry in the cranial skeleton remains unknown, however it correlates well with an asymmetric distribution of the cranial sensory neuromasts (Gross et al., 2016). Neuromasts are volcano-shaped organs composed of sensory hair-cells distributed through the fish body that are part of the lateral line system, involved in mechanosensation. Whereas the organization or density of neuromasts on the trunk or tail is similar in the two *Astyanax* morphs, there is an increased density of neuromasts on the cavefish head, particularly in the suborbital region (Yoshizawa et al., 2010). Moreover, the fine anatomy of the cavefish neuromasts themselves is different, conferring them with heightened sensitivity (Yoshizawa et al., 2014)—which also poses interesting questions in terms of development and morphogenesis of neuromasts cells. These changes are associated with the emergence of a behavioral adaptation denominated "vibration attraction behavior" (VAB), in which cavefish but not surface fish swim toward an object vibrating at 35Hz (Yoshizawa et al., 2012). This behavior may be relevant for food localization in the caves, as a way of sensory compensation. Finally, early embryonic manipulations (*Shh* mRNA injections) to decrease eye size in surface fish do not affect VAB or superficial neuromast numbers, indirectly suggesting that the cavefish lateral line phenotype is independent from eye loss (Yoshizawa et al., 2012).

In addition to morphofunctional changes in the mechanosensory system, the chemosensory modalities of taste and olfaction are also modified in cavefish. They possess more taste buds with a broader distribution on the ventral mouth. This difference is thought to be responsible for the morph-specific food seeking behavior: whereas in total darkness, surface fish feed perpendicular to the substrate, the blind cavefish adopt a 45 degrees angle (Schemmel, 1967). Numeric differences in taste buds become evident in larvae about 12 days old and again, this change may be linked developmentally to eye loss and jaw enlargement (Varatharasan et al., 2009; Yamamoto et al., 2009). However, taste bud numbers in surface fish are not increased following induced eye regression by lens removal (Dufton et al., 2012).

In cave-adapted *Astyanax* the loss of the visual system is also accompanied with an enlargement of the olfactory epithelia, the organs involved in the detection of odorant molecules dissolved in the surrounding environment (Bibliowicz et al., 2013). This anatomical change is correlated with enhanced olfactory skills: cavefish have a threshold for detection of the amino acid alanine 100.000 times lower than surface fish ($10^{-10}$ M vs. $10^{-5}$ M) (Hinaux et al., 2016). Interestingly, the enlargement of the olfactory organs in cavefish is not due to eye degeneration, since the organ primordium, the olfactory placodes, are already larger compared to surface fish before the first signs of apoptosis in the lens arise (Hinaux et al., 2016). In addition, surface fish that are visually deprived after early embryonic lens ablation show enhanced responses to olfactory cues compared to normal surface fish. This surgical procedure mimics cavefish eye degeneration and produces strongly reduced eyes but does not affect the size of the olfactory epithelium in juveniles (Blin et al., 2018) —although a slight enlargement has been reported in adults (Yamamoto et al., 2003). Improved olfaction in eyeless surface fish apparently results from developmental plasticity, since eyed surface fish raised in total darkness also show similar improvements in olfactory performance. However, the threshold detection levels of the cavefish were never reached (Blin et al., 2018). The neural mechanisms involved in olfactory developmental plasticity remain unknown. Finally, with the exception of the size of the olfactory organ, the developmental basis of improved olfaction in cavefish may also originate from the neuronal composition of the olfactory epithelium, which emerges during the development of olfactory sensory neurons (OSNs). In fish there are three main types of OSN (ciliated, microvillous and crypt), each with distinctive morphology, signaling transduction pathways and connections to the olfactory bulb in the brain. Cavefish have a higher density of microvillous OSNs (Blin et al., 2018), which are those involved in the detection of amino acids food cues.

In sum, the trophic and sensory apparatuses on the cavefish head are markedly enhanced, and all these morphological traits arise during embryonic and larval development. Interestingly, the changes can be divided into eye-dependent and eye-independent processes. The former corresponds to developmental plasticity mechanisms and the later are probably genetically encoded, highlighting the intricate processes which underlie adaptive developmental evolution.

## Developmental constraints for cavefish evolution

An intriguing question regarding cavefish developmental evolution concerns the development of an eye primordium altogether, even though it will later be eliminated. The answer comes from mechanical constraints imposed during forebrain neurulation, as formation of the optic vesicles is a required step for the whole morphogenetic process which cannot be circumvented (Fig. 15.2A). At the neural plate stage, the precursors of the optic cups are confined to a single morphogenetic domain within the ANP called the eye-field, which is surrounded anteriorly by the precursors of the prospective telencephalon and posteriorly by precursors of the future hypothalamus and diencephalon. The prospective telencephalon is surrounded by the preplacodal field, from where cells will segregate, giving rise to the olfactory and lens placodes.

As cellular boundaries of the eye-field are shared with the prospective telencephalon and hypothalamus, movements during optic vesicle evagination must be coordinated with cell movement of adjacent tissues. In particular, the eye-field splits in two domains by lateral movements forming the two optic vesicles, whereas hypothalamic precursors move anteriorly, ventral to the dividing eye-field (Varga et al., 1999; England et al., 2006; Giger and Houart, 2018; Devos et al., 2021). At the same time, telencephalic cells converge toward the midline dorsally (Werner et al., 2021) (Fig. 15.2A). Thus, if optic vesicles were not formed, the presumptive telencephalic and hypothalamic cells would not end up in a correct position, and the condition would be embryonic-lethal.

At neural plate stage, the regionalization of the ANP into discrete morphogenetic domains begins to be refined by restricted expression of developmental position-coding or identity-coding genes such as *rx3* (eyefield), *nkx2.1a* (hypothalamus), and *emx3* (telencephalon) (Agnès et al., 2022). Changes in the size of these gene expression domains could modify the identity of cells at the boundaries, expanding the size of a prospective morphogenetic domain at the expense of another (Stigloher et al., 2006; Menuet et al., 2007; Bielen and Houart, 2012; Agnès et al., 2022). Indeed, a mechanistic "trade-off" between adjacent fields (eye-field vs. prospective hypothalamus) explains the reduction of ventral retina as hypothalamic tissue gains in size (Pottin et al., 2011; Devos et al., 2019).

Although the whole prospective hypothalamus is larger in cavefish compared to surface fish (Menuet et al., 2007), there are some subdivisions within the hypothalamus that are particularly enlarged relative to others. For example, the transcription factor *lhx9* is expressed in the rostral part of the hypothalamus, in a domain that is larger in cavefish embryos. Cells expressing *lhx9* constitute a pool of progenitors that will differentiate into peptidergic neurons expressing *hypocretin*, which are consequently more numerous in cavefish (Alié et al., 2018). A similar increase in neuropeptidergic lineages occur for NPY-expressing neurons in the rostral-most hypothalamus, and this is due to a larger domain of precursors expressing the transcription factor *lhx7*. Conversely, cavefish have less pro-opio-melanocortin neurons. Peptidergic systems in the hypothalamus control several physiological aspects of body homeostasis such as food intake, energetic balance, locomotion, and sleep. The evolutionary or adaptive relevance of developmental modifications in peptidergic neuronal networks have been studied, showing that enhanced hypocretinergic neurotransmission leads to increased locomotion and reduction in sleep, two functions that could be important during food searching in caves (Alié et al., 2018; Jaggard et al., 2018).

## Embryonic origins of cavefish evolution

The development of the vertebrate nervous system requires precise temporal and spatial coordination of multiple events including the activation of signaling pathways, gene expression and cell movements (Fig. 15.3A). The patterning of embryonic territories is dependent on the localized production and secretion of morphogens, generating gradients of signaling activity throughout the embryo, setting the three-dimensional axes. Among them, a strong repression of the BMP signals produced ventrally and an activation of the FGF pathway are necessary for the induction of the prospective neural tissue in the dorsal pole of the embryo (Piccolo et al., 1996; Kudoh et al., 2004). In addition, an anterior-low to posterior-high WNT signaling gradient patterns the neural tissue in its antero-posterior dimension (Hashimoto et al., 2000; Kiecker and Niehrs, 2001).

Patterning of the prospective nervous system in the dorso-ventral axis in its entirety comes from signals produced by the underlying axial mesoderm. The axial mesoderm is composed of the prechordal plate and the notochord, rostrally and caudally, respectively. Both structures have distinctive signaling properties and specific inductive capabilities in the anterior and posterior neural plate. SHH signaling from the axial mesoderm has an important role in ventral patterning of both the forebrain and posterior neural tube (Aoto et al., 2009) (Fig. 15.3A). In the prospective forebrain, prechordal SHH activity is necessary for the induction of the ventral hypothalamus, and absence of SHH has a dorsalizing effect on ventral structures (Barth and Wilson, 1995; Varga et al., 2001).

In *A. mexicanus* cavefish, SHH overactivation in the midline is involved in many developmental phenotypes, including the enlarged hypothalamus (Menuet et al., 2007; Rétaux et al., 2008), the reduction of the ventral retina during optic development (Yamamoto et al., 2004; Pottin et al., 2011), and lens apoptosis (Yamamoto et al., 2004). Subtle modifications in other

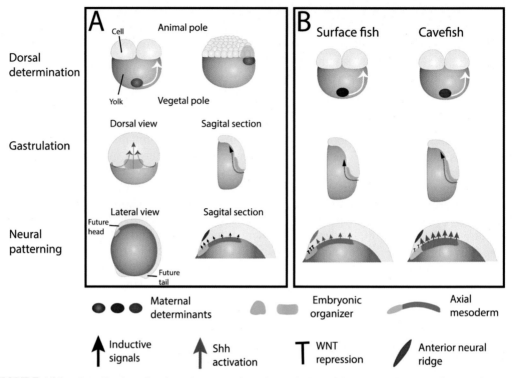

**FIGURE 15.3**    **Contribution of early embryogenesis to the evolution of the cavefish brain.** In fish development, maternal determinants in the vegetal pole of the fertilized embryo are transported toward the future dorsal side (white arrow, top left). The maternal determinants induce the expression of a particular set of genes in the dorsal side, leading to the induction of the embryonic organizer (green). During gastrulation (center), cells in the organizer migrate from the margin toward the animal side of the embryo (black arrows with dashed lines) where they will form the anterior extreme of the axial mesoderm, the prechordal plate. Once in the animal side of the embryo, the prechordal plate (light and dark green, bottom) will induce and pattern the anterior brain (black arrows, bottom right). Prechordal signals induce the development of the secondary signaling center at the anterior neural ridge (lilac). **B.** In *Astyanax* morphotypes, maternal determinants have a different composition (top). As a consequence, differences occur during gastrulation. Migration of prechordal plate cells is more advanced in cavefish embryos compared to surface fish (center). In addition, the prechordal plate has different inductive properties in the two morphs (bottom). Compared to the surface fish, the posterior part of the prechordal plate in cavefish produces higher levels of SHH (green arrows), whereas the anterior part leads to a lower repression of the WNT pathway (blunt-end black arrows). Higher levels of SHH signaling lead to an earlier onset of *fgf8* expression in the ANP of cavefish embryos (dark lilac) compared to surface fish (pale lilac).

prechordal signals also contribute to the cavefish phenotype. This includes changes in the dynamics of BMP signaling affecting sensory placode development and leads to a larger olfactory epithelia and a reduced lens in cavefish (Hinaux et al., 2016), as well as a reduction of WNT inhibition producing morphogenetic defects in optic development such as reduced optic cups and abnormal optic fissure (Torres-Paz et al., 2019).

In addition to the axial mesoderm, secondary signaling centers develop within the neural tube, orchestrating the morphogenesis and differentiation of neighboring neural tissues (Vieira et al., 2010). One of them, the anterior neural ridge, is involved in organizing the development of the forebrain through local secretion of FGF8 (Shimamura and Rubenstein, 1997; Houart et al., 1998). In *Astyanax* cavefish, *fgf8* expression in this signaling center starts earlier than in surface fish embryos. This temporal shift occurs in response to enhanced SHH activation in the axial mesoderm and contributes to developmental defects in the optic tissue (Pottin et al., 2011).

In sum, subtle variations in the intensity or timing of signaling from embryonic organizing centers that induce and pattern the ANP have (directly or indirectly) profound consequences on the morphogenesis of the cavefish head and brain. Such variations typically correspond to part of the developmental evolutionary "toolkit," at work in cavefish embryos, and have an important role in the generation of brain diversity through evolution (Sylvester et al., 2010; Cavodeassi and Houart, 2012; Rétaux et al., 2013).

In fact, investigating earlier stages of embryogenesis has revealed extremely precocious origins of cavefish morphological evolution. The axial mesoderm is considered the main embryonic signaling center due to its inductive properties on the neural ectoderm. It is a derivative of the embryonic organizer (known as the Spemann-Mangold organizer in amphibians and "shield" in teleost fishes). The organizer is a group of cells responsible for neural induction, capable of inducing ectopic secondary axes when transplanted in host embryos (i.e., embryos with two bifurcated heads). From the organizer, prechordal plate precursor cells migrate in close intimacy with neural precursors during a process called gastrulation, with its definitive position underneath the anterior neural plate (Fig. 15.3A).

Comparative studies in *A. mexicanus* have shown morphotype-specific prechordal plate features, not only at the molecular level, but also morphologically. Cavefish axial mesodermal structures are wider compared to surface fish (Yamamoto et al., 2004), and these differences in prechordal plate organization and morphology are attributed to heterochronic migratory events during gastrulation (Ren et al., 2018; Torres-Paz et al., 2019). During their journey from posterior to anterior, signaling from organizer-derived cells starts to pattern the developing neural plate (García-Calero et al., 2008). Thus, timing differences observed during gastrulation would have an impact on regional neural specification, due to differential exposure of neural cells to morphogens (Fig. 15.3A and B).

The development of the organizer occurs during the establishment of the embryonic dorso-ventral polarity. In fish, the dorsal pole is set by maternal factors, deposited in the egg during oogenesis, which are actively transported from the vegetal side (Fig. 15.3A and B). Organizer properties, and therefore later developmental events, fully depend on these maternal factors. Owing to the interfertility between *Astyanax* morphotypes, it is possible to test maternal contribution to phenotypic evolution. By comparing F1 hybrid embryos obtained by reciprocal crosses (breeding surface fish female with cavefish male and vice versa) it was found that differences observed during gastrulation are fully due to a maternal genetic effect (Torres-Paz et al., 2019). As development advances, phenotypes in reciprocal hybrids become more similar to each other (intermediate between "pure" surface fish and cavefish). However, the maternal effect persists particularly in eye phenotypes and hypothalamic patterning (Ma et al., 2018; Torres-Paz et al., 2019). Transcriptomic analyses of maternal mRNAs deposited in eggs have demonstrated a distinctive morphotype-specific signature that could prepattern a determined developmental program (Torres-Paz et al., 2019). In conclusion, the cavefish "natural mutant" may well represent an extreme case of developmental evolution, whereby the variation in the developmental program is set during oogenesis, in the mother's gonads.

# Conclusions

The advancement of *Astyanax* as a genetic and developmental model of evolution has been marked by consistent interest in this remarkable creature since its discovery. Technical advances such as the ability to hybridize cave and surface forms, have catalyzed important discoveries over the last 9 decades. Early classical genetic studies have been revisited in the context of contemporary QTL analyses, which in turn have improved with the development of genome-level resources today. Similarly, developmental studies have started from insightful classical embryonic manipulations, were followed with candidate gene approaches and expression studies, and have currently turned toward more functional and mechanistic studies, thanks to genome-editing and gene expression manipulation tools. The expansion of interest in *Astyanax* as a model for genetic and developmental studies has been accompanied by an expansion of techniques and resources. The future for this leading model of troglomorphic evolution is very bright and promises to provide new insight to the developmental and genetic basis for extreme environmental adaptation.

# Acknowledgments

We thank C.L. Peichel for constructive comments that improved this chapter.

# References

Agnès, F., Torres-Paz, J., Michel, P., Rétaux, S., 2022. A 3D molecular map of the cavefish neural plate illuminates eye-field organization and its borders in vertebrates. Development 149, dev199966. https://doi.org/10.1242/dev.199966.

Alié, A., Devos, L., Torres-Paz, J., Prunier, L., Boulet, F., Blin, M., Elipot, Y., Rétaux, S., 2018. Developmental evolution of the forebrain in cavefish, from natural variations in neuropeptides to behavior. Elife 7, e32808. https://doi.org/10.7554/eLife.32808.

Alunni, A., Menuet, A., Candal, E., Pénigault, J.B., Jeffery, W.R., Rétaux, S., 2007. Developmental mechanisms for retinal degeneration in the blind cavefish *Astyanax mexicanus*. Journal of Comparative Neurology 505 (2), 221–233. https://doi.org/10.1002/cne.21488.

Aoto, K., Shikata, Y., Imai, H., Matsumaru, D., Tokunaga, T., Shioda, S., Yamada, G., Motoyama, J., 2009. Mouse Shh is required for prechordal plate maintenance during brain and craniofacial morphogenesis. Developmental Biology 327 (1), 106–120. https://doi.org/10.1016/j.ydbio.2008.11.022.

Avise, J.C., Selander, R.K., 1972. Evolutionary genetics of cave-dwelling fishes of the genus *Astyanax*. Evolution 26 (1), 1–19. https://doi.org/10.1111/j.1558-5646.1972.tb00170.x.

Barth, K.A., Wilson, S.W., 1995. Expression of zebrafish *nk2.2* is influenced by sonic hedgehog/vertebrate hedgehog-1 and demarcates a zone of neuronal differentiation in the embryonic forebrain. Development 121 (6), 1755–1768. https://doi.org/10.1242/dev.121.6.1755.

Behrens, M, Langecker, T.G., Wilkens, H., Schmale, H., 1997. Comparative analysis of *Pax-6* sequence and expression in the eye development of the blind cave fish *Astyanax fasciatus* and its epigean conspecific. Molecular Biology and Evolution 14 (3), 299–308.

Bibliowicz, J., Alié, A., Espinasa, L., Yoshizawa, M., Blin, M., Hinaux, H., Legendre, L., Père, S., Rétaux, S., 2013. Differences in chemosensory response between eyed and eyeless *Astyanax mexicanus* of the Rio Subterráneo cave. Evodevo 4 (1), 25. https://doi.org/10.1186/2041-9139-4-25.

Bielen, H., Houart, C., 2012. BMP signaling protects telencephalic fate by repressing eye identity and its Cxcr4-dependent morphogenesis. Developmental Cell 23 (4), 812–822. https://doi.org/10.1016/j.devcel.2012.09.006.

Bilandžija, H., Hollifield, B., Steck, M., Meng, G., Ng, M., Koch, A., Gračan, R., Ćetković, H., Porter, M., Renner, K., Jeffery, W.R., 2020. Phenotypic plasticity as an important mechanism of cave colonization and adaptation. eLife 9, e51830.

Blin, M., Tine, E., Meister, L., Elipot, Y., Bibliowicz, J., Espinasa, L., Rétaux, S., 2018. Developmental evolution and developmental plasticity of the olfactory epithelium and olfactory skills in Mexican cavefish. Developmental Biology 441 (2), 242–251. https://doi.org/10.1016/j.ydbio.2018.04.019.

Borowsky, R., Wilkens, H., 2002. Mapping a cave fish genome: polygenic systems and regressive evolution. Journal of Heredity 93 (1), 19–21.

Burchards, H., Dölle, A., Parzefall, J., 1985. Aggressive behaviour of an epigean population of Astyanax mexicanus (Characidae, Pisces) and some observations of three subterranean populations. Behavioural Processes 11 (3), 225–235.

Carlson, B.M., Onusko, S.W., Gross, J.B., 2015. A high-density linkage map for Astyanax mexicanus using genotyping-by-sequencing technology. Genes, Genomes, Genetics 5 (2), 241–251. https://doi.org/10.1534/g3.114.015438.

Carroll, S.B., 2008. Evo-devo and an expanding evolutionary synthesis: a genetic theory of morphological evolution. Cell 134 (1), 25–36. https://doi.org/10.1016/j.cell.2008.06.030.

Casane, D., Rétaux, S., 2016. Evolutionary genetics of the cavefish Astyanax mexicanus. Advances in genetics 95, 117–159.

Cavodeassi, F., Houart, C., 2012. Brain regionalization: of signaling centers and boundaries. Developmental Neurobiology 72 (3), 218–233. https://doi.org/10.1002/dneu.20938.

De Robertis, E.M., Moriyama, Y., Colozza, G., 2017. Generation of animal form by the Chordin/Tolloid/BMP Gradient: 100 Years after D'Arcy Thompson. Development Growth and Differentiation 59, 580–592. https://doi.org/10.1111/dgd.12388.

Devos, L., Klee, F., Edouard, J., Simon, V., Legendre, L., Khallouki, N.E., Barbachou, S., Sohm, F., Rétaux, S., 2019. Morphogenetic and patterning defects explain the coloboma phenotype of the eye in the Mexican cavefish. bioRxiv 698035. https://doi.org/10.1101/698035.

Devos, L., Agnès, F., Edouard, J., Simon, V., Legendre, L., El Khallouki, N., Barbachou, S., Sohm, F., Rétaux, S., 2021. Eye morphogenesis in the blind Mexican cavefish. Biol Open 10 (10), bio059031.

Dufton, M., Hall, B.K., Franz-odendaal, T.A., 2012. Early lens ablation causes dramatic long-term effects on the shape of bones in the craniofacial skeleton of Astyanax mexicanus. Plos One 7 (11), e50308. https://doi.org/10.1371/journal.pone.0050308.

Elliott, W.R., 2018. The Astyanax Caves of Mexico. Cavefishes of Tamaulipas, San Luis Potosí, and Guerrero, Bulletin 26. Association for Mexican Cave Studies, Austin, Texas, pp. 1–326.

England, S.J., Blanchard, G.B., Mahadevan, L., Adams, R.J., 2006. A dynamic fate map of the forebrain shows how vertebrate eyes form and explains two causes of cyclopia. Development 133 (23), 4613–4617. https://doi.org/10.1242/dev.02678.

Gal, A., Orth, U., Baehr, W., Schwinger, E., Rosenberg, T., 1994. Heterozygous missense mutation in the rod cGMP phosphodiesterase β–subunit gene in autosomal dominant stationary night blindness. Nature Genetics 7 (1), 64–68. https://doi.org/10.1038/ng0594-64.

García-Calero, E., Fernández-Garre, P., Martínez, S., Puelles, L., 2008. Early mammillary pouch specification in the course of prechordal ventralization of the forebrain tegmentum. Developmental Biology 320 (2), 366–377. https://doi.org/10.1016/j.ydbio.2008.05.545.

Giger, F.A., Houart, C., 2018. The birth of the eye vesicle: when fate decision equals morphogenesis. Frontiers in Neuroscience 12, 87. https://doi.org/10.3389/fnins.2018.00087.

Gould, S.J., 1977. Ontogeny and Phylogeny. Harvard University Press, p. 501.

Gross, J.B., 2012. The complex origin of Astyanax cavefish. BMC Evolutionary Biology 12 (1), 105. https://doi.org/10.1186/1471-2148-12-105.

Gross, J.B., Borowsky, R., Tabin, C.J., 2009. A novel role for Mc1r in the parallel evolution of depigmentation in independent populations of the cavefish Astyanax mexicanus. PLoS Genetics 5 (1), e1000326.

Gross, J.B., Furterer, A., Carlson, B.M., Stahl, B.A., 2013. An integrated transcriptome-wide analysis of cave and surface dwelling Astyanax mexicanus. PloS One 8 (2), e55659.

Gross, J., Gangidine, A., Powers, A., 2016. Asymmetric facial bone fragmentation mirrors asymmetric distribution of cranial neuromasts in blind Mexican cavefish. Symmetry 8 (11), 118. https://doi.org/10.3390/sym8110118.

Hashimoto, H., Itoh, M., Yamanaka, Y., Yamashita, S., Shimizu, T., Solnica-Krezel, L., Hibi, M., Hirano, T., 2000. Zebrafish Dkk1 functions in forebrain specification and axial mesendoderm formation. Developmental Biology 217 (1), 138–152. https://doi.org/10.1006/dbio.1999.9537.

Herman, A., Brandvain, Y., Weagley, J., Jeffery, W.R., Keene, A.C., Kono, T.J., Bilandžija, H., Borowsky, R., Espinasa, L., O'Quin, K., Ornelas-García, C.P., 2018. The role of gene flow in rapid and repeated evolution of cave-related traits in Mexican tetra, *Astyanax mexicanus*. Molecular Ecology 27 (22), 4397–4416.

Hinaux, H., Devos, L., Blin, M., Elipot, Y., Bibliowicz, J., Alié, A., Rétaux, S., 2016. Sensory evolution in blind cavefish is driven by early embryonic events during gastrulation and neurulation. Development 143 (23), 4521–4532. https://doi.org/10.1242/dev.141291.

Hinaux, H., Pottin, K., Chalhoub, H., Pere, S., Elipot, Y., Legendre, L., Rétaux, S., 2011. A developmental staging table for *Astyanax mexicanus* surface fish and Pachón cavefish. Zebrafish 8 (4), 155–165.

Hinaux, H., Poulain, J., Da Silva, C., Noirot, C., Jeffery, W.R., Casane, D., Rétaux, S., 2013. De novo sequencing of *Astyanax mexicanus* surface fish and Pachón cavefish transcriptomes reveals enrichment of mutations in cavefish putative eye genes. PLoS One 8 (1), e53553. https://doi.org/10.1371/journal.pone.0053553.

Hinaux, H., Blin, M., Fumey, J., Legendre, L., Heuzé, A., Casane, D., Rétaux, S., 2015. Lens defects in *Astyanax mexicanus* cavefish: evolution of crystallins and a role for alphaA-crystallin. Developmental Neurobiology 75 (5), 505–521.

Houart, C., Westerfield, M., Wilson, S.W., 1998. A small population of anterior cells patterns the forebrain during zebrafish gastrulation. Nature 391 (6669), 788–792. https://doi.org/10.1038/35853.

Hubbs, C.L., Innes, W.T., 1936. The first known blind fish of the family Characidae: a new genus from Mexico. Occasional Papers of the Museum of Zoology University of Michigan 342, 1–7.

Hüppop, K., 1989. Genetic analysis of oxygen consumption rate in cave and surface fish of *Astyanax fasciatus* (Characidae, Pisces). Further support for the neutral mutation theory. Mémoires de Biospéologie 16, 163–168.

Jaggard, J.B., Stahl, B.A., Lloyd, E., Prober, D.A., Duboue, E.R., Keene, A.C., 2018. Hypocretin underlies the evolution of sleep loss in the Mexican cavefish. Elife 7, e32637. https://doi.org/10.7554/eLife.32637.

Jeffery, W.R., 2001. Cavefish as a model system in evolutionary developmental biology. Developmental Biology 231 (1), 1–12.

Jeffery, W.R., Strickler, A.G., Guiney, S., Heyser, D.G., Tomarev, S.I., 2000. *Prox1* in eye degeneration and sensory organ compensation during development and evolution of the cavefish *Astyanax*. Development Genes and Evolution 210 (5), 223–230.

Keene, A.C., Yoshizawa, M., McGaugh, S.E., 2016. Biology and Evolution of the Mexican Cavefish. Academic Press.

Kiecker, C., Niehrs, C., 2001. A morphogen gradient of Wnt/β-catenin signalling regulates anteroposterior neural patterning in Xenopus. Development 124, 4189–4201.

Klaassen, H., Wang, Y., Adamski, K., Rohner, N., Kowalko, J.E., 2018. CRISPR mutagenesis confirms the role of oca2 in melanin pigmentation in *Astyanax mexicanus*. Developmental Biology 441 (2), 313–318.

Kowalko, J.E., Rohner, N., Linden, T.A., Rompani, S.B., Warren, W.C., Borowsky, R., Tabin, C.J., Jeffery, W.R., Yoshizawa, M., 2013. Convergence in feeding posture occurs through different genetic loci in independently evolved cave populations of *Astyanax mexicanus*. Proceedings of the National Academy of Sciences 110 (42), 16933–16938.

Kudoh, T., Concha, M.L., Houart, C., Dawid, I.B., Wilson, S.W., 2004. Combinatorial Fgf and Bmp signalling patterns the gastrula ectoderm into prospective neural and epidermal domains. Development 131 (15), 3581–3592. https://doi.org/10.1242/dev.01227.

Lande, R., 1981. The minimum number of genes contributing to quantitative variation between and within populations. Genetics 99 (3–4), 541–553.

Langecker, T.G., Neumann, B., Hausberg, C., Parzefall, J., 1995. Evolution of the optical releasers for aggressive behavior in cave-dwelling *Astyanax fasciatus* (Teleostei, Characidae). Behavioural Processes 34 (2), 161–167.

Litt, M., Kramer, P., Lamorticella, D.M., Murphey, W., Lovrien, E.W., Weleber, R.G., 1998. Autosomal dominant congenital cataract associated with a missense mutation in the human alpha crystallin gene CRYAA. Human Molecular Genetics 7 (3), 471–474.

Ma, L., Ng, M., Weele, C.M., Yoshizawa, M., Jeffery, W.R., 2020. Dual roles of the retinal pigment epithelium and lens in cavefish eye degeneration. Journal of Experimental Zoology Part B: Molecular and Developmental Evolution 334, 438–449. https://doi.org/10.1002/jez.b.22923.

Ma, L., Parkhurst, A., Jeffery, W.R., 2014. The role of a lens survival pathway including Sox2 and α A-crystallin in the evolution of cavefish eye degeneration. EvoDevo 5, 1–14. https://doi.org/10.1186/2041-9139-5-28.

Ma, L., Strickler, A.G., Parkhurst, A., Yoshizawa, M., Shi, J., Jeffery, W.R., 2018. Maternal genetic effects in *Astyanax* cavefish development. Developmental Biology 441 (2), 209–220. https://doi.org/10.1016/J.YDBIO.2018.07.014.

McGaugh, S.E., Gross, J.B., Aken, B., Blin, M., Borowsky, R., Chalopin, D., Hinaux, H., Jeffery, W.R., Keene, A., Ma, L., Minx, P., 2014. The cavefish genome reveals candidate genes for eye loss. Nature Communications 5 (1), 1–10.

McGaugh, S.E., Passow, C.N., Jaggard, J.B., Stahl, B.A., Keene, A.C., 2020. Unique transcriptional signatures of sleep loss across independently evolved cavefish populations. Journal of Experimental Zoology Part B: Molecular and Developmental Evolution 334 (7–8), 497–510.

McLaughlin, M.E., Sandberg, M.A., Berson, E.L., Dryja, T.P., 1993. Recessive mutations in the gene encoding the β–subunit of rod phosphodiesterase in patients with retinitis pigmentosa. Nature Genetics 4 (2), 130–134. https://doi.org/10.1038/ng0693-130.

Menuet, A., Alunni, A., Joly, J., Jeffery, W.R., Rétaux, S., 2007. Expanded expression of Sonic Hedgehog in *Astyanax* cavefish: multiple consequences on forebrain development and evolution. Development 134, 845–855. https://doi.org/10.1242/dev.02780.

Mitchell, R., Russell, W., Elliott, W., 1977. Mexican eyeless characin fishes, genus *Astyanax*: environment, distribution, and evolution. KIP Monographs 17, 1–89.

O'Quin, K., McGaugh, S., 2016. Mapping the genetic basis of troglomorphy in *Astyanax*: how far we have come and where do we go from here. In: Keene, A., Yoshizawa, M., McGaugh, S. (Eds.), Biology and Evolution of the Mexican Cavefish. Academic Press, pp. 111–135.

Parzefall, J., Senkel, S., 1986. Schooling behavior in cavernicolous fish and their epigean conspecifics, 2. 9th Congres International de Espeleologica, Barcelona, Spain, pp. 107–109.

Pfeiffer, W., 1966. Die verbreitung der schreckreaktion bei kaulquappen und die herkunft des schreckstoffes. Zeitschrift für vergleichende Physiologie 52 (1), 79–98.

Piccolo, S., Sasai, Y., Lu, B., De Robertis, E.M., 1996. Dorsoventral patterning in Xenopus: inhibition of ventral signals by direct binding of chordin to BMP-4. Cell 86 (4), 589–598. https://doi.org/10.1016/S0092-8674(00)80132-4.

Policarpo, M., Fumey, J., Lafargeas, P., Naquin, D., Thermes, C., Naville, M., Dechaud, C., Volff, J.N., Cabau, C., Klopp, C., Møller, P.R., Bernatchez, L., García-Machado, E., Rétaux, S., Casane, D., 2021. Contrasted gene decay in subterranean vertebrates : insights from cavefishes and fossorial mammals. Molecular Biology and Evolution 38 (2), 589–605. https://doi.org/10.1093/molbev/msaa249.

Pottin, K., Hinaux, H., Rétaux, S., 2011. Restoring eye size in *Astyanax mexicanus* blind cavefish embryos through modulation of the Shh and Fgf8 forebrain organising centres. Development 138 (12), 2467–2476. https://doi.org/10.1242/dev.054106.

Powers, A.K., Davis, E.M., Kaplan, S.A., Gross, J.B., 2017. Cranial asymmetry arises later in the life history of the blind Mexican cavefish, *Astyanax mexicanus*. PLOS One 12, e0177419. https://doi.org/10.1371/journal.pone.0177419.

Powers, A.K., Kaplan, S.A., Boggs, T.E., Gross, J.B., 2018. Facial bone fragmentation in blind cavefish arises through two unusual ossification processes. Scientific Reports 8 (1), 7015. https://doi.org/10.1038/s41598-018-25107-2.

Protas, M., Conrad, M., Gross, J.B., Tabin, C., Borowsky, R., 2007. Regressive evolution in the Mexican cave tetra, *Astyanax mexicanus*. Current Biology 17 (5), 452–454.

Protas, M.E., Hersey, C., Kochanek, D., Zhou, Y., Wilkens, H., Jeffery, W.R., Zon, L.I., Borowsky, R., Tabin, C.J., 2006. Genetic analysis of cavefish reveals molecular convergence in the evolution of albinism. Nature Genetics 38 (1), 107–111. https://doi.org/10.1038/ng1700.

Protas, M., Tabansky, I., Conrad, M., Gross, J.B., Vidal, O., Tabin, C.J., Borowsky, R., 2008. Multi-trait evolution in a cave fish, *Astyanax mexicanus*. Evolution & Development 10 (2), 196–209.

Ren, X., Hamilton, N., Müller, F., Yamamoto, Y., 2018. Cellular rearrangement of the prechordal plate contributes to eye degeneration in the cavefish. Developmental Biology 441 (2), 221–234. https://doi.org/10.1016/J.YDBIO.2018.07.017.

Rétaux, S., Bourrat, F., Joly, J.-S., Hinaux, H., 2013. Perspectives in evo-devo of the vertebrate brain. In: Streelman, J.T. (Ed.), Advances in Evolutionary Developmental Biology. John Wiley & Sons, Inc, Hoboken, NJ, pp. 151–172. https://doi.org/10.1002/9781118707449.ch8.

Rétaux, S., Casane, D., 2013. Evolution of eye development in the darkness of caves: adaptation, drift, or both? Evo-Devo 4 (1), 26. https://doi.org/10.1186/2041-9139-4-26.

Rétaux, S., Pottin, K., Alunni, A., 2008. Shh and forebrain evolution in the blind cavefish *Astyanax mexicanus*. Biology of the Cell 100 (3), 139–147. https://doi.org/10.1042/bc20070084.

Sadoğlu, P., 1957. A Mendelian gene for albinism in natural cave fish. Experientia 13, 394. https://doi.org/10.1007/BF02161111.

Sadoğlu, P., McKee, A., 1969. A second gene that affects eye and body color in Mexican blind cave fish. Journal of Heredity 60 (1), 10–14.

Schemmel, C., 1967. Vergleichende Untersuchungen an den Hautsinnesorganen ober- und unterirdisch lebender *Astyanax*-Formen - Ein Beitrag zur Evolution der Cavernicolen. Zeitschrift Für Morphologie Der Tiere 61 (2), 255–316. https://doi.org/10.1007/BF00400988.

Schemmel, C., 1974. Genetische Untersuchungen zur Evolution des Geschmacksapparates bei cavernicolen Fischen. Journal of Zoological Systematics and Evolutionary Research 12 (1), 196–215.

Schemmel, C., 1980. Studies on the genetics of feeding behaviour in the cave fish *Astyanax mexicanus* f. Anoptichthys: an example of apparent monofactorial inheritance by polygenes. Zeitschrift für Tierpsychologie 53 (1), 9–22.

Shimamura, K., Rubenstein, J.L., 1997. Inductive interactions direct early regionalization of the mouse forebrain. Development 124 (14), 2709–2718. https://doi.org/10.1242/dev.124.14.2709.

Simon, N., Fujita, S., Porter, M., Yoshizawa, M., 2019. Expression of extraocular opsin genes and light-dependent basal activity of blind cavefish. PeerJ 7, e8148.

Stahl, B.A., Gross, J.B., 2017. A comparative transcriptomic analysis of development in two *Astyanax* cavefish populations. Journal of Experimental Zoology Part B: Molecular and Developmental Evolution 328 (6), 515–532.

Stahl, B.A., Sears, C.R., Ma, L., Perkins, M., Gross, J.B., 2018. *Pmela* and *Tyrp1b* contribute to melanophore variation in Mexican cavefish. In: Pontarotti, P. (Ed.), Origin and Evolution of Biodiversity. Springer International Publishing, pp. 3–22.

Stigloher, C., Ninkovic, J., Laplante, M., Geling, A., Tannhäuser, B., Topp, S., Kikuta, H., Becker, T.S., Houart, C., Bally-Cuif, L., 2006. Segregation of telencephalic and eye-field identities inside the zebrafish forebrain territory is controlled by Rx3. Development 133 (15), 2925–2935. https://doi.org/10.1242/dev.02450.

Strickler, A.G., Yamamoto, Y, Jeffery, W.R., 2001. Early and late changes in *Pax6* expression accompany eye degeneration during cavefish development. Development Genes & Evolution 211 (3), 138–144. https://doi.org/10.1007/s004270000123.

Strickler, A.G., Yamamoto, Y., Jeffery, W.R., 2007. The lens controls cell survival in the retina: evidence from the blind cavefish *Astyanax*. Developmental Biology 311 (2), 512–523. https://doi.org/10.1016/j.ydbio.2007.08.050.

Sylvester, J.B., Rich, C.A., Loh, Y.-H.E., van Staaden, M.J., Fraser, G.J., Streelman, J.T., 2010. Brain diversity evolves via differences in patterning. Proceedings of the National Academy of Sciences of the United States of America 107 (21), 9718–9723. https://doi.org/10.1073/pnas.1000395107.

Torres-Paz, J., Leclercq, J., Rétaux, S., 2019. Maternally-regulated gastrulation as a source of variation contributing to cavefish forebrain evolution. eLife 8, e50160. https://doi.org/10.7554/eLife.50160.

Varatharasan, N., Croll, R.P., Franz-Odendaal, T., 2009. Taste bud development and patterning in sighted and blind morphs of *Astyanax mexicanus*. Developmental Dynamics 238 (12), 3056–3064. https://doi.org/10.1002/dvdy.22144.

Varga, Z.M., Amores, a, Lewis, K.E., Yan, Y.L., Postlethwait, J.H., Eisen, J.S., Westerfield, M., 2001. Zebrafish smoothened functions in ventral neural tube specification and axon tract formation. Development 128 (18), 3497–3509. https://doi.org/10.1242/dev.128.18.3497.

Varga, Z.M., Wegner, J., Westerfield, M., 1999. Anterior movement of ventral diencephalic precursors separates the primordial eye field in the neural plate and requires cyclops. Development 126 (24), 5533–5546. https://doi.org/10.1242/dev.126.24.5533.

Vieira, C., Pombero, A., García-Lopez, R., Gimeno, L., Echevarria, D., Martínez, S., 2010. Molecular mechanisms controlling brain development: an overview of neuroepithelial secondary organizers. International Journal of Developmental Biology 54 (1), 7–20. https://doi.org/10.1387/ijdb.092853cv.

Warren, W.C., Boggs, T.E., Borowsky, R., Carlson, B.M., Ferrufino, E., Gross, J.B., Hillier, L., Hu, Z., Keene, A.C., Kenzior, A., Kowalko, J.E., Tomlinson, C., Kremitzki, M., Lemieux, M.E., Graves-Lindsay, T., McGaugh, S.E., Miller, J.T., Mommersteeg, M.T.M., Moran, R.L., Peuß, R., Rice, E.S., Riddle, M.R., Sifuentes-Romero, I., Stanhope, B.A., Tabin, C.J., Thakur, S., Yamamoto, Y., 2021. A chromosome-level genome of *Astyanax mexicanus* surface fish for comparing population-specific genetic differences contributing to trait evolution. Nature Communications 12 (1), 1447. https://doi.org/10.1038/s41467-021-21733-z.

Yamamoto, Y., Jeffery, W.R., 2000. Central role for the lens in cave fish eye degeneration. Science (New York, N.Y.) 289 (5479), 631–633. https://doi.org/10.1126/SCIENCE.289.5479.631.

Werner, J.M., Negesse, M.Y., Brooks, D.L., Caldwell, A.R., Johnson, J.M., Brewster, R., 2021. Hallmarks of primary neurulation are conserved in the zebrafish forebrain. Communications Biology 4 (1), 147. https://doi.org/10.1038/s42003-021-01655-8.

Wilkens, H., 1970. Beiträge zur Degeneration des Auges bei Cavernicolen, Genzahl und Manifestationsart. Journal of Zoological Systematics and Evolutionary Research 8 (1), 1—47.

Wilkens, H., 1970. Beiträge zur Degeneration des Melaninpigments bei cavernicolen Sippen des *Astyanax mexicanus* (Filippi) (Characidae, Pisces). Journal of Zoological Systematics and Evolutionary Research 8 (1), 173—199.

Wilkens, H., 1972. Zur phylogenetischen Rückbildung des Auges Cavernicoler: Untersuchungen an *Anoptichthys jordani* (*Astyanax mexicanus*), Characidae, Pisces. Annales de Spéléologie 27, 411—432.

Wilkens, H., 1976. Genotypic and phenotypic variability in cave animals. Studies on a phylogenetically young cave population of *Astyanax mexicanus* (Filippi) (Characidae, Pisces). Annales de Spéléologie 31, 137—148.

Wilkens, H., 1988. Evolution and genetics of epigean and cave *Astyanax fasciatus* (Characidae, Pisces). In: Hecht, M.K., Wallace, B. (eds) Evolutionary Biology. Springer, Boston, MA, pp. 271—367. https://doi.org/10.1007/978-1-4613-1043-3_8.

Yamamoto, Y., Byerly, M.S., Jackman, W.R., Jeffery, W.R., 2009. Pleiotropic functions of embryonic sonic hedgehog expression link jaw and taste bud amplification with eye loss during cavefish evolution. Developmental Biology 330 (1), 200—211. https://doi.org/10.1016/j.ydbio.2009.03.003.

Yamamoto, Y., Espinasa, L., Stock, D.W., Jeffery, W.R., 2003. Development and evolution of craniofacial patterning is mediated by eye-dependent and -independent processes in the cavefish *Astyanax*. Evolution and Development 5, 435—446. https://doi.org/10.1046/j.1525-142X.2003.03050.x.

Yamamoto, Y., Stock, D.W., Jeffery, W.R., 2004. Hedgehog signalling controls eye degeneration in blind cavefish. Nature 431 (October), 844—847. https://doi.org/10.1038/nature02864.

Yokoyama, R., Yokoyama, S., 1990. Convergent evolution of the red- and green-like visual pigment genes in fish, *Astyanax fasciatus*, and human. Proceedings of the National Academy of Sciences 87 (23), 9315—9318.

Yokoyama, R., Yokoyama, S., 1990. Isolation, DNA sequence and evolution of a color visual pigment gene of the blind cave fish *Astyanax fasciatus*. Vision Research 30 (6), 807—816.

Yoshizawa, M., Gorički, Š., Soares, D., Jeffery, W.R., 2010. Evolution of a behavioral shift mediated by superficial neuromasts helps cavefish find food in darkness. Current Biology 20 (18), 1631—1636. https://doi.org/10.1016/j.cub.2010.07.017.

Yoshizawa, M., Jeffery, W.R., Netten, S. M. Van, Mchenry, M.J., 2014. The sensitivity of lateral line receptors and their role in the behavior of Mexican blind cavefish (*Astyanax mexicanus*). Journal of Experimental Biology 217 (6), 886—895. https://doi.org/10.1242/jeb.094599.

Yoshizawa, M., Yamamoto, Y., O'Quin, K.E., Jeffery, W.R., 2012. Evolution of an adaptive behavior and its sensory receptors promotes eye regression in blind cavefish. BMC Biology 10 (1), 108. https://doi.org/10.1186/1741-7007-10-108.

# 16

# Ecological and evolutionary perspectives on groundwater colonization by the amphipod crustacean *Gammarus minus*

## Daniel W. Fong and David B. Carlini

Department of Biology, American University, Washington, DC, United States

## Introduction

Earth is facing a sixth mass extinction, resulting from increasing human populations that demand accelerating rates of extractive economic activities to support rising levels of consumption (Kolbert, 2014; Ceballos et al., 2015). Effective conservation and management of Earth's rapidly diminishing biodiversity are essential tools to mitigate the deleterious consequences of biodiversity loss in this Anthropocene Epoch (Lewis and Maslin, 2015). However, such efforts are hobbled by inadequate to complete lack of knowledge of species richness at regional and local levels as well as the basic biology of many component species (Hortal et al., 2015). This is especially acute for noncharismatic invertebrates inhabiting subterranean ecosystems (Culver and Pipan, 2019). Studies of subterranean species face unique challenges. Much of subterranean spaces are too small for human access or have no human-sized natural openings or both (Curl, 1986), and exploration of accessible subterranean voids is often technically challenging. Consequently, the geographic ranges and potentials for dispersal for most subterranean species are poorly known (see Hutchins et al., 2010). In addition, the lengthy life spans and small population sizes of most subterranean species also render basic ecological studies impractical. Aquatic subterranean species, or stygobionts, face the additional challenge of increasing demand for groundwater withdrawal and likelihood of groundwater pollution due to expanding urban development and agricultural practices. In addition, the continuing increase in the average global temperature poses a potentially severe threat especially to stygobionts. Although there may be a lag time between warming of

*Groundwater Ecology and Evolution, Second Edition*
https://doi.org/10.1016/B978-0-12-819119-4.00016-0

surface waters and corresponding warming of subterranean waters, the result is inevitable. Subterranean habitats are generally limited in food resources compared to surface habitats, and most stygobionts are adapted to low food resources with lowered metabolic rates compared to surface-dwelling relatives (see Culver and Pipan, 2019). Because all stygobionts are ectothermic, warming of their habitats will result in an increase in their metabolic rates, which will require greater quantities of food resources to maintain routine physiological functions, and this may be highly detrimental in a low food resource environment. Because water is a good disperser of heat relative to air, thermal refuges from microhabitat variation in temperature are more limited for stygobionts compared to terrestrial troglobionts. Yet, even with such obstacles, there is growing awareness of a need to protect subterranean taxa and their habitats, driven by the continuing discovery of new species, the utility of some subterranean species as indicators of environmental quality and as models for the study of morphological evolution, and an increasing literature on the vulnerability of these species to anthropogenic disturbances (see Culver and Pipan, 2019).

In this chapter, we summarize the current knowledge of one stygobiont that inhabit groundwater resurgences, or karst springs, and migrated upstream to colonize subterranean vadose streams, the freshwater amphipod crustacean *Gammarus minus*. This species is an excellent model to understand upstream colonization of subterranean waters and its subsequent evolution (Fong, 2019). It exists as one species with multiple morphologically differentiated nontroglomorphic surface and troglomorphic cave populations, and different cave populations are independently derived from ancestral surface populations, representing multiple naturally replicated units for comparative studies. Knowledge on the *G. minus* system may inform on many other subterranean species of *Gammarus*, such as *G. cohabitus* (Holsinger et al., 2008), *G. acherondytes* (Wilhelm et al., 2006) and the *G. pecos* species complex (see Cannizzaro et al., 2017) in North America, *G. lacustris* (Østbye et al., 2018) and *G. balcanicus* (Pacioglu et al., 2020) in Europe, and the multitude of species in China (e.g., Hou and Li, 2010), as well as on many subterranean amphipods in the genus *Hyalella* in Central and South America (e.g., Rodrigues et al., 2014).

## Ecological setting and morphological variation

*Gammarus minus* was first described from specimens obtained in a spring run in southern Pennsylvania, USA (Say, 1818). The species has a wide geographic range, spanning much of the east central United States to west across the Mississippi River (Fig. 16.1, Culver et al., 1995). Within its range, it is ubiquitous in cold, carbonate springs in karst areas. The karst spring habitats of *G. minus* are characterized by hard, alkaline water of pH six or higher, conductivity greater than 100 μS/cm (Glazier et al., 1992), and temperatures that vary little from near the average annual ambient temperature for the latitude, about 8−12°C, over its range. Karst springs with appropriate physical and chemical properties but without *G. minus* are rare, at least in the mid-Atlantic region of the eastern USA. Within its spring habitat, *G. minus* is usually the dominant macroinvertebrate in terms of density, reaching 200 individuals/m$^2$. and probably biomass.

**FIGURE 16.1**  Distribution of *Gammarus minus* (shaded area) in the eastern United States. Troglomorphic cave populations located in Greenbrier Valley (GV) in the state of West Virginia and Wards Cove (WC) in the state of Virginia.

Sexually mature *Gammarus minus* of spring populations have large compound eyes with about 40 ommatidia, first pair of antennae reaching about 45%—50% of body length, and brownish body pigmentation (Fig. 16.2). Body size is highly variable among spring populations, with length of mature males ranging from 6 to 12 mm. The variation in body size among spring populations may be inversely correlated with the intensity of size-selective predation by small fish predators such as sculpins (*Cottus* sp.) (See below). Interestingly, eye size of *G. minus* appears larger in springs with higher density of sculpins (Glazier and Deptola, 2011).

*Gammarus minus* also inhabits caves streams throughout its range. Water from cave streams within a subterranean drainage basin resurge onto the surface at a spring, and the morphology of most cave populations of *G. minus* is identical to or only slightly different from that of hydrologically connected spring populations. However, populations of *G. minus* occurring in some caves in two small geographic areas, Greenbrier Valley of West Virginia, and Wards Cove of Virginia (Fig. 16.1), exhibit the classic troglomorphic syndrome common to species that are highly adapted to the subterranean environment. They differ from spring populations and nontroglomorphic cave populations in having consistently large body size with mature males attaining 10—12 mm in length, pale to bluish

**FIGURE 16.2**    Specimen of *Gammarus minus* from a surface spring (top) and from a cave (bottom). Both are mature males. The cave specimen is about 10 mm in body length. The two specimens are shown at the same scale. *Photos by Michael E. Slay. Used with permission.*

body coloration, long first antennae that are at least 65% of body length, and greatly reduced compound eyes with only a few to no discernible ommatidia (Culver et al., 1995: Fig. 16.2). The densities of nontroglomorphic populations in cave streams are highly variable and may reach similarly high values as spring populations, but densities of troglomorphic populations are at most an order of magnitude lower or less.

Typical of most amphipods in the family Gammaridae, *Gammarus minus* is sexually dimorphic in size, the body length of mature females is about two-thirds of mature males. During reproduction males and females form precopulatory pairs in amplexus. Although pairs in amplexus are observed throughout the year, their numbers tend to peak between December and March in spring populations, but do not show a consistent temporal pattern among troglomorphic cave populations, with some showing a peak in the winter months and some not. Fertilization is external. Each female deposits her eggs into a ventral marsupium formed by overlapping oostegites. Fertilized eggs develop directly within the marsupium into miniature amphipods, each about 1 mm in length. Brood size of spring populations is highly dependent on female body size, ranging from 5 to 16. Brood size of troglomorphic cave populations is much smaller and range from only two to six. The duration of embryonic development is about one to 2 months in spring populations, as the marsupia of the majority of ovigerous females contain dark egg masses in December and January but contain cream-colored, well-developed, embryonic miniature amphipods in February and March. The newly hatched amphipods are released directly from the marsupium. In the laboratory, the duration from when a female releases her first offspring to her last is variable, ranging from 1 to 8 days, and may or may not be associated with molting of the female, for both troglomorphic and nontroglomorphic populations. The life cycle is about one to 2 years in spring populations and probably also in nontroglomorphic cave populations but may be longer in troglomorphic cave populations. Specimens collected from any of the habitats can live for two to three additional years in the laboratory.

*Gammarus minus* in surface and subterranean habitats feed on organic detritus and derive nutrients from the associated microbial fauna and flora (Kostalos and Seymour, 1976). Troglomorphic cave populations also graze on the biofilm coating the substratum of cave streams and have been observed to prey on other organisms of smaller or similar sizes, such as annelids, isopods, and other amphipods, and show a tendency toward cannibalism, especially on smaller or injured individuals (Fong, 2011). These observations are confirmed by results from analysis of stable isotopes by MacAvoy et al. (2016) showing that a troglomorphic cave population obtained its nitrogen from animal sources, in addition to detritus associated microbial sources, but not in the hydrologically connected spring population. They also showed that nitrogen was not obtained from animal sources in another spring population of similar body size as the cave population, indicating the tendency to prey on other species was not solely because it attained larger body size in caves. Broadening of food niches as an adaptation to the cave environment has also been demonstrated for subterranean amphipods in the Edwards Aquifer (Hutchins et al., 2014) and is consistent with the general theory of niche expansion in resource poor environments.

## Upstream colonization of subterranean waters by *Gammarus minus*

There is evidence that *Gammarus minus* has a strong tendency to migrate upstream, indicating that troglomorphic cave populations likely originated through colonization of cave streams by *G. minus* dwelling at the spring where the cave waters resurge. An analysis by Fong and Culver (1994) of the distribution of five aquatic crustacean species, including *G. minus*, within the complex Organ Cave system (Stevens, 1998) indicated that *G. minus* colonized subterranean streams from the downstream direction. They found that while three crangonyctid amphipods, one in the genus *Crangonyx* and two in the genus *Stygobromus*, as well as one asellid isopod of the genus *Caecidotea*, were abundant in smaller upper-level streams and rapidly decreased in abundance at larger lower-level streams, the opposite was true for *G. minus*. They suggested that the pattern was explained by different routes of colonization of the cave system, with the other species colonizing from the epikarst and *G. minus* colonizing from downstream originating from the resurgence.

This idea was supported by data from Carlini et al. (2009), who, as part of a larger study, analyzed sequence variation in the cytochrome *c* oxidase I (COI) gene between *Gammarus minus* specimens from the downstream most section of Organ Cave, designated OCB, which was hydrologically nearest to the resurgence, and specimens from an upper-level head-water stream, called OCM, which was far from the resurgence. The average nucleotide diversity of COI at OCB (0.0185), although low, was as high or higher than those at 13 other cave and spring *G. minus* populations, while at OCM there was no variation at all, with all specimens fixed for the same haplotype. Within a cave drainage system, larger lower-level streams more proximal to its resurgence are fed by more tributaries, and thus are more stable habitats than are smaller upper-level streams that may dry out from occasional regional droughts causing episodic local extinctions. In fact, one of us (DWF) has personally observed complete drying of the stream at the OCM section at least four times over a 30-year span, but never at the OCB location. They suggested that *G. minus* at OCB and OCM represented classic source and sink populations, respectively. In this scenario, founder individuals from OCB are continuously probing upstream to colonize upper-level sections, such as OCM. Therefore, periodic extinction followed by recolonization by founder specimens, combined with genetic drift, could account for the absence of COI nucleotide variation in *G. minus* at OCM.

## Impetus for colonizing cave streams

A long-standing dogma is the greatly reduced amplitude of variation in environmental parameters, especially temperature, of subterranean habitats compared to surface habitats (Poulson and White, 1969). Avoidance of seasonal extremes of temperature and associated conditions is often proposed as part of the impetus driving surface organisms into subterranean space (see Cooper et al., 2007; Danielopol and Rouch, 2012). *Gammarus minus*, however, does not fit well with this paradigm. Like most freshwater *Gammarus* species, it is a cold stenotherm, seldom occurring where temperature exceeds 15°C and absent from waters exceeding 20°C (Marchant, 1981). It is likely a climate relict, which previously occupied cold streams south of the last glacial maximum. Upon warming of surface streams since, however, it became restricted to karst springs serving as consistently cold refuges. Temperatures of karst springs are generally stable (Carroll and Thorp, 2014), showing yearly variation of less than 2–3°C in the mid-Atlantic region of the USA (Man, 1991; Gooch and Glazier, 1991; Glazier et al., 1992; Fong, unpublished data). Temperatures of cave streams that resurge at these springs, however, show much greater seasonal as well as short-term temporal variation, including rapid spikes in response to storm events, especially at small upper-level passages near insurgences fed by sinking surface streams (Zeit, 1993; Culver et al., 1995). Therefore, within a subterranean drainage basin, there exists a gradient of wide temperature variation at upper-level headwater streams to less variation at lower-level large streams to almost no variation at the resurgence. Thus, the impetus by *G. minus* to colonize cave streams is not to avoid extremes of fluctuations in temperature. Contrarily, cave streams may expose

G. *minus* to greater fluctuations in temperature and associated chemical parameters, such as calcium and magnesium concentrations, compared to the resurgence. Furthermore, compared to cave streams, spring runs have richer and more stable food resources such as decaying aquatic macrophytes and allochthonous coarse particulate organic matter for G. *minus*, a detritivore. Instead, the selective agent driving colonization of cave streams by G. *minus* may partially involve ecological release from avoidance of predators at karst springs.

## Size-selective predation, sexual selection and cave colonization

As indicated above, the body lengths of sexually mature *Gammarus minus* individuals vary tremendously among karst springs in the mid-Atlantic region of the eastern US, with males ranging from 6 to 12 mm among springs in West Virginia and Virginia (Culver et al., 1995). Cloud (2019) and Cloud and Fong (in prep.) examined the potential causes of body size variation among these populations. They hypothesized that the body size at any one spring is a balance between size-selective predation by the sculpin fish predator, *Cottus carolinae*, driving it down, and sexual selection among males competing for access to larger females carrying more eggs, driving it up. Therefore, populations of G. *minus* in karst spring where C. *carolinae* is absent should have larger body sizes than ones in karst springs where the fish predator is present. The effects of size-selective predation on life history traits of prey are well documented and much work have focused on effects of fish predators, especially sculpins, on amphipods, resulting in decreased body size (Newman and Waters, 1984; Mac-Neil et al., 1999; Kinzler and Maier, 2006; Berezina, 2011), shifted ontogenetic metabolic scaling (Glazier et al., 2011), and altered mating behavior (Cothran, 2004; Dunn et al., 2008; Lewis and Loch-Mally, 2010).

The main results of Cloud's work are summarized in Table 16.1. The mean and the minimum body sizes of sexually mature males and females and the mean body sizes of ovigerous females were significantly smaller in eight spring populations where *Cottus carolinae* was present than in nine spring populations where the sculpin was absent, providing strong indirect evidence that G. *minus* evolved toward smaller body size in response to predation by the sculpin. The cost of smaller body size appears to be lower fecundity. Ovigerous females carried significantly fewer eggs in springs where C. *carolinae* was present than in springs without sculpins. Sculpin predation also affected the relationship between brood size and female body size. The magnitudes of the slopes of regression of egg number on female body size were, on average, significantly lower in springs where the sculpin was present than where the fish was absent. The correlation of body size between amplexing pairs of males and females were significantly lower in springs with sculpins than in springs without. These results suggest that in springs with sculpins, competition among males for larger females was weak, partially because fertilizing the slightly higher number of eggs in larger females did not balance the increased risk of predation. Therefore, individuals that migrated upstream from springs with sculpins and colonized subterranean streams where sculpins were absent may incurred higher fitness, not only because they were released from predation but also because release from predation favored larger females with higher fecundity, which then favored larger males competing for access to larger females. Interestingly, maximum body sizes of

**TABLE 16.1**    Range of mean values of measurements of *Gammarus minus* from eight springs where sculpins were present (Sculpin +) and from nine springs where sculpins were absent (Sculpin −). Body size is given as the head capsule length (HL) in ocular micrometer units (at 42 units equals 1.0 mm).

|  | Sculpin + [8] | Sculpin − [9] |
| --- | --- | --- |
| Amp M | 37.4−45.4 | 49.7−55.4 |
| Amp F | 27.2−33.8 | 36.5−45.0 |
| Min M | 30−38 | 39−47 |
| Min F | 21−29 | 26−34 |
| Corr | 0.15−0.57 | 0.25−0.73 |
| Ovi F | 28.8−34.5 | 37.0−44.6 |
| Egg # | 5.0−8.3 | 9.1−16.4 |
| Slope | 0.29−0.78 | 0.28−1.20 |

*Amp M and Amp F*, mean HL of males and females in amplexus, respectively; *Min M and Min F*, minimum HL of males and females in amplexus, respectively; *Corr*, Pearson's correlation coefficients between HL of males and females in amplexus; *Ovi F*, mean HL of ovigerous females; *Egg #*, mean numbers of eggs per ovigerous female; Slope, range of slopes of linear regressions of number of eggs on ovigerous female HL.

all troglomorphic and some nontroglomorphic cave populations, in the absence of sculpin predation, are about 11−12 mm, the same as maximum body sizes of nontroglomorphic populations from springs where sculpins are absent. Thus, avoidance of sculpin predation may be a strong driving force behind initial colonization of subterranean vadose habitats by *G. minus*.

## Multiple independent colonization of cave streams

There is strong evidence that different troglomorphic populations of *Gammarus minus* resulted from multiple independent colonization of subterranean drainage basins, each from an ancestral population dwelling in a karst spring. Groundwater resurging at karst springs flow as spring runs for varying distance before emptying into their respective base level streams. The mean values and especially the temporal variation in physical and chemical parameters, such as temperature and cation concentrations, are usually very different at karst springs and at their base level streams (see Carroll and Thorp, 2014). Thus, migration among *G. minus* populations inhabiting even different karst springs that empty into the same base level stream is rare, and populations in different subterranean basins are effectively isolated from each other. Therefore, populations in different subterranean basins had likely evolved troglomorphy independently multiple times.

## Allozyme variation

A population genetic survey of eight polymorphic allozyme loci demonstrated that, despite the substantial morphological divergence of troglomorphic cave and nontroglomorphic spring populations of *Gammarus minus*, cave populations were most closely related to hydrologically proximate surface populations than to other cave populations (Kane et al., 1992). This supports the notion that troglomorphy evolved independently in different cave populations, and that each cave population originated via colonization by individuals of the surface population dwelling at the spring where cave water resurges. Cluster analysis of the allozyme data did not always group populations hydrologically. To explain this, Kane et al. (1992) and Culver et al. (1995) suggested that the genetic relationship between a cave population and its hydrologically connected surface population has been diluted or lost resulting from periodic, local extinction of spring populations when some springs cease flowing during draughts, followed by recolonization by individuals from other spring populations. Furthermore, the allozyme survey of *G. minus* did not reveal any significant differences in the level of genetic variation in cave versus spring populations, and over half of the loci (10 of 18) were invariant within populations. These results could be attributed to several factors, including random sampling error due to few detectable alleles at allozyme loci, the modest number of polymorphic allozyme loci surveyed, gene flow between seemingly disparate populations, or to direct selection on alleles of the allozyme loci. Coelho et al. (2003) also found very low intra- and interpopulation allozyme variability in the widespread European species *Gammarus locusta* and suggested that this may be a common feature among continuously distributed amphipods with extensive ranges.

## COI and ITS sequences

In an effort to further characterize the genetic relationships of the populations examined in the Kane et al. (1992) allozyme study, Carlini et al. (2009) conducted a population genetic survey of 15 *Gammarus minus* populations through analysis of DNA sequences from one mitochondrial gene, cytochrome *c* oxidase I (COI), and one nuclear gene, internal transcribed spacer 1 (ITS-1). Standing variation at both loci was low within populations (<2% for COI, <1% for ITS-1), but in accordance with the allozyme data a significant degree of divergence and spatial structuring of populations was observed. In general, divergence values reflected the hydrological distances among the populations, but hydrologically proximate nontroglomorphic spring populations and troglomorphic cave population pairs exhibited low levels of divergence despite significant morphological differentiation. This result supports the idea of multiple independent colonization of subterranean habitats by hydrologically proximate spring populations.

In addition, levels of genetic variation at the COI locus were significantly lower within cave populations than within their corresponding sister spring populations, suggesting that cave populations were initially colonized by only a few individuals, or that they are more prone to bottlenecks, possibly a consequence of larger temperature fluctuations or higher frequency of drought conditions associated with these cave stream habitats. Also consistent with a higher incidence of bottlenecks in cave populations was the finding that the COI sequences of cave populations exhibited less codon bias, as measured by the effective number of codons (ENC), than their sister populations (Wright, 1990) (Fig. 16.3). Codon bias arises due to weak selection at synonymous sites, and the strength of weak selection is inversely proportional to the effective population size ($N_e$). Thus, selection at synonymous sites is more efficient in larger populations than in smaller populations, where genetic drift plays a more prominent role in determining the fate of synonymous variants. Therefore, the finding that cave populations harbor significantly lower levels of codon bias than that of their surface sister populations suggests smaller $N_e$ in cave populations. These results indicate that initial colonization of subterranean habitats was difficult, and few individuals could adapt and form new cave populations; yet, such cave populations were successfully established multiple times independently, suggesting strong selection to invade subterranean streams from karst springs, possibly to avoid fish predation (see Size-selective Predation, Sexual Selection and Cave Colonization).

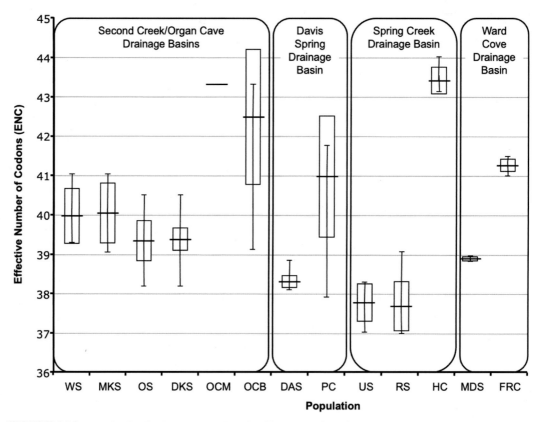

**FIGURE 16.3**    Levels of codon bias measured as the effective number of codons (ENC) in COI sequences from 13 populations of *Gammarus minus*. For each population, the averages (thick horizontal lines), standard deviation (boxes), and range (vertical bars) of ENC are indicated. Low ENC values indicate high codon bias, and vice versa. Populations are grouped by drainage basin and habitat type (surface springs vs. caves). *WS*, Ward Spring; *MKS*, Mike Spring; *OS*, Organ Spring; *DKS*, Dickson Spring; *OCM*, Organ Cave Main; *OCB*, Organ Cave Bowen Canyon; *DAS*, Davis Spring; *PC*, Persinger Cave; *US*, US219 Spring; *RS*, Rock Spring; *HC*, The Hole Cave; *MDS*, Maiden Spring; *FRC*, Fallen Rock Cave.

# Evolutionary perspectives

Upon colonization of cave streams, *Gammarus minus* populations must simultaneously adapt to the subterranean environment. We have discussed above the broadened food niche and a possible reason for larger body size in cave compared to surface *G. minus* populations. Although our current understanding is that the subterranean environment consists of a diversity of habitat types with different physical parameters and biological components, the one overriding constant is the absence of light and its consequences (Culver and Pipan, 2019). The most obvious morphological differences of cave compared to surface *G. minus*, as is true for many other cave and surface species, are reductions in eyes and body pigment, both directly related to adapting to the aphotic environment. In this section, we summarize recent studies using molecular techniques to explore adaptation of *G. minus* to the subterranean environment.

## Eye development gene expression

Aspiras et al. (2012) compared the levels of expression of four genes, *hedgehog, pax6, sine oculus* and *dachshund,* involved in the developmental pathway of arthropod eyes in three sister-pairs of spring and cave populations of *Gammarus minus.* They found a parallel reduction in expressions of only one of the genes, *hedgehog,* in the cave populations compared to surface populations but not in the other three genes upstream of *hedgehog* in the pathway. Their results mirror that of Yamamoto et al. (2004), who showed that *hedgehog* related genes were also involved in eye reduction in the cavefish *Astyanax mexicanus,* although *hedgehog* expression was upregulated in the cavefish compared to surface fish (see Gross et al., this volume). These results indicate that relaxed selection may operate on similar genes governing eye development in a vertebrate and an invertebrate following colonization of the subterranean environment. The implication is that the genetic mechanism behind convergent morphological adaptation among diverse species may be due to constraints on other morphogens involved in eye development because they could lead to other deleterious pleiotropic effects, whereas any pleiotropic effects of altered *hedgehog* expression may not be deleterious.

## Opsins show No loss of functional constraints in cave populations

Carlini et al. (2013) compared DNA sequence variation and levels of expression in two paralogs of the gene for opsin, a protein that functions in phototransduction and is responsible for photosensitivity, also in three sister-pairs of nontroglomorphic surface and troglomorphic cave populations of *Gammarus minus.* Comparisons of orthologous opsin nucleotide sequences revealed low levels of sequence divergence, both within and between populations, for both paralogs. Levels of divergence within cave populations were not appreciably different from those within surface populations, nor were there any consistent differences in cave and surface pairwise $dN/dS$ ratios ($dN$ = substitution rate at nonsynonymous sites, $dS$ = substitution rate at synonymous sites). Given that the individuals from cave populations have reduced or entirely absent eyes, one might expect a neutral $dN/dS$ ratio within cave populations than within surface populations, reflecting some relaxation of functional constraint. However, this was not the case for both opsin genes, where the *dN/dS* ratio for cave populations was often equal to or less than that of corresponding sister population from the surface (Fig. 16.4). These results are consistent with previous work on opsins in crayfish, where there was no evidence for loss of functional constraint in cave species compared to surface species (Crandall and Hillis, 1997). For both paralogs, the average relative expressions were significantly higher, with at least a fivefold difference, in the spring compared to cave populations. For both genes there was also substantial variation in expression from individual to individual within populations. These results suggest that these opsin genes may be differentially regulated in the surface and cave populations, but the genetic basis of that difference remains unknown. Probably, functions of the opsins in darkness were maintained by selection through pleiotropy because opsins served important but unknown functions unrelated to photosensitivity. Thus, reduction in opsin expression may be advantageous in darkness up to a point, beyond which further reduction is selected against because of unknown pleiotropic effects. On the other hand, it is possible that the persistence of opsin expression may be consequence of there not having been enough time since colonization for both genes to have incurred mutations that completely inactivate expression.

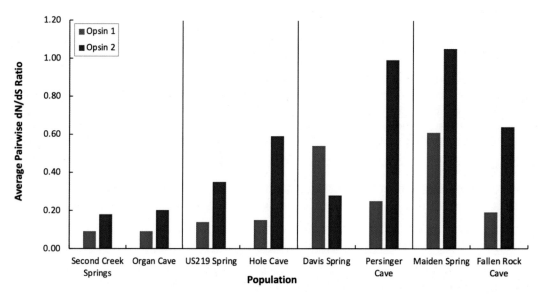

**FIGURE 16.4**    Average dN/dS ratios of opsin one and opsin two for pairwise comparisons within populations. We used modified dN/dS, where dN/dS = dN/((dS+1)/S). Using this modification, dN/dS remains defined when dS = 0. Vertical lines separate populations grouped by drainage basin and habitat type (surface springs vs. caves).

## Changes in gene expression associated with adaptation to the cave habitat

Carlini and Fong (2017) conducted RNA-Seq on four individuals each from a pair of morphologically distinct sister populations inhabiting Ward Spring (WS) and Organ Cave (OC) to identify genes that were differentially expressed in the two populations, as well as to compare levels and patterns of genetic variation within and between populations. They found that 57% of transcripts had higher average levels of expression in the cave population and that there was an almost three-fold enrichment of cave-upregulated genes. This general tendency toward upregulation of gene expression in a subterranean population or species is difficult to explain given that, in general, previous studies have documented reduction in the expression of sets of related genes in cave populations/species, such as those related to vision or circadian rhythms. However, there were some interesting transcripts that were significantly downregulated in the OC population, including homologs of vision related genes such as *opsin*, *arrestin*, and the *Bardet-Biel syndrome protein*, as well as a reduction in expression in the gene for the DNA repair protein photolyase. The downregulation of two opsin homologs in the OC population, as well as the downregulation of *hedgehog* expression in the OC population relative to the WS population were already discussed in the sections above. However, there were no obvious candidates for significantly downregulated genes involved in pigmentation pathways.

### Evidence for positive selection

Carlini and Fong (2017) also found that levels of genetic variation in the transcripts were generally lower in the OC population, and the average level of nucleotide diversity across all transcripts was significantly lower in OC. Within OC, there was nearly 1.5-fold reduction in levels of genetic variation in cave-downregulated transcripts relative to cave-upregulated transcripts, a highly significant difference. In contrast, within the WS population the nucleotide diversities of cave-downregulated and cave-upregulated transcripts was virtually identical (Fig. 16.5). Three lines of evidence suggested that the reduced variation in cave downregulated transcripts was due to positive selection in the cave population. First, the average neutrality index (NI), the odds ratio of fixed nonsynonymous to fixed synonymous substitutions ($F_N/F_S$) over nonsynonymous to synonymous polymorphisms ($P_N/P_S$), was less than one for cave downregulated transcripts, consistent with positive selection, and significantly less than that of cave-upregulated genes. This could result from positive selection on the regulatory regions of cave-downregulated transcripts. A second line of evidence in favor of positive selection was that Tajima's $D$, a test statistic for positive selection that is independent of NI (where $D < 1$ indicates positive selection), was positively correlated with the cave:surface (C:S) expression ratio. In other words, Tajima's D was generally less than one in transcripts with a low C:S expression ratio and greater than one in transcripts with a high C:S expression ratio. The third line of evidence suggesting that cave-downregulated transcripts are the target of positive selection is that they were more highly diverged from their surface homologs than cave-upregulated transcripts. This follows from the fact that, if positive selection acted on the cave-downregulated transcripts in the OC population then, on average, those transcripts should exhibit a higher degree of sequence divergence from those of the WS population than the corresponding comparisons between the cave-upregulated transcripts, resulting in an enrichment of highly diverged cave downregulated transcripts. There was an eightfold enrichment of very highly diverged cave-downregulated transcripts over cave-upregulated transcripts, and an over twofold enrichment of highly diverged cave-downregulated transcripts over cave-upregulated transcripts, consistent with what would be expected if the cave-downregulated transcripts were the targets of positive selection.

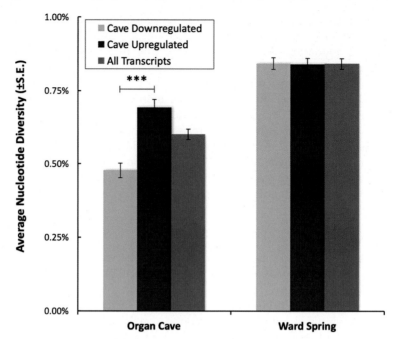

**FIGURE 16.5** Average nucleotide diversities of cave downregulated, cave upregulated, and all unigene transcripts within the Organ Cave and Ward Spring populations. In Ward Spring, the average nucleotide diversities of cave downregulated and cave upregulated genes were nearly identical at 0.840% and 0.843%, respectively. In Organ Cave, the nucleotide diversity of cave downregulated genes, 0.479%, was significantly less than that of cave upregulated genes, 0.693% ($P < 10^{-25}$ Welch's t-test, $P < 10^{-118}$, paired t-test, as indicated by the asterisks).

### Photolyase

In addition, Carlini and Fong (2017) discovered that among the five transcripts that had a fixed premature termination codon (PTC) in the OC population (all four individuals were homozygous for the PTC) was a transcript, which encodes photolyase, a protein that repairs UV-induced DNA damage using short-wavelength visible light. There is strong evidence for the relaxation of functional constraint in this light-dependent protein in the subterranean OC population. Because the downregulation and relaxed functional constraint of photolyase in the OC population made evolutionary sense given the absence of UV and visible light in the cave environment, they assayed an additional 10 individuals each from the OC and WS populations for photolyase gene expression and DNA sequence. They confirmed that the photolyase expression was downregulated in the OC population, although the level of downregulation, while remaining statistically significant, was less pronounced than the results from the RNA-seq data on only four individuals from each population. Sequencing also confirmed that the PTC was fixed in and restricted to the OC population: all 10 individuals from the OC population were homozygous for the PTC (TAA), whereas all 10 individuals from the WS population were homozygous for the TAC tyrosine codon. Given the absence of both UV and visible light in the OC habitat, it makes sense that a specific mechanism to repair UV-induced pyrimidine dimers may not be required in that population. Therefore, any mutations, which negatively affect the function of photolyase would not be subject to purifying selection and could drift to fixation. Results from UV tolerance experiments indicate that individuals from the OC population seemed less tolerant of UV exposure in the lab, suggesting that the photoreactivation repair mechanism is compromised due to the PTC in the photolyase gene of OC individuals (Carlini, unpublished data).

Photoreactivation is a highly conserved and ancient DNA repair mechanism, present in Eubacteria, Archaea, and Eukaryotes. Photoreactivation has been independently lost in multiple lineages within all three groups including, among the Eukaryotes, placental mammals. Given that placental mammals are completely dependent on the alternative pathway of nucleotide excision repair (NER) to remove pyrimidine dimers, it potentially renders them, and other groups in which photoreactivation has been lost, more prone to UV-induced DNA damage. The loss of photoreactivation has recently been documented in a second cave dwelling species, the Somalian blind cavefish *Phreatichthys andruzzii* (Zhao et al., 2018), thus loss of photoreactivation may be a common feature of subterranean organisms.

## Melanin pigment loss and innate immunity

In addition to eye loss, loss of body pigmentation is another hallmark feature of organisms adapted to subterranean environments. A longstanding idea is that relaxed selection for pigmented traits in darkness allows for accumulation of mutations that eventually eliminate pigment production. Indeed, functions of pigments, such as protection from harmful UV radiation, camouflage, aposematic coloration, attracting potential mates, etc., are unnecessary in aphotic environments. In a variety of organisms, melanin is a major component of body pigmentation. The melanin synthesis pathway in arthropods is well categorized in insects (True, 2003). It starts with conversion of tyrosine into DOPA, catalyzed by tyrosine hydroxylase or phenoloxidase (PO) or both. Further downstream processing of DOPA results in the end products dopamine melanin and dopamelanin, and PO also catalyzes critical parts of this downstream processing. Thus, PO is involved in multiple parts of the pathway of conversion of the precursor molecule to the final product during melanin synthesis. Bilandžjia et al. (2012) discovered that a species each in two lineages of albino cave cixiid planthoppers, one from Hawaii and one from an island off the coast of Croatia, were convergent for disruption of melanin production at the same initial step, between tyrosine and DOPA, because application of DOPA rescued melanin production, while application of DOPA and concomitant inhibition of enzymes involved in downstream processing of DOPA, including PO, again stopped melanin production. They suggested that disruption of this initial step, leading to albinism from loss of melanin, is selectively advantageous because tyrosine is also a precursor to synthesis of catecholamines, molecules involved in stress responses. Assuming the subterranean habitat is a more stressful environment than surface habitats, catecholamines levels should be higher in cave than in surface populations. Bilandžjia et al. (2013) did show that catecholamine levels were much higher in the cave forms compared to surface forms of the fish *Astyanax mexicanus*. This result does not exclude possible neutral mutations affecting the expression levels of stress response molecules in subterranean organisms.

Disruption of melanin synthesis may, however, incur a significant cost because melanin and intermediate molecules during PO mediated melanin synthesis play important roles in innate immunity and wound healing in invertebrates (Sugumaran, 2002; Cerenius and Söderhäll, 2004; Christensen et al., 2005), and thus the melanin synthesis pathway in albino subterranean invertebrates may only be inactivated and not genetically disrupted. Bilandžjia et al. (2017) tested this idea by amputating appendages of surface and cave arthropods belonging to most of the higher taxonomic groups. In almost all groups, including two sister-pairs of surface and cave *Gammarus minus*, melanin was produced during wound healing at the amputation sites but was absent when exposed to an inhibitor of PO. Furthermore, the speed of wound melanization was not different between cave and surface *G. minus* and the troglomorphic *G. cohabitus*, with different metabolic rates. Thus, the PO mediated melanin synthesis pathway in *G. minus* and in most arthropods is highly regulated and is only activated in response to an immune challenge, such as wounding.

Tyrosine is a critical component in the early steps of the PO mediated melanin synthesis pathway, and it is derived from phenylalanine, which is obtainable only from ingested food, while melanin is itself a nitrogen rich compound and requires significant input of protein rich resources for its synthesis (Lee et al., 2005). Therefore, melanin synthesis is costly in terms of quantity and quality of food resources and this cost is especially substantial in subterranean habitats. Furthermore, subterranean environments may harbor lower abundance and diversity of parasites and pathogens (Tobler et al., 2007). Thus, it is conceivable that cave organisms may experience relaxed selection to maintain the costly melanin synthesis pathway, but clearly this idea is not supported by the results.

## Future directions

Amphipods and isopods, among other crustaceans, constitute major components of the stygobiotic fauna in groundwaters, and the potential routes and processes of their colonization of subterranean waters are varied and complex (see Culver and Pipan, 2019). Obviously, colonization of subterranean waters starts with nontroglomorphic, nonstygobiotic surface populations. Yet, the occurrence, abundance, and diversity of nontroglomorphic taxa in subterranean systems are understudied (Fong and Culver, 2018). Furthermore, the basic ecology in their respective native habitats of most of surface taxa with subterranean relatives are also understudied. As illustrated by the example of *Gammarus minus*, a simple study of life-history variation among surface spring populations informed on a hypothesis on the impetus for colonization of subterranean waters, and this hypothesis may apply to many other amphipods with related cave-adapted populations or species, including the many amphipod species of *Gammarus* in the USA, Europe and especially China, and of *Hyalella* in the Americas. Specific to *G. minus*, future work on variations in intensity of size-selective predation and in intensity of sexual-selection within and among springs with and without sculpin predators are needed to further examine why spring populations started to colonize subterranean waters. Clearly, the dogma of avoidance of variations in environmental conditions or the idea of avoidance of predation as the impetus for colonization of subterranean environments does not apply to all taxa in all subterranean habitats. More ecological studies of surface taxa with subterranean relatives should generate a diverse set of hypotheses on not only how, but why surface taxa colonize subterranean waters, including not only vadose streams but hypotelminorheic, epikarstic and phreatic waters.

There is often a focus on diversity in terms of species richness of troglomorphic taxa in major subterranean systems, usually in the form of a species list. Yet, data on the distribution within such systems are lacking for most of the taxa. As illustrated by Fong and Culver (1994), a fine-grained analysis of the distribution of aquatic taxa such as *Gammarus minus* within a subterranean drainage basin may be informative on putative routes of colonization. Such microgeographic distribution data are clearly needed in future studies of subterranean taxa, and such data should be easier to obtain for aquatic than for terrestrial organisms as they are confined to where water occurs in the subterranean systems.

Finally, the rapidly accelerating advancement of molecular techniques should allow for discovery of mechanisms of evolution of specific phenotypes associated with the subterranean environment. Comparison of subterranean taxa with closely related surface taxa is critical for evolutionary studies. *Gammarus minus* is among only a handful of species with morphologically differentiated surface and subterranean populations where multiple subterranean populations have been repeatedly and independently derived from ancestral surface of populations, and it should continue to play an important role in studies on adaptations to the subterranean realm, such as illustrated in this chapter.

## Acknowledgments

We thank M.E. Bichuette and M. Policarpo for constructive comments that improved this chapter.

## References

Aspiras, A.C., Prasad, R., Fong, D.W., Carlini, D.B., Angelini, D.R., 2012. Parallel reduction in expression of the eye development gene *hedgehog* in separately derived cave populations of the amphipod *Gammarus minus*. Journal of Evolutionary Biology 25, 995–1001. https://doi.org/10.1111/j.1420-9101.2012.02481.x.

Berezina, N.A., 2011. Perch-mediated shifts in reproductive variables of *Gammarus lacustris* (Amphipoda, Gammaridae) in lakes of northern Russia. Crustaceana 84 (5–6), 523–542. https://doi.org/10.1163/001121611X577909.

Bilandžjia, H., Cetkovic, H., Jeffery, W.R., 2012. Evolution of albinism in cave planthoppers by a convergent defect in the first step of melanin biosynthesis. Evolution and Development 14 (21), 196–203. https://doi.org/10.1111/j.1525-142X.2012.00535.x.

Bilandžjia, H., Ma, L., Parkhurst, A., Jeffery, W.R., 2013. A potential benefit of albinism in *Astyanax* cavefish: downregulation of the oca2 gene increases tyrosine and catecholamine levels as an alternative to melanin synthesis. PLoS One 8 (1), e80823. https://doi.org/10.1371/journal.pone.0080823.

Bilandžjia, H., Laslo, M., Porter, M.L., Fong, D.W., 2017. Melanization in response to wounding is ancestral in arthropods and conserved in albino cave species. Scientific Reports 7, 17148. https://doi.org/10.1038/s41598-017-17471-2.

Cannizzaro, A.G., Walters, A.D., Berg, D.J., 2017. A new species of freshwater *Gammarus* Fabricius, 1773 (Amphipoda: Gammaridae) from a desert spring in Texas, with a key to the species of the genus *Gammarus* from North America. Journal of Crustacean Biology 37 (6), 709–722. https://doi.org/10.1093/jcbiol/rux088.

Carlini, D.B., Fong, D.W., 2017. The transcriptomes of cave and surface populations of *Gammarus minus* (Crustacea: Amphipoda) provide evidence for positive selection on cave downregulated transcripts. PLoS One 12 (10), e0186173. https://doi.org/10.1371/journal.pone.0186173.

Carlini, D.B., Satish, S., Fong, D.W., 2013. Parallel reduction in expression, but no loss of functional constraint, in two opsin paralogs within cave populations of *Gammarus minus* (Crustacea: Amphipoda). BMC Evolutionary Biology 2013 (13), 89. http://www.biomedcentral.com/1471-2148/13/89.

Carlini, D.B., Manning, J., Sullivan, P.G., Fong, D.W., 2009. Molecular genetic variation and population structure in morphologically differentiated cave and surface *Gammarus minus* (Crustacea, Amphipoda). Molecular Ecology 18, 1932–1945. https://doi.org/10.1111/j.1365-294X.2009.04161.x.

Carroll, T.M., Thorp, J.H., 2014. Ecotonal shifts in diversity and functional traits in zoobenthic communities of karst springs. Hydrobiologia 738 (1), 1–20. https://doi.org/10.1007/s10750-014-1907-4.

Ceballos, G., Ehrlich, P.R., Barnosky, A.D., Garcia, A., Pringle, R.M., Palmer, T.M., 2015. Accelerated modern human-induced species losses: entering the sixth mass extinction. Science Advances 1 (5), e1400253. https://doi.org/10.1126/sciadv.1400253.

Cerenius, L., Söderhäll, K., 2004. The prophenoloxidase-activating system in invertebrate immunity. Immunological Reviews 198, 116–126. https://doi.org/10.1111/j.0105-2896.2004.00116.x.

Christensen, B.M., Li, J., Chen, C.-C.C., Nappi, A.J., 2005. Melanization immune responses in mosquito vectors. Trends in Parasitology 21 (4), 192–199. https://doi.org/10.1016/j.pt.2005.02.007.

Cloud, M., 2019. Size-Selective Predation and Sexual Selection on Body Size Variation Among Karst Spring Populations of the Amphipod *Gammarus minus*. M.S. thesis. American University, Washington, DC.

Coelho, H., Costa, F.O., Costa, M.H., Coelho, M.M., 2003. Low genetic variability of the widespread amphipod *Gammarus locusta*, as evidenced by allozyme electrophoresis of southern European populations. Crustaceana 75 (11), 1335–1348. https://www.jstor.org/stable/20105523.

Cooper, S.J.B., Bradbury, J.H., Saint, K.M., Leys, R., Austin, A.D., Humphreys, W.F., 2007. Subterranean archipelago in the Australian arid zone: mitochondrial DNA phylogeography of amphipods from central Western Australia. Molecular Ecology 16, 1533–1544. https://doi.org/10.1111/j.1365-294X.2007.03261.x.

Cothran, R.D., 2004. Precopulatory mate guarding affects predation risk in two freshwater amphipod species. Animal Behaviour 68 (5), 1133–1138. https://doi.org/10.1016/j.anbehav.2003.09.021.

Crandall, K.A., Hillis, D.M., 1997. Rhodopsin evolution in the dark. Nature 387, 667–668. https://www.nature.com/articles/42628.

Culver, D.C., Pipan, T., 2019. The Biology of Cave and Other Subterranean Habitats, second ed. Oxford University Press, New York, NY.

Culver, D.C., Kane, T.C., Fong, D.W., 1995. *Adaptation and Natural Selection in Caves: The Evolution of* Gammarus Minus. Harvard University Press, Cambridge, MA.

Curl, R.L., 1986. Fractal dimensions and geometrics of caves. Mathematical Geology 18 (8), 765–783. https://doi.org/10.1007/BF00899743.

Danielopol, D.L., Rouch, R., 2012. Invasion, active versus passive. In: White, W.B., Culver, D.C. (Eds.), Encyclopedia of Caves, second ed. Elsevier Academic Press, Amsterdam, pp. 404–409.

Dunn, A.M., Dick, J.T.A., Hatcher, M.J., 2008. The less amorous *Gammarus*: predation risk affects mating decisions in *Gammarus duebeni* (Amphipoda). Animal Behaviour 76 (4), 1289–1295. https://doi.org/10.1016/j.anbehav.2008.06.013.

Fong, D.W., 2011. Management of subterranean fauna in karst. In: Beynen, P.E. (Ed.), Karst Management. Springer, Dordrecht, pp. 201–224.

Fong, D.W., 2019. *Gammarus minus*: a model system for the study of adaptation to the cave environment. In: White, W.B., Culver, D.C., Pipan, T. (Eds.), Encyclopedia of Caves, third ed. Elsevier Academic Press, Cambridge, pp. 451–458.

Fong, D.W., Culver, D.C., 1994. Fine-scale biogeographic differences in the crustacean fauna of a cave system in West Virginia, USA. Hydrobiologia 287, 29–37. https://doi.org/10.1007/BF00006894.

Fong, D.W., Culver, D.C., 2018. The subterranean aquatic fauna of the Greenbrier Karst. In: White, W.B. (Ed.), Cave and Karst of the Greenbrier Valley in West Virginia, Cave and Karst Systems of the World. Springer, Cham, pp. 385–397. https://doi.org/10.1007/978-3-319-65801-8_19.

Glazier, D.S., Deptola, T.J., 2011. The amphipod *Gammarus minus* has larger eyes in freshwater springs with numerous fish predators. Invertebrate Biology 130 (1), 60–67. https://doi.org/10.1111/j.1744-7410.2010.00220.x.

Glazier, D.S., Horne, M.T., Lehman, M.E., 1992. Abundance, body composition and reproductive output of *Gammarus minus* (Amphipoda, Gammaridae), in ten cold springs differing in pH and ionic content. Freshwater Biology 28, 149–163. https://doi.org/10.1111/j.1365-2427.1992.tb00572.x.

Glazier, D.S., Butler, E.M., Lombardi, S.A., Deptola, T.J., Reese, A.J., Satterthwaite, E.V., 2011. Ecological effects on metabolic scaling: amphipod responses to fish predators in freshwater springs. Ecological Monographs 81, 599–618. https://doi.org/10.1890/11-0264.1.

Gooch, J.L., Glazier, D.S., 1991. Temporal and spatial patterns in mid-Appalachian springs. Memoirs of the Entomological Society of Canada 123 (S155), 29—49. https://doi.org/10.4039/entm123155029-1.

Holsinger, J.R., Shafer, J., Fong, D.W., Culver, D.C., 2008. *Gammarus cohabitus*, a new species of subterranean amphipod crustacean (Gammaridae) from groundwater habitats in central Pennsylvania, USA. Subterranean Biology 6, 41—51.

Hortal, J., de Bello, F., Diniz-Filho, J.A.F., Lewinsohn, T.M., Lobo, J.M., Ladle, R.J., 2015. Seven shortfalls that beset large-scale knowledge of biodiversity. Annual Review of Ecology and Systematics 46, 523—549. https://doi.org/10.1146/annurev-ecolsys-112414-054400.

Hou, Z., Li, S., 2010. Intraspecific or interspecific variation: delimitation of species boundaries within the genus *Gammarus* (Crustacea, Amphipoda, Gammaridae), with description of four new species. Zoological Journal of the Linnean Society 160, 215—253. https://doi.org/10.1111/j.1096-3642.2009.00603.x.

Hutchins, B., Fong, D.W., Carlini, D.B., 2010. Genetic population structure of the Madison cave isopod, *Anrolana lira* (Cymothoida: Cirolanidae) in the Shenandoah Valley of the eastern United States. Journal of Crustacean Biology 30, 312—322. https://doi.org/10.1651/09-3151.1.

Hutchins, B.T., Schwartz, B.F., Nowlin, W.H., 2014. Morphological and trophic specialization in a subterranean amphipod assemblage. Freshwater Biology 59, 2447—2461. https://doi.org/10.1111/fwb.12440.

Kane, T.C., Culver, D.C., Jones, R.T., 1992. Genetic structure of morphologically differentiated populations of the amphipod *Gammarus minus*. Evolution 46, 272—278. https://www.jstor.org/stable/2409822.

Kinzler, W., Maier, G., 2006. Selective predation by fish: a further reason for the decline of native gammarids in the presence of invasives? Journal of Limnology 65 (1), 27—34. https://doi.org/10.4081/jlimnol.2006.27.

Kolbert, E., 2014. The Sixth Extinction: An Unnatural History. Henry Holt & Co, New York, NY.

Kostalos, M.S., Seymour, R.L., 1976. Role of microbial enriched detritus in the nutrition of *Gammarus minus* (Amphipoda). Oikos 27 (3), 512—516. https://www.jstor.org/stable/pdf/3543471.pdf.

Lee, K.P., Simpson, S.J., Wilson, K., 2005. Dietary protein-quality influences melanization and immune function in an insect. Functional Ecology 22, 1052—1061. https://doi.org/10.1111/j.1365-2435.2008.01459.x.

Lewis, S.E., Loch-Mally, A.M., 2010. Ovigerous female amphipods (*Gammarus pseudolimnaeus*) face increased risks from vertebrate and invertebrate predators. Journal of Freshwater Ecology 25 (3), 395—402. https://doi.org/10.1080/02705060.2010.9664382.

Lewis, S.L., Maslin, M.A., 2015. Defining the Anthropocene. Nature 519, 171—180. https://doi.org/10.1038/nature14258.

MacAvoy, S.E., Braciszewski, A., Tengi, E., Fong, D.W., 2016. Trophic plasticity among spring vs. cave populations of *Gammarus minus*: examining functional niches using stable isotope and C/N ratios. Ecological Research 31, 589—595. https://doi.org/10.1007/s11284-016-1359-6.

MacNeil, C., Dick, J.T.A., Elwood, R.W., 1999. The dynamics of predation on *Gammarus spp* (Crustacea, Amphipods). Biological Reviews 74 (4), 375—395. https://doi.org/10.1111/j.1469-185X.1999.tb00035.x.

Man, Z., 1991. Life History Variation Among Spring-Dwelling Populations of *Gammarus minus* Say. M.S. thesis. American University, Washington, DC.

Marchant, R., 1981. The ecology of *Gammarus* in running water. In: Lock, M.A., Williams, D.D. (Eds.), Perspectives in Running Water Ecology. Plenum Press, New York, NY, pp. 225—249.

Newman, R.M., Waters, T.F., 1984. Size-selective predation on *Gammarus pseudolimnaeus* by trout and sculpins. Ecology 65 (5), 1535—1545. https://doi.org/10.2307/1939133.

Østbye, K., Østbye, E., Lien, A.M., Lee, L.R., Lauritzen, S.-E., Carlini, D.B., 2018. Morphology and life history divergence in cave and surface populations of *Gammarus lacustris* (L.). PLoS One 13 (10), e0205556. https://doi.org/10.1371/journal.pone.0205556.

Pacioglu, O., Sturngaru, S.-A., Ianovici, N., Filimon, M.N., Sinitean, A., Jacob, G., Barabas, H., Acs, A., Muntean, H., Plăvan, G., Schulz, R., Zubrod, J.P., Pârvulescu, L., 2020. Ecophysiological and life-history adaptations of *Gammarus balconicus* (Schäferna, 1922) in a sinking-cave stream from Western Carpathians (Romania). Zoology 1, 125754. https://doi.org/10.1016/j.zool.2020.125754.

Poulson, T.L., White, W.B., 1969. The cave environment. Science 165, 971—981. https://www.jstor.org/stable/1727057.

Rodrigues, S.G., Bueno, A.A., Ferreira, R.L., 2014. A new troglobiotic species of *Hyalella* (Crustacea, Amphipoda, Hyalellidae) with a taxonomic key for the Brazilian species. Zootaxa 3815 (2), 200—214. https://doi.org/10.11646/zootaxa.3815.2.2.

Say, T., 1818. An account of the Crustacea of the United States. Journal of the Academy of Natural Sciences of Philadelphia 1, 37–401.

Stevens, P.J., 1998. Caves of the organ cave plateau, Greenbrier County, West Virginia. West Virginia Speleological Survey Bulletin 9, 1–200.

Sugumaran, M., 2002. Comparative biochemistry of eumelanogenesis and the protective roles of phenoloxidase and melanin in insects. Pigment Cell Research 15, 2–9. https://doi.org/10.1034/j.1600-0749.2002.00056.x.

Tobler, M., Schlupp, I., Garcia De León, F.J., Glaubrecht, M., Plath, M., 2007. Extreme habitats as refuge from parasite infections? Evidence from an extremophile fish. Acta Oecologica 31, 270–275. https://doi.org/10.1016/j.actao.2006.12.002.

True, J.R., 2003. Insect melanism: the molecules matter. Trends in Ecology & Evolution 18, 640–647. https://doi.org/10.1016/j.tree.2003.09.006.

Wilhelm, F.M., Taylor, S.J., Adams, G.L., 2006. Comparison of routine metabolic rates of the stygobite, *Gammarus acherondytes* (Amphipoda: Gammaridae) and the stygophile, *Gammarus troglophilus*. Freshwater Biology 51, 1162–1174. https://doi.org/10.1111/j.1365-2427.2006.01564.x.

Wright, F., 1990. The 'effective number of codons' used in a gene. Gene 87, 23–29. https://doi.org/10.1016/0378-1119(90)90491-9.

Yamamoto, Y., Stock, D.W., Jeffery, W.R., 2004. Hedgehog signalling controls eye degeneration in blind cavefish. Nature 431, 844–847. https://www.nature.com/articles/nature02864.

Zeit, L.B., 1993. The Effect of Temperature and Water Level Variation on the Distribution and Abundance of Aquatic Invertebrates in Two Cave Streams and Their Resurgence. M.S. thesis. American University, Washington, DC.

Zhao, H., Di Mauro, G., Lungu-Mitea, S., Negrini, P., Guarino, A.M., Frigato, E., Braunbeck, T., Ma, H., Lamparter, T., Vallone, D., Bertolucci, C., Foulkes, N.S., 2018. Modulation of DNA repair systems in blind cavefish during evolution in constant darkness. Current Biology 28 (20), 3229–3243. https://doi.org/10.1016/j.cub.2018.08.039.

# Evolutionary genomics and transcriptomics in groundwater animals

*Didier Casane*[1,2], *Nathanaelle Saclier*[3], *Maxime Policarpo*[1], *Clémentine François*[4] *and Tristan Lefébure*[4]

[1]Université Paris-Saclay, CNRS, IRD, UMR Évolution, Génomes, Comportement et Écologie, Gif-sur-Yvette, France; [2]Université Paris Cité, UFR Sciences du Vivant, Paris, France; [3]ISEM, CNRS, Univ. Montpellier, IRD, EPHE, Montpellier, France; [4]Univ Lyon, Université Claude Bernard Lyon 1, CNRS, ENTPE, UMR 5023 LEHNA, Villeurbanne, France

## Introduction

The main goal of evolutionary genomics and transcriptomics is to understand how the architecture of the genome and its expression evolve. This is mainly based on comparisons of genomes and transcriptomes. Beyond the mere description of differences in the organization of genomes and changes in gene expression, it also seeks to relate mutations to variations in gene function and/or level of expression, and ultimately to phenotypic changes. Most morphological, physiological and behavioral traits are under stabilizing selection, but an environmental shift can produce either an optimum displacement leading to directional selection toward this new optimum, or relaxed selection on traits that are less or no longer necessary (Rétaux and Casane, 2013). Comparative studies may be conducted at the intra-species level, with data from either individuals from a single population or different populations, or at a larger evolutionary scale, where genomes and transcriptomes from individuals of different species are compared. Within this larger evolutionary framework, an environmental shift from surface water to groundwater is an outstanding model because the environmental shift is extreme in many biotic and abiotic components. With such an extreme shift, large phenotypic changes might be expected, underpinned by many mutations. A major challenge for molecular evolutionists is to determine the relative importance of different evolutionary processes, which drive the fixation of mutations in genomes, in particular positive selection and genetic drift. Differences in gene expression can result from either environmental or genomic differences and disentangling these different factors is another major challenge. Similarly, identifying gene expression changes that are adaptive can be difficult. During the last 10 years, genomes and transcriptomes from groundwater organisms have been accumulating at an ever increasing rate, allowing us to compare groundwater organisms with other groundwater organisms or with close surface relatives. Below we present evolutionary genomic and transcriptomic studies in groundwater vertebrates and arthropods.

## Evolution of genes and genome architecture

Genomes for groundwater animals are currently available for a limited number of taxonomic groups, and of these, cave teleosts (hereafter called cavefishes) and asellid isopods have been most thoroughly studied. Nevertheless, comparisons of independently evolving groundwater species with closely related surface species has revealed some general trends in the evolution of genome architecture and protein coding genes. It has been shown that eye-specific genes and nonvisual opsins are under relaxed purifying selection in cavefishes, but as the cave colonization events are relatively recent, only a small fraction of these genes are pseudogenized (Policarpo et al., 2021a, b). An analysis of asellid isopods has shown that the small population size of cave species has resulted in a higher dN/dS and genome expansion compared to surface species, a signature of relaxed selection on nearly neutral mutations such as slightly deleterious nonsynonymous mutations and transposon insertions in noncoding sequences (Lefébure et al., 2017; Saclier et al., 2018).

### Cavefish genomics and gene decay

The first cavefish genome sequenced was that of the Mexican *Astyanax mexicanus* (McGaugh et al., 2014). This species includes surface and cave populations that have very different morphotypes but can interbreed in the wild and the laboratory. They can also be easily maintained in the laboratory. For these reasons, most genetic and behavorial studies of cave animals have used *A. mexicanus*. Soon after, the genomes of other cavefish and related surface fish became available. The genomes of three Chinese *Sinocyclocheilus* species were sequenced (Yang et al., 2016): one species, *S. grahami* with a surface morphotype, *S. rhinocerous*, which has small eyes and *S. anshuensis*, which is blind and depigmented. These fishes are recent tetraploids and therefore have large genomes (Yang et al., 2016). The genome of an American blind and depigmented amblyopsid cavefish, *Typhlichthys subterraneus* (Malmstrøm et al., 2017) and the genomes of two Cuban *Lucifuga* species (Policarpo et al., 2021a) were published more recently. The *L. dentata* specimen was blind and depigmented while the *L. gibarensis* specimen was small-eyed and pigmented. Also, the genome of a blind and depigmented cichlid, *Lamprologus lethops*, was published (Aardema et al., 2020). This fish is not a cavefish, but it thrives in the deep freshwater of the Congo River, an environment that presents several constraints that are similar to those in groundwater. The phylogenetic relationships between these fishes and their close surface species for which genomes are available are shown in Fig. 17.1. The availability of these genomes has allowed to carry out a

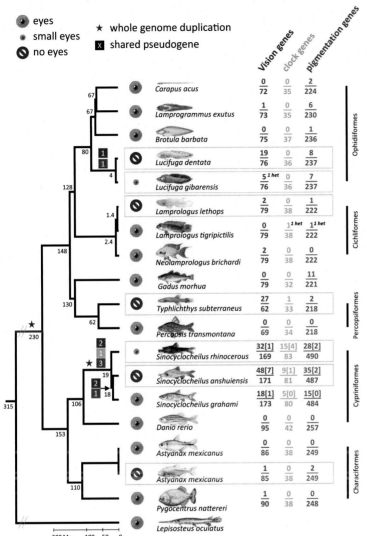

**FIGURE 17.1** Phylogeny and pseudogene mapping. For each gene set, the number of pseudogenes identified and the number of genes examined in a species are given to the right of the species name. Cavefishes and the blind cichlid living in deep-water (*L. lethops*) are framed. no eyes: no external eyes and highly degenerate remnants under the skin in cavefish. het: the specimen examined was heterozygous. *Redrawn from Policarpo et al. (2021a, b).*

comparative analysis of relaxed selection on vision, circadian clock and pigmentation genes (Policarpo et al., 2021a, b). These three gene sets have been examined in cavefish, looking for loss of function (LoF) mutations such as premature stop codons and frameshifts in coding sequences, which are at the origin of nonfunctional genes (pseudogenes). The set of vision genes is composed of eye-specific genes that are involved in visual phototransduction, eye-specific crystallins and nonvisual opsins. These genes are assumed to be dispensable in a dark environment. The circadian clock and pigmentation genes are large sets of genes, most are pleiotropic and therefore are not dispensable in an aphotic environment. Among circadian clock and pigmentation genes, very few pseudogenes were found in cavefishes (Fig. 17.1). Interestingly, only four circadian clock genes were found pseudogenized in cavefishes, and two of them, *dash* and *per2*, were found pseudogenized independently in three species. These two genes appeared as mutation hot spots involved in the circadian clock loss. Among the pigmentation genes, 18 pseudogenes were found and seven were identified as independently pseudogenized in two species. Most genes belonging to this gene set are also under strong purifying selection against the fixation of LoF mutations. However, a few genes seem to be recurrent targets for fixation of LoF mutations associated with depigmentation, indicating that these genes are not involved in other essential functions.

In sharp contrast to the circadian and pigmentation genes, the number of pseudogenes among vision genes was very high (Fig. 17.1 and Fig. 17.2). A similar percentage of pseudogenes (about 50%) was found among the different types of vision genes, that is, phototransduction, crystallins and nonvisual opsins. However, the percentage of vision pseudogenes varied widely between species. In *Astyanax mexicanus*, only one pseudogene was identified in 85 genes (1%) while 27 were identified in 62 genes (44%) in *Typhlichthys subterraneus* and 19 in 76 genes (25%) in the Cuban blind cavefish *Lucifuga dentata*. In the African deepwater cichlid, *Lamprologus lethops*, which is blind, only two pseudogenes were identified out of 79 genes (3%). The highly variable number of vision pseudogenes indicates that eye loss is very recent in some lineages while it is much more ancient in others.

The small-eyed cavefish, *Lucifuga gibarensis*, has less pseudogenes (7%) than the blind *Lucifuga dentata* but many more than *Astyanax mexicanus*. The lower level of vision gene decay in *L. gibarensis* indicates that some small-eyed cavefishes may be much more ancient than blind cavefishes and that therefore the level of eye decay is a poor proxy for the time of cave adaptation. There is strong evidence that most if not all genes that were classified as vision genes in this study are dispensable in blind fishes, while most of them are indispensable in eyed fishes. We therefore opine that the small eyes in *L. gibarensis* could be a stable state rather than an intermediary state between large eyes and blindness. In the tetraploid *Synocyclocheilus*, one cannot necessarily associate the loss of a vision gene with cave settlement since all genes have been recently duplicated and therefore an ohnolog of any gene may be lost without any loss of function, as in *S. grahami*, which doesn't have regressed eyes. Nevertheless, there are more vision pseudogenes in cavefishes, and more pseudogenes were identified in the blind cavefish *S. anshuensis* than in the small-eyed *S. rhinocerous* (Fig. 17.1). Moreover, if we assume that the function of a gene is lost only if both ohnologs are pseudogenized, then there is only one gene loss in *S. grahami* and *S. rhinocerous*, while seven genes have been lost in *S. anshuensis*. The level of vision gene decay is therefore higher in the blind cavefish.

Two other methods have been used to estimate relaxed selection in vision genes. The first method is based on changes in ω (the ratio of the mean number of nonsynonymous substitutions per nonsynonymous site to the mean number of synonymous substitutions per synonymous site, also known as d$N$/d$S$ or $K_a$/$K_s$). This ratio is expected to be lower than one under purifying selection, equal to one under neutral evolution, and larger than one under adaptive selection (Boxes 17.1 and 17.2). Policarpo et al. (2021a) found a shift of ω toward one in the terminal branches leading to cavefishes, which is the result of an initial period under purifying selection before groundwater colonization (ω < 1), followed by a period of neutral evolution (ω = 1) after cave colonization. The second method is based on a machine

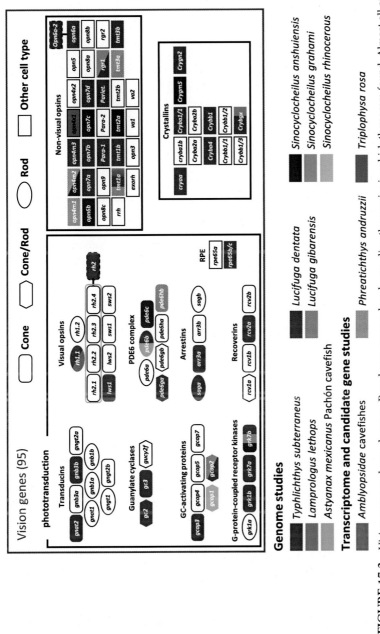

**FIGURE 17.2** Vision genes and pseudogenes. Pseudogenes are colored according the species in which they were found. Almost all of the genes were examined in the genome analyses, while only a few of the genes were examined in the transcriptomic and candidate gene studies. Gene copy number variations: there are four copies of *Rh2* in zebrafish whereas there is only one *Rh2* gene in *T. subterraneus*; there is one *opn6a* gene in zebrafish whereas there are two *opn6a* genes in *T. subterraneus*. Gene expression patterns in zebrafish are from the ZFIN database (https://zfin.org/). Redrawn from Policarpo et al. (2021a, b).

## BOX 17.1

### S e l e c t i o n   i n t e n s i t y

In a diploid population of effective size $N_e$, when a new allele appears, its frequency is $1/2N_e$. If this allele is subject to selection with a selection coefficient $s$, the probability that the mutant is ultimately fixed is:

$$P_{fix} = \frac{2s}{1 - e^{-4N_e s}}$$

The selection coefficient $s$ is negative if the new allele is deleterious, it is equal to zero if the new allele is neutral and is positive if the new allele is advantageous.

When $s \to 0$, $P_{fix} \to 1/2N_e$, the probability of fixation of a new and neutral allele is lower in a large population than in a small population.

When a new allele is subject to selection, its probability of reaching fixation depends on the selection intensity, $N_e s$, in a quite complicated manner shown in the figures below.

and independent of the population size (figure on the left). However, for mutations that are nearly neutral, that is with a weak advantageous or a weak deleterious effect, the fate of the mutation depends also on the population size. Hence, a slightly deleterious or slightly advantageous mutation ($-0.002 < s < 0.002$) has a probability of reaching fixation similar to that of a neutral mutation, that is the ratio of the probability of fixation is close to one if the population size is small (figure on the right). However, in larger populations this ratio is not close to one, in particular, slightly deleterious mutations are effectively removed. In other words, in large populations selection intensity is high and genetic drift is low, and selection is effective

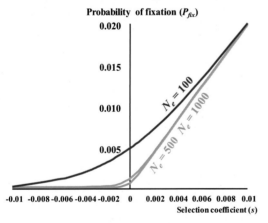

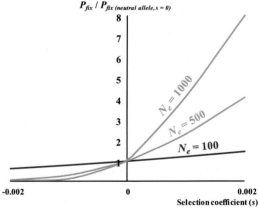

When $s$ is very small or very large, that is, the new allele is highly deleterious or highly advantageous, the probability of it being, respectively, eliminated or fixed is very high

on nearly neutral mutations. In small populations, such as many cave populations, slightly deleterious mutations will therefore be more likely to reach fixation.

---

## BOX 17.2

### Measures of selection on coding sequences

The evolutionary rate ($\lambda$) of a nucleotide at a particular position in a DNA sequence depends on the probability of the occurrence of a mutation ($\mu$) and the probability of fixation of the mutation, which is shown in Box 17.1. In a population of effective size $N_e$, $2N_e\mu$ mutations occur at each generation, thus the evolutionary rate is:

$$\lambda = 2N_e\mu \frac{2s}{1 - e^{-4N_e s}} \quad \text{and} \quad \frac{\lambda}{\mu} = \frac{4N_e s}{1 - e^{-4N_e s}}$$

Interestingly, when $s \to 0$, $\lambda/\mu \to 1$, that is the rate of fixation of new mutations in a population, known as the substitution rate, is equal to the mutation rate of a sequence and does not depend on the population size. Thus, the substitution rate at neutral sites can be used to estimate the mutation rate. If new alleles are advantageous, then $s > 0$ and $\lambda/\mu > 1$, and if new alleles are deleterious, then $s < 0$ and $\lambda/\mu < 1$.

This theoretical result is used to analyze the evolution of protein coding genes. Differences between DNA sequences are classified as synonymous or nonsynonymous. Synonymous differences do not change the protein sequence and are assumed to be neutral. Nonsynonymous differences change the protein sequence. Several methods have been proposed to estimate the number of synonymous substitutions per synonymous site ($d_S$), the number of nonsynonymous substitutions per nonsynonymous site ($d_N$), and $d_N/d_S$, which is an estimate of $\lambda/\mu$. A $d_N/d_S = 1$ indicates neutral evolution of the amino acid sequence, a $d_N/d_S < 1$ indicates negative (or purifying) selection for removal of nonsynonymous mutations and a

$d_N/d_S > 1$ indicates positive (or adaptive) selection for nonsynonymous mutations. For most protein coding genes, $d_N/d_S < 1$ because most protein sequences have been optimized over a long evolutionary time and most amino acid changes are deleterious. Very few amino acid changes are advantageous. In the transition from a surface to a groundwater environment, if a protein coding gene was under strong purifying selection in surface water but is dispensable in groundwater, one would expect a shift of $d_N/d_S$ from a value less than 1 to 1.

Another approach relies on the measurement of the deleterious effect of an amino acid substitution. Under purifying selection, most substitutions would be expected to have no or a low deleterious effect. However, if purifying selection becomes relaxed after an environmental shift, such as in the shift from surface water to groundwater, mutations that were deleterious in surface water and that would have been eliminated, are now neutral in groundwater and can reach fixation. The fixation of mutations classified as highly deleterious in the original environment suggests that the environmental shift relaxed purifying selection on proteins in which such substitutions are found. The most difficult problem with this approach is to estimate accurately the deleterious effect of a mutation in the original environment. There are several machine learning methods dedicated to this task, which are improving rapidly, which could become a valuable approach in the near future to measure changes in selection on protein coding genes.

learning approach to identify deleterious mutations. Vision genes in cavefishes accumulated more deleterious mutations than in surface fishes (Policarpo et al., 2021a).

## Isopods genomics

### *Genome size evolution*

The simplest characteristic that defines a genome - its size - remains a puzzling enigma: why does this genome characteristic vary so much across taxa and what forces drive this huge variability? Two opposite views initially emerged: genome size variations are adaptive and evolve under positive selection, or they are nonadaptive and are the result of the interaction between mutation and drift (Blommaert, 2020). The adaptationist hypothesis is rooted in countless evidence of covariation between genome size and phenotypic traits such as body size, basal metabolism, generation time or growth rate. Based on these numerous correlations, a general hypothesis was proposed, named the bulk hypothesis (Gregory, 2005), which integrates genome size as a central trait in interaction with many phenotypic traits. Under this hypothesis the size of the genome controls the cellular growth rate and metabolism, which themselves impact other traits in a web of complex interactions. The size of the genome, in addition to the genetic information it carries, is thus expected to be central in defining an organism's fitness and is therefore tightly controlled by natural selection. Interestingly, with the discovery that genome size is mainly dependent on the amount of repetitive DNA present, this hypothesis remained valid in the sense that while the main mutational force generating genome size variation quickly is the rate of insertion and deletion of repetitive elements, it is ultimately natural selection that will select organisms with smaller or larger genomes depending on their respective fitness. While variations are mainly generated by the activity of transposable elements and the direction of evolution is controlled by natural selection, genome size plays a central role between these two forces. In parallel, the discovery of the large role played by transposable elements and their frequent deleterious impact on a genome functioning, led to the emergence of a totally different hypothesis where positive selection plays no role. A first neutral hypothesis postulated that genome size has little to no impact on the fitness, and that genome size is mainly under the control of the rate of insertion and deletion (Petrov, 2002). Combined with mechanisms such as ectopic recombination or transposable element inactivation by host responses, this model can generate complex genome size evolutionary trajectories (Bourgeois and Boissinot, 2019). To accommodate for the numerous documented covariations between genome size and phenotypic traits, another nonadaptive hypothesis was proposed in the 2000s, named the mutational-hazard (MH) hypothesis (Lynch, 2007). This hypothesis integrates the effective population size ($N_e$) as a central parameter, which modulates the intensity of genetic drift and ultimately generates a noncausal covariation between genome size and many phenotypic traits. The efficacy of selection depends on genetic drift, in particular for slightly deleterious and slightly advantageous mutations, which can become effectively neutral in small populations (Ohta, 1992). The MH hypothesis postulates that many genomic characteristics are slightly deleterious but are more likely to be fixed in small populations where they become effectively neutral (Box 17.1). This includes the insertion of transposable elements, which are expected to be more frequent in small populations. In the long run, populations with small $N_e$ should therefore evolve larger genomes. Since population size is also correlated to many traits such as generation time or body size, the MH hypothesis suggests that the observed correlation between genome size and many phenotypic traits is noncausal, and is the result of the confounding effect of $N_e$. While rooted in population genetic theory, the MH is difficult to test and lacks empirical support.

One of the most important characteristics of most subterranean habitats is the low availability of trophic resources (Venarsky et al., 2014). When considering two closely related taxa, one a surface and the other a subterranean dweller, the latter is therefore more likely to have lower population densities and smaller effective population sizes. Many counterexamples are likely to be found, but provided that many pairs of such taxa are studied, we can hypothesize that on average subterranean taxa will have lower $N_e$. If this is correct and if many closely related surface/subterranean pairs of species can be examined, then we have at hand an original and powerful model to test the influence of $N_e$ on the evolution of genome size. Unfortunately, while there is a huge diversity of subterranean species, many of them don't have a closely related surface taxon. However, for a few groups, like the Asellidae waterlouse, many such pairs can be found (Fig. 17.3) (Morvan et al., 2013). By using 11 pairs of subterranean/surface Asellidae waterlouse species, Lefébure et al., 2017 tested if $N_e$ was indeed lower in the subterranean waterlouse species. They estimated the efficacy of selection against slightly deleterious substitutions in protein coding genes extracted from whole transcriptomes. Under low $N_e$, slightly deleterious nonsynonymous substitutions are more likely to reach fixation, and the ratio of nonsynonymous over synonymous substitution rates ($dN/dS$) is expected to increase (Box 17.2). On average subterranean asellids had higher $dN/dS$, demonstrating that the efficacy of selection was lower in these taxa because they had smaller $N_e$. In the same study, they estimated genome sizes of the same species and found that on average subterranean asellids have larger genomes (25% average increase). The largest genome size difference was observed between the surface species *Proasellus karamani* (1.1 Gb) and the subterranean species *Proasellus hercegovinensis* (2.7 Gb). Subterranean taxa while evolving smaller $N_e$ may also develop other traits such as lower metabolism and larger body size. These traits can play a confounding role, masking the influence of positive selection in the evolution of genome size in subterranean asellids. However, body size and growth rate were not correlated with genome size, but $dN/dS$ was significantly correlated with genome size. An analogous genome size increase was also observed in two pairs of surface/subterranean decapods and one pair of gastropods, suggesting that this might be a general tendency in subterranean taxa. Altogether, subterranean organisms offered the first solid empirical support to the MH hypothesis.

## Molecular evolution rate in subterranean habitats

The rate of molecular evolution is the rate at which DNA sequences accumulate substitutions (i.e., mutations shared by all the individuals of a population). This rate depends on the mutation rate ($\mu$) and on the probability of a mutation spreading in the population (called the probability of fixation $p$). Thus, the substitution rate is the mutation rate multiplied by the probability of fixation ($\mu \times p$). For neutral mutations, the substitution rate equals the mutation rate (Boxes 17.1 and 17.2). One way to estimate the mutation rate, therefore, is to collect a large comparative dataset and estimate the substitution rate for a class of substitutions that is considered to have no impact on the fitness. One commonly used substitution class are synonymous

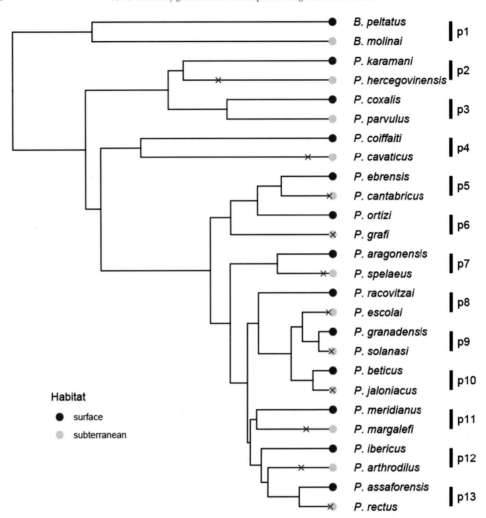

**FIGURE 17.3** Phylogeny of 13 pairs of Asellidae species inhabiting either surface or ground water. *P: Proasellus; B: Bragasellus;* a red cross indicates the estimated time of groundwater colonization. Redrawn from Saclier et al. (2018). The Asellidae group contains numerous endemic species, both inhabiting surface and ground water. Starting from a large phylogeny of the group with their corresponding habitat (Morvan et al., 2013), it is possible to delineate several pairs of species each composed of one surface and one subterranean species. In this figure, 13 pairs have been selected. Given that the colonization of surface habitat by a subterranean species is very unlikely to succeed, each pair corresponds to an independent habitat shift from a surface to a subterranean habitat. This offers an unprecedented level of replication of the same evolutionary process, which has been instrumental to describe the impact of groundwater colonization on the evolution of the genome (e.g., François et al., 2016; Lefébure et al., 2017; Saclier et al., 2018). Thanks to the convergent regression of the eye in subterranean asellids, it is possible to estimate the time of colonization of a subterranean species by looking for the amount of nonsynonymous substitution one opsin gene has accumulated. Indeed, once useless, opsin genes will slowly become pseudogenes and gradually accumulate nonsense and nonsynonymous mutations.

substitutions in protein coding genes. The mutation rate is known to vary tremendously among species, by almost 1000-fold, from $10^{-11}$ mutations per nucleotide site per generation in some unicellular organisms to approximately $10^{-8}$ in primates (Lynch, 2015), and is the subject of a long-standing debate about the factors driving this variation. The colonization of subterranean habitats can lead to variation in the mutation rate for two main reasons. First, a major cause of variation is thought to be attributable to differences in life history traits among species, as the colonization of subterranean habitats is associated with changes in life history traits, mutation rates can be expected to vary as well. Second, mutation rate can vary because of differences in mutagenic factors between surface and subterranean habitats.

### Life history traits

Many life history traits (LHT) have been proposed to impact the rate of molecular evolution, the most widely accepted is generation time (GT). Every time DNA is duplicated during gamete production, replication errors occur, resulting in new mutations. The GT hypothesis (Li and Tanimura, 1987) proposes that in species with longer generation times the genome is copied less often, resulting in less mutations per unit of time. However, many other traits have also been proposed to explain variations in substitution rate between species. It has been proposed that parental investment could impact the balance between selection and drift on mutation rate (Britten, 1986). A species producing few offspring should transmit less mutations to its offspring to avoid deleterious mutations, which would significantly impact their fitness. There is also a large body of literature linking LHT and substitution rate based on the oxygen metabolism in the mitochondria (Lanfear et al., 2007). Oxygen metabolism generates reactive oxygen species (ROS), which are mutagenic. The metabolic rate hypothesis (Martin et al., 1992) proposes that species with a higher metabolic rate generate more ROS and subsequently have a higher mutation rate. As this metabolic rate is strongly correlated with body size, it is also correlated with many LHT. However, the amount of ROS also depends on the efficiency of the mitochondrial respiratory chain. In addition, the mutational impact of these ROS depends on the efficiency of protection mechanisms (Galtier et al., 2009). It has been suggested that the link between ROS and substitution rate does not depend on the metabolic rate but on the balance between the cost of mechanisms avoiding mutations and the cost of deleterious mutations (Nabholz et al., 2008). They propose that the point of balance between these two costs depends on the organism's longevity. As ROS can accelerate aging, natural selection would lead to a decrease of ROS production or enhance cell protection against ROS in long-lived species.

While the literature on the link between substitution rate and LHT is large, there is no consensus on which traits are predominant in controlling the substitution rate, except for the impact of GT. Groundwater species offer a unique opportunity to test these hypotheses. Empirical tests carried out so far examining these hypotheses (1) were mostly done on mammals, calling into question the universality of the observed correlations, (2) often used only mitochondrial genes and, (3) were based on few genes (between 1 and 15). If, as outlined in the previous sections of this chapter, LHT changed predictably during the transition from surface to groundwater habitats, then groundwater organisms offer useful case studies to test theories linking LHT and rate of molecular evolution. Indeed, the comparison between groundwater and surface species provides us with replicates of the same changes in LHT, giving us an ideal experimental design to disentangle the aforementioned hypotheses.

To date, a single study has examined the above hypotheses using groundwater organisms (Saclier et al., 2018). By computing nuclear and mitochondrial substitution rates on a large number of genes (382 and 13, respectively) for 26 species of Asellid isopods, forming 13 pairs of groundwater-surface species (Fig. 17.3), the authors showed that groundwater isopods have lower nuclear rate of molecular evolution, suggesting a generation time effect. However, the mitochondrial rate of molecular evolution was not impacted by the colonization of groundwater and the important LHT modifications that are associated with this habitat shift. This was a totally unexpected result, which challenges our view of the impact of LHT on the mitochondrial mutation rate. This first example demonstrates that groundwater species are an ideal model to study the link between LHT and molecular evolution.

### Environmental radiation

The mutation rate is also impacted by environmental mutagenic factors such as ionizing radiation. The main source of ionizing radiation is ultra-violet (UV) or bedrock radiation. Because individuals move and disperse they are exposed to very low but variable doses of natural radiation, thus it is very difficult to study the impact of radiation in most natural populations. Thanks to their limited dispersal ability, subterranean species offer a great opportunity to study the impact of the ionizing environment on the mutation rate. Indeed, a subterranean population is likely to have been exposed to the same radiation over many generations, while this is very unlikely for a surface species. Moreover, unlike surface species, subterranean species are not exposed to UV radiation, offering the opportunity to (1) study the impact of UV radiation on the mutation rate by comparing surface and subterranean species, and (2) to discard UV radiation as a confounding factor to study the impact of natural bedrock radioactivity on mutation rate. UV radiation leaves a particular mutational signature: it causes a photo-excitation of pyrimidine dimers, which twists the DNA molecules. The damaged base is replaced by two thymines (Brash, 2015), causing a mutation to a TpT dinucleotide. Looking for variation in the frequency of pyrimidine dinucleotides, no evidence of differential impact of UV on the mutation rate between surface and subterranean Asellidae was identified (Saclier et al., 2020). This result suggests that surface species have protection mechanisms to avoid mutations from UV radiation in germ cells. Different geological formations are characterized by different levels of radioactive radiation. For example, limestone formations emit less radiation than igneous formations. By studying closely related endemic subterranean species inhabiting different geological environments, it is possible to test the influence of bedrock radioactivity on the mutation rate. Using 14 species of subterranean water-lice living in environments with contrasting levels of radioactivity, it was found that the mitochondrial and nuclear mutation rates increased by 60% and 30%, respectively (Saclier et al., 2020). They also found that bedrock radioactivity modifies not only the mutation rate but also increases the probability of G to T mutations. This mutation is a hallmark of oxidative stress, suggesting that radioactive environments influence the mutation rate, possibility by generating free radicals, in particular in the mitochondria.

### Elemental composition of genomes, transcriptomes and proteomes in subterranean habitats

One of the consequences of the transition from surface to subterranean environments is commonly a critical decrease in nutrients and energy available for groundwater metazoan

(Venarsky et al., 2014; Francois, 2015). Numerous studies have investigated the potential adaptive strategies to such limitations. A novel theoretical framework has been recently developed on the influence of nutrient availability on the elemental composition of the genome, transcriptome and proteome (Elser et al., 2011). The 20 amino acids, as well as the four nucleotides, differ in the number of nutrients they contain (e.g., carbon or nitrogen). Thus, changes in the composition of RNAs or proteins may reduce nutrient requirements and increase the fitness of an organism facing environmental limitations. This hypothesis conceptually applies to any nutrient, but is particularly relevant to nitrogen ($N$), as it accounts for a large proportion of biological macromolecules (proteins and nucleic acids) (Sterner and Elser, 2002). Thus, low $N$ availability could theoretically select for changes in proteins or RNA composition through the preferential use of $N$-poor amino acids or nucleotides. Such $N$-saving adaptive mechanisms have been reliably shown in marine microorganisms evolving in $N$-poor environments (Luo et al., 2015).

As groundwater environments typically have extremely low nutrients (including $N$) availability (Venarsky et al., 2014; Francois, 2015; Francois et al., 2016), subterranean organisms could have selected $N$-saving mutations since the transition to a groundwater environment. A comparative study (Francois et al., 2016) tested this hypothesis by analyzing the transcriptomes (and the corresponding proteomes) of 13 pairs of surface and subterranean species. Neither global analyses (total $N$-budget, average $N$-cost) nor analyses restricted to orthologous gene families ($N$-cost of the lineage-specific substitutions) showed any evidence of $N$-savings in the transcripts and proteins of subterranean species. The only pattern compatible with $N$-savings selection was a lower $N$-usage at third codon positions in subterranean species. However, a careful analysis of the transcriptomic data showed that the mechanisms responsible for this adaptation-like pattern were actually neutral, and not linked to a selective pressure linked to $N$ availability. Two hypotheses could explain why $N$-saving mutations have not been selected in the transcriptome and proteome of these subterranean isopods evolving for thousands or millions of years under extremely low $N$ availability. First, even when occurring in a highly expressed transcript or protein, an $N$-saving point-mutation may have a too small adaptive effect to be efficiently selected for. This holds especially for small $N_e$ populations, such as subterranean isopods, characterized by low natural selection efficacy. Second, other nutrients-saving adaptations may have been selected more quickly and efficiently following groundwater colonization. For example, subterranean isopods have lower growth rates than their surface counterparts (Lefébure et al., 2017). Such adaptive response might have alleviated the selective pressure for further decreasing nutrient requirements.

## Evolution of gene expression in groundwater

In parallel with genome sequencing, transcriptomes have been obtained for several cave-fishes and cave arthropods. The comparison of transcriptomes of different tissues, at different developmental stages, in different environments and from individuals belonging to different populations or species has allowed to identify gene expression changes depending on the ontogenetic stages, environmental factors and genome divergence. Variations of gene expression according to the environment can be involved in phenotypic plasticity, i.e., variations in the phenotype for a given genotype, which could be involved in the first steps of adaptation of some surface species to groundwater. Like differences in genomes, variations in gene expression are numerous, but their association with phenotypic changes is often a very difficult task. It is even more difficult to relate these changes to adaptation.

## Cavefish transcriptomics

Most studies of cavefish gene expression have been in the Mexican characid *A. mexicanus* and a few species belonging to the Chinese genus *Sinocyclocheilus*. Transcriptomes have also been analyzed in the Chinese cave cypriniform *Oreonectes daqikongensis, Oreonectes jiarongensis* and *Triplophysa rosa*, and different populations of the Mexican poeciliid *Poecilia mexicana*. The main results and references are given below.

The first high throughput differential gene expression analysis used RNA extracted from 3-day postfertilization *A. mexicanus* surface fish from Texas and cavefish from Pachón cave (Mexico). Surface and cavefish cRNA was hybridized to Affymetrix Zebrafish Genome Array chips (Strickler and Jeffery, 2009). A total of 67 differentially expressed genes were found, six upregulated and 61 downregulated in cavefish relative to surface fish. Many of these genes are involved either in eye development and/or maintenance, or in programmed cell death. Another study compared RNA from adult males and females with the same provenance as the first study (Gross et al., 2013). Sequencing of cDNA followed by de novo transcriptome assembly and annotation identified many genes involved in visual system maintenance expressed in surface fish, but not in the adult Pachón cavefish. Conversely, several metabolism-related genes expressed in cavefish were not detected in surface fish. Further sequencing of *A. mexicanus* surface fish and Pachón cavefish transcriptomes showed enrichment of radical mutations in cavefish eye genes that may be evidence of relaxed selection for vision in the absence of light (Hinaux et al., 2013). Differential gene expression has also been analyzed in two geographically distinct cave-dwelling populations, Pachón and Tinaja (Stahl and Gross, 2017). Gene ontology (GO) analysis showed that similar functional profiles evolved in the two cave lineages. However, enrichment studies indicated that the similar GO profiles were in some cases mediated by different genes. Certain "master" regulators, such as *Otx2* and *Mitf*, may be important loci for cave adaptation. Shared expression profiles may reflect a common origin, common environmental pressures or parallel expression drift, while unique expression in a cave may reflect adaptation specific to this cave or divergent expression drift in the caves.

In order to detect a maternal effect on early gene expression in surface and cavefish, four crosses were performed (surface ♀ × surface ♂; surface ♀ × cave ♂; cave ♀ × cave ♂; cave ♀ × surface ♂) and transcriptomes of 2-cell stage embryos were obtained (Torres-Paz et al., 2019). A sample-to-sample distance analysis and principal component analyses (PCA) clustered these samples into two well-separated groups, that is hybrid embryos together with those from their maternal morphotype, demonstrating a strong maternal effect on gene expression at this very early developmental stage. Among the 20,730 genes that were expressed at the 2-cell stage, close to a third (32%) were differentially expressed between embryos with mothers of respectively surface and cave origins. A similar proportion was up- or down-regulated (17.3% and 14.7%, respectively). A gene ontology (GO) enrichment analysis was carried out on differentially expressed genes, and "cell adhesion" (7.1%) and "signaling" (6.5%) were among the significantly enriched biological processes. They might be the most relevant to understand the effect of early differences in gene expression on adult surface and cavefish phenotypes.

Several recent studies examined phenotypic plasticity, that is gene expression differences due to environmental variations. Comparative transcriptome analysis of wild and lab populations of *A. mexicanus* identified differential effects of environment and genetics on gene expression in the liver (Krishnan et al., 2020). A PCA plot showed clustering of individual samples within each population and the relative effect of the environment on gene expression (Fig. 17.4A). Another study examined transcriptome-wide changes associated with sleep deprivation in surface fish and cavefish (McGaugh et al., 2020). The first axis (PC1) of a multi-dimensional scaling (MDS) plot clearly separates the three cave populations (Pachón, Tinaja and Molino) from the surface population, while sleep deprivation has a relatively weak effect

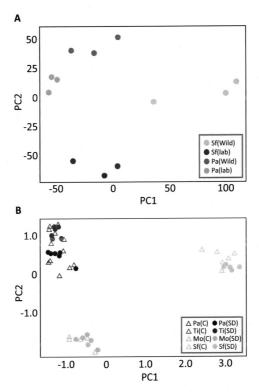

**FIGURE 17.4** Gene expression variation among different populations and environments. (A) Comparison of liver transcriptomes from lab-raised and wild-caught *A. mexicanus*. The principal component analysis (PCA) plot shows the clustering of individual samples within each population and environment. Sf: surface fish; Pa: Pachón cavefish (redrawn from Krishnan et al., 2020). (B) Comparison of whole 30-dpf fry transcriptomes from lab-raised surface and cave fish after a nighttime sleep deprivation. Multidimensional scaling (MDS) plot of 500 top genes. PC1 clearly demarcates population of origin. C: control; SD: sleep deprivation; Pa: Pachón cavefish; Ti: Tinaja cavefish; Mo: Molino cavefish; Sf: surface fish. *(Redrawn from McGaugh et al., 2020).*

on expression of most genes (Fig. 17.4B). In two separate studies surface fish and cavefish were reared under constant darkness of light/dark cycle to analyze the effect of darkness on gene expression (Bilandžija et al., 2020; Sears et al., 2020). Both studies found a rearing condition effect.

Several studies have focused on transcriptome evolution in *Sinocyclocheilus* cavefishes. The blind cavefish *Sinocyclocheilus anophthalmus* retains small eyes that are buried deeply within adipose tissue and are covered with skin, the eyes still have the basic vertebrate eye structure, with fully formed lens, cornea, iris and neural retina. A comparison of the eye transcriptome of *S. anophthalmus* with that of the surface fish *S. angustiporus* (Meng et al., 2013) showed that of 9649 genes, 1658 genes were differentially regulated, in particular, photoreceptor genes were downregulated in the cavefish. Four genes (*rho*, *gnat1*, *gnat2* and *cish*) have a reduced transcription in both *S. anophthalmus* and *A. mexicanus* cavefish. This analysis was extended with the addition of the transcriptome of *S. tileihornes*, another cavefish with small eyes (Huang et al., 2019). Out of 9958 genes, 891 were differentially expressed (395 up and 496 down) in *S. tileihornes* cavefish compared with the surface fish *S. angustiporus*, 889 were differentially expressed (401 up and 488 down) in the cavefish *S. anophthalmus* compared with the surface fish *S. angustiporus*, and 605 were differentially expressed (266 up and 339 down) in the cavefish *S. tileihornes* compared with the cavefish *S. anophthalmus*. The comparison of the eye transcriptome of the cave loach *Triplophysa rosa* and the surface loach *Triplophysa bleekery* (Zhao et al., 2020a) identified 2049 differentially expressed genes. A comparison of the transcriptome of the brain of *S. angustiporus* and *S. anophthalmus* revealed that out of 11,471 genes, 1080 were downregulated and 1067 upregulated in the brain of the cavefish (Meng et al., 2018). A similar result was obtained in a comparison of the brain transcriptome of the surface *S. malacopterus* and the semicave-dwelling *S. rhinocerous* (Zhao et al., 2020b). Comparison of the skin transcriptome of four *Sinocyclocheilus* surface species and four cave species (Li et al., 2020) identified thousands of differentially expressed genes. Eleven of these genes, which are involved in key pigment regulation pathways, were downregulated in the cavefish species and may be involved in pigmentation loss in *Sinocyclocheilus* cavefish species. Skin transcriptomes were also obtained for the depigmented cavefish *Oreonectes dakikongensis* and the pigmented *Oreonectes jiarongensis* (Liu et al., 2019). Thousands of differentially expressed genes were identified, some involved in melanogenesis.

The role of different environmental conditions (nonsulfidic surface streams, sulfidic surface streams, a nonsulfidic cave and a sulfidic cave) on gene expression has been characterized in the small live-bearing fish *Poecilia maxicana* (Passow et al., 2017). In these different habitats, differentially expressed genes were observed in the brain, gill and liver. Cavefish have smaller eyes than surface fish. The comparison of the eye transcriptome of surface fish and cavefish *Poecilia maxicana* showed that 20 genes with eye-related functions were down regulated in cavefish. On the other hand, gene expression differences were almost absent in comparisons of populations inhabiting sulfidic and nonsulfidic habitats (McGowan et al., 2019).

## Arthropod transcriptomics

Apart from the cavefish studies described above, there have been few studies analyzing the transcriptomes of groundwater animals. There have been three publications on the isopod

*Asellus aquaticus*, two on cambarid crayfish, one on the amphipod *Niphargus hrabei*, one on a second amphipod, *Gammarus minus*, one on the syncarid *Allobathynella bangokensis* and one on cave diving beetles. The main results and references are given below.

*Asellus aquaticus*, an isopod crustacean found throughout Europe, has the potential to be an excellent genetic model. Similar to *Astyanax mexicanus*, this species has both cave and surface dwelling populations that can interbreed and produce fertile offspring. RNA was extracted from the heads of a surface dwelling male, a cave dwelling male, a hybrid male, and from pooled surface embryos and hatchlings (Stahl et al., 2015). In a more recent study, RNA was extracted from pooled embryos of surface, cave and hybrid origins (Gross et al., 2020). RNA was also extracted from whole surface and cave specimens (Pérez-Moreno et al., 2018). Differential RNA-seq analysis between embryos of cave and surface morphs identified genes involved in eye and pigmentation underexpressed in the cave morph, which makes biological sense. Genes overexpressed in the cave morph include those involved in metabolism and a gene expressed in stripes in each limb bud segment that could be a candidate for differential antennal characteristics in the cave form (Gross et al., 2020).

Cambaridae, a North American family of crayfish, with about 45 obligate cave-dwellers is another interesting model (Stern et al., 2017). RNA was extracted from both eyes of one to three individuals of eight blind species and six sighted species with close phylogenetic relationships to the blind species. Two of the sighted species were collected from caves. The analysis of expression variation of 3560 genes suggested convergence in transcriptome evolution in independently blind animals. Moreover, modeling expression evolution suggested that there is an increase in evolutionary rates in the blind lineages, consistent with a relaxation of selective constraint maintaining optimal expression levels, and highlighting the importance of gene expression drift (Stern and Crandall, 2018a). Using these 14 transcriptomes, the expression of 17 genes with functions putatively related to rhabdomeric phototransduction were analyzed to investigate molecular and gene expression evolution in caves (Stern and Crandall, 2018b). Signatures of positive and relaxed selection on some gene sequences of cave animals were found. Analyses of gene expression evolution revealed a pervasive signal of convergent and possibly adaptive downregulation of these genes in blind lineages.

The eyeless amphipod *Niphargus hrabeihas* has successfully colonized surface environments despite belonging to an almost exclusively cave-dwelling genus. Total RNA was extracted from two whole specimens. The expression of putative functional visual opsins and other phototransduction genes was maintained, suggesting that this species may be capable of extraocular photoreception (Pérez-Moreno et al., 2018). Comparison of the transcriptomes of cave and surface whole specimens of the amphipod *Gammarus minus* (Carlini and Fong, 2017) identified, out of 104,630 transcripts, 1517 and 551 transcripts were significantly upregulated and downregulated, respectively, in the cave population. Out of five transcripts with fixed premature termination codons in the cave population, one codes for a photolyase, which is a light-dependent enzyme that repairs UV-induced DNA damage.

In the transcriptome of the eyeless syncarid *Allobathynella bangokensis*, which lives in groundwater, no visual or nonvisual opsin sequences were found, suggesting loss of expression of genes with light-related functions in this species (Kim et al., 2017).

A comparison of the head transcriptomes of two surface and three blind subterranean diving beetles from two tribes of Dytiscidae (Tierney et al., 2015) showed the absence of transcription and downregulation of several visual and nonvisual opsin genes in subterranean beetles.

# Conclusion

The recent application of comparative genomics and transcriptomics to groundwater animals has much improved our understanding of the evolution of gene sequence and expression and genome architecture associated with surface to groundwater transition. However, few studies used explicit evolutionary models, particularly in analyses of gene expression differences between surface and groundwater species or populations. Future studies should rely on larger samples and thorough evolutionary analyses to better understand the evolutionary mechanisms at the origin of the extraordinary phenotypes of groundwater animals.

## Acknowledgments

We thank M. Protas and W. Warren for constructive comments that improved this chapter.

## References

Aardema, M.L., Stiassny, M.L.J., Alter, S.E., 2020. Genomic analysis of the only blind cichlid reveals extensive inactivation in eye and pigment formation genes. Genome Biology and Evolution 12, 1392–1406.

Bilandžija, H., Hollifield, B., Steck, M., Meng, G., Ng, M., Koch, A.D., Gračan, R., Ćetković, H., Porter, M.L., Renner, K.J., Jeffery, W., 2020. Phenotypic plasticity as a mechanism of cave colonization and adaptation. eLife 9, e51830.

Blommaert, J., 2020. Genome size evolution: towards new model systems for old questions. Proceedings of the Royal Society B: Biological Sciences 287 (1933), 20201441.

Bourgeois, Y., Boissinot, S., 2019. On the population dynamics of junk: a review on the population genomics of transposable elements. Genes 10 (6), 419.

Brash, D.E., 2015. UV signature mutations. Photochemistry and Photobiology 91, 15–26.

Britten, R.J., 1986. Rates of DNA sequence evolution differ between taxonomic groups. Science 231, 1393–1398.

Carlini, D.B., Fong, D.W., 2017. The transcriptomes of cave and surface populations of *Gammarus minus* (Crustacea: Amphipoda) provide evidence for positive selection on cave downregulated transcripts. PLoS One 12, e0186173.

Elser, J.J., Acquisti, C., Kumar, S., 2011. Stoichiogenomics: the evolutionary ecology of macromolecular elemental composition. Trends in Ecology & Evolution 26, 38–44.

Francois, C.M., 2015. Évaluation des stratégies adaptatives des métazoaires aux faibles disponibilités en nutriments: couplage d'approches d'écologie isotopique et de transcriptomique chez des isopodes épigés et hypogés. PhD thesis. Université Lyon 1, Lyon, France.

Francois, C.M., Duret, L., Simon, L., Mermillod-Blondin, F., Malard, F., Konecny-Dupré, L., Planel, R., Penel, S., Douady, C.J., Lefébure, T., 2016. No evidence that nitrogen limitation influences the elemental composition of isopod transcriptomes and proteomes. Molecular Biology and Evolution 33, 2605–2620.

Galtier, N., Jobson, R.W., Nabholz, B., Glémin, S., Blier, P.U., 2009. Mitochondrial whims: metabolic rate, longevity and the rate of molecular evolution. Biology Letters 5, 413–416.

Gregory, T.R., 2005. Genome size evolution in animals. In: Gregory, T.R. (Ed.), The Evolution of the Genome. Academic Press, Burlington, VT, pp. 3–87.

Gross, J.B., Furterer, A., Carlson, B.M., Stahl, B.A., 2013. An integrated transcriptome-wide analysis of cave and surface dwelling *Astyanax mexicanus*. PLoS One 8, e55659.

Gross, J.B., Sun, D.A., Carlson, B.M., Brodo-Abo, S., Protas, M.E., 2020. Developmental transcriptomic analysis of the cave-dwelling crustacean, *Asellus aquaticus*. Genes 11 (1), 42.

Hinaux, H., Poulain, J., Da Silva, C., Noirot, C., Jeffery, W.R., Casane, D., Rétaux, S., 2013. De novo sequencing of *Astyanax mexicanus* surface fish and Pachon cavefish transcriptomes reveals enrichment of mutations in cavefish putative eye genes. PLoS One 8, e53553.

Huang, Z., Titus, T., Postlethwait, J.H., Meng, F., 2019. Eye degeneration and loss of *otx5b* expression in the cavefish *Sinocyclocheilus tileihornes*. Journal of Molecular Evolution 87, 199–208.

Kim, B.-M., Kang, S., Ahn, D.-H., Kim, J.-H., Ahn, I., Lee, C.-W., Cho, J.-L., Min, G.-S., Park, H., 2017. First insights into the subterranean crustacean bathynellacea transcriptome: transcriptionally reduced opsin repertoire and evidence of conserved homeostasis regulatory mechanisms. PLoS One 12, e0170424.

Krishnan, J., Persons, J.L., Peuß, R., Hassan, H., Kenzior, A., Xiong, S., Olsen, L., Maldonado, E., Kowalko, J.E., Rohner, N., 2020. Comparative transcriptome analysis of wild and lab populations of *Astyanax mexicanus* uncovers differential effects of environment and morphotype on gene expression. Journal of Experimental Zoology Part B: Molecular and Developmental Evolution 334, 530–539.

Lanfear, R., Thomas, J.A., Welch, J.J., Brey, T., Bromham, L., 2007. Metabolic rate does not calibrate the molecular clock. Proceedings of the National Academy of Sciences 104 (39), 15388–15393.

Lefébure, T., Morvan, C., Malard, F., François, C., Konecny-Dupré, L., Guéguen, L., Weiss-Gayet, M., Seguin-Orlando, A., Ermini, L., Der Sarkissian, C., Charrier, N.P., Eme, D., Mermillod-Blondin, F., Duret, L., Vieira, C., Orlando, L., Douady, C.J., 2017. Less effective selection leads to larger genomes. Genome Research 27 (6), 1016–1028.

Li, C., Chen, H., Zhao, Y., Chen, S., Xiao, H., 2020. Comparative transcriptomics reveals the molecular genetic basis of pigmentation loss in *Sinocyclocheilus* cavefishes. Ecology and Evolution 10, 14256–14271.

Li, W.-H., Tanimura, M., 1987. The molecular clock runs more slowly in man than in apes and monkeys. Nature 326, 93–96.

Liu, Z., Wen, H., Hailer, F., Dong, F., Yang, Z., Liu, T., Han, L., Shi, F., Hu, Y., Zhou, J., 2019. Pseudogenization of *Mc1r* gene associated with transcriptional changes related to melanogenesis explains leucistic phenotypes in *Oreonectes* cavefish (Cypriniformes, Nemacheilidae). Journal of Zoological Systematics and Evolutionary Research 57, 900–909.

Luo, H., Thompson, L.R., Stingl, U., Hughes, A.L., 2015. Selection maintains low genomic GC content in marine SAR11 lineages. Molecular Biology and Evolution 32, 2738–2748.

Lynch, M., 2015. Feedforward loop for diversity. Nature 523, 414–416.

Lynch, M., 2007. The Origins of Genome Architecture. Sinauer, Sunderland, MA.

Malmstrøm, M., Matschiner, M., Tørresen, O.K., Jakobsen, K.S., Jentoft, S., 2017. Whole genome sequencing data and de novo draft assemblies for 66 teleost species. Scientific Data 4, 160132.

Martin, A.P., Naylor, G.J.P., Palumbi, S.R., 1992. Rates of mitochondrial DNA evolution in sharks are slow compared with mammals. Nature 357, 153–155.

McGaugh, S.E., Gross, J.B., Aken, B., Blin, M., Borowsky, R., Chalopin, D., Hinaux, H., Jeffery, W.R., Keene, A., Ma, L., Minx, P., Murphy, D., O'Quin, K.E., Rétaux, S., Rohner, N., Searle, S.M., Stahl, B.A., Tabin, C., Volff, J.N., Yoshizawa, M., Warren, W.C., 2014. The cavefish genome reveals candidate genes for eye loss. Nature Communications 5, 5307.

McGaugh, S.E., Passow, C.N., Jaggard, J.B., Stahl, B.A., Keene, A.C., 2020. Unique transcriptional signatures of sleep loss across independently evolved cavefish populations. Journal of Experimental Zoology Part B: Molecular and Developmental Evolution 334 (7–8), 497–510.

McGowan, K.L., Passow, C.N., Arias-Rodriguez, L., Tobler, M., Kelley, J.L., 2019. Expression analyses of cave mollies (*Poecilia mexicana*) reveal key genes involved in the early evolution of eye regression. Biology Letters 15 (10), 20190554.

Meng, F., Braasch, I., Phillips, J.B., Lin, X., Titus, T., Zhang, C., Postlethwait, J.H., 2013. Evolution of the eye transcriptome under constant darkness in *Sinocyclocheilus* cavefish. Molecular Biology and Evolution 30, 1527–1543.

Meng, F., Zhao, Y., Titus, T., Zhang, C., Postlethwait, J.H., 2018. Brain of the blind: transcriptomics of the golden-line cavefish brain. Current Zoology 64, 765–773.

Morvan, C., Malard, F., Paradis, E., Lefébure, T., Konecny-Dupré, L., Douady, C.J., 2013. Timetree of aselloidea reveals species diversification dynamics in groundwater. Systematic Biology 62, 512–522.

Nabholz, B., Glémin, S., Galtier, N., 2008. Strong variations of mitochondrial mutation rate across mammals—the longevity hypothesis. Molecular Biology and Evolution 25, 120–130.

Ohta, T., 1992. The nearly neutral theory of molecular evolution. Annual Review of Ecology and Systematics 23, 263–286.

IV. Principles of evolution in groundwater

Passow, C.N., Brown, A.P., Arias-Rodriguez, L., Yee, M.-C., Sockell, A., Schartl, M., Warren, W.C., Bustamante, C., Kelley, J.L., Tobler, M., 2017. Complexities of gene expression patterns in natural populations of an extremophile fish (*Poecilia mexicana*, Poeciliidae). Molecular Ecology 26, 4211–4225.

Pérez-Moreno, J.L., Balázs, G., Bracken-Grissom, H.D., 2018. Transcriptomic insights into the loss of vision in Molnár János Cave's Crustaceans. Integrative and Comparative Biology 58, 452–464.

Petrov, D.A., 2002. Mutational equilibrium model of genome size evolution. Theoretical Population Biology 61, 531–544.

Policarpo, M., Fumey, J., Lafargeas, P., Naquin, D., Thermes, C., Naville, M., Dechaud, C., Volff, J.N., Cabau, C., Klopp, C., Møller, P.R., Bernatchez, L., García-Machado, E., Rétaux, S., Casane, D., 2021a. Contrasting gene decay in subterranean vertebrates: insights from cavefishes and fossorial mammals. Molecular Biology and Evolution 38 (2), 589–605.

Policarpo, M., Laurenti, P., García-Machado, E., Metcalfe, C., Rétaux, S., Casane, D., 2021b. Genomic evidence that blind cavefishes are not wrecks of ancient life. bioRxiv. https://doi.org/10.1101/2021.06.02.446701.

Rétaux, S., Casane, D., 2013. Evolution of eye development in the darkness of caves: adaptation, drift, or both? Evo-Devo 4 (1), 26.

Saclier, N., Chardon, P., Malard, F., Konecny-Dupré, L., Eme, D., Bellec, A., Breton, V., Duret, L., Lefebure, T., Douady, C.J., 2020. Bedrock radioactivity influences the rate and spectrum of mutation. eLife 9, e56830.

Saclier, N., François, C.M., Konecny-Dupré, L., Lartillot, N., Guéguen, L., Duret, L., Malard, F., Douady, C.J., Lefébure, T., 2018. Life history traits impact the nuclear rate of substitution but not the mitochondrial rate in isopods. Molecular Biology and Evolution 35, 2900–2912.

Sears, C.R., Boggs, T.E., Gross, J.B., 2020. Dark-rearing uncovers novel gene expression patterns in an obligate cave-dwelling fish. Journal of Experimental Zoology 334, 518–529.

Stahl, B.A., Gross, J.B., 2017. A comparative transcriptomic analysis of development in two *Astyanax* cavefish populations. Journal of Experimental Zoology Part B: Molecular and Developmental Evolution 328, 515–532.

Stahl, B.A., Gross, J.B., Speiser, D.I., Oakley, T.H., Patel, N.H., Gould, D.B., Protas, M.E., 2015. A transcriptomic analysis of cave, surface, and hybrid isopod crustaceans of the species *Asellus aquaticus*. PLoS One 10, e0140484.

Stern, D.B., Breinholt, J., Pedraza-Lara, C., López-Mejía, M., Owen, C.L., Bracken-Grissom, H., Fetzner Jr., J.W., Crandall, K.A., 2017. Phylogenetic evidence from freshwater crayfishes that cave adaptation is not an evolutionary dead-end. Evolution 71, 2522–2532.

Stern, D.B., Crandall, K.A., 2018a. The evolution of gene expression underlying vision loss in cave animals. Molecular Biology and Evolution 35, 2005–2014.

Stern, D.B., Crandall, K.A., 2018b. Phototransduction gene expression and evolution in cave and surface crayfishes. Integrative and Comparative Biology 58, 398–410.

Sterner, R.W., Elser, J.J., 2002. Ecological Stoichiometry: The Biology of Elements from Molecules to the Biosphere. Princeton University Press, Princeton, NJ.

Strickler, A.G., Jeffery, W.R., 2009. Differentially expressed genes identified by cross-species microarray in the blind cavefish *Astyanax*. Integrative Zoology 4, 99–109.

Tierney, S.M., Cooper, S.J., Saint, K.M., Bertozzi, T., Hyde, J., Humphreys, W.F., Austin, A.D., 2015. Opsin transcripts of predatory diving beetles: a comparison of surface and subterranean photic niches. Royal Society Open Science 2 (1), 140386.

Torres-Paz, J., Leclercq, J., Rétaux, S., 2019. Maternally regulated gastrulation as a source of variation contributing to cavefish forebrain evolution. eLife 8, e50160.

Venarsky, M.P., Huntsman, B.M., Huryn, A.D., Benstead, J.P., Kuhajda, B.R., 2014. Quantitative food web analysis supports the energy-limitation hypothesis in cave stream ecosystems. Oecologia 176, 859–869.

Zhao, Q., Zhang, R., Xiao, Y., Niu, Y., Shao, F., Li, Y., Peng, Z., 2020a. Comparative transcriptome profiling of the Loaches *Triplophysa bleekeri* and *Triplophysa rosa* reveals potential mechanisms of eye degeneration. Frontiers in Genetics 10, 1334.

Yang, J., Chen, X., Bai, J., Fang, D., Qiu, Y., Jiang, W., Yuan, H., Bian, C., Lu, J., He, S., Pan, X., Zhang, Y., Wang, X., You, X., Wang, Y., Sun, Y., Mao, D., Liu, Y., Fan, G., Zhang, H., Chen, X., Zhang, X., Zheng, L., Wang, J., Cheng, L., Chen, J., Ruan, Z., Li, J., Yu, H., Peng, C., Ma, X., Xu, J., He, Y., Xu, Z., Xu, P., Wang, J., Yang, H., Wang, J., Whitten, T., Xu, X., Shi, Q., 2016. The *Sinocyclocheilus* cavefish genome provides insights into cave adaptation. BMC Biology 14 (1).

Zhao, Y., Chen, H., Li, C., Chen, S., Xiao, H., 2020b. Comparative transcriptomics reveals the molecular genetic basis of cave adaptability in *Sinocyclocheilus* fish species. Frontiers in Ecology and Evolution 8, 589039.

# Biological traits in groundwater

# Dissolving morphological and behavioral traits of groundwater animals into a functional phenotype

Cene Fišer[1], Anton Brancelj[2,3], Masato Yoshizawa[4], Stefano Mammola[5,6] and Žiga Fišer[1]

[1]University of Ljubljana, Biotechnical Faculty, Department of Biology, Ljubljana, Slovenia; [2]Université Paris-Saclay, CNRS, IRD, UMR Évolution, Génomes, Comportement et Écologie, Gif-sur-Yvette, France; [3]Université de Paris, UFR Sciences du Vivant, Paris, France; [4]University of Hawai'i at Mānoa, School of Life Sciences, Honolulu, HI, United States; [5]Molecular Ecology Group (dark-MEG), Water Research Institute (IRSA), National Research Council (CNR), Verbania-Pallanza, Italy; [6]University of Helsinki, Finnish Museum of Natural History (LUOMUS), Helsinki, Finland

## Introduction

Animal's fitness critically depends on detection of food and mates, finding appropriate shelter as well as predator or pathogen avoidance (Odum, 1971). Organisms interact with their environment through morphological structures, behavior and physiology. Receptors and related accessory structures detect and filter the relevant information from the background noise, convey it to the central nervous system, where it is processed and translated into an appropriate motor response that is realized via morphological traits (Alcock, 2005). The performance of both sensory input and motor response can be maximized through fine-tuning of sensors, morphologies and corresponding behaviors (Irschick and Higham, 2014; Aiello et al., 2017; De Meyer et al., 2019). Hence, behavior and morphology jointly comprise functional traits. Organism fitness relates to performance of functional traits, which could be theoretically maximized through morphology, behavior or both. Consequently, both morphology and functionally related behaviors are subject to natural selection. In ecologically heterogeneous environments, different environmental factors lead to divergence in functional traits (Schluter, 2000; Rundle and Nosil, 2005). Functional links between morphology, behavior and environment have been studied in surface organisms (Dickinson et al., 2000; Foster et al., 2015; Aiello et al., 2017), but less so in subterranean ones. With rare exceptions (Christiansen, 1961, 1965; Kralj-Fišer et al., 2020), studies of morphology of subterranean organisms were decoupled from behavioral research.

The research effort and the amount of knowledge between morphology and behavior differs. Spectacular morphology of subterranean species was recognized earlier. The first records of olm (*Proteus anguinus* Laurenti, 1768), a century prior its scientific description, mentioned the species as a baby dragon (Valvasor, 1689). This pale, reptile-looking animal that was frequently washed out from the Carniolian caves (nowadays Central Slovenia) drew attention by local people who did not recognize connections with the salamanders inhabiting the neighboring forests (Aljančič, 2019). Few centuries later leading authorities in Europe and America, including Charles Darwin, started explaining the unique morphologies of cave animals such as lack of eyes and elongated appendages as products of evolutionary changes in the dark cave environments (Romero, 2009). Subterranean species have retained their appeal since then, being peculiar enough to be repeatedly represented in yearly lists of the top 10 most outstanding newly described species (e.g., https://www.esf.edu/top10/past.htm). Morphological traits of subterranean organisms have led to the creation of jargon (e.g., troglomorphy) broadly accepted by speleobiologists (Martínez and Mammola, 2021, Culver et al., this volume). Patterns of morphological variations are relatively well known, linked to evolutionary theory, and knowledge gaps are well recognized (Mammola et al., 2020a). By contrast, behavior of subterranean animals has been less systematically studied (Parzefall, 1982; Hüppop, 2000; Langecker, 2000; Friedrich, 2013; Kowalko, 2019). This can be attributed to technical challenges such as maintenance of long-lasting populations in the laboratory and lack of appropriate infrastructure to conduct experiments (Mammola et al., 2021b). Despite the different research efforts and taxa studied, researchers in both fields would agree that the patterns of morphological and behavioral variation are not perfectly consistent across taxa (e.g., appendage lengths differ drastically between cave organisms). The reason may be the heterogeneity of processes involved in functional trait evolution, which may favor multiple solutions to achieve the same goal, i.e., the so-called "many-to-one relationship between the form and function" paradigm (Losos, 2009).

In this chapter, we identify links between morphology and behavior with the aim to improve our understanding of phenotype functionality. We reconsider these links from an organismal perspective, by emphasizing how an organism directly or indirectly interacts with its environment. In other words, we evaluate how the functional phenotype (morphology and/or behavior) corresponds to environmental variation. We find this an essential step that precedes further ecological and evolutionary studies, which, e.g., quantify trait heritability or link trait variations to individual's fitness (Schluter, 2000). The chapter is organized into three main sections. We first define habitat templates as selective pressures and make predictions about which factors dominate the selective environment of these habitats. The central section reconsiders six general aspects of the functional phenotype, namely sensory input, locomotion, feeding, (micro)habitat choice, reproduction and antipredation response. We conclude with a synthesis and some perspectives.

# Habitat template

Understanding phenotypic variation within the eco-evolutionary context requires a definition of the selective environment, i.e., environmental variation needs to be framed into a habitat template serving as a basis for hypotheses construction. Subterranean researchers generally emphasized three key properties in which subterranean habitats differ from surface ones, i.e., darkness, reduced environmental variation and oligotrophy (Culver and Pipan, 2019). Over the past decades it became clear that what we call the subterranean environment is an aggregate of numerous subterranean habitats differing from each other in physical, chemical and biological properties (see Robertson et al., this volume; Brancelj, 2002; Trontelj et al., 2012; Brancelj et al., 2016; C. Fišer et al., 2019; Mammola et al., 2020b). For the needs of this chapter, we summarize differences among subterranean aquatic habitats in Table 18.1 (we refer the reader to Robertson et al., this volume, for a detailed description and discussion). This table unveils high heterogeneity of subterranean habitats that share only one property — deprivation of light (Pipan and Culver, 2012). This finding leads to an important prediction: functional traits associated with the lack of light will tend to converge in all subterranean animals across all habitats, whereas functional traits that are associated with other habitat properties will evolve in a habitat-specific manner (Pipan and Culver, 2012).

# Morphological-behavioral functional phenotype

We selected six aspects of the functional phenotype essential for species' fitness, namely sensory input, locomotion, feeding, reproduction, (micro)habitat choice and predator response. Each of these aspects is presented in its own section. Some structures are used for multiple functions and thus contribute to different functional traits. Multi-functionality is potentially linked with different trade-offs. We intentionally omitted them from this review, and we briefly return to them in the last section.

## Sensory input

Individual organisms collect information from the environment through different sensory systems. The most common modalities that animals rely on are chemicals, light, mechanical, as well as magnetic and electric fields. Each modality and/or their combinations translate into meaningful signals (Stevens, 2013; Jordan and Ryan, 2015). The origin of the particular sensory modalities for a given species can be traced in their phylogenetic root, and these are often elaborated to adapt to their habitat (Stevens, 2013).

In subterranean habitats, no visual information can be transmitted by light. Therefore, eyes and pigments (both in eyes and body) are no longer needed and thought as relaxed from selection. These structures became reduced through diverse mechanisms (Gross et al., this volume; Protas et al., this volume). To compensate the loss of visual information, many subterranean organisms rearranged nonvisual sensory systems. The magnitude of such rearrangement depends on developmental and phylogenetic constraints. For instance, an eye cannot be transformed into another antenna in arthropods. Furthermore, modification of sensory equipment may be indistinguishable in taxa that receive little visual information, such as

**TABLE 18.1** An overview of subterranean habitats with key properties acting on functional phenotypes. Note high high diversity except in single common property, light conditions. Brief characterization of less well-known habitats: hypotelminorheic — shallow subterranean habitat of small patches, collecting water from soil at impermeable layers at slopes of low inclinations; epikarst — water-filled fissure system at the boundary with soil; hygropetric — water film flowing across the rock; epiphreatic — upper layer of phreatic zone; phreatic zone — permanently flooded zone, the most stable and the most isolated subterranean habitat.

| Habitat | Light conditions | Water current | Discharge variation | Living space (diameter of voids) | Food availability[c] | Mode of feeding[d] | Physical variation[e] | Other | Main selective combination |
|---|---|---|---|---|---|---|---|---|---|
| Hypotelminorheic | no light | none[a] to slow | small | 1 mm – few mm | moderate – high | 1,2?,3,4 | moderate – high | NA | abundant food supply; high variation of temperature; limiting space |
| Epikarst | no light | none to fast | high | 0.1 mm – few mm | low | 1,2?,3,4 | moderate | NA | high discharge and temperature variation; limiting space; low food availability |
| Vadose zone (incl. hygropteric zone) | no light | stagnant[b] to fast | high | 1 m – several m | low | 1,2,3,4 | low | NA | high discharge variation; limiting food |
| Epiphreatic zone | no light | stagnant to fast | high | 1 m – several m | low – high | 1,2,3,4 | low | NA | seasonal variation in chemical and physical parameters |
| Phreatic zone – freshwater | no light | stagnant to very slow | very small | 1 m – several m | low | 1,2,3,4 | low | NA | low variation in physical and chemical parameters |
| Phreatic zone – anchialine | no light | stagnant to very slow | very small | 1 m – several m | low – high | 1,2,3,4 | low | freshwater and marine water separated by halocline | moderate stability in physical and chemical parameters; salinity gradient |
| Phreatic zone – sulfidic[f] | no light | stagnant to very slow | very small | 1 m – several m | high | 1,2,3,4 | high | high level of sulfide; chemoautotrophy concentration of sulfide | high stability in physical and chemical parameters; high |
| Sinking rivers (ponors) | no light – twilight | slow to fast | high | 1 m – several m | moderate – high | 1,2,3,4 | high | NA | low and fast water currents |

| | | | | | | | | | |
|---|---|---|---|---|---|---|---|---|---|
| Springs – temporary | no light – twilight | none to fast | high | 1 m – several m | low | | 2,3,4, | high | NA | ecotone; high fluctuations in environmental parameters |
| Springs – permanent | no light – twilight | slow to fast | moderate to high | <1 m – several m | low – moderate | 1,2,3,4 | low | NA | ecotone; significant variation in environmental parameters; contact between surface and subterranean inhabitants; refugium |
| Springs – thermal | no light – twilight | slow to fast | small | <1 m – several m | low | 1?,2,3,4 | low | above mean annual surface temperature | elevated/high temperature; specific water chemistry; refugium |
| Interstitial – hyporheic | no light | slow | small | <0.1 mm – < 1 cm | moderate – high | 1,2,3,4 | moderate | NA | limiting space; low water flow; contact between surface and subterranean inhabitants |
| Interstitial – phreatic | no light | very slow | very small | <0.1 mm – < 1 cm | low | 1,2,3,4 | low | NA | limiting space; low water flow; low food; low variation in physical and chemical parameters |

[a] *water present as capillary water.*
[b] *water present in pools.*
[c] *this is a heuristic estimation; proper estimations mainly lacking.*
[d] *based on presence of species; 1-filtration, 2-scraping, 3-scavenging, 4-predation.*
[e] *physical variation except discharge, e.g., temperature, water conductivity, and dissolved oxygen.*
[f] *oxygen present only in a thin upper most layer of water.*

V. Biological traits in groundwater

nocturnal species or inhabitants of deep and/or murky waters. In the following subsections we review: (a) morphological sensory rearrangements, (b) behavioral sensory rearrangements and their interaction with morphology and (c) how environmental factors other than darkness act on sensory components of morphology and behavior. We recognize that the absence of light also has a significant influence on animals' circadian physiology/homeostasis (Friedrich, 2013; Abhilash et al., 2017), that is however beyond the scope of this chapter.

## Morphology

Eyes and pigments are reduced or entirely lost in all subterranean species studied, even in phyla with simple ocelli, like flatworms or simple eyes, like snails (Culver et al., 1995; Klaus et al., 2013; Konec et al., 2015; Leal-Zanchet and Marques, 2018). At the first glance, the loss of light-processing structures and pigments seems a universal evolutionary response to permanent darkness. However, photoreception is not lost as universally as eyes (see section Habitat choice).

Rearrangements of sensory systems seem to be common among subterranean animals and manifest in diverse combinations. Most studies reported enhanced chemo- and mechanoreceptors. These sensory systems elaborated through the increase of the size of sensory organs (e.g., accessory structures like antennae), more numerous sensory units (e.g., taste buds, mechanosensory lateral line), and/or more numerous receptor cells per each sensory unit, although in some rare occasions number of sensory units are reduced. Antennae of different crustacean groups, barbells or parts of a head are accessory structures bearing chemoreceptors. These antennae and/or their accessory structures are often enlarged providing more space for more receptors that reach further out from the body (Soares and Niemiller, 2013; Klaus et al., 2013; Ramm and Scholtz, 2017). These modifications likely advance detection of chemical cues. In Mexican tetra *Astyanax mexicanus*, the size of the chemosensing olfactory epithelium (Hinaux et al., 2016), the taste bud number (Boudriot and Reutter, 2001; Varatharasan et al., 2009) and the number of the mechanosensing lateral line neuromasts (Schemmel, 1967) are more enhanced than in their surface-dwelling conspecifics. The enlarged olfactory epithelium allows detection of amino acids concentrations 100–1000 times lower than in surface-dwelling individuals (Hinaux et al., 2016). On a sensory unit level, the lateral line neuromasts of subterranean fishes have larger mucus dome or rod structure covering hair cells within a neuromast and more hair cells than their surface relatives (Soares and Niemiller, 2013). These facilitate the detection of lower frequencies of disturbance such as those generated by crawling crustaceans (Lang, 1980; Yoshizawa et al., 2014). In parallel, many subterranean copepods (Crustacea: Copepoda) have antennal olfactory organs, termed aesthetacs, much larger compared to surface species, especially in highly specialized representatives of subterranean families Gelyellidae and Parastenocarididae and the genus *Stygepactophanes* (Huys and Boxshall, 1991).

Much less is known about electroreceptors. Surface and subterranean species of glass knifefish genus *Eigenmannia* (Actinopterygii: Sternopygidae) from the Amazon basin communicate through energetically costly electric organ discharges. The strength of electric discharges in subterranean species is greater than in surface fish, implying an increased reliance on electro-sensory perception (Fortune et al., 2020). Magnetoreception is another potentially important yet understudied modality. It has been documented in olm (Schlegel et al., 2009) and can be expected in other salamanders and fish. However, it is unclear whether and how it differs from magnetoreception of related surface species.

The elaboration or reduction of peripheral sensory structures corresponds to changes in the respective parts of the central nervous system. Fishes and crustaceans have reduced optic and enlarged olfactory lobes (Soares and Niemiller, 2013; Moran et al., 2015; Stegner et al., 2015; Ramm and Scholtz, 2017).

Few studies imply that rearrangements of different sensory systems are interlinked. The best known is antagonistic regulation of taste-buds and eye during ontogeny of *A. mexicanus*, in which sonic hedgehog (*shh*) genes control development of both, taste-buds and eye (Yamamoto et al., 2009). Overexpression of *shh* interrupts the development of the eye but widens the jaw and increases the number of taste buds. A shift in regulation corresponds to cave colonization and a consequent change of selective regime, i.e., directional selection for olfacto-reception coined with relaxed stabilizing selection for the eye (Gross et al., this volume). Moreover, in the skin covering eyeless orbits, subterranean populations develop additional neuromasts, aiding in vibration attraction behavior (see below) (Yoshizawa et al., 2010, 2012). An interesting analogy was reported from three different families of shrimps, where tactile hairs developed on rudiments of the eye peduncle (Mejía-Ortíz et al., 2006). It would be worth to explore whether these cases of apparently interlinked sensory rearrangements are more generally present across different phyla.

## Behavior and its interaction with morphology

Behavior alone may provide some degree of sensory compensation in the absence of morphological changes of sensory systems. Surface populations of *A. mexicanus* communicate using sharp clicks in response to visual or olfactory stimuli, respectively. Hearing properties of surface and subterranean populations are similar, but the frequencies of clicking and their meaning differ between the two populations: while the onset of clicking in surface fish corresponds with social aggressive and hierarchic behavior, in subterranean fish they correlate to foraging behavior (Hyacinthe et al., 2019).

More commonly, behavior is functionally intertwined with morphological changes in order to maximize performance of an organism in a specific environment. One of the good examples for the morphological shift and its associated behavior was presented again in *A. mexicanus*. Subterranean populations enhanced vibration attraction behavior (VAB), where fish are attracted to and attack a vibrating object in the dark. VAB is advantageous for foraging in the dark (Yoshizawa et al., 2010). This behavior is promoted by the enhancement of the neuromasts, the sensory units of the mechanosensory lateral line, particularly at the eye orbit region. Accordingly, there is a negative correlation between the eye size and the neuromasts number at the eye orbit, as well as between eye size and VAB (Yoshizawa et al., 2012). The mechanical ablation of the neuromasts at the eye orbit also attenuates the VAB (Yoshizawa et al., 2010). Quantitative trait locus mapping unveiled that eye size, neuromasts number and VAB shared one locus driving subterranean phenotypes in these three traits (Yoshizawa et al., 2010, 2012, 2013). Thus, the morphological-behavioral functional traits may share genetic bases.

Sometimes, the links between morphology and behavior are less obvious. During prey detection, spring salamander *Gyrinophilus porphyriticus* raises on its legs to expose and increase efficiency of its lateral line system, linking morphology of legs, sensory system and behavior (Culver, 1973). Other functional traits, such as locomotor appendages seem to be fine-tuned to sensory input. Longer legs of salamanders and fish fins are energetically more efficient for exploring wider areas, but also decrease water disturbance to minimize interference with the lateral line (Hüppop, 2000).

In many cases, it remains unclear whether compensation for visual information relies on elaborated sensory systems, behavior or both. The functional links between behavior and morphology were rarely explored and remain unsupported even in the most cited cases. No study so far explicitly tested whether long antennae of subterranean arthropods indeed improve mechano- and/or chemoreception. Likewise, some behavioral studies provided experimental evidence for sensory compensation in darkness, but receptors involved remain unknown. For example, *P. anguinus* detects and finds prey faster than the Pyrenean brook salamander *Euproctus asper* (Dugès, 1852), a facultative cave-dweller, although receptors mediating this response are still unknown (Uibleinet al., 1992). Subterranean males and females of mollies *Poecillia mexicana* Steindachner, 1863 preferentially mate with larger individuals of the opposite sex. In light, both surface and subterranean individuals select larger mates using visual information. In darkness, only subterranean individuals can discriminate mates by size. The modality or modalities they rely on are not known (Plath et al., 2004, 2007); apparently, at least mechanoreception of the lateral line system can be ruled out (Rüschenbaum and Schlupp, 2013). Despite these gaps, it seems reasonable to assume that morphology and behavior mediating sensory input are functionally intertwined.

### *Beyond the darkness*

Ecological differences between subterranean habitats add an additional layer of complexity to the overall selective regime. The result are different tradeoffs, and numerous examples of apparently contradictory results, i.e., where the same nonoptic sensor becomes elaborated or reduced. As mentioned above, some groundwater copepods have enlarged chemoreceptors called aesthetascs. By contrast, groundwater copepods in the order Calanoida have reduced number and size of aesthetascs compared to surface species (Petkovski, 1978; Brancelj, 1991; Tran Duc and Brancelj, 2017). Apparent reduction of chemoreceptors was reported also from decapod species of the genus *Orconectes*, where subterranean species have fewer yet longer aesthetascs than surface species (Ziemba et al., 2003). Even more striking is the example of waterlice *Asellus aquaticus* (Linnaeus, 1758), an isopod that independently colonized several caves in Europe. A comparison of two surface-subterranean population pairs unveiled only 18 (29%) of the total 62 morphological traits show a pattern that could be attributed to convergence (Konec et al., 2015).

These observations may have several explanations, one of them being the effect of another environmental factor, water movements (Table 18.1). Lentic and lotic water bodies are substantially different. From the sensory point of view, in a running water chemical and mechanical signals are modified all the time (Webster and Weissburg, 2009). Thus, the sensory apparatus of lentic and lotic species should differ. Habitat disturbance (e.g., water level and speed) may add additional constraints on species' morphology affecting especially accessory structures such as antennae (Townsend et al., 1997). In subterranean habitats, water flow affects morphology of receptors as well as their accessory structures. Decapod antennae have fewer aesthetascs in running water than in stagnant water (Ziemba et al., 2003). Likewise, the amphipod antennae are shorter in running water than in stagnant water (Trontelj et al., 2012; Delić et al., 2016, Fig. 18.1). Thus, diversity of sensory rearrangements can be to some extent attributed to differences among subterranean habitats.

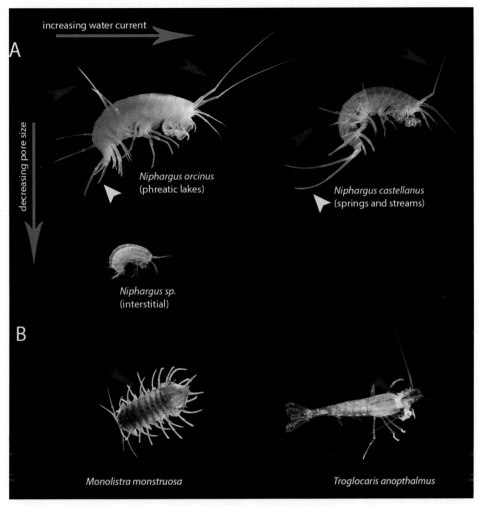

**FIGURE 18.1** An overview of the crustacean diversity from the Dinaric Karst. (A) Three *Niphargus* amphipod species illustrate three aspects of morphological variation in relation to habitat properties. Species from open subterranean waters, like streams (e.g., *N. castellanus*) and phreatic lakes (e.g., *N. orcinus*), are larger than interstitial or epikarstic species (unknown niphargid). Water currents affect the length of appendages (red arrowheads), and consequently locomotor behavior. Species from shallower habitats, such as springs or streams from unsaturated zone (e.g., *N. castellauns*), exhibit sexually dimorphic uropods, whereas phreatic species (*N. orcinus*) are mainly sexually monomorphic (yellow arrowheads). (B) Community composition may further affect the functional morphology of cave crustaceans. Predation of olm (*Proteus anguinus*) may affect rostrum length of *Troglocaris* shrimps (e.g., *T. anophthalmus*), and we hypothesize that cuticular spines of isopods *Monolistra* (e.g., *M. monstruosa*) may have the same function (blue arrowheads). Animals are not to the scale. *All photos courtesy of Teo Delić.*

It is hard to estimate which other environmental factors interfere with sensory input. In theory, at least available living space (e.g., unconsolidated vs. karstic) could have effects on accessory structures, such as antennae, but the firm evidence for this assumption is lacking.

## Locomotion

Locomotion could be defined as any movement of an animal between two places. Appendage length and shape, as well as the overall body shape and size, determine locomotion performance such as speed and energetic costs (Hüppop, 2000; Dickinson et al., 2000; Belanger, 2013; Foster et al., 2015).

It could be expected that oligotrophic environments favor longer appendages, which achieve higher speed (Belanger, 2013) and at the same time exploration of larger areas per unit of expended energy (Poulson, 1963; Langecker, 2000). Indeed, relatively long locomotion appendages are commonly observed among subterranean animals, reported across diverse taxa including long fish fins, salamander and arthropod legs, polychaeta parapodi and even brittle star arms (Langecker, 2000; Hou and Li, 2002; Soares and Niemiller, 2013; Konec et al., 2015; Flammang et al., 2016; Gonzalez et al., 2018; Márquez-Borrás et al., 2020).

The phenomenon, however, is not ubiquitous and seems to be under the influence of at least two additional environmental factors: available living space and water flow. The effect of water flow was studied in the genus *Niphargus* (Amphipoda, Crustacea). Long legs are more common among inhabitants of subterranean slow-running or lentic waters (Fig. 18.1), while species of lotic waters have short appendages (Trontelj et al., 2012; Delić et al., 2016), even shorter than distantly related surface species of genus *Gammarus* (unpublished). The effects of limiting space are most obvious in species from fissure systems and unconsolidated sediments. Several crustacean groups like Ingolfiellidae (Amphipoda), Calanoida, Cyclopoida, Harpacticoida (Copepoda) or Syncarida (Bathynelaceae) have a reduced number of segments in their swimming legs (Coineau, 2000; Brancelj and Karanovic, 2015). Surface representatives of Copepoda have four pairs of swimming legs normally with 3-segmented exopodite and 3-segmented endopodite. In subterranean relatives, this number of segments may be reduced to 1-segmented exopodite and completely reduced endopodite. In the genus *Gelyella*, the fourth pair of swimming legs is absent. Epikarstic Copepoda have reduced antennules and swimming legs but also strong spines/claws on anterior legs, which enable them moving through narrow spaces. In addition, some species have strong spiniform setae on furcal rami preventing them being washed-down from epikarst (Brancelj, 2009; Boonyanusith et al., 2018). Representatives of the genus *Phreatalona* (Cladocera) from the hyporheic zone have reduced antennules, antenna, exopodal lobes on swimming legs, and spines on postabdomen as well as fusiform body shape to facilitate specimens to move among sand particles (Van Damme et al., 2009).

Only few studies addressed the functional significance of locomotion related behavior and morphology in a given ecological context, but they consistently suggest coevolution of morphology and behavior. The relative length of appendages positively correlates with speed, but also with the mode of locomotion, i.e., behavior. Representatives of the genus *Niphargus* from phreatic lakes no longer crawl on a side, i.e., the most common locomotion mode in amphipods, but switch to walking in an upright position (Kralj-Fišer et al.,

2020). Fish with longer fins can glide through water and minimize the disturbance of sensory input (Poulson, 1963; Schemmel, 1980). Clearly, locomotion deserves to be studied more often, given that it bridges all aspects of the functional phenotype (sensory input, feeding, breeding, habitat choice, antipredation response) and may be involved in multiple trade-offs acting on these aspects of the phenotype.

## Feeding

### Energy-saving mechanisms

Many subterranean habitats are energy-poor (see Table 18.1) and prompted evolution of energy-saving mechanisms and specific foraging behaviors (Fišer, 2019). Energy-saving mechanisms encompass mainly physiological and life-history adaptations presented by Di Lorenzo et al. (this volume) and Venarsky et al. (chapter 19, this volume), respectively. Morphological responses on the anatomic level to energy saving are lipid reserves that supply organism during long periods of starvation (Vogt and Štrus, 1992; Vogt, 1999; Hüppop, 2000). Many authors argued that reductions of traits that in darkness became nonfunctional, such as eye and pigment, may be part of energy-saving mechanisms. The evidence for this hypothesis is contradictory. A study on *A. mexicanus* indeed implies that loss of eyes and corresponding parts of brain saves energy (Moran et al., 2015). The result, however, cannot explain loss of eyes in invertebrates living in food-rich habitats, such as hypotelminorheic (Culver and Pipan, 2014). However, behavioral adaptations may additionally contribute to energy saving mechanism, such as energetically economic locomotion (Poulson, 1963), decreased locomotor activity during starvation (Hervant et al., 1997), increased sedentary (Balázs et al., 2020) and loss of circadian rhythms (Moran et al., 2014).

Darkness can reinforce food limitations. Joint effects of both factors were studied in fish. Surface fish evolved antipredator aggregation behaviors, shoaling and schooling. In subterranean environments without visual predators, these are no longer under selection. Moreover, fish flocks in subterranean environment experience increased intraspecific competition for food. Perhaps, to decrease competition for food and save energy-costly injuries, aggregation and aggressive behaviors are reduced (Parzefall, 2001; Timmermann et al., 2004; Kowalko et al., 2013). To prolong periods of feeding, *A. mexicanus* shortened periods of sleep (Duboué et al., 2011), but also evolved specific behaviors, such as VAB, which presumably guide fish to food droppings or crawling crustaceans in water (Yoshizawa et al., 2010). Some of these behaviors have a genetic basis (Yoshizawa et al., 2015), though plasticity should not be ruled out (Espinasa et al., 2021). Most of these studies were made on a single species, *A. mexicanus*. Hence, generalizations need to be drawn with care. It is noteworthy that all these food-finding behaviors seem to be correlated with morphological changes. Loss of aggregating behavior correlates with loss of eyes (Kowalko et al., 2013). Reduction of sleep may be associated with a hypertrophied lateral line system (Jaggard et al., 2017). At last, vibration attraction behavior is linked to the development of superficial neuromasts from the region of the eyeless orbit (Yoshizawa et al., 2012).

## Differentiation with respect to type of food

Species within a community differ in their trophic niches either within the same trophic level (i.e., type of food), or among trophic levels. This differentiation is parallel to surface communities, and it is hard to link it to any habitat properties in Table 18.1. An implicit yet untested assumption is that more productive subterranean habitats support more structured trophic webs (Schneider et al., 2011; Saccò et al., 2019).

The evidence for the trophic differentiation has been inferred mainly from correlational analyses between stable isotopes and morphology or behavior. Edwards Aquifer in Texas harbors rich amphipod fauna, counting at least 19 species. Of these, seven species were studied in detail. These species differ in stable isotopes with respect to food type (detritus, chemolithoautotrophy) and morphology of mouthparts (Hutchins et al., 2014, 2016). In parallel, coexisting amphipods of *Niphargus* from caves on Dinaric Karst differ in both, stable isotopes and morphology of gnathopods they use for handling food (Premate et al., 2021a). The shape of gnathopods weakly correlates with differentiation within a trophic level, whereas the size correlates with trophic position. Morphological specializations were detected also in other groups, like predatory Copepoda with elongated maxillipeds (Brancelj, 2000). Behavioral studies such as food-choice experiments are mostly lacking and badly needed.

The structure of groundwater feeding groups is incompletely understood. There is growing evidence for detritivores and predators, but most commonly, we cannot tell them apart using morphology alone. However, one group, suspension feeders, seems to be rare in subterranean habitats (Fig. 18.2). Some species-rich taxa that in surface habitats feed on suspended particles are completely absent in groundwater, while some, such as sedentary clams, polychaetes and sponges or planktonic crustaceans, like Cladocera and Calanoida, are absent or underrepresented in groundwater (Stoch and Galassi, 2010; Bilandžija et al., 2013). Some subterranean benthic species, like representatives of Cladocera of the genus *Brancelia* switched from detritivory to scraping and sediment feeding, as inferred from finely

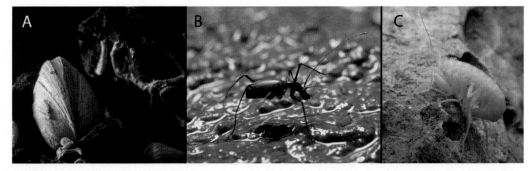

FIGURE 18.2   Environmental constraints to subterranean fauna. Suspension feeders seem to be rare in subterranean waters, presumably due to limited availability of suspended organic material. The clam *Congeria kusceri* and the tube worm *Marifugia cavatica* (A) are among the few filter feeders. By contrast, cave hygropetric, i.e., water flowing across the limestone rock and bringing organic material, seems to be nutrient-rich and was colonized by several species, including the beetles *Hadeisa* (B) and the amphipod *Typhlogammarus mrazeki* (C). Water currents and food type select for convergence in climbing claws and mouthparts shaped for filter-feeding, which allow both beetles and amphipods to enter and forage within the hygropetric. *All photos courtesy of Teo Delić.*

serrated and elongated scrapers on the second trunk limb (Brancelj and Dumont, 2007; Van Damme and Sinev, 2011). By contrast, none of few known obligate groundwater Calanoida (predominantly planktonic group) species possess specific adaptations in food-gathering structures on mouthparts or swimming legs (Brancelj and Dumont, 2007). An exception is the polychaete worm *Megadrilus pelagicus* Martínez, Kvindebjerg, Iliffe & Worsaee (2017) from anchialine caves. This species evolved long ciliated palps, an autoapomorphic dorsal ciliated keel and several longitudinal and transverse ciliary bands, which are used in suspension feeding (Martínez et al., 2017).

## Reproduction

Reproduction is a vital function securing species' survival. Reproductive biology is under the influence of sexual selection that may yield opposing effects on males and females and thus lead to phenomena of sexual dimorphism, mate-choice and sex-specific agonistic behaviors. An important question in subterranean species is how natural selection interacts with sexual selection and modifies all these phenomena. This question received relatively little attention and has been rarely systematically examined or rigorously tested. Moreover, in many species, reproductive biology is insufficiently explored, thus it is hard to make assumptions, which habitat properties from Table 18.1 interact with sexual selection.

Sexual dimorphism has been reported from some taxa. In the amphipod genus *Niphargus*, some articles of caudal appendages (uropods) are more elongated in males than in females and sometimes males are larger than females (Fišer et al., 2008, Fig. 18.1). Interestingly, a comparative analysis of sexually dimorphic surface and subterranean *A. aquaticus* isopods revealed that during the transition to the subterranean environment some traits change sex-specifically, resulting in divergent sexual dimorphism between both environments (Balázs et al., 2021). Finally, although there are no clear differences in sexual dimorphism between surface and cave populations in *A. mexicanus*, its mating behavior has diversified in subterranean water (Simon et al., 2019). Ecological context and functional significance of these patterns have never been explored, but they could be linked to sex ratio and/or reduced seasonality. In some groups, like cladoceran genus *Brancellia*, no males were found (Brancelj and Dumont, 2007; Jeong et al., 2017). In *Niphargus* species living in cave streams and shallow subterranean habitats, sex ratio approximates 1:1, whereas in species from phreatic lakes sex ratio is female-biased (Premate et al., 2021b). If males are common, they need to compete for females. Male-male competition may favor larger and more aggressive males, i.e., enhanced sexual dimorphism (Fišer et al., in preparation). Reduced seasonality might have similar effects onto selective regime as sex ratio. Seasonality in surface waters or shallow subterranean habitats may favor synchronous receptiveness of females, which temporarily increases male-male competition yielding sexual dimorphism. When seasonality is strongly buffered such as in more isolated subterranean habitats, receptive females may occur asynchronously and thus be rare to encounter. In such cases males may gain fitness with efficient mate-finding strategies rather than being superior in agonistic interactions.

Mate choice has been explicitly explored in subterranean fish, and results imply that subterranean individuals retained ancestral selection for larger and presumably more fecund mate (see section Sensory input and Plath et al., 2006). A presumably opposing pattern has been reported in some crustaceans. Males in many surface peracarid and some copepod crustaceans guard females prior to copulation. Mate guarding has not been recorded and is presumably absent in some groups of subterranean isopods (genus *Stenasellus*), amphipods (genus *Niphargus*) (Ginet, 1967; Magniez, 1978) and subterranean copepods (genera *Elaphoidella, Morariopsis*; Brancelj, pers. obs.). Loss of mate guarding may relate to female-biased sex ratio and presumably reduced male-male competition.

## Habitat choice and adaptation to specific habitats

Several studies acknowledged that subterranean species segregate in space (Sket, 1996; Brancelj, 2002; Trontelj et al., 2012; Fišer et al., 2015). These species differ in morphology, which might point to that they perform optimally only in a subset of all possible subterranean habitats (see previous sections). If this is true, these species should have evolved habitat choice mechanisms that help them detect ecological boundaries of the habitats they are adapted to. These mechanisms should rely on characteristic environmental factors or their combinations and are closely related to sensory input. However, habitat choice mechanisms among subterranean organisms are poorly understood. We are aware of a single study that explicitly addressed habitat choice (Fišer et al., 2016). We, however, list also some examples that hint to specific habitat specializations, where habitat choice mechanisms could be expected.

The only study that explicitly addressed habitat choice mechanisms studied amphipods of subterranean genus *Niphargus* and surface *Gammarus* living in springs. Subterranean species were consistently more photophobic than surface ones, suggesting that light might help them discriminate between the surface and subterranean environment (Fišer et al., 2016). An unsolved question is how these species detect light. Nevertheless, response to light was studied in few other eyeless subterranean species. Some subterranean species do not react to light stimuli, while others are strongly photophobic. Some authors suggested that maintenance of photoreception has been retained as a habitat-choice mechanism in species that interact with the surface through springs, and has been lost in species inhabiting deeper subterranean habitats (Vawter et al., 1987; Langecker, 2000; Friedrich, 2013; Fišer et al., 2016). The generality of this hypothesis remains untested.

Hypoxia regularly occurs in some subterranean water bodies or parts of these bodies. Resistance to or avoidance of hypoxia might be part of habitat choice mechanisms. The effects of hypoxia were studied mainly for physiological traits (Hervant et al., 1998; Malard and Hervant, 1999). A special case of hypoxia are sulfide-rich water bodies, where organisms apparently developed specific morphological and behavioral adaptations. Sulfide is toxic as it binds to the electron transportation system in mitochondria, thus species need to find an appropriate solution of oxygen uptake. Species living in sulfide-rich waters, like *Poecilia mexicana,* compensate the deficiency in oxygen with an increased number of gill filaments but also by swimming to the upper-most water layer that is best oxygenated (Tobler et al., 2008). It seems that some species of the genus *Niphargus* living in sulfidic caves exploit a similar technique — clinging to the water surface and hanging upside down. In such position they expose gills on the ventral body side to the upper-most layer of water with higher oxygen concentrations (Borko et al., 2019). The resistance to hypoxia may be a habitat specific adaptation, a prerequisite for habitat choice mechanisms to evolve.

The last specific habitat we discuss is cave hygropetric, water streaming across the carbonate rock, carrying bacteria and organic debris (Table 18.1, Fig. 18.2). Many organisms would be washed from this habitat, but few species seem to be specialized to it, including unrelated crustacean species from the families Typhlogammaridae and Niphargidae, as well as leiodid beetles of the genus *Hadesia* Müeller 1911 and erpodbelid leech of the genus *Croatobranchus*. These species have evolved claws and specific behavior, helping them to enter and resist strong currents, where they filter organic particles (Dorigo et al., 2017). Moreover, all above mentioned arthropod species, albeit unrelated, have mouthparts specialized for feeding in the rapids (Sket, 2004; Polak et al., 2016). Their specialized morphologies and climbing in strong currents could be considered as joint morphological-behavioral part of a habitat choice mechanism.

## Antipredation mechanisms

The predator-prey arms race is a common phenomenon in surface habitats, but it has been rarely studied in subterranean ones. Therein, antipredation mechanisms manifest in two ways. On the one hand, reduced number of predatory species led to loss of these mechanisms. For example, a subterranean population of *A. aquaticus* exhibits reduced thigmotaxis and sheltering behavior, presumably due to the lack of visual predators in its habitat (Fišer et al., 2019). Likewise, relaxation of the same selective driver in combination with competition for food probably also contributed to the loss of aggregative behaviors in cavefish of Mexican tetra (see section on Feeding). By contrast, presence of predators may lead to protective mechanisms. *Proteus anguinus* is a top predator in Dinaric Karst. Shrimps from the genus *Troglocaris* and amphipods from the genus *Niphargus* constitute much of its diet. Shrimp lineages living in syntopy with *Proteus* have a longer rostrum than lineages inhabiting areas where this predator is absent. The pattern was replicated across different phylogenetic lineages of shrimps. Preliminary observations indicated that consumption of shrimps with long rostra took longer, and that, when attacked frontally, the olm spat out long-rostrated shrimps, which survived (Jugovic et al., 2010, Fig. 18.1). Morphological spines might therefore act against predation. Similar conclusions were drawn from a correlative study of *Niphargus* species living in phreatic lakes that have pleon armed with spines. Analyses of spatial distribution, co-occurrences of predator and prey as well as the co-evolution patterns imply that spines on *Niphargus* pleon may act as an antipredation trait (Premate et al., 2021c). Nevertheless, many other crustacean species with less obvious protective structures live in Dinaric Karst as well, and it is possible that these species evolved predator avoidance behaviors instead.

This evidence for a predator-prey arms race paves at least two interesting venues for future research. The first question is whether habitat properties modify predatory attack and antipredation response. For example, vibrational and chemical signals spread differently through lotic and lentic systems. It can be expected that the suite of modalities that predators and prey use for mutual detection differs in either environment and yields environment-specific foraging and avoidance strategies. Moreover, the strategy of escape may differ too, e.g., animals in lakes and streams may preferably hide in crevices or drift away in currents, respectively. The second question relates to predator-prey coevolution, i.e., whether and how the evolution of antipredatory mechanisms elicits evolution of foraging strategies, and thus affects the feeding aspect of the functional phenotype.

## Synthesis and perspectives

A perspective with an emphasis on the functional phenotype composed of morphological and behavioral components leads to four important conclusions.

First, the above presented studies, albeit fragmented, support the idea that response to environmental factors can be either morphological, behavioral, physiological or combined, so called "many-to-one relationship between the form and function" (Fig. 18.3). For instance, the second pair of antennae is elongated in at least five subterranean populations of *Asellus aquaticus*. In four populations, this elongation corresponds to an increase in article number, whereas in the fifth population the elongation emerges from elongation of individual articles (Protas et al., this volume). This is a clear example of the "many-to-one relationship" that can be easily extended to nonmorphological traits. For instance, animals can cope with food scarcity by more effective food finding, more economic locomotion, lipid accumulation, reduced metabolism, adjusted reproductive strategy — or any combination thereof (Fišer, 2019). Applying the many-to-one hypothesis might explain the absence of convergent morphologies in some populations, which colonized subterranean habitats on several independent occasions (Konec et al., 2015).

Second, subterranean habitats are more diverse than generally appreciated. The only ubiquitous environmental factor is darkness (Pipan and Culver, 2012, see also Table 18.1). Thus, the only convergences observed across nearly all subterranean species, i.e., reduction of eyes and pigments, likely evolved in response to darkness. All other factors like water flow, available living space, availability of food and chemical conditions vary across subterranean habitats just as they do on the surface. Therefore, there is no theoretical ground to expect other convergences inherent to all subterranean habitats. This is an important notion justifying the need that subterranean biology better integrates with the rest of biology (Mammola et al., 2021b). Many subterranean (micro)habitats are more alike their analogs among the surface (micro)habitats, than they are to other subterranean ones. We thus expect that aligning research of sinking rivers and phreatic lakes with research of surface rivers and lakes or even deep sea should pinpoint general principles ruling transversally across aquatic systems. That said, we can hardly emphasize enough that subterranean habitats are an ideal eco-evolutionary model system. Despite high diversity of subterranean habitats, subterranean communities are simpler than surface ones and consequently more suitable for investigations of more complex questions, like trade-offs (Mammola, 2019). This chapter highlighted two possible trade-offs that deserve further attention. The first trade-off result from the interaction of opposing abiotic components, e.g., darkness and water current yield opposing effects on antenna length. The second trade-off arises due to multifunctional traits, where performance of two different functions depends on different trait optima. For example, while large invertebrates are energetically more efficient, move faster and exhibit higher fecundity (synergistic effects), they may be more attractive to predators (antagonistic effect).

Third, it should be emphasized that our future approach requires a holistic consideration of animal biology within their specific environmental context. Any functional phenotype results from the interactions of morphology, behavior, physiology and life histories. The latter two phenotype components (see Di Lorenzo et al., this volume and Venarsky et al., chapter 19, this volume) could be easily integrated into our scheme (Fig. 18.3). Both, physiology and

A

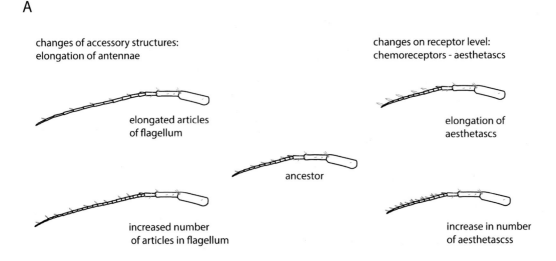

changes of accessory structures:
elongation of antennae

changes on receptor level:
chemoreceptors - aesthetascs

elongated articles
of flagellum

elongation of
aesthetascs

ancestor

increased number
of articles in flagellum

increase in number
of aesthetascss

B

FUNCTION
(feeding)

life history
(allocation strategy)

morphology
(long antenna for
food detection)

behavior
(food finding strategy,
fast locomotion)

physiology
(low metabolism,
starvation)

**FIGURE 18.3** Schematic example of the "many-to-one relationship of form and function". The principle applies on different levels of organismal organization. (A) An example on the morphological level. Chemoreception can be improved by elaboration of accessory structures (left) or receptors (right). Flagellum of antenna bearing aesthetascs can double its length by doubling the lengths of individual flagellar articles (left, upper) or doubling the number of articles of unchanged length (left, lower). Likewise, chemoreception can be improved by elongation of aesthetasc receptors (right, upper), or by an increase in their number. Combinations of these solutions may further increase chemoreception. (B) The "many-to-one" principle can involve relationships of morphology, behavior, physiology and life history. Each trait can individually lead to similarly enhanced performance. Additional enhancement can again be achieved via combinations of different traits, as discussed in text.

life histories are intertwined with behavior and/or morphology (Hervant, 2012; Fišer et al., 2013). Study design should therefore consider all possible phenotype components that contribute to a specific function in a specific subterranean habitat and control them.

The fourth and final conclusion reaches beyond the scope of this chapter. Understanding how species function within their habitat is a prerequisite for species' protection, but also for conservation of all functions they perform in a given habitat (Cadotte et al., 2011; Kosman et al., 2019). Hence, functional traits are in the heart of basic eco-evolutionary research, nature conservation and applied ecology (Mammola et al., 2021a). In the evermore threatened natural world, where freshwater-dependent ecosystems are subjected to mounting anthropogenic pressures, knowledge of functional ecology is highly needed (Mammola et al., 2019).

Integration of morphology and behavior into the functional phenotype unveiled deep gaps in our knowledge. We foresee two main directions of research to fill them. Above all, we urgently need to identify additional model systems to be studied in detail. We propose seeking and developing study systems on a species as well as population level that could be studied in breadth and depth, using comparative methods and experimental approaches, respectively (Mammola et al., 2021b). It is also important to fill the gaps in basic biology of these organisms (Alfred and Baldwin, 2015). This will require development of breeding programs, a serious technical challenge that is nevertheless a prerequisite for assessments of heritability, experimental manipulation and for linking phenotypic variation with fitness (McGaugh et al., 2020).

These proposed research steps seem little rewarding in the contemporary molecular "−omics" era that stampedes over much of scientific efforts that are not lightningly fast in producing big data. We plead not to get discouraged, though. After all, we can benefit from the −omics studies only if such big data can be linked to well-documented, measurable and observable biological functions. On the more positive note, technical advances, such as those in computer vision and machine learning, as well as their accessibility via cheap or open-source software and hardware, will eventually find their way underground and transform the traditionally slow quantification of the subterranean phenotype into high throughput phenotyping as well.

## Acknowledgments

We thank Rüdiger Riesch and Florian Malard for enlightening suggestions that substantially improved the earlier version of the manuscript. Teo Delić kindly provide us with photographs of cave animals. CF, AB and ŽF were supported by the Slovenian Research Agency (SRA) through core funding programs P1-0184 and P1-0255 as well as projects N1-0069, N1-0096 and J1-2464. SM was supported by the European Commission through Horizon 2020 Marie Skłodowska-Curie Actions (MSCA) individual fellowships (Grant no. 882221). MY was supported by the United States National Institute of Health NIGMS grant (P20GM125508).

## References

Abhilash, L., Shindey, R., Sharma, V.K., 2017. To be or not to be rhythmic? A review of studies on organisms inhabiting constant environments. Biological Rhythm Research 48 (5), 677−691.
Aiello, B.R., Westneat, M.W., Hale, M.E., 2017. Mechanosensation is evolutionarily tuned to locomotor mechanics. Proceedings of the National Academy of Sciences 114 (17), 4459−4464.
Alcock, J., 2005. Animal Behavior, eight ed. Sinauer Associates, Sunderland, MA.

Alfred, J., Baldwin, I.T., 2015. The natural history of model organisms: new opportunities at the wild frontier. eLife 4, e06956.

Balázs, G., Lewarne, B., Herczeg, G., 2020. Extreme site fidelity of the olm (*Proteus anguinus*) revealed by a long-term capture—mark—recapture study. Journal of Zoology 311 (2), 99—105.

Aljančič, G., 2019. History of research on *Proteus anguinus* Laurenti 1768 in Slovenia / Zgodovina raziskovanja človeške ribice (*Proteus anguinus* Laurenti 1768) v Sloveniji. Folia Biologica et Geologica 60 (1), 39—69.

Balázs, G., Biró, A., Fišer, Ž., Fišer, C., Herczeg, G., 2021. Parallel morphological evolution and habitat-dependent sexual dimorphism in cave- vs. surface populations of the *Asellus aquaticus* (Crustacea: Isopoda: Asellidae) species complex. Ecology and Evolution 11 (21), 15389—15403.

Belanger, J., 2013. Appendage diversity and mode of locomotion: walking. In: Watling, L., Thiel, M. (Eds.), The Natural History of Crustacea: Functional Morphology & Diversity, vol. 1. Oxford University Press, New York, NY, pp. 261—275.

Bilandžija, H., Morton, B., Podnar, M., Cetković, H., 2013. Evolutionary history of relict *Congeria* (Bivalvia: Dreissenidae): unearthing the subterranean biodiversity of the Dinaric Karst. Frontiers in Zoology 10 (1), 5.

Boonyanusith, C., Sanoamuang, L.O., Brancelj, A., 2018. A new genus and two new species of cave-dwelling cyclopoids (Crustacea, Copepoda) from the epikarst zone of Thailand and up-to-date keys to genera and subgenera of the *Bryocyclops* and *Microcyclops* groups. European Journal of Taxonomy 431, 1—30.

Borko, Š., Collette, M., Brad, T., Zakšek, V., Flot, J.-F., Vaxevanopoulos, M., Sarbu, S.M., Fišer, C., 2019. Amphipods in a Greek cave with sulphidic and non-sulphidic water: phylogenetically clustered and ecologically divergent. Systematics and Biodiversity 17 (6), 558—572.

Boudriot, F., Reutter, K., 2001. Ultrastructure of the taste buds in the blind cave fish *Astyanax jordani* ("Anoptichthys") and the sighted river fish *Astyanax mexicanus* (Teleostei, Characidae). Journal of Comparative Neurology 434 (4), 428—444.

Brancelj, A., 1991. Stygobitic Calanoida (Crustacea: Copepoda) from Yugoslavia with the description of a new species — *Stygodiaptomus petkovski* from Bosnia and Herzegovina. Stygologia 6 (3), 165—176.

Brancelj, A., 2000. *Parastenocaris andreji* n. sp. (Crustacea ; Copepoda) — the first record of the genus in Slovenia (SE Europe). Hydrobiologia 437, 235—239.

Brancelj, A., 2002. Microdistribution and high diversity of Copepoda (Crustacea) in a small cave in central Slovenia. Hydrobiologia 477, 59—72.

Brancelj, A., 2009. Fauna of an unsaturated karstic zone in Central Slovenia: two new species of Harpacticoida (Crustacea: Copepoda), *Elaphoidella millennii* n. sp. and *E. tarmani* n. sp., their ecology and morphological adaptations. Hydrobiologia 621 (1), 85—104.

Brancelj, A., Dumont, H.J., 2007. A review of the diversity, adaptations and groundwater colonization pathways in Cladocera and Calanoida (Crustacea), two rare and contrasting groups of stygobionts. Fundamental and Applied Limnology 168 (1), 3—17.

Brancelj, A., Karanovic, T., 2015. A new subterranean *Maraenobiotus* (Crustacea : Copepoda) from Slovenia challenges the concept of polymorphic and widely distributed harpacticoids. Journal of Natural History 49 (45—48), 2905—2928.

Brancelj, A., Žibrat, U., Jamnik, B., 2016. Differences between groundwater fauna in shallow and in deep intergranular aquifers as an indication of different characteristics of habitats and hydraulic connections. Journal of Limnology 75 (2), 248—261.

Cadotte, M.W., Carscadden, K., Mirotchnick, N., 2011. Beyond species: functional diversity and the maintenance of ecological processes and services. Journal of Applied Ecology 48 (5), 1079—1087.

Christiansen, K., 1961. Convergence and parallelism in cave Entomobryinae. Evolution 15 (3), 288—301.

Christiansen, K., 1965. Behavior and form in the evolution of cave Collembola. Evolution 19 (4), 529—537.

Coineau, N., 2000. Adaptations to interstitial groundwater life. In: Wilkens, H., Culver, D.C., Humphreys, W.F. (Eds.), Subterranean Ecosystems. Elsevier Academic Press, Amsterdam, pp. 189—210.

Culver, D.C., 1973. Feeding behavior of the salamander *Gyrinophilus porphyriticus* in caves. International Journal of Speleology 5 (3/4), 369—377.

Culver, D.C., Pipan, T., 2014. Shallow Subterranean Habitats: Ecology, Evolution, and Conservation, first ed. Oxford University Press, New York.

Culver, D.C., Pipan, T., 2019. The Biology of Caves and Other Subterranean Habitats. Oxford University Press, New York.

Culver, D.C., Kane, T.C., Fong, D.W., 1995. Adaptation and Natural Selection in Caves: The Evolution of Gammarus Minus. Harward University Press, Cambridge, MA; London.

Delić, T., Trontelj, P., Zakšek, V., Fišer, C., 2016. Biotic and abiotic determinants of appendage length evolution in a cave amphipod. Journal of Zoology 299 (1), 42−50.

De Meyer, J., Irschick, D.J., Vanhooydonck, B., Losos, J.B., Adriaens, D., Herrel, A., 2019. The role of bite force in the evolution of head shape and head shape dimorphism in *Anolis* lizards. Functional Ecology 22 (11), 2191−2202.

Dickinson, M.H., Farley, C.T., Full, R.J., Koehl, M., Kram, R., Lehman, S., 2000. How animals move: an integrative view. Science 288 (5463), 100−106.

Dorigo, L., Squartini, A., Toniello, V., Dreon, A.L., Pamio, A., Concina, G., Simonutti, V., Ruzzier, E., Perreau, M., Engel, A.S., Gavinelli, F., Martinez-Sañudo, I., Mazzon, L., Paoletti, M.G., 2017. Cave hygropetric beetles and their feeding behaviour, a comparative study of *Cansiliella servadeii* and *Hadesia asamo* (Coleoptera, Leiodidae, Cholevinae, Leptodirini). Acta Carsologica 46 (2−3), 317−328.

Duboué, E.R., Keene, A.C., Borowsky, R.L., 2011. Evolutionary convergence on sleep loss in cavefish populations. Current Biology 21 (8), 671−676.

Espinasa, L., Heintz, C., Rétaux, S., Yoshisawa, M., Agnès, F., Ornelas-Garcia, P., Balogh-Robinson, R., 2021. Vibration attraction response is a plastic trait in blind Mexican tetra (*Astyanax mexicanus*), variable within subpopulations inhabiting the same cave. Journal of Fish Biology 98 (1), 304−316.

Fišer, C., Zagmajster, M., Zakšek, V., 2013. Coevolution of life history traits and morphology in female subterranean amphipods. Oikos 122, 770−778.

Fišer, C., Bininda-Emonds, O.R.P., Blejec, A., Sket, B., 2008. Can heterochrony help explain the high morphological diversity within the genus *Niphargus* (Crustacea: Amphipoda)? Organisms, Diversity and Evolution 8 (2), 146−162.

Fišer, C., Delić, T., Luštrik, R., Zagmajster, M., Altermatt, F., 2019. Niches within a niche: ecological differentiation of subterranean amphipods across Europe's interstitial waters. Ecography 42 (6), 1212−1223.

Fišer, C., Luštrik, R., Sarbu, S.M., Flot, J.-F., Trontelj, P., 2015. Morphological evolution of coexisting amphipod species pairs from sulfidic caves suggests competitive interactions and character displacement, but no environmental filtering and convergence. PLoS One 10 (4), e0123535.

Fišer, Ž., 2019. Adaptation to low food. In: White, W.B., Culver, D.C., Pipan, T. (Eds.), Encyclopaedia of Caves. Elsevier Academic Press, Amsterdam, pp. 1−7.

Fišer, Ž., Novak, L., Luštrik, R., Fišer, C., 2016. Light triggers habitat choice of eyeless subterranean but not of eyed surface amphipods. Naturwissenschaften 103 (1−2), 7.

Fišer, Ž., Prevorčnik, S., Lozej, N., Trontelj, P., 2019. No need to hide in caves: shelter-seeking behavior of surface and cave ecomorphs of *Asellus aquaticus* (Isopoda: Crustacea). Zoology 134, 58−65.

Flammang, B.E., Suvarnaraksha, A., Markiewicz, J., Soares, D., 2016. Tetrapod-like pelvic girdle in a walking cavefish. Scientific Reports 6, 23711.

Fortune, E.S., Andanar, N., Madhav, M., Jayakumar, R.P., Cowan, N.J., Bichuette, M.E., Soares, D., 2020. Spooky interaction at a distance in cave and surface dwelling electric fishes. Frontiers in Integrative Neuroscience 14, 561524.

Foster, K.L., Collins, C.E., Higham, T.E., Garland, T., 2015. Determinants of lizard escape performance: decision, motivation, ability, and opportunity. In: Cooper Jr., W., Blumstein, D.T. (Eds.), Escaping from Predators: An Integrative View of Escape Decisions. Cambridge University Press, Cambridge, pp. 287−321.

Friedrich, M., 2013. Biological clocks and visual systems in cave-adapted animals at the dawn of speleogenomics. Integrative and Comparative Biology 53 (1), 50−67.

Ginet, R., 1967. Compartement sexuel de *Niphargus virei* (Crustacé hypogé). Comparaison avec les autres Amphipodes. Revue du Comportement Animal 4 (45), 56.

Gonzalez, B.C., Worsaae, K., Fontaneto, D., Martínez, A., 2018. Anophthalmia and elongation of body appendages in cave scale worms (Annelida: Aphroditiformia). Zoologica Scripta 48 (1), 106−121.

Hervant, F., 2012. Starvation in subterranean species versus surface-dwelling species: crustaceans, fish, and salamanders. In: McCue, M.D. (Ed.), Comparative Physiology of Fasting, Starvation, and Food Limitation. Springer, Heidelberg, pp. 91−102.

Hervant, F., Mathieu, J., Messana, G., 1998. Oxygen consumption and ventitaltion in declining oxygen tension and recovery in epigean and hypogean crustaceans. Journal of Crustacean Biology 18 (4), 717−727.

Hervant, F., Mathieu, J., Barre, H., Simon, K., Pinon, C., 1997. Comparative study on the behavioral, ventilatory and respiratory responses of hypogean and epigean crustaceans to long-term starvation and subsequent feeding. Comparative Biochemistry and Physiology - A Molecular and Integrative Physiology 118 (4), 1277–1287.

Hinaux, H., Devos, L., Blin, M., Elipot, Y., Bibliowicz, J., Alié, A., Rétaux, S., 2016. Sensory evolution in blind cavefish is driven by early embryonic events during gastrulation and neurulation. Development 143 (23), 4521–4532.

Hou, Z.-E., Li, S., 2002. A new cave amphipod of the genus *Sinogammarus* from China. Crustaceana 75 (6), 815–825.

Hutchins, B.T., Schwartz, B.F., Nowlin, W.H., 2014. Morphological and trophic specialization in a subterranean amphipod assemblage. Freshwater Biology 59, 2447–2461.

Hutchins, B.T., Summers Engel, A., Nowlin, W.H., Schwartz, B.F., 2016. Chemolithoautotrophy supports macroinvertebrate food webs and affects diversity and stability in groundwater communities. Ecology 97 (6), 1530–1542.

Hüppop, K., 2000. How do animals cope with the food scarcity in caves? In: Wilkens, H., Culver, D.C., Humphreys, F., William (Eds.), Ecosystems of the World 30: Subterranean Ecosystems, first ed. Elsevier, Amsterdam, pp. 159–188.

Huys, R., Boxshall, G.A., 1991. Copepod Evolution. The Natural History Museum, London.

Hyacinthe, C., Attia, J., Rétaux, S., 2019. Evolution of acoustic communication in blind cavefish. Nature Communications 10 (4231), 1–12.

Irschick, D.J., Higham, T., 2014. Locomotion unplugged: how movement in animals is influenced by the environment. Functional Ecology: Virtual Issues. https://besjournals.onlinelibrary.wiley.com/pb-assets/hub-assets/besjournals/1365-2435_FE/VI_Locomotion%20unplugged-1560334868717.pdf.

Jaggard, J., Robinson, B.G., Stahl, B.A., Oh, I., Masek, P., Yoshizawa, M., Keene, A.C., 2017. The lateral line confers evolutionarily derived sleep loss in the Mexican cavefish. Journal of Experimental Biology 220 (2), 284–293.

Jeong, H.G., Sinev, A.Y., Brancelj, A., Chang, K.-H., Kotov, A.A., 2017. A new blind groundwater- dwelling genus of the Cladocera (Crustacea: Branchiopoda) from the Korean Peninsula. Zootaxa 4341 (4), 451–474.

Jordan, L.A., Ryan, M.J., 2015. The sensory ecology of adaptive landscapes. Biology Letters 11 (5), 20141054.

Jugovic, J., Prevorčnik, S., Aljančič, G., Sket, B., 2010. The atyid shrimp (Crustacea: Decapoda: Atyidae) rostrum: phylogeny versus adaptation, taxonomy versus trophic ecology. Journal of Natural History 44 (41–42), 2509–2533.

Klaus, S., Mendoza, C.E.J., Liew, J.H., Plath, M., Meier, R., Yeo, D.C.J., 2013. Rapid evolution of troglomorphic characters suggests selection rather than neutral mutation as a driver of eye reduction in cave crabs. Biology Letters 9, 20121098.

Konec, M., Prevorčnik, S., Sarbu, S.M., Verovnik, R., Trontelj, P., 2015. Parallels between two geographically and ecologically disparate cave invasions by the same species, *Asellus aquaticus* (Isopoda, Crustacea). Journal of Evolutionary Biology 28 (4), 864–875.

Kosman, E., Burgio, K.R., Scheiner, S.M., Presley, S.J., Willig, M.R., 2019. Conservation prioritization based on trait - based metrics illustrated with global parrot distributions. Diversity and Distributions 25 (7), 1156–1165.

Kowalko, J.E., 2019. Adaptations: behavioral. In: White, W.B., Culver, D.C., Pipan, T. (Eds.), Encyclopaedia of Caves, third ed. Academic Press, Boston, MA, pp. 24–32.

Kowalko, J.E., Rohner, N., Rompani, S.B., Peterson, B.K., Linden, T.A., Yoshizawa, M., Kay, E.H., Weber, J., Hoekstra, H.E., Jeffery, W.R., Borowsky, R., Tabin, C.J., 2013. Loss of schooling behavior in cavefish through sight-dependent and sight-independent mechanisms. Current Biology 23 (19), 1874–1883.

Kralj-Fišer, S., Premate, E., Copilas-Ciocianu, D., Volk, T., Fišer, Ž., Balazs, G., Herzeg, G., Delić, T., Fišer, C., 2020. The interplay between habitat use, morphology and locomotion in subterranean crustaceans of the genus *Niphargus*. Zoology 139, 125742.

Lang, H.H., 1980. Surface wave discrimination between prey and nonprey by the back swimmer *Notonecta glauca* L. (Hemiptera, Heteroptera). Behavioral Ecology and Sociobiology 6 (3), 233–246.

Langecker, T.G., 2000. The effects of continuous darkness on cave ecology and cavernicolous evolution. In: Wilkens, H., Humphreys, W.F., Culver, D.C. (Eds.), Ecosystems of the World 30: Subterranean Ecosystems, first ed. Elsevier, Amsterdam, pp. 135–157.

Leal-Zanchet, A.M., Marques, A.D., 2018. Coming out in a harsh environment: a new genus and species for a land flatworm (Platyhelminthes: Tricladida) occurring in a ferruginous cave from the Brazilian savanna. PeerJ 6, e6007.

V. Biological traits in groundwater

Losos, J.B., 2009. Lizards in an Evolutionary Tree: Ecology and Adaptive Radiation of Anoles. University of California Press, Berkeley, CA.

Magniez, G., 1978. Précopulation et vie souterraine chez quelques Peracarides (Crustacea: Malacoraca). Archives de Zoologie Expérimentale et Générale 119, 471–478.

Malard, F., Hervant, F., 1999. Oxygen supply and adaptation of animals in groundwater. Freshwater Biology 41 (1), 1–30.

Mammola, S., 2019. Finding answers in the dark: caves as models in ecology fifty years after Poulson and White. Ecography 42, 1331–1351.

Mammola, S., Amorim, I.R., Bichuette, M.E., Borges, P.A.V., Cheeptham, N., Cooper, S.J.B., Culver, D.C., Deharveng, L., Eme, D., Ferreira, R.L., Fišer, C., Fišer, Ž., Fong, D.W., Griebler, C., Jeffery, W.R., Jugovic, J., Kowalko, J.E., Lilley, T.M., Malard, F., Manenti, R., Martínez, A., Meierhofer, M.B., Niemiller, M.L., Northup, D.E., Pellegrini, T.G., Pipan, T., Protas, M., Reboleira, A.S.P.S., Venarsky, M.P., Wynne, J.J., Zagmajster, M., Cardoso, P., 2020a. Fundamental research questions in subterranean biology. Biological Reviews 95 (6), 1855–1872.

Mammola, S., Arnedo, M.A., Fišer, C., Cardoso, P., Dejanaz, A.J., Isaia, M., 2020b. Environmnetal filtering and convergent evolution determine the ecological specialisation of subterranean spiders. Functional Ecology 34 (5), 1064–1077.

Mammola, S., Cardoso, P., Culver, D.C., Deharveng, L., Ferreira, R.L., Fišer, C., Galassi, D.M.P., Griebler, C., Halse, S., Humphreys, W.F., Isaia, M., Malard, F., Martinez, A., Moldovan, O.T., Niemiller, M.L., Pavlek, M., Reboleira, A.S.P.S., Souza-Silva, M., Teeling, E.C., Wynne, J.C., Zagmajster, M., 2019. Scientists' warning on the conservation of subterranean ecosystems. BioScience 69 (8), 641–650.

Mammola, S., Carmona, C.P., Guillerme, T., Cardoso, P., 2021a. Concepts and applications in functional diversity. Functional Ecology 35 (9), 1869–1885.

Mammola, S., Lunghi, E., Bilandžija, H., Cardoso, P., Grimm, V., Schmidt, S.I., Hesselberg, T., Martínez, A., 2021b. Collecting eco-evolutionary data in the dark: Impediments to subterranean research and how to overcome them. Ecol Evol. 11 (11), 5911–5926.

Márquez-Borrás, F., Solís-Marín, F.A., Mejía-Ortiz, L.M., 2020. Troglomorphism in the brittle star *Ophionereis commutabilis* Bribiesca-Contreras et al., 2019 (Echinodermata, Ophiuroidea, Ophionereididae). Subterranean Biology 108, 87–108.

Martínez, A., Mammola, S., 2021. Specialized terminology limits the reach of new scientific knowledge. Proceedings of the Royal Society B 288, 20202581.

Martínez, A., Kvindebjerg, K., Iliffe, T.M., Worsaae, K., 2017. Evolution of cave suspension feeding in Protodrilidae (Annelida). Zoologica Scripta 46 (2), 214–226.

McGaugh, S.E., Kowalko, J.E., Rohner, N., Duboué, E., Gross, J.B., Lewis, P., Odendaal, T.A.F., Rohner, N., Gross, J.B., Keene, A.C., 2020. Dark world rises: the emergence of cavefish as a model for the study of evolution, development, behavior, and disease. Journal of Experimental Zoology Part B: Molecular and Developmental Evolution 334 (7–8), 397–404.

Mejía-Ortíz, L.M., Hartnoll, R.G., Mejía, M.L., 2006. Progressive troglomorphism of ambulatory and sensory appendages in three Mexican cave decapods. Journal of Natural Histroy 40 (5–6), 225–264.

Moran, D., Softley, R., Warrant, E.J., 2014. Eyeless Mexican cavefish save energy by eliminating the circadian rhythm in metabolism. PLoS One 9 (9), e107877.

Moran, D., Softley, R., Warrant, E.J., 2015. The energetic cost of vision and the evolution of eyeless Mexican cavefish. Science Advances 1 (8), e1500363.

Odum, E.P., 1971. Fundamentals of Ecology, third ed. WB Saunders Company, Philadelphia, PA.

Parzefall, J., 1982. Changement of behaviour during the evolution of cave animals. Mémoires de Biospéologie 8, 55–62.

Parzefall, J., 2001. A review of morphological and behavioural changes in the cave molly, *Poecilia mexicana*, from Tabasco, Mexico. Environmental Biology of Fishes 62, 263–275.

Petkovski, T.K., 1978. *Troglodiaptomus sketi* n. gen., n. sp., ein neuer Höhlen-Calanoide vom Karstgelände Istriens (Crustacea, Copepoda). Acta Musei Macedonici Scientiarum Naturalium 25 (7), 151–165.

Pipan, T., Culver, D.C., 2012. Convergence and divergence in the subterranean realm: a reassessment. Biological Journal of the Linnean Society 107 (1), 1–14.

Plath, M., Parzefall, J., Körner, K.E., Schlupp, I., 2004. Sexual selection in darkness? Female mating preferences in surface- and cave-dwelling Atlantic mollies, *Poecilia mexicana* (Poeciliidae, Teleostei). Behavioral Ecology and Sociobiology 55 (6), 596—601.

Plath, M., Schlupp, I., Parzefall, J., Riesch, R., 2007. Female choice for large body size in the cave molly, *Poecilia mexicana* (Poeciliidae, Teleostei): influence of species- and sex-specific cues. Behaviour 144 (10), 1147—1160.

Plath, M., Rohde, M., Schroder, T., Taebel-Hellwig, A., Schlupp, I., 2006. Female mating preferences in blind cave tetras *Astyanax fasciatus* (Characidae, Teleostei). Behaviour 143 (1), 15—32.

Polak, S., Delic, T., Kostanjšek, R., Trontelj, P., 2016. Molecular phylogeny of the cave beetle genus *Hadesia* (Coleoptera: Leiodidae: Cholevinae: Leptodirini), with a description of a new species from Montenegro. Arthropod Systematics and Phylogeny 74 (3), 241—254.

Poulson, T.L., 1963. Cave adaptation in Amblyopsid fishes. The American Midland Naturalist 70, 257—290.

Premate, E., Zagmajster, M., Fišer, C., 2021c. Inferring predator—prey interaction in the subterranean environment: a case study from Dinaric caves. Scientific Reports 11 (1), 1—9.

Premate, E., Borko, Š., Delić, T., Malard, F., Simon, L., Fišer, C., 2021a. Cave amphipods reveal co-variation between morphology and trophic niche in a low-productivity environment. Freshwater Biology 66, 1876—1888.

Premate, E., Borko, Š., Kralj-Fišer, S., Fišer, Ž., Jennions, M., Balazs, G., Biro, A., Bračko, G., Copilas-Ciocianu, D., Hrga, N., Herzeg, G., Rexhepi, B., Zagmajster, M., Zakšek, V., Fromhage, L., Fišer, C., 2021b. No room for males in caves: female biased sex ratio in subterranean amphipods of the genus *Niphargus*. Journal of Evolutionary Biology 34 (10), 1653—1661.

Ramm, T., Scholtz, G., 2017. No sight, no smell? — brain anatomy of two amphipod crustaceans with different lifestyles. Arthropod Structure & Development 46 (4), 537—551.

Romero, A., 2009. Cave Biology Life in Darkness, first ed. Cambridge University Press, New York, NY.

Rundle, H.D., Nosil, P., 2005. Ecological speciation. Ecology Letters 8 (3), 336—352.

Rüschenbaum, S., Schlupp, I., 2013. Non-visual mate choice ability in a cavefish (*Poecilia mexicana*) is not mechanosensory. Ethology 119, 368—376.

Saccò, M., Blyth, A., Bateman, P.W., Hua, Q., Mazumder, D., White, N., Humphreys, W.F., Laini, A., Griebler, C., Grice, K., 2019. New light in the dark - a proposed multidisciplinary framework for studying functional ecology of groundwater fauna. Science of the Total Environment 662, 963—977.

Schemmel, C., 1967. Vergleichende Untersuchungen an den Hautsinnesorganen ober-und unterirdisch lebender Astyanax-Formen. Zeitschrift für Morphologie der Tiere 61, 255—316.

Schemmel, C., 1980. Studies on the genetics of feeding behaviour in the cave fish *Astyanax mexicanus f. anoptichthys*. Zeitschrift für Tierpsychologie 53 (1), 9—22.

Schlegel, P.A., Steinfartz, S., Bulog, B., 2009. Non-visual sensory physiology and magnetic orientation in the blind cave salamander, *Proteus anguinus* (and some other cave-dwelling urodele species). Review and new results on light-sensitivity and non-visual orientation in subterranean urodeles (Amphibia). Animal Biology 59 (3), 351—384.

Schluter, D., 2000. Ecological character displacement in adaptive radiation. The American Naturalist 156, S4—S16.

Schneider, K., Christman, M.C., Fagan, W.F., 2011. The influence of resource subsidies on cave invertebrates: results from an ecosystem-level manipulation experiment. Ecology 92 (3), 765—776.

Simon, V., Hyacinthe, C., Rétaux, S., 2019. Breeding behavior in the blind Mexican cavefish and its river-dwelling conspecific. PLoS One 14 (2), 1—16.

Sket, B., 1996. The ecology of anchihaline caves. Trends in Ecology & Evolution 11 (5), 221—225.

Sket, B., 2004. The cave hygropetric — a little known habitat and its inhabitants. Archiv für Hydrobiologie 160 (3), 413—425.

Soares, D., Niemiller, M.L., 2013. Sensory adaptations of fishes to subterranean environments. BioScience 63 (4), 274—283.

Stegner, M.E.J., Stemme, T., Iliffe, T.M., Richter, S., Wirkner, C.S., 2015. The brain in three crustaceans from cavernous darkness. BMC Neuroscience 16, 19.

Stevens, M., 2013. Sensory Ecology, Behavior, and Evolution, first ed. Oxford University Press, Glasgow.

Stoch, F., Galassi, D.M.P., 2010. Stygobiotic crustacean species richness: a question of numbers, a matter of scale. Hydrobiologia 653 (1), 217—234.

Timmermann, M., Schlupp, I., Plath, M., 2004. Shoaling behaviour in a surface-dwelling and a cave-dwelling population of a barb *Garra barreimiae* (Cyprinidae, Teleostei). Acta Ethologica 7 (2), 59—64.

V. Biological traits in groundwater

Tobler, M., DeWitt, T.J., Schlupp, I., Garcia de León, F.J., Herrmann, R., Feulner, P.G.D., Tiedemann, R., Plath, M., 2008. Toxic hydrogen sulfide and dark caves: phenotypic and genetic divergence across two abiotic gradients in *Poecilia mexicana*. Evolution 62 (10), 2643–2659.

Townsend, C.R., Dolédec, S., Scarsbrook, M.R., 1997. Species traits in relation to temporal and spatial heterogeneity in streams: a test of habitat templet theory. Freshwater Biology 37 (2), 367–387.

Tran Duc, L., Brancelj, A., 2017. Amended diagnosis of the genus *Nannodiaptomus* (Copepoda, Calanoida), based on redescription of *N. phongnhaensis* and description of a new species from caves in central Vietnam. Zootaxa 4221 (4), 457–476.

Trontelj, P., Blejec, A., Fišer, C., 2012. Ecomorphological convergence of cave communities. Evolution 66 (12), 3852–3865.

Uiblein, F., Durand, J.P., Juberthie, C., Parzefall, J., 1992. Predation in caves: the effects of prey immobility and darkness on the foraging behaviour of two salamanders, *Euproctus asper* and *Proteus anguinus*. Behavioral Processes 28 (1–2), 33–40.

Valvasor, J.V., 1689. Die Ehre Dess Hertzogthums Crain. Nürnberg.

Van Damme, K., Sinev, A.Y., 2011. A new genus of cave-dwelling microcrustaceans from the Dinaric Region (southeast Europe): adaptations of true stygobitic Cladocera (Crustacea: Branchiopoda). Zoological Journal of the Linnean Society 161 (1), 31–52.

Van Damme, K., Brancelj, A., Dumont, H.J., 2009. Adaptations to the hyporheic in Aloninae (Crustacea: Cladocera): allocation of *Alona protzi* Hartwig, 1900 and related species to *Phreatalona* gen. nov. Hydrobiologia 618 (1), 1–34.

Varatharasan, N., Croll, R.P., Franz-Odendaal, T., 2009. Taste bud development and patterning in sighted and blind morphs of *Astyanax mexicanus*. Developmental Dynamics 238 (12), 3056–3064.

Vawter, T.A., Fong, D.W., Culver, D.C., 1987. Negative phototaxis in surface and cave populations of the amphipod *Gammarus minus*. Stygologia 3 (1), 83–88.

Vogt, G., 1999. Hypogean life-style fuelled by oil. Naturwissenschaften 86, 43–45.

Vogt, G., Štrus, J., 1992. Oleospheres of the cave-dwelling shrimp *Troglocaris schmidtii*: a unique mode of extracellular lipid storage. Journal of Morphology 211, 31–39.

Webster, D.R., Weissburg, M.J., 2009. The hydrodynamics of chemical cues among aquatic organisms. Annual Review of Fluid Mechanics 41, 73–90.

Yamamoto, Y., Byerly, M.S., Jackman, W.R., Jeffery, W.R., 2009. Pleiotropic functions of embryonic sonic hedgehog expression link jaw and taste bud amplification with eye loss during cavefish evolution. Developmental Biology 330 (1), 200–211.

Yoshizawa, M., O'Quin, K.E., Jeffery, W.R., 2013. QTL clustering as a mechanism for rapid multi-trait evolution. Communicative & Integrative Biology 6 (4), 3–5.

Yoshizawa, M., Gorički, Š., Soares, D., Jeffery, W.R., 2010. Evolution of a behavioral shift mediated by superficial neuromasts helps cavefish find food in darkness. Current Biology 20, 1631–1636.

Yoshizawa, M., Jeffery, W.R., Van Netten, S.M., McHenry, M.J., 2014. The sensitivity of lateral line receptors and their role in the behavior of Mexican blind cavefish (*Astyanax mexicanus*). Journal of Experimental Biology 217 (6), 886–895.

Yoshizawa, M., Robinson, B.G., Duboué, E.R., Masek, P., Jaggard, J.B., Quin, K.E.O., Borowsky, R.L., Jeffery, W.R., Keene, A.C., 2015. Distinct genetic architecture underlies the emergence of sleep loss and prey-seeking behavior in the Mexican cavefish. BMC Biology 13, 15.

Yoshizawa, M., Yamamoto, Y., O'Quin, K.E., Jeffery, W.R., 2012. Evolution of an adaptive behavior and its sensory receptors promotes eye regression in blind cavefish. BMC Biology 10 (1), 108.

Ziemba, R.E., Simpson, A., Hopper, R., Cooper, R.L., 2003. A comparison of antennule structure in a surface- and a cave-dwelling crayfish, genus *Orconectes* (Decapoda: Astacidea). Crustaceana 76 (7), 859–869.

# 19

# Life histories in groundwater organisms

*Michael Venarsky*[1,2], *Matthew L. Niemiller*[3], *Cene Fišer*[4], *Nathanaelle Saclier*[5] *and Oana Teodora Moldovan*[6]

[1]Department of Biodiversity Conservation and Attractions, Kensington, WA, Australia; [2]Australian Rivers Institute, Griffith University, Nathan, QLD, Australia; [3]Department of Biological Sciences, The University of Alabama in Huntsville, Huntsville, AL, United States; [4]University of Ljubljana, Biotechnical Faculty, Department of Biology, Ljubljana, Slovenia; [5]Univ Lyon, Université Claude Bernard Lyon 1, CNRS, ENTPE, UMR 5023 LEHNA, Villeurbanne, France; [6]Emil Racovitza Institute of Speleology, Cluj-Napoca, Romania

## Introduction

Studies of "life history" span multiple disciplines of biology. Ecosystem ecologists incorporate life history information into models describing the flow of materials and energy through food webs and across landscapes (e.g., Mims and Olden, 2013; Massol et al., 2017), while evolutionary biologists examine life histories to better understand how natural selection shapes phenotypes (e.g., Reznick et al., 1990; Riesch et al., 2010). In conservation biology, a comprehensive understanding of life history is paramount to predict how anthropogenic stressors will influence population demographics (e.g., survival or population size) (e.g., Garcia et al., 2008; Pearson et al., 2014). The study of groundwater ecosystems is no exception to this trend, with the interactions between species life history traits and environmental conditions commonly used to explain patterns in groundwater communities (e.g., Huntsman et al., 2011) and speciation (e.g., Riesch et al., 2010, 2016). Furthermore, groundwater and other subterranean ecosystems represent an endpoint along several environmental gradients (i.e., light, resource availability, and environmental stability) and are thus commonly viewed as model systems in ecology and evolutionary biology, including the study of life history evolution (Poulson and White, 1969; Culver et al., 1995; Niemiller and Poulson, 2010; Fišer, 2019a; Mammola, 2019; Mammola et al., 2020).

A large body of life history research has resulted in varying definitions as to what constitutes a "life history trait." We define a life history trait as a trait that affects the survival and reproduction of an organism, that is, fitness. While some authors define functional traits in general as traits that influence individual fitness (e.g., Violle et al., 2007; Sobral 2021), we focus our discussion in this chapter on a classic set of life history traits (Box 19.1), such as size at hatching/birth, number of eggs/offspring per reproductive episode, and age and size at sexual maturity (Stearns, 1992; Flatt and Heyland, 2011). Note that we do not include behavioral, morphological, or physiological traits in this chapter, as these traits are covered in other chapters of this volume (Fišer et al., this volume; Di Lorenzo et al., this volume). Importantly though, life history traits cannot be discussed in isolation, as life history, behavioral, morphological, and physiological traits interact to influence one another (Stearns, 1989; Ricklefs, 2000). Therefore, while our discussion is focused on life history traits, we explicitly acknowledge interactions with other traits as appropriate.

---

## BOX 19.1

### Life history traits and life table variables in studies of groundwater fauna

**Growth rate** refers to the increase of body size (e.g., mass, body length, head capsule width) per unit time (day, month, year). Studies on groundwater species have not differentiated between the growth of somatic and reproductive tissues, and generally report that growth rates of groundwater species are slower than surface relatives. See Cooper (1975), Venarsky et al. (2007), Huntsman et al. (2011), Venarsky et al. (2012), Niemiller et al. (2016), Carpenter (2021), among others.

**Lifespan** (e.g., longevity) refers to the duration of life between birth and death of an organism. However, reports of lifespan are not always consistent in the literature and may refer to one of several similar aspects, such as minimum lifespan, maximum lifespan, life expectancy, and reproductive lifespan. Lifespan exhibits considerable variation both within and among species and is related to several factors, such as body size (Lindsetdt and Calder, 1981; Ricklefs, 2010). In

general, groundwater taxa are longer-lived compared to surface relatives. See Heuts (1951), Magniez (1975), Ginet and Decou (1977), Cooper (1975), Turquin and Barthelemy (1985), Trajano (1991), Mathieu and Turguin (1992), Huntsman et al. (2011), Voituron et al. (2011), Venarsky et al. (2012), Niemiller et al. (2016), and Carpenter (2021).

**Age and size at sexual maturity** is the age and body size at which a species can first reproduce, which can differ between males and females. Currently there is little consensus in the literature regarding age and size at sexual maturity in groundwater fauna, with some studies reporting that sexual maturity is delayed and occurs at larger body sizes in groundwater species when compared to surface species (Poulson, 1963; Niemiller and Poulson, 2010; Venarsky et al., 2012; Trochet et al., 2014; Reisch et al., 2016; Østbye et al., 2018; Pacioglu et al., 2020; Carpenter, 2021), is similar between groundwater and surface species (Fenolio et al., 2014a;

---

## BOX 19.1   *(cont'd)*

Niemiller et al., 2016), or that groundwater species mature earlier and at smaller body sizes than surface species (Reisch et al., 2016; Simons et al., 2017).

**Parity** refers to the number of reproductive episodes in an organism's lifetime. Some organisms reproduce once followed shortly by death (semelparity), whereas others reproduce multiple times throughout their lives (iteroparity). Most groundwater species appear to be iteroparous (Poulson, 1963; Ginet and Decou, 1977; Wilson and Ponder, 1992; Niemiller and Poulson, 2010; Voituron et al., 2011; Reisch et al., 2016).

**Fecundity** is the number of offspring (eggs or young) produced during a reproductive episode or over the lifetime of an organism, while **size of offspring** refers to the size of eggs or young at birth of hatching. In general, groundwater species produce fewer but larger eggs per reproductive episode than related surface species. For examples, see Poulson (1963), Jegla (1969), Hobbs and Barr (1972), Hüppop and Wilkens (1991), Culver et al. (1995), Pouilly and Miranda (2003), Fong (2009), Zigler and Cooper (2011), Reisch et al. (2012), Fišer et al. (2013), Fenolio et al. (2014b), Trochet et al. (2014), Østbye et al. (2018), Paciouglu et al. (2020), and Carpenter (2021).

**Parental care** is the amount of time and energy invested by parents in the caring for or rearing of eggs and offspring and may refer to any behavior that contributes to the survival of offspring, such as provisioning of offspring or defending eggs/young from predators. Parental care is intricately linked to fecundity, with species that produce large numbers of eggs making smaller investments in parental care than species that produce fewer offspring. However, parental care and the trade-off with fecundity have not been well studied in groundwater species. Examples of parental care exhibited in groundwater fauna include mouthbrooding of eggs in amblyopsid cavefishes (Poulson, 1963; Niemiller and Poulson, 2010) and egg/young brooding in cave crustaceans (Zigler and Cooper, 2011; Fenolio et al., 2014b; Carpenter, 2021).

**Life tables** summarize birth (fecundity) and death (mortality) rates of organisms at different stages or ages and are the core metrics that dictate the survival or extinction of a population. Because birth and death rates can differ among age, size, and life history stages in a population, we are often interested in accounting for variation in schedules of births and deaths associated with these factors. Life tables are constructed based on data collected from individuals within a population. One approach is to collect data from the same cohort of individuals throughout their lives (i.e., a group of individuals all born during the same time interval and tracked until death). However, these **cohort** or **age-specific** life tables are rarely possible with natural populations of groundwater species as reproduction is rarely synchronized, meaning that recruitment of young can occur at any time of the year, which makes it difficult to identify cohorts and thus calculate rates of fecundity or mortality. An alternative approach is to collect data from the population at a particular point in time or for a short period of time, which produces a **static** or **time-specific** life table, respectively. This approach assumes a cohort was followed through time. Finally, the two approaches can be combined in a **composite** life table in which data are gathered over several time periods (e.g., years) and generations using cohort and

*Continued*

V. Biological traits in groundwater

static approaches allowing for variation in rates of fecundity and mortality to be assessed. We are unaware of any studies that have generated life tables for a groundwater species to date; however, some studies have estimated specific life table variables from demographic data, such as survivorship (e.g., Fenolio et al., 2014a; Niemiller et al., 2016; Huntsman et al., 2020).

Common life table variables include:

- **Class (x)**—the age, size class, or life history stage
- **Survivorship ($l_x$)**—the proportion of individuals surviving to the next life stage or age-class.
- **Survival ($n_x$)**—the number of individuals surviving at the start of a life stage or age-class.
- **Mortality ($d_x$)**—the number or proportion of individuals that die during each life stage or age-class.
- **Finite rate of mortality ($q_x$)**—Calculated as $dx$ divided by $n_x$ during the age interval $x$ to $x+1$. This variable is often used for comparisons within and between species.

- **Fecundity rate ($m_x$)**—the average number of female offspring produced per female in a population over a period of time.
- **Net reproductive rate per generation ($R_0$)**—the mean number of female offspring produced by a female during her lifetime; also known as the replacement rate. Calculated by taking the sum of survivorship multiplied by fecundity rate ($l_x * m_x$) for each life stage or age-class then dividing by the survivorship of the initial life stage or age-class ($L_0$). This variable can be used to infer trends in size of population: $R0 < 1$ indicates a population is declining in size, $R0 > 1$ indicates a population is increasing in size, and $R0 = 0$ indicates a population is stable.
- **Generation time (G)**—the mean length of a generation often calculated as the amount of time between the birth of a female and the mean time (age) of production of her offspring, or as the average age that adult females produce offspring.

We begin with a brief discussion of life history evolution, life history traits, and life table variables. Our intent is not to provide an in-depth explanation of life history evolution as this can be found in other sources (e.g., Stearns, 1992; Roff, 1993, 2002; Flatt and Heyland, 2011), but merely put the life history studies of subterranean organism into a more general context. Next, we outline the current view of life history evolution in groundwater species and review the support for this conceptual model. Lastly, we discuss how various areas of research relate to life histories.

## A brief overview of life history evolution, life history traits, and life table variables

Discussions of life history often conflate two types of life history information: life history traits (e.g., somatic growth rates and life span) and life table variables (e.g., rates of survival and reproduction, mean number of offspring produced per individual) (Ricklefs and Wikelski, 2002; see Box 19.1). While life history traits and life table variables are related to and interact with one another, they provide distinctly different types of information on the life history of a species. Life history traits are shaped by selection, meaning that the gene frequencies associated with these traits change in response to the environment, which in-turn influences the distribution of life history phenotypes in a population. In contrast, life table variables measure how a population is responding to contemporary environmental conditions, such as density-dependent factors (competition, consumer—resource interactions) or measures of environment quality in relation to the biology and ecology of the species (e.g., temperature variation, water chemistry, or toxic compounds) (Ricklefs, 2000; Ricklefs and Wikelski, 2002). Thus, life history traits represent the adaptations of individuals to the environment, while life table variables are a feature of populations that measure how life history traits, or more accurately combinations thereof, perform within the context of contemporary environmental conditions.

Life history traits and life table variables are complemented with two other concepts, namely life history plasticity and trade-offs. The environment can strongly influence the phenotypic response of life history traits. A life history trait is considered "plastic" when an individual genotype has a variable phenotypic response under different environmental conditions (i.e., phenotypic plasticity; Roff, 1997; Pigliucci, 2001; DeWitt and Scheiner, 2004). Note that natural populations are generally comprised of multiple genotypes, each of which potentially may have a distinct phenotypic response (i.e., genotype by environmental interaction; Box 19.2). Trade-offs are an interaction among two or more traits and

---

### BOX 19.2

## Interactions between life history traits and life table variables under different environmental conditions

In this example, we illustrate life history adaptation by showing the interactions between life history traits and life table variables under different environmental conditions in a conceptualized common garden transplantation experiment. We use food availability as the environmental driver, number of eggs per reproductive event (assuming one event per year) and size of eggs as the life history traits, and survival rate of young as the life table variable (Fig. 19.1). We do not use fecundity or fertility as the life history trait because: (i) the definitions for these variables varies among authors and subdisciplines (i.e., human population demographics vs. population

ecology) and (ii) these variables are often the product of multiple life history traits (e.g., fecundity is the product of the number of young per reproductive event and the number of reproductive events per unit time). Moreover, we use the survival rate of young as the life table variable because this metric determines if genotypes are carried into future generations and is, therefore, a measure of both contemporary ecological performance as well as evolutionary fitness (Ricklefs, 2000).

We divide the original population (triangles) into three groups with identical genotype frequencies and then transplant each group into a new habitat with different

*Continued*

---

V. Biological traits in groundwater

## BOX 19.2 *(cont'd)*

amounts of food: high-food (squares), low-food (circles), and very low-food (diamonds). Note that average food availability (solid horizontal lines) in the high- and low-food environments differs from the original habitat, but the variability overlaps among habitats, indicating that the populations transplanted to the high- and low-food environments are not experiencing completely "novel" environmental conditions. In contrast, food availability in the very low-food habitat is quite different from the original habitat, which resulted in failed reproductive events and ultimately extinction as the transplanted population did not contain genotypes capable of surviving and reproducing under the new environmental conditions.

The transplanted populations in the high- and low-food environments survived, and after the initial transplant we see that the number of eggs per reproductive episode responds in accordance with the new habitat conditions, with individuals in the high- and low-food environments producing more and fewer eggs, respectively. In contrast, we see no change in the size of eggs produced per reproductive event. These patterns in the number and size of eggs following transplantation are measures of the phenotypic response for each life history trait (*sensu* reaction norm; Ricklefs and Wikelski, 2002). In this example, phenotypic plasticity in life history traits can be seen immediately after relocation, with the "number of eggs per reproductive event" changing in response to the new environmental conditions, while the size of eggs showed no change. Our example also illustrates the advantages of phenotypic

plasticity in life history traits as the survival of young correlates with quantity of food in the environment, meaning that populations can "bend rather than break" under changing environmental conditions.

Over time, we see changes in the life history traits in the low-food population, indicating the evolution of new genotypes and phenotypes that are better adapted to the low-food environment. Individuals in the low-food population now produce fewer, larger eggs per reproductive episode, and the performance of these life history traits has improved as the survival rate increased to levels similar to the original population. The negative relationship between the number and size of eggs illustrates trade-offs life history traits. Note that trade-offs often occur among life history, behavioral, physiological, and morphological traits. In other words, an organism can invest resources into growth, maintenance, or reproduction, and investment in one trait will likely result in changes in one or more other traits. Moreover, trade-offs can occur during the same time in the life cycle or between different life stages (e.g., juvenile vs. adult; reproductive vs. nonreproductive time periods) (Chippindale et al., 1996; Zera and Harshman, 2001). Many potential trade-offs exist among life history traits, such as number versus size of offspring (Smith and Fretwell, 1974), current versus future reproduction (Williams, 1966), current reproduction versus survival (Hirshfield and Tinkle, 1975; Stearns, 1989), and current reproduction versus growth (Gadgil and Bossert, 1970; Stearns, 1989, 1992).

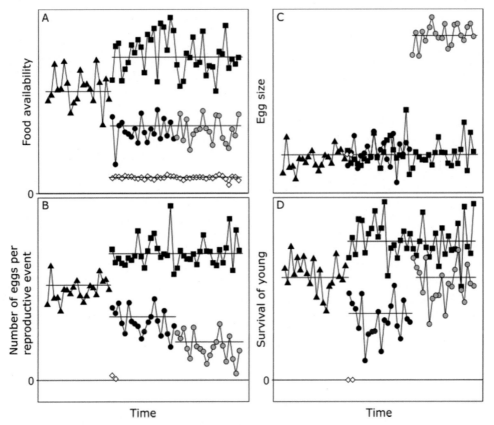

**FIGURE 19.1**   Illustration of the interactions between food availability, number of eggs per reproductive event, egg size, and survival rate of young in a conceptualized common garden transplantation experiment. Horizontal lines are the means for each variable in each population.

are measured through both field and laboratory experiments (Stearns, 1989, 1992; Roff, 1993; Zera and Harshman, 2001; Flatt and Heyland, 2011). For example, increasing the size of eggs generally reduces the number of eggs per reproductive event. See Box 19.2 for a conceptualized common garden transplantation experiment that illustrates the interactions among these concepts.

## The current conceptual model of life history evolution in groundwater species

The "habitat template" approach (Southwood, 1977, 1988; Poff and Ward, 1990) is effective at explaining life history evolution because habitat characteristics change spatially and temporally, and this variability in the habitat template provides the fuel for evolution to shape life history strategies (Townsend and Hildrew, 1994). The habitat template of groundwater ecosystems is generally characterized as "environmentally stable" but "energy limited" when compared to neighboring surface ecosystems, and the characterization appears to be consistent among many groundwater ecosystems across the planet (Pipan and Culver, 2012; Marmonier et al., this volume). These characteristics are the primary factors shaping the evolution of life history traits of groundwater organisms (e.g., Racovitza, 1907; Poulson, 1963; Poulson and White, 1969; Ginet and Decou, 1977; Culver, 1982; Howarth and Moldovan, 2018; Culver and Pipan, 2019).

Strong selective forces associated with the groundwater habitat template, such as permanent darkness and food scarcity, have resulted in the convergence of similar characteristics among disparate groups of organisms (e.g., Racovitza, 1907; Poulson, 1963; Ginet and Decou, 1977; Culver et al., 1995; Pipan and Culver, 2012). This includes life history traits which follow the classic, albeit somewhat dated, description of *k*- and *r*-selected life history strategies (i.e., combinations of life history traits; Reznick et al., 2002). Species that only complete their life cycle in groundwater habitats are "obligate groundwater species" (i.e., stygobionts) and often exhibit "*k*-selected" life history strategies, including long life span, slow somatic growth rates, maturity at older ages and larger body sizes, low numbers of large eggs/offspring per reproductive episode, and few or nonregular reproductive events over their lifetime (e.g., Poulson, 1963; Poulson and Lavoie, 2000; Niemiller and Poulson, 2010; Voituron et al., 2011; Østbye et al., 2018). This contrasts with surface species that are found in groundwater habitats, hereafter referred "non-obligate groundwater taxa" (i.e., stygophiles and stygoxenes). Nonobligate groundwater taxa often exhibit more "*r*-selected" life history traits, including shorter life spans, faster somatic growth rates, maturity at younger and smaller size classes, more smaller eggs/offspring per reproductive event, and regular reproductive events (Poulson and White, 1969; Venarsky and Huntsman, 2018). Importantly though, some nonobligate groundwater taxa can complete their life cycle in groundwater habitats (i.e., stygophiles), meaning obligate and nonobligate groundwater taxa can directly compete for habitat and resources in groundwater habitats (e.g., Eberly, 1960; Venarsky et al., 2018).

## Support for the current conceptual model of life history evolution in groundwater species

The current conceptual model of life history evolution in groundwater fauna is generally accepted, and, as we show below, is supported by the available data. However, life history traits are poorly described in many groundwater species, and comparative analyses of life history traits between groundwater species and surface relatives have rarely been conducted, indicating that the conceptual model of life history evolution in groundwater species still necessitates rigorous study. In this section, we review the currently available data on life history traits and trade-offs in groundwater species, with the intent of highlighting areas for future research, rather than providing a rigorous test of the conceptual model. We present a brief description of common life history traits and life table variables in Box 19.1, including exemplar studies of groundwater fauna. We also direct readers to other recent discussions of these traits and variables (Fišer, 2019b).

Natural selection favors the optimal strategy that maximizes lifetime reproductive output within the context of many intrinsic and extrinsic factors, including sex, age, environmental conditions, resource availability, predation, and competition (Zera and Harshman, 2001). Reproduction often involves important trade-offs, such as whether to allocate resources for current reproduction or delaying reproduction and investing in somatic growth and maintenance to reproduce at an older age with increased risk of mortality (Stearns, 1992; Roff, 1993). This trade-off is especially relevant to females because: (i) eggs are more energetically demanding to produce when compared to male sperm and (ii) the number of eggs per reproductive event is often positively correlated with body size (Fisher, 1930; Williams, 1966; Trivers, 1972; Stearns, 1992). In general, age at first reproduction is linked to lifespan, and thus a common suite of factors increases measures in both of these life history traits, including lower adult mortality rates, higher juvenile mortality rates, higher variability in juvenile reproductive rates among reproductive episodes, and lower variability in adult mortality rates among reproductive episodes (Stearns, 2000).

While delaying reproduction comes with the increased risk of mortality, this life history strategy does come with several advantages. Longer-lived species generally invest less energy per-reproductive episode than short-lived species and have multiple reproductive episodes during their lifetime (i.e., iteroparity; Ginet and Decou, 1977; Stearns, 1992; Hamel et al., 2010; Howarth and Moldovan, 2018). Furthermore, iteroparous species can delay or skip reproduction (i.e., intermittent breeding; Hamel et al., 2010; Shaw and Levin, 2013) based on intrinsic and extrinsic factors, including body condition, past reproductive success, population density, and contemporary habitat quality (e.g., Hamel et al., 2009, 2010; Desprez et al., 2011). This strategy appears to be adaptive under low-energy environmental conditions, which includes groundwater ecosystems (e.g., Forcada et al., 2008; Huntsman et al., 2020). Delayed reproduction provides a specific advantage to females as females have a finite amount of energy to allocate to reproduction and can maximize either the number of offspring per reproductive episode or the size of offspring, but not both—the offspring size—fecundity trade-off (Smith and Fretwell, 1974). The number of offspring produced is often negatively correlated with egg size, which also suggests a negative correlation with female fitness. However, offspring survival generally increases with increasing size at birth/hatching, meaning that fewer larger eggs can increase female fitness (Smith and Fretwell, 1974; Pianka, 1976; Roff, 2002). Importantly though, the offspring size that maximizes offspring fitness may not coincide to that which also maximizes parental fitness and can lead to conflict between parent and offspring due to asymmetries in the relatedness between parents and offspring (Trivers, 1974). In species that lack parental care, theory predicts that selection should maximize parental fitness (Trivers, 1974; Godfray, 1995).

Groundwater species are thought to be longer-lived compared to their surface relatives, but studies of natural populations are very few (see Box 19.1). At the extreme is the olm *Proteus anguinus*, which has been kept in captivity for over 60 years and is estimated to live 100 years or more (Voituron et al., 2011). Related to longer lifespan in groundwater species are concomitant delays in sexual maturity and multiple reproductive episodes. In general, reproduction occurs at older ages and larger body sizes when compared to related surface species. For example, sexual maturity is delayed in the cave crayfish *Orconectes inermis* and *O. australis* as much as 2—4 times longer than surface crayfishes in the family Cambaridae (Venarsky et al., 2012). Groundwater amphipods (Turquin, 1984) and amblyopsid fishes (Poulson, 1963; Niemiller and Poulson, 2010) reach sexual maturity at an older age compared to their surface relatives. Likewise, groundwater species have multiple reproductive episodes but have lower reproductive effort per episode (Poulson, 1963; Fišer et al., 2013). Two factors are thought to drive these reproductive strategies. First, because many groundwater environments are resource-limited, fewer resources are available for females to invest in reproduction. However, the relative stability of groundwater environments suggests that the chances of survival are higher. Thus, females can delay reproduction to an older age, and in doing so, they are larger at the time of reproduction, which can reduce the stress of reproduction while also allowing them to potentially produce more eggs per reproductive event (body size—fecundity relationship; Roff, 2002).

Offspring size and the number of offspring produced by females appears to be consistent with predictions of the offspring size—fecundity trade-off in groundwater species that have been studied. Groundwater species or populations tend to produce fewer but larger eggs per reproductive episodes compared to surface species or populations. For example, amblyopsid cavefishes tend to have fewer but larger eggs compared to their surface-dwelling relatives (Poulson, 1963; Niemiller and Poulson, 2010). Populations of cave mollies (*Poecilia mexicana*) produce fewer offspring per reproductive episode, but the young are larger and have higher survival compared to surface molly offspring, indicating that female cave mollies have adapted to increase their fitness (Riesch et al., 2010, 2012). Similarly, females in cave populations of the amphipods *Gammarus minus* and *Niphargus* sp. produce fewer but larger eggs that populations of surface relatives (Culver et al., 1995; Fišer et al., 2013). Greater energetic investment in offspring size may increase juvenile survival through improved ability to cope with environmental challenges, such as resistance to starvation and accommodating multiple food sources (Hüppop, 2000). However, it is important to note that our knowledge of life history in groundwater organisms is a synthesize of information from studies of relatively few taxa and no studies have comprehensively examined life history traits and evolution in groundwater fauna.

## Expanding the conceptual model of life history evolution in groundwater species

The discussion in the previous section outlined our current understanding of life history evolution in groundwater species. While theory and data generally appear to align, more studies are needed because our perspective on the life histories of groundwater species has advanced little beyond the generalized characterization of "groundwater species exhibit *k*-selected life history strategies." In this section, we discuss how to expand our understanding of life history traits in groundwater fauna, and how we can use life history studies to explore both life history evolution while also better explaining the assembly of groundwater communities under various environmental conditions.

## The environment and life history plasticity

The environment is a primary driver of life history evolution in ecosystems (Stearns, 1992; Roff, 2002), and groundwater ecosystems are no exception to this rule. In surface ecosystems, the drivers of life history evolution, as well as behavioral and physiological traits, have been identified among different types of habitats (e.g., temperate vs. tropical; Sarma et al., 2005; Martin, 2015) and among systems within the same habitat (e.g., lakes with and without predators; Fisk et al., 2007; Garcia et al., 2007). However, our understanding of life history evolution in groundwater taxa is largely limited to comparisons between groundwater and surface habitats, with less emphasis on variation in life history traits among groundwater ecosystems.

A driver of this bias appears to be the view that groundwater ecosystems are broadly similar from the perspective of environmental (e.g., temperature and light) and biodiversity metrics. While the degree of intrasystem variability in these metrics will be lower in groundwater habitats when compared to surface habitats, intrasystem variability still exists and the role of this variability in driving life history evolution in groundwater species remains largely unexplored. One aspect of intrasystem variability that has not been included in life history studies of groundwater taxa is that of superficial subterranean habitats (Culver and Pipan, 2008), which are groundwater habitats close to the surface (epikarst, vadose zones, hypotelminorheic seeps) that can experience relatively high energy availability and temporal variability in many environmental conditions (Culver et al., 2006; Gabrovsek and Peric, 2006; Culver and Pipan, 2008; Engel, 2010; Pipan and Culver, 2013; Kogovsek and Petric, 2014; Moldovan et al., 2018). Furthermore, these superficial habitats also differ in biotic factors that can affect life history traits, namely, reduced predation (e.g., Tobler et al., 2006; Plath and Schlupp, 2008; Howarth and Moldovan, 2018).

As we discussed above, phenotypic plasticity in life history traits is a key mechanism by which organisms can cope with changes in the environment. Phenotypic plasticity has been explored in some groundwater species, but most of this research has focused on morphological (e.g., pigmentation and eye development), physiological, and behavioral traits (e.g., Bilandžija et al., 2020; Espinasa et al., 2020), rather than life history traits (e.g., Reisch et al., 2009). Furthermore, much of this research has focused on comparisons between surface and cave populations of a few model species, such as the cavefishes *Astyanax mexicanus* (e.g., Bilandžija et al., 2020) and *Poecilia mexicana* (Reisch et al., 2009, 2010) and the amphipod *Gammarus minus* (e.g., Culver et al., 1995). While these studies have certainly provided important insights into our understanding how species colonized groundwater habitats from surface environments, they provide limited information on phenotypic plasticity in life history traits of groundwater species.

To date, we are unaware of any studies related to groundwater fauna that have directly explored intraspecific variation in life history traits, but there are several studies that have explored this issue using indirect methods. Reisch et al. (2010) examined number and size of eggs in female cave mollies between sulfidic and nonsulfidic aquatic cave habitats, finding that number of eggs was lower, and size was larger in the sulfidic cave habitats than in the nonsulfidic cave habitats. Huntsman et al. (2011) estimated the growth rates of two populations of stygobitic salamanders, while Venarsky et al. (2012) estimated the growth rates and age at sexual maturity for three populations of a cave crayfish. Both studies were conducted in multiple cave systems with varying levels of basal energy resources, and neither study found that growth rates or age at sexual maturity differed among cave populations.

Using among population variation in life history traits is not an ideal test of plasticity because these *post hoc* studies cannot attribute phenotypic variation in life history traits to either plasticity (i.e., genotypes that produce multiple phenotypes) or adaptation (i.e., genotypes that have evolved to produce phenotypes for local conditions). Consequently, such studies provide limited information on how populations of groundwater species may respond to changing environmental conditions. Among the primary ways to examine plasticity within a population are common garden experiments where individuals are transferred from their parent habitat to a habitat with novel environmental conditions. These experiments have been conducted in aquatic and terrestrial surface species under both field and laboratory conditions (e.g., Johnson, 2001; De Block and Stoks, 2004; Dybdahl and Kane, 2005), but we are unaware of any studies that have conducted such experiments on groundwater species.

## Trade-offs among life history, morphological, and physiological traits

As we have already discussed, a change in one life history trait often leads to a corresponding change in another trait. For example, increasing the size of eggs generally is correlated with a decrease in the number of eggs (Smith and Fretwell, 1974). However, trade-offs can occur among other different trait groups. For example, body size has a large influence on reproductive rates in groundwater species, with larger females producing more larger eggs (e.g., Fišer et al., 2013). However, if growth rate is constant and females decide to wait to reproduce until they reach a larger body size, then the age at sexual maturity will also likely increase. This can reduce the potential for an individual to reproduce due to an increased chance of death. Furthermore, the response in life history traits could be the result of how environmental factors influence other biology systems within an individual, which in-turn influence the expression of life history traits.

## Life history evolution and population demography

There is a growing body of literature illustrating how interactions between species traits (life history, behavioral, physiological) and population demographics dictate the capacity of populations to adapt to changing environmental conditions or to rebound from population declines (Kokko and Lopez-Sepulcre, 2007; Bolnick et al., 2011; Lowe et al., 2017). However, our understanding of population demographics in groundwater species is even more limited than our knowledge of life history traits. To date, we are aware of only a handful of studies that have attempted to quantity population demographic variables in groundwater species. Estimates of population size of and survival rates are available for cave salamanders (Huntsman et al. 2011, 2020; Fenolio et al., 2014a; Niemiller et al., 2016), cave crayfish (Cooper, 1975; Venarsky et al., 2014), and cave isopods and amphipods (Simon and Buikema, 1997; Carpenter, 2021).

The lack of demographic data on groundwater species is understandable given the logistical challenges of working within groundwater environments. Acquiring measures of population demographics often requires capture—mark—recapture approaches that uniquely mark individuals, such as toe or uropod clips, visible implant elastomers, and alphanumeric or genetic tags (Huntsman et al., 2011; Venarsky et al., 2014; Niemiller et al., 2016). Furthermore, many groundwater species are encountered in relatively low densities and occupy an environment that can be sampled only through small windows, such as boreholes, springs, or cave systems that intersect the groundwater table when human access is permittable. Thus, capture-mark-recapture studies generally need to be conducted over multiple years to ensure enough individuals are encountered to estimate population demographic parameters.

While population demographic studies will provide useful information on groundwater species population sizes and growth rates, this demographic information can be combined with results from genetic methods to provide a more holistic understanding of population dynamics of groundwater species. For example, population genetic studies use measures of genetic diversity to estimate effective population size, with higher genetic diversity and effective population sizes generally correlating with both the ability of a population to persist as well as adapt to changing environmental conditions. However, the relationship between effective population size and actual measures of population size from demographic studies has not been the subject of study in groundwater species. Furthermore, placing population genetics within the context of population demographics can also allow researchers to better understand how genetic drift, selection, and gene flow are influenced by measures of population size, population density, and emigration and immigration rates (i.e., source-sink dynamics; Lowe et al., 2017).

## Linking life history traits to molecular evolution

Evolutionary adaptation is driven by the accumulation of mutations and life history traits can influence the rate at which DNA sequences accumulate substitutions, which in-turn influences the rate of molecular evolution and thus evolutionary adaptation (Britten, 1986; Li et al., 1987; Nabholz et al., 2008). Many traits have been proposed to impact the rate of molecular evolution, including, generation time (Li et al., 1987), parental investment Britten (1986), and metabolic rate (Martin et al., 1992). While there is a rich literature surrounding this topic, there is little consensus on which traits are predominant in controlling substitution rate. However, groundwater species offer useful case studies to test theories linking life history traits and rates of molecular evolution. Indeed, the comparison between groundwater and surface species allows to obtain replicates of the same changes in life history traits, and thus, build a powerful experimental design to disentangle hypotheses concerning generation time (Li et al., 1987), metabolic rate (Martin et al., 1992), and longevity (Nabholz et al., 2008). To date, a single study has examined these hypotheses using surface and groundwater isopods (Saclier et al., 2018). This study found that groundwater isopods have lower nuclear rates of molecular evolution, but mitochondrial rates of molecular evolution were similar between surface and groundwater taxa. Thus, the nuclear rates of molecular evolution support the generation time hypothesis, while the mitochondrial rates do not. This example suggests that groundwater species could be an ideal model to test the link between life history traits and molecular evolution.

## Conclusions

Life histories are central to the evolution of behavioral, physiological, and morphological traits on one hand, and the structuring of communities and functioning of ecosystems on the other. Groundwater organisms are interesting models for studying life histories. They exist in a well-defined habitat template that occurs on an endpoint of several ecological gradients from surface habitats. However, despite this habitat template and strong theoretical expectations about life history traits derived from it, our knowledge of life history in groundwater fauna remains limited and largely derived from studies of a few model taxa. Life history traits are poorly described in most groundwater species and comparative studies of life history traits between groundwater species and surface are few. Consequently, the conceptual model of life history evolution in groundwater species warrants thorough inquiry. We are unaware of studies that have examined intraspecific variation in life history traits and plasticity in groundwater fauna using direct approaches, such as common garden experiments. Likewise, few studies on demography and rates of molecular evolution have been conducted, and no studies to date have generated life tables for groundwater species. Such studies are greatly needed not only for informing conservation initiatives but also for advancing our knowledge of population and community dynamics and life history evolution.

## Acknowledgments

MLN was supported by the National Science Foundation (award no. 2047939). CF was supported by Slovenian research Agency through long-term funding P1-0184 and project grant N1-0069. OTM was supported by a grant of the Romanian Ministry of Research and Innovation CNCS-UEFISCDI, project number PN-III-P4-ID-PCCF-2016-0016, DARKFOOD. We thank S. Mammola and F. Malard for constructive comments that improved this chapter.

## References

Bilandžija, H., Hollifield, B., Steck, M., Meng, G., Ng, M., Koch, A.D., Gračan, R., Ćetković, H., Porter, M.L., Renner, K.J., Jeffery, W., 2020. Phenotypic plasticity as a mechanism of cave colonization and adaptation. Elife 21 (9), e51830.

Britten, R.J., 1986. Rates of DNA sequence evolution differ between taxonomic groups. Science 231, 1393—1398.

Bolnick, D.I., Amarasekare, P., Araújo, M.S., Bürger, R., Levine, J.M., Novak, M., Rudolf, V.H., Schreiber, S.J., Urban, M.C., Vasseur, D.A., 2011. Why intraspecific trait variation matters in community ecology. Trends in Ecology & Evolution 26 (4), 183—192.

Carpenter, J.H., 2021. Forty-year natural history study of *Bahalana geracei* Carpenter, 1981, an anchialine cave-dwelling isopod (Crustacea, Isopoda, Cirolanidae) from San Salvador Island, Bahamas: reproduction, growth, longevity, and population structure. Subterranean Biology 37, 105—156.

Chippindale, A.K., Chu, T.J., Rose, M.R., 1996. Complex trade-offs and the evolution of starvation resistance in *Drosophila melanogaster*. Evolution 50, 753—766.

Cooper, J.E., 1975. Ecological and Behavioral Studies in Shelta Cave, Alabama, with Emphasis on Decapod Crustaceans. PhD Thesis,. University of Kentucky, Lexington, KY, USA.

Culver, D.C., 1982. Cave Life: Evolution and Ecology. Harvard University Press.

Culver, D.C., Kane, T.C., Fong, D.W., 1995. Adaptation and Natural Selection in Caves: The Evolution of Gammarus Minus. Harvard University Press.

Culver, D.C., Pipan, T., 2008. Superficial subterranean habitats — gateway to the subterranean realm. Cave and Karst Science 35 (1—2), 5—12.

Culver, D.C., Pipan, T., 2019. The Biology of Caves and Other Subterranean Habitats. Oxford University Press.

Culver, D.C., Pipan, T., Gottstein, S., 2006. Hypotelminorheic — a unique freshwater habitat. Subterranean Biology 4, 1—8.

De Block, M., Stoks, R., 2004. Life-history variation in relation to time constraints in a damselfly. Oecologia 140, 68—75.

Desprez, M., Pradel, R., Cam, E., Monnat, J.Y., Gimenez, O., 2011. Now you see him, now you don't: experience, not age, is related to reproduction in kittiwakes. Proceedings of the Royal Society B: Biological Sciences 278, 3060—3066.

DeWitt, T.J., Scheiner, S.M. (Eds.), 2004. Phenotypic Plasticity: Functional and Conceptual Approaches. Oxford University Press.

Dybdahl, M.F., Kane, S.L., 2005. Adaptation vs. phenotypic plasticity in the success of a clonal invader. Ecology 86, 1592—1601.

Eberly, W.R., 1960. Competition and evolution in cave crayfishes of southern Indiana. Systematic Zoology 9 (1), 29—32.

Engel, A.S., 2010. Microbial diversity of cave ecosystems. In: Barton, L.L., Mandl, M., Loy, A. (Eds.), Geomicrobiology: Molecular and Environmental Perspective. Springer, Dordrecht 219—238.

Espinasa, L., Heintz, C., Rétaux, S., Yoshisawa, M., Agnès, F., Ornelas-Garcia, P., Balogh-Robinson, R., 2020. Vibration Attraction Response (VAB) is a plastic trait in Blind Mexican tetra (*Astyanax mexicanus*), variable within subpopulations inhabiting the same cave. Journal of Fish Biology 98 (1), 304—316.

Fenolio, D.F., Niemiller, M.L., Bonett, R.M., Graening, G.O., Collier, B.A., Stout, J.F., 2014a. Life history, demography, and the influence of cave-roosting bats on a population of the Grotto Salamander (*Eurycea spelaea*) from the Ozark Plateaus of Oklahoma (Caudata: Plethodontidae). Herpetological Conservation and Biology 9 (2), 394—405.

Fenolio, D.B., Niemiller, M.L., Martinez, B., 2014b. Observations of reproduction in captivity by the Dougherty plain cave crayfish, *Cambarus cryptodytes* (Decapoda: Astacoidea: Cambaridae). Speleobiology Notes 6, 14—26.

Fišer, C., 2019a. *Niphargus*—a model system for evolution and ecology. In: White, W.B., Culver, D.C., Pipan, T. (Eds.), Encyclopedia of Caves, third ed. Academic Press 746—755.

Fišer, C., 2019b. Life histories. In: White, W.B., Culver, D.C., Pipan, T. (Eds.), Encyclopedia of Caves, third ed. Academic Press 652—657.

Fišer, C., Zagmajster, M., Zakšek, V., 2013. Coevolution of life history traits and morphology in female subterranean amphipods. Oikos 122, 770—778.

Fisher, R.A., 1930. The Genetical Theory of Natural Selection. Oxford University Press.

Fisk, D.L., Latta, L.C., Knapp, R.A., Pfrender, M.E., 2007. Rapid evolution in response to introduced predators I: rates and patterns of morphological and life-history trait divergence. BMC Evolutionary Biology 7, 1—11.

Flatt, T., Heyland, A., 2011. Mechanisms of Life History Evolution. The Genetics and Physiology of Life History Traits and Trade-Offs. Oxford University Press Inc, New York, NY.

Fong, D.W., 2009. Brood size of the stygobiotic isopod *Caecidotea pricei* from a spring run in West Virginia, USA. Speleobiology Notes 1, 1—2.

Forcada, J., Trathan, P.N., Murphy, E.J., 2008. Life history buffering in Antarctic mammals and birds against changing patterns of climate and environmental variation. Global Change Biology 14, 2473—2488.

Gabrovšek, F., Peric, B., 2006. Monitoring the flood pulses in the epiphreatic zone of karst aquifers: the case of Reka river system, Karst plateau, SW Slovenia. Acta Carsologica 35 (1), 35—45.

Gadgil, M., Bossert, W.H., 1970. Life historical consequences of natural selection. The American Naturalist 104 (935), 1—24.

Garcia, C.E., De Jesús Chaparro-Herrera, D., Nandini, S., Sarma, S.S.S., 2007. Life-history strategies of *Brachionus havanaensis* subject to kairomones of vertebrate and invertebrate predators. Chemistry and Ecology 23, 303—313.

García, V.B., Lucifora, L.O., Myers, R.A., 2008. The importance of habitat and life history to extinction risk in sharks, skates, rays and chimaeras. Proceedings of the Royal Society B: Biological Sciences 275 (1630), 83—89.

Ginet, R., Decou, V., 1977. Initiation à la biologie et à l'écologie souterraines. Delarge, Paris.

Godfray, H.C.J., 1995. Evolutionary theory of parent—offspring conflict. Nature 376 (6536), 133—138.

Hamel, S., Gaillard, J.M., Festa-Bianchet, M., CôTE, S.D., 2009. Individual quality, early-life conditions, and reproductive success in contrasted populations of large herbivores. Ecology 90, 1981—1995.

Hamel, S., Gaillard, J.M., Yoccoz, N.G., Loison, A., Bonenfant, C., Descamps, S., 2010. Fitness costs of reproduction depend on life speed: empirical evidence from mammalian populations. Ecology Letters 13, 915—935.

Heuts, M.J., 1951. Ecology, variation and adaptation of the blind fish *Caecobarbus geertsi* Blgr. Annales de la Société Royale Zoologique de Belgique 82, 155—230.

Hirshfield, M.F., Tinkle, D.W., 1975. Natural selection and the evolution of reproductive effort. Proceedings of the National Academy of Sciences 72, 2227—2231.

Hobbs Jr., H.H., Barr Jr., T.C., 1972. Origins and affinities of troglobitic crayfishes of North America (Decapoda: Astacidae) II. Genus *Orconectes*. Smithsonian Contributions to Zoology 105, 1—84.

Howarth, F.G., Moldovan, O.T., 2018. Where cave animals live. In: Moldovan, O.T., Kovac, L., Halse, S. (Eds.), Cave Ecology. Springer 23—37.

Huntsman, B.M., Venarsky, M.P., Abadi, F., Huryn, A.D., Kuhajda, B.R., Cox, C.L., Benstead, J.P., 2020. Evolutionary history and sex are significant drivers of crayfish demography in resource-limited cave ecosystems. Evolutionary Ecology 34, 235–255.

Huntsman, B.M., Venarsky, M.P., Benstead, J.P., Huryn, A.D., 2011. Effects of organic matter availability on the life history and production of a top vertebrate predator (Plethodontidae: *Gyrinophilus palleucus*) in two cave streams. Freshwater Biology 56 (9), 1746–1760.

Hüppop, K., 2000. How do animals cope with the food scarcity in caves? In: Wilkens, H., Culver, D.C., Humphreys, W.F. (Eds.), Ecosystems of the World: Subterranean Ecosystems. Elsevier, Amsterdam 159–188.

Hüppop, K., Wilkens, H., 1991. Bigger eggs in subterranean *Astyanax fasciatus* (Characidae, Pisces)—their significance and genetics. Zeitschrift für Zoologische Systematik und Evolutionsforschung 29, 280–288.

Jegla, T.C., 1969. Cave crayfish: annual periods of molting and reproduction. in Actes du IV Congres International de Speleologie en Yugoslavie (12–26 IX 1965) 135–137.

Johnson, J.B., 2001. Adaptive life-history evolution in the livebearing fish *Brachyrhaphis rhabdophora*: genetic basis for parallel divergence in age and size at maturity and a test of predator-induced plasticity. Evolution 55, 1486–1491.

Kogovsek, J., Petric, M., 2014. Solute transport processes in a karst vadose zone characterized by long-term tracer tests (the cave system of Postojnska Jama, Slovenia). Journal of Hydrology 519, 1205–1213.

Kokko, H., Lopez-Sepulcre, A., 2007. The ecogenetic link between demography and evolution: can we bridge the gap between theory and data? Ecology Letters 10, 773–782.

Li, W.-H., Tanimura, M., Sharp, P.M., 1987. An evaluation of the molecular clock hypothesis using mammalian DNA sequences. Journal of Molecular Evolution 25, 330–342.

Lindstedt, S., Calder III, W., 1981. Body size, physiological time, and longevity of homeothermic animals. Quarterly Review of Biology 56, 1–16.

Lowe, W.H., Kovach, R.P., Allendorf, F.W., 2017. Population genetics and demography unite ecology and evolution. Trends in Ecology & Evolution 32 (2), 141–152.

Magniez, G., 1975. Observations sur la biologie de *Stenasellus virei* (Crustacea Isopoda Asellota des eaux souterraines). International Journal of Speleology 7 (1), 79–228.

Mammola, S., 2019. Finding answers in the dark: caves as models in ecology fifty years after Poulson and White. Ecography 42 (7), 1331–1351.

Mammola, S., Amorim, I.R., Bichuette, M.E., Borges, P., Cheeptham, N., Cooper, S.J.B., Culver, D.C., Deharveng, L., Eme, D., Ferreira, R.L., Fiser, C., Fiser, Z., Fong, D.W., Griebler, C., Jeffery, W.R., Kowalko, J.E., Jugovic, J., Lilley, T.M., Malard, F., Manenti, R., Martinez, A., Meierhofer, M.B., Northup, D.E., Pellegrini, T.G., Protas, M., Niemiller, M.L., Reboleira, A.S., Pipan, T., Venarsky, M.P., Wynne, J.J., Zagmajster, M., Cardoso, P., 2020. Fundamental research questions in subterranean biology. Biological Reviews 95, 1855–1872.

Martin, T.E., 2015. Age-related mortality explains life history strategies of tropical and temperate songbirds. Science 349, 966–970.

Martin, A.P., Naylor, G.J.P., Palumbi, S.R., 1992. Rates of mitochondrial DNA evolution in sharks are slow compared with mammals. Nature 357, 153–155.

Massol, F., Altermatt, F., Gounand, I., Gravel, D., Leibold, M.A., Mouquet, N., 2017. How life-history traits affect ecosystem properties: effects of dispersal in meta-ecosystems. Oikos 126 (4), 532–546.

Mathieu, J., Turquin, M.J., 1992. Biological processes at the population level. II. Aquatic populations: *Niphargus* (stygobiont amphipod) case. In: Camacho, A.I. (Ed.), The Natural History of Biospeleology. Monografias. Museo Nacional de Ciencias Naturales, Madrid 263–293.

Mims, M.C., Olden, J.D., 2013. Fish assemblages respond to altered flow regimes via ecological filtering of life history strategies. Freshwater Biology 58 (1), 50–62.

Moldovan, O.T., Constantin, S., Cheval, S., 2018. Drip heterogeneity and the impact of decreased flow rates on the vadose zone fauna in Ciur-Izbuc Cave, NW Romania. Ecohydrology 11, e2028.

Nabholz, B., Glémin, S., Galtier, N., 2008. Strong variations of mitochondrial mutation rate across mammals—the longevity hypothesis. Molecular Biology and Evolution 25, 120–130.

Niemiller, M.L., Glorioso, B.M., Fenolio, D.B., Reynolds, R.G., Taylor, S.J., Miller, B.T., 2016. Growth, survival, longevity, and population size of the Big Mouth Cave Salamander (*Gyrinophilus palleucus necturoides*) from the type locality in Grundy County, Tennessee, USA. Copeia 104 (1), 35–41.

Niemiller, M.L., Poulson, T.L., 2010. Subterranean fishes of North America: Amblyopsidae. In: Trajano, E., Bichuette, M.E., Kapoor, B.G. (Eds.). Biology of Subterranean Fishes. CRC Press 169–280.

Østbye, K., Østbye, E., Lien, A.M., Lee, L.R., Lauritzen, S.E., Carlini, D.B., 2018. Morphology and life history divergence in cave and surface populations of *Gammarus lacustris* (L.). PLoS One 13 (10), e0205556.

Pacioglu, O., Strungaru, S.-A., Ianovici, N., Filimon, M.N., Sinitean, A., Iacob, G., Barabas, H., Acs, A., Muntean, H., Plăvan, G., Schulz, R., Zubrod, J.P., Pârvulescu, L., 2020. Ecophysiological and life-history adaptations of *Gammarus balcanicus* (Schäferna, 1922) in a sinking-cave stream from Western Carpathians (Romania). Zoology 139, 125754.

Pearson, R.G., Stanton, J.C., Shoemaker, K.T., Aiello-Lammens, M.E., Ersts, P.J., Horning, N., Fordham, D.A., Raxworthy, C.J., Ryu, H.Y., McNees, J., Akçakaya, H.R., 2014. Life history and spatial traits predict extinction risk due to climate change. Nature Climate Change 4 (3), 217–221.

Pianka, E.R., 1976. Natural selection of optimal reproductive tactics. American Zoologist 16 (4), 775–784.

Pigliucci, M., 2001. Phenotypic Plasticity: Beyond Nature and Nurture. John Hopkins University Press.

Pipan, T., Culver, D.C., 2012. Convergence and divergence in the subterranean realm: a reassessment. Biological Journal of the Linnean Society 107 (1), 1–14.

Pipan, T., Culver, D.C., 2013. Organic carbon in shallow subterranean habitats. Acta Carsologica 42 (2–3), 291–300.

Plath, M., Schlupp, I., 2008. Parallel evolution leads to reduced shoaling behavior in two cave dwelling populations of Atlantic mollies (*Poecilia mexicana*, Poeciliidae, Teleostei). Environmental Biology of Fishes 82 (3), 289–297.

Poff, N.L., Ward, J.V., 1990. Physical habitat template of lotic systems: recovery in the context of historical pattern of spatiotemporal heterogeneity. Environmental Management 14 (5), 629–645.

Pouilly, M., Miranda, G., 2003. Morphology and reproduction of the cavefish *Trichomycterus chaberti* and the related epigean *Trichomycterus cf. barbouri*. Journal of Fish Biology 63, 490–505.

Poulson, T.L., 1963. Cave adaptation in amblyopsid fishes. The American Midland Naturalist 70, 257–290.

Poulson, T.L., Lavoie, K.H., 2000. The trophic basis of subsurface ecosystems. In: Wilkens, H., Culver, D.C., Humphreys, W.F. (Eds.), Ecosystems of the World, Subterranean Ecosystems, vol 30 231–249.

Poulson, T.L., White, W.B., 1969. The cave environment. Science 165 (3897), 971–981.

Racovitza, E.G., 1907. Essai sur les Problèmes biospéologiques. Archives de Zoologie Expérimentale et Générale 6, 371–488.

Reznick, D., Bryant, M.J., Bashey, F., 2002. r-and K-selection revisited: the role of population regulation in life-history evolution. Ecology 83 (6), 1509–1520.

Reznick, D.A., Bryga, H., Endler, J.A., 1990. Experimentally induced life-history evolution in a natural population. Nature 346 (6282), 357–359.

Ricklefs, R.E., 2000. Density dependence, evolutionary optimization, and the diversification of avian life histories. The Condor 102 (1), 9–22.

Ricklefs, R.E., 2010. Life-history connections to rates of aging in terrestrial vertebrates. Proceedings of the National Academy of Sciences 107, 10314–10319.

Ricklefs, R.E., Wikelski, M., 2002. The physiology/life-history nexus. Trends in Ecology & Evolution 17 (10), 462–468.

Riesch, R., Plath, M., Schlupp, I., 2010. Toxic hydrogen sulfide and dark caves: life-history adaptations in a livebearing fish (*Poecilia mexicana*, Poeciliidae). Ecology 91 (5), 1494–1505.

Riesch, R., Plath, M., Schlupp, I., 2012. The offspring size/fecundity trade-off and female fitness in the Atlantic molly (*Poecilia mexicana*, Poeciliidae). Environmental Biology of Fishes 94 (2), 457–463.

Riesch, R., Tobler, M., Plath, M., Schlupp, I., 2009. Offspring number in a livebearing fish (*Poecilia mexicana*, Poeciliidae): reduced fecundity and reduce plasticity in a population of cave mollies. Environmental Biology of Fishes 84, 89–94.

Riesch, R., Reznick, D.N., Plath, M., Schlupp, I., 2016. Sex-specific local life-history adaptation in surface-and cave-dwelling Atlantic mollies (*Poecilia mexicana*). Scientific Reports 6 (1), 1–13.

Roff, D.A., 1993. Evolution of Life Histories: Theory and Analysis. Springer.

Roff, D.A., 1997. Phenotypic Plasticity and Reaction Norms. Springer, Boston, MA.

Roff, D.A., 2002. Life History Evolution. Sinauer Associates, Sunderland.

Saclier, N., François, C.M., Konecny-Dupré, L., Lartillot, N., Guéguen, L., Duret, L., Malard, F., Douady, C.J., Lefébure, T., 2018. Life history traits impact the nuclear rate of substitution but not the mitochondrial rate in isopods. Molecular Biology and Evolution 35, 2900–2912.

Sarma, S.S.S., Nandini, S., Gulati, R.D., 2005. Life history strategies of cladocerans: comparisons of tropical and temperate taxa. Hydrobiologia 542, 315–333.

Shaw, A.K., Levin, S.A., 2013. The evolution of intermittent breeding. Journal of Mathematical Biology 66, 685–703.

V. Biological traits in groundwater

Simon, K.S., Buikema Jr., A.L., 1997. Effects of organic pollution on an Appalachian cave: changes in macroinverte-brate populations and food supplies. The American Midland Naturalist 138, 387—401.

Simon, V., Elleboode, R., Mahé, K., Legendre, L., Ornelas-Garcia, P., Espinasa, L., Retaux, S., 2017. Comparing growth in surface and cave morphs of the species *Astyanax mexicanus*: insights from scales. EvoDevo 8, 23.

Smith, C.C., Fretwell, S.D., 1974. The optimal balance between size and number of offspring. The American Naturalist 108 (962), 499—506.

Sobral, M., 2021. All traits are functional: an evolutionary viewpoint. Trends in Plant Science 26 (7), 674—676.

Southwood, T.R.E., 1977. Habitat, the templet for ecological strategies? Journal of Animal Ecology 46, 337—365.

Southwood, T.R.E., 1988. Tactics, strategies and templets. Oikos 52, 3—18.

Stearns, S.C., 1989. Trade-offs in life-history evolution. Functional Ecology 3 (3), 259—268.

Stearns, S.C., 1992. The Evolution of Life Histories. Oxford University Press, Oxford.

Stearns, S.C., 2000. Life history evolution: successes, limitations, and prospects. Naturwissenschaften 87, 476—486.

Tobler, M., Schlupp, I., Heubel, K.U., Riesch, R., García de León, F.J., Giere, O., Plath, M., 2006. Life on the edge: hydrogen sulfide and the fish communities of a Mexican cave and surrounding waters. Extremophiles 10, 577—585.

Townsend, C.R., Hildrew, A.G., 1994. Species traits in relation to a habitat templet for river systems. Freshwater Biology 31 (3), 265—275.

Trajano, E., 1991. Population ecology of *Pimelodella kronei*, troglobitic catfish from Southeastern Brazil (Siluriformes, Pimelodidae). Environmental Biology of Fishes 30, 407—421.

Trivers, R.L., 1972. Parental investment and sexual selection. In: Campbell, B. (Ed.), Sexual Selection and the Descent of Man, 1871—1971. Aldine 136—179.

Trivers, R.L., 1974. Parent-offspring conflict. Integrative and Comparative Biology 14 (1), 249—264.

Trochet, A., Moulherat, S., Calvez, O., Stevens, V.M., Clobert, J., Schmeller, D.S., 2014. A database of life-history traits of European amphibians. Biodiversity Data Journal 2 (2), e4123.

Turquin, M.J., 1984. Age et croissance de *Niphargus virei* (amphipode perennant) dans le systeme karstique de Drom: méthodes d'estimation. Mémoires de Biospéologie 11, 37—49.

Turquin, M.J., Barthelemy, D., 1985. The dynamics of a population of the troglobitic amphipod *Niphargus virei* Chevreux. Stygologia 1, 109—117.

Venarsky, M.P., Huryn, A.D., Benstead, J.P., 2012. Re-examining extreme longevity of the cave crayfish *Orconectes australis* using new mark—recapture data: a lesson on the limitations of iterative size-at-age models. Freshwater Biology 57 (7), 1471—1481.

Venarsky, M.P., Benstead, J.P., Huryn, A.D., Huntsman, B.M., Edmonds, J.W., Findlay, R.H., Wallace, J.B., 2018. Experimental detritus manipulations unite surface and cave stream ecosystems along a common energy gradient. Ecosystems 21 (4), 629—642.

Venarsky, M.P., Huntsman, B.M., 2018. Food webs in caves. In: Moldovan, O.T., Kováč, L., Halse, S. (Eds.), Cave Ecology. Springer, Cham 309—328.

Venarsky, M.P., Huntsman, B.M., Huryn, A.D., Benstead, J.P., Kuhajda, B.R., 2014. Quantitative food web analysis supports the energy-limitation hypothesis in cave stream ecosystems. Oecologia 176 (3), 859—869.

Venarsky, M.P., Wilhelm, F.M., Anderson, F.E., 2007. Conservation strategies supported by non-lethal life history sampling of the US federally listed Illinois cave amphipod, *Gammarus acherondytes*. Journal of Crustacean Biology 27 (2), 202—211.

Violle, C., Navas, M.L., Vile, D., Kazakou, E., Fortunel, C., Hummel, I., Garnier, E., 2007. Let the concept of trait be functional. Oikos 116 (5), 882—892.

Voituron, Y., de Fraipont, M., Issartel, J., Guillaume, O., Clobert, J., 2011. Extreme lifespan of the human fish (*Proteus anguinus*): a challenge for ageing mechanisms. Biology Letters 7 (1), 105—107.

Williams, G.C., 1966. Natural selection, the costs of reproduction, and a refinement of Lack's principle. The American Naturalist 100, 687—690.

Wilson, G.D.F., Ponder, W.F., 1992. Extraordinary new subterranean isopods (Peracarida: Crustacea) from the Kimberley region, Western Australia. Records of the Australian Museum 44, 279—298.

Zera, A.J., Harshman, L.G., 2001. The physiology of life history trade-offs in animals. Annual Review of Ecology and Systematics 32 (1), 95—126.

Zigler, K.S., Cooper, G.M., 2011. Brood size of the stygobiotic asellid isopod *Caecidotea bicrenata bicrenata* from Franklin County, Tennessee, USA. Speleobiology Notes 3, 1—3.

# Physiological tolerance and ecotoxicological constraints of groundwater fauna

*Tiziana Di Lorenzo[1,2,9,11], Maria Avramov[3], Diana Maria Paola Galassi[4], Sanda Iepure[2,5], Stefano Mammola[6,7], Ana Sofia P.S. Reboleira[8,9] and Frédéric Hervant[10]*

[1]Research Institute on Terrestrial Ecosystems of the National Research Council of Italy (IRET-CNR), Florence, Italy; [2]Emil Racovita Institute of Speleology, Cluj-Napoca, Romania; [3]Helmholtz Zentrum München, German Research Center for Environmental Health, Institute of Groundwater Ecology, Neuherberg, Germany; [4]Department of Life, Health and Environmental Sciences, University of L'Aquila, L'Aquila, Italy; [5]Institutul Român de Știință și Tehnologie, Cluj-Napoca, Romania; [6]Molecular Ecology Group (dark-MEG), Water Research Institute (IRSA), National Research Council (CNR), Verbania-Pallanza, Italy; [7]University of Helsinki, Finnish Museum of Natural History (LUOMUS), Helsinki, Finland; [8]Natural History Museum of Denmark, University of Copenhagen, Copenhagen, Denmark; [9]Centre for Ecology, Evolution and Environmental Changes (cE3c), Departamento de Biologia Animal, Faculdade de Ciências, Universidade de Lisboa, Lisbon, Portugal; [10]Univ Lyon, Université Claude Bernard Lyon 1, CNRS, ENTPE, UMR 5023 LEHNA, Villeurbanne, France; [11]National Biodiversity Future Center (NBFC), Palermo, Italy

## Introduction

Groundwater has long been perceived as a stable environment, with evolutionary and geological changes occurring at a timescale of hundreds or thousands of years (Gibert et al., 1994). Accordingly, obligate groundwater organisms have long been considered major examples of highly adapted animals with restricted tolerance to changing environmental conditions (Vandel, 1965; Ginet and Decou, 1977; Botosaneanu and Holsinger, 1991; Danielopol et al., 1994). This perception changed in the 90s when individual studies underlined that groundwater environments were much less stable than initially thought and that, besides groundwater species with a narrow physiological tolerance, others were there too with a much broader tolerance. Accordingly, Danielopol et al. (1994) demonstrated that some groundwater crustacean species, such as *Proasellus slavus*, are "regulators," that is, able to maintain their activities more or less unaltered, independently of the variable environmental conditions. Apart from natural environmental variability, human activities induce considerable variation in environmental conditions due to interventions causing habitat destruction, climate modification, or contamination by many kinds of chemicals (Mammola et al., 2019a; Boulton, 2020). This new scenario of groundwater environmental variability induced by anthropic activities represents additional selective pressures acting over groundwater species. In consequence, the assessment of the physiological tolerance of groundwater species is important to predict the effect of human stressors on the structure and functioning of groundwater ecosystems.

We devoted this chapter to explore the physiological tolerance of groundwater species, understood as the ability of groundwater organisms to cope with human-generated stress, encompassing temperature changes (Section Physiological tolerance of groundwater invertebrates to changing thermal conditions), chemical contamination scenarios (Section Physiological tolerance of groundwater organisms to chemical stress), and constraints to the laboratory research that can elucidate responses of groundwater species to these stressors (Section Physiological tolerance of groundwater organisms to light, food and oxygen variations: indications for ecotoxicological protocols). Each section of this chapter consists of a small introductory paragraph, a discussion of the main specific aspects of physiological tolerance, and a conclusive paragraph wrapping up key points and providing future prospects. We conclude the chapter with final remarks about the expected tolerance of groundwater fauna regarding (i) potential future scenarios of climate change and (ii) chemical pollution, with some recommendations for ecological risk assessment.

## Physiological tolerance of groundwater invertebrates to changing thermal conditions

Temperature plays a critical ecological role in animal life, determining physiological limits, distribution range and, ultimately, phenology (Parmesan, 2006; Colinet et al., 2015; Muñoz and Bodensteiner, 2019). The Intergovernmental Panel on Climate Change denounced global warming of 1.5°C above preindustrial levels and reported that a temperature increase between 2 and 6°C is expected by the next 90 years (IPPC, 2018). Satellite-derived multiannual measurements of groundwater temperatures (GWT) in shallow (<60 m depth) groundwater bodies in 29 countries and 2 overseas territories worldwide revealed that the shallow GWT has a limited range of variation globally, namely from 0 to 40°C (Benz et al., 2017). The GWT in shallow Quaternary aquifers is projected to rise by up to 5°C at 45° latitude in the northern hemisphere within the next century (Taylor and Stefan, 2009). There are, however, also other anthropogenic causes for temperature fluctuations in groundwater, which can, at a local scale, be much more significant than the ones resulting from climate change. For example, the GWT underneath big cities is already 2–7°C higher than the temperature of the surrounding aquifers due to continuous heat discharge from district heating, public sewer networks, insufficiently insulated power plants, landfills or open geothermal systems (Taniguchi et al., 2005; Menberg et al., 2013; Zhu and Grathwohl, 2015). In addition, when heat is being actively stored in aquifers (e.g., for heating purposes), temperatures $\geq 30°C$ are regularly observed (Griebler et al., 2016). The increase in GWT due to anthropogenic causes should concern us today and in the near future. Therefore, we can argue that GWT increases due to climate change occurring at a regional scale will likely add up to local GWT increases due to urbanization. Since many physical and chemical processes are strongly related to temperature, such a global warming scenario would significantly modify the living conditions in groundwater ecosystems, potentially posing major threats to the groundwater microbiome and fauna (Retter et al., 2020).

Two main approaches have been mostly used for the investigation of groundwater species' response to thermal change so far. The first one, not tackled in this chapter, is based on correlative models, whereby biodiversity, presence/absence of individual species and distributional ranges are correlated with current and past environmental conditions to foresee survival capabilities in future climate change scenarios (e.g., Iepure et al., 2016; Mammola and Leroy, 2018). The second approach is based on experiments in which the physiological tolerance of groundwater organisms to changing thermal conditions is tested in the laboratory so as to minimize the effect of other variables such as food availability, interspecific interactions and sublethal concentrations of contaminants. This section summarizes the outcomes of the latter approach.

## Tolerance to cold

Due to specific physiological characteristics, groundwater species can exhibit a high level of cold hardiness. For instance, the groundwater amphipod *Niphargus rhenorhodanensis* varies the amount of cryoprotective substances (mainly polyols, free amino acids, and sugars), which are considerably higher at low temperatures, differently from the epigean amphipod *Gammarus fossarum* (Issartel et al., 2006). In detail, *N. rhenorhodanensis* shows a significant increase in trehalose content (a sugar identified as a common membrane and protein protectant at cold temperatures) during thermal stress, as well as in amino acids (mainly glutamine, proline, alanine, lysine, and glycine), which act as cryoprotectants by stabilizing enzymes and preserving their catalytic activity (Issartel et al., 2006; Colson-Proch et al., 2009). The crystallization temperature of this species is lower than that of *G. fossarum*, as well as the percentage of body water transformed to ice at $-2°C$. Finally, cold-acclimated individuals of *N. rhenorhodanensis* are able to survive at subzero temperatures for 15 days (90%–100% of survival), while the individuals of *G. fossarum* die much earlier (Issartel et al., 2006). Over a temperature range of 0–5°C, the groundwater amphipod and isopod species *Niphargus inopinatus* and *Proasellus cavaticus* frequently exhibited a cold-rigor from which, however, almost all specimens recovered as soon as temperatures returned to above 5°C (Brielmann et al., 2011).

## Tolerance to heat

Groundwater species have a low tolerance to heat. Increased temperatures have been shown to cause stress and even death to several groundwater species. For example, the groundwater amphipod *N. inopinatus* is not able to withstand temperatures much higher than those of the GWT at its collection site (12°C) for a long time. In an experiment carried out by Brielman et al. (2011) in a chamber with a temperature gradient of 2−35°C, *N. inopinatus* was found positioned at temperatures between 8 and 16°C in 77% of the observations (one observation every 30 min for a total of 9 h), with a mean residence temperature of 11.7 ± 3.4°C. Repeated tests with the groundwater isopod *P. cavaticus* provided a similar result (mean residence temperature of 11.4 ± 5°C). Control experiments in a chamber at 12−13°C (i.e., without temperature gradient) showed a more or less even distribution of the individuals of both species. Brielmann et al. (2011) also performed classical temperature-response tests by measuring the LT50 (i.e., the temperature that causes the death of 50% of exposed individuals) at different exposure times (Fig. 20.1). LT50 values decreased for both species with the protraction of exposure time (Fig. 20.1), but *P. cavaticus* was more sensitive to temperature than *N. inopinatus* (e.g., *N. inopinatus*: LT50 at 48h = 23.3 ± 2.9°C; *P. cavaticus*: 18.8 ± 0.4°C). Individuals of both species experienced a heat-rigor at temperatures in the range of 23−25°C. Only some individuals recovered when put back to a cooler temperature and, however, frequently died a few hours or days later (Brielmann et al., 2011).

In terms of physiological response to elevated temperatures, several variables have been studied in groundwater fauna. Among them are the heat shock proteins (HSP). There are various groups of HSPs classified by their molecular weight (e.g., HSP90, HSP70, HSP60). They are produced in response to exposure to stressful conditions in order to refold proteins that have been damaged. To study the effect of temperature on HSP70 gene expression, individuals of *N. rhenorodanensis*, previously acclimated at 10°C (=GWT at the collection site), were abruptly exposed to 25°C for 12 h (+15°C heat shock) and then promptly returned

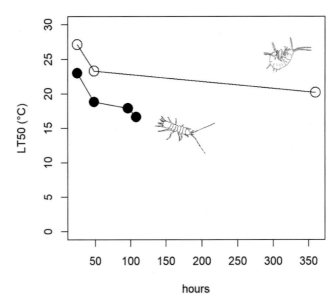

**FIGURE 20.1**    Trends of LT50 (i.e., the temperature that causes the death of 50% of exposed individuals) values at different exposure time for the species *Niphargus inopinatus* (*white dots*) and *Proasellus cavaticus* (*black dots*). Data from Brielmann et al. (2011).

to the control temperature (10°C) for a period of 48 h. The heat shock induced a significant induction of the HSP70 transcription in *N. rhenorhodanensis* compared to the controls (Colson-Proch et al., 2010). A heat shock of +2°C did not produce a significant induction of HSP70 gene transcription whatever the duration of the thermal stress was, while a heat shock of +6°C induced a significant induction of the amount of HSP70 transcripts after 1 and 2 months but not after 3 months (Colson-Proch et al., 2010). Other examples are catecholamines, such as adrenaline, noradrenaline and dopamine, biogenic amines that are known to be involved in the stress response of vertebrates and invertebrates. A sudden heat shock (prompt temperature increase of either +6 or +12°C with respect to the GWT at the collection site) induced an increase in the adrenaline content in *N. inopinatus* (Avramov et al., 2013a).

Another tool to assess the effects of temperature is the $Q_{10}$ coefficient, which expresses the change of a physiological process rate as a result of a temperature increase by 10°C. In general, the value of $Q_{10}$ for physiological processes equals to a 2-3 fold increase in process rate, although it is not uncommon to observe values both higher and lower than these for specific processes such as, for example, oxygen consumption (Willmer et al., 2009). Intraspecific or even intrapopulation variability is often observed in $Q_{10}$ values related to respiration rates. If $Q_{10} = 1$, the rate of a physiological process—say respiration rate, for instance - is considered totally independent of temperature. If, on the other hand, $Q_{10} > 1$, then the rate must be considered temperature-dependent. For instance, a $Q_{10} = 2$ indicates a doubling of the respiration rates when temperature increases from T to T+10°C. Finally, when $Q_{10} < 1$, the respiration rates are considered inversely or negatively dependent of temperature, meaning that they tend to decrease with increasing temperature. This is observed for many species at high temperatures and is considered an adaptive means of "metabolic homeostasis" (Willmer et al., 2009). The respiration rates of the epigean amphipod *G. fossarum* and of the two groundwater amphipods *N. rhenorhodanensis* and *N. virei* are highly temperature-dependent in a range between −2 and 14°C ($Q_{10} = 5.33, 7.13$, and 16.13, respectively; Issartel et al., 2005). *N. virei* increases its respiration rate more than the two other species do. Since a high $Q_{10}$ value means that temperature is impacting the physiological functions of an individual (Hochachka and Somero, 2002), it can be concluded that a large temperature variation has a strong impact on the metabolism of the three species, and on *N. virei* in particular, suggesting a lower capacity to maintain optimal enzymatic activities in this species. However, all three species show inversely temperature-dependent respiration rates in the range 21−26°C ($Q_{10} = 0.66, 0.68$, and 0.36, respectively; Issartel et al., 2005). This suggests that all three species control the energy expenditure in this temperature range, although *N. virei* does so less efficiently than the others. Similar results were observed for the hypogean *N. stygius* and the epigean *N. zagrebensis* (Simčič and Sket, 2019, 2021)

The individuals of *N. rhenorhodanensis*, *N. virei* and *G. fosssarum* are able to survive (100% survival) in the temperature range of 3−14°C for 3 months at least (Issartel et al., 2005). At higher temperatures, the survival rate decreases for all three species, however, the epigean *G. fossarum* shows significantly higher survival percentages than *N. virei* and *N. rhenorhodanensis* in the thermal range 17−28°C (Issartel et al., 2005). Oxygen consumption curves of the three species follow the classical bell-shaped profile with a peak in the range of 20−24°C. However, the mean values of oxygen consumption of *G. fossarum* are significantly higher than those of *N. rhenorhodanensis* (+30%) and *N. virei* (+143%), respectively (Issartel et al., 2005). A temperature rising from −2 to 28°C has also a strong effect on the ventilatory activity of the three amphipod species (Issartel et al., 2005). In contrast, the locomotory activity varies very little with temperature rising in the two hypogean species, while *G. fossarum* shows a rapid change in activity (Issartel et al., 2005). The performance breadths at 80% (B80), that is, the range of environmental temperature over which the physiological performances, such as oxygen consumption, ventilatory or movement, are $\geq$80% of the respective maximum performance values, are narrower for *N. virei* than in the other two species (Issartel et al., 2005). *G. fossarum* shows the largest performance breadth in terms of locomotory and ventilatory activities (B80 = 14 and B80 = 12.7°C, respectively), but not in terms of oxygen consumption (B80 = 13.2°C). Conversely, *N. rhenorhodanensis* can maintain 80% of the maximum oxygen consumption over a thermal variation of about 16°C, thus showing evident eurythermal traits, albeit it lives in heavily buffered habitats. Overall, *N. rhenorhodanensis* seems to tolerate temperature rising significantly better than *N. virei*. This result is unexpected because both species dwell in thermally stable groundwater habitats where annual temperature variation is < 1°C (Issartel et al., 2005).

The groundwater isopod *Proasellus valdensis* is surprisingly tolerant to high temperatures while two other species belonging to the same genus (*Proasellus* n. sp. 1 and *Proasellus* n. sp. 2) are extremely sensitive even to moderate changes in temperature ($\pm$2°C) with respect to their thermal optimum (Mermillod-Blondin et al., 2013). After 2 months, the survival of both *Proasellus* n. sp.1 and *Proasellus* n. sp. 2 is < 90% at temperature <9°C or >12°C, suggesting a very narrow thermal tolerance. Conversely, *P. valdensis* shows an ampler thermal breadth, with 90% of individuals surviving in the temperature range of 5.2–16.6°C (Mermillod-Blondin et al., 2013). Temperature variation also induces an immune response and an effect on the concentrations of total free amino acids in two *Proasellus* species out of three with a decrease of the phenoloxidase activity and the concentrations of several free amino acids at 4 and 16°C as compared to 10°C (Mermillod-Blondin et al., 2013).

## Insights and future perspectives for global warming scenarios

The relationship between environmental temperature and the physiological functions of groundwater organisms, such as oxygen consumption, ventilatory activity and movement, can be described by a typical asymmetric bell-shaped curve (Fig. 20.2), where the performance of a physiological trait is maximized at the thermal optimum which is usually an intermediate environmental temperature (Angilletta et al., 2002). However, in comparison to surface water species, groundwater species maximize their physiological performance over a narrower temperature range (Fig. 20.2). Outside the optimal temperature, groundwater organisms undergo a decrease in the performance of physiological processes that they try to counteract with overexpression of protective proteins and production of catecholamines at high temperatures, and with the production of cryoprotective substances at low temperatures. These defense mechanisms are efficient at low temperatures at which groundwater species seem to perform better than the surface water ones. On the other hand, at temperatures higher than the thermal optimum, the defense mechanisms of groundwater species are less efficient so that the survival of some species is seriously compromised already at 18°C.

In addition, differences in the physiological performance vary considerably among congeneric groundwater species, indicating that the thermal physiology of groundwater fauna is not solely shaped by the ambient temperature in groundwater habitats. Overall, the studies

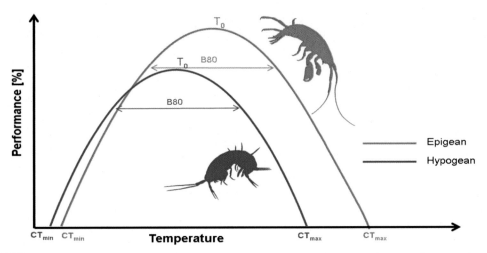

**FIGURE 20.2**    Relationship between temperature and physiological performance of a groundwater and surface water invertebrate species (ectotherms). The thermal optimum (To) is the temperature that maximizes the performance; the performance breadth B80 is the range of temperatures over which the performance is $\geq$80% of maximum. The critical thermal minimum ($CT_{min}$) and critical thermal maximum ($CT_{max}$) are the minimum and maximum temperatures, respectively, at which the performance is possible. B80, To, $CT_{max}$ and $CT_{min}$ of groundwater species are narrower/lower than those of surface water species. Based on Angilletta et al. (2002).

available are still limited in their taxonomic and geographic coverage. Conclusions about how differently these species respond, compared to their surface water relatives, could imply a bias due to the different evolutionary and ecological traits (Castaño-Sánchez et al., 2020). Most studies focused on species dwelling in groundwater habitats in temperate areas, especially in Europe, leaving unresolved if subtropical/tropical species, or species of extreme groundwater habitats such as ice caves fauna, display similar thermal patterns. Several species inhabiting ice caves (Iepure, 2017), or the hyporheic zone of the glacier-fed rivers (Camacho et al., 2019), have to cope with permanently low temperatures ranging from 0°C down to −14°C, but very little is known about their tolerance to temperature increase. Finally, experiments were typically performed over short timeframes (hours to a few months), a very short period for these long-living animals, some of them having a life span of decades (e.g., Voituron et al., 2010). This last is a critical aspect because groundwater biotopes are often fragmented (see Malard et al., this volume), and groundwater species cannot easily avoid temperature alterations through dispersal. In fact, even if they can endure low thermal stress for a few months, we have no evidence that they can do it with unchanged efficiency for longer periods of time, namely over timeframes that are congruent with those of climate change. Moreover, elevated temperatures inevitably lead to an increase in enzymatic activity, and metabolic rates, and thus also to an increase in energetic costs. In groundwater ecosystems, where food availability is typically low, this could become a critical factor for survival.

We can conclude that groundwater invertebrate species in alluvial aquifers beneath the big cities are already at high risk of disappearance due to the groundwater heat-ups related to geothermal energy usage and urban heat islands. In addition, in the perspective of global warming (perceptible, though less intensively, even in groundwater environments: Green et al., 2011; Taylor et al., 2013), the low tolerance to heat is expected to pose a major disadvantage hindrance (potentially more severe than expected based on the few available laboratory studies on survival; Pallarés et al. (2020)) to the survival of the most specialized/stenothermic species in the long term (Mammola et al., 2019b).

## Physiological tolerance of groundwater organisms to chemical stress

Groundwater pollution is a problem of growing concern. Anthropogenic activities, including intensive agriculture, improper discharge of industrial and municipal wastewater, as well as insufficiently sealed landfill sites (to name just a few), are exposing groundwater to a variety of organic and inorganic pollutants. Comprehensive reviews about groundwater pollution sources have pointed out that the main organic contaminants in groundwater are pesticides, herbicides, mineral oil, polycyclic aromatic hydrocarbons, pharmaceuticals, and perfluorooctanoic acids, while contamination due to inorganic compounds mainly involves arsenic and other metals, but also fluoride and nitrate (Meffe and de Bustamante, 2014; Kurwadkar, 2019). Quarry wastes, slurries, and microplastics have been recently included in the list (e.g., Veado et al., 2006; Mintenig et al., 2018; Piccini et al., 2019).

### Pattern of tolerance to chemicals

The tolerance of groundwater fauna to chemicals is variable and no taxon is consistently more or less sensitive than the others, according to the current data (Gerhardt, 2019; Grimm and Gerhardt, 2019; Hose et al., 2019; Castaño-Sanchez et al., 2020). This is, in part, due to the variability of the test conditions and the lack of standard protocols (Di Lorenzo et al., 2019a), as well as to the individual physiological characteristics of the species. Nevertheless, a tolerance pattern to toxicants can be inferred considering the available data from test duration $\leq 4$ days based on 50% survival or immobility endpoints ($LC_{50}/EC_{50}$). Pesticides (insecticides and fungicides) seem to be the most toxic chemicals to groundwater invertebrates, with $LC_{50}/EC_{50}$ values in the range of 6 ng/L—2.9 mg/L (Castaño-Sanchez et al., 2020). Groundwater invertebrates seem to better tolerate metals, meeting the 50% lethal/effect endpoints at concentrations in the range of 150 µg/L—424 mg/L (from the most toxic to the least toxic: Cu > Cd > Zn > Cr > Pb > As > Ni). The β-blocker propranolol, a widely used pharmaceutical compound, is toxic at concentrations <10 mg/L ($LC_{50} = 5$ mg/L for the groundwater copepod *Diacyclops belgicus* at 96 h; Di Lorenzo et al., 2019b). On the other hand, diclofenac, the nonsteroidal drug most widely used for the treatment of musculoskeletal and systemic inflammatory states in humans and animals widespread, is toxic at concentrations >10 mg/L ($LC_{50} = 12$ mg/L for the groundwater harpacticoid *Nitocrella achaiae* at 96 h; Di Lorenzo et al., 2021). BPA and volatile organic compounds (VOCs), such as toluene, are highly tolerated ($LC_{50-96h} = 64$ mg/L and $LC_{50-24h} = 6$ mg/L, respectively) by groundwater invertebrates (Avramov et al., 2013b; Gerhardt, 2019). Widely used fertilizers, such as ammonium nitrates, can lead groundwater species to death at concentrations of about 14 mg/L $NH_4^+$ (Di Lorenzo et al., 2014; Di Marzio et al., 2018). When fertilizers, such as ammonium nitrate, are mixed up with herbicides (a combination that often occurs in agricultural practice), the action of the compounds in the mixture is often synergic, causing groundwater species to die at concentrations lower than those tolerated when the compounds are tested individually (Di Marzio et al., 2018). Organic fertilizers based on nitrates deserve separate consideration. At present, there are no $LC_{50}/EC_{50}$ data of nitrates computed for groundwater species. However, the results of two experiments suggest that the $LC_{50}/EC_{50}$ values at 96 h could be in the range of 50–100 mg/L $NO_3^-$ (Gerhardt, 2020). The tolerance of groundwater species could be even lower if nitrates were administered in the form of manure associated with antibiotic substances, such as sulfamethoxazole, and with other substances, such as nitrites (Gerhardt et al., 2020). Based on the knowledge available so far, the tolerance of groundwater invertebrates toward anthropogenic chemicals can be rated as follows (from the least tolerated to the most tolerated): pesticides > metals > pharmaceuticals > ammonia-based fertilizers > bisphenol A > VOCs > $NO_3^-$.

## Pattern of environmental risk

Environmental risk assessment (ERA) is an approach that evaluates the likelihood that adverse ecological effects may occur, or are occurring, as a result of exposure of organisms and communities to one or more chemical compounds (e.g., EMA, 2018a,b). From a regulatory perspective, the environmental risk (R) is usually calculated as R = MEC/PNEC, where PNEC is the predicted no effect concentration of the chemical compound and MEC is its measured environmental concentration (EMA, 2018a,b). An environmental risk must be assumed if R $\geq$ 1 (EMA, 2018a,b; Di Lorenzo et al., 2018, 2019b). In the last few years, many countries have diligently compiled MEC databases for their groundwater bodies (EEA, 2019). To depict a first preliminary scenario of ERA in groundwater, we calculated the ratios of the maximum MECs of contaminants detected in European groundwater bodies since 1960 (EEA, 2019) to the PNEC for groundwater fauna for 15 chemical compounds (Table 20.1). We calculated the PNEC by applying an assessment factor of 1000 to the lowest short-term $LC_{50}$ or $EC_{50}$ value available for groundwater species for each compound, as

**TABLE 20.1**   Maximum measured environmental concentrations (MECs in µg/L) of 15 contaminants detected in European groundwater bodies, the lowest short-term endpoints ($LC_{50}$ or $EC_{50}$ in µg/L) of groundwater species for individual compounds, predicted no effect concentrations (PNECs in µg/L), ecological risk values and interpretation. n indicates the number of endpoints available from literature. Tests' duration <10% of the species life span. All MECs were obtained from WATERBASE (EEA, 2019) with the exception of bisphenol A (BPA) MEC that corresponded to 25% of the highest BPA concentration detected in German streams (EMA, 2018b).

| Name | Class | MEC | $LC_{50}$ or $EC_{50}$ | PNEC | Risk | Interpretation | PNEC references |
|------|-------|-----|-----------|------|------|----------------|------------------|
| $Zn^{2+}$ | Metals | 17,500 | 450 (n = 17) | 0.45 | 38,888 | At risk | Boutin et al. (1995) |
| $Cu^{2+}$ | Metals | 9530 | 150 (n = 5) | 0.15 | 63,533 | At risk | Boutin et al. (1995) |
| $Ni^{2+}$ | Metals | 9000 | 75,550 (n = 1) | 75.5 | 119 | At risk | Barr (1976) |

*(Continued)*

| Name | Class | MEC | LC$_{50}$ or EC$_{50}$ | PNEC | Risk | Interpretation | PNEC references |
|------|-------|-----|-----------------------|------|------|----------------|-----------------|
| Pb$^{2+}$ | Metals | 5000 | 800 ($n = 2$) | 0.80 | 6250 | At risk | Boutin et al. (1995) |
| As$^{3+}$ | Metals | 1170 | 250 ($n = 8$) | 0.25 | 4680 | At risk | Hose et al. (2019) |
| Cd$^{2+}$ | Metals | 818 | 290 ($n = 19$) | 0.29 | 2820 | At risk | Hose et al. (2019) |
| Cr(VI) | Metals | 289 | 500 ($n = 8$) | 0.51 | 566 | At risk | Hose et al. (2019) |
| S-Metolachlor | Pesticide-herbicide | 0.9 | 36,900 ($n = 2$) | 36.9 | 0.02 | No risk | Maazouzi et al. (2016) |
| Chlorpyrifos | Pesticide-insecticide | 1.4 | 57 ($n = 1$) | 0.057 | 24 | At risk | Notenboom and Boessenkool (1992) |
| Pentachlorophenol | Pesticide-insecticide | 0.4 | 11 ($n = 4$) | 0.01 | 36 | At risk | Notenboom et al., 1991 |
| NH$_4^+$ | Fertilizer | 767,000 | 1400 ($n = 4$) | 1.4 | 540,873 | At risk | Di Marzio et al., 2018 |
| Toluene | VOCs | 478,000 | 23,300 ($n = 1$) | 23.3 | 20,515 | At risk | Avramov et al. (2013b) |
| Propranolol | Pharmaceuticals | 0.2 | 5000 ($n = 1$) | 5.0 | 0.04 | No risk | Di Lorenzo et al. (2019a) |
| Diclofenac | Pharmaceuticals | 0.00255 | 12,000 ($n = 1$) | 12.0 | 0.0002 | No risk | Di Lorenzo et al., 2021 |
| BPA | Diphenylmethane deriv. | 2.4 | 6300 ($n = 1$) | 6.3 | 0.38 | No risk | Gerhardt (2019) |

indicated in the guidelines (ECHA, 2017). The lowest LC$_{50}$ or EC$_{50}$ values in Table 20.1 were obtained from Castaño-Sanchez et al. (2020).

The herbicide S-Metolachlor, the pharmaceutical compounds propranolol and diclofenac and the organic compound bisphenol A, do not induce a significant environmental risk in groundwater as opposed to heavy metals, some insecticides, some VOCs such as toluene, and ammonium-based synthetic fertilizers, which generate a serious environmental risk in groundwater with R $>>$ 1. Since there are no standard ecotoxicological data on nitrate, it is not possible to calculate the R value for this contaminant. However, assuming that the LC$_{50}$/EC$_{50}$ values are in the range of 50–100 mg/L NO$_3$ for short-term exposures of up to 96 h as suggested by Gerhardt (2020), and given the high nitrate concentrations recorded in European groundwater bodies (up to 300 mg/L; EEA, 2019), we should expect the environmental risk induced in groundwater by nitrate to be significant.

The risk pattern could be even worse than that shown in Table 20.1. First, groundwater is almost always contaminated with mixtures of pollutants (Loos et al., 2010). For instance, ammonium contamination is often associated with nitrite and nitrate pollution (Di Lorenzo et al., 2019c) and contamination by VOCs often includes BTEX (benzene, toluene, ethylbenzene and p-xylene) as well as chlorinated aliphatic hydrocarbons, such as chloroform and tetrachloroethylene (Di Lorenzo et al., 2015). Current guidelines suggest calculating R by adding up the risks associated with the individual contaminants in the mixture (EMA 2018a,b; Strona et al., 2019). However, synergistic effects should be considered, since they worsen the outcomes of risk scenarios. A further problem lies in the use of short-term endpoints since groundwater species have a long life span and their low metabolism may mask the toxic effects under short exposures. Di Lorenzo et al. (2020) analyzed the structure of groundwater copepod assemblages in a shallow unconsolidated European aquifer subjected to long-lasting nitrate contamination (>50 mg/L) for over 15 years. Unexpectedly, the copepod assemblages were highly diversified and represented by 17 groundwater species. However, the analyses of functional traits showed signs of alteration of these assemblages, indicating that chronic exposure to nitrate impaired the copepod populations. The guidelines suggest using sublethal endpoints instead of mortality/effect-based ones (EMA, 2018b). However, sublethal endpoints are only available for a very limited number of groundwater species at the moment (Di Lorenzo et al., 2019b; Hose et al., 2019; Castaño-Sanchez et al., 2020). Nevertheless, using sublethal concentrations as PNEC proxies in the future would represent an improvement of ERA toward establishing feasible regulatory risk benchmarks. For instance, when individuals of the groundwater copepod *D. belgicus* are exposed to 2 mg/L propranolol for 4 days, although they do not die, they move significantly slower than the control individuals and travel a shorter distance in the same unit of time, covering a different convex hull (Fig. 20.3). Such physiological alterations can be fatal in the long run, impairing the search for food and mating partners and the avoidance of predators/stressful conditions.

The current groundwater ERA guidelines (e.g., EMA, 2018a,b) suggest computing the PNECs for groundwater ($PNEC_{GW}$) using surface water standard taxa (i.e., algae, *Daphnia* and fish) as proxies by applying an additional assessment factor equal to 10, so that $PNEC_{GW} = PNEC_{SW}/10$. Although this approach has been much criticized because it does not reflect the traits of groundwater species, it proved to be highly protective nonetheless (Di Lorenzo et al., 2019b). Alternative approaches for risk assessment in groundwater have been suggested . For instance, Di Lorenzo et al. (2018) and Strona et al. (2019) suggested using $PNEC_{SW}$ of the sole crustacean taxon, and not those of algae and fish, as proxies of $PNEC_{GW}$ values in alluvial aquifers because these ecosystems are largely dominated by crustaceans (mostly, copepods, ostracods, syncarids, amphipods, and isopods). The debate about which approach leads to more realistic results is still open, the difference in sensitivity to chemicals of groundwater and surface water species being the burning point of discussion. Are groundwater species more sensitive to chemicals than surface water species? This question is difficult to answer. Overall, the answer seems to be "yes" if one considers closely related groundwater and surface water species belonging to the same order or family, as terms of comparison (e.g., Mösslacher, 2000; Di Lorenzo et al., 2014; Di Marzio et al., 2018; Gerhardt, 2019). However, the answer might be "no" when considering standard test species, such as freshwater cladocerans,

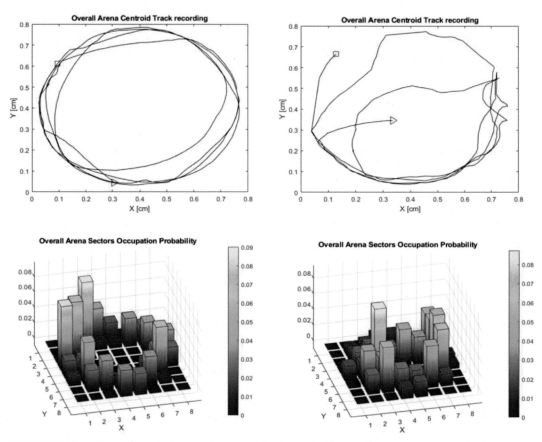

**FIGURE 20.3** Movement tracks (top) and convex hulls (bottom) obtained from the frequencies of occurrence in specific points of an experimental arena (diameter: 8 mm) of an individual of the groundwater copepod *Diacyclops belgicus* in the control (left) and after exposure to a concentration of propranolol equal to 2 mg/L for 96 h (right). 1-min videos per specimen were recorded with an HD digital microscope camera integrated into a stereomicroscope at 7.5x magnification. Live images with up to 30 frames per second and a standard capture resolution of 5 megapixels were recorded in real-time on a solid-state disk. The videos were examined using the software TrAQ (Di Censo et al., 2018). All the methodological details about the experimental protocol and software are provided in Di Lorenzo et al. (2019b).

in comparison (e.g., Avramov et al., 2013b; Reboleira et al., 2013; Hose et al., 2016; Gerhardt, 2019; Grimm and Gerhardt, 2019). For instance, concerning zinc, it is not possible to give an unambiguous answer to the question (Fig. 20.4). In fact, based on the available data, the cladoceran *Ceriodaphnia dubia* is the most sensitive species to zinc, while the acute sensitivities of other species, for instance of the standard test species *Daphnia magna* and *D. lumholtzi*, are lower than those of several groundwater species (Fig. 20.4).

The assessment of the ecological risk of metals could benefit from Biotic Ligand Models (BLM). BLMs are an effective means for assessing how water chemistry affects the speciation and biological availability of metals in aquatic systems. However, there are many challenges that make the application of BLMs difficult in groundwater. The main problem is the

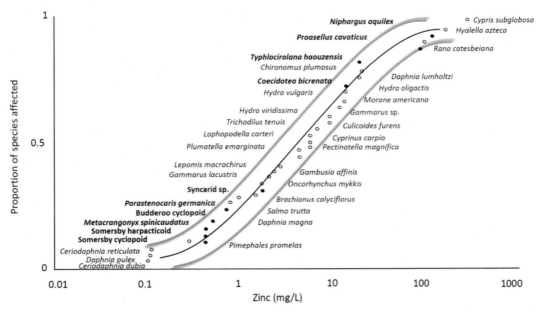

**FIGURE 20.4** Species sensitivity distribution curves (solid black line: main trend; blurred gray lines: 95% CI) of zinc based on short-term endpoints assessed for groundwater (bold) and surface water species (normal type) retrieved from the U.S. EPA ECOTOX database. The data have not been corrected for pH and hardness and, therefore, the actual trend of the curve may be slightly different.

derivation of the stability constants of the BLMs. These constants are calculated through multiple toxicity tests (typically 4 to 7). Large-scale toxicity testing must be conducted simultaneously using a large number of test organisms. Although alternative methods for calculating these constants have been proposed in recent years (Liang et al., 2021), the number of test organisms still remains very high (e.g., in the case of *Daphnia*, about 2000 individuals are needed). The low specific abundances of groundwater species and their low fertility are an insurmountable problem for BLMs' application in groundwater ecosystems.

## Insights and future perspectives

In this section, we provided an overview of the tolerance of groundwater fauna to chemical pollution and also tried to unveil the issues of the guidelines on ERA in groundwater. From a regulatory perspective, the ERA in groundwater is currently based on the tolerance of surface water species that do not possess any of the distinctive traits of the true inhabitants of subterranean environments (Hose et al., 2022). The European guidelines suggest estimating the sensitivity of groundwater species simply by applying an arbitrary correction factor of 10 to the endpoint of the most sensitive surface water species, without taking into account further corrective factors related to the differences in biomass, body size, uptake processes and pathways and experimental conditions. On the other hand, currently, there is no real alternative, given the lack of ecotoxicological data on true groundwater species. Although the use of surface water standard species as proxies is attractive, a realistic scenario of ecological risks in groundwater can be likely achieved only by testing the tolerance of the obligate dwellers of groundwater habitats to sublethal toxicant concentrations.

## Physiological tolerance of groundwater organisms to light, food and oxygen variations: indications for ecotoxicological protocols

The ecotoxicological studies on groundwater species that have been carried out so far (see Section Physiological tolerance of groundwater organisms to chemical stress in this chapter) have been conducted under different experimental protocols based on the optimal conditions of the species under study (Di Lorenzo et al., 2019a). In this section, we focus on three experimental factors that affect the tolerance of groundwater species in the laboratory and, hence, may lead to poor reliability of test results: light, oxygen, and food.

### Tolerance to light

Groundwater fauna live in complete darkness. However, it is difficult to carry out ecotoxicological and physiological tests under the complete absence of light. Handling groundwater species in the laboratory must necessarily be carried out under some visible light and, for some taxa like groundwater copepods, also with the aid of a stereomicroscope at 12x magnification at least. Even an expert operator cannot manage to reduce exposure to visible light to less than 10 min for the overall test phases. Is light harmful to groundwater species? Based on the few specific experiments carried out so far, the answer seems to be "yes." As an indication, oxygen consumption of the groundwater amphipod *Niphargus stygius* was observed to significantly increase during 1h exposure to white light (spectral range: 300–700 nm) with intensity >720 lx at 10°C, with consumption levels remaining elevated even after up to 3h of recovery in the dark (Simčic and Brancelj, 2007). In addition, the ETS/R ratio (Electron Transport System activity/basal metabolic rate) of this species significantly decreased during exposure to high-intensity light (4700 lx), indicating that the animals spend much of their potential energy for breathing during this exposure to light (Simčic and Brancelj, 2007). Specimens of 10 different *Niphargus* species displayed a distinct photophobic behavior, showing a tendency to stay longer in the dark spaces of a Petri dish, avoiding illuminated areas, while eyed surface amphipods did not appear to orient themselves with light cues (Fišer et al., 2016). The photophobia of *Niphargus* could be associated with the need to distinguish the border between surface and subterranean environments, so as to avoid surface habitats where competition and predation may be higher and UV rays dangerous for depigmented animals (Borowsky, 2011; Fišer et al., 2016; Manenti and Barzaghi, 2021). Finally, blind cave fish have a pineal organ that is still photosensitive and display a clear phototactic behavior (Green and Romero, 1997). Light also significantly altered the social behavior of the groundwater crayfish *Orconectes australis packardi* (Li and Cooper, 2002).

### Tolerance to oxygen depletion

A common requirement in carrying out ecotoxicological tests with aquatic animals is to maintain a well-oxygenated environment so that the animals are not forced to switch to anaerobic metabolism with consequent accumulation of end-products (mainly lactate) and depletion of energy stores. In the trials carried out with surface water fauna, it is preferred to oversaturate the test medium. Is this necessary in tests with groundwater species? Likely, it is not, even in the case of long-term experiments and in sealed test vials. Oxygen concentrations in groundwater ecosystems are generally lower than in surface environments, sometimes hardly exceeding 3–4 mg/L (Winograd and Robertson, 1982; Malard and Hervant, 1999). Moreover, groundwater species can tolerate even severely hypoxic environments much better than their surface relatives (Issartel et al., 2009). For instance, when individuals of N. rhenorhodanensis and G. fossarum were placed in an oxygen-depleted environment (anoxic water) at a temperature of 11°C, half of the groundwater individuals survived for 2 days while half of the individuals of the epigean species died within 6 h (Hervant et al., 1996).

The high tolerance to low environmental oxygen concentrations of both groundwater invertebrates and vertebrates is likely based on several factors. First, the primary energy reserves, stored in the form of glycogen and phosphagen (i.e., arginine phosphate or creatine phosphate), are much more abundant in groundwater than in epigean organisms and this is a functional advantage in terms of the availability of fermentable resources (to fuel the anaerobic metabolism) during oxygen depletion (Hervant et al., 1996, 1997a, 1998). During severe hypoxia, groundwater amphipods must shift from aerobic to anaerobic metabolism in an attempt to produce sufficient ATP to meet primary energetic needs. Thus, they use the coupled fermentation of amino acids and glycogen (with lactate, alanine and succinate as end-products), significantly increasing the ATP production during severe hypoxia (Hervant et al., 1996, 1997a, 1998). Finally, during oxygen deprivation, groundwater amphipods and isopods display a drastic reduction in locomotory and ventilatory activities (Hervant et al., 1996, 1997a). The anaerobic metabolism is promptly abandoned when the oxygen concentrations return to favorable levels. In this event, the lactate produced during the anaerobic phase is remetabolized in groundwater invertebrates while their epigean relatives simply eliminate it by excretion, therefore losing energy-rich compounds (Hervant et al., 1996, 1997a,b). This strategy allows groundwater species to rapidly restore (without any food source) their glycogen stock (Hervant et al., 1997a,b, 1998). However, reoxygenation does not occur without a metabolic cost. Overshoot in oxygen consumption is observed during the time course of posthypoxic recovery, inducing a significant increase in reactive oxygen species (ROS) production. Overall, antioxidant mechanisms (mainly enzymatic) are overexpressed during reoxygenation in groundwater species, to protect cells from damage from oxygen radicals, as observed in N. rhenorhodanensis (Lawniczak et al., 2013). There is, at least, a known exception to this rule in groundwater ecosystems: the cave salamander Proteus anguinus presents a very moderate antioxidant response during reoxygenation (Issartel et al., 2009).

## Tolerance to starvation

The food factor represents an additional issue in ecotoxicological studies with groundwater animals. Very little is known about their diet and, currently, there is no standardized food that can be used throughout long-term tests (Di Lorenzo et al., 2019a). The best survival conditions in the test controls are usually achieved by not supplying any additional food to the vials, allowing the animals to feed on the bacteria contained in the groundwater of the collection site (Di Lorenzo et al., 2019a). In some experiments, the researchers have even carried out the tests using reconstituted ultrafiltered water (free of bacteria), observing no mortality in the control group after 7 days (Di Lorenzo et al., 2014; Di Marzio et al., 2018). Similarly, 85% of the test individuals of *N. rhenorhodanensis*, *N. virei*, and the isopod *Stenasellus virei* survived 200 days of starvation while 100% of the test individuals of *G. fossarum* died after 100 days (Hervant et al., 1997b; Hervant and Renault, 2002) or even earlier according to Mezek et al. (2010). Moreover, the dry mass and water content of *N. rhenorhodanensis* and *N. virei* show no significant changes under starvation, at least up to 180 days, in contrast to what is observed for *G. fossarum* (Hervant et al., 1999).

To cope with food scarcity for a long time, groundwater fauna induces several types of physiological responses. During food deprivation, both the locomotory and ventilatory activities and, to a lesser extent, also the oxygen consumption, decrease considerably, already after about 2 weeks of starvation (Hervant et al., 1997b). Moreover, groundwater species display a sequential energy strategy during starvation, consisting in successive periods of glucidic, lipidic and finally lipido-proteic-dominant catabolism, while epigean species show a monophasic response with an immediate, linear, and substantial decrease of all the energy reserves (Hervant et al., 1997b, Hervant et al., 1999, 2001; Issartel et al., 2010; Mezek et al., 2010; Nair et al., 2020). Accordingly, the groundwater amphipod *Stygobromus pecki* mobilizes mainly proteins as an energy reserve during 90 days of starvation (Nair et al., 2020). Interspecific differences in the utilization of lipids (mainly fatty acids) during fasting were also highlighted by Mezek et al. (2010) who observed that the groundwater amphipod *N. stygius* tends to preserve all of its fatty acids during 50 days of starvation, while the epigean *G. fossarum* displays a tendency to use monounsaturated and saturated fatty acids preferentially, retaining polyunsaturated lipids. During fasting, the ratio ETS/R does not change in *N. stygius* while it increases in *G. fossarum*, indicating that the groundwater species modulates its respiration according to its energy storage (Mezek et al., 2010).

As a general rule, by the time groundwater animals are refed after a prolonged starvation period, all the body stores and physiological activities related to locomotion, ventilation and oxygen consumption are efficiently restored (even though they are altered compared to the control in the first few days of recovery), to then return to normal values (Hervant et al., 1997b, Hervant et al., 1999, 2001). The exception to the rule has been, up to now, only found in a particular genus of tropical fish that has both epigean and subterranean species coexisting in tropical caves. The cave species *Astyanax mexicanus* catabolizes slightly more energy stores than the epigean *Astyanax fasciatus* during the experimental starvation period, conversely to the results of the comparative studies with nontropical species (Salin et al., 2010). In this case, the observations made on *Astyanax* might indicate that adaptations to starvation are not necessarily shared by all cave animals but are rather correlated to the "energetic state" of each ecosystem.

## Insights and future perspectives

The tolerance of groundwater organisms to light, low oxygen concentrations and food scarcity suggests that the ecotoxicological tests should not be carried out using the standard experimental protocols issued for surface water species. The experimental conditions suitable for surface water species (that include, for instance, the oversaturation of oxygen in water, the administration of artificial food and a photoperiod of 18 h of light and 6 h of darkness) are not adequate for groundwater species. Light affects the physiology and behavior of some groundwater species. If, as stated above, it is not possible to fully exclude light from all handling and preparation procedures, it seems conceivable to limit the exposure of the test animal to light as much as possible and then perform the actual testing in the dark. It is not advisable to oxygenate the vials during acclimation and testing, especially if the tests are carried out at low temperatures. It is recommended to keep oxygen values close to the concentrations of the collection site. For a test to be valid, appropriate controls, including any solvent or carrier control, should be designed and the dissolved oxygen concentration at the end of the test should be within 20% of the value at the beginning of the test while <20% of the individuals in the control should be dead or immobilized (Di Lorenzo et al., 2019a). Finally, based on the revised data, providing groundwater species with food is not necessary for short-term tests (Di Lorenzo et al., 2019a). As for chronic tests, the addition of food is advisable. Which kind of artificial food is suited the best represents a research gap awaiting to be addressed (Di Lorenzo et al., 2019a).

## Conclusions

Due to the limited number of studies, it is not possible to make sound predictions on possible future changes in the distribution pattern of groundwater fauna according to the expected scenarios of climate change, groundwater warming, chemical contamination, and other ecological risks. Life-history traits, such as reproduction rates and growth, should be investigated, and laboratory data must be interpreted in the light of current distribution patterns of groundwater fauna (Kotta et al., 2019), to better infer the distribution impacts of climate change and urban heat discharge on groundwater species and to set-up effective conservation strategies in the long run (Mammola et al., 2019a,b). Even more uncertain are the scenarios of groundwater ERA related to anthropogenic stressors, because they are not based on the tolerance of stygobites and troglobites (= true groundwater and cave water dwelling invertebrates). It is necessary to increase the number of experimental tests with these animals, using laboratory techniques that are specifically designed to suit their physiology, as described in the last section of this chapter. Based on the increasing societal awareness for environmental issues (e.g., as recently expressed through the high popularity of initiatives such as *Fridays For Future* and postpandemic intergovernmental recovery plans such as the European Green Deal), we anticipate an increased scientific interest related to studying the effects of global warming, emerging contaminants, as well as micropollutants, toxicant mixtures, and micro- and nanoplastics. Nevertheless, groundwater ecosystems remain largely "unseen" by society—both literally and in terms of societal perception. Therefore, we conclude with an exhortation to scientists to participate as much as possible in the dialogue between research, decision-makers and society, and to disseminate the results of their studies to the wide public. As the world's scientists have recently emphasized in countless warnings to humanity regarding the collapse of ecosystems and the unprecedented loss in biodiversity on Earth (e.g., Cavicchioli et al., 2019; Finlayson et al., 2019; Cardoso et al., 2020), "[…] *soon it will be too late to shift course away from our failing trajectory*" (Ripple et al., 2017), and this also holds true for groundwater ecosystems (Mammola et al., 2019a; Ferreira et al., 2022).

## Acknowledgments

We thank J.L. Pereira, W.D. Di Marzio and C. Griebler for constructive comments that improved the chapter. TLD acknowledges support from the National Biodiversity Future Center (NBFC; CN_00000033) of the National Recovery and Resilience Plan (NRRP) of Italian Ministry of University and Research, funded by the European Union (NextGenerationEU). DL, DMPG and SM acknowledge support from Biodiversa+ 2021—2022 program (project DarCo; BIODIV21_0006). SI was supported by a grant from the Romanian Ministry of Education and Research, CNCS - UEFISCDI (project number: PN-III-P4-PCE-2020-2843; title: Evo-Devo-Cave), within PNCDI III.

# References

Angilletta, M.J., Niewiarowski, P.H., Navas, C.A., 2002. The evolution of thermal physiology in ectotherms. Journal of Thermal Biology 27, 249—268.

Avramov, M., Rock, T.M., Pfister, G., Karl-Werner Schramm, K.-W., Schmid, S.I., Griebler, C., 2013. Catecholamine levels in groundwater and stream amphipods and their response to temperature stress. General and Comparative Endocrinology 194, 110—117.

Avramov, M., Schmidt, S.I., Griebler, C., 2013b. A new bioassay for the ecotoxicological testing of VOCs on groundwater invertebrates and the effects of toluene on *Niphargus inopinatus*. Aquatic Toxicology 130—131, 1—8.

Barr, T.C., 1976. Ecological Effects of Water Pollutants in Mammoth Cave: Final Technical Report to the National Park Service. University of Kentucky.

Benz, S.A., Bayer, P., Blum, P., 2017. Global patterns of shallow groundwater temperatures. Environmental Research Letters 12, 034005.

Borowsky, B., 2011. Responses to light in two eyeless cave dwelling amphipods (*Niphargus ictus* and *Niphargus frasassianus*). Journal of Crustacean Biology 31 (4), 613—616.

Botosaneanu, L., Holsinger, J., 1991. Some aspects concerning colonization of the subterranean realm - especially subterranean waters: a response to Rouch and Danielopol, 1987. Stygologia 6 (1), 11—39.

Boutin, C., Boulanouar, M., Yacoubi—Khebiza, M., 1995. Un test biologique simple pour apprécier la toxicité de l'eau et des sédiments d'un puits. Toxicité comparée, in vitro, de quelques métaux lourds et de l'ammonium, vis-à-vis de trois genres de crustacés de la zoocénose. Hydroécologie Appliquée 7 (1—2), 91—109.

Boulton, A.J., 2020. Conservation of groundwaters and their dependent ecosystems: Integrating molecular taxonomy, systematic reserve planning and cultural values. Aquatic Conservation: Marine and Freshwater Ecosystems 30 (1), 1—7.

Brielmann, H., Lueders, T., Schreglmann, K., 2011. Shallow geothermal energy usage and its potential impacts on groundwater ecosystems. Grundwasser 16, 77—91.

Camacho, A., Mas-Peinado, P., Iepure, S., Perina, G., Dorda, B.A., Casado, A., Rey, I., 2019. Novel sexual dimorphism in a new genus of Bathynellidae from Russia, with a revision of phylogenetic relationships. Zoologica Scripta 49 (1), 47—63.

Cardoso, P., Barton, P., Birkhofer, K., Chichorro, F., Deacon, C., Fartmann, T., Fukushima, C.S., Gaigher, R., Habel, J.C., Hallmann, C.A., Hill, M.J., Hochkirchi, A., Kwak, M.L., Mammola, S., Noriega, J.A., Orfingern, A.B., Pedraza, F., Pryke, J.S., Roque, F.O., Settele, J., Simaika, J.P., Stork, N.E., Suhling, F., Vorsterd, C., Samwaysd, M.J., 2020. Scientists' warning to humanity on insect extinctions. Biological Conservation 242, 108426.

Castaño-Sánchez, A., Hose, G.C., Reboleira, A.S.P.S., 2020. Ecotoxicological effects of anthropogenic stressors in subterranean organisms: a review. Chemosphere 244, 125422.

Cavicchioli, R., Ripple, W.J., Timmis, K.N., Azam, F., Bakken, L.R., Baylis, M., Behrenfeld, M.J., Boetius, A., Boyd, P.W., Classen, A.T., Crowther, T.W., Danovaro, R., Foreman, C.M., Huisman, J., Hutchins, D.A., Jansson, J.K., Karl, D.M., Koskella, B., Welch, D.B.M., Martiny, J.B.H., Moran, M.A., Orphan, V.J., Reay, D.S., Remais, J.V., Rich, V.I., Singh, B.K., Stein, L.Y., Stewart, F.J., Sullivan, M.B., van Oppen, M.J.H., Weaver, S.C., Webb, E.A., Webster, N.S., 2019. Scientists' warning to humanity: microorganisms and climate change. Nature Reviews Microbiology 17, 569—586.

Colinet, H., Sinclair, B.J., Vernon, P., Renault, D., 2015. Insects in fluctuating thermal environments. Annual Review of Entomology 60, 123—140.

Colson-Proch, C., Morales, A., Hervant, F., Konecny, L., Moulin, C., Douady, C.J., 2010. First cellular approach of the effects of global warming on groundwater organisms: a study of the HSP70 gene expression. Cell Stress & Chaperones 15 (3), 259—270.

Colson-Proch, C., Renault, D., Gravot, A., Douady, C.J., Hervant, F., 2009. Do current environmental conditions explain physiological and metabolic responses of subterranean crustaceans to cold? Journal of Experimental Biology 212 (12), 1859–1868.

Danielopol, D.L., Creuzé des Châtelliers, M., Moeszlacher, F., Pospisil, P., Popa, R., 1994. Adaptation of Crustacea to interstitial habitats: a practical agenda for ecological studies. In: Gibert, J., Danielopol, D.L., Stanford, J.A. (Eds.), Groundwater Ecology. Academic Press, pp. 217–243.

Di Lorenzo, T., Di Marzio, W.D., Sáenz, M.E., Baratti, M., Dedonno, A.A., Iannucci, A., Cannicci, S., Messana, G., Galassi, D.M.P., 2014. Sensitivity of hypogean and epigean freshwater copepods to agricultural pollutants. Environmental Science and Pollution Research 21 (6), 4643–4655.

Di Lorenzo, T., Fiasca, B., Di Cicco, M., Galassi, D.M.P., 2020. The impact of nitrate on the groundwater assemblages of European unconsolidated aquifers is likely less severe than expected. Environmental Science and Pollution Research 28 (9), 11518–11527.

Di Lorenzo, T., Di Marzio, W.D., Fiasca, B., Galassi, D.M.P., Korbel, K., Iepure, S., Pereira, J.L., Reboleira, A.S.P.S., Schmidt, S.I., Hose, G.C., 2019a. Recommendations for ecotoxicity testing with stygobiotic species in the framework of groundwater environmental risk assessment. Science of the Total Environment 681 (1), 292–304.

Di Lorenzo, T., Cifoni, M., Fiasca, B., Di Cioccio, A., Galassi, D.M.P., 2018. Ecological risk assessment of pesticide mixtures in the alluvial aquifers of central Italy: toward more realistic scenarios for risk mitigation. Science of the Total Environment 644, 161–172.

Di Lorenzo, T., Di Cicco, M., Di Censo, D., Galante, A., Boscaro, F., Messana, G., Galassi, D.M.P., 2019b. Environmental risk assessment of propranolol in the groundwater bodies of Europe. Environmental Pollution 255, 113189.

Di Lorenzo, T., Murolo, A., Fiasca, B., Tabilio Di Camillo, A., Di Cicco, M., Galassi, D.M.P., 2019c. Potential of a trait-based approach in the characterization of an N-contaminated alluvial aquifer. Water 11, 2553.

Di Lorenzo, T., Borgoni, R., Ambrosini, R., Cifoni, M., Galassi, D.M.P., Petitta, M., M., Petitta, M., 2015. Occurrence of volatile organic compounds in shallow alluvial aquifers of a Mediterranean region: baseline scenario and ecological implications. Science of the Total Environment 538, 712–723.

Di Lorenzo, T., Cifoni, M., Baratti, M., Pieraccini, G., Di Marzio, W.D., Galassi, D.M.P., 2021. Four scenarios of environmental risk of diclofenac in European groundwater ecosystems. Environmental Pollution 287, 117315.

Di Censo, D., Florio, T.M., Rosa, I., Ranieri, B., Scarnati, E., Alecci, M., Galante, A., 2018. A novel, versatile and automated tracking software (TrAQ) for the characterization of rodent behavior. Book of Abstracts, 11th FENS, July 07-11-2018, Berlin, Germany, p. F18eF4686.

Di Marzio, W.D., Cifoni, M., Sáenz, M.E., Galassi, D.M.P., Di Lorenzo, T., 2018. The ecotoxicity of binary mixtures of Imazamox and ionized ammonia on freshwater copepods: implications for environmental risk assessment in groundwater bodies. Ecotoxicology and Environmental Safety 149, 72–79.

ECHA (European Chemicals Agency), 2017. Guidance on Biocidal Products Regulation: Volume IV Environment - Assessment and Evaluation (Parts B+C), ISBN 978-92-9020-151-9. https://echa.europa.eu/documents/10162/23036412/bpr_guidance_ra_vol_iv_part_b-c_en.pdf/e2622aea-0b93-493f-85a3-f9cb42be16ae (Accessed on 19 March 2020).

EEA (European environmental Agency), 2019. Waterbase — Water Quality. Available at: https://www.eea.europa.eu/data-and-maps/data/waterbase-water-quality-2 (Accessed on 19 March 2020).

EMA (European Medicines Agency), 2018a. Guideline on Assessing the Environmental and Human Health Risks of Veterinary Medicinal Products in Groundwater. EMA/CVMP/ERA/103555/2015. In: http://www.ema.europa.eu/docs/en_GB/document_library/Regulatory_and_procedural_guideline/2018/04/WC500248219.pdf (Accessed on 19 March 2020).

EMA (European Medicines Agency), 2018b. Guideline on the Environmental Risks Assessment of Medicinal Products for Human Use, 1. https://www.ema.europa.eu/en/environmental-risk-assessment-medicinal-products-human-use-scientific-guideline (Accessed on 19 March 2020).

Ferreira, R.L., Bernard, E., da Cruz Júnior, F.W., Piló, L.B., Calux, A., Souza-Silva, M., Barlow, J., Pompeu, P.S., Cardoso, P., Frick, W.F., et al., 2022. Brazilian cave heritage under siege. Science 375 (6586), 1238–1239.

Finlayson, C.M., Davies, G.T., Moomaw, W.R., Chmura, G.L., Natali, S.M., Perry, J.E., Roulet, N., Sutton-Grier, A.E., 2019. The second warning to humanity — providing a context for wetland management and policy. Wetlands 39, 1–5.

Gerhardt, A., 2020. Sensitivity towards nitrate: comparison of groundwater versus surface water crustaceans. Journal of Soil and Water Science 4 (1), 112–121.

Gerhardt, A., Badouin, N., Weiler, M., 2020. *In situ* online biomonitoring of groundwater quality using freshwater amphipods exposed to organic fertilizer and rainfall events. Current Topics in Toxicology 16, 13−23.

Fišer, Z., Novak, L., Luštrik, R., Fišer, C., 2016. Light triggers habitat choice of eyeless subterranean but not of eyed surface amphipods. Naturwissenschaften 103 (1−2), 7.

Gerhardt, A., 2019. Plastic additive Bisphenol A: toxicity in surface- and groundwater crustaceans. Journal of Toxicology and Risk Assessment 5 (1), 5017.

Gibert, J., Danielopol, D.L., Stanford, J.A., 1994. Groundwater Ecology. Academic Press.

Ginet, R., Decou, V., 1977. Initiation à la biologie et à l'écologie souterraines. Jean-Pierre Delarge, Paris.

Green, S.M., Romero, A., 1997. Responses to light in two blind cave fishes (*Amblyopsis spelaea* and *Typhlichthys subterraneus*) (Pisces: Amblyopsidae). Environmental Biology of Fishes 50, 167−174.

Green, T.R., Taniguchi, M., Kooi, H., Gurdak, J.J., Allen, D.M., Hiscock, K.M., Treidel, H., Aureli, A., 2011. Beneath the surface of global change: impacts of climate change on groundwater. Journal of Hydrology 405, 532−560.

Griebler, C., Brielmann, H., Haberer, C.M., Kaschuba, S., Kellermann, C., Stumpp, C., Hegler, F., Kuntz, D., Walker-Hertkorn, S., Lueders, T., 2016. Potential impacts of geothermal energy use and storage of heat on groundwater quality, biodiversity, and ecosystem processes. Environmental Earth Sciences 75, 1391.

Grimm, C., Gerhardt, A., 2019. Sensitivity towards copper: comparison of stygal and surface water species' biomonitoring performance in water quality surveillance. International Journal of Scientific Research in Environmental Science and Toxicology 31 (1), 15.

Hervant, F., Renault, D., 2002. Long-term fasting and realimentation in hypogean and epigean isopods: a proposed adaptive strategy for groundwater organisms. Journal of Experimental Biology 205, 2079−2087.

Hervant, F., Mathieu, J., Barre, H., 1999. Comparative study on the metabolic responses of subterranean and surface-dwelling amphipod crustaceans to long-term starvation and subsequent refeeding. Journal of Experimental Biology 202, 3587−3595.

Hervant, F., Mathieu, J., Durand, J.P., 2001. Behavioural, physiological and metabolic responses to long-term starvation and refeeding in a blind cave-dwelling salamander (*Proteus anguinus*) and a facultative cave-dwelling newt (*Euproctus asper*). Journal of Experimental Biology 204, 269−281.

Hervant, F., Mathieu, J., Messana, G., 1998. Oxygen consumption and ventilation in declining oxygen tension and posthypoxic recovery in epigean and hypogean aquatic crustaceans. Journal of Crustacean Biology 18, 717−727.

Hervant, F., Mathieu, J., Garin, D., Freminet, A., 1996. Behavioral, ventilatory, and metabolic responses of the hypogean amphipod *Niphargus virei* and the epigean isopod *Asellus aquaticus* to severe hypoxia and subsequent recovery. Physiological Zoology 69, 1277−1300.

Hervant, F., Mathieu, J., Messana, G., 1997a. Locomotory, ventilatory and metabolic responses of the subterranean *Stenasellus virei* (Crustacea: Isopoda) to severe hypoxia and subsequent recovery. Comptes Rendus de l'Académie des Sciences (Paris), Life Sciences 320, 139−148.

Hervant, F., Mathieu, J., Barre, H., Simon, K., Pinon, C., 1997. Comparative study on the behavioral, ventilatory and respiratory responses of hypogean and epigean crustaceans to long-term starvation and subsequent feeding. Comparative Biochemistry and Physiology 118A, 1277−1283.

Hochachka, P., Somero, G., 2002. Biochemical Adaptation, Mechanism and Physiological Evolution. Oxford University Press, New York.

Hose, G.C., Chariton, A.A., Daam, M.A., Di Lorenzo, T., Galassi, D.M.P., Halse, S.A., Reboleira, A.S.P.S., Robertson, A.L., Schmidt, S.I., Korbel, K.L., 2022. Invertebrate traits, diversity and the vulnerability of groundwater ecosystems. Functional Ecology 38 (9), 2200−2214.

Hose, G.C., Symington, K., Lategan, M.J., Siegele, R., 2019. The toxicity and uptake of As, Cr and Zn in a stygobitic syncarid (Syncarida: Bathynellidae). Water 11 (12), 2508.

Hose, G.C., Symington, K., Lott, M.J., Lategan, M.J., 2016. The toxicity of arsenic (III), chromium(VI) and zinc to groundwater copepods. Environmental Science and Pollution Research 23, 18704−18713.

Iepure, S., 2017. Ice caves fauna. In: Persoiu, A., Lauritzen, S.E. (Eds.), Ice Caves, first ed. Elsevier Science and Technology Book, pp. 163−171.

Iepure, S., Feurdean, A.N., Bădălută, C., Nagavciuc, V., Perşoiu, A., 2016. Pattern of richness and distribution of groundwater Copepoda (Cyclopoida: Harpacticoida) and Ostracoda in Romania: an evolutionary perspective. Biological Journal of the Linnean Society 119, 593−608.

IPCC, 2018. Global Warming of 1.5°C. An IPCC Special Report on the impacts of global warming of 1.5 °C above pre-industrial levels and related global greenhouse gas emission pathways. In: Zhai, V.P., Pörtner, H.-O., Roberts, D.,

Skea, J., Shukla, P.R., Pirani, A., Moufouma-Okia, W., Péan, C., Pidcock, R., Connors, S., Matthews, J.B.R., Chen, Y., Zhou, X., Gomis, M.I., Lonnoy, E., Maycock, T., Tignor, M., Waterfield, T. (Eds.), The Context of Strengthening the Global Response to the Threat of Climate Change, Sustainable Development, and Efforts to Eradicate Poverty [Masson-Delmotte. World Meteorological Organization, Geneva, Switzerland, p. 32.

Issartel, J., Hervant, F., De Fraipont, M., Clobert, J., Voituron, Y., 2009. High anoxia tolerance in the subterranean salamander *Proteus anguinus* without oxidative stress nor activation of antioxidant defenses during reoxygenation. Journal of Comparative Physiology B 179, 543–551.

Issartel, J., Hervant, F., Voituron, Y., Renault, D., Vernon, P., 2005. Behavioural, ventilatory and respiratory responses of epigean and hypogean crustaceans to different temperatures. Comparative Biochemistry and Physiology A 141 (1), 1–7.

Issartel, J., Voituron, Y., Guillaume, O., Clobert, J., Hervant, F., 2010. Selection of physiological and metabolic adaptations to food deprivation in the Pyrenean newt *Calotriton asper* during cave colonisation. Comparative Biochemistry and Physiology A 155, 77–83.

Issartel, J., Voituron, Y., Odagescu, V., Baudot, A., Guillot, G., Ruaud, J.-P., Renault, D., Vernon, P., Hervant, F., 2006. Freezing or supercooling: how does an aquatic subterranean crustacean survive exposures at subzero temperatures? Journal of Experimental Biology 209, 3469–3475.

Kotta, J., Vanhatalo, J., Jänes, H., Orav-Kotta, H., Rugiu, L., Jormalainen, V., Bobsien, I., Viitasalo, M., Virtanen, E., Nyström Sandman, A., Isaeus, M., Leidenberger, S., Jonsson, P.R., Johannesson, K., 2019. Integrating experimental and distribution data to predict future species patterns. Scientific Reports 9 (1), 1821.

Kurwadkar, S., 2019. Occurrence and distribution of organic and inorganic pollutants in groundwater. Water Environment Research 91 (10), 1001–1008.

Lawniczak, M., Romestaing, C., Roussel, D., Maazouzi, C., Renault, D., Hervant, F., 2013. Preventive antioxidant responses to extreme oxygen level fluctuation in a subterranean crustacean. Comparative Biochemistry and Physiology A 165, 299–303.

Li, H., Cooper, R.L., 2002. The effect of ambient light on blind cave crayfish: social interactions. Journal of Crustacean Biology 22 (2), 449–458.

Liang, W.-Q., Xie, M., Tan, Q.-G., 2021. Making the Biotic Ligand Model kinetic, easier to develop, and more flexible for deriving water quality criteria. Water Research 188, 116548.

Loos, R., Locoro, G., Comero, S., Contini, S., Schwesig, D., Werres, F., Balsaa, P., Gans, O., Weiss, S., Blaha, L., Bolchi, M., Gawlik, B.M., 2010. Pan-European survey on the occurrence of selected polar organic persistent pollutants in ground water. Water Research 44 (14), 4115–4126.

Maazouzi, C., Coureau, C., Piscart, C., Saplairoles, M., Baran, N., Marmonier, P., 2016. Individual and joint toxicity of the herbicide S-metolachlor and a metabolite, deethylatrazine on aquatic crustaceans: difference between ecological groups. Chemosphere 165, 118e125.

Malard, F., Hervant, F., 1999. Oxygen supply and the adaptations of animals in groundwater. Freshwater Biology 41, 1–30.

Mammola, S., Pedro, C., Culver, D.C., Deharveng, L., Ferreira, R.L., Fišer, C., Galassi, D.M.P., Griebler, C., Halse, S., Humphreys, W.F., Isaia, M., Malard, F., Martinez, A., Moldovan, O.T., Niemiller, M.L., Pavlek, M., Reboleira, A.S.P.S., Souza-Silva, M., Teeling, E.C., Wynne, J.C., Zagmajster, M., 2019a. Scientists' warning on the conservation of subterranean ecosystems. Bioscience 69 (9), 641–650.

Mammola, S., Piano, E., Cardoso, P., Vernon, P., Domínguez-Villar, D., Culver, D.C., Pipan, T., Isaia, M., 2019b. Climate change going deep: the effects of global climatic alterations on cave ecosystems. The Anthropocene Review 6 (1–2), 98–116.

Mammola, S., Leroy, B., 2018. Applying species distribution models to caves and other subterranean habitats. Ecography 41 (7), 1194–1208.

Manenti, R., Barzaghi, B., 2021. Diel activity of *Niphargus* amphipods in spring habitats. Crustaceana 9 (6), 705–721.

Meffe, R., de Bustamante, I., 2014. Emerging organic contaminants in surface water and groundwater: a first overview of the situation in Italy. Science of the Total Environment 15 (481), 280–295.

Menberg, K., Bayer, P., Zosseder, K., Rumohr, S., Blum, P., 2013. Subsurface urban heat islands in German cities. Science of the Total Environment 44, 123–133.

Mermillod-Blondin, F., Lefour, C., Lalouette, L., Renault, D., Malard, F., Simon, L., Douady, C.J., 2013. Thermal tolerance breadths among groundwater crustaceans living in a thermally constant environment. Journal of Experimental Biology 216 (9), 1683–1694.

Mezek, T., Simčič, T., Arts, M., Brancelj, A., 2010. Effect of fasting on hypogean (*Niphargus stygius*) and epigean (*Gammarus fossarum*) amphipods: a laboratory study. Aquatic Ecology 44, 397–408.

Mintenig, S.M., Löder, M.G.J., Primpke, S., Gerdts, G., 2018. Low numbers of microplastics detected in drinking water from ground water sources. Science of the Total Environment 648, 631–635.

Mösslacher, F., 2000. Sensitivity of groundwater and surface water crustaceans to chemical pollutants and hypoxia: implication for pollution management. Archiv für Hydrobiologie 149 (1), 51–66.

Muñoz, M.M., Bodensteiner, B.L., 2019. Janzen's hypothesis meets the Bogert effect: Connecting climate variation, thermoregulatory behavior, and rates of physiological evolution. Integrative Organismal Biology 1 (1), 1–11.

Nair, P., Huertas, M., Nowlin, W.H., 2020. Metabolic responses to long-term food deprivation in subterranean and surface amphipods. Subterranean Biology 33, 1–15.

Notenboom, J., Boessenkool, J.-J., 1992. Acute toxicity testing with the groundwater copepod *Parastenocaris Germanica* (Crustacea). First International Conference of Ground Water Ecology, U.S. Environmental Protection Agency, American Water Resources Association, Tampa, Florida, pp. 301–309. April 26-29, 1992.

Notenboom, J., Cruys, K., Hoekstra, J., van Beelen, P., 1991. Effect of ambient oxygen concentration upon the acute toxicity of chlorophenols and heavy metals to the groundwater copepod *Parastenocaris germanica* (Crustacea). Ecotoxicology and Environmental Safety 24 (2), 131–143.

Pallarés, S., Sanchez-Hernandez, J.C., Colado, R., Balart-García, P., Comas, J., Sánchez-Fernández, D., 2020. Beyond survival experiments: using biomarkers of oxidative stress and neurotoxicity to assess vulnerability of subterranean fauna to climate change. Conservation Physiology 8 (1), coaa067.

Parmesan, C., 2006. Ecological and evolutionary responses to recent climate change. Annual Review of Ecology, Evolution, and Systematics 37, 637–669.

Piccini, L., Di Lorenzo, T., Costagliola, P., Galassi, D.M.P., 2019. Marble slurry's impact on groundwater: the case study of the Apuan Alps karst aquifers. Water 11, 2462.

Reboleira, A.S., Abrantes, N., Oromì, P., Gonçalves, F., 2013. Acute toxicity of copper sulfate and potassium dichromate on stygobiont *Proasellus*: general aspects of groundwater ecotoxicological and future perspective. Water Air and Soil Pollution 224, 1550.

Retter, A., Karwautz, C., Griebler, C., 2020. Groundwater microbial communities in times of climate change. Current Issues in Molecular Biology 41, 509–538.

Ripple, W.J., Wolf, C., Newsome, T.M., Galetti, M., Alamgir, M., Crist, E., Mahmoud, M.I., Laurance, W.F., and 15,364 scientist signatories from 184 countries, 2017. World scientists' warning to humanity: a second notice. BioScience 67 (12), 1026–1028.

Salin, K., Voituron, Y., Mourin, J., Hervant, F., 2010. Cave colonization without fasting capacities: an example with the fish *Astyanax fasciatus mexicanus*. Comparative Biochemistry and Physiology A 156, 451–457.

Simčič, T., Brancelj, A., 2007. The effect of light on oxygen consumption in two amphipod crustaceans - the hypogean *Niphargus stygius* and the epigean *Gammarus fossarum*. Marine and Freshwater Behaviour and Physiology 40 (2), 141–150.

Simčič, T., Sket, B., 2019. Comparison of some epigean and troglobitic animals regarding their metabolism intensity. Examination of a classical assertion. International Journal of Speleology 48 (2), 2.

Simčič, T., Sket, B., 2021. Ecophysiological responses of two closely related epigean and hypogean *Niphargus* species to hypoxia and increased temperature: Do they differ?. International Journal of Speleology 50 (2), 1.

Strona, G., Fattorini, S., Fiasca, B., Di Lorenzo, T., Di Cicco, M., Lorenzetti, W., Boccacci, F., Galassi, D.M.P., 2019. AQUALIFE software: a new tool for a standardized ecological assessment of groundwater dependent ecosystems. Water 11 (12), 2574.

Taniguchi, M., Uemura, T., Sakura, Y., 2005. Effects of urbanization and groundwater flow on subsurface temperature in three megacities in Japan. Journal of Geophysics and Engineering 2, 320–325.

Taylor, R.G., Scanlon, B., Döll, P., Rodell, M., van Beek, R., Wada, Y., Longuevergne, L., Leblanc, M., Famiglietti, J.S., Edmunds, M., Konikow, L., Green, T.R., Chen, J., Taniguchi, M., Bierkens, M.F.P., MacDonald, A., Fan, Y., Maxwell, R.M., Yechieli, Y., Gurdak, J.J., Allen, D.M., Shamsudduha, M., Hiscock, K., Yeh, P.H.-F., Holman, I., Treidel, H., 2013. Ground water and climate change. Nature Climate Change 3, 322–329.

Taylor, C.A., Stefan, H.G., 2009. Shallow groundwater temperature response to climate change and urbanization. Journal of Hydrology 375, 601–612.

Vandel, A., 1965. Biospeleology. The Biology of Cavernicolous Animals. Pergamon Press.

Veado, M.A.R.V., Arantes, I.A., Oliveira, A.H., Almeida, M.R.M.G., Miguel, R.A., Severo, M.I., Cabaleiro, H.L., 2006. Metal pollution in the environment of Minas Gerais state—Brazil. Environmental Monitoring and Assessment 117 (1—3), 157—172.

Winograd, I., Robertson, F., 1982. Deep oxygenated groundwater. Anomaly or common occurrence? Science 216, 1227—1229.

Voituron, Y., de Frapoint, M., Issartel, J., Guillaume, O., Clobert, J., 2010. Extreme lifespan of the human fish (*Proteus anguinus*): a challenge for ageing mechanisms. Biology Letters 7, 105—107.

Willmer, P., Stone, G., Johnston, I., 2009. Environmental Physiology of Animals. Blackwell Publishing Ltd, UK.

Zhu, K., Grathwohl, P., 2015. Groundwater temperature evolution in the subsurface urban heat island of Cologne, Germany. Hydrological Processes 29 (6), 965—978.

# Biodiversity and ecosystem management in groundwater

# 21

# Global groundwater in the Anthropocene

*Daniel Kretschmer[1], Alexander Wachholz[2] and Robert Reinecke[1]*

[1]Institute of Environmental Science and Geography, University Potsdam, Potsdam, Germany;
[2]Helmholtz Center for Environmental Research (UFZ), Department for Aquatic Ecosystem Analysis and Management, Magdeburg, Germany

## Introduction

Billions of people rely on groundwater as a reliable source of freshwater, accounting for up to 40% of all human water abstractions (Döll et al., 2014). At the same time, it is an essential source for freshwater biota in rivers, lakes, wetlands, and an often overlooked habitat (Ficetola et al., 2019). Despite its importance to the ecosystems, we do not successfully protect this hidden resource (Gleeson et al., 2020b). The depletion of fossil groundwater worldwide, for example, displays clear marks of the Anthropocene.

Groundwater is stored in aquifers, a layer of water carrying permeable material, which can be hundreds of meters deep and extend for hundreds of kilometers. Connected to water bodies like streams and wetlands, groundwater can be a steady freshwater supply even without rain. Aquifers are refilled by a complex process called groundwater recharge driven by precipitation and various other factors. With climate change, the importance of groundwater as an accessible source of drinking water and irrigation will increase (Taylor et al., 2013). As droughts and other extremes become more frequent, longer, and more intense (Chiang et al., 2021), access to a reliable source of freshwater is pivotal. This access is becoming less reliable as groundwater availability is impacted by continued overuse, leading to widespread lowered water tables and depletion (Wada et al., 2012; Konikow, 2015). The overuse of groundwater has a widespread impact on global food security and the health of freshwater ecosystems worldwide (Kløve et al., 2014). Depleted aquifers no longer supply water to surface water bodies or the ocean, a threat, especially in ecological and economically critical periods. Especially groundwater-dependent ecosystems (GDEs), for example in coastal lagoons, can be highly affected (Erostate et al., 2020).

While extraction impacts availability, various factors affect the quality of groundwater. Agriculture, urbanization, and industry are the main driving factors that lead to degraded quality. For example, in the U.S., over 100,000 cancer cases are attributed to contaminants in drinking water alone (McDonough et al., 2020). This degradation of groundwater quality not only affects surface water bodies like rivers and wetlands but may even reach coastal ecosystems through submarine groundwater discharge (SGD; Luijendijk et al., 2020). Humans have already altered terrestrial fluxes with a lasting impact through dams and river regulations that affect flow regimes and, thus, freshwater biota (Biemans et al., 2011). This is also the case for the less visible but essential resource groundwater. In this chapter, we take a global perspective on groundwater and discuss it in three aspects: (1) availability and distribution, (2) sustainability frameworks, and (3) threats to availability and quality. While doing so, we touch on related aspects of interactions to other components in the hydrologic cycle or the impacts of climate change and rising sea levels.

## Groundwater availability and distribution

Groundwater is the largest available freshwater resource on our planet, but its distribution varies. It makes up 97% of the world's available (i.e., nonfrozen) freshwater (Clarke, 1996; Guppy et al., 2018). Only 1.5% of this storage is modern groundwater, replenished in the past 50 years, the remaining 98.5% are old groundwater (see Fig. 21.1; Gleeson et al., 2016).

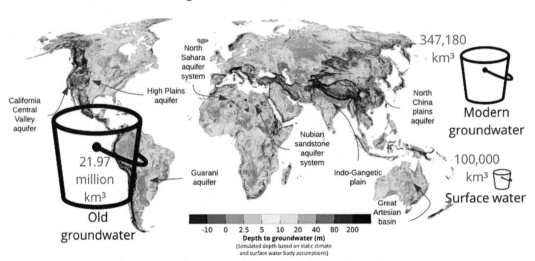

FIGURE 21.1  **Global groundwater abundance and distribution.** Arrows mark the location of major aquifers. The depth to groundwater map is based on a steady-state simulation of Reinecke et al. (2019). Volumes of groundwater and surface water are based on Gleeson et al. (2016), where modern groundwater is considered to be recharged over the past 50 years.

Modern groundwater is the largest freshwater store in the hydrologic cycle, with a volume three times larger than surface water (Fig. 21.1; Gleeson et al., 2016).

Groundwater interacts with various fluxes in the hydrologic cycle. The exchange of water between groundwater and the other, not necessarily fresh or liquid stores of the hydrologic cycle (i.e., atmosphere, surface water, soil, vegetation, snow/ice, oceans) takes part in regulating the Earth's climate system (Gleeson et al., 2020b). Furthermore, groundwater resources help aquatic ecosystems overcome short-term climatic variability impacts (Gleeson et al., 2020b). Shallow groundwater reserves influence over a fifth of the global land area by supplying water to surface water bodies and plants (Fan et al., 2013). Billions of people depend on groundwater, but millions of wells are at risk of running dry (Jasechko and Perrone, 2021).

This section describes the global distribution of groundwater, its interactions with both surface water and the oceans, how groundwater recharge is impacted by climate and highlights the necessity of sustainable groundwater use and management.

## Global distribution

More than 22 million km$^3$ of groundwater is estimated to be stored in the upper 2 km of continental crust (Gleeson et al., 2016), and a comparable amount is likely stored in the deep zone from 2 to 10 km depth (Ferguson et al., 2021). Thus, groundwater in the zone up to 10 km depth makes up more than 10 times the volume of the Mediterranean Sea (Eakins and Sharman, 2010). However, deep storages are far out of reach to be used economically as pumping costs increase with every meter of well depth. Fresh groundwater below the seabed within 200 km off the ocean shoreline is estimated to be one million km$^3$ and thus could be a vital resource for coastal regions (Zamrsky et al., 2022). The exact distribution of groundwater storage across the continents is uncertain, as installed observation networks may cover specific areas but are insufficient for reliable global assessments (Lall et al., 2020). Though climate change and population growth raise new questions on how we can use these vast but highly unequally distributed resources, observation networks, for example, in the Mediterranean, are degrading (Leduc et al., 2017).

Due to its importance as a global resource and compartment that interacts with various water cycle components, complex computer simulations modeling groundwater on a global scale have been developed in recent years (Graaf et al., 2015; Reinecke et al., 2019). Uncertainty is high in all global estimates of groundwater distribution, mainly because (1) global hydrogeological data is missing or uncertain, (2) the amount of groundwater recharge is uncertain (Reinecke et al., 2021), just as the interactions of groundwater with surface water (Reinecke et al., 2019) and (3) the spatial resolution of the used models is coarse. Suggestions have been made on improving the evaluation of such large-scale models (Gleeson et al., 2021), and calls for global groundwater platforms to promote interdisciplinary collaboration have been brought forward (Condon et al., 2021). A typical modeling parameter in large-scale groundwater models is the water table depth below the surface (Fan et al., 2013; Graaf et al., 2015; Reinecke et al., 2019). The map in Fig. 21.1 shows the simulated depth to groundwater table for steady-state conditions if climate and groundwater interactions with surface water balanced out over millions of years and if no anthropogenic impact or abstraction occurred. As groundwater flow follows gravitational gradients to the lowest elevation (Gleeson et al., 2020a), the strong effect of topography is displayed in the figure.

## Interactions

Shallow groundwater influences up to 32% of the global land area by flowing into surface waters and by directly supplying plants with water (Fan et al., 2013). Ecosystems may depend on groundwater flow in areas where groundwater discharges, providing a stable flow into rivers, wetlands, and lakes in dry periods (Gleeson et al., 2020a). Groundwater is an important contributor to streamflow (Fig. 21.2A; Hare et al., 2021; Reinecke et al., 2019). Conditions of groundwater flowing into a stream typically occur in (semi-)humid climates and wet periods, when groundwater levels are closer to the Earth's surface (Bierkens and Wada, 2019) and if the streambed permeability is large enough (Jasechko et al., 2021). On the other hand, streams may also recharge groundwater by losing water at low groundwater levels, likely caused by low groundwater recharge amounts or groundwater pumping (Jasechko et al., 2021). On a global scale, groundwater contribution to streamflow outweighs groundwater recharge from streams by far (Reinecke et al., 2019), but continued pumping of groundwater can decrease this contribution (Fig. 21.2C; Graaf et al., 2019).

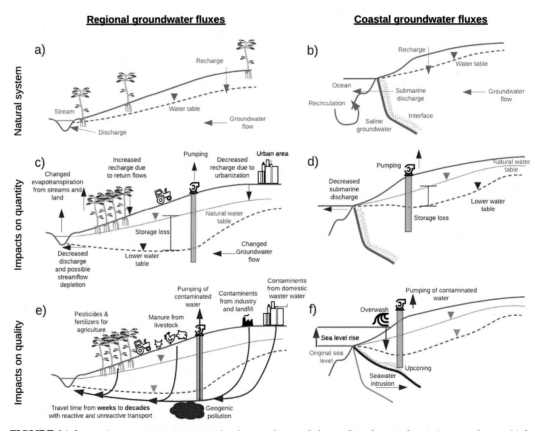

**FIGURE 21.2** Anthropogenic impacts on inland groundwater (left panel) and groundwater in coastal areas (right panel). Natural and impacted systems do not include focused recharge or preferential flow.

Groundwater is not only interacting with streams, wetlands, and lakes but also with the oceans. The exact magnitude of global SGD is highly uncertain (Burnett et al., 2003). Recent modeling efforts show that SGD likely makes up less than 2% of global river discharge to the oceans (Zhou et al., 2019; Luijendijk et al., 2020). Orders of magnitude separate these estimations from model results in an earlier study, estimating SGD into the Atlantic and Indo-Pacific Oceans to be three times greater than river discharge (Kwon et al., 2014). SGD is known to be highly variable along the global coastlines and transports substantial amounts of solutes to the coast (Zhou et al., 2019; Luijendijk et al., 2020). SGD significantly impacts coastal ecosystems by introducing bacteria, nutrients, and other dissolved compounds (Alorda-Kleinglass et al., 2021). As the hydraulic gradient (slope of the water table) determines the flow direction, pumping fresh groundwater in coastal regions can increase the risk of seawater intrusion (see also subsection "Seawater intrusion" and Fig. 21.2F; Burnett et al., 2003; van Camp et al., 2014). Sea level rise may cause ocean water to flow over the beach crests and intrude freshwater aquifers vertically (Befus et al., 2020). Coastal fresh groundwater resources are threatened by both horizontal and vertical (due to overwash) seawater intrusion (see Fig. 21.2F; Post et al., 2018). However, coastal aquifers are more vulnerable to groundwater extraction than sea-level rise (Ferguson and Gleeson, 2012). Lowering the coastal land surface, e.g., due to excessive groundwater extraction, can trigger seawater intrusion and may lead to a loss in the structural support of material in the subsurface (Post et al., 2018).

## Climate

At regional scales, climate is likely the most important driver of water table depth (Fan et al., 2013; Cuthbert et al., 2019a). In observation data, high groundwater recharge rates were found for high amounts of rainfall (Moeck et al., 2020). Not only the amount of rainfall is important. High annual groundwater recharge rates are also associated with high-intensity rainfall and flooding events, potentially making groundwater resilient against climate variability even for years of overall low precipitation (Cuthbert et al., 2019b). Seasonality of rainfall and temperatures influence groundwater recharge as well (Santoni et al., 2018; Moeck et al., 2020). As shown in Fig. 21.3, regions with arid climates (e.g., the Sahara in North Africa) experience low groundwater recharge per precipitated water, while regions with a humid climate (e.g., the Amazon) receive high groundwater recharge per precipitated amount (Reinecke et al., 2021).

High groundwater recharge rates have been observed in the Amazon and Indonesia, while very low groundwater recharge was found, amongst others, in North Africa, Australia, and Pakistan (Mohan et al., 2018). Nonrenewable groundwater resources, i.e., having low groundwater recharge rates, are found in arid climate conditions, where the aquifer reservoir is thick and beneath confining layers (Margat et al., 2006). Groundwater recharge is highly variable. In the Mediterranean, over 90% of groundwater recharge occurs in the northern and less than 10% in eastern and southern countries (Fader et al., 2020). In Africa, the distribution of groundwater storage and groundwater recharge supports water security as regions with low groundwater recharge have high storage and vice versa (MacDonald et al., 2021).

Climate change and the subsequent variability in precipitation are expected to increase the importance of groundwater in future decades (Taylor et al., 2013). Groundwater could act as an essential buffer in periods of water scarcity (Fader et al., 2020), and it is likely to react slower to changing climates than surface water bodies (Gamvroudis et al., 2017). However,

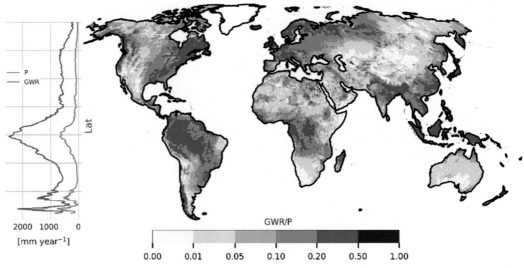

**FIGURE 21.3**   Groundwater recharge ratio (diffuse groundwater recharge (GWR) relative to precipitation (P)) at preindustrial $CO_2$ concentrations based on an ensemble of eight global hydrological models (Reinecke et al., 2021).

shallow groundwater resources can be particularly vulnerable to increased temperatures (Hare et al., 2021). As precipitation and temperature are important for groundwater recharge (Kløve et al., 2014; Moeck et al., 2020), a changing climate can lead to fundamentally altered groundwater recharge rates (National Research Council, 2004). Large groundwater reservoirs located in today's arid regions were recharged under more humid conditions in the past (National Research Council, 2004). While groundwater recharge will be reduced in some world regions, others may experience higher groundwater recharge rates (Reinecke et al., 2021). Groundwater fluxes in humid regions are expected to respond to climate change more strongly than in arid regions (Cuthbert et al., 2019a). However, anthropogenic changes of the land cover, combined with increased pumping and discharge, are expected to change groundwater dynamics more than climate change in the Mediterranean (Gamvroudis et al., 2017; Leduc et al., 2017).

## Sustainable groundwater use and management

Groundwater is distributed unevenly globally, and, in some regions, it might be the only source of freshwater. Sustainable use of groundwater is essential for over two billion people around the globe (Gleeson et al., 2020a; Jasechko and Perrone, 2021). Groundwater sustainability means that storage and high-quality groundwater flow are maintained in the long term (Gleeson et al., 2020a). It depends on physical factors (stable storage and water quality) and a long-term management balancing environmental, economic, and social interests (Gleeson et al., 2020a). The abstraction of renewable and nonrenewable groundwater can be sustainable if the aquifer water table is not persistently lowered (Bierkens and Wada, 2019). Thus, neither the observation of declining groundwater levels over a few years nor the pumping of old groundwater resources is sufficient evidence of nonsustainable groundwater use (Ferguson et al., 2020; Gleeson et al., 2020a).

Estimating the (dynamically) stable range of sustainability and long-term goals is essential to sustainably manage an aquifer, as an existing deficit may be recharged in a future period (Gleeson et al., 2020a). The complex dynamics of groundwater are still difficult to assess, as a lack of observations limits our knowledge and process understanding (Leduc et al., 2017; Lall et al., 2020). Anthropogenic action can alter the natural system of both land and coastal groundwater fluxes and induce severe storage losses (Fig. 21.2C and D; Konikow, 2015; Gleeson et al., 2020a). A drop in formerly long-term stable groundwater levels or deterioration of groundwater quality from anthropogenic action can cause severe damage to subterranean biodiversity and groundwater-dependent ecosystems (Hancock et al., 2005; Boulton, 2009; Ianella et al., 2021). Damaging or disturbing the groundwater ecosystem may cause decreased groundwater quality (Hancock et al., 2005). The sustainability of groundwater is at risk, as nonsustainable groundwater use has been enabled by technical advances and economic improvements (Margat et al., 2006; Bierkens and Wada, 2019). Current trends like population growth and increased living standards accelerate extraction rates and threaten global groundwater resources (Fader et al., 2020; Ashraf et al., 2021).

## Frameworks for sustainable use of groundwater in the Anthropocene

Roughly 1.7 billion people live in regions of nonsustainable groundwater use (Gleeson et al., 2012). Abstractions have led to groundwater quantity and quality reduction worldwide (Aeschbach-Hertig and Gleeson, 2012; Lall et al., 2020). Anthropogenic actions cause the most significant changes to the water cycle, which is vital to provide food and income to billions (Gleeson et al., 2020b). In regions that rely on groundwater, sustainable extraction is essential to sustain human welfare (Gleeson et al., 2020a). Modifications to water cycle components may accumulate to planetary-scale threats by amplifying feedback in interconnected systems (Gleeson et al., 2020b). Alterations at a local scale may trigger large-scale transitions; for example, large amounts of anthropogenic water consumption can induce changes in ecosystems by decreasing water availability (Rockström et al., 2009). On the other hand, global-scale forcing changes may trigger local changes, such as an altered freshwater cycle, which may lower the resilience of ecosystems by decreasing biodiversity (Rockström et al., 2009). In the following, we thus discuss groundwater in the context of two frameworks for sustainability, the United Nations (U.N.) Sustainable Development Goals (SDGs) and the planetary boundary concept.

## Groundwater and its link to the United Nations Sustainable Development Goals

Approaching sustainability globally while developing economically and socially will require action both on global and local scales. The United Nations have defined 17 SDGs to pursue by 2030, addressing human needs with goals like ending hunger and poverty, and calling for action to combat climate change. Each goal is separated into several targets. The underlying agenda is intended to stimulate action protecting our planet from degradation while ensuring economic progress and supporting peaceful societies (United Nations, 2015).

The importance of groundwater supplying water to urban, agricultural, and industrial purposes has increased over the 20th century, making it an essential resource supporting economic development and human livelihoods (Velis et al., 2017; Guppy et al., 2018). Only target 6.6 on water-related ecosystems contains an explicit reference to groundwater or its importance for sustainable development (Guppy et al., 2018). However, direct links exist between ground-water and SDGs of clean water availability and sanitation for all (SDG 6), sustainable consumption and production (SDG 12), and to combat climate change (SDG 13) (Guppy et al., 2018; Velis et al., 2017). Consequently, some SDG targets can have adverse impacts on groundwater, e.g., increased agricultural activities intending to end hunger (SDG 2) may increase ground-water abstractions and deteriorate groundwater quality (Guppy et al., 2018).

## Planetary boundary of freshwater use

The exponential growth of anthropogenic activities caused a discussion about the resilience of the Earth System. As a result, the framework of planetary boundaries, defining a "planetary playing field" for humanity, was introduced by Rockström et al. (2009). The idea is that human activities are intended to stay within a safe operating space, thus maintaining a functioning Earth System and avoiding undesired sudden or irreversible changes. Therefore, the boundaries are set at a certain distance from thresholds, commonly known as tipping points. Crossing these in a subsystem of the Earth could cause a destabilized system in which it is not possible to revert that change in a reasonable time scale. The uncertainty of the defined threshold determines the distance between boundaries and thresholds. For a highly uncertain threshold, the planetary boundary is located further away than for a less uncertain threshold. Planetary boundaries were introduced for nine processes. Among others, they include Climate Change, Biodiversity Loss, and Freshwater Use (Rockström et al., 2009). Ecosystem-related tipping points of groundwater subsystems include (1) supplying stable baseflow to rivers preventing streamflow depletion and (2) delivering moisture to or near the root zone (Gleeson et al., 2020b). Other tipping points related to groundwater are seawater intrusion (Mazi et al., 2013) and land subsidence (Minderhoud et al., 2020).

The planetary boundary for freshwater use allows the assessment of fair water consumption at sub-global scales and compares water management across watersheds or nations (Gleeson et al., 2020b). However, there is no evidence that regional shifts in water cycles caused by freshwater consumption could scale up to global-scale failures. Even if such regional-scale boundaries exist, they cannot be reflected by a planetary boundary of freshwater use (Heistermann, 2017). Further, the simple sum of freshwater use in different regions does not represent the complexity of the underlying processes, and the current global metrics for water are inadequate (Gleeson et al., 2020c). Although groundwater is the largest store of accessible freshwater in the global water cycle (Gleeson et al., 2016), it is rarely addressed in the planetary boundaries framework. Thus, six sub-boundaries for water were proposed to improve the representation of the water boundary (Gleeson et al., 2020c): (1) atmospheric water (for hydroclimatic regulation), (2) atmospheric water (for hydro-ecological regulation), (3) soil moisture, (4) surface water, (5) groundwater, and (6) frozen water.

## Anthropogenic threats to groundwater

Groundwater availability and quality face various threats worldwide (Gleeson et al., 2012; Burri et al., 2019). Fig. 21.2 shows the anthropogenic impacts on inland (left panel) and coastal groundwater (right panel). The natural systems (Fig. 21.2A and B) are disturbed by continued pumping of groundwater through wells (Fig. 21.2C and D), which has led to depleted aquifers globally (Wada et al., 2010). Hazards from sea-level rise—induced by global ice sheet melting due to a warmer climate—include salinization through increased coastal flooding and saltwater intrusion (Fig. 21.2E; Befus et al., 2020). Saltwater intrusion threatens 32% of coastal metropolises (Fig. 21.2F; Cao et al., 2021). Hence, population density strongly impacts groundwater quality at the coast. Additionally, groundwater quality is deteriorated by anthropogenic land development, land use, waste production, and wastewater (Burri et al., 2019). This section highlights the main anthropogenic threats to groundwater by pointing out the drivers and global state of groundwater depletion. It then highlights factors of groundwater quantity reduction in the context of human-induced seawater intrusion and the other causes of anthropogenic groundwater quality deterioration.

## Depletion

Groundwater withdrawal can alter the underground flow regime and may lead to declining groundwater levels if the withdrawal rate is larger than the groundwater recharge rate (Bierkens and Wada, 2019). Groundwater is depleted when a persistent drop in groundwater levels is caused by overexploitation (Margat et al., 2006; Scanlon et al., 2012). As depletion makes water available for anthropogenic use, it may lead to an intermediate social and economic development until the wells run dry (Margat et al., 2006; Gleeson et al., 2020a). Anthropogenic impacts on quantity, i.e., storage loss leading to a lower water table and changed groundwater flow, are shown in Fig. 21.2C and D.

Depletion of groundwater is a demanding issue around the globe (Wada et al., 2010). In arid regions, the climatic conditions limit surface water occurrence and groundwater recharge. The result is a small margin of groundwater extraction to avoid depleting nonrenewable groundwater resources (Döll, 2009; Bierkens and Wada, 2019). Currently, India has the largest groundwater depletion, followed by the USA, Saudi Arabia, and China (Döll et al., 2014). North-western parts of India are the region with the most pronounced depletion worldwide (Aeschbach-Hertig and Gleeson, 2012). An analysis of depletion trends in the High Plains aquifer in the USA (from 1950 to 2007) showed a strongly localized depletion of 35% in 4% of the aquifer area (Scanlon et al., 2012). In the southern Mediterranean, abstraction rates exceed renewable rates by 24% (Fader et al., 2020).

The primary use of pumped groundwater is irrigation (Wada et al., 2014). Estimated global groundwater depletion embedded in food production has increased by about 24% from 2000 to 2010 (Dalin et al., 2017). While depletion could theoretically be reduced by developing and installing highly efficient irrigation systems, in practice, it merely permits expansions of irrigation (Scanlon et al., 2012). Global climate change may increase irrigated agricultural land in arid regions (Fader et al., 2016). Anthropogenic demand for water is the main driver of depletion (e.g., due to population rise and economic developments), but it is limited by the capability or cost of drilling more and deeper wells (Margat et al., 2006; Konikow, 2015; Fader et al., 2020; Ashraf et al., 2021). Many groundwater wells worldwide already have depths beyond 100 m below the surface, and many are not much deeper than the depth to groundwater table (Jasechko and Perrone, 2021).

Most of the pumped water will eventually end up in the ocean, either by runoff or by evaporation followed by precipitation over the sea. Wada et al. (2010) found that only 3% of the extracted water returns to the groundwater storage, 97% of it evaporates and precipitates over the ocean, leading to a substantial addition of 25% (0.8 mm/year) onto the global rate of sea-level rise. While the water quality of coastal aquifers may suffer from seawater intrusion caused by sea-level rise, most coastal aquifers are more vulnerable to groundwater extraction (see subsection "Seawater intrusion"; Ferguson and Gleeson, 2012). Depletion of coastal groundwater can lead to seawater intrusion deteriorating groundwater quality (Werner et al., 2013).

## Seawater intrusion

Seawater intrusion can occur in horizontal and vertical directions (see Fig. 21.2F). Horizontal seawater intrusion is caused by persistent or episodic changes in coastal groundwater levels (Werner et al., 2013). In a natural system, various terrestrial and marine factors influence the mixing of salty ocean water and meteoric groundwater. Among these are groundwater flow toward the ocean, hydrogeological setup of the aquifer, tidal effects, wave forcing, and density differences between seawater and coastal groundwater (Werner et al., 2013; Post et al., 2018). When anthropogenic abstraction rates are larger than the renewal of the coastal groundwater, the water table may drop, and a horizontal landward flow of seawater compensates the residual amount (Fig. 21.2F), potentially leading to contaminations of well locations kilometers away from the shoreline (Post et al., 2018). Upcoming, the upward motion of saltwater into a well, is caused by groundwater wells pumping freshwater in proximity to salty groundwater and can contaminate the pumped water quickly (see Fig. 21.2f; Werner et al., 2013; Post et al., 2018). Another anthropogenic cause of vertical seawater intrusion is land drainage increasing the risk of land subsidence and overwash (see Fig. 21.2F; Post et al., 2018).

High population density and large-scale groundwater-irrigated agriculture increase the risk of seawater intrusion (Post et al., 2018). Globally, about 187 million people live in coastal areas affected by seawater intrusion (van Weert and van der Gun, 2012). High pumping rates cause a high fraction of anthropogenic saltwater intrusion. Alterations to the natural state, such as reduced groundwater recharge and thus decreased groundwater outflow from aquifers, lead to a landward shift of saline groundwater (Oude Essink, 2001). Countermeasures to limit seawater intrusion are manifold, including injection of fresh groundwater at the coast, extraction of contaminated saline coastal groundwater, optimizing pumping rates and locations, reclaiming land, increasing groundwater recharge inland, and creating physical barriers at the coast (Oude Essink, 2001).

As ocean water typically has a density 2.5% larger than freshwater (Oude Essink, 2001), coastal groundwater models implement the density difference between the water masses to improve the representation of seawater intrusion compared to conventional constant density groundwater models (Okuhata et al., 2021). Simplified analytical approximations of the mixing area between groundwater and ocean water assume a sharp boundary, or interface, located at the mean density between the water bodies adjacent to it (Burnett et al., 2003; Werner et al., 2013). Computationally, interface models approximate the location of one or more interfaces in simulations with relatively low computational effort (e.g., in large-scale models). In contrast, variable density models discretizing the transition zone into high-resolution grid cells can approximate the actual salinity distribution (e.g., at a specific site) (Werner et al., 2013).

Sea level rise will have a lower impact on underground seawater intrusion than anthropogenic activities like extraction (Ferguson and Gleeson, 2012). However, coastal surface water deterioration from increasing storm surges, wave attacks, flooding, and increased saltwater intrusion into rivers and estuaries, pose a direct threat to coastal settlements and coastal groundwater quality (Oude Essink, 2001; McGranahan et al., 2007). The population at risk from coastal flooding is projected to increase from 267 million in 2020 to at least 410 million by 2100 (Hooijer and Vernimmen, 2021). Sea water intrusion due to sea level rise is projected to reach far inland in aquifers where the groundwater table is very close to the land surface (Michael et al., 2013). In such topographically-limited aquifers, the hydraulic gradient cannot be maintained when sea levels are rising, and seawater can intrude the aquifer bottom (Michael et al., 2013). Less intrusion is expected in recharge-limited aquifers as the groundwater table can lift in response to the changing conditions, maintaining the hydraulic gradient (Michael et al., 2013). It is critical to limit the intrusion of seawater to a minimum, no matter what the cause is, as a mixture of freshwater with only 1% of seawater can make it unsuitable for drinking water use (Post et al., 2018).

## Global groundwater quality

The quality of groundwater does not only concern abstractions for drinking water or irrigation. Ecosystems in rivers, lakes, and coastal zones receive and depend not only on a certain quantity but also on the quality of groundwater inflow (Murray et al., 2003). Groundwater quality, which refers to its chemical and physical properties, can be deteriorated by various human activities ranging from agricultural and industrial practices to the inevitable handling of domestic wastewater (Burri et al., 2019). This large variety of pollution sources is reflected by an even larger spectrum of pollutants (Burri et al., 2019). Some examples of pollutants from different human activities are given in Fig. 21.4.

Groundwater pollution is often classified according to its origin from point and non-point (or diffuse) sources (Talabi and Kayode, 2019). Point source pollution stems from a small spatial extent, e.g., the discharge pipe of a wastewater treatment plant, and is therefore easily identifiable and manageable, e.g., with technical solutions at the so-called "end-of-pipe." Non-point or diffuse sources, on the other hand, such as nitrate from fertilizers, are distributed across the landscape and, therefore, more difficult to control (Talabi and Kayode, 2019).

Once they enter the groundwater system, pollutants are transported at different velocities (Fig. 21.2E) and different degrees of chemical alteration (Burri et al., 2019). The transport of contaminants through an aquifer to a receiving water body can take days to decades (Lerner and Harris, 2009), which means that pollutants that entered the groundwater system decades ago can still cause harm when they discharge into receiving surface water bodies today. During their transport, biochemical reactions performed by microorganisms can reduce pollutant loads, e.g., nitrogen removal via denitrification (Rivett et al., 2008) or the attenuation of hydrocarbons stemming from oil spills (Scow and Hicks, 2005). Apart from anthropogenic contamination, geogenic contamination is widespread. It occurs in areas in which, e.g., heavy metals are enriched in soils or aquifers due to natural processes and are then mobilized and transported within the hydrologic cycle (Liu et al., 2021). Common geogenic pollutants are cadmium, zinc, and arsenic, threatening plant, human, and animal health (Srivastava et al., 2017). Globally, it was estimated that between 94 and 220 million people are potentially exposed to elevated arsenic concentrations via their domestic water supply (Podgorski and Berg, 2020).

# Contaminant sources

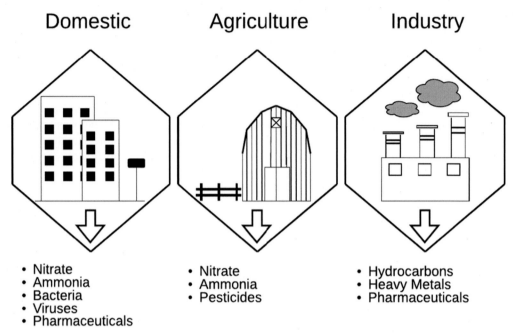

FIGURE 21.4    Exemplary groundwater pollutants and their sources, summarized from Burri et al. (2019).

On the global scale, consistent estimation of the extent of groundwater pollution is hindered by the lack of available data (Gleeson et al., 2020a), which further complicates the assessment of overall available water resources (van Vliet et al., 2021). However, groundwater quality is under increasing pressure from human development and climate change, with grave consequences for humans and ecosystems (World Water Quality Alliance, 2021).

## Outlook

In 2022, the UNESCO has called out the year of groundwater to draw attention to this hidden but vital resource. While groundwater is the largest accessible resource of freshwater that will be essential in adapting to the challenges of climate change, its quantity and quality are in peril. It is yet uncertain to what extent humans have already altered this slow-changing system, but it is evident that past and present overuse and degradation of quality will continue to affect humans and ecosystems alike. Additional research is necessary to investigate multiple aspects of how groundwater is distributed, used, and connected. Current modeling efforts largely ignore the importance of preferential flow, and our knowledge about exchange fluxes between surface water and groundwater is limited. This lack of expertise also extends to how groundwater-dependent ecosystems are affected by quantity and quality changes on large scales.

The Anthropocene has led to changes in Earth's geology, ecosystems, and climate, and its markings are also clearly visible in modifications of the water cycle. Groundwater depletion contributes to rising sea levels, and quality degradation through pollutants leaves a long-lasting legacy. Groundwater is hidden but vital for humans and ecosystems. A continued investigation of the interconnections of groundwater within the water cycle, its quality, and its role for ecosystems on the global scale are necessary to tackle future challenges.

## Glossary

**Aquifer** A layer of (ground-) water carrying permeable materials
**Baseflow** Streamflow that is sustained by groundwater inflow
**GDEs** Groundwater-dependent ecosystems
**Grid cell** Discretization of the spatial domain of a model into parts
**Groundwater** Water that exists beneath the Earth's surface in saturated zones
**Groundwater depletion** A shortage of groundwater supply
**Groundwater recharge** Water moving from the Earth's surface or soil to the groundwater
**Groundwater sustainability** Maintaining long-term, dynamically stable storage and flows of high-quality groundwater using inclusive, equitable, and long-term governance and management
**GWAAEs** Groundwater associated aquatic ecosystems
**Land subsidence** Lowering of the ground due to a lack of support (e.g., due to missing groundwater)
**Meteoric groundwater** Groundwater derived from precipitation
**Modern groundwater** Groundwater that is less than 50 years old
**Nonrenewable groundwater** Groundwater resources that are replenished in periods exceeding a human lifetime (>100 years)
**Nonsustainable groundwater use** Prolonged (multi-annual) withdrawal of groundwater from an aquifer in quantities exceeding average annual replenishment, leading to a persistent decline in groundwater levels and reduction of groundwater volumes
**Permeability** Ability of porous material to let fluids pass through
**Planetary boundaries** The boundaries for humanity if we want to be sure of avoiding major human-induced environmental change on a global scale
**Renewable groundwater** Groundwater resources that are replenished in periods within human time scales (<100 years)
**Seawater Intrusion** The landward incursion of seawater
**Stygofauna** Animal species living in groundwater systems or aquifers
**Submarine Groundwater Discharge** Groundwater flowing from continental margins through the seabed to the coastal ocean
**Submarine Groundwater Recharge** Ocean water flowing from the coastal ocean through the seabed to continental margins
**Water table** Boundary between water-saturated and unsaturated zone underground

## Acknowledgments

We thank C. Griebler and F. Malard for editing an earlier version of this chapter.

## References

Aeschbach-Hertig, W., Gleeson, T., 2012. Regional strategies for the accelerating global problem of groundwater depletion. Nature Geoscience 5, 853–861. https://doi.org/10.1038/ngeo1617.

Alorda-Kleinglass, A., Ruiz-Mallén, I., Diego-Feliu, M., Rodellas, V., Bruach-Menchén, J.M., Garcia-Orellana, J., 2021. The social implications of Submarine Groundwater Discharge from an Ecosystem Services perspective: a systematic review. Earth-Science Reviews 221, 103742. https://doi.org/10.1016/j.earscirev.2021.103742.

Ashraf, S., Nazemi, A., AghaKouchak, A., 2021. Anthropogenic drought dominates groundwater depletion in Iran. Scientific Reports 11, 9135. https://doi.org/10.1038/s41598-021-88522-y.

Befus, K.M., Barnard, P.L., Hoover, D.J., Finzi Hart, J.A., Voss, C.I., 2020. Increasing threat of coastal groundwater hazards from sea-level rise in California. Nature Climate Change 10, 946—952. https://doi.org/10.1038/s41558-020-0874-1.

Biemans, H., Haddeland, I., Kabat, P., Ludwig, F., Hutjes, R.W.A., Heinke, J., Bloh, W von, Gerten, D., 2011. Impact of reservoirs on river discharge and irrigation water supply during the 20th century. Water Resources Research 47, 3. https://doi.org/10.1029/2009WR008929.

Bierkens, M.F.P., Wada, Y., 2019. Non-renewable groundwater use and groundwater depletion: a review. Environmental Research Letters 14, 63002. https://doi.org/10.1088/1748-9326/ab1a5f.

Boulton, A.J., 2009. Recent progress in the conservation of groundwaters and their dependent ecosystems. Aquatic Conservation: Marine and Freshwater Ecosystems 19, 7. https://doi.org/10.1002/aqc.1073.

Burnett, W.C., Bokuniewicz, H., Huettel, M., Moore, W.S., Taniguchi, M., 2003. Groundwater and pore water inputs to the coastal zone. Biogeochemistry 66, 3—33. https://doi.org/10.1023/B:BIOG.0000006066.21240.53.

Burri, N.M., Weatherl, R., Moeck, C., Schirmer, M., 2019. A review of threats to groundwater quality in the anthropocene. Science of the Total Environment 684, 136—154. https://doi.org/10.1016/j.scitotenv.2019.05.236.

Cao, T., Han, D., Song, X., 2021. Past, present, and future of global seawater intrusion research: a bibliometric analysis. Journal of Hydrology 603, 126844. https://doi.org/10.1016/j.jhydrol.2021.126844.

Chiang, F., Mazdiyasni, O., AghaKouchak, A., 2021. Evidence of anthropogenic impacts on global drought frequency, duration, and intensity. Nature Communications 12, 2754. https://doi.org/10.1038/s41467-021-22314-w.

Clarke, R., 1996. Groundwater: A Threatened Resource. UNEP Environment Library No 15. UNEP, United Nations Environment Programme, Nairobi, Kenya. https://www.ircwash.org/sites/default/files/212.0-96GR-13834.pdf. (Accessed 15 March 2022).

Condon, L.E., Kollet, S., Bierkens, M.F., Fogg, G.E., Maxwell, R.M., Hill, M.C., Hendricks Fransen, H.-J., Verhoef, A., van Loon, A.F., Sulis, M., Abesser, C., 2021. Global groundwater modeling and monitoring: opportunities and challenges. Water Resources Research 57, 12. https://doi.org/10.1029/2020WR029500.

Cuthbert, M.O., Gleeson, T., Moosdorf, N., Befus, K.M., Schneider, A., Hartmann, J., Lehner, B., 2019a. Global patterns and dynamics of climate—groundwater interactions. Nature Climate Change 9, 137—141. https://doi.org/10.1038/s41558-018-0386-4.

Cuthbert, M.O., Taylor, R.G., Favreau, G., Todd, M.C., Shamsudduha, M., Villholth, K.G., MacDonald, A.M., Scanlon, B.R., Kotchoni, D.O.V., Vouillamoz, J.-M., Lawson, F.M.A., Adjomayi, P.A., Kashaigili, J., Seddon, D., Sorensen, J.P.R., Ebrahim, G.Y., Owor, M., Nyenje, P.M., Nazoumou, Y., Goni, I., Ousmane, B.I., Sibanda, T., Ascott, M.J., Macdonald, D.M.J., Agyekum, W., Koussoubé, Y., Wanke, H., Kim, H., Wada, Y., Lo, M.-H., Oki, T., Kukuric, N., 2019b. Observed controls on resilience of groundwater to climate variability in sub-Saharan Africa. Nature 572, 230—234. https://doi.org/10.1038/s41586-019-1441-7.

Dalin, C., Wada, Y., Kastner, T., Puma, M.J., 2017. Groundwater depletion embedded in international food trade. Nature 543, 700—704. https://doi.org/10.1038/nature21403.

Döll, P., 2009. Vulnerability to the impact of climate change on renewable groundwater resources: a global-scale assessment. Environmental Research Letters 4, 3. https://doi.org/10.1088/1748-9326/4/3/035006.

Döll, P., Müller-Schmied, H., Schuh, C., Portmann, F.T., Eicker, A., 2014. Global-scale assessment of groundwater depletion and related groundwater abstractions: combining hydrological modeling with information from well observations and GRACE satellites. Water Resources Research 50, 5698—5720. https://doi.org/10.1002/2014WR015595.

Eakins, B.W., Sharman, G.F., 2010. Volumes of the World's Oceans from ETOPO1. https://ngdc.noaa.gov/mgg/global/etopo1_ocean_volumes.html. (Accessed 15 March 2022).

Erostate, M., Huneau, F., Garel, E., Ghiotti, S., Vystavna, Y., Garrido, M., Pasqualini, V., 2020. Groundwater dependent ecosystems in coastal Mediterranean regions: characterization, challenges and management for their protection. Water Research 172, 115461. https://doi.org/10.1016/j.watres.2019.115461.

Fader, M., Giupponi, C., Burak, S., Dakhlaoui, H., Koutroulis, A., Lange, M.A., Llasat, M.C., Pulido-Velazquez, D., Sanz-Cobena, A., 2020. Water. In: Cramer, W., Guiot, J., Marini, K. (Eds.), Climate and Environmental Change in the Mediterranean Basin - Current Situation and Risks for the Future. First Mediterranean Assessment Report. Union for the Mediterranean, Plan Bleu, UNEP/MAP, Marseille, France, pp. 182—235.

Fader, M., Shi, S., Bloh, W von, Bondeau, A., Cramer, W., 2016. Mediterranean irrigation under climate change: more efficient irrigation needed to compensate for increases in irrigation water requirements. Hydrology and Earth System Sciences 20, 953—973. https://doi.org/10.5194/hess-20-953-2016.

Fan, Y., Li, H., Miguez-Macho, G., 2013. Global patterns of groundwater table depth. Science 339, 940–943. https://doi.org/10.1126/science.1229881.

Ferguson, G., Gleeson, T., 2012. Vulnerability of coastal aquifers to groundwater use and climate change. Nature Climate Change 2, 342–345. https://doi.org/10.1038/nclimate1413.

Ferguson, G., Cuthbert, M.O., Befus, K., Gleeson, T., McIntosh, J.C., 2020. Rethinking groundwater age. Nature Geoscience 13, 592–594. https://doi.org/10.1038/s41561-020-0629-7.

Ferguson, G., McIntosh, J.C., Warr, O., Sherwood Lollar, B., Ballentine, C.J., Famiglietti, J.S., Kim, J.-H., Michalski, J.R., Mustard, J.F., Tarnas, J., McDonnell, J.J., 2021. Crustal groundwater volumes greater than previously thought. Geophysical Research Letters 48, 16. https://doi.org/10.1029/2021GL093549.

Ficetola, G.F., Canedoli, C., Stoch, F., 2019. The Racovitzan impediment and the hidden biodiversity of unexplored environments. Conservation Biology 33, 214–216. https://doi.org/10.1111/cobi.13179.

Gamvroudis, C., Dokou, Z., Nikolaidis, N.P., Karatzas, G.P., 2017. Impacts of surface and groundwater variability response to future climate change scenarios in a large Mediterranean watershed. Environmental Earth Sciences 76, 385. https://doi.org/10.1007/s12665-017-6721-7.

Guppy, L., Uyttendaele, P., Villholth, K.G., 2018. Groundwater and Sustainable Development Goals: Analysis of Interlinkages. United Nations University - Institute for Water, Hamilton, ON, CA. https://inweh.unu.edu/wp-content/uploads/2018/12/Groundwater-and-Sustainable-Development-Goals-Analysis-of-Interlinkages.pdf. (Accessed 15 March 2022).

Gleeson, T., Befus, K.M., Jasechko, S., Luijendijk, E., Cardenas, M.B., 2016. The global volume and distribution of modern groundwater. Nature Geoscience 9, 161–167. https://doi.org/10.1038/ngeo2590.

Gleeson, T., Cuthbert, M., Ferguson, G., Perrone, D., 2020a. Global groundwater sustainability, resources, and systems in the anthropocene. Annual Review of Earth and Planetary Sciences 48, 431–463. https://doi.org/10.1146/annurev-earth-071719-055251.

Gleeson, T., Wada, Y., Bierkens, M.F.P., van Beek, L.P.H., 2012. Water balance of global aquifers revealed by groundwater footprint. Nature 488, 197–200. https://doi.org/10.1038/nature11295.

Gleeson, T., Wang-Erlandsson, L., Porkka, M., Zipper, S.C., Jaramillo, F., Gerten, D., Fetzer, I., Cornell, S.E., Piemontese, L., Gordon, L.J., Rockström, J., Oki, T., Sivapalan, M., Wada, Y., Brauman, K.A., Flörke, M., Bierkens, M.F.P., Lehner, B., Keys, P., Kummu, M., Wagener, T., Dadson, S., Troy, T.J., Steffen, W., Falkenmark, M., Famiglietti, J.S., 2020b. Illuminating water cycle modifications and Earth system resilience in the Anthropocene. Water Resources Research 56, 4. https://doi.org/10.1029/2019WR024957.

Gleeson, T., Wang-Erlandsson, L., Zipper, S.C., Porkka, M., Jaramillo, F., Gerten, D., Fetzer, I., Cornell, S.E., Piemontese, L., Gordon, L.J., Rockström, J., Oki, T., Sivapalan, M., Wada, Y., Brauman, K.A., Flörke, M., Bierkens, M.F., Lehner, B., Keys, P., Kummu, M., Wagener, T., Dadson, S., Troy, T.J., Steffen, W., Falkenmark, M., Famiglietti, J.S., 2020c. The water planetary boundary: interrogation and revision. One Earth 2, 223–234. https://doi.org/10.1016/j.oneear.2020.02.009.

Gleeson, T., Wagener, T., Döll, P., Zipper, S.C., West, C., Wada, Y., Taylor, R., Scanlon, B., Rosolem, R., Rahman, S., Oshinlaja, N., Maxwell, R., Lo, M.-H., Kim, H., Hill, M., Hartmann, A., Fogg, G., Famiglietti, J.S., Ducharne, A., Graaf, I de, Cuthbert, M., Condon, L., Bresciani, E., Bierkens, M.F.P., 2021. GMD perspective: the quest to improve the evaluation of groundwater representation in continental- to global-scale models. Geoscientific Model Development 14, 7545–7571. https://doi.org/10.5194/gmd-14-7545-2021.

Graaf, IEM de, Sutanudjaja, E.H., van Beek, L.P.H., Bierkens, M.F.P., 2015. A high-resolution global-scale groundwater model. Hydrology and Earth System Sciences 19, 823–837. https://doi.org/10.5194/hess-19-823-2015.

Graaf, IEM de, Gleeson, T., van Rens Beek, L.P.H., Sutanudjaja, E.H., Bierkens, M.F.P., 2019. Environmental flow limits to global groundwater pumping. Nature 574, 90–94. https://doi.org/10.1038/s41586-019-1594-4.

Hancock, P.J., Boulton, A.J., Humphreys, W.F., 2005. Aquifers and hyporheic zones: towards an ecological understanding of groundwater. Hydrogeology Journal 13, 1. https://doi.org/10.1007/s10040-004-0421-6.

Hare, D.K., Helton, A.M., Johnson, Z.C., Lane, J.W., Briggs, M.A., 2021. Continental-scale analysis of shallow and deep groundwater contributions to streams. Nature Communications 12, 1450. https://doi.org/10.1038/s41467-021-21651-0.

Heistermann, M., 2017. HESS Opinions: a planetary boundary on freshwater use is misleading. Hydrology and Earth System Sciences 21, 3455–3461. https://doi.org/10.5194/hess-21-3455-2017.

Hooijer, A., Vernimmen, R., 2021. Global LiDAR land elevation data reveal greatest sea-level rise vulnerability in the tropics. Nature Communications 12, 3592. https://doi.org/10.1038/s41467-021-23810-9.

Iannella, M., Fiasca, B., Di Lorenzo, T., Di Cicco, M., Biondi, M., Mammola, S., Galassi, D.M.P., 2021. Getting the 'most out of the hotspot' for practical conservation of groundwater biodiversity. Global Ecology and Conservation 31, e01844. https://doi.org/10.1016/j.gecco.2021.e01844.

Jasechko, S., Perrone, D., 2021. Global groundwater wells at risk of running dry. Science 372, 418—421. https://doi.org/10.1126/science.abc2755.

Jasechko, S., Seybold, H., Perrone, D., Fan, Y., Kirchner, J.W., 2021. Widespread potential loss of streamflow into underlying aquifers across the USA. Nature 591, 391—395. https://doi.org/10.1038/s41586-021-03311-x.

Kløve, B., Ala-Aho, P., Bertrand, G., Gurdak, J.J., Kupfersberger, H., Kværner, J., Muotka, T., Mykrä, H., Preda, E., Rossi, P., Uvo, C.B., Velasco, E., Pulido-Velazquez, M., 2014. Climate change impacts on groundwater and dependent ecosystems. Journal of Hydrology 518, 250—266. https://doi.org/10.1016/j.jhydrol.2013.06.037.

Konikow, L.F., 2015. Long-term groundwater depletion in the United States. Ground Water 53, 2—9. https://doi.org/10.1111/gwat.12306.

Kwon, E.Y., Kim, G., Primeau, F., Moore, W.S., Cho, H.M., DeVries, T., Sarmiento, J.L., Charette, M.A., Cho, Y.K., 2014. Global estimate of submarine groundwater discharge based on an observationally constrained radium isotope model. Geophysical Research Letters 41, 8438—8444. https://doi.org/10.1002/2014GL061574.

Lall, U., Josset, L., Russo, T., 2020. A snapshot of the world's groundwater challenges. Annual Review of Environment and Resources 45, 171—194. https://doi.org/10.1146/annurev-environ-102017-025800.

Leduc, C., Pulido-Bosch, A., Remini, B., 2017. Anthropization of groundwater resources in the Mediterranean region: processes and challenges. Hydrogeology Journal 25, 1529—1547. https://doi.org/10.1007/s10040-017-1572-6.

Lerner, D.N., Harris, B., 2009. The relationship between land use and groundwater resources and quality. Land Use Policy 26, 265—273. https://doi.org/10.1016/j.landusepol.2009.09.005.

Liu, Y., Xiao, T., Zhu, Z., Ma, L., Li, H., Ning, Z., 2021. Geogenic pollution, fractionation and potential risks of Cd and Zn in soils from a mountainous region underlain by black shale. Science of the Total Environment 760, 143426. https://doi.org/10.1016/j.scitotenv.2020.143426.

Luijendijk, E., Gleeson, T., Moosdorf, N., 2020. Fresh groundwater discharge insignificant for the world's oceans but important for coastal ecosystems. Nature Communications 11, 1260. https://doi.org/10.1038/s41467-020-15064-8.

MacDonald, A.M., Lark, R.M., Taylor, R.G., Abiye, T., Fallas, H.C., Favreau, G., Goni, I.B., Kebede, S., Scanlon, B., Sorensen, J.P.R., Tijani, M., Upton, K.A., West, C., 2021. Mapping groundwater recharge in Africa from ground observations and implications for water security. Environmental Research Letters 16, 34012. https://doi.org/10.1088/1748-9326/abd661.

Margat, J., Foster, S., Droubi, A., 2006. Concept and importance of non-renewable resources. In: Foster, S., Loucks, D.P. (Eds.), Non-Renewable Groundwater Resources: A Guidebook on Socially-Sustainable Management for Water-Policy Makers. UNSESCO, Paris, pp. 13—24. https://unesdoc.unesco.org/ark:/48223/pf0000146997. (Accessed 15 March 2022).

Mazi, K., Koussis, A.D., Destouni, G., 2013. Tipping points for seawater intrusion in coastal aquifers under rising sea level. Environmental Research Letters 8, 14001. https://doi.org/10.1088/1748-9326/8/1/014001.

McDonough, L.K., Santos, I.R., Andersen, M.S., O'Carroll, D.M., Rutlidge, H., Meredith, K., Oudone, P., Bridgeman, J., Gooddy, D.C., Sorensen, J.P.R., Lapworth, D.J., MacDonald, A.M., Ward, J., Baker, A., 2020. Changes in global groundwater organic carbon driven by climate change and urbanization. Nature Communications 11, 1279. https://doi.org/10.1038/s41467-020-14946-1.

McGranahan, G., Balk, D., Anderson, B., 2007. The rising tide: assessing the risks of climate change and human settlements in low elevation coastal zones. Environment and Urbanization 19, 17—37. https://doi.org/10.1177/0956247807076960.

Michael, H.A., Russoniello, C.J., Byron, L.A., 2013. Global assessment of vulnerability to sea-level rise in topography-limited and recharge-limited coastal groundwater systems. Water Resources Research 49, 2228—2240. https://doi.org/10.1002/wrcr.20213.

Minderhoud, P.S.J., Middelkoop, H., Erkens, G., Stouthamer, E., 2020. Groundwater extraction may drown mega-delta: projections of extraction-induced subsidence and elevation of the Mekong delta for the 21st century. Environmental Research Communications 2, 11005. https://doi.org/10.1088/2515-7620/ab5e21.

Moeck, C., Grech-Cumbo, N., Podgorski, J., Bretzler, A., Gurdak, J.J., Berg, M., Schirmer, M., 2020. A global-scale dataset of direct natural groundwater recharge rates: a review of variables, processes and relationships. Science of the Total Environment 717, 137042. https://doi.org/10.1016/j.scitotenv.2020.137042.

Mohan, C., Western, A.W., Wei, Y., Saft, M., 2018. Predicting groundwater recharge for varying land cover and climate conditions — a global meta-study. Hydrology and Earth System Sciences 22, 2689—2703. https://doi.org/10.5194/hess-22-2689-2018.

Murray, B.B.R., Zeppel, M.J.B., Hose, G.C., Eamus, D., 2003. Groundwater-dependent ecosystems in Australia: it's more than just water for rivers. Ecological Management and Restoration 4, 110—113. https://doi.org/10.1046/j.1442-8903.2003.00144.x.

National Research Council, 2004. Groundwater Fluxes across Interfaces. The National Academies Press, Washington, D.C. https://www.nap.edu/catalog/10891/groundwater-fluxes-across-interfaces. (Accessed 15 March 2022).

Okuhata, B.K., El-Kadi, A.I., Dulai, H., Lee, J., Wada, C.A., Bremer, L.L., Burnett, K.M., Delevaux, J.M.S., Shuler, C.K., 2021. A density-dependent multi-species model to assess groundwater flow and nutrient transport in the coastal Keauhou aquifer, Hawai'i, USA. Hydrogeology Journal 30, 231—250. https://doi.org/10.1007/s10040-021-02407-y.

Oude Essink, G.H., 2001. Improving fresh groundwater supply—problems and solutions. Ocean and Coastal Management 44, 429—449. https://doi.org/10.1016/S0964-5691(01)00057-6.

Podgorski, J., Berg, M., 2020. Global threat of arsenic in groundwater. Science 368, 845—850. https://doi.org/10.1126/science.aba1510.

Post, V.E.A., Eichholz, M., Brentführer, R., 2018. Groundwater Management in Coastal Zones. Bundesanstalt für Geowissenschaften und Rohstoffe (BGR), Hannover, Germany. https://www.bgr.bund.de/EN/Themen/Wasser/Produkte/Downloads/groundwater_management_in_coastal_zones.pdf?__blob=publicationFile&v=3. (Accessed 15 March 2022).

Reinecke, R., Foglia, L., Mehl, S., Trautmann, T., Cáceres, D., Döll, P., 2019. Challenges in developing a global gradient-based groundwater model ($G^3M$ v1.0) for the integration into a global hydrological model. Geoscientific Model Development 12, 2401—2418. https://doi.org/10.5194/gmd-12-2401-2019.

Reinecke, R., Müller-Schmied, H., Trautmann, T., Andersen, L.S., Burek, P., Flörke, M., Gosling, S.N., Grillakis, M., Hanasaki, N., Koutroulis, A., Pokhrel, Y., Thiery, W., Wada, Y., Yusuke, S., Döll, P., 2021. Uncertainty of simulated groundwater recharge at different global warming levels: a global-scale multi-model ensemble study. Hydrology and Earth System Sciences 25, 787—810. https://doi.org/10.5194/hess-25-787-2021.

Rivett, M., Buss, S., Morgan, P., Smith, J.W.N., Bemment, C.D., 2008. Nitrate attenuation in groundwater: a review of biogeochemical controlling processes. Water Research 16, 4215—4232. https://doi.org/10.1016/j.watres.2008.07.020.

Rockström, J., Steffen, W., Noone, K., Persson, Å., Chapin, F.S.I., Lambin, E., Lenton, T.M., Scheffer, M., Folke, C., Schellnhuber, H.J., Nykvist, B., de Wit, C.A., Hughes, T., van der Leeuw, S., Rodhe, H., Sörlin, S., Snyder, P.K., Costanza, R., Svedin, U., Falkenmark, M., Karlberg, L., Corell, R.W., Fabry, V.J., Hansen, J., Walker, B., Liverman, D., Richardson, K., Crutzen, P., Foley, J., 2009. Planetary boundaries: exploring the safe operating space for humanity. Ecology and Society 14 (2), 32. https://doi.org/10.5751/ES-03180-140232.

Santoni, S., Huneau, F., Garel, E., Celle-Jeanton, H., 2018. Multiple recharge processes to heterogeneous Mediterranean coastal aquifers and implications on recharge rates evolution in time. Journal of Hydrology 559, 669—683. https://doi.org/10.1016/j.jhydrol.2018.02.068.

Scanlon, B.R., Faunt, C.C., Longuevergne, L., Reedy, R.C., Alley, W.M., McGuire, V.L., McMahon, P.B., 2012. Groundwater depletion and sustainability of irrigation in the U.S. High Plains and central valley. Proceedings of the National Academy of Sciences of the United States of America 109, 9320—9325. https://doi.org/10.1073/pnas.1200311109.

Scow, K., Hicks, K., 2005. Natural attenuation and enhanced bioremediation of organic contaminants in groundwater. Current Opinion in Biotechnology 16, 246—253. https://doi.org/10.1016/J.COPBIO.2005.03.009.

Srivastava, V., Sarkar, A., Singh, S., Singh, P., Araujo, ASF de, Singh, R.P., 2017. Agroecological responses of heavy metal pollution with special emphasis on soil health and plant performances. Frontiers in Environmental Science 5, 64. https://doi.org/10.3389/fenvs.2017.00064.

Talabi, A.O., Kayode, T.J., 2019. Groundwater pollution and remediation. Journal of Water Resource and Protection 11, 1. https://doi.org/10.4236/jwarp.2019.111001.

Taylor, R.G., Scanlon, B., Döll, P., Rodell, M., van Beek, R., Wada, Y., Longuevergne, L., Leblanc, M., Famiglietti, J.S., Edmunds, M., Konikow, L., Green, T.R., Chen, J., Taniguchi, M., Bierkens, M.F.P., MacDonald, A., Fan, Y., Maxwell, R.M., Yechieli, Y., Gurdak, J.J., Allen, D.M., Shamsudduha, M., Hiscock, K., Yeh, P.J.-F., Holman, I.,

Treidel, H., 2013. Ground water and climate change. Nature Climate Change 3, 322−329. https://doi.org/10.1038/nclimate1744.

United Nations, 2015. Transforming Our World: The 2030 Agenda for Sustainable Development − A/RES/70/1. United Nations. https://sdgs.un.org/2030agenda. (Accessed 15 March 2022).

van Camp, M., Mtoni, Y., Mjemah, I.C., Bakundukize, C., Walraevens, K., 2014. Investigating seawater intrusion due to groundwater pumping with schematic model simulations: the example of the Dar es Salaam coastal aquifer in Tanzania. Journal of African Earth Sciences 96, 71−78. https://doi.org/10.1016/j.jafrearsci.2014.02.012.

van Vliet, M.T.H., Jones, E.R., Flörke, M., Franssen, W.H.P., Hanasaki, N., Wada, Y., Yearsley, J.R., 2021. Global water scarcity including surface water quality and expansions of clean water technologies. Environmental Research Letters 16, 24020. https://doi.org/10.1088/1748-9326/abbfc3.

van Weert, F., van der Gun, J., 2012. Saline and Brackish Groundwater at Shallow and Intermediate Depths: Genesis and World-wide Occurrence. International Groundwater Resources Assessment Centre, Delft The Netherlands. http://www.un-igrac.org/sites/default/files/resources/files/van%20Weert%20and%20van%20der%20Gun,%20IAH%202012%20Congress.pdf. (Accessed 15 March 2022).

Velis, M., Conti, K.I., Biermann, F., 2017. Groundwater and human development: synergies and trade-offs within the context of the sustainable development goals. Sustainability Science 12, 1007−1017. https://doi.org/10.1007/s11625-017-0490-9.

Wada, Y., van Beek, L.P.H., van Kempen, C.M., Reckman, J.W.T.M., Vasak, S., Bierkens, M.F.P., 2010. Global depletion of groundwater resources. Geophysical Research Letters 37, L20402. https://doi.org/10.1029/2010GL044571.

Wada, Y., van Beek, L.P.H., Sperna Weiland, F.C., Chao, B.F., Wu, Y.-H., Bierkens, M.F.P., 2012. Past and future contribution of global groundwater depletion to sea-level rise. Geophysical Research Letters 39, L09402. https://doi.org/10.1029/2012GL051230.

Wada, Y., Wisser, D., Bierkens, M.F.P., 2014. Global modeling of withdrawal, allocation and consumptive use of surface water and groundwater resources. Earth System Dynamics 5, 15−40. https://doi.org/10.5194/esd-5-15-2014.

Werner, A.D., Bakker, M., Post, V.E., Vandenbohede, A., Lu, C., Ataie-Ashtiani, B., Simmons, C.T., Barry, D.A., 2013. Seawater intrusion processes, investigation and management: recent advances and future challenges. Advances in Water Resources 51, 3−26. https://doi.org/10.1016/j.advwatres.2012.03.004.

World Water Quality Alliance, 2021. Assessing groundwater quality: a global perspective: importance, methods and potential data sources. In: Assessing Groundwater Quality: A Global Perspective. A report by the Friends of Groundwater in the World Water Quality Alliance. Information Document Annex for display at the 5th Session of the United Nations Environment Assembly, Nairobi 2021. https://groundwater-quality.org/sites/default/files/2021-01/Assessing%20Groundwater%20Quality_A%20Global%20Perspective.pdf. (Accessed 15 March 2022).

Zamrsky, D., Essink, G.H.P.O., Sutanudjaja, E.H., van Beek, L.P.H., Bierkens, M.F.P., 2022. Offshore fresh groundwater in coastal unconsolidated sediment systems as a potential fresh water source in the 21st century. Environmental Research Letters 17, 14021. https://doi.org/10.1088/1748-9326/ac4073.

Zhou, Y., Sawyer, A.H., David, C.H., Famiglietti, J.S., 2019. Fresh submarine groundwater discharge to the near-global coast. Geophysical Research Letters 46, 5855−5863. https://doi.org/10.1029/2019GL082749.

# Assessing groundwater ecosystem health, status, and services

Grant C. Hose[1], Tiziana Di Lorenzo[2,3,4], Lucas Fillinger[5],
Diana Maria Paola Galassi[6], Christian Griebler[5],
Hans Juergen Hahn[7], Kim M. Handley[8], Kathryn Korbel[1],
Ana Sofia Reboleira[4,9], Tobias Siemensmeyer[7],
Cornelia Spengler[7], Louise Weaver[10] and
Alexander Weigand[11]

[1]School of Natural Sciences, Macquarie University, Sydney, Australia; [2]Research Institute on
Terrestrial Ecosystems of the National Research Council of Italy (IRET-CNR), Florence, Italy;
[3]Emil Racovita Institute of Speleology, Cluj-Napoca, Romania; [4]Center for Ecology, Evolution
and Environmental Changes (cE3c), Departamento de Biologia Animal, Faculdade de Ciências,
Universidade de Lisboa, Lisbon Portugal; [5]University of Vienna, Department of Functional &
Evolutionary Ecology, Vienna, Austria; [6]Department of Life, Health and Environmental
Sciences, University of L'Aquila, L'Aquila, Italy; [7]Institute for Environmental Sciences,
University of Koblenz-Landau, Landau, Germany; [8]School of Biological Sciences, The
University of Auckland, Auckland, New Zealand; [9]Natural History Museum of Denmark,
University of Copenhagen, Copenhagen, Denmark; [10]Institute of Environmental Science and
Research (ESR) Christchurch, Canterbury, New Zealand; [11]National Museum of Natural
History Luxembourg, Luxembourg

## Introduction

The notion of ecosystem health has become well established as a means of communicating
the status or condition of a natural environment (or part thereof) to a diverse audience. The
concept is embedded in environmental policies globally, where the objectives are frequently

to achieve or maintain good health or status. In this context, ecosystem health is considered analogous to human health, with a healthy ecosystem being one that is free from stress or disease, with its component parts present and functioning appropriately (Karr, 1999).

The concept of ecosystem health has for decades underpinned river assessments, with major programs such as the River Invertebrate Prediction and Classification System (RIVPACS) in the UK and the River Assessment System (AUSRIVAS) in Australia (Wright et al., 2000). Building on the conceptual platform that river health provided (Vugteveen et al., 2006), Korbel and Hose (2011, 2017) defined ecosystem health specifically for groundwaters as *"an expression of an aquifer's ability to sustain its ecological functioning (vigor and resilience) in accordance with its organisation while maintaining the provision of ecosystem goods and services."* This was the first time that the concept of ecosystem health had been explicitly defined for subterranean systems, however, just as the concept evolved for river management, the application to groundwaters was a natural progression from growing interest in the use of bioindicators and ecosystem assessment in groundwater since the mid-1990s (e.g., Malard et al., 1996; Mösslacher and Notenboom 1999; Steube et al., 2009; Griebler et al., 2010).

Inherent in definitions of ecosystem health is that a healthy ecosystem should be providing ecosystem goods and services (collectively "ecosystem services") that are of benefit to humans (Karr, 1999). The ecosystem services provided by groundwater organisms are increasingly being recognized and valued (Griebler and Avramov, 2015; Griebler et al., 2019, Fillinger et al., Mermillod-Blondin et al., this volume). Microbes improve water quality by breaking down organic compounds and have been used widely in aquifer remediation (Griebler et al., 2019). The ecological function and services provided by stygofauna have until recently been based largely on perceived parallels to the activities of invertebrates in surface aquatic sediments (see Mermillod-Blondin et al., this volume). However, there is growing evidence of the role of stygofauna in maintaining the hydraulic properties of aquifers through burrowing (Hose and Stumpp, 2019) and maintaining microbial communities through grazing (Griebler et al., 2019). The recognition of ecosystem services, although an anthropocentric ideal, is what separates ecosystem health from other related concepts of ecosystem condition, status or integrity, for which the provision of ecosystem services is not specifically defined (Boulton, 1999).

Critical discussion of the ecosystem health concept, including a definition, has been ongoing since the early 1990s (Scrimgeour and Wicklum, 1996; Boulton, 1999; Karr, 1999; Vugteveen et al., 2006). Some authors have argued that ecosystem health is undefinable and impossible to adequately benchmark (Scrimgeour and Wicklum, 1996). Other researchers have suggested the use of benchmarks that encompass the best available contemporary conditions (Bailey et al., 2004). It has been further argued that health is a property of species and is not a measurable ecological property (Suter, 1993) and that the provision of services for humans is a questionable basis for assessing ecosystem status (Wilkins, 1999). Nevertheless, the ecosystem health concept has merit in being readily interpretable by politicians, managers and stakeholders, such that it is now commonplace in public policy and dialogue (Vugteveen et al., 2006).

The aim of this chapter is to provide a critical review of methods for the monitoring and assessment of health in groundwater ecosystems. These include conventional groundwater assessment methods in the fields of community ecology, functional ecology (measurement of ecosystem processes) and ecotoxicology and how these methods are being enhanced by integrating molecular tools (e.g., DNA metabarcoding of groundwater communities, qPCR on targeted functional genes, metagenomics, transcriptomics, DNA-stable isotope probing, ecotoxicogenomics). We conclude the discussion by considering the future directions, knowledge gaps and need to progress the understanding of ecosystem health, and the provision of ecosystem services in groundwaters (Box 22.1).

## Assessing ecosystem health and condition

### Current approaches in surface and groundwaters

Assessments of ecosystem health are fundamentally based on recording the biophysical attributes of a site and comparing those to the values expected in the absence of disturbance. Existing approaches for assessing ecosystem health can be divided broadly into predictive models and multimetric indices, which differ in how the "expected" attributes are determined. Predictive models, such as the RIVPACS and AUSRIVAS models for river health, predict the biota based on a suite of environmental variables recorded at a site. Underpinning the predictions are large data sets containing the biophysical attributes of undisturbed or reference sites (see Wright et al., 2000). The expected species can be predicted using multivariate models, Bayesian models (e.g., Adriaenssens et al., 2004), or environmental filters (*sensu* Chessman and Royal, 2004), in which site environmental attributes are matched to the known distributions or preferences of taxa, leading to, by process of elimination, the suite of taxa expected to be present. Deviation of the observed taxa lists from the expected taxa lists are used

---

### BOX 22.1

#### Definitions of ecosystem health and indicator types

**Ecosystem health** is an expression of an aquifer's ability to sustain its ecological functioning (vigor and resilience) in accordance with its organization while maintaining the provision of ecosystem goods and services (Korbel and Hose, 2011).

**Ecosystem services** are goods and services provided by the environment that are of value or benefit to humans (Griebler and Avramov, 2015).

**Functional indicators** are those that reflect and or quantify the processes (functions)

undertaken within an ecosystem, and which may serve as surrogates for the provision of ecosystem services.

**Organizational indicators** reflect the structure and composition of the biota and physicochemical attributes of the ecosystem.

**Stressors** are factors or pressures that potentially or actually disrupt the natural state of an ecosystem.

---

to indicate changes in condition. Although described as measuring ecosystem health, such models often fail to incorporate the ecosystem services perspective, and as such, are instead typically measures of ecosystem condition based on community composition (see below).

As the name implies, multimetric models require measurements of multiple biological, physical and chemical attributes (metrics) at a site. These are compared to the "normal" or expected range of those attributes and are used together to provide a picture of the state or condition of a system (e.g., Barbour et al., 1999). Where attributes are measured across a large number of undisturbed or "reference sites," the ranges of observed values are used to represent the range of acceptable conditions (Bailey et al., 2004).

Both predictive and multimetric approaches have been applied with some success in groundwater ecosystems (Castellarini et al., 2007; Steube et al., 2009). Each approach has advantages and disadvantages in the information contained (or lost) and the subsequent interpretation of the final metric (see Bonada et al., 2006). However, the development of predictive models is complicated by the often heterogeneous distributions of groundwater biota over space and time, and current poor understanding of relationships between biota and environmental gradients in aquifers. The application of deep-learning models may be useful here, but these require large datasets, and their development is in its infancy (see Christin et al., 2019).

The concept of assessing groundwater using bioindicators began in the mid-1990s, with a number of early publications describing the sensitivities of groundwater organisms (mostly one, or a small number of invertebrate taxa) to natural factors such as physical habitats, and anthropogenic impacts (e.g., Malard et al., 1996; Mösslacher and Notenboom, 1999). First attempts to use groundwater faunal composition for groundwater ecosystem assessments emerged in the mid-2000s, prompted by the development of the European Union Groundwater Directive (EC-GWD, 2006). Hahn (2006) proposed the Groundwater Fauna Index based on abiotic indicators, which was followed by efforts to predict groundwater diversity (Stoch et al., 2009) based on the richness of specific taxonomic groups. The EU Groundwater Directive was also the stimulus for Steube et al. (2009) to propose a structured scheme for ecological assessment of groundwater ecosystems using microbes, fauna and physicochemical conditions as indicators. From these studies, it was evident that there was a need to develop not only an inventory of biological and ecological indicators, but also to establish the natural or "reference" ecological conditions within aquifers.

Growing recognition of the importance of groundwater to the sustainability of agricultural industries prompted the development of the Groundwater Health Index (GHI) in Australia as a method to assess the ecological status and health of groundwaters using a combination of biotic and abiotic measures (Korbel and Hose, 2011). The GHI uses indicators of functions (e.g., metabolism), ecosystem organisation (e.g., biotic composition) and stressors (natural and anthropogenic) (Fig. 22.1), combining these in a multimetric approach to generate a single measure of ecosystem health. This method provides a tiered approach, in which a preliminary assessment of health can be made based on a list of standard criteria based on the expected attributes of groundwater ecosystems (Tier 1, Fig. 22.2). A second tier allows for rigorous analysis and benchmark setting where outcomes of the Tier 1 assessments indicate potential impairment or are uncertain.

Griebler et al. (2014) adopted a similar approach to the GHI in Europe. This approach aims to provide an ecological assessment scheme for groundwater systems (FMER, 2020) to support the European Union Water Framework Directive (EC-WFD, 2000) and the EC-GWD

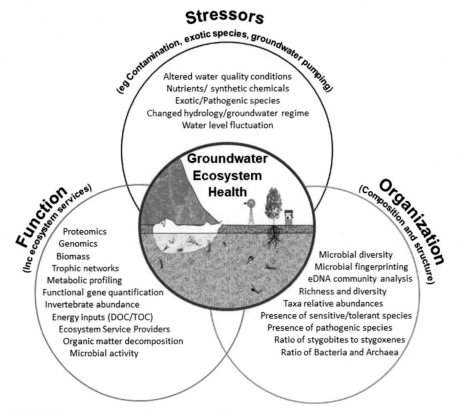

**FIGURE 22.1**   Types of indicators required for assessment of groundwater ecosystem health.

(2006). Just like the GHI approach, microbial, and faunal indicators are used for the assessment of ecological status. This project developed standardized biological and ecological criteria and methods for monitoring groundwater quality, ecosystem status, and services in a practical manner. Resulting from this project, a microbial (Density-Activity-(Carbon)) index (Fillinger et al., 2019a) and a faunal index (Hahn et al., in prep.) for monitoring groundwater health and status were developed.

While most approaches have been developed to assess nonspecific impacts to groundwaters, tools have been developed for some specific anthropogenic threats. Spengler and Hahn (2018) proposed the "Thermo-Index," which combines temperature threshold values for groundwater with an index based on sensitive indicator species. Di Lorenzo et al. (2020) refined the GHI of Korbel and Hose (2017) to assess the compliance of unconsolidated aquifers in Italy with the nitrate requirements of the European Directives. The GHI has also been applied in a simplified form as part of a citizen science program (Fig. 22.3; Korbel pers. comm.). This work demonstrated the GHI to be a globally relevant tool that could be applied to assess compliance with regional regulations, albeit with some modification of the thresholds and metrics to meet local conditions.

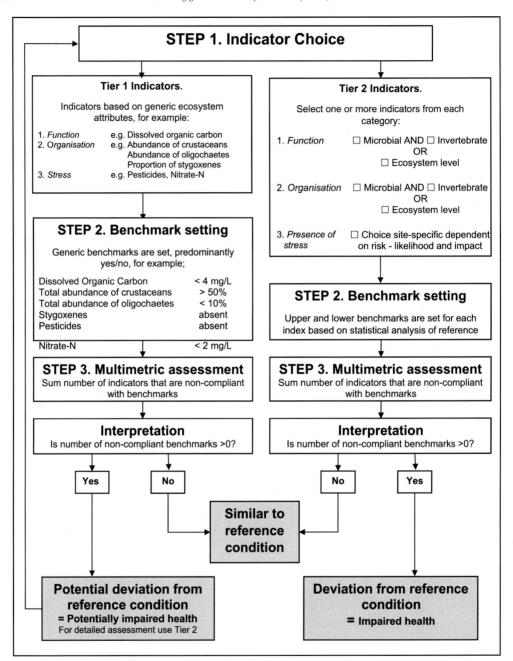

**FIGURE 22.2**    Tiered framework for assessing groundwater ecosystem health assessment. *Adapted from Korbel and Hose (2011, 2017).*

**FIGURE 22.3** Approaches to sampling groundwater ecosystems. (A) Motorised inertia pump (K Korbel). (B) Net sampling for stygofauna (G Hose, inset C Spengler). (C) Hyporheic sampling in a cave stream (D Galassi). (D) Fauna sampling in a cave pool (D Galassi). (E) Artificial substrates (cages with clay beads) and water quality probe for deployment in a well (L Volatier). (F) Farmers collecting from a production well as part of a citizen science program (K Korbel). (G) Net sampling in a cave stream (C Spengler).

## Indicators of ecosystem health and condition

Health assessments should consider the condition of both the physical and biological components of an ecosystem and include the presence of stressors as a potential early warning indicator (Vugteveen et al., 2006). The attributes of ecosystem condition can be further categorized as those reflecting ecosystem organization (composition and structure) and those reflecting function (Fig. 22.1). Within the latter, specific indicators may reflect or serve as surrogates for the provision of ecosystem services. Ideally, indicators should cover a range of spatial and temporal scales and include microbes that may turn over quickly and vary over small spatial scales (Fig. 22.4), through to invertebrates which have longer life spans and broader distributions, to hydrological and chemical variables that cover to regional and millennial scales (Fig. 22.4).

Most approaches to assess groundwater quality, ecological status and health have relied on groundwater samples, which can be collected in a variety of ways (Fig. 22.3), and the quantification of biota or chemistry therein. In a few cases, substrates deployed in wells have been used, particularly to assess microbial communities and activity, but also invertebrates. Voisin et al. (2016) identified clay balls as a suitable substrate for microbial

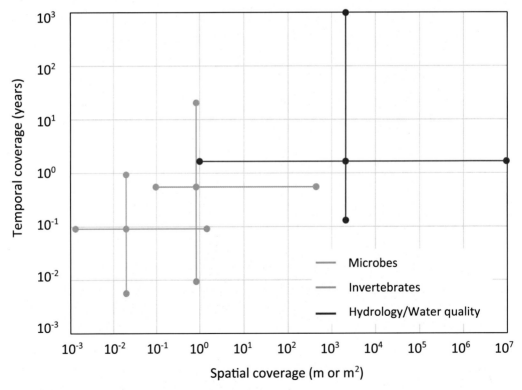

FIGURE 22.4 Spatial and temporal coverage in microbial, invertebrate, and physicochemical attributes of groundwater ecosystems. *Adapted from Innis et al. (2000).*

colonization which were collected and analyzed using DNA-based approaches (e.g., Mermillod-Blondin et al., 2019). The tool clearly separated sites in the metropolitan area of Lyon, France, that were impacted by urban inputs and artificial recharge from those outside the recharge plume. Fillinger et al. (2019b) deployed and analyzed sterile river sediments in wells in a similar way. Lategan et al. (2010) deployed cotton strips into wells, with the loss of strength of the strip correlated with and used as a surrogate for microbial activity. This cost-effective tool has been used to distinguish between climatic conditions and nitrate contamination (Korbel and Hose, 2015) and may reflect redox processes active in the subsurface. The use of traps for groundwater biota was established by Hahn (2005), whose simple method allows simultaneous sampling of hydro-chemical, faunal, and microbial samples. Samples from traps had similar composition as those from the aquifer (Bork et al., 2008), demonstrating their potential for use in ongoing monitoring. Marmonier et al. (2018) deployed live invertebrates (surface and stygobitic) and measured mortality and condition in reference and stormwater-impacted wells. However, one challenge that in situ approaches face is the differences in biotic and abiotic conditions in wells compared to the surrounding aquifer (Korbel et al., 2017).

In the following section, we list a select suite of indicators, categorized into organisational, functional and stressors, and review their application in, or suitability for groundwater monitoring, with a consideration of the likely challenges to implementation and setting of benchmark values. The list is far from exhaustive, and practitioners are encouraged to explore a diversity of potential indicators beyond those listed here.

## Organizational indicators for assessing health

Compared to surface waters, the "truncated biodiversity" of groundwaters (Gibert and Deharveng, 2002) means that there are relatively fewer organisms available for use as bioindicators, particularly macrophytes, algae, and fish, which are actuely sensitive to disturbances, but which are often absent from groundwater ecosystems (see Marmonier et al., this volume). Despite this, there is growing evidence that groundwater microbes and stygofauna provide a sufficiently diverse and sensitive array of metrics to underpin bioassessment frameworks (e.g., Griebler et al., 2010; Stein et al., 2010; Korbel and Hose, 2011; Spengler and Hahn, 2018; Fillinger et al., this volume).

As in rivers, collections of macro- and meio-invertebrates have been the primary tool for groundwater bioassessment (Korbel and Hose, 2011; Griebler et al., 2014). However, the application of invertebrate-based assessments in groundwaters has been hampered by the often low abundance of animals collected, high spatial variability (patchiness) of stygofauna and challenges associated with species and genus-level taxonomy. Nevertheless, the abundance and the diversity and composition of fauna are sensitive to changes in environmental conditions (e.g., Sinton, 1984; Korbel et al., 2013). The biomass of fauna can reflect changes in the nutrient condition of the aquifer (e.g., Sinton, 1984), and the proportions of stygobites/stygoxenes and taxa such as crustaceans and oligochaetes may reflect the degree of surface-water exchange or disturbance (Hahn, 2006; Griebler et al., 2014; Korbel and Hose, 2017). Increases in stygobiont abundance following a disturbance have been reported only rarely in the literature (e.g., Sinton, 1984). More frequently, declines in abundance occur either indirectly, such as through competition, or directly, such as through toxicity or habitat modification.

Analysis of the morphological and physiological traits of invertebrates can reveal compositional, functional and structural changes in assemblages, and provide an alternative to taxonomic-based assessments of community structure. Trait-based analyses of stygofauna assemblages are sensitive to groundwater contamination (Di Lorenzo et al., 2019a), with the added potential of improving biological assessments through inference of causality (Culp et al., 2011). Hose et al. (2022) provide a list of traits for stygofauna and discuss opportunities for trait-based bioassessment of groundwaters.

While invertebrates have historically been the primary indicators for aquatic ecosystem assessments, microbiota (Archaea, Bacteria, protists, and fungi) have been relatively rarely used, apart from detection of specific (often pathogenic) taxa. Microbiota have not been widely used because of difficulties in culturing and identifying the vast majority of *Bacteria*, *Archaea* and fungi and a lack of understanding of their ecology. The advent of next-generation DNA sequencing has allowed larger-scale analyses of microbiota, and an exponential increase in the use and understanding of microbial assemblages as indicators of environmental change. While analysis of individual taxa is still uncommon (notwithstanding pathogenic screening), microbial community composition can provide an informative indicator of groundwater quality and anthropogenic impacts (e.g., Mermillod-Blondin et al., 2019; Korbel et al., 2022). Diversity metrics can be used to summarize the community-level changes. Multivariate and network analyses can also be used to highlight deviation of assemblages from reference condition (e.g., Pearce et al., 2011). The relative proportions of specific microbial taxa within a community (e.g., Santoro et al., 2008; Korbel et al., 2022) provide meaningful metrics reflecting possible causality and ecosystem functions (see below).

## Functional indicators for assessing health

Functional indicators reflect the activity, metabolism, or primary production of the ecosystem and include ecosystem services and resilience (Korbel and Hose, 2011). In the absence of light in groundwater ecosystems, organic carbon infiltrating from the surface is generally the primary energy source, and the quality and quantity of carbon in groundwater can be a key functional indicator. Changes to carbon inputs, such as from sewage or stormwater (Foulquier et al., 2011), strongly influence microbial assemblages and ecosystem functions. Aquifer productivity, which can be measured by rates of heterotrophic metabolism and aerobic respiration and determined by $CO_2$ and $CH_4$ generation and the depletion of dissolved oxygen (Baker et al., 2000) is a valuable indicator of ecosystem functioning.

Microbial cell abundance has been used as an indicator of disturbance in groundwater ecosystems (Fillinger et al., 2019a; Retter et al., 2021) and can be measured by a variety of techniques such as flow cytometry, microscopic cell counts, or protein concentrations. Such metrics integrate biological responses across taxonomic and functional groups of microorganisms, and thus can be applied as general measures of disturbances without the need for detailed knowledge on specific target organisms.

Microbial activity provides a measure of ecosystem function and has been used as an indicator of disturbance in groundwater ecosystems (Mermillod-Blondin et al., 2019). Lategan et al. (2010) showed that the degradation of cellulose in groundwater was correlated with overall microbial activity (measured by hydrolysis of fluorescein diacetate) and made for a cheap and effective indicator that varied in response to a range of environmental conditions (Korbel and Hose, 2015). Measurements of the microbial activity such as intracellular ATP concentrations or activities of general enzymes such as dehydrogenase and hydrolases (Fillinger et al., 2019a) also show potential as indicators of ecosystem function.

Functional indicators have proven useful for monitoring the stability of drinking water distribution systems (Vital et al., 2012), and for detecting changes in planktonic and surface-attached microorganisms in response to aquifer recharge (Foulquier et al., 2011) and contamination with hydrocarbons (Herzyk et al., 2017). Although microbial cell abundances and activity are often correlated (Foulquier et al., 2011; Fillinger et al., 2019a), they are not necessarily mutually redundant as they can be affected differently and over different timescales by different types of disturbances. However, combining both indicators in a multivariate analysis can enable detection of various disturbances (e.g., eutrophication, pollution, warming) and can provide more sensitive and reliable results than analysis of the individual parameters alone (Fillinger et al., 2019a).

The use of specific carbon sources, such as BIOLOG Ecoplates, provides a community metabolic profile that reflects the function of the community. Such fingerprinting analyzed either by multivariate methods or summary metrics such as the number of different carbon sources used can provide a metric of change in groundwaters (Korbel et al., 2013; Melita et al., 2019).

The physiological response of microorganisms to groundwater conditions, and their metabolic impact on groundwater chemistry, can also be effectively determined through the abundance or expression of characteristic genes or proteins (e.g., Wilkins et al., 2011; Wilson et al., 2019), or where functional traits are phylogenetically highly conserved, by the 16S rRNA gene-based identification of characteristic taxa (e.g., Ivanova et al., 2000; Ritalahti et al., 2006). Microbial changes can therefore be measured as shifts in taxa, genetic potential (in terms of gene content), and gene or protein expression. Taken together, prokaryotic molecular biomarkers are useful for determining changes in groundwater conditions, and the increase or decrease of biogeochemical processes.

Quantification of genes (DNA) or gene transcript (RNA) copies in a given volume of groundwater provides a means to rapidly screen groundwaters for genetic potential or activity and are particularly useful where metabolic processes can be linked to one or few characteristic genes. In this case, quantitative polymerase chain reaction (qPCR) assessment of taxa or functional genes has been widely used for evaluating biological activity in groundwater. Approaches include using primers that target various functional genes, for example, the 16S rRNA genes of *Dehalococcoides* involved in the degradation of chlorinated ethenes (Ritalahti et al., 2006; Wilson et al., 2019), and the 16S rRNA or hydrazine synthase genes of anaerobic ammonium oxidizing (anammox) planctomycetes, which convert ammonium to $N_2$ gas (Wang et al., 2020; Mosley et al., 2022a). Functional marker genes have further been used as indicators of the anaerobic degradation of aromatic compounds like benzene, toluene, ethylbenzene and xylene (BTEX) by quantifying genes involved in their metabolism (Kuntze et al., 2011). Conversely, functional genes can be used to indicate increased groundwater toxicity, such as through the reductive mobilization of naturally occurring arsenic in aquifer sediments due to groundwater abstraction (Héry et al., 2010). Similarly, human fecal contamination of groundwater can be tracked by quantification of bacterial antibiotic resistance genes and human pathogens using taxa specific 16S rRNA gene primers (Böckelmann et al., 2009).

While targeted gene approaches, such as qPCR, provide an efficient method for screening particular functional genes, "omics" approaches (metagenomics, metatranscriptomics, metaproteomics, metabolomics) are more effective for identifying biological processes that are undertaken by multiple genes or pathways. A number of studies have used metaproteomics (shotgun proteomics), coupled with reference genome databases for peptide annotation, to explore microbial activity in groundwater (Wrighton et al., 2014). For example, protein biomarkers have been used to identify and track strain-level changes in uranium bioremediation by U(VI)-reducing *Geobacter* (Wilkins et al., 2011; Chourey et al., 2013). Metaproteomics and metatranscriptomics have also been used to identify microbial-mediated nitrogen removal by anammox and/or heterotrophic denitrification (via nitrous oxide reduction to $N_2$) (Chourey et al., 2013; Handley et al., 2013; Mosely et al., 2022a,b), and sulfide production and reoxidation (Handley et al., 2013; Bell et al., 2018) in aquifers. Although cutting-edge "omics" techniques can provide a wealth of information on microbial metabolic capacity and processes in aquifers, they are currently too costly, time-consuming, and dependent on a high level of expertise in bioinformatics, for routine groundwater health surveys. Furthermore, understanding of the natural individual variability in the omics outputs is limited but critical for environmental monitoring programs (Bahamonde et al., 2016). However, detailed omics analyses can be used to identify gene markers of interest for PCR-based screens.

The transfer of energy through an ecosystem via food webs is a critical ecosystem function. Saccò et al. (2019) detected, using stable isotopes, shifts in groundwater food webs attributable to rainfall events, but these same tools could logically be applied to other environmental changes. A recent study quantifying Ribulose-1,5-bisphosphate carboxylase-oxygenase (RuBisCO)-encoding genes in a karstic aquifer suggests that complex eukaryotic food webs are underpinned by microbial chemolithoautotrophy (Herrmann et al., 2020). However, the intricacies of microbial food webs in groundwater are still largely enigmatic. Current "omics" methodologies allow detailed reconstructions of microbial food webs in aquifers and have shown the partitioning of resources, and the production and consumption of substrates in subsurface microbial communities. This includes identifying community fractions that supply or utilize fermentation products, rely on exogenous amino acids or their derivatives (Wrighton et al., 2014; Borton et al., 2018), and acquire nutrients via carbon and nitrogen fixation (Wrighton et al., 2012; Handley et al., 2013). These and other comprehensive "omic" studies provide a growing inventory of important microbial metabolic processes (Anantharaman et al., 2016) and functional genes (Wegner et al., 2019) in aquifers, which can be developed as metrics for community assessment and used for future comparisons over space and time.

Measuring ecological resilience, that is, the capacity of an ecosystem to "absorb" disturbance and still maintain its structure, processes, and functioning, is a challenge in any ecosystem (Quinlan et al., 2016), and may be particularly challenging with groundwaters. While the ability of groundwater ecosystems to recover from disturbance is low because of the immense challenges of remediating impacts (particularly contamination) to aquifers and the likely slow recovery of biota due to low capacity for recruitment through dispersal and reproduction and low energy of the system (Boulton, 2020; Hose et al., 2022), the ability to resist disturbance has been little considered. Resilience can be considered in terms of the traits of the community and there is a range of potential trait-based, and other metrics that could be applied to explore and quantify the resilience of groundwater ecosystems (see Baho et al., 2017). However, with functional redundancy and adaptive capacity among biota likely to be low, resilience in groundwater ecosystems may be limited (Mammola et al., 2019), at least for invertebrates, but perhaps not for microbes (Weaver et al., 2015). Greater understanding of the thresholds of change is still needed for resilience metrics to be useful for indicating ecosystem health.

## Stress indicators for assessing health

The presence of stressors in the groundwater environment has alone been the principal means of quality assessment and remains the underpinning metric of the EC-GWD and similar policies elsewhere. A stressor may be present prior to observable ecological change but will not be present in a healthy system.

The list of potential stressors is endless and may include chemical (e.g., Castaño-Sanchez et al., 2020), physical (e.g., thermal (Griebler et al., 2016)), and biological (e.g., exotic species (Mazza et al., 2014) and pathogenic (e.g., Weaver et al., 2016; Fout et al., 2017) pressures. The choice of stress indicators should be tailored to those likely to be present at a site (EC-WFD, 2000; Korbel and Hose, 2011). Monitoring of stressors should consider the intensity, extent, timing, and duration of the exposure to the stressor and allow a prediction of risk or an assessment of impact.

## Defining the reference condition for groundwater ecosystems

Most biological assessment schemes are based on a comparison of current condition to a natural, undisturbed state, which is generally referred to as the "reference condition" (Stoddard et al., 2006). A first step to defining reference conditions at a broad spatial scale can be the classification of groundwater ecosystems (Robertson et al., this volume) based on specific geological units and a defined spatial scale (e.g., Weitowitz et al., 2017; Iannella et al., 2020). Stein et al. (2012) examined the biotic composition of aquifers in Germany and identified stygofauna bioregions ("Stygoregions"), which could identify sites that are similar, and inform the expected biotic attributes of sites within a region. Weitowitz et al. (2017) provide a similar classification for the UK and Wales. At a regional scale, Korbel and Hose (2011, 2017) identified a suite of relatively undisturbed sites as a point of reference to set benchmark values. However, the most suitable spatial scale for reference sites is still debated (Robertson et al., this volume).

Following classification by type or region, pristine and unpolluted representative sites are then selected within each unit and a number of abiotic and biotic parameters (indicators) is recorded. Reference thresholds for indicators are defined by the range of values recorded at the reference sites. Importantly, few if any parts of the planet remain unaffected by human activities, such that "pristine" sites may be unavailable, and reference sites may be based on relatively undisturbed or "best available" sites (Stoddard et al., 2006).

Robust classifications require adequate assessment of the natural variation within and between units and may require intensive sampling of a large number of sites to adequately identify site "types" and characterize the reference condition state. First attempts of classifying reference conditions for stygofauna were made by Illies (1978), Botosaneanu (1986) and Stein et al. (2012). Due to the ubiquitous nature of bacteria in groundwater (Fillinger et al., this volume), the definition of reference conditions seems to be more feasible, with Fillinger et al. (2019a) proposing the Density—Activity—Carbon (D-A-C) concept.

The alternate approach is to define values of indicators that reflect (or are expected in) "pristine" or undisturbed groundwater ecosystems. Groundwater bodies with similar values (Korbel and Hose, 2011) and that are not affected by natural (e.g., low oxygen concentrations for stygofauna) or anthropogenic stressors (e.g., pesticides) can be considered as a suitable reference site (Griebler et al., 2014). This approach underpinned the Tier 1 indicators set by Korbel and Hose (2011).

Complicating the definition of reference conditions are natural factors that limit biotic distributions. Among the hydrochemical parameters, oxygen is probably the most important natural stressor with respect to fauna, with metazoans requiring on the long term $\geq 1$ mg/L of dissolved oxygen (Hahn, 2006). Regional temperature must also be considered since many stygobites have narrow temperature tolerance ranges (Spengler and Hahn, 2018; Di Lorenzo et al., this volume). Temperature variation also indicates surface water influences, and potential organic matter supply. The Groundwater Fauna Index summarizes these three abiotic parameters and reflects the hydrological exchange and the degree of surface water intrusion (Hahn, 2006). Additionally, the porosity of the aquifer matrix and the sediment grain size composition impact the distribution of biota (Fiasca et al., 2014; Korbel et al., 2019), with fractured aquifers and fine sand alluvial deposits characterized by an impoverished fauna (Hahn and Fuchs, 2009).

In principle, reference habitats for metazoans should comply with the following requirements: oxygen concentration $>1$ mg/L, matrix pore size $>200$ µm, sufficient food (organic matter) supply (Groundwater Fauna Index $>2$) and no fine sediments, however these limitations do not appear to apply to microbial indicators. Outside of these conditions, general reference condition thresholds are problematic, for example, a faunal assessment of fine sandy aquifers in Europe incorrectly indicated a pristine aquifer as disturbed (Robertson et al., this volume) due to naturally low oxygen concentrations (Stein et al., 2012). In such cases, a different approach, using local reference habitats to establish catchment-specific reference condition thresholds may be more applicable (e.g., Korbel and Hose, 2017). These reference conditions can be established using national standards and guidelines (e.g., nitrate concentrations) or by collating data from multiple sampling events and reference sites, thus accounting for temporal variation, to establish reference thresholds for each indicator (e.g., biotic diversity). A variety of statistical methods can be used to establish these reference values, including the use of means, ranges or presence/absence of specific taxa. It is also possible to account for natural biotic variation in weighted frameworks, meaning that sites with environmental factors conducive to particular taxa (e.g., high dissolved oxygen) or detrimental to taxa (e.g., high electrical conductivity) are accounted for in any individual assessment of groundwater health.

## Combining indicators into summary indices

As described in the previous sections, there is an array of individual indicators that can be grouped under the broad categories of ecosystem function, organization, and stressors. One of the most common and difficult issues within multimetric frameworks in ecological studies is the question of how different indicators should be combined to give an overall summary index, particularly as indicators consist of a variety of quantitative and qualitative data, for example, biotic diversity, species sensitivities, water chemistry, and environmental data. However, when there is such diversity of data combined with limited understanding of groundwater ecosystems by managers and the general public, it is vital that a clear, easy-to-understand process, producing a simple suite of summary results or one overall health index is provided. Furthermore, summary indices must be transparent and able to be decomposed into their components for a closer assessment. This ensures that monitoring, assessment and evaluation of groundwater ecosystem health can be understood and aid management objectives.

Fig. 22.5 describes one method by which indicators can be compared to reference conditions (or predicted reference conditions), summarized into indices, and then aggregated into a singular index describing overall ecosystem health. This approach is common in the assessment of surface waters (Barbour et al., 1999; Wright et al., 2000), and has been adopted in the GHI frameworks (Korbel and Hose, 2011, 2017). Another approach available is to use

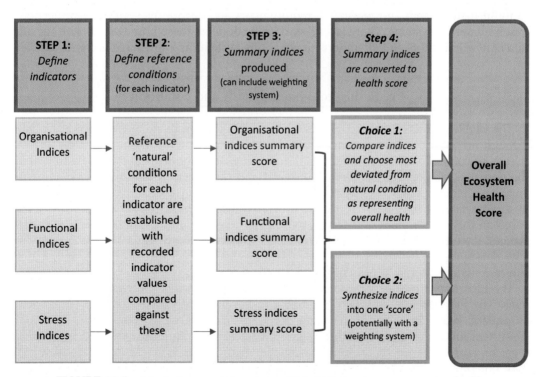

FIGURE 22.5   A basic framework for using multiple indicators to assess ecosystem health.

individual indices (Fig. 22.5 - Step 3) which are each compared to the thresholds and the index that deviates most from reference condition is used to indicate the overall ecosystem health. Such approaches are used in the EC-WFD (2000) where the most deviated summary index is considered the result of the multi-indexed assessment.

The simplest form of this framework assumes that all indicators are equally important in the overall determination of ecosystem health. Weighting systems have also been used to recognize the relative importance of different indicators, account for the cumulative impact of multiple stressors, give value to credibility of data, and to incorporate natural variation into ecological assessments (Jørgensen et al., 2010; Suter and Cormer, 2011). "Weighted" indicator values are then combined into one overall "score." For example, the weighted-GHI (Korbel and Hose, 2017) used a simple multiplication system to ensure that biotic and abiotic factors had equal influence in final ecosystem health scores.

The use of weighting evidence in ecological assessments does not come without criticism, as many such systems use degrees of expert judgment for combining indicators. However, other methods can be used, including criteria-guided assessments, normalization of data and numerical indices, and combined with expert judgment to devise rigorous and transparent weighting systems which allow for the combination of multiple variables into one overall summary index (Suter and Cormer, 2011).

This entire process of assessing ecosystem health, from selection of indicators through to the final assessment of indices and classification of health, provides an important decision-making tool for groundwater managers. Such tools allow monitoring of groundwater health to fulfill management and legislative requirements and ensure the continued protection of groundwater ecosystems. There is still work to be done on these frameworks with the addition of new indicators and alterations to weighting systems as knowledge becomes available.

## Predicting ecosystem health and condition

The likelihood that human activities adversely affect groundwater ecosystems is evaluated through ecological risk assessment (ERA). ERA consists of identifying hazards (e.g., the release of contaminants in groundwater) and using measurements and models to quantify the potential relationship between the hazard and the effect on individuals, communities, and ecosystems as a whole. Several conceptual frameworks for ERA of chemical substances in aquatic habitats have been developed worldwide (US EPA, 1998; ANZG, 2018). They are based on the identification of the measured environmental concentrations (MEC) of the substances, the estimation of a Predicted No Effect Concentration (PNEC) derived from ecotoxicity data and, finally, the risk (R), calculated as the ratio of MEC to PNEC. If $R < 1$, the substance is considered to present no risk to the ecosystem.

The ERA approach has some important limitations, particularly for groundwater ecosystems. The first issue is related to the difficulty in obtaining toxicity data on all species in an aquatic ecosystem. The current practice is to test representatives of the three major ecosystem functional groups, namely algae, crustaceans (mainly *Daphnia*) and fish, and use these as surrogates for the whole ecosystem. This method is questionable because (1) it may not protect the most sensitive species and (2) two of these major functional groups are not appropriate for groundwater ecosystems, in which photosynthetic organisms and fish are rare or absent. Finally, the paucity of toxicity data on groundwater organisms has prompted the use of data for surface water species as surrogates. This practice may lead to unreliable ERA results due to the many physiological differences between epigean and groundwater species (Hose, 2007; Di Lorenzo et al., 2018, 2019b).

To be truly realistic, the ERA scenarios in groundwater must be based on PNEC values obtained from ecotoxicological tests on groundwater fauna. However, the standard protocols for determining PNECs are not entirely applicable to groundwater species, which occur naturally at low abundance, and are difficult to collect and rear in the laboratory (Castaño-Sánchez et al., 2020). For this reason, researchers have used their own testing protocols to suit the traits of the tested species. This is certainly a sound approach but prevents the comparison of results between different laboratories. In order to overcome these difficulties, Di Lorenzo et al. (2019b) have recommended approaches for ecotoxicological studies using stygofauna.

Ideally the PNEC calculation should use Hazardous Concentration 5% (HC5) values, which represent threshold concentrations of "no concern" and are derived from specific sensitivity distribution (SSD) curves (see ANZG, 2018). SSD curves should be constructed using sensitivity data of representatives of different functional groups. In the absence of sufficient data for groundwater biota, SSD curves may be created using sensitivity data for taxonomically similar surface water species as surrogates, after applying a correction factor that takes into account the higher sensitivity of groundwater species compared to their epigean relatives (Di Lorenzo et al., 2018). While common, the application of correction and safety factors (see ANZG, 2018) is not ideal and often results in PNEC values that are overly conservative and well below nontoxic concentrations (Di Lorenzo et al., 2021). Many PNEC values are freely available or can be calculated online by entering the MEC values of the chemicals (Strona et al., 2019).

Notwithstanding costs and demanding computations, toxicology frameworks based on "omics" methods may help in revealing the complex suite of stressors and pressures to which groundwater ecosystems may be exposed (Martins et al., 2019). In particular, "omics" techniques allow characterization of changes to ecosystem structure and function instead of using single species (Oziolor et al., 2017). Changes in species abundance, species richness, genetic diversity, and shifts in gene expression may serve as useful indicators of ecosystem health (Oziolor et al., 2016). Meta-analyses have shown that evolutionary adaptations to environmental contaminants may be observed at pollutant concentrations below regulatory benchmarks (Oziolor et al., 2016). For further details on groundwater ERA and ecotoxicological data with groundwater species, the reader can refer to Di Lorenzo et al. (this volume).

## Future directions

Promising emerging technologies and concepts should be tested whenever possible, as they can provide new indicators and analytical frameworks for groundwater health assessment. The ultimate goal must be to assess groundwater health based on rich, holistic and reliable data. Here, innovative and data-driven approaches can lead to more robust science-informed management decisions. For instance, niche-partitioning models, building on the assumption that relative biomass is a surrogate for the manner in which resources are distributed among species, is a powerful approach to assess the impact of stressors on a community but has been little tested in groundwater ecosystems (Di Lorenzo et al., 2019; Reiss et al., 2019). In another direction, the biological component of groundwater can be accessed by genetic sequence variants, covering diverse taxa such as microorganisms, fungi, protists, and metazoans. Due to a variable degree of coverage of genetic sequences in public data repositories, taxonomic lists might remain incomplete for certain taxa. However, genetic sequence variants can be investigated directly as a taxonomic surrogate and correlated with environmental parameters of interest via machine-learning approaches. As such, "responsive" and hence "informative" genetic sequence variants (or taxa) can be identified and find their way into large-scale biomonitoring programs (Pearce et al., 2011; Pawlowski et al., 2018).

Communities of practice should continually discuss, develop and evaluate good practice approaches within a shared vision framework. Well-understood reference sites could be selected to ground-truth innovative tools and concepts. Promising outcomes then should be picked up by teams involving stakeholders from all sectors (i.e., science, water management, politics, industry), which can i) formulate a consensus based on the demands and needs from all involved parties, ii) mediate the results for target group-specific purposes, iii) seek further steps for the implementation of promising approaches, and iv) generate a culture of a shared vision of groundwater health (Leese et al., 2018). This process can be best supported by continuous interdisciplinary trainings organized by and provided to protagonists from all sectors, thereby engaging long-lasting partnerships and offering opportunities for innovative products and technologies accompanied by the development of new markets.

Future scientific directions should emphasize better understanding of the temporal and spatial variability of indicators and the interconnectivity of groundwaters with adjoining ecosystems (European Union, 2015; Bugnot et al., 2019), and disentangling the effects of multiple stressors. Interconnectivity and subsequent variability of parameters and conditions have an impact on groundwater health assessments and the robustness of assessments must be tested under different scenarios. Importantly, new scientific knowledge and practice must be integrated into policy—something which remains challenging for practitioners on both sides of the science-policy interface (Quevauviller, 2007).

Given the interconnected nature of groundwater ecosystems, a major step for groundwater health assessment would be to achieve legal treatment (i.e., regulatory protection and monitoring) equal with surface waters. Aquatic groundwater-dependent ecosystems, such as springs, wetlands, streams and rivers, hyporheic zones, lakes, coastal lagoons, estuarine and marine discharge zones, all need to be appropriately included in groundwater characterization and risk assessment (Strona et al., 2019; Boulton, 2020).

## Acknowledgments

We thank Florian Malard and an anonymous referee for constructive comments that improved this chapter. We thank Laurence Volatier and Florian Mermillod-Blondin for access to photos of field sampling. Alexander Weigand was supported by DNAqua-Net COST Action CA15219. Christian Griebler was supported by the German Federal Ministry for Education and Research (BMBF) in the framework of the project consortium 'GroundCare' (033W037A).

# References

Adriaenssens, V., Goethals, P.L.M., Charles, J., De Pauw, N., 2004. Application of Bayesian Belief Networks for the prediction of macroinvertebrate taxa in rivers. Annales de Limnologie-International Journal of Limnology 40, 181–191.

Anantharaman, K., Brown, C.T., Hug, L.A., Sharon, I., Castelle, C.J., Probst, A.J., Thomas, B.C., Singh, A., Wilkins, M.J., Karaoz, U., Brodie, E.L., Williams, K.H., Hubbard, S.S., Banfield, J.F., 2016. Thousands of microbial genomes shed light on interconnected biogeochemical processes in an aquifer system. Nature Communications 7, 13219.

ANZG (Australian and New Zealand Governments), 2018. Australian and New Zealand Guidelines for Fresh and Marine Water Quality. Australian and New Zealand Governments and Australian State and Territory Governments, Canberra, ACT, Australia (Accessed 06 January 2023). https://www.waterquality.gov.au/guidelines/anz-fresh-marine.

Bahamonde, P.A., Feswick, A., Isaacs, M.A., Munkittrick, K.R., Martyniuk, C.J., 2016. Defining the role of omics in assessing ecosystem health: perspectives from the Canadian Environmental Monitoring Program. Environmental Toxicology and Chemistry 35, 20–35.

Baho, D.L., Allen, C.R., Garmestani, A.S., Fried-Petersen, H.B., Renes, S.E., Gunderson, L., Angeler, D.G., 2017. A quantitative framework for assessing ecological resilience. Ecology and Society 22 (3), 17.

Bailey, R.C., Norris, R.H., Reynoldson, T.B., 2004. Bioassessment of Freshwater Ecosystems: Using the Reference Condition Approach. Springer, New York.

Baker, M.A., Valett, H.M., Dahm, C.N., 2000. Organic carbon supply and metabolism in a shallow groundwater ecosystem. Ecology 81 (11), 3133–3148.

Barbour, M.T., Gerritsen, J., Snyder, B.D., Stribling, J.B., 1999. Rapid Bioassessment Protocols for Use in Streams and Wadeable Rivers: Periphyton, Benthic Macroinvertebrates and Fish, second ed. U.S. Environmental Protection Agency, Office of Water, Washington. EPA 841-B-99-002.

Bell, E., Lamminmaki, T., Alneberg, J., Andersson, A.F., Qian, C., Xiong, W., Hettich, R.L., Balmer, L., Frutschi, M., Sommer, G., Bernier-Latmani, R., 2018. Biogeochemical cycling by a low-diversity microbial community in deep groundwater. Frontiers in Microbiology 9, 2129.

Böckelmann, U., Dorries, H.H., Ayuso-Gabella, M.N., Salgot de Marcay, M., Tandoi, V., Levantesi, C., Masciopinto, C., Van Houtte, E., Szewzyk, U., Wintgens, T., Grohmann, E., 2009. Quantitative PCR monitoring of antibiotic resistance genes and bacterial pathogens in three European artificial groundwater recharge systems. Applied and Environmental Microbiology 75 (1), 154–163.

Bonada, N., Prat, N., Resh, V.H., Statzner, B., 2006. Developments in aquatic insect biomonitoring: a comparative analysis of recent approaches. Annual Review of Entomology 51 (1), 495–523.

Bork, J., Bork, S., Berkhoff, S.E., Hahn, H.J., 2008. Testing unbaited stygofauna traps for sampling performance. Limnologica 38 (2), 105–115.

Borton, M.A., Hoyt, D.W., Roux, S., Daly, R.A., Welch, S.A., Nicora, C.D., Purvine, S., Eder, E.K., Hanson, A.J., Sheets, J.M., Morgan, D.M., Wolfe, R.A., Sharma, S., Carr, T.R., Cole, D.R., Mouser, P.J., Lipton, M.S., Wilkins, M.J., Wrighton, K.C., 2018. Coupled laboratory and field investigations resolve microbial interactions that underpin persistence in hydraulically fractured shales. Proceedings of the National Academy of Sciences 115 (28), E6585–E6594.

Botosaneanu, L. (Ed.), 1986. Stygofauna Mundi: a faunistic, distributional, and ecological synthesis of the World fauna inhabiting subterranean waters. Brill, Leiden.

Boulton, A.J., 2020. Editorial: conservation of groundwaters and their dependent ecosystems: integrating molecular taxonomy, systematic reserve planning and cultural values. Aquatic Conservation: Marine and Freshwater Ecosystems 30, 1–7.

Boulton, A.J., 1999. An overview of river health assessment: philosophies, practice, problems and prognosis. Freshwater Biology 41, 469–479.

Bugnot, A.B., Hose, G.C., Walsh, C., Floerl, O., French, K., Dafforn, K.A., Hanford, J., Lowe, L., Hahs, A., 2019. Urban impacts across realms: making the case for inter-realm monitoring and management. Science of the Total Environment 648, 711–719.

Castaño-Sánchez, A., Hose, G.C., Reboleira, A.S.P.S., 2020. Salinity and temperature increases impact groundwater crustaceans. Scientific Reports 10, 12328.

Castellarini, F., Malard, F., Dole-Olivier, M.J., Gibert, J., 2007. Modelling the distribution of stygobionts in the Jura Mountains (eastern France). Implications for the protection of ground waters. Diversity and Distributions 13, 213–224.

Chessman, B., Royal, M., 2004. Bioassessment without reference sites: use of environmental filters to predict natural assemblages of river macroinvertebrates. Journal of the North American Benthological Society 23, 599–615.

Chourey, K., Nissen, S., Vishnivetskaya, T., Shah, M., Pfiffner, S., Hettich, R.L., Loffler, F.E., 2013. Environmental proteomics reveals early microbial community responses to biostimulation at a uranium- and nitrate-contaminated site. Proteomics 13 (18–19), 2921–2930.

Christin, S., Hervet, E., Lecomte, N., 2019. Applications for deep learning in ecology. Methods in Ecology and Evolution 10 (10), 1632–1644.

Culp, J.M., Armanini, D.G., Dunbar, M.J., Orlofske, J.M., Poff, N.L., Pollard, A.I., Yates, A.G., Hose, G.C., 2011. Incorporating traits in aquatic biomonitoring to enhance causal diagnosis and prediction. Integrated Environmental Assessment and Management 7 (2), 187–197.

Di Lorenzo, T, Fiasca, B, Di Cicco, M, Cifoni, M, Galassi, D.M.P., 2021. Taxonomic and functional trait variation along a gradient of ammonium contamination in the hyporheic zone of a Mediterranean stream. Ecological Indicators 132, 108268.

Di Lorenzo, T., Murolo, A., Fiasca, B., Tabilio Di Camillo, A., Di Cicco, M., Galassi, D.M.P., 2019a. Potential of a trait-based approach in the characterization of an N-contaminated alluvial aquifer. Water 11 (12), 2553.

Di Lorenzo, T., Cifoni, M., Fiasca, B., Di Cioccio, Galassi, D.M.P., 2018. Ecological risk assessment of pesticide mixtures in the alluvial aquifers of central Italy: toward more realistic scenarios for risk mitigation. Science of the Total Environment 644, 161–172.

Di Lorenzo, T., Di Marzio, W.D., Fiasca, B., Galassi, D.M.P., Korbel, K., Iepure, S., Pereira, J., Reboleira, A.S.P.S., Schmidt, S., Hose, G.C., 2019b. Recommendations for ecotoxicity testing with stygobiotic species in the framework of groundwater environmental risk assessment. Science of the Total Environment 681 (1), 292–304.

Di Lorenzo, T., Fiasca, B., Di Camillo Tabilio, A., Murolo, A., Di Cicco, M., Galassi, D.M.P., 2020. The weighted Groundwater Health Index (wGHI) by Korbel and Hose (2017) in European groundwater bodies in nitrate vulnerable zones. Ecological Indicators 116, 106525.

EC-GWD, 2006. Directive 2006/118/EC of the European Parliament and of the Council of 12 December 2006 on the protection of groundwater against pollution and deterioration. Official Journal of the European Union L 372 (19).

EC-WFD, 2000. Directive 2000/60/EC of the European Parliament and of the Council of 23 October 2000 establishing a framework for Community action in the field of water policy. Official Journal of the European Commission L327 (1). Brussels, Belgium.

European Union, 2015. Technical Report on Groundwater Associated Aquatic Ecosystems. Publications Office of the European Union, Luxembourg. Technical Report No. 9.

Fiasca, B., Stoch, F., Olivier, M.-J., Maazouzi, C., Petitta, M., Di Cioccio, A., Galassi, D.M.P., 2014. The dark side of springs: What drives small-scale spatial patterns of subsurface meiofaunal assemblages? Journal of Limnology 73, 55–64.

Fillinger, L., Hug, K., Trimbach, A.M., Wang, H., Kellermann, C., Meyer, A., Bendinger, B., Griebler, C., 2019a. The D-A-(C) index: a practical approach towards the microbiological-ecological monitoring of groundwater ecosystems. Water Research 163, 114902.

Fillinger, L., Zhou, Y., Kellermann, C., Griebler, C., 2019b. Non-random processes determine the colonization of groundwater sediments by microbial communities in a pristine porous aquifer. Environmental Microbiology 21 (1), 327–342.

FMER, 2020. GroundCare - Parameterisation and Quantification of Ecosystem Services as a Basis for Sustainable Groundwater Management (Accessed 06 January 2023). https://bmbf.nawam-rewam.de/en/projekt/groundcare/.

Foulquier, A., Mermillod-Blondin, F., Malard, F., Gibert, J., 2011. Response of sediment biofilm to increased dissolved organic carbon supply in groundwater artificially recharged with stormwater. Journal of Soils and Sediments 11 (2), 382–393.

Fout, G.S., Borchardt, M.A., Kieke, B.A., Karim, M.R., 2017. Human virus and microbial indicator occurrence in public-supply groundwater systems: meta-analysis of 12 international studies. Hydrogeology Journal 25 (4), 903–919.

Gibert, J., Deharveng, L., 2002. Subterranean ecosystems: a truncated functional biodiversity. BioScience 52 (6), 473−481.

Griebler, C., Avramov, M., Hose, G., 2019. Groundwater ecosystems and their services - current status and potential risks. In: Schröter, M., Bonn, A., Klotz, S., Seppelt, R., Baessler, C. (Eds.), Atlas of Ecosystem Services - Drivers, Risks, and Societal Responses. Springer, Cham, Switzerland, pp. 197−203.

Griebler, C., Stein, H., Hahn, H.J., Steube, C., Kellermann, C., Fuchs, A., Berkhoff, S.E., Brielmann, H., 2014. In: Entwicklung biologischer Bewertungsmethoden und -kriterien für Grundwasserökosysteme (Development of biological assessent schemes and criteria for groundwater ecosystems). UFOPLAN, FKZ 3708 23 200, pp. 153. https://www.umweltbundesamt.de/sites/default/files/medien/378/publikationen/uba_bericht_grundwasser _web.pdf.

Griebler, C., Avramov, M., 2015. Groundwater ecosystem services: a review. Freshwater Science 34 (20), 355−367.

Griebler, C., Brielmann, H., Haberer, C.M., Kaschuba, S., Kellermann, C., Stumpp, C., Hegler, F., Kuntz, D., Walker-Hertkorn, S., Lueders, T., 2016. Potential impacts of geothermal energy use and storage of heat on groundwater quality, biodiversity, and ecosystem processes. Environmental Earth Sciences 75 (20), 1391.

Griebler, C., Stein, H., Kellermann, C., Berkhoff, S., Brielmann, H., Schmidt, S., Selesi, D., Steube, C., Fuchs, A., Hahn, H.J., 2010. Ecological assessment of groundwater ecosystems − vision or illusion? Ecological Engineering 36 (9), 1174−1190.

Hahn, H.J., 2006. A first approach to a quantitative ecological assessment of groundwater habitats: the GW-Fauna-Index. Limnologica 36 (2), 119−137.

Hahn, H.J., Fuchs, A., 2009. Distribution patterns of groundwater communities across aquifer types in southwestern Germany. Freshwater Biology 54, 848−860.

Hahn, H.J., 2005. Unbaited phreatic traps: a new method of sampling stygofauna. Limnologica 35, 248−261.

Handley, K.M., VerBerkmoes, N.C., Steefel, C.I., Williams, K.H., Sharon, I., Miller, C.S., Frischkorn, K.R., Chourey, K., Shah, M.B., Long, P.E., Hettich, R.L., Banfield, J.F., 2013. Biostimulation induces syntrophic interactions that impact C, S and N cycling in a sediment microbial community. The ISME Journal 7 (4), 800−816.

Herrmann, M., Geesink, P., Yan, L., Lehmann, R., Totsche, K.U., Kusel, K., 2020. Complex food webs coincide with high genetic potential for chemolithoautotrophy in fractured bedrock groundwater. Water Research 170, 115306.

Héry, M., Van Dongen, B.E., Gill, F., Mondal, D., Vaughan, D.J., Pancost, R.D., Polya, D.A., Lloyd, J.R., 2010. Arsenic release and attenuation in low organic carbon aquifer sediments from West Bengal. Geobiology 8 (2), 155−168.

Herzyk, A., Fillinger, L., Larentis, M., Qiu, S., Maloszewski, P., Hünniger, M., Schmidt, S.I., Stumpp, C., Marozava, S., Knappett, P.S., Elsner, M., 2017. Response and recovery of a pristine groundwater ecosystem impacted by toluene contamination − a meso-scale indoor aquifer experiment. Journal of Contaminant Hydrology 207, 17−30.

Hose, G.C., Chariton, A., Di Lorenzo, T., Galassi, D.M.P., Halse, S.A., Reboleira, A.S.P.S., Robertson, A.L., Schmidt, S.I., Korbel, K., 2022. Invertebrate traits, diversity and the vulnerability of groundwater ecosystems. Functional Ecology 36, 2200−2214.

Hose, G.C., Stumpp, C., 2019. Architects of the underworld: Bioturbation by groundwater invertebrates influences aquifer hydraulic properties. Aquatic Sciences 81, 20.

Hose, G.C., 2007. A response to comments on Assessing the need for groundwater quality guidelines using the species sensitivity distribution approach. Human and Ecological Risk Assessment 13, 241−246.

Iannella, M., Fiasca, B., Di Lorenzo, T., Biondi, M., Di Cicco, M., Galassi, D.M.P., 2020. Jumping into the grids: mapping biodiversity hotspots in groundwater habitat types across Europe. Ecography 43, 1825−1841.

Illies, J., 1978. Limnofauna Europaea. Fischer, Stuttgart.

Innis, S.A., Naiman, R.J., Elliott, S.R., 2000. Indicators and assessment methods for measuring the ecological integrity of semi-aquatic terrestrial environments. Hydrobiologia 422/423, 111−131.

Ivanova, I.A., Stephen, J.R., Chang, Y.-J., Brüggemann, J., Long, P.E., McKinley, J.P., Kowalchuk, G.A., White, D.C., Macnaughton, S.J., 2000. A survey of 16S rRNA and amoA genes related to autotrophic ammonia-oxidizing bacteria of the β- subdivision of the class proteobacteria in contaminated groundwater. Canadian Journal of Microbiology 46, 1012−1020.

Jørgensen, S., Xu, L., Marques, J.C., Salas, F., 2010. Application of indicators for the assessment of ecosystem health. In: Jørgensen, S., Xu, L., Costanza, R. (Eds.), Handbook of Ecological Indicators for Assessment of Ecosystem Health. CRC Press, Boca Raton, USA, pp. 9−76.

Karr, J.R., 1999. Defining and measuring river health. Freshwater Biology 41, 221−234.

Korbel, K., Greenfield, P., Hose, G.C., 2022. Agricultural practices linked to shifts in groundwater microbial structure and denitrifying bacteria. Science of the Total Environment 807, 150870.

Korbel, K., Hancock, P.J., Serov, P., Lim, R.P., Hose, G.C., 2013. Groundwater ecosystems change with landuse across a mixed agricultural landscape. Journal of Environmental Quality 42, 380—390.

Korbel, K., Hose, G.C., 2015. Habitat, water quality, seasonality or site? Identifying environmental correlates of the distribution of groundwater biota. Freshwater Science 34, 329—343.

Korbel, K.L., Hose, G.C., 2011. A tiered framework for assessing groundwater ecosystem health. Hydrobiologia 661 (1), 329—349.

Korbel, K.L., Hose, G.C., 2017. The weighted groundwater health index: improving the monitoring and management of groundwater resources. Ecological Indicators 75, 164—181.

Korbel, K., Stephenson, S., Hose, G.C., 2019. Sediment size influences habitat selection and use by groundwater macrofauna and meiofauna. Aquatic Sciences 81, 39.

Kuntze, K., Vogt, C., Richnow, H.H., Boll, M., 2011. Combined application of PCR-based functional assays for the detection of aromatic-compound-degrading anaerobes. Applied and Environmental Microbiology 77 (14), 5056—5061.

Lategan, M.J., Korbel, K., Hose, G.C., 2010. Is cotton-strip tensile strength a surrogate for microbial activity in groundwater? Marine and Freshwater Research 61, 351—356.

Leese, F., Bouchez, A., Abarenkov, K., Altermatt, F., Borja, Á., Bruce, K., Ekrem, T., Ciampor Jr., F., Ciamporova-Zatovicova, Z., Costa, F.O., Elbrecht, V., Fontaneto, D., Franc, A., Geiger, M.F., Hering, D., Kahlert, M., Kalamujic Stroil, B., Weigand, A.M., 2018. Why we need sustainable networks bridging countries, disciplines, cultures and generations for aquatic biomonitoring 2.0: a perspective derived from the DNAqua-Net COST action. Advances in Ecological Research 58, 63—99.

Malard, F., Plenet, S., Gibert, J., 1996. The use of invertebrates in ground water monitoring: a rising research field. Ground Water Monitoring and Remediation 16 (2), 103—113.

Mammola, S., Cardoso, P., Culver, D.C., Deharveng, L., Ferreira, R.L., Fišer, C., Galassi, D.M.P., Griebler, C., Halse, S., Humphreys, W.F., Isaia, M., Malard, F., Martinez, A., Moldovan, O.T., Niemiller, M.L., Pavlek, M., Reboleira, A.S.P.S., Souza-Silva, M., Teeling, E.C., Wynne, J.J., Zagmajster, M., 2019. Scientists' warning on the conservation of subterranean ecosystems. BioScience 69, 641—650.

Martins, C., Dreij, K., Costa, P.M., 2019. The state-of-the art of environmental toxicogenomics: challenges and perspectives of "Omics" approaches directed to toxicant mixtures. International Journal of Environmental Research and Public Health 16, 4718.

Mazza, G., Reboleira, A.S., Gonçalves, F., Aquiloni, L., Inghiles, I., Spigoli, D., Stoch, F., Taiti, S., Gherardi, F., Tricarico, E., 2014. A new threat to groundwater ecosystems: first occurrences of the invasive crayfish *Procambarus clarkii* (Girard, 1852) in European caves. Journal of Cave and Karst Studies 76, 62—65.

Melita, M., Amalfitano, S., Preziosi, E., Ghergo, S., Frollini, E., Parrone, D., Zoppini, A., 2019. Physiological profiling and functional diversity of groundwater microbial communities in a municipal solid waste landfill area. Water 11, 2624.

Mermillod-Blondin, F., Voisin, J., Marjolet, L., Marmonier, P., Cournoyer, B., 2019. Clay beads as artificial trapping matrices for monitoring bacterial distribution among urban stormwater infiltration systems and their connected aquifers. Environmental Monitoring and Assessment 191, 58.

Marmonier, P., Maazouzi, C., Baran, N., Blanchet, S., Ritter, A., Saplairoles, M., Dole-Olivier, M.-J., Galassi, D.M.P., Eme, D., Dolédec, S., Piscart, C., 2018. Ecology-based evaluation of groundwater ecosystems under intensive agriculture: a combination of community analysis and sentinel exposure. Science of the Total Environment 613—614, 353—1366.

Mosley, O.E., Gios, E., Weaver, L., Close, M., Daughney, C., van der Raaij, R., Martindale, H., Handley, K.M., 2022a. Metabolic diversity and aero-tolerance in anammox bacteria from geochemically distinct aquifers. mSystems 7, e01255—21.

Mosley, O.E., Gios, E., Close, M., Weaver, L., Daughney, C., Handley, K.M., 2022b. Nitrogen cycling and microbial cooperation in the terrestrial subsurface. The ISME Journal 16 (11), 2561—2573.

Mösslacher, F., Notenboom, J., 1999. Groundwater biomonitoring. In: Gerhardt, A. (Ed.), Biomonitoring of Polluted Water, Environmental Science Forum Vol. 96, Trans Tech Publications, Switzerland, pp. 119—140.

Oziolor, E.M., Bickham, J.W., Matson, C.W., 2017. Evolutionary toxicology in an omics world. Evolutionary Applications 10 (8), 752—761.

Oziolor, E.M., De Schamphelaere, K., Matson, C.W., 2016. Evolutionary toxicology: meta-analysis of evolutionary events in response to chemical stressors. Ecotoxicology 25, 1—9.

Pawlowski, J., Kelly-Quinn, M., Altermatt, F., Apothéloz-Perret-Gentil, L., Beja, P., Boggero, A., Borja, A., Bouchez, A., Cordier, T., Domaizon, I., Feio, M.J., Filipe, A.F., Fornaroli, R., Graf, W., Herder, J., van der Hoorn, B., Jones, J.I., Sagova-Mareckova, M., Moritz, C., Barquín, J., Piggott, J.J., Pinna, M., Rimet, F., Rinkevich, B., Sousa-Santos, C., Specchia, V., Trobajo, R., Vasselon, V., Vitecek, S., Zimmerman, J., Weigand, A., Leese, F., Kahlert, M., 2018. The future of biotic indices in the ecogenomic era: integrating (e) DNA metabarcoding in biological assessment of aquatic ecosystems. Science of the Total Environment 637, 1295—1310.

Pearce, A.R., Rizzo, D.M., Mouser, P.J., 2011. Subsurface characterization of groundwater contaminated by landfill leachate using microbial community profile data and a nonparametric decision-making process. Water Resources Research 47, W06511.

Quevauviller, P., 2007. Water protection against pollution. conceptual framework for a science policy interface. Environmental Science and Pollution Research 14 (5), 297—307.

Quinlan, A.E., Berbés-Blázquez, M., Haider, L.J., Peterson, G.D., 2016. Measuring and assessing resilience: broadening understanding through multiple disciplinary perspectives. Journal of Applied Ecology 53, 677—687.

Reiss, J., Perkins, D.M., Fussmann, K.E., Krause, S., Canhoto, C., Romeijn, P., Robertson, A.L., 2019. Groundwater flooding: ecosystem structure following an extreme recharge event. Science of the Total Environment 625, 1252—1260.

Retter, A., Griebler, C., Haas, J., Birk, S., Stumpp, C., Brielmann, H., Fillinger, L., 2021. Application of the D-A-(C) index as a simple tool for microbial-ecological characterization and assessment of groundwater ecosystems—a case study of theMur River Valley, Austria. Österreichische Wasser- und Abfallwirtschaft 73, 455—467.

Ritalahti, K.M., Amos, B.K., Sung, Y., Wu, Q., Koenigsberg, S.S., Loffler, F.E., 2006. Quantitative PCR targeting 16S rRNA and reductive dehalogenase genes simultaneously monitors multiple Dehalococcoides strains. Applied and Environmental Microbiology 72 (4), 2765—2774.

Saccò, M., Blyth, A.J., Humphreys, W.F., Kuhl, A., Mazumder, D., Smith, C., Grice, K., 2019. Elucidating stygofaunal trophic web interactions via isotopic ecology. PLoS One 14 (10), e0223982.

Santoro, A.E., Francis, C.A., De Sieyes, N.R., Boehm, A.B., 2008. Shifts in the relative abundance of ammonia-oxidizing bacteria and archaea across physicochemical gradients in a subterranean estuary. Environmental Microbiology 10 (4), 1068—1079.

Scrimgeour, G.J., Wicklum, D., 1996. Aquatic ecosystem health and integrity: problems and potential solutions. Journal of the North American Benthological Society 15, 254—261.

Sinton, L.W., 1984. The macroinvertebrates in a sewage polluted aquifer. Hydrobiologia 119, 161—169.

Spengler, C., Hahn, H.J., 2018. Thermostress: Ökologisch begründete, thermische Schwellenwerte und Bewertungsansätze für das Grundwasser. Korrespondenz Wasserwirtschaft 11 (9), 525—525.

Stein, H., Griebler, C., Berkhoff, S., Matzke, D., Fuchs, A., Hahn, H.J., 2012. Stygoregions - a promising approach to a bioregional classification of groundwater systems. Scientific Reports 2, 673.

Stein, H., Kellermann, C., Schmidt, S.I., Brielmann, H., Steube, C., Berkhoff, S.E., Fuchs, A., Hahn, H.J., Thulin, B., Griebler, C., 2010. The potential use of fauna and bacteria as ecological indicators for the assessment of groundwater quality. Journal of Environmental Monitoring 12 (1), 242—254.

Steube, C., Richter, S., Griebler, C., 2009. First attempts towards an integrative concept for the ecological assessment of groundwater ecosystems. Hydrogeology Journal 17 (1), 23—35.

Stoch, F., Artheau, M., Brancelj, A., Galassi, D., Malard, F., 2009. Biodiversity indicators in European ground waters; towards a predictive model of stygobiotic species richness. Freshwater Biology 54, 745—755.

Stoddard, J.L., Larsen, D.P., Hawkins, C.P., Johnson, R.K., Norris, R.H., 2006. Setting expectations for the ecological condition of streams: the concept of reference condition. Ecological Applications 16, 1267—1276.

Strona, G., Fattorini, S., Fiasca, B., Di Lorenzo, T., Di Cicco, M., Lorenzetti, W., Boccacci, F., Galassi, D.M.P., 2019. AQUALIFE software: a new tool for a standardized ecological assessment of groundwater dependent ecosystems. Water 11, 2574.

Suter, G.W., 1993. A critique of ecosystem health concepts and indexes. Environmental Toxicology and Chemistry 12, 1533—1539.

Suter, G.W., Cormer, S.M., 2011. What and how to combine evidence in environmental assessments: weighing evidence and building cases. Science of The Total Environment 409, 1406—1417.

VI. Biodiversity and ecosystem management in groundwater

US EPA, 1998. Guidelines for Ecological Risk Assessment. https://www.epa.gov/sites/production/files/2014-11/documents/eco_risk_assessment1998.pdf (Accessed 06 January 2023).

Vital, M., Dignum, M., Magic-Knezev, A., Ross, P., Rietveld, L., Hammes, F., 2012. Flow cytometry and adenosine triphosphate analysis: alternative possibilities to evaluate major bacteriological changes in drinking water treatment and distribution systems. Water Research 46 (15), 4665–4676.

Voisin, J., Cournoyer, B., Mermillod-Blondin, F., 2016. Assessment of artificial substrates for evaluating groundwater microbial quality. Ecological Indicators 71, 577–586.

Vugteveen, P., Leuven, R., Huijbregts, M., Lenders, H., 2006. Redefinition and elaboration of river ecosystem health: perspective for river management. Hydrobiologia 565, 289–308.

Wang, S., Zhu, G., Zhuang, L., Li, Y., Liu, L., Lavik, G., Berg, M., Liu, S., Long, X.-E., Guo, J., Jetten, M.S.M., Kuypers, M.M.M., Li, F., Schwark, L., Yin, C., 2020. Anaerobic ammonium oxidation is a major N-sink in aquifer systems around the world. The ISME Journal 14 (1), 151–163.

Weaver, L., Karki, N., Mackenzie, M., Sinton, L., Wood, D., Flintoft, M., Havelaar, P., Close, M., 2016. Microbial transport into groundwater from irrigation: comparison of two irrigation practices in New Zealand. Science of the Total Environment 543, 83–94.

Weaver, L., Webber, J.B., Hickson, A.C., Abraham, P.M., Close, M.E., 2015. Biofilm resilience to desiccation in groundwater aquifers: a laboratory and field study. Science of the Total Environment 514, 281–289.

Wegner, C.E., Gaspar, M., Geesink, P., Herrmann, M., Marz, M., Kusel, K., 2019. Biogeochemical regimes in shallow aquifers reflect the metabolic coupling of the elements nitrogen, sulfur, and carbon. Applied and Environmental Microbiology 85 (5), e02346.

Weitowitz, D.C., Maurice, L., Lewis, M., Bloomfield, J.P., Reiss, J., Robertson, A.L., 2017. Defining geo-habitats for groundwater ecosystem assessments: an example from England and Wales (UK). Hydrogeology Journal 25, 2453–2466.

Wilkins, D.A., 1999. Assessing ecosystem health. Trends in Ecology & Evolution 14, 69.

Wilkins, M.J., Callister, S.J., Miletto, M., Williams, K.H., Nicora, C.D., Lovley, D.R., Long, P.E., Lipton, M.S., 2011. Development of a biomarker for *Geobacter* activity and strain composition; proteogenomic analysis of the citrate synthase protein during bioremediation of U(VI). Microbial Biotechnology 4 (1), 55–63.

Wilson, J.T., Mills, J.C., Wilson, B.H., Ferrey, M.L., Freedman, D.L., Taggart, D., 2019. Using qPCR assays to predict rates of cometabolism of TCE in aerobic groundwater. Groundwater Monitoring & Remediation 39, 53–63.

Wright, J.F., Sutcliffe, D.W., Furse, M.T. (Eds.), 2000. Assessing the Biological Quality of Fresh Waters: RIVPACS and Other Techniques. Freshwater Biological Association, Ambleside.

Wrighton, K.C., Castelle, C.J., Wilkins, M.J., Hug, L.A., Sharon, I., Thomas, B.C., Handley, K.M., Mullin, S.W., Nicora, C.D., Singh, A., Lipton, M.S., Long, P.E., Williams, K.H., Banfield, J.F., 2014. Metabolic interdependencies between phylogenetically novel fermenters and respiratory organisms in an unconfined aquifer. The ISME Journal 8 (7), 1452–1463.

Wrighton, K.C., Thomas, B.C., Sharon, I., Miller, C.S., Castelle, C.J., VerBerkmoes, N.C., Wilkins, M.J., Hettich, R.L., Lipton, M.S., Williams, K.H., Long, P.E., Banfield, J.F., 2012. Fermentation, hydrogen, and sulfur metabolism in multiple uncultivated bacterial phyla. Science 337 (6102), 1661–1665.

# Recent concepts and approaches for conserving groundwater biodiversity

*Andrew J. Boulton[1], Maria Elina Bichuette[2],*
*Kathryn Korbel[3], Fabio Stoch[4], Matthew L. Niemiller[5],*
*Grant C. Hose[3] and Simon Linke[6]*

[1]School of Environmental and Rural Science, University of New England, Armidale, NSW, Australia; [2]Laboratory of Subterranean Studies, Federal University of São Carlos, São Carlos, Brazil; [3]School of Natural Sciences, Macquarie University, Sydney, Australia; [4]Evolutionary Biology & Ecology, Université libre de Bruxelles, Brussels, Belgium; [5]Department of Biological Sciences, The University of Alabama in Huntsville, Huntsville, AL, United States; [6]CSIRO, Dutton Park, Brisbane, QLD, Australia

## Introduction

Long considered devoid of life, groundwater ecosystems are increasingly recognized for their unique, unusual, and sometimes surprisingly biodiverse fauna. Many of these species are not found in surface ecosystems but, being specially adapted to subterranean conditions. They are likely to be vulnerable to external pressures that cause even small changes in one or more of these conditions. This vulnerability is particularly relevant considering the hydrological connectivity of many groundwater ecosystems to surface ones. Therefore, human activities and global climate change that affect surface ecosystems likely have repercussions on groundwater ecosystems and must be considered when conserving subterranean biodiversity. As scientific and public recognition grows of this biotic (including microbial) diversity and its values, the focus of groundwater conservation is shifting from solely protection as a water source toward policies and practices that preserve subterranean biodiversity and ecological processes.

Broadly, groundwater ecosystems are defined as saturated subterranean voids occurring in unconsolidated alluvial or colluvial sediments, in pores or fractures of igneous, sedimentary or metamorphic rocks, and in saturated soils and organic matter (e.g., hypotelminorheic habitats, Culver et al., 2006). These ecosystems include flooded caves, alluvial and fractured-rock aquifers and deep artesian aquifers and associated springs, as well as lesser-known ecosystems such as groundwater-fed bogs and coastal anchialine systems (Fig. 23.1). For most

*Groundwater Ecology and Evolution, Second Edition*
https://doi.org/10.1016/B978-0-12-819119-4.00001-9

525

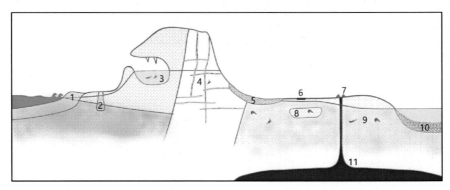

**FIGURE 23.1**    Groundwater ecosystems across the landscape. 1. Coastal sandbeds, 2. Anchialine systems in karst, 3. Flooded caves, 4. Fractured rock aquifers, 5. Groundwater-fed springs, bogs and swamps, 6. Hypotelminorheic habitat, 7. Artesian springs, 8. Karst calcretes, 9. Alluvial and colluvial aquifers, 10. Hyporheic zones, 11. Artesian aquifers. *Credit: Grant Hose.*

subterranean ecosystems, recharge of water, organic matter, other nutrients and energy from overlying surface ecosystems is crucial. Therefore, conservation of their biodiversity must be suitably holistic (e.g., Linke et al., 2019) and adequately protect surface-subterranean hydrological and ecological connectivity.

Groundwater invertebrates and vertebrates, collectively termed stygofauna, co-occur with subterranean microbial communities. The diverse microbial communities present provide valuable ecosystem services (Fillinger et al., this volume). Despite this, they are seldom considered in biodiversity conservation, except for several extreme communities found in some deep caves (e.g., Holmes et al., 2001). Instead, conservation has focused on stygofauna because these taxa are better known, morphologically distinctive and often inhabit narrow geographical ranges. Although the ecosystem services of some groundwater invertebrates have been quantified (e.g., enhancing water storage and transmission—Hose and Stumpp, 2019), those of cave fish, salamanders and other subterranean vertebrates remain largely unknown. However, some species have considerable cultural values for particular societies (e.g., the olm to Slovenia, Kostanjšek et al., this volume) and, as flagship species in Europe and the United States, have attracted most of the conservation efforts for groundwater ecosystems to date (e.g., Niemiller et al., 2018a).

As the science of conservation biology has evolved, it has remained underpinned by the central tenets that diversity of organisms, ecological complexity and evolution is desirable, and that biodiversity has an intrinsic value. Thus, the goal of biodiversity conservation is to maintain organismal (and genetic) diversity, ecological complexity, evolutionary processes and the ecological and ecosystem service values of biodiversity. However, achieving these goals in groundwaters is complicated by a lack of fundamental biological knowledge, the predominant conservation focus on surface ecosystems, and the limited tools available to protect what we often cannot see. In the face of global threats to groundwater (Kretschmer et al., this volume) such as overextraction, contamination, introduced species and climate change, effective strategies must be immediately enacted to conserve our beleaguered groundwater biodiversity. Protection is especially crucial because many groundwater ecosystems can seldom be restored once damaged, especially if hydrological connectivity is lost or impaired (Boulton, 2020).

This chapter reviews past and current concepts about groundwater biodiversity and conservation, focusing on present and potential tools and approaches to minimize the risk of subterranean biodiversity loss and to protect diverse groundwater ecosystems at multiple spatial scales. We start by describing how scientific understanding of subterranean biodiversity and its drivers has changed, especially since this book's first edition in 1994. We then focus on how groundwater biodiversity is currently measured and conserved, including several recent developments that show great promise. Specific challenges to conservation and management of groundwater ecosystems are discussed, and we suggest how these might be addressed in a future that will be particularly demanding for groundwater ecosystems and their biota.

## Past concepts and approaches in groundwater biodiversity conservation

### Past concepts of groundwater biodiversity

The prevailing "classic" view (see references in Marmonier et al., 1993) was that subterranean biodiversity was predicted to be much lower than surface biodiversity because groundwaters were considered to have less habitat diversity, less physical and chemical variability, and apparently no primary production (Gibert and Deharveng, 2002), although chemoautotrophic primary production was known in several caves (Sarbu et al., 1996). This perception of an impoverished subterranean biodiversity within an "extreme environment" rested heavily on the conceptual basis that spatial and temporal homogeneity limits biodiversity. Over time, awareness increased about the diversity of groundwater habitats at multiple spatial scales. For example, at a fine scale, variations in direction and magnitude of surface-groundwater exchange in hyporheic sediments (Fig. 23.1) interact with spatial heterogeneity in particle and pore sizes to create diverse microhabitats that support rich subterranean biodiversity (e.g., Rouch, 1995). Biogeochemical gradients in, for example, redox conditions and nutrient concentrations further increase this biodiversity by making extra energy available. At broader scales, especially where subterranean ecotones between surface waters and groundwaters occur (e.g., spring-fed wetlands, hyporheic zones, Tonina et al., this volume), biodiversity is enhanced by gradients of habitat complexity and groundwater influence. More attention has also been paid to groundwater-dependent ecosystems (GDEs) such as terrestrial vegetation communities, submarine seepage areas and river baseflows, all of which harbor further biodiversity promoted by spatial and temporal heterogeneity in habitat structure and groundwater regime.

The "classic" view also perceived that groundwater biodiversity would be constrained by the apparent lack of primary producers in a lightless environment, described as a "truncated functional biodiversity" by Gibert and Deharveng (2002). The lack of photosynthetic basal trophic levels in subterranean food webs not only limits overall biodiversity but also that of upper trophic levels (predators) because of food and energy shortages (Gibert and Deharveng, 2002). Of course, these bioenergetic constraints seldom apply to surface GDEs where primary production occurs or to groundwater ecosystems hydrologically connected to surficial sources of organic matter (e.g., caves fed by streams).

Ecologically, groundwater biodiversity (Marmonier et al., this volume) is primarily determined by physical habitat availability, water physicochemistry, hydrological connectivity to sources of colonists, nutrients and energy, and biotic interactions such as competition and predation (Fig. 23.2). These primary drivers interact, and disruption of one or more of them usually leads to a decline in biodiversity of native subterranean species. Therefore, conservation strategies include those that seek to prevent, constrain or remediate human activities that potentially disrupt the primary drivers of subterranean biodiversity (Fig. 23.2). Strategies also include protecting recharge zones as well as areas deemed of particular importance, including those with especially high biodiversity, endemism, or both.

## Past approaches to groundwater conservation

Although the first edition of this book had only four index references to groundwater "conservation," it was already recognized that "management instruments used to deal with the protection of the intrinsic ecological values of groundwater are barely developed"

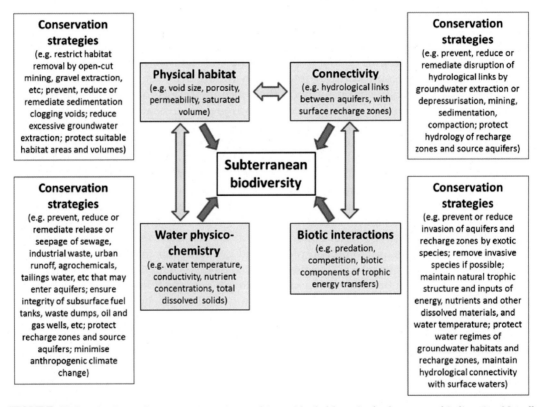

FIGURE 23.2 The four, often interacting primary drivers (shaded boxes) of subterranean biodiversity. Not all interactions (double-headed arrows) among the drivers are shown (e.g., connectivity can affect water physicochemistry). Potential conservation strategies (unshaded boxes) are matched with each driver but, of course, are complementary and many strategies apply to more than one of the primary drivers. *Credit: Andrew Boulton.*

(Notenboom et al., 1994: 498) and that groundwater conservation was likely to benefit surface biota as well as subterranean biodiversity (Stanford et al., 1994). The book's epilogue also stated that because most groundwater organisms evolved in stable environmental conditions, they are unlikely to be resistant to environmental change (Stanford and Gibert, 1994). However, the relevance of this observation to promoting strategies for groundwater conservation was not highlighted.

By the early 2000s, concerns were emerging about the impacts of human activities on groundwater ecosystems worldwide and there were calls for more effective groundwater protection and conservation (Sket, 1999; Boulton et al., 2003; Danielopol et al., 2003). At that time, conservation of groundwater fauna consisted of either nominating particular species as "threatened" or identifying and protecting "hotspots" such as single caves or cave systems that harbored numerous rare or endemic species (e.g., Juberthie and Juberthie-Jupeau, 1975; Malard et al., 1997). Initially, protection of caves and their groundwaters was motivated more by their geological, archeological and esthetic attractiveness than their biodiversity. For example, Vjetrenica (the largest cave in Bosnia and Herzegovina, Fig. 23.3A) was declared a protected natural monument in 1952 and designated a "special geological reserve" in 1965. It is also one of the richest hotspots of subterranean biodiversity (Culver and Sket, 2000), with almost 100 animal species of which 37 were first described from this cave. Another cave, the Postojna-Planina system in Slovenia (Fig. 23.3B), is the world's richest subterranean biodiversity hotspot with 84 obligate subterranean species, 48 of them living in groundwaters (Culver

**FIGURE 23.3** (A) The huge flooded galleries of Vjetrenica cave, a hotspot of subterranean biodiversity protected since 1952; (B) A spectacular hall with rimstone pools in the Postojna cave, the most biodiverse cave in the world; (C and D) mound springs of the Dalhousie Springs complex, central Australia, conserved in 1984. *Credit: (A) Roman Ozimec; (B) Peter Gedei; (C and D) Fran Sheldon.*

and Sket, 2000). Full legal protection of Postojna cave dates back to 1948. Since 1999, like all Slovenian caves, it has been protected by the nation's Nature Conservation Act.

Prior to 2000, there were also several groundwater-fed surface waters, especially springs and their associated wetlands, whose biodiversity was deemed worthy of conservation. For instance, Ponder (1986) made a strong case for conservation of mound springs of the Great Artesian Basin, Australia, based on their diverse endemic fauna, their scientific value as "natural laboratories" for ecological and evolutionary studies, and their cultural value for Indigenous and other Australians. In 1984, the pastoral property of Mt Dare station was purchased by the South Australian National Parks and Wildlife agency to protect and rehabilitate Dalhousie Springs (Fig. 23.3C and D), a group of over 60 natural artesian springs. Another example is the Edwards Aquifer that supplies the two largest freshwater springs in Texas (Comal Springs and San Marcos Springs) which are inhabited by several species of endemic fish, amphibians, crustaceans, insects and plants. Largely in response to a lawsuit filed by the Sierra Club in 1991, a conservation plan was developed to protect seven species listed under the U.S. *Endangered Species Act* that live in the Edwards Aquifer and its springs (National Academy of Sciences, 2015; Griebler et al. this volume). This initial conservation plan focused heavily on controlling groundwater extraction from the aquifer to protect spring flows.

## Initial impediments to groundwater conservation

Early efforts to conserve subterranean habitats were severely hampered by limited knowledge of their biodiversity and distribution (Danielopol et al., 2003). Traditional taxonomic studies are time-consuming and based on morphological characters that are usually highly homoplastic (i.e., arise independently due to convergence) in obligate groundwater species. Taxonomic uncertainty and the widespread presence of cryptic species (Eme et al., 2018) led to unreliable species richness estimations (a by-product of the "Linnean shortfall": Lomolino, 2004).

In addition to the lack of taxonomic knowledge, little was known about species' distributions. Many groundwater habitats are challenging to sample and often inaccessible (the "Racovitzan shortfall": Ficetola et al., 2008). As early sampling was spatially limited, it was unclear whether the high rate of endemism often found was real or an artifact of undersampling. Nevertheless, it was expected that many groundwater species would have much smaller geographical ranges than equivalent surface species (Gibert and Deharveng, 2002) because of their limited powers of dispersal. Logically, this implied that most stygofauna should be considered of conservation concern—a major impediment to develop protection plans because, as Culver et al. (2000) argued "there are simply too many subterranean species at risk to deal with them one at a time." There had long been calls for protection of whole catchment areas to protect groundwaters (e.g., de Marsily, 1992) and GDEs such as alluvial aquifers (e.g., Danielopol and Pospisil, 2001). However, catchment- or landscape-scale protection was hampered by the lack of regional information at suitably broad scales. Early efforts to describe general patterns of distribution and diversity of stygofauna on a continental scale (e.g., Europe: Thienemann, 1950) or globally (Botosaneanu, 1986) were not amenable to guiding conservation measures because they relied on data that varied widely in quality, collection method, and taxonomic resolution.

A final impediment was the limited public appreciation of groundwaters as ecosystems in their own right whose biodiversity and ecological processes were valuable, often threatened and worthy of conservation. Up until the last decade of last century, even most scientists were unaware that many groundwater ecosystems harbored diverse stygofaunal assemblages and rich microbial communities. Understanding of ecosystem services was still in its infancy (Daily, 1997), and most groundwaters were viewed as almost limitless supplies of water for exploitation where surface waters were sparse or of poor water quality. In the next section, we describe how many of these conceptual, technological, and social impediments to effective groundwater conservation have been addressed since the early 2000s.

## Recent concepts and approaches in groundwater biodiversity conservation

### Introduction

In the first two decades of the 21st century, there have been dramatic technological advances, several of which (e.g., molecular approaches, remote sensing) have great potential to address one or more of the impediments to groundwater conservation described above. Meanwhile, societal awareness of the vulnerability of groundwater ecosystems and their biodiversity is growing swiftly. Although there have been no "new" concepts about the drivers of biodiversity, molecular methods such as the use of environmental DNA (eDNA, see next section) seem particularly suited for resolving impediments such as the limited knowledge of species richness and distribution, particularly for cryptic and very small species, as well as estimating subterranean microbial diversity and functioning.

Of course, successful conservation relies on more than the knowledge of species richness and distribution. As indicated earlier, previous groundwater conservation has focused on species or on single habitats such as caves (e.g., the Habitats Directive in Europe) (Griebler et al. this volume). Groundwater habitat maps such as the one for Europe (Cornu et al., 2013) coupled with large-scale species occurrence data sets (e.g., the European Groundwater Crustacean Database, Zagmajster et al., 2014) allow preliminary identification of particular habitats that harbor stygofauna of major conservation significance (Iannella et al., 2020, 2021). However, recent developments in hydrological and ecological modeling (e.g., using biomarkers and species as indicators of subterranean water connections: Brancelj et al., 2020) confirm that most groundwater ecosystems are complex interconnected networks of subterranean habitats spanning aquifers, karstic and nonkarstic groundwaters and often the soil water itself. Surface-expression GDEs are variably connected to these habitats (Kath et al., 2018) and provide crucial conduits for water, nutrients and fauna. Thus, effective conservation of groundwater biodiversity relies on preserving the integrity of these conduits in this connected network of surface and subterranean ecosystems (a "holistic" approach, Linke et al., 2019).

Integrating the improved information on subterranean species richness and distribution from molecular data with remotely sensed spatial data and recent systematic reserve planning methods (see later) provides an excellent basis for a "landscape network approach" to conserve connected groundwater habitats (including surface GDEs) and their recharge zones. Mapping these landscape networks would encompass land-use information and potential pathways of contaminants, invasive species and other threats to subterranean diversity, as well as biogeographic and phylogenetic data (both especially important in groundwaters). In the following sections, we review the application and limitations of several recent advances in groundwater conservation: molecular methods, vulnerability mapping, assessing hotspots of biodiversity and endemism, conservation biogeography, and systematic conservation planning (SCP).

## Environmental DNA and metabarcoding

As mentioned earlier, effective conservation of subterranean biodiversity is limited by a lack of knowledge of species' distributions. This problem is exacerbated for species that are rare and endangered, small, and requiring collection of whole specimens for taxonomic identification from habitats that are exceptionally difficult to access or survey. The analysis of DNA shed in the environment, termed "environmental DNA" (eDNA), is a powerful, rapid, noninvasive, and potentially cost-efficient tool for addressing many of these challenges.

Coupling eDNA techniques with high-throughput sequencing enables detection of DNA from many different species within a single environmental sample (eDNA metabarcoding, Fig. 23.4A). This approach allows the characterization of entire ecological communities (eukaryotes and prokaryotes) and, combined with bioinformatics, greatly increases our ability to understand biotic functions within ecosystems (reviews in Deiner et al., 2017; Ruppert et al., 2019). In recent years, the ability to detect trace amounts of eDNA (i.e., sensitivity) has improved while costs have plummeted, making the technique feasible for routine assessment and monitoring (Fig. 23.4B). Future eDNA metabarcoding studies show exciting potential for rapidly advancing our understanding of groundwater ecosystems and improving their conservation (Fig. 23.4C). Even more exciting is the possible use of eRNA as a supplement to eDNA, assessed using the same procedures shown in Fig. 23.4. In the environment, eRNA persists for a much shorter time after cell death so it provides a more accurate representation of the living biotic communities (e.g., Mengoni et al., 2005). However, eRNA samples have added costs and transportation difficulties.

These eDNA-based approaches have been used to aid biodiversity conservation in many aquatic habitats (reviewed in Thomsen and Willerslev, 2015). Because the technique provides information on the occurrence of focal organisms with a high degree of efficiency and sensitivity yet without destructive sampling, it has proved valuable for targeted searches of taxa considered rare or endangered (Thomsen et al., 2012; Niemiller et al., 2018b), identifying species new to science and tracking invasive species (Goldberg et al., 2016; Ruppert et al., 2019). However, it has seldom been used in groundwater ecosystems (Saccò et al., 2022). The few examples include using eDNA to detect populations of the salamander *Proteus anguinus* in Europe (Gorički et al., 2017; Voros et al., 2017), microbial communities in Australia (Korbel et al., 2017), and amphipods in the United States (Niemiller et al., 2018b). This latter study demonstrated the efficiency of using eDNA approaches for screening rare groundwater invertebrates, with the benefit of avoiding the removal of rare species as required by traditional methods.

As well as species distribution data, effective conservation requires information on the functional roles of groundwater biota, their delivery of ecosystem services, trophic interactions, and hydrological and ecological connectivity with surface waters, all of which can be studied using molecular biology. For example, trophic interactions have been studied by metabarcoding

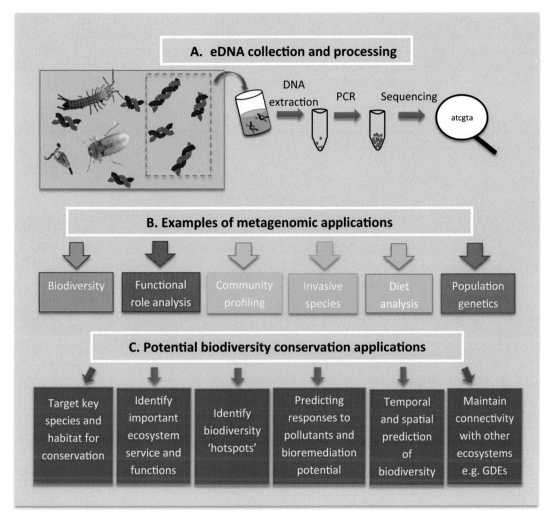

**FIGURE 23.4** Workflow and examples of conservation applications associated with environmental DNA sampling of prokaryotic and eukaryotic biota within groundwater. GDEs = groundwater-dependent ecosystems; PCR = polymerase chain reaction, a technique for amplifying targeted DNA. *Credit: Kathryn Korbel.*

the gut contents of stygofauna (Saccò et al., 2019, Venarsky et al. this volume), eDNA has identified metabolic functions of microorganisms within aquifers (e.g., detection of methanogens, Korbel et al., 2017), and qPCR has been used to quantify biogeochemical processes such as nitrification in groundwaters (e.g., De Vet et al., 2011). Although only at a preliminary stage, the application of molecular methods shows promise in identifying groundwater connectivity (Segawa et al., 2015) and faunal exchanges with the surface. Given the importance of hydrological and ecological connectivity to subterranean biodiversity (Fig. 23.2), the information gleaned from eDNA is crucial when delineating protected areas that incorporate sites of groundwater recharge and connections among groundwater habitats in the landscape.

For monitoring biota, eDNA has the potential to aid conserving biodiversity "hotspots" and endemism, protecting rare species, protecting species with important functional roles and preventing invasive species more rapidly, comprehensively and cheaply than traditional approaches (e.g., Thomsen et al., 2012). However, it is also crucial to acknowledge the pitfalls in the use of eDNA methods (Box 23.1) and results should be interpreted cautiously in their application to conservation.

---

### BOX 23.1

## Pitfalls of using eDNA techniques in groundwater conservation

Despite their great promise, eDNA-based approaches have several methodological and bioinformatic limitations that produce challenges in groundwater ecosystems (see Goldberg et al., 2016; Deiner et al., 2017; Ruppert et al., 2019). Firstly, the origin and transport of eDNA are largely unknown in groundwater. In surface waters, eDNA can travel several kilometers, obscuring the precise origin of source biota (Barnes and Turner, 2016), and the same is likely true for groundwaters. Thus, a positive detection from a water sample collected at a spring could represent DNA of organisms just a few meters upstream of the spring or several kilometers away.

Secondly, the persistence of eDNA in groundwater is largely unknown. In most surface waters, DNA degrades within hours to days (e.g., Thomsen et al., 2012) and so its presence is usually interpreted as coming from contemporary sources. However, eDNA can settle out and become incorporated in sediment (Turner et al., 2015) for long periods of time, perhaps hundreds of years, which may be problematic where groundwater sampling disturbs sediments and "old" eDNA. The degradation of DNA depends on factors such as UV radiation, temperature, pH, salinity, and microbial activity (Barnes et al., 2014; Strickler et al., 2015) and may be slower in groundwater than in surface habitats because of the typically stable, dark groundwater environment. Preliminary results from a mesocosm experiment in a cave in Alabama suggest that eDNA may persist for months or even longer (Niemiller, unpublished data). Therefore, understanding rates of DNA degradation in different groundwater habitats is crucial if eDNA is to be used to monitor species of conservation concern.

Finally, eDNA studies in groundwater ecosystems are severely hampered by the lack of reference DNA sequence data for most taxonomic groups (Zagmajster et al., 2022). Comprehensive reference sequence databases are necessary for designing species-specific assays and for accurate taxonomic assignment in metabarcoding studies. Poor representation in reference databases of intraspecific variation among populations within a target species can lead to high rates of false negatives. Conversely, poor representation of cooccurring taxa can result in false positives if nontarget species' DNA is amplified. The lack of sequences from specimens collected from type localities can impede correct species identification and introduce "misidentifications" within sequence databases.

---

These molecular methods should be combined with traditional ones to fully describe ecological communities, especially for eukaryotes (Korbel et al., 2017). In addition to the lack of sequence data from type localities (Box 23.1), species delimitation can yield very different results between methods (Eme et al., 2018), especially if based on single loci (Delli-cour and Flot, 2018). Some groups such as crustaceans (common in groundwaters) are especially challenging to barcode (see Asmyhr and Cooper, 2012), and primers need further development to obtain reliable eDNA for cryptic and small species that are harder to identify using traditional methods. Thus, eDNA and eRNA approaches should be viewed as a promising tool in groundwater biodiversity assessment for conservation but their current limitations must be acknowledged.

## Vulnerability mapping

Assessing the vulnerability of groundwater habitats to pollution and other threats is essential for their effective conservation. There are many approaches (reviews in Wachniew et al., 2016; Iván and Mádl-Szőnyi, 2017; Machiwal et al., 2018), most of which model the susceptibility of groundwater aquifers to contamination conveyed as a vulnerability risk map. Vulnerability is usually expressed as a function of the geological, hydrological, and hydrogeological properties of the groundwater-flow system (intrinsic vulnerability) and factors related to specific contaminants (specific vulnerability) such as proximity to a contaminant source and physical and biogeochemical attenuation processes (Focazio et al., 2002; Machiwal et al., 2018).

Groundwater vulnerability models vary widely in scope, complexity and computational approach. They broadly fall into three classes: (i) index-based (qualitative) methods, (ii) process-based (quantitative) methods, and (iii) statistical methods. Index-based models can be subdivided further by aquifer media (e.g., porous vs. karst) and generally aim at assessing intrinsic vulnerability. Process-based methods consider or simulate groundwater flow and transport processes of water and contaminants. Statistical methods attempt to predict the concentrations or probabilities of contamination based on relationships with empirical data (Iván and Mádl-Szőnyi, 2017) and can include regression and artificial intelligence approaches.

Statistical and index-based approaches gained popularity in the 1990s with advances in computer technology and GIS. Currently, the most popular index-based model is DRASTIC (Aller et al., 1987), which uses seven hydrogeological parameters to assess groundwater vulnerability of porous aquifers: depth to water table ($D$), net recharge ($R$), aquifer media ($A$), soil media ($S$), topography ($T$), impact of the vadose zone ($I$), and hydraulic conductivity of the aquifer ($C$). These parameters are assigned weights and relative ratings to generate a vulnerability index for a given region. DRASTIC has been modified over time for specific assessment needs (e.g., KARSTIC for karst aquifers; Davis et al., 2002). Other popular index-based methods, particularly for karst and fractured aquifers, include SINTACS (Civita and De Maio, 2004), GOD (Foster, 1987), and EPIK (Doerfliger et al., 1999). Most of these methods can distinguish degrees of vulnerability at regional scales across different rock-types but are less effective at assessing vulnerability in carbonate aquifers. For these aquifers, another method (COP, Vías et al., 2006) is recommended that considers karst's special hydrogeological properties and can be applied to different climatic conditions and types of carbonate aquifers.

No single vulnerability assessment model consistently outperforms the others in all hydrogeological settings. For example, although process-based models can be powerful, many are complex and limited by the availability and quality of current spatial and temporal datasets (Machiwal et al., 2018). Index-based approaches are straightforward and easily integrated into a GIS framework, but there is considerable subjectivity in assigning weights and ratings for model parameters. To address these problems, hybrid models combining multiple approaches have been developed (e.g., Neshat and Pradhan, 2015). For instance, the subjective selection, weighting, and rating of parameters in many index-based models can be diminished using a sensitivity analysis based on quantitative data to eliminate redundant parameters. Methods for assessing vulnerability of aquifer systems are still being developed and will continue to improve in the future.

Groundwater vulnerability assessments underpin conservation and ecological risk assessments for many groundwater species and ecosystems. One good example is the modified DRASTIC model for karst groundwater in northern Arkansas, USA, implemented at the watershed scale (Fig. 23.5A, Inlander et al., 2011). This GIS-based model, DRASTIK,

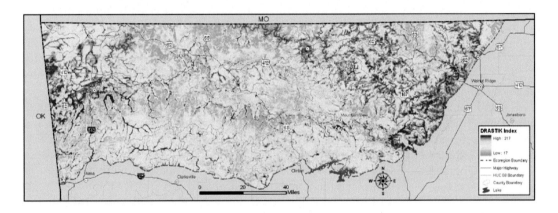

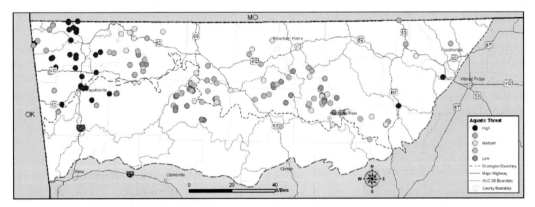

**FIGURE 23.5**   A: Heat map of the DRASTIK index for karst groundwater in northern Arkansas, USA at the watershed scale; B: Threat ratings for sites occupied by 18 groundwater species of state conservation concern. *Credit: Michael Slay.*

included [K]arst lineaments as a parameter. The vulnerability model was combined with an assessment of surface water quality and quantity as well as rates of human visitation to derive a groundwater sensitivity model for sites occupied by 18 groundwater species of state conservation concern (Fig. 23.5B). This assessment suggested that all species are experiencing some level of threat, with species such as Ozark Cavefish (*Troglichthys rosae*) and the Benton Cave Crayfish (*Cambarus aculabrum*) in urban areas experiencing the greatest impacts.

## Conserving hotspots of groundwater biodiversity and endemism

Hotspots of subterranean biodiversity and high levels of endemism (most large-bodied stygofauna species have a range <200 km in length: Trontelj et al., 2009) occur worldwide but are poorly known in many areas. Further, as biodiversity includes phylogenetic and genetic diversities, we need models predicting the most probable location of hotspots of all forms of groundwater biodiversity to help us identify and then conserve appropriate hotspots. To achieve this goal, historical, spatial and ecological determinants driving patterns of subterranean species distribution (e.g., paleobiogeographical, climatic, geological, topographic, and hydrological factors) must all be taken in account.

When subterranean biodiversity data are scarce, a useful approach is to map potential distributions based on known preferences of stygofauna and using whatever spatial data are available. In the previous section, we saw how maps and models can reveal vulnerability, and similar approaches can be used to predict areas that are potentially rich in subterranean species, relicts and/or endemic taxa, and using this information to guide conservation plans. Phylogenetic diversity (measuring species relatedness) and functional diversity (based on functional traits) has proved to be particularly sensitive for detecting responses of communities to environmental changes (Cianciaruso et al., 2009). The presence of relictual taxa increases phylogenetic diversity and may be more relevant than taxonomic diversity for many conservation purposes (Arponen, 2012).

Functional, phylogenetic and "dark" (defined as all species in a determined region that can potentially inhabit those particular conditions— Pärtel et al., 2011) diversities are crucial attributes of subterranean biodiversity (e.g., Gallão and Bichuette, 2015; Fernandes et al., 2016). In many subterranean communities, the absence of some taxa is conspicuous and ecologically relevant, resulting in increased phylogenetic and "dark" diversities. The implications for conservation are clear: phylogenetic, functional and dark diversity may be useful complements to species richness, and the relative contribution of an individual species, including its putative position in a phylogeny, whether it is relictual, its degree of specialization, endemism and genetic diversification, and its behavior and physiology must all be considered in groundwater conservation strategies.

In one of the few examples assessing subterranean functional and dark diversity in the Neotropics, Fernandes et al. (2016) showed that functional diversity of terrestrial oniscidean isopods (crustaceans) within caves was greater than in surface habitats. This implies phenotypic overdispersion (Weiher and Keddy, 1995), adding another strong reason for the conservation of such fragile environments. Extending this work, Fernandes et al. (2019) assessed the dark diversity of these cave isopods to detect patterns of co-occurrence among phylogenetically related species and predict the theoretical species pool for comparison with the observed richness. Except for one cave, dark diversity equaled or exceeded observed richness, suggesting that those caves with higher completeness represented a valuable sample of the regional subterranean species pool and should be conserved as biodiversity hotspots. Extending these parallels to the abundant aquifers in the Neotropics (e.g., 181 aquifers in Brazil) allows maps of hotspots of stygofaunal biodiversity and endemicity to be derived and compared with land-use intensity, revealing that many of these hotspots unfortunately coincide with high rates of groundwater extraction for irrigation and domestic use (Fig. 23.6).

## Conservation biogeography of subterranean species

Conservation biogeography is a recent discipline involving "the application of biogeographical principles, theories and analyses to problems concerning the conservation of biodiversity" (Richardson and Whittaker, 2010). Older methods that relied on the analysis of current-day biodiversity patterns are now being replaced by methods that incorporate the ecological and evolutionary dynamic processes that generated and shaped biodiversity patterns. The advent of rapid DNA-sequencing technologies has led to reconstruction of many groundwater species phylogenies, enabling assessment of the evolutionary value of species and subterranean habitats. Analyzing the currently available data with recently developed statistical techniques like geographical weighted regressions and spatial autocorrelation analysis allows us to predict broad-scale groundwater diversity patterns and locations of biodiversity hotspots. These predictions also need data about current or past climatic conditions, subterranean habitat heterogeneity and surface productive energy (Zagmajster et al., 2014). Understanding multicausality (i.e., that geographic variation in species richness is more likely shaped by multiple factors) and spatial nonstationarity (i.e., that the relative contribution of factors may vary among regions) provides a broader perspective of groundwater biodiversity determinants (Eme et al., 2015) but disentangling their effects remains challenging.

Despite the increase in available data and the power of new analytical techniques, we are far from a theory capable of predicting species distribution in groundwaters based on complex environmental and historical factors acting at different scales in space and time. Most practical decisions in biodiversity conservation are taken at regional or local scales whereas the predictive models mentioned above have been tested at broader scales. More detailed species distribution models (SDMs, see next section) may help to fill distributional "gaps" at smaller scales but these models are hampered by high endemism (see Mammola and Leroy, 2018 for possible options).

Although the predictive power of ecological and spatial analyses has been satisfactory at some scales, accounting for historical biogeography is more challenging because of the uncertainty of paleogeographical scenarios. Several novel techniques may help in understanding the history of present-day species distributions to guide conservation planning. For example, Bayesian phylogeographic inference may be used to reveal colonization dynamics of stygobionts based on molecular phylogenies (Eme et al., 2013). Paleogeographical modeling using software packages like BioGeoBEARS (Matzke, 2013) applied to multilocus phylogenetic

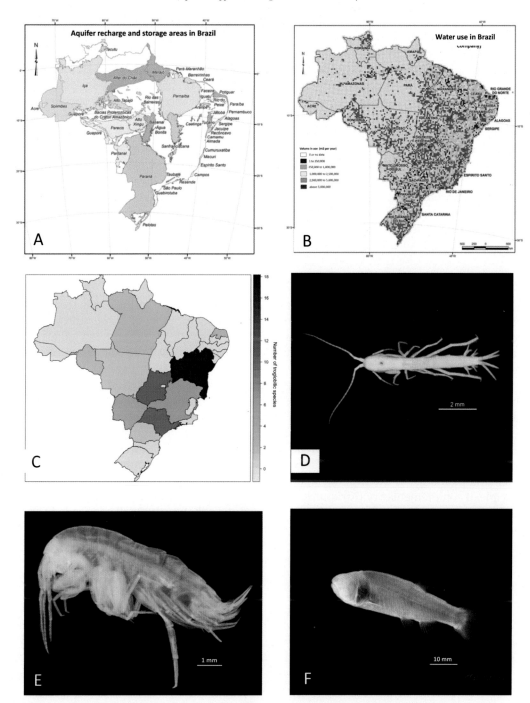

**FIGURE 23.6** Map of Brazil showing aquifer regions (A), water use (B) and regional numbers of amphibious and aquatic stygofauna (C) that include endemic species such as the spelaeogriphacean *Potiicoara brasiliensis* (D), the amphipod *Spelaeogammarus spinilacertus* (E), and the "piaba" *Stygichthys typhlops* (F). *Credit: (A and B) CPRM Mineral Resources Research Company; (C) Maria Elina Bichuette; (D and E) Luciana B. R. Fernandes; (F) Adriano Gambarini.*

trees can infer patterns of dispersal and vicariance, and has been applied to groundwaters (Delič et al., 2020). However, dispersal and distribution of single species or clades may not be representative of large areas, and the uneven species distribution and high degree of endemism may require the use of different indicators in different areas (Stoch et al., 2009).

Conservation biogeographers have also focused recently on heterogeneity in species distribution (i.e., "beta" diversity) and the phylogenetic resemblance of different species assemblages (phylobetadiversity). Partitioning beta diversity into nestedness and turnover (Baselga, 2010) is relevant for conservation because the degree of nestedness correlates with the probability of extinction, indicating areas where conservation is more urgent. Analysis of species turnover decay with distance may help calculate the extension of areas which may be considered "conservation units," acknowledging that turnover in groundwaters varies with latitude (Zagmajster et al., 2014). The species turnover component of beta diversity underpins a new technique called Generalized Dissimilarity Modeling (GDM) which has been recently applied to groundwater monitoring programs and conservation planning (Mokany et al., 2019b).

## Systematic conservation planning approaches

Systematic Conservation Planning is a technique that emerged in the 1980s (Kirkpatrick, 1983) to guide objective decisions about the nature and location of conservation actions. Most approaches allow the operator to compile a "shopping list" of conservation values which is then fulfilled using a prioritization algorithm that trades off attributes for elimination from the list with a measure of cost.

Most SCP approaches require a comprehensive list of conservation attributes before embarking on a systematic planning exercise. Objectives must be stated directly; whether targets are species, ecosystems or surrogates (e.g., landscape measures representing biodiversity), an exact statement of quantities is needed which can then be recalibrated in the early stages of the plan. These SCP approaches also rely on a selection process, usually a complementarity-based optimization algorithm. This is what is commonly associated with systematic planning—it attempts to minimize the impact on stakeholders by invoking the principle of complementarity (Kirkpatrick, 1983). In the example in Fig. 23.7, although species richness is higher in aquifers A and B, the only solution that protects all species is D and E. This logic underlies all systematic conservation plans.

| Aquifer | | | | | | | | Species richness |
|---------|---|---|---|---|---|---|---|------------------|
| A | x | x | x | x | | | x | 5 |
| B | x | x | x | x | | | x | 5 |
| C | x | x | | | | | x | 3 |
| D | | | x | x | x | | | 3 |
| E | x | x | | | | x | x | 4 |

FIGURE 23.7    The principle of complementarity. Despite Aquifers A and B containing more species, protecting D and E would conserve all seven species of stygofauna. *Credit: Simon Linke, Grant Hose and Kath Korbel.*

Although SCP had been developed in the 1980s, its application to surface freshwaters such as rivers was hindered by the challenges of dealing with hydrological connectivity, both upstream and onto floodplains. The first systematic conservation plan in freshwater systems had no explicit spatial component in the algorithm so longer river stretches were flagged for protection (e.g., Roux et al., 2002). Several years later, software for SCP had advanced to include explicit spatial components, and flexible upstream protection was devised for the leading packages Marxan (Hermoso et al., 2011) and Zonation (Moilanen et al., 2008). Despite the impediments (e.g., taxonomic difficulties, challenges with mapping and spatial frameworks) described in previous sections, SCP in groundwater ecosystems started not long after the first river studies were published. Drawing on the European PASCALIS database, Michel et al. (2009) assessed 10,000 records of >1000 taxa for SCP, using multiple approaches to prioritization, including Marxan, which can trade off conservation priorities with both ecological condition and socio-economic costs. In this project, taxonomic records were used as surrogates.

Asmyhr et al. (2014) developed a completely novel approach to include phylogenetic data (see previous sections) from stygofauna in prioritizations. Instead of using taxa as input to the complementarity matrix, they developed a phylogenetic tree from genetically sequenced samples of stygofauna and then used the branches as "pseudospecies." This has interesting consequences in the conservation planning algorithm because it goes beyond just counting the number of species and the complementarity between species by considering phylogenetic similarity. For example, if two sites have the same level of rarity, the one that adds more phylogenetic diversity to the overall reserve system is chosen. Although developed for applications to stygofauna protection, this approach has found prominence in mammal conservation (Rosauer et al., 2017).

Conservation planning is always a tradeoff between spatial resolution and taxonomic completeness. Typically, datasets comprise opportunistically collected data that are partly boosted by expert assessments, but are not spatially complete. This can lead to cases where areas are flagged as hotspots because they have been more heavily sampled rather than actually having the highest richness, endemism or complementarity. The opposite effect—high completeness but low taxonomic resolution—arises when using surrogates. The earliest landscape-scale study simply used aquifers in a standard conservation plan in China, retrofitted onto surface units (Li et al., 2017). A later study (Linke et al., 2019), done in the Hunter Valley, Australia, attempted to account for the spatial complexities of aquifers, combining conventional longitudinal connectivity in rivers with lateral connectivity to floodplains and vertical connectivity to the aquifer (Fig. 23.8A). Aquifer surrogates were simply subterranean ecoregions, defined by similar geology and stygofauna. Connectivity rules were modified so that aquifers were vertically integrated with wetlands and rivers above, resulting in steady increases in reserved area up to 28.3% of the catchment needed to protect connected rivers, wetlands and 75% of aquifers (Fig. 23.8). This example shows that integration of surface and groundwaters in SCP is feasible. At least for coarse surrogates, protecting aquifers does not necessarily increase cost substantially but does change spatial priorities. In other words, planning to protect aquifers can come as an extra and relatively inexpensive benefit to conserving surface waters as long as it is included a priori.

Several challenges currently hinder the application of SCP for conserving groundwater biodiversity. Despite new analytical prioritization methods (e.g., Linke et al., 2019), functional

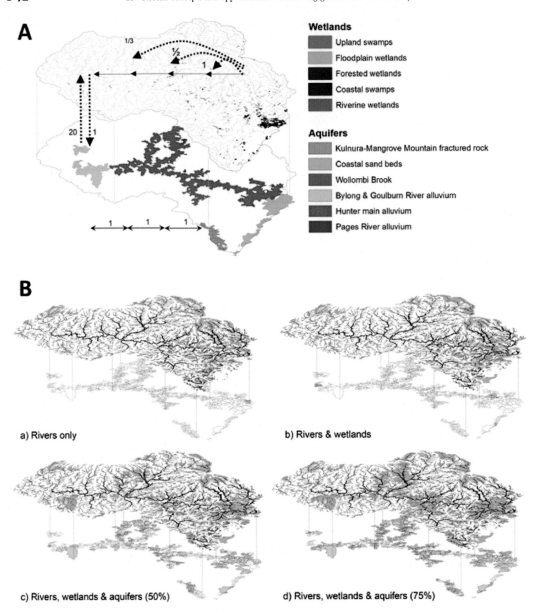

**FIGURE 23.8** A: Connected rivers, wetlands and groundwater features in the spatial framework of the Hunter Valley. Australia (numbers above arrows indicate connection strength in Marxan); B: conservation areas (pale green) needed to protect only rivers (a, 14% of total area), rivers and wetlands (b, 17%), rivers, wetlands, and 50% of aquifers (c, 21.5%) and rivers, wetlands, and 75% of aquifers (d, 28.3%). *Credit: Linke et al. (2019).*

connections between aquifers and above-ground features are not yet routinely included, leading to inadequate estimates of connectivity. Further, the lack of comprehensive data severely limits the quality of the plans although this may be partly resolved using species distribution models (SDMs) to extrapolate species records across the landscape by using environmental correlates. Theoretically, this should be possible as recent research has shown that subterranean faunal assemblages respond to depth and sediment (Hose et al., 2017; Korbel et al., 2019) and salinity (Shapouri et al., 2015)—classic predictor variables in conservation SDMs.

Such SDMs have not been used in groundwater planning because 5–10 observations of the same taxon are usually needed for acceptable landscape models (but see Smith et al., 2019). This can be circumvented by a new statistical technique termed GDM that uses turnover of biodiversity (how similar sampled sites are and what environmental drivers are responsible for the similarity/dissimilarity). Recently, this technique has been used in the Western Australian Pilbara region to derive links between stygofauna identified in standard monitoring programs and landscape drivers (Mokany et al., 2019a, 2019b). These models produce a stratified list of environmental drivers which can be used as surrogates in planning software packages like Marxan. The input requirements for GDM models are extremely flexible and could use phylogenetic information in which genetic dissimilarity is measured instead of species dissimilarity. This seems a logical step forward, combining phylogenetic monitoring planning approaches (Asmyhr et al., 2014) with landscape prioritization, and building on the development of conservation biogeography to preserve hotspots of subterranean biodiversity and endemism.

## Conclusion and future directions

In the two decades since this book's first edition, our knowledge of the diversity, distribution, and vulnerability of subterranean fauna and microbial assemblages has increased dramatically, along with an increased awareness of the urgent need to effectively conserve groundwater ecosystems and their surface connections. However, there is still much that we do not know and many impediments to groundwater conservation remain. A "taxonomy crisis" exists where there are not enough taxonomists globally with the expertise to describe let alone identify species in widespread subterranean taxonomic groups such as flatworms, gastropods and some crustaceans. Many groundwater taxa verging on extinction may disappear as they await formal description (Niemiller et al., 2013; Trajano et al., 2016).

Although novel technological advances such as eDNA and metabarcoding are likely to help overcome some of these impediments, these methods cannot wholly replace traditional taxonomy and there are still constraints to their use. Systematic reserve planning and judicious spatial modeling can guide efficient allocation of resources to conserve connected surface and groundwater ecosystems (e.g., Linke et al., 2019) but these methods depend on reliable empirical data currently only available for restricted areas globally. We cannot protect what we do not know.

Currently, conservation targets for groundwater ecosystems are seldom proposed on ecological or biodiversity grounds. Instead, policies concerning groundwater sustainability are deeply rooted in hydrogeology, with the premise that maintaining water supply and not exceeding recharge leads to sustainability (e.g., Alley et al., 1999; Zhou et al., 2009). In many countries, groundwater sustainability is perceived solely in terms of sustainable pumping yields for water supply, and defined in terms of water balances, recharge, and abstraction; there is no acknowledgment of other ecosystem services such as biodiversity, enhancing water quality, and supporting GDEs. Although several countries have included these values in their groundwater legislation (e.g., the EU's Groundwater Daughter Directive (Directive, 2006/118/EC); Australia's National Water Initiative (NWI, 2014), severe data limitations constrain effective enforcement (Griebler et al. this volume). Most conservation effort focuses on surface-expressed GDEs, whereas the protection of aquifers and subterranean ecosystems remains largely elusive in government policies, management strategies, and routine monitoring.

This ignorance of the importance of conserving subterranean ecosystems appears to arise because they are "out of sight, out of mind." However, the importance of conserving subterranean ecosystems is increasingly promoted in terms of ecosystem services provided by intact groundwater ecosystems and their biodiversity. Public awareness is greatest about "provisioning" services for water. "Cultural" services are also highly valued in some parts of the world, so there is scope to better harness this value by conserving groundwater as "cultural groundwater" (Boulton, 2020). Groundwater ecosystems perform numerous crucial "regulating services" such as nutrient recycling and purifying water (Griebler and Avramov, 2015). Groundwater biologists and conservationists need to be more active in public outreach exercises that highlight these ecosystem services because if we do not effectively conserve our groundwater ecosystems, many of these services will be impaired or lost.

What does the future hold for conservation of groundwater biodiversity, especially in the application of novel approaches and concepts? We suspect that major new concepts in biodiversity are unlikely but that there will be growing awareness of the importance of genetic diversity to complement organismal biodiversity, as well as its causes and mechanisms. Better understanding of the mechanisms that generate biodiversity could make future conservation strategies more effective by focusing on protecting the processes that generate species richness (e.g., speciation, extinction and dispersal) rather than basing conservation plans solely on the output of these processes (e.g., species richness and other biodiversity metrics). Various metrics, including phylogenetic diversity indexes, enable planners to map out regions where there is a low extinction rate and diversity has been preserved over time ("museums," Jablonski et al., 2006), there are high rates of recent speciation in situ ("cradles of diversity," Jablonski et al., 2006) and/or where lineages have been preserved through time and received phylogenetically distant migrants from neighboring regions ("Noah's Arks," Zagmajster et al., 2018). Conservation priorities may then be specified by using this knowledge of the main processes and associated factors (e.g., spatial heterogeneity, long-term environmental stability) that contribute to subterranean biodiversity.

Increasing use of molecular methods will continue, supplemented with traditional taxonomic approaches for eukaryotic species. Meanwhile, spatial and hydrogeological maps will be refined to better identify where groundwater ecosystems occur and how they are connected, while concurrent surveys and taxonomic assessments incorporating metabarcoding and traditional taxonomy provide publicly accessible data on subterranean biodiversity. This information, along with data from geodatabases such as the European Groundwater Crustacean Database (Zagmajster et al., 2014) and the Subterranean Biodiversity Database (Bregović et al., 2019), will increasingly be analyzed and modeled to generate SCP for reserves that explicitly include surface connections to groundwater recharge and discharge areas. However, databases like these should be made more accessible to promote their efficient use (Fišer, 2019).

Finally, as scientific and public awareness grows of the threats facing the continued provision of ecosystem services by groundwaters (Mammola et al., 2019), there will be greater impetus to develop more powerful methods to use strategic databases and sophisticated models to produce conservation plans that integrate surface and subterranean waters to protect hotspots of biodiversity and endemism. These plans must actively involve citizen scientists, relevant land and water managers and administrators, local regional councils, Indigenous peoples, and other stakeholders to be involved in restoration programs (e.g., Manenti et al., 2019) so that "out of sight" will no longer be "out of mind" and facing extinction in our lifetimes.

## Acknowledgments

We thank Slavko Polak, Peter Gedei, Roman Ozimec, Fran Sheldon, Luciana Fernandes and Adriano Gambarini for providing photos, Michael Slay for the images in Fig. 23.5, and Florian Malard, Christian Griebler and Stefano Mammola for constructive comments on an earlier draft. We are especially grateful to Florian Malard for suggesting that groundwater conservation may be more effective by focusing on the processes generating species richness. MEB thanks J.E. Gallão for help in the conception of ideas and projects, and FAPESP, CNPQ and Capes for financial support.

## References

Aller, L., Bennett, T., Lehr, J.H., Petty, R.G., Hackett, G., 1987. DRASTIC: A Standardized System for Evaluating Ground Water Pollution Potential Using Hydrogeologic Settings. Doc. EPA/600/2-87/035. US Environmental Protection Agency, Washington, DC.

Alley, W.M., Reilly, T.E., Franke, O.L., 1999. Sustainability of Ground-Water Resources. US Geological Survey Circular 1186. US Geological Survey, Denver, Colorado.

Arponen, A., 2012. Prioritizing species for conservation planning. Biodiversity & Conservation 21, 875—893.

Asmyhr, M.G., Cooper, S.J.B., 2012. Difficulties barcoding in the dark: the case of crustacean stygofauna from eastern Australia. Invertebrate Systematics 26, 583—591.

Asmyhr, M.G., Linke, S., Hose, G., Nipperess, D.A., 2014. Systematic conservation planning for groundwater ecosystems using phylogenetic diversity. PLoS One 9, e115132.

Barnes, M.A., Turner, C.R., 2016. The ecology of environmental DNA and implications for conservation genetics. Conservation Genetics 17, 1—17.

Barnes, M.A., Turner, C.R., Jerde, C.L., Renshaw, M.A., Chadderton, W.L., Lodge, D.M., 2014. Environmental conditions influence eDNA persistence in aquatic systems. Environmental Science and Technology 48, 1819—1827.

Baselga, A., 2010. Partitioning the turnover and nestedness components of beta diversity. Global Ecology and Biogeography 19, 134—143.

Botosaneanu, L., 1986. Stygofauna Mundi. A Faunistic, Distributional and Ecological Synthesis of the World Fauna Inhabiting Subterranean Waters (Including the Marine Interstitial). Brill, Leiden.

Boulton, A.J., 2020. Conservation of groundwaters and their dependent ecosystems: integrating molecular taxonomy, systematic reserve planning and cultural values. Aquatic Conservation: Marine and Freshwater Ecosystems 30, 1—7.

Boulton, A.J., Humphreys, W.F., Eberhard, S.M., 2003. Imperilled subsurface waters in Australia: biodiversity, threatening processes and conservation. Aquatic Ecosystem Health 6, 41—54.

Brancelj, A., Mori, N., Treu, F., Stoch, F., 2020. The groundwater fauna of the Classical Karst: hydrogeological indicators and descriptors. Aquatic Ecology 54, 205—224.

Bregović, P., Fišer, C., Zagmajster, M., 2019. Contribution of rare and common species to subterranean species richness patterns. Ecology and Evolution 9, 11606—11618.

Cianciaruso, M.V., Silva, I.A., Batalha, M.A., 2009. Diversidades filogenética e funcional: novas abordagens para a Ecologia de comunidades. Biota Neotropica 9, 93—103.

Civita, M., De Maio, M., 2004. Assessing and mapping groundwater vulnerability to contamination: the Italian "combined" approach. Geofísica Internacional 43, 513–532.

Cornu, J.-F., Eme, D., Malard, F., 2013. The distribution of groundwater habitats in Europe. Hydrogeology Journal 21, 949–960.

Culver, D.C., Sket, B., 2000. Hotspots of subterranean biodiversity in caves and wells. Journal of Cave and Karst Studies 62, 11–17.

Culver, D.C., Master, L.L., Christman, M.C., Hobbs, H.H., 2000. Obligate cave fauna of the 48 contiguous United States. Conservation Biology 14, 386–401.

Culver, D.C., Pipan, T., Gottstein, S., 2006. Hypotelminorheic — a unique freshwater habitat. Subterranean Biology 4, 1–8.

Daily, G.C., 1997. Nature's Services: Societal Dependence on Natural Ecosystems. Island Press, Washington, DC.

Danielopol, D.L., Pospisil, P., 2001. Hidden biodiversity in the groundwater of the Danube Floodplain National Park (Austria). Biodiversity & Conservation 10, 1711–1721.

Danielopol, D.L., Griebler, C., Gunatilaka, A., Notenboom, J., 2003. Present state and future prospects for groundwater ecosystems. Environmental Conservation 30, 104–130.

Davis, A., Long, A., Wireman, M., 2002. KARSTIC: a sensitivity method for carbonate aquifers in karst terrain. Environmental Geology 42, 65–72.

De Marsily, G., 1992. Creation of 'Hydrological Nature Reserves': a plea for the defence of ground water. Ground Water 30, 658–659.

De Vet, W.W.J.M., Kleerebezem, R., Van der Wielen, P.W.J.J., Rietveld, L.C., van Loosdrecht, M.C.M., 2011. Assessment of nitrification in groundwater filters for drinking water production by qPCR and activity measurement. Water Research 45, 4008–4018.

Deiner, K., Bik, H.M., Machler, E., Seymour, M., Lacoursiere-Roussel, A., Altermatt, F., Creer, S., Bista, I., Lodge, D.M., de Vere, N., Pfrender, M.E., Bernatchez, L., 2017. Environmental DNA metabarcoding: transforming how we survey animal and plant communities. Molecular Ecology 26, 5872–5895.

Delić, T., Stoch, F., Borko, Š., Flot, J.-F., Fišer, C. 2020. How did subterranean amphipods cross the Adriatic Sea? Phylogenetic evidence for dispersal–vicariance interplay mediated by marine regression–transgression cycles. Journal of Biogeography 47, 1875–1887.

Dellicour, S., Flot, J.-F., 2018. The hitchhiker's guide to single-locus species delimitation. Molecular Ecology Resources 18, 1234–1246.

Doerfliger, N., Jeannin, P.-Y., Zwahlen, F., 1999. Water vulnerability assessment in karst environments: a new method of defining protection areas using a multi-attribute approach and GIS tools (EPIK method). Environmental Geology 39, 165–176.

Eme, D., Malard, F., Konecny-Dupré, L., Lefébure, T., Douady, C.J., 2013. Bayesian phylogeographic inferences reveal contrasting colonization dynamics among European groundwater isopods. Molecular Ecology 22, 5685–5699.

Eme, D., Zagmajster, M., Fišer, C., Galassi, D., Marmonier, P., Stoch, F., Cornu, J.-F., Oberdorff, T., Malard, F., 2015. Multi-causality and spatial non-stationarity in the determinants of groundwater crustacean diversity in Europe. Ecography 38, 531–540.

Eme, D., Zagmajster, M., Delić, T., Fišer, C., Flot, J.-F., Konecny-Dupré, L., Pálsson, S., Stoch, F., Zakšek, V., Douady, C.J., Malard, F., 2018. Do cryptic species matter in macroecology? Sequencing European groundwater crustaceans yields smaller ranges but does not challenge biodiversity determinants. Ecography 41, 424–436.

Fernandes, C.S., Batalha, M.A., Bichuette, M.E., 2016. Does the cave environment reduce functional diversity? PLoS One 11, e0151958.

Fernandes, C.S., Batalha, M.A., Bichuette, M.E., 2019. Dark diversity in the dark: a new approach to subterranean conservation. Subterranean Biology 32, 69–80.

Ficetola, G.F., Miaud, C., Pompanon, F., Taberlet, P., 2008. Species detection using environmental DNA from water samples. Biology Letters 23, 423–425.

Fišer, C., 2019. Collaborative databasing should be encouraged. Trends in Ecology & Evolution 34, 184–185.

Focazio, M.J., Reilly, T.E., Rupert, M.G., Helsel, D.R., 2002. Assessing Ground-Water Vulnerability to Contamination: Providing Scientifically Defensible Information for Decision Makers. US Geological Survey Circular No. 1224. US Department of Interior and US Geological Survey, Reston, VA.

Foster, S.S.D., 1987. Fundamental concepts in aquifer vulnerability, pollution risk and protection strategy. In: van Duijvenbooden, W., van Waegeningh, H.G. (Eds.), Proceedings and Information in Vulnerability of Soil and Ground-Water to Pollutants, vol 38. The Hague: TNO Committee on Hydrological Research, pp. 69–86.

Gallão, J.E., Bichuette, M.E., 2015. Taxonomic distinctness and conservation of a new high biodiversity subterranean area in Brazil. Anais da Academia Brasileira de Ciências 87, 209–217.

Gibert, J., Deharveng, L., 2002. Subterranean ecosystems: a truncated functional biodiversity. BioScience 52, 473–481.

Goldberg, C.S., Turner, C.R., Deiner, K., Klymus, K.E., Thomsen, P.F., Murphy, M.A., Spear, S.F., McKee, A., Oyler-McCance, S.J., Cornman, R.S., Laramie, M.B., Mahon, A.R., Lance, R.F., Pilliod, D.S., Strickler, K.M., Waits, L.P., Fremier, A.K., Takahara, T., Herder, J.E., Taberlet, P., 2016. Critical considerations for the application of environmental DNA methods to detect aquatic species. Methods in Ecology and Evolution 7, 1299–1307.

Gorički, Š., Stanković, D., Snoj, A., Kuntner, M., Jeffery, W.R., Trontelj, P., Pavićević, M., Grizelj, Z., Năpăruş-Aljančič, M., Aljančič, G., 2017. Environmental DNA in subterranean biology: range extension and taxonomic implications for *Proteus*. Scientific Reports 7, 45054.

Griebler, C., Avramov, M., 2015. Groundwater ecosystem services: a review. Freshwater Science 34, 355–367.

Hermoso, V., Linke, S., Prenda, J., Possingham, H.P., 2011. Addressing longitudinal connectivity in the systematic conservation planning of fresh waters. Freshwater Biology 56, 57–70.

Holmes, A.J., Tujula, N.A., Holley, M., Contos, A., James, J.M., Rogers, P., Gillings, M.R., 2001. Phylogenetic structure of unusual aquatic microbial formations in Nullarbor caves, Australia. Environmental Microbiology 3, 256–264.

Hose, G.C., Stumpp, C., 2019. Architects of the underworld: bioturbation by groundwater invertebrates influences aquifer hydraulic properties. Aquatic Sciences 81, 20.

Hose, G.C., Fryirs, K.A., Bailey, J., Ashby, N., White, T., Stumpp, C., 2017. Different depths, different fauna: habitat influences on the distribution of groundwater invertebrates. Hydrobiologia 797, 145–157.

Iannella, M., Fiasca, B., Di Lorenzo, T., Biondi, M., Di Cicco, M., Galassi, D.M.P., 2020. Jumping into the grids: mapping biodiversity hotspots in groundwater habitat types across Europe. Ecography 43, 1825–1841.

Iannella, M., Fiasca, B., Di Lorenzo, T., Di Cicco, M., Biondi, M., Mammola, S., Galassi, D.M.P., 2021. Getting the 'most out of the hotspot' for practical conservation of groundwater biodiversity. Global Ecology and Conservation 31, e01844.

Inlander, E., Gallipeau, C., Slay, M., 2011. Mapping the Distribution, Habitat, and Threats for Arkansas' Species of Greatest Conservation Need. Technical Report to Arkansas Game and Fish Commission. Little Rock, Arkansas, USA.

Iván, V., Mádl-Szőnyi, J., 2017. State of the art karst vulnerability assessment: overview, evaluation and outlook. Environmental Earth Sciences 76, 112.

Jablonski, D., Roy, K., Valentine, J.W., 2006. Out of the tropics: evolutionary dynamics of the latitudinal diversity gradient. Science 314, 102–106.

Juberthie, C., Juberthie-Jupeau, L., 1975. La réserve biologique du laboratoire souterrain du CNRS à Sauve (Gard). Annales de Spéléologie 30, 539–551.

Kath, J., Boulton, A.J., Harrison, E., Dyer, F., 2018. A conceptual framework for ecological responses to groundwater regime alteration (FERGRA). Ecohydrology 11, e2010.

Kirkpatrick, J.B., 1983. An iterative method for establishing priorities for the selection of nature reserves: an example from Tasmania. Biological Conservation 25, 127–134.

Korbel, K., Chariton, A., Stephenson, S., Greenfield, P., Hose, G.C., 2017. Wells provide a distorted view of life in the aquifer: implications for sampling monitoring assessment of groundwater ecosystems. Scientific Reports 7, 40702.

Korbel, K., Stephenson, S., Hose, G.C., 2019. Sediment size influences habitat selection and use by groundwater macrofauna and meiofauna. Aquatic Sciences 81, 39.

Li, X.W., Shi, J.B., Song, X.L., Ma, T.T., Man, Y., Cui, B.S., 2017. Integrating within-catchment and interbasin connectivity in riverine and nonriverine freshwater conservation planning in the North China Plain. Journal of Environmental Management 204, 1–11.

Linke, S., Turak, E., Asmyhr, M.G., Hose, G., 2019. 3D conservation planning: including aquifer protection in freshwater plans refines priorities without much additional effort. Aquatic Conservation: Marine and Freshwater Ecosystems 29, 1063–1072.

Lomolino, M.V., 2004. Introduction. In: Lomolino, M.V., Heaney, L.R. (Eds.), Conservation Biogeography. Frontiers of Biogeography: New Directions in the Geography of Nature. Sinauer Associates, Sunderland, Massachusetts, pp. 293–296.

Machiwal, D., Jha, M.K., Singh, V.P., Mohan, C., 2018. Assessment and mapping of groundwater vulnerability to pollution: current status and challenges. Earth-Science Reviews 185, 901–927.

Malard, F., Gibert, J., Laurent, R., 1997. L'aquifère de la source du Lez: un réservoir d'eau et de biodiversité. Karstologia 30, 49–54.

Manenti, R., Barzaghi, B., Tonni, G., Ficetola, G.F., Melotto, A., 2019. Even worms matter: cave habitat restoration for a planarian species increased environmental suitability but not abundance. Oryx 53, 216–221.

Mammola, S., Leroy, B., 2018. Applying species distribution models to caves and other subterranean habitats. Ecography 41, 1194–1208.

Mammola, S., Cardoso, P., Culver, D.C., Deharveng, L., Ferreira, R.L., Fišer, C., Galassi, D.M.P., Griebler, C., Halse, S., Humphreys, W.F., Isaia, M., Malard, F., Martinez, A., Moldovan, O.T., Niemiller, M.L., Pavlek, M., Reboleira, A.S.P.S., Souza-Silva, M., Teeling, E.C., Wynne, J.J., Zagmajster, M., 2019. Scientists' warning on the conservation of subterranean ecosystems. BioScience 69, 641–650.

Marmonier, P., Vervier, P.H., Gibert, J., Dole-Olivier, M.-J., 1993. Biodiversity in ground waters. Trends in Ecology & Evolution 8, 392–395.

Matzke, N.J., 2013. Probabilistic historical biogeography: new models for founder-event speciation, imperfect detection, and fossils allow improved accuracy and model-testing. Frontiers of Biogeography 5, 242–248.

Mengoni, A., Tatti, E., Decorosi, F., Viti, C., Bazzicalupo, M., Giovannetti, L., 2005. Comparison of 16S rRNA and 16S rDNA T-RFLP approaches to study bacterial communities in soil microcosms treated with chromate as perturbing agent. Microbial Ecology 50, 375–384.

Michel, G., Malard, F., Deharveng, L., Di Lorenzo, T., Sket, B., De Broyer, C., 2009. Reserve selection for conserving groundwater biodiversity. Freshwater Biology 54, 861–876.

Moilanen, A., Leathwick, J., Elith, J., 2008. A method for spatial freshwater conservation prioritization. Freshwater Biology 53, 577–592.

Mokany, K., Harwood, T.D., Ferrier, S., 2019a. Improving links between environmental accounting and scenario-based cumulative impact assessment for better-informed biodiversity decisions. Journal of Applied Ecology 56, 2732–2741.

Mokany, K., Harwood, T.D., Halse, S.A., Ferrier, S., 2019b. Riddles in the dark: assessing diversity patterns for cryptic subterranean fauna of the Pilbara. Diversity and Distributions 25, 240–254.

National Academy of Sciences, 2015. Review of the Edwards Aquifer Habitat Conservation Plan. Water Science Report, Water Science and Technology Board, Washington, DC.

Neshat, A., Pradhan, B., 2015. An integrated DRASTIC model using frequency ratio and two new hybrid methods for groundwater vulnerability assessment. Natural Hazards 76, 543–563.

Niemiller, M.L., Graening, G.O., Fenolio, D.B., Godwin, J.C., Cooley, J.R., Pearson, W.R., Near, T.J., Fitzpatrick, B.M., 2013. Doomed before they are described? The need for conservation assessments of cryptic species complexes using an amblyopsid cavefish (Amblyopsidae: *Typhlichthys*) as a case study. Biodiversity & Conservation 22, 1799–1820.

Niemiller, M.L., Bichuette, E., Taylor, S.J., 2018a. Conservation of cave fauna in Europe and the Americas. In: Moldovan, O.T., Kovac, L., Halse, S. (Eds.), Ecological Studies: Cave Ecology. Springer, Dordrecht, pp. 451–478.

Niemiller, M.L., Porter, M.L., Keany, J., Gilbert, H., Fong, D.W., Culver, D.C., Hobson, C.S., Kendall, K.D., Davis, M.A., Taylor, S.J., 2018b. Evaluation of eDNA for groundwater invertebrate detection and monitoring, a case study with endangered *Stygobromus* (Amphipoda, Crangonyctidae). Conservation Genetics Resources 10, 247–257.

Notenboom, J., Plénet, S., Turquin, M.-J., 1994. Groundwater contamination and its impact on groundwater animals and ecosystems. In: Gibert, J., Danielopol, D.L., Stanford, J.A. (Eds.), Groundwater Ecology. Academic Press, San Diego, pp. 477–504.

NWI (National Water Initiative), 2014. Intergovernmental Agreement on a National Water Initiative. Australian Government, Canberra. https://www.pc.gov.au/inquiries/completed/water-reform/national-water-initiative-agreement-2004.pdf [Accessed 7 January 2023].

Pärtel, M., Szava-Kovats, R., Zobel, M., 2011. Dark diversity: shedding light on absent species. Trends in Ecology & Evolution 26, 124–128.

Ponder, W.F., 1986. Mound springs of the Great Artesian Basin. In: De Deckker, P., Williams, W.D. (Eds.), Limnology in Australia. Springer, Dordrecht, pp. 403–420.

Richardson, D.M., Whittaker, R.J., 2010. Conservation biogeography — foundations, concepts and challenges. Diversity and Distributions 16, 313—320.

Rosauer, D.F., Pollock, L.J., Linke, S., Jetz, W., 2017. Phylogenetically informed spatial planning is required to conserve the mammalian tree of life. Proceedings of the Royal Society B: Biological Sciences 284, 2017-2627.

Rouch, R., 1995. Peuplement des Crustacés dans la zone hyporhéique d'un ruisseau des Pyrénées. Annales de Limnologie 31, 9—28.

Roux, D., de Moor, F., Cambray, J., Barber-James, H., 2002. Use of landscape-level river signatures in conservation planning: a South African case study. Conservation Ecology 6 (2), 6.

Ruppert, K.M., Kline, R.J., Rahman, M.S., 2019. Past, present, and future perspectives of environmental DNA (eDNA) metabarcoding: a systematic review in methods, monitoring, and applications of global eDNA. Global Ecology and Conservation 17, e00547.

Saccò, M., Blyth, A., Bateman, P.W., Hua, Q., Mazumder, D., White, N., Humphreys, W.F., Laini, A., Griebler, C., Grice, K., 2019. New light in the dark — a proposed multidisciplinary framework for studying functional ecology of groundwater fauna. Science of the Total Environment 662, 963—977.

Saccò, M., Guzik, M.T., van der Heyde, M., Nevill, P., Cooper, S.J.B., Austin, A.D., Coates, P.J., Allentoft, M.E., White, N.E., 2022. eDNA in subterranean ecosystems: applications, technical aspects, and future prospects. Science of the Total Environment 820, 153223.

Sarbu, S.M., Kane, T.C., Kindle, B.K., 1996. A chemoautotrophically based cave ecosystem. Science 272, 1953—1955.

Shapouri, M., Cancela da Fonseca, L., Iepure, S., Stigter, T., Ribeiro, L., Silva, A., 2015. The variation of stygofauna along a gradient of salinization in a coastal aquifer. Hydrology Research 47, 89—103.

Segawa, T., Sugiyama, A., Kinoshita, T., Sohrin, R., Nakano, T., Nagaosa, K., Greenidge, D., Kato, K., 2015. Microbes in groundwater of a volcanic mountain, Mt. Fuji; 16S rDNA phylogenetic analysis as a possible indicator for the transport routes of groundwater. Geomicrobiology Journal 32, 677—688.

Sket, B., 1999. High biodiversity in hypogean waters and its endangerment - the situation in Slovenia, the Dinaric karst, and Europe. Crustaceana 72, 767—779.

Smith, A.B., Godsoe, W., Rodríguez-Sánchez, F., Wang, H.H., Warren, D., 2019. Niche estimation above and below the species level. Trends in Ecology & Evolution 34, 260—273.

Stanford, J.A., Gibert, J., 1994. Conclusions and perspective. In: Gibert, J., Danielopol, D.L., Stanford, J.A. (Eds.), Groundwater Ecology. Academic Press, San Diego, pp. 543—547.

Stanford, J.A., Ward, J.V., Ellis, B.K., 1994. Ecology of the alluvial aquifers of the Flathead River, Montana. In: Gibert, J., Danielopol, D.L., Stanford, J.A. (Eds.), Groundwater Ecology. Academic Press, San Diego, pp. 367—390.

Stoch, F., Artheau, M., Brancelj, A., Galassi, D.M.P., Malard, F., 2009. Biodiversity indicators in European groundwaters: towards a predictive model of stygobiotic species richness. Freshwater Biology 54, 745—755.

Strickler, K.M., Fremier, A.K., Goldberg, C.S., 2015. Quantifying effects of UV-B, temperature, and pH on eDNA degradation in aquatic microcosms. Biological Conservation 183, 85—92.

Thienemann, A., 1950. Die Verbreitungsgeschichte der Süßwassertierwelt Europas. Versuch einer historischen Tiergeographie (The history of the distribution of the European freshwater fauna. An attempt at historical zoogeography). Die Binnengewässer 18, 1—809.

Thomsen, P.F., Willerslev, E., 2015. Environmental DNA: an emerging tool in conservation for monitoring past and present biodiversity. Biological Conservation 183, 4—18.

Thomsen, P.F., Kielgast, J., Iversen, L.L., Wiuf, C., Rasmussen, M., Gilbert, M.T.P., Orlando, L., Willerslev, E., 2012. Monitoring endangered freshwater biodiversity using environmental DNA. Molecular Ecology 21, 2565—2573.

Trajano, E., Gallão, J.E., Bichuette, M.E., 2016. Spots of high diversity of troglobites in Brazil: the challenge of measuring subterranean diversity. Biodiversity & Conservation 25, 1805—1828.

Trontelj, P., Douady, C.J., Fišer, C., Gibert, J., Gorički, Š., Lefébure, T., Sket, B., Zakšek, V., 2009. A molecular test for cryptic diversity in ground water: how large are the ranges of macro-stygobionts? Freshwater Biology 54, 727—744.

Turner, C.R., Uy, K.L., Everhart, R.C., 2015. Fish environmental DNA is more concentrated in aquatic sediments than surface water. Biological Conservation 183, 93—102.

Vías, J.M., Andreo, B., Perles, M.J., Carrasco, F., Vadillo, I., Jiménez, P., 2006. Proposed method for groundwater vulnerability mapping in carbonate (karstic) aquifers: the COP method. Hydrogeology Journal 14, 912—925.

Vörös, J., Marton, O., Schmidt, B.R., Gal, J.T., Jelic, D., 2017. Surveying Europe's only cave-dwelling chordate species (Proteus anguinus) using environmental DNA. PLoS One 12, e0170945.

Wachniew, P., Zurek, A.J., Stumpp, C., Gemitzi, A., Gargini, A., Filippini, M., Rozanski, K., Meeks, J., Kværner, J., Witczak, S., 2016. Toward operational methods for the assessment of intrinsic groundwater vulnerability: a review. Critical Reviews in Environmental Science and Technology 46, 827–884.

Weiher, E., Keddy, P.A., 1995. The assembly of experimental wetland plant communities. Oikos 73, 323–335.

Zagmajster, M., Borko, Š., Delić, T., Douady, C.J., Eme, D., Malard, F., Trontelj, P., Fišer, C., 2022. Availability of DNA barcodes in subterranean amphipods of Europe. In: Gauchon, C., Jaillet, S. (Eds.), Proceedings of the 18th IUS Congress, Heritage & Ecology, vol 1, Karstologia Mémoires 21. French Federation of Speleology, pp. 361–364.

Zagmajster, M., Eme, D., Fišer, C., Galassi, D., Marmonier, P., Stoch, F., Cornu, J.-F., Malard, F., 2014. Geographic variation in range size and beta diversity of groundwater crustaceans: insights from habitats with low thermal seasonality. Global Ecology and Biogeography 23, 1135–1145.

Zagmajster, M., Malard, F., Eme, D., Culver, D.C., 2018. Subterranean biodiversity patterns from global to regional scales. In: Moldovan, O.T., Kovac, L., Halse, S. (Eds.), Ecological Studies: Cave Ecology. Springer, Dordrecht, pp. 195–227.

Zhou, Y., 2009. A critical review of groundwater budget myth, safe yield and sustainability. Journal of Hydrology 370, 207–213.

# Legal frameworks for the conservation and sustainable management of groundwater ecosystems

*Christian Griebler[1], Hans Juergen Hahn[2],*
*Stefano Mammola[3], Matthew L. Niemiller[4], Louise Weaver[5],*
*Mattia Saccò[6], Maria Elina Bichuette[7] and Grant C. Hose[8]*

[1]University of Vienna, Department of Functional & Evolutionary Ecology, Vienna, Austria;
[2]Institute for Environmental Sciences, University of Koblenz-Landau, Landau, Germany;
[3]Molecular Ecology Group (dark-MEG), Water Research Institute (IRSA), National Research
Council (CNR), Verbania-Pallanza, Italy; [4]Department of Biological Sciences, The University
of Alabama in Huntsville, Huntsville, AL, United States; [5]Institute of Environmental Science
and Research (ESR) Christchurch, Canterbury, New Zealand; [6]Subterranean Research and
Groundwater Ecology (SuRGE) Group, Trace and Environmental DNA (TrEnD) Laboratory,
School of Molecular and Life Sciences, Curtin University, Perth, WA, Australia; [7]Laboratory
of Subterranean Studies, Federal University of São Carlos, São Carlos, Brazil; [8]School of
Natural Sciences, Macquarie University, Sydney, Australia

## Introduction

Groundwater is everywhere. It is present on all continents, where it is found hundreds of meters below the ground up to the land surface where it sustains wetlands and other groundwater-dependent ecosystems. Besides occupying space in unconsolidated sediments, groundwater is frequently contained in fissured and karstified rocks and caves. As a result, groundwater-filled subterranean spaces together constitute the largest continental aquatic biome. While groundwater ecosystems may substantially differ in size, structure (e.g., geologic formation and interstitial space), and environmental conditions (e.g., depth, salinity, oxygen saturation, and temperature), they share many characteristics that differentiate them from surface limnic environments. Most obvious, groundwater ecosystems lack light, and thus lack phototrophic primary production, which has huge consequences for the energetic status of the ecosystem, consigning groundwaters to being typically low productivity (oligotrophic) environments. Groundwater habitats exhibit features in common with the deep sea (e.g., Overholt et al., 2022), namely, a pronounced stability in physicochemical conditions and a high degree of isolation of habitats and communities. Moreover, water residence time is high when compared to surface inland waters (Danielopol et al., 2003). These peculiarities strongly shape microbial and animal communities in groundwater ecosystems (see also Fillinger et al., this volume; Marmonier et al., this volume).

Considering these environmental conditions, it may be surprising that there is a vast diversity of microbial and metazoan species, of which some are shared with adjoining surface ecosystems, but most are unique to the subsurface aquatic ecosystems (Malard et al., this volume; Zagmajster et al., this volume). Some 25,000 (or even more) groundwater-dwelling species (stygobites) are estimated to live exclusively in cave waters and karstic, fissured, and porous aquifers (Culver and Holsinger, 1992; Martinez et al., 2018). While most groundwater biodiversity is found in shallow alluvial and karst aquifers, occurrence of organisms including fauna extends to >1000 m below the ground (e.g., Essafi et al., 1998). Microorganisms have even been found several kilometers below the land surface (Krumholz, 2000).

Typically, abundance and biomass of organisms in groundwater habitats are orders of magnitude lower than in surface waters. Additionally, many groundwater species are rare and endemic, with numerous cryptic and relict species, as well as some living fossils (Gibert and Deharveng, 2002; Hahn, 2006; Griebler and Lueders, 2009; Hahn and Fuchs, 2009; Fillinger et al. this volume). Due to the energy-limited environment, organisms often have low metabolic rates. Moreover, groundwater metazoans typically reproduce later in life (i.e., delayed sexual maturity), exhibit low reproduction rates and a low fecundity per reproductive episode (Ginet and Decou, 1977). All of these characteristics make groundwater organisms, communities, and ecosystems vulnerable to the many potential threats, mainly related to anthropogenic activities (Hose et al., 2022).

## Conservation of groundwater ecosystems and species at risk

Strategies and programs dedicated to the protection and conservation of individual endangered species, selected habitats, entire ecosystems or large areas of land and water have a long history. However, subsurface ecosystems suffer from their invisibility (Niemiller et al., 2018; Ficetola et al., 2019). Groundwater ecosystems and their impressive biodiversity have traditionally been overlooked in global conservation agendas and multilateral agreements (Mammola et al., 2019; Wynne et al., 2021; Sánchez-Fernández et al., 2021; Fišer et al., 2022; Mammola et al., 2022a; Saccò et al., 2022a; Boulton et al., this volume). In fact, there are only a few examples of targeted protection and conservation of subterranean habitats and/or groundwater species globally. Protection of exclusive areas of land and surface water, in some cases, also protects the underlying subterranean habitats. To date, efforts to conserve subterranean ecosystems have been dominated by problem-based studies focused on identifying the main drivers associated with declines in subterranean biodiversity (Mammola et al.,

2019; Gerovasileiou and Bianchi, 2021; Mammola et al., 2022a). Worth mentioning, when it comes to political and stakeholder negotiations, the perception is that there is still insufficient knowledge and tools (e.g., bioindicators and other ecological criteria) to routinely implement cost-effective assessment and monitoring schemes for groundwater ecosystems, and where needed conservation interventions (Mammola et al., 2022a; Hose et al., this volume). The same applies to individual rare and/or endemic species that are frequently at risk of extinction. There is often a lack of fundamental understanding of their distribution and abundance, let alone their autecology and role in terms of ecosystem services. And yet, the field of groundwater ecology has grown consistently in the past 3 decades, accumulating considerable knowledge and experience that still awaits application to its full potential. It is thus a "chicken and egg" problem. Will we make use of what we already know while accepting serious knowledge gaps, or will we continue waiting for some more clues to come? If we agree to treat groundwater ecosystems in a similar way to their surface counterparts, additional effort is needed.

## Why study, assess, and protect groundwater ecosystems?

There are several answers to this question. As chapters of this book attest, groundwater ecosystems may be considered extremely sensitive to anthropogenic threats, such as contamination from agriculture and mining activities, landfills, accidental spills, and urban polluted waters and areas, warming, and overexploitation. The potentially low resilience of groundwater communities against environmental perturbations is heavily related to the poor energetic status and productivity of groundwater environments (Saccò et al., 2022b; Venarsky et al., Chapter 10, this volume). Consequently, we argue that because of the comparably long residence times of water in the subsurface, groundwater habitats, once stressed, take a disproportionately long time to recover. Moreover, while it is commonly agreed that surface terrestrial and aquatic ecosystems and biodiversity deserve protection and sustainable management, there is no valid argument for treating subterranean ecosystems and biodiversity differently. Today's efforts to protect the environment, and in particular biodiversity, are based on the awareness that humans depend on processes and services delivered by intact communities living in healthy ecosystems. This is likewise true for groundwater ecosystems. Several important ecosystem services, such as the purification of water and biodegradation of contaminants, can be attributed to groundwater communities (Griebler and Avramov, 2015). Finally, life in groundwater deserves respectful treatment and conservation regardless of economic values and interests. As highlighted in several chapters of this book and many scientific papers and reports, groundwater ecosystems are home to an impressive biodiversity of microorganisms (Griebler and Lueders, 2009; Fillinger et al., this volume) and metazoans (Marmonier et al., this volume; Zagmajster et al., this volume). It is estimated that >50% of the metazoan biodiversity in aquifers is not yet discovered and described, and that species regularly go extinct before even being described (e.g., Niemiller et al., 2013). In fact, typical features of groundwater fauna, i.e., rareness, endemism, low competitiveness, low reproduction rate, highly fragmented populations, and low dispersal capacity (Hose et al., 2022), strongly argue for their protection. On the other hand, taking the immense number of rare and endemic species in groundwater habitats into consideration will push future conservation efforts readily to its practical limits. In consequence, prioritization in the protection of sites and the conservation of endangered species is needed (read below about "future challenges"). In the following section, we will briefly review legal frameworks dedicated to the ecological assessment, monitoring, and conservation of groundwater ecosystems and communities already in place.

## Legal frameworks related to groundwater ecosystems

Legislation with respect to groundwater ecosystems can be viewed from several different angles. First, there are the international conventions—such as the Convention on Biological Diversity, the Ramsar Convention on Wetlands of International Importance, the World Heritage Convention (WHC), and the Convention on International Trade in Endangered Species of Wild Fauna and Flora (CITES)—that focus on biodiversity issues and are directly or indirectly relevant to the protection and conservation of groundwater ecosystems and its biodiversity. Many countries have adapted and combined these conventions into national legal frameworks. Second, from the perspective of endangered species, there is the International Union for Conservation of Nature (IUCN) Red List of Threatened Species, established in 1964 and widely regarded as the most objective source of information on the conservation status of species and their extinction threats. The IUCN Red List is complemented by numerous national and regional Red Lists and Red Data Books. Although often based on criteria slightly different from that of the IUCN, these regional tools contribute to the effective conservation and recovery of threatened species on a smaller scale. Third, there is a plethora of regional regulations dedicated to maintain the sustainable use of groundwater in a qualitative and quantitative manner.

## International and national conventions for the protection of groundwater ecosystems

For the protection of groundwater ecosystems and its inhabitants, four previously outlined major international conventions that focus on biodiversity issues are of relevance. Each of these biodiversity-related conventions aims to implement conservation actions at the international, national, and regional levels. However, only a few groundwater ecosystems and/or species are currently protected under the auspices of these conventions. Niemiller et al. (2018) and Niemiller and Taylor (2019) recently produced a detailed overview of legislation in place dedicated to the conservation of cave fauna with an emphasis on Europe and the Americas. Where applicable to aquatic subterranean environments, selected information was extracted from these reviews.

The Convention on Biological Diversity (CBD, 1993), an international treaty among 196 countries which came into force in 1993, was developed for the conservation and sustainable use of biodiversity. The convention requires countries to adopt a national biodiversity strategy, called National Biodiversity Strategie and Action Plan (NBSAP), and to ensure that NBSAPs are implemented into all relevant planning and activities that may have a positive or negative impact on biodiversity. However, the focus on subterranean aquatic fauna and ecosystems in NBSAPs is highly variable among countries. Cave and groundwater biodiversity and ecosystems are specifically addressed in the NBSAPs of only a few countries. For example, the Slovenian NBSAP has a specific objective pertaining to cave habitat types "to maintain subterranean habitat types in ecologically important areas, and the entire subterranean fauna, at favorable conservation status."

The Convention on Wetlands of International Importance (Ramsar Convention, 1975) is the oldest global international environmental agreement which entered into force in 1975. Its mission is the conservation of all wetlands through local and national actions and international cooperation. Wetlands are frequently groundwater-dependent ecosystems. Many Ramsar Sites are karst wetlands, including notable cave systems such as Skocjanske Jame in Slovenia, the Caves of the Demänová valley in Slovakia, or the Anillo de Cenotes in the Yucatan, Mexico. There are also Ramsar sites that include alluvial aquifer systems characterized by an extraordinarily high groundwater fauna biodiversity, i.e., the Lobau wetland as part of the Danube Floodplain National Park, Austria (Danielopol and Pospisil, 2001). Considering South America, which is the wettest continent on Earth and has wetlands covering ~20% of its area, there are only a few sites proposed in karst areas, such as the Canyon of Peruaçu karst (located in eastern Brazil). Most of the Ramsar sites in South America are located in the Amazon region, and on a representative area of the Guarani aquifer, which underlies parts of Brazil and southern South American countries. Globally, Ramsar sites are continuously increasing in both numbers and area, today covering 113 sites, totaling an area of ~373,000 km$^2$.

The World Heritage Convention (WHC, 1975) was adopted by the UNESCO General Conference in 1972 and came into force in 1975. The WHC aims to promote cooperation among nations to protect outstanding cultural and natural heritage globally. Several sites included in the UNESCO World Heritage List contain significant cave and karst systems. Of these sites, seven are specifically recognized for their outstanding biodiversity value under the biodiversity criteria (Niemiller et al., 2018). Gunn (2021) estimated the number of karst sites and its area being part of the four UNESCO protected area categories: Biosphere Reserves (BR), Ramsar Sites (RS), UNESCO Global Geoparks (UGGp) and World Heritage Properties (WHP). His conclusion is that there are 86 countries in which there is at least one UNESCO protected area with karst, potentially harboring groundwater fauna.

The Convention on International Trade in Endangered Species of Wild Fauna and Flora (CITES, 1975) is an international agreement signed in 1973 and entered into force in 1975 between governments to ensure that international trade of wild animals and plants of conservation concern does not further threaten their continued survival. To our best knowledge, no stygobites or troglobites (true cave animals) are listed so far, despite the fact that some subterranean species (e.g., cavefishes and cave salamanders from China) are becoming increasingly popular in the trade (e.g., Cunningham et al., 2016).

Some countries that are member parties to the conventions mentioned above have enacted national endangered species legislation, despite worldwide threats and decline in biodiversity. Even when enacted, only a few groundwater taxa are directly protected under endangered species legislation. Moreover, there is considerable variation among countries in the level of protection given to subterranean biodiversity and ecosystems (Huppert, 1995, 2006; Juberthie, 1995; Lamoreaux et al., 1997; Tercafs, 2001; Niemiller et al., 2018).

Caves, rather than alluvial aquifers, have more frequently received particular conservation protection (Moldovan, 2019). There are many countries that have set up specific Cave Protection Acts. An early example is the U.S. Federal Cave Resources Protection Act (USFCRPA, 1988), a United States federal law that aims "to secure, protect, and preserve significant caves on Federal lands for the perpetual use, enjoyment, and benefit of all people; and to foster increased cooperation and exchange of information between governmental authorities and those who utilize caves located on Federal lands for scientific, education, or recreational purposes." A stronger environmental focus has the Canadian Cave Protection Act (CCPA, 2019). Caves that are uncommon and unique environments that can harbor rare and threatened species, unique mineralogy and sediments shall be protected. Moreover, protection is needed where caves are nonrenewable, site-specific landscape features with natural, cultural, spiritual, esthetic, and scientific value and constitute sensitive ecosystems that can underlie developed landscapes and as such are vulnerable to pollution, destruction by quarrying, vandalism, mismanagement, species extinction, and general degradation caused by human activities. Already mentioned above, the Slovenian Cave Conservation Act includes besides monitoring of interventions also restoration measures (Ravbar and Šebela, 2015). In the United States, there are national parks where the protection of caves is the primary focus, such as Mammoth Cave National Park in Kentucky, Carlsbad Caverns National Park in New Mexico, and Wind Cave National Park in South Dakota.

The central legislation tool for safeguarding biodiversity in the European Union is the Directive on flora, fauna, and habitats, the so-called Habitats Directive (EUHD, 1992). The Habitats Directive supports the conservation of rare, threatened, or endemic animal and plant species. Together with the Birds Directive, it forms the cornerstone of Europe's nature conservation policy and is the basis for the creation of a network of Special Areas of Conservation (SACs), called the Natura 2000 Network of protected areas. Over 26,000 Natura 2000 sites have been designated across Europe to date. The EUHD recommends the delineation of SACs specifically designated to subterranean biodiversity. To date, neither groundwater species nor groundwater habitats are mentioned in the EUHD. This has far reaching consequences, since impact assessment and regulation is not required for groundwater systems, and there is currently no legal background at the EU level for the protection of groundwater habitats and species. However, individual biodiversity directives and acts have been developed at national levels that do consider subterranean and groundwater environments. For example, in Slovenia, dozens of Olm (*Proteus anguinus*) localities have been protected within 26 SACs (Hudoklin, 2011) and selected subterranean environments are protected by the national Cave Protection Act, with a specific focus on biodiversity. A similar situation is found in Croatia, where many stygobionts and troglobionts are strictly protected under the Regulation on Protection of Wild Species. It is particularly caves that are considered by national legislation including the establishment of national parks, nature reserves, and Natura 2000 sites (Juberthie, 1995).

The Council of Europe's Convention on the Conservation of European Wildlife and Natural Habitats of 1979, also known as the Bern Convention (BC, 1979), was the first international treaty to protect species and habitats in Europe. In recommendation no. 36 of the Bern Convention (1992), it is emphasized that national inventories of subterranean invertebrates and subterranean habitats shall be compiled and that species of conservation concern shall be identified. Unfortunately, this recommendation remains to be fulfilled (Haslett, 2007).

To date, the only aquatic troglobiont species listed in the Bern Convention is the Olm (*P. anguinus*).

In contrast to Europe, there are Australian laws in place that specifically target groundwater ecosystems and biodiversity. The protection of groundwater ecosystems in Australia is achieved through legislation at both national and state levels. At the national level, the Environmental Protection of Biodiversity and Conservation Act (EPBC, 1999) is the primary piece of legislation that provides legal protection for plants, animals, habitats, and places such as heritage sites, marine areas, and wetlands that are considered matters or assets of national significance. The Act also covers lands owned or managed by the national government. Specifically, the Act requires that any developments that potentially impact listed communities (i.e., those considered as being assets of national significance) are required to undergo a detailed assessment, which is then approved (or not) by the relevant government minister. The EPBC provides specific protection for habitats, species, or communities that are listed as being threatened, endangered, or significant, which can include subterranean species or habitats. Indeed, nearly 2000 species and ecological communities have been identified nationally as threatened and at risk of extinction under the Act and are protected. While the EPBC Act has the potential for protecting groundwater ecosystems, it is in practice limited because few groundwater ecosystems or species are currently listed under the Act. Species, communities, or threatening processes may be listed by way of public nomination but listing of groundwater ecosystems or species is difficult because the status of the species or ecosystem must be well characterized, including data to demonstrate threats and changes to populations.

The EPBC Act in Australia does recognize and protect wetlands of national and international significance, and the inclusion of subterranean wetlands under the RAMSAR classification does provide an avenue for protection of subterranean biota (Hose et al., 2015). While several groundwater dependent wetland ecosystems in Australia are RAMSAR listed (such as the Piccaninnie Ponds Karst wetlands in South Australia, Fig. 24.1A), no aquifers are yet listed for protection.

In 2013, the EPBC Act was modified to provide protection for water-dependent ecosystems (including groundwaters) that may be affected by coal mining and coal seam gas developments. Termed the "Water Trigger," this change to the Act legislated that water resources are a matter of national environmental significance in relation to coal seam gas and large coal mining activities, and thus such developments would necessarily require governmental approval. Effectively, this change made water resources a matter of national significance, but *only* with respect to coal and coal seam gas development types. Other developments (including other mines or extractive industries) are not subject to the EPBC act and thus groundwater and other water dependent ecosystems have no legislated protection from those developments. The protection of groundwater ecosystems in Australia is complicated by the division of powers between national and state governments under the Australian Constitution, in which states have primary responsibility for environmental protection. As a consequence, environmental protection policies and legislation vary between jurisdictions.

There are a number of other organizations that deserve being mentioned for their efforts in protecting aquatic subterranean habitats and species, i.e., speleological societies, NGOs and community groups; however, their recommendations are not legally binding, and as such not the focus of this compilation.

FIGURE 24.1    (A) Piccaninnie Ponds (©Rae Young), (B) gudgeon *Milyeringa* veritas (©Douglas E., Western Australian Museum), (C) blind cave eel *Ophisternon candidum* (©Allen M. and G. Moore, Western Australian Museum), and (D) remipede *Kumonga exleyi* (©Douglas E., Western Australian Museum).

In the context of deliberate ignorance of groundwater ecosystems and biodiversity in EU legislation, it is surprising that the European Medicines Agency (EMA) developed guidelines for assessing the environmental and human health risks of veterinary medicinal products in groundwater (EMA, 2018). In these guidelines, it is assumed that groundwater invertebrates are 10 times more sensitive to pollutants than their surface water counterparts.

## Legislation focused on the protection of endangered species

Switching from the habitat-level to the species-level, further legislation is of relevance for the conservation of groundwater biodiversity. In North and Central America, the primary legislation for species protection is the Endangered Species Act of 1973 (ESA, 1973) in the United States and the Species at Risk Act of 2002 (SARA, 2002) in Canada. Here, species are the primary focus. Of the many groundwater species known for the United States, only 17 aquatic cave species are listed as federally endangered or threatened. Similar endangered species legislation has been passed in Mexico (General Wildlife Act of 2000), Costa Rica (Biodiversity Law of 1998), in Bermuda (Protected Species Act, 2003), and other countries (see Table 22.1 in Niemiller et al., 2018). In South America, Brazil is the country with most studies focusing on conservation purposes, and at the time of writing, about 40 of more than 70 known groundwater species, are listed as federally endangered (Gallão and Bichuette, 2018; Bichuette and Gallão, unpubl. data). Due to large knowledge gaps with groundwater fauna in South American countries, groundwater biodiversity is largely neglected in conservation acts.

A well-known instrument for highlighting species at risk is the "Red List." The IUCN Global Species Program, in conjunction with the IUCN Species Survival Commission

(SSC), developed and maintains the IUCN Red List of Threatened Species. The IUCN was founded in 1948 and is the world's oldest international environmental and conservation organization (IUCN, 1948). It assists governments and other conservation organizations with national and international biodiversity policies and initiatives. Overall, groundwater fauna is poorly represented in the IUCN Red List (Fig. 24.2). One reason is that it is rather difficult to gather the necessarily comprehensive information needed for assessing the risk of extinction of subterranean species under the IUCN Red List criteria. Furthermore, the Red List criteria and thresholds are notoriously insufficient for invertebrates (Cardoso et al., 2011, 2012). Consequently, most groundwater taxa would be considered "data deficient."

IUCN Red List categories and criteria also are applied at smaller spatial scales, the so-called Regional Red List Assessments, which are more practical for management and conservation planning in specific countries and regions. Indeed, a widespread species may be assessed as "Least Concern" across its global distribution, but considered "Vulnerable" in a regional assessment focusing on a location where specific threats are present. Beyond the IUCN Red List, Regional and National Red Lists that do not strictly follow the IUCN criteria (although there may be some overlaps) have been developed in more than 100 countries. A database of these Regional and National Red Lists is maintained by the IUCN National Red List project (http://www.nationalredlist.org). An outstanding example is the Croatian Red Book of Cave Fauna. It is the first Red List assessment of troglobionts and stygobionts of its kind in the world, covering almost 200 taxa of which 35% are assessed as "Critically Endangered" (Ozimec, 2011). Similar Red Lists containing groundwater animal species are found in Slovenia, France, and Germany, to name a few examples (Baillie and Groombridge, 1996; Burmeister, 2003; Arntzen et al., 2008; Allanic, 2012; Weber and Flot, 2019). IUCN is currently developing categories and criteria for a Red List of Ecosystems (RLE) as a global standard for ecosystem risk assessment at multiple scales (Rodriguez et al., 2011, 2012; Keith et al., 2013, 2015).

As discussed, the Australian EPBC Act provides protection for listed habitats and species. Under this Act, the listing of stygobitic species is challenging because the taxonomy and spatial distribution of the species must be well known, which is a particular challenge for groundwater taxa and ecosystems in Australia, given the declining state level of taxonomic expertise and the high incidence of short-range endemism among groundwater species. Currently, the remipede *Kumonga exleyi*, the cave gudgeon *Milyeringa veritas* Whitley, 1945,

FIGURE 24.2    The red list species *Proteus anguinus* (Ravbar and Pipan, 2022).

and the blind cave eel *Ophisternon candidum* (Mees, 1962) all from the Cape Range Peninsula (Western Australia) (Fig. 24.1) are the only three subterranean aquatic species listed under the EPBC Act.

## Groundwater ecosystem protection by water laws

Groundwater without doubt is an essential (re)source of drinking water, water for irrigation, and water for industrial purposes. Unlike surface waters, where ecological aspects have been implemented in monitoring, assessment, and protection schemes decades ago, groundwater is still treated like an abiotic raw material. This perception is mirrored in water legislation. Although we observe a slow mind switch in some areas of the world (e.g., individual countries in Europe and Australia), strong efforts for an equitable consideration of groundwater systems in terms of ecosystem health (Danielopol et al., 2004, 2007) are, although noticed, deliberately ignored and delayed. Nevertheless, it is also these laws targeting the quantity and (physical-chemical) quality of groundwater that essentially contribute directly and indirectly to the protection of subterranean habitats and species.

### *Groundwater laws in Europe*

An early and often cited example for the intended and targeted protection of groundwater, including its biocenosis, is the Swiss Water Protection Ordinance released in 1998 (GSchV, 1998), which includes ecological objectives with respect to ecological water quality: "the biocoenosis should be in a natural state adapted to the habitat and characteristic of water that is not or only slightly polluted" (Goldscheider et al., 2006). Only very recently, the natural reference status in terms of communities has been defined and methods for the assessment of the ecological status of an aquifer developed (see Stein et al., 2010; Korbel and Hose, 2011, 2017; Fillinger et al., 2019).

Within the European Union, the main legal instruments for the protection and management of waters are the Water Framework Directive (EC-WFD, 2000) and the affiliated Groundwater Directive (EC-GWD, 2006). While the EC-WFD comprises the general policy framework for the management of both surface waters and groundwater, the EC-GWD defines assessment criteria, threshold values, and measures particular for groundwaters. In preparation of the EC-GWD, a lively discussion about the necessity and perspectives of considering "ecological status criteria" in future groundwater monitoring schemes started (Danielopol et al., 2004; Quevauviller, 2005; Cunningham et al., 2006, Danielopol et al., 2007). Released in December 2006, the EC-GWD finally only included a brief statement in its preamble hinting at the importance of protective measures for groundwater ecosystems and it further states: "Research should be conducted in order to provide better criteria for ensuring groundwater ecosystem quality." Besides, it does not contain any obligatory measures related to biocoenosis and/or other ecological criteria. While the "good ecological status" of surface waters is key to the EC-WFD, something similar still is missing for European groundwater systems 17 years later.

With respect to individual European national water laws, the absence or deliberate ignorance of groundwater ecological measures becomes obvious. One example is the German Water Act (WHG, Wasserhaushaltsgesetz), which translates the EC-WFD into national law. The

WHG first states that "all" types of natural waters are essential part of the natural environment and deserve protection as habitats of plants and animals. Their economic use must be sustainable for the collective good, circumventing avoidable impairment of ecological functions and capacities. It further states, the goal is to sustain and improve aquatic habitats, in particular via protection against adverse changes of environmental conditions. However, the specific use and protection of groundwater is treated in a later section of the Water Act and lacks any cross reference to biocoenosis or a good ecological status (Hahn et al., 2018). Nevertheless, the German Ministry for Education and Research (BMBF) and the German Environment Agency (UBA) in the past decade financially supported numerous research projects dedicated to the development of ecological criteria and assessment schemes for groundwater (e.g., Griebler et al., 2014; BMBF-Project GroundCare, 2015). Similarly, research for a better understanding of groundwater biodiversity as well as the development of methods for an ecologically sound sampling and assessment of groundwater environments have been funded at EU and national levels (e.g., PASCALIS, 2002, GENESIS, 2009, Aqualife, 2013). The current revision of the EC-WFD provides a unique chance for the first implementation of ecological criteria into European and national groundwater laws.

### Groundwater laws in the United States

Historically, water management and allocation, including groundwater, has been decentralized in the United States. While the federal government has established standards for water quality—i.e., the U.S. Clean Water Act regulates water quality standards—it has deferred implementation of those standards to the states (Abrams, 2012; Megdal et al., 2015), and, consequently, there is considerable variation in governance of groundwater across the United States. Megdal et al. (2015) surveyed state agencies about the extent and scope of use, laws and regulations, and tools and strategies related to groundwater in the United States. While all states have legislation that includes groundwater, specific legislation exclusively related to groundwater varies. Some states have explicit legislation, such as Nebraska's Groundwater Management and Protection Act, while other states indirectly protect groundwater in a more piecemeal fashion through specific statutes and regulations associated with various activities (Megdal et al., 2015). While it has been recognized that legislation that integrate surface water and groundwater are more responsive and efficient (Hoffman and Zellmer, 2013), only half of all U.S. states have legislation that recognizes the connection between surface and subsurface water. Moreover, there is considerable variation among states in the recognition of groundwater quality, conservation, and groundwater-dependent ecosystems in legislation (Megdal et al., 2015). Top groundwater governance priorities among states include water quality and contamination, conflicts between water users, and declining water levels (Megdal et al., 2015). While these and other governance priorities can impact groundwater biodiversity, most states do not directly prioritize groundwater ecosystems.

### Groundwater laws in Central and South America

In Central America, only little legislation directly targets the conservation and protection of groundwater and subterranean ecosystems. An exception is the National Law for Aquifer Protection (Law 6938) in Brazil, dating back to the year 1981. In addition, there are regional regulations dedicated to the protection and management of aquifers and springs. The Guarani Aquifer is an example. It is with an area of 1.2 million $km^2$, the second largest aquifer in the World, of which 70% is located below Brazil. In total the Guarani Aquifer extends through eight states. With its quantitative relevance, the Guarani Aquifer in Brazil is, together with surface waters, protected by the Water Law. This law regulates that groundwater is in the public domain and therefore cannot be exploited by the private sector. In consequence, the Government of Brazil cannot grant the exploration of the aquifer to the private sector. In addition, the Guarani Aquifer is protected by international agreements. Four countries sharing the Guarani Aquifer, i.e., Argentina, Brazil, Paraguay, and Uruguay, signed a document establishing the Guarani Aquifer Project (SAG). In the past decade, changes to existing laws, due to pressure from the mining and agricultural sectors, significantly weakened the few protection measures in place (e.g., Ferreira et al., 2022). Major threats to groundwater ecosystems in Central and South America include pollution of aquifer by (agro)chemicals, mining, and inadequate resource use (e.g., unauthorized installation of artesian wells). Because of the lack of specific national policies, limited financial and human resources related to a general lack of infrastructure and narrow monitoring capacity, threats are ongoing.

### Groundwater laws in Australia

Water resource management in Australia is the dual responsibility of states and national agencies. States, as signatories of the National Water Initiative (NWI, 2004), are tasked with its implementation. This is achieved through independent state-level legislation that allows States the authority to identify and set ecological objectives, quantify environmental water needs, and determine water allocations. In the most populous state, New South Wales, the *Water* Management Act (WMA, 2000) protects water dependent ecosystems, and requires equitable sharing of water across stakeholders, including the environment, through the implementation of Water Sharing Plans. In this way, the water needs of groundwater ecosystems may be protected. The largest state (by area), Western Australia (WA), has the most comprehensive legislation with specific regards to subterranean fauna. The broader legislative framework in WA is the Wildlife Conservation Act (WCA, 1950) and the Environmental Protection Act (EPA, 1986), where the protection of unique habitats and individual taxa (species and subspecies) is promoted and regulated. More recently, the WA Biodiversity Conservation Act (BCA, 2016) expanded the state listed fauna taxa and threatened ecological communities. Under the WA Environmental Protection Act 1986, proponents of significant activities, such as mining and associated infrastructure, are required to survey for subterranean species where subterranean habitat exists and/or subterranean fauna are known to occur. Most Australian states have equivalent legislation for water resource and environmental protection, but different mechanisms for implementation and management, meaning that the actual and potential protection of groundwater ecosystems varies considerably.

### Further international and national groundwater laws

In New Zealand, the National Policy Statement for Freshwater Management (NPSFM, 2020) has stated that freshwater biodiversity must be both measured and maintained. It also states that included in freshwater is groundwater but, as mentioned in previous sections there are no specific laws in place at present to uphold this legislation. In fact, the National Environmental Standards for sources of human drinking water (SR, 2007/396) include groundwater as a source of drinking water that requires protection. The monitoring criteria resulting from these standards only consider groundwater as a resource for drinking water and as such do not consider the protection of groundwater ecosystems present, rather limits are set for contaminant presence. There is evidence that future regulations in New Zealand will include consideration of groundwater ecosystems. The current National Policy Statement for Freshwater Management has Te Mana o te wai[1] as its central concept. Te Mana o te wai upholds water as a living entity and prioritizes the health and well-being of water bodies and freshwater ecosystems as the first obligation. This concept is a pivotal shift in thinking for freshwater management in New Zealand and will mean regulatory bodies must consider groundwater ecosystems as part of management plans for freshwater. There are, however, currently no limits set for groundwater ecosystems, bringing into question the ability of regulatory bodies to protect the ecosystems when managing freshwater. At regional level, councils must ensure the policy statement is met through their management plans. At present, there are wide ranging approaches in terms of groundwater ecosystems due to the lack of fundamental information on the diversity and sensitivity of groundwater ecosystems. Recently, a compilation of existing knowledge was commissioned by one of the regional councils in New Zealand (Fenwick et al., 2018). Since this time, however, there has not been substantive improvement and as such many regions are not able to monitor or set limits for contaminants to protect groundwater ecosystems. Nevertheless, some of the regional councils are moving forward to improve the foundational knowledge to be able to start setting standards (Bolton and Weaver, 2021). There is still a way to go, however, before the nationwide legislation will provide a solid framework.

There is currently a water conservation order ongoing that includes the assessment of the status and sensitivity of groundwater ecosystems to proposed land use change. The outcome of this conservation order is being closely watched by other regions as it is the first such case in New Zealand. The fact that it is ongoing (currently 3 years) is related, however, to the difficulties faced when trying to assess the impact of land use on groundwater ecosystems when there is a lack of knowledge on the species present and their likely sensitivity to contaminants and nutrient levels. Worth mentioning, already in 1998 a report on "Environmental Performance Indicators for Groundwater" prepared by Bright and colleagues for the New Zealand Ministry for the Environment contained a first discussion on the use of groundwater invertebrates as bioindicators (Bright et al., 1998).

## Current challenges and the future of groundwater conservation

As we have discussed in this and other chapters (Boulton et al., this volume), groundwater conservation faces herculean challenges which have prevented us to fully implement these systems into legal frameworks. Foremost, in many parts of the world, the conservation of groundwater ecosystems is restricted to the protection of groundwater resources for its economic use or the unintended overlap between valuable groundwater ecosystems and protected areas established for surface species or habitats. A recent global estimation indicates

[1] "Te Mana o te Wai is a concept that refers to the fundamental importance of water and recognizes that protecting the health of freshwater protects the health and well-being of the wider environment. It protects the mauri* of the wai. Te Mana o te Wai is about restoring and preserving the balance between the water, the wider environment, and the community." National Policy Statement for Freshwater Management 2020.

that currently, only 6.9% of known subterranean ecosystems overlap with protected areas (Sánchez-Fernández et al., 2021). Importantly, since most of these protected areas were designed to protect surface species and/or ecosystems, they may not be optimally effective to target specific threats and conservation needs of groundwater systems. Second, direct protection, restoration, and management of groundwater ecosystems on the basis of their ecological values are exceedingly rare. To our knowledge, there are just a handful of examples of direct conservation intervention, often restricted to local systems or taxa (e.g., Manenti et al., 2019). Third, groundwater systems are riddled by extensive knowledge gaps about their communities and ecosystem functions, susceptibility to anthropogenic impacts, and effective conservation measures. As a consequence, risk assessment schemes for groundwater ecosystems and monitoring schemes including biological and ecological criteria, as well as ecologically sound thresholds, are lacking in most countries of the world and in their legislations. Lastly, there is still poor awareness about the importance of groundwater ecosystems across policymakers, stakeholders, and the general public; consequently, these systems are systematically overlooked in general legislations, conservation agendas, and biodiversity targets (Sánchez-Fernández et al., 2021; Fišer et al., 2022). In the light of these and other challenges, moving forward we will have to confront a pressing question: "How can we change this unwanted situation and have groundwater ecosystems more comprehensively represented into legal systems?"

A large consortium of subterranean biologists recently published a *"conservation roadmap for the subterranean biome"* (Wynne et al., 2021) delineating future strategies for subterranean conservation globally, from the filling of knowledge gaps, along with the inclusion of these systems into legislation, to their effective protection. Building upon this roadmap and adapting it to the specific case of groundwater ecosystems, we foresee five main points that need to be improved toward the goal of achieving a comprehensive legal account of the biotic component of groundwaters:

1. It is critical to continue acquiring basic knowledge about groundwater species across multiple dimensions of diversity (taxonomy, distribution patterns, interspecific interactions, trait, and genetic diversity) and the anthropogenic impacts affecting them. Such information is fundamental to support conservation and justify the incorporation of groundwater ecosystems into national and international legislations (see next points). Importantly, the collection of these diverse data is complicated by several impediments (Hortal et al., 2015; Ficetola et al., 2019; Mammola et al., 2021). We foresee that the exploitation of emerging technologies and monitoring tools will be key. This is indirectly confirmed by the recent upsurge of groundwater-based studies relying on environmental DNA for mapping species distribution and monitoring their trends (Saccò et al., 2022c), the emergence of citizen science as a reliable tool to map different features of groundwaters systems (Little et al., 2016; Alther et al., 2021), the growing use of species distribution modeling to infer distributions from partial data and environmental variables (Mammola and Leroy, 2018), and many more (Besson et al., 2022).
2. Armed with the information above, we should strive to improve on the mapping of conservation priorities, the assessment of species extinction risks, the study of sensitivity of groundwater fauna to anthropogenic stressors, and the establishment of sound biological thresholds that can be implemented into national and international legislations (e.g.,

maximum exposure concentration for major contaminants). In the face of major uncertainties regarding the distribution of groundwater systems and their biodiversity, special attention should be paid into the mapping of conservation priorities, given the proven effectiveness of protected area in maintaining biodiversity and ecosystem services and their simple implementability into legal frameworks (see, e.g., the legal effectiveness of the Natura 2000 network in Europe). In times of limited resources, a valid question is then how to select priority sites for establishing protected areas (Mammola et al., 2022a). Due to data limitation, most studies to date were restricted to small geographic areas, narrow taxonomic scopes, and/or they focused exclusively on taxonomic diversity (Michel et al., 2009; Rabelo et al., 2018; Linke et al., 2019; Fattorini et al., 2020; Iannella et al., 2021; Saccò et al., 2022c). However, as the knowledge increases (point one), it should be possible to achieve a regional to global picture about sound protected areas targets, as recently shown for cave-roosting bats (Tanalgo et al., 2022), as today, a unique case in the panorama of subterranean studies.

3. Despite the fact that knowledge about conservation actions targeting groundwater ecosystems is growing, quantifications of their effectiveness remain scarce (Mammola et al., 2022a). This knowledge gap calls for renewed effort to testing conservation intervention while simultaneously improving monitoring standards to quantify progress of protection measures. The latter point will require both a standardization of sampling methods and protocols for effective monitoring (Wynee et al., 2021) and the implementation of emerging monitoring technologies (see point one above). That said, going back to the "chicken and egg" dilemma mentioned at the beginning of this chapter, we cannot postpone for too long the implementation of direct conservation measures and the establishment of legal standards. Just like epidemiology or medicine, biological conservation is regarded as a "crisis discipline" (Soulé, 1985), often requiring practitioners to make decisions within restricted time windows and without possessing all the information. This is especially true in areas where threats to subterranean ecosystems are escalating and where there is complacency and inaction from the local governments toward the protection and recovery of the environment (see, e.g., the recent change to Brazilian legislation concerning caves; Ferreira et al., 2022).

4. Concerning strictly legal aspects, it is fundamental to establish a direct dialogue with policymakers and stakeholders aiming to achieve legal protection and recognition for the biological component of groundwater ecosystems. Concretely, it would be important to obtain: (i) legal equality for groundwater and surface water ecosystems; (ii) explicit implementation of the terms "groundwater ecosystems" and "good ecological status" in the laws pertaining to water and to conservation inclusive impact regulation; (iii) definition and legal consideration of biological references, indicator parameters and threshold values for the monitoring of groundwater ecosystems, (iv) implementation of groundwater ecological indicator parameters and threshold values into groundwater management plans.

5. Finally, we must reiterate that all of this must be paralleled by an increase in people's awareness, including policymakers, about environmental issues and values. Concerning groundwaters, education, and outreach has been a central tenet of the 2021–22 International Year of Caves and Karst (http://iyck2021.org/) initiative. Yet, even if awareness about subterranean ecosystems has been growing as a result of these and other initiatives, achieving

environmental education remains challenging because groundwaters are so foreign to human experience. Going forward in educating about the importance of these *"out of sight"* ecosystems, we must therefore be creative in our communication efforts. Examples include the use of art to showcase subterranean biota (Danielopol, 1998), citizen science (Alther et al., 2021), guided tours of karst areas (North and van Beynen, 2016), and publications intended for the general public, including kids (Mammola et al., 2022b). All these activities are of the utmost importance: with education comes awareness and, ultimately, an empowerment of local communities to stand for their biodiversity and natural resources. This will ultimately ensure that legislation will be concretely implemented and respected.

We here scratched the surface of possible measures to achieve a better recognition of groundwater ecosystems into legal frameworks. So far groundwater ecosystems, although highly sensitive to perturbations and inhabited by diverse microbial and metazoan communities, have been treated in a different way than surface terrestrial and aquatic ecosystems. While, in the past, limited knowledge might have been the major counterargument, today the anticipation of high labor and costs slow down the trend reversal. On a positive note, the road is clear on how to change the status quo. It is time to act.

## Acknowledgments

We thank S. Rétaux and F. Malard for editing an earlier version of this chapter.

## References

Abrams, R.H., 2012. Legal convergence of east and west in contemporary American water law. Environmental Law 42, 65–91.

Allanic, Y., 2012. Crustacés d'eau douce de France métropolitaine. In: La Liste rouge des espèces menacées en France. IUCN Comité Français, Muséum national d'Histoire naturelle. https://uicn.fr/wp-content/uploads/2012/06/Liste_rouge_France_Crustaces_d_eau_douce_de_metropole.pdf. (Accessed 28 September 2019).

Alther, R., Bongni, N., Borko, Š., Fišer, C., Altermatt, F., 2021. Citizen science approach reveals groundwater fauna in Switzerland and a new species of *Niphargus* (Amphipoda, Niphargidae). Subterranean Biology 39, 1–31.

AquaLife, 2013. The Aqualife project. https://www.researchgate.net/project/AQUALIFE-project-LIFE12-BIO-IT-000231-AQUALIFE. (Accessed 12 January 2023).

Arntzen, J.W., Denoël, M., Miaud, C., Andreone, F., Vogrin, M., Edgar, P., Isailovic, J.C., Ajtic, R., Corti, C., 2008. *Proteus anguinus*. IUCN Red List of Threatened Species, IUCN. Available at: http://www.iucnredlist.org.

Baillie, J., Groombridge, B., 1996. 1996 IUCN Red List of Threatened Animals (Hrsg.). https://portals.iucn.org/library/sites/library/files/documents/RL-1996-001.pdf. (Accessed 7 January 2023).

BC, 1979. The Bern Convention - The Council of Europe's Convention on the Conservation of European Wildlife and Natural Habitats. https://www.coe.int/en/web/bern-convention. (Accessed 12 January 2023).

BCA, 2016. Western Australia Biodiversity Conservation Act. https://www.legislation.wa.gov.au/legislation/statutes.nsf/main_mrtitle_13811_homepage.html. (Accessed 12 January 2023).

Besson, M., Alison, J., Bjerge, K., Gorochowski, T., Høye, T., Jucker, T., Clements, C., 2022. Towards the fully automated monitoring of ecological communities. Ecology Letters 25, 2753–2775.

BMBF-Project GroundCare, 2015. https://bmbf.nawam-rewam.de/projekt/groundcare/. (Accessed 12 January 2023).

Bolton, A., Weaver, L., 2021. Preliminary Assessment of Groundwater Dependent Ecosystems: Invertebrate Groundwater Fauna, Takaka, Golden Bay, Tasman. Client Report CSC20026a, Report Number 2110-TSDC172-1, Prepared by ESR Ltd., for Tasman District Council under Envirolink Contract C03X2002-1.

Bright, J., Bidwell, V., Robb, C., Ward, J., 1998. Environmental Performance Indicators for Groundwater. Technical Paper No. 38, Freshwater, Ministry for the Environment, Wellington, New Zealand.

Burmeister, E.-G., 2003. Rote Liste gefährdeter wasserbewohnender Krebse, exkl. Kleinstkrebse (limn. Crustacea) Bayerns. In: Rote Liste der gefährdeten Tiere und Gefäßpflanzen Bayerns, pp. 328–330.

Cardoso, P., Borges, P.A., Triantis, K.A., Ferránd, M.A., Martín, J.L., 2011. Adapting the IUCN red list criteria for invertebrates. Biological Conservation 144 (10), 2432–2440.

Cardoso, P., Borges, P.A., Triantis, K.A., Ferránd, M.A., Martín, J.L., 2012. The underrepresentation and misrepresentation of invertebrates in the IUCN Red List. Biological Conservation 149 (1), 147–148.

CBD, 1993. The Convention on Biological Diversity. https://www.cbd.int/. (Accessed 12 January 2023).

CCPA, 2019. Canadian Cave Protection Act. https://www.ubcm.ca/convention-resolutions/resolutions/resolutions-database/cave-protection-act. (Accessed 12 January 2023).

CITES, 1975. The Convention on International Trade in Endangered Species f Wild Fauna and Flora. https://cites.org/eng/disc/text.php. (Accessed 12 January 2023).

Culver, D.C., Holsinger, J.R., 1992. How many species of troglobites are there? National Speleological Society Bulletin 54, 79–80.

Cunningham, R., Scheuer, S., Eberhardt, D., Schweer, C., 2006. A Critical Assessment of Europe's Groundwater Quality Protection under the New Groundwater Directive. Europäisches Umweltbüro (EEB). http://www.wrrl-info.de/docs/wrrl_grundwasser_critical_assessment.pdf. (Accessed 7 January 2023).

Cunningham, A.A., Turvey, S.T., Zhou, F., Meredith, H.M., Guan, W., Liu, X., Wu, M., 2016. Development of the Chinese giant salamander *Andrias davidianus* farming industry in Shaanxi Province, China: conservation threats and opportunities. Oryx 50 (2), 265–273.

Danielopol, D.L., 1998. Conservation and protection of the biota of karst: assimilation of scientific ideas through artistic perception. Journal of Cave and Karst Studies 60, 67.

Danielopol, D.L., Gibert, J., Griebler, C., Gunatilaka, A., Hahn, H.J., Messana, G., Notenboom, J., Sket, B., 2004. Incorporating ecological perspectives in European groundwater management policy. Environmental Conservation 31 (3), 185–189.

Danielopol, D.L., Griebler, C., Gunatilaka, A., Hahn, H.J., Gibert, J., Mermillod-Blondin, F., Messana, G., Notenboom, J., Sket, B., 2007. Incorporation of roundwater Ecology in Environmental Policy. In: Quevauviller, P. (Ed.), Groundwater Science and Policy: An International Overview. Royal Society of Chemistry, London, pp. 671–689.

Danielopol, D.L., Griebler, C., Gunatilaka, A., Notenboom, J., 2003. Present state and future prospects for groundwater ecosystems. Environmental Conservation 30, 104–130.

Danielopol, D.L., Pospisil, P., 2001. Hidden biodiversity in the groundwater of the Danube flood plain national park (Austria). Biodiversity & Conservation 10, 1711–1721.

EC-GWD, 2006. Directive 2006/118/EC—groundwater Directive of the European Parliament and of the council on the protection of groundwater against pollution and deterioration. Official Journal of the European Union L372 (19).

EC-WFD, 2000. Directive 2000/60/EC—water framework directive of the European parliament and of the council. Official Journal of the European Commission L327 (1).

EPA, 1986. Environment Protection Act of Western Australia. https://www.legislation.wa.gov.au/legislation/statutes.nsf/main_mrtitle_304_homepage.html. (Accessed 12 January 2023).

EPBC, 1999. Environmental Protection of Biodiversity and Conservation Act. https://www.legislation.gov.au/Details/C2021C00182. (Accessed 12 January 2023).

ESA, 1973. U.S. Endangered Species Act. https://www.fws.gov/law/endangered-species-act. (Accessed 12 January 2023).

Essafi, K., Mathieu, J., Berrady, I., Chergui, H., 1998. Qualité de l'eau et de la faune au niveau de forages artesiens dans la Plaine de Fès et la Plaine des Beni-Sadden. Premiers resultats.—Mèmoires de Biospéologie 25, 157–166.

EUHD, 1992. European Directive on flora, fauna, and habitats. https://ec.europa.eu/environment/nature/legislation/habitatsdirective/index_en.htm. (Accessed 12 January 2023).

European Medicines Agency, 2018. Guideline on Assessing the Toxicological Risk to Human Health and Groundwater Communities from Veterinary Pharmaceuticals in Groundwater. EMA/CVMP/ERA/103555/2015, Committee for Medicinal Products for Veterinary Use (CVMP) London.

Fattorini, S., Fiasca, B., Di Lorenzo, T., Di Cicco, M., Galassi, D.M.P., 2020. A new protocol for assessing the conservation priority of groundwater-dependent ecosystems. Aquatic Conservation 30, 1483–1504.

Fenwick, G., Greenwood, M., Williams, E., Milne, J., Watene-Rawiri, E., 2018. Groundwater Ecosystems: Functions, Values, Impacts and Management. NIWA Client Report 2018184CH Prepared for Horizons Regional Council.

Ferreira, R.L., et al., 2022. Brazilian cave heritage under siege. Science 375 (6586), 1238–1239.

Ficetola, G.F., Canedoli, C., Stoch, F., 2019. The Racovitzan impediment and the hidden biodiversity of unexplored environments. Conservation Biology 33, 214–216.

Fišer, C., Borko, S., Delić, T., Kos, A., Premate, E., Zagmajster, M., Zakšek, V., Altermatt, F., 2022. The European green deal misses Europe's subterranean biodiversity hotspots. Nature Ecology and Evolution 6, 1403–1404.

Fillinger, L., Hug, K., Trimbach, A.M., Wang, H., Kellermann, C., Meyer, A., Bendinger, B., Griebler, C., 2019. The D-A-(C) index: a practical approach towards the microbiological-ecological monitoring of groundwater ecosystems. Water Research 163, 114902.

Gallão, J.E., Bichuette, M.E., 2018. Brazilian obligatory subterranean fauna and the threats to hypogean environment. ZooKeys 746, 1–23.

GENESIS, 2009. Groundwater and dependent ecosystems: new scientific basis on climate change and land-use impacts for the update of the EU groundwater directive. https://cordis.europa.eu/project/id/226536/reporting. (Accessed 12 January 2023).

Gerovasileiou, V., Bianchi, C.N., 2021. Mediterranean marine caves: a synthesis of current knowledge. Oceanography and Marine Biology an Annual Review 59, 1–88.

Gibert, J., Deharveng, L., 2002. Subterranean ecosystems: a truncated functional biodiversity. BioScience 52, 473–481.

Ginet, R., Decou, V., 1977. Initiation à la Biologie et à l'Écologie Souterraine, Jean Pierre Delarge, 345, Paris.

Goldscheider, N., Hunkeler, D., Rossi, P., 2006. Review: microbial biocenoses in pristine aquifers and an assessment of investigative methods. Hydrogeology Journal 14, 926–941.

Griebler, C., Avramov, M., 2015. Groundwater ecosystem services—a review. Freshwater Science 34, 355–367.

Griebler, C., Hahn, H.J., Stein, H., Kellermann, C., Fuchs, A., Steube, C., Berkhoff, S., Brielmann, H., 2014. Entwicklung biologischer Bewertungsmethoden und -kriterien für Grundwasserökosysteme (Development of Biological Assessment Methods and Criteria for Groundwater Ecosystems). UFOPLAN, FKZ 3708 23 200, pp. 153. https://www.umweltbundesamt.de/sites/default/files/medien/378/publikationen/uba_bericht_grundwasser_web.pdf.

Griebler, C., Lueders, T., 2009. Microbial biodiversity in groundwater ecosystems. Freshwater Biology 54, 649–677.

GSchV, 1998. Gewasserschutzverordnung (Swiss Water Ordinance) 814.201. Der Schweizer Bundesrat, Bern, Switzerland.

Gunn, J., 2021. Karst groundwater in UNESCO protected areas: a global overview. Hydrogeology Journal 29 (1), 297–314.

Hahn, H.J., 2006. A first approach to a quantitative ecological assessment of groundwater habitats: the GW-Fauna-Index. Limnologica 36 (2), 119–137.

Hahn, H.J., Fuchs, A., 2009. Distribution patterns of groundwater communities across aquifer types in Southwestern Germany. Freshwater Biology 54, 848–860.

Hahn, H.J., Schweer, C., Griebler, C., 2018. Grundwasserökosysteme im Recht?—Eine kritische Betrachtung zur rechtlichen Stellung von Grundwasserökosystemen (Groundwater ecosystems rights acknowledged?—a critical evaluation of the legal status of groundwater ecosystems). Grundwasser 23, 209–218.

Haslett, J.R., 2007. European Strategy for the Conservation of Insects. No. 145, Convention on the Conservation of European Wildlife and Natural Habitats. Council of Europe Publishing.

Hoffman, C., Zellmer, S., 2013. Assessing institutional ability to support adaptive, integrated water resources management. Nebraska Law Review 91, 805–865.

Hortal, J., de Bello, F., Diniz-Filho, J.A.F., Lewinsohn, T.M., Lobo, J.M., Ladle, R.J., 2015. Seven shortfalls that beset large-scale knowledge of biodiversity. Annual Review of Ecology, Evolution, and Systematics 46 (1), 523–549.

Hose, G.C., Asmyhr, M.G., Cooper, S.J.B., Humphreys, W.F., 2015. Down under down under: austral groundwater life. In: Stow, A., Maclean, N., Holwell, G.I. (Eds.), Austral Ark. Cambridge University Press, pp. 512–536.

Hose, G.C., Chariton, A., Daam, M., Di Lorenzo, T., Galassi, D.M.P., Halse, S.A., Reboleira, A.S.P.S., Robertson, A.L., Schmidt, S.I., Korbel, K., 2022. Invertebrate traits, diversity and the vulnerability of groundwater ecosystems. Functional Ecology 36 (9), 2200–2214.

Hudoklin, A., 2011. Are we guaranteeing the favourable status of the *Proteus anguinus* in the Natura 2000 Network in Slovenia? In: Prelovšek, M., Zupan Hajna, N. (Eds.), Pressures and Protection of the Underground Karst: Cases from Slovenia and Croatia. Postojna, Inštitut Za Raziskovanje Krasa ZRC SAZU. Postojna/Karst Research Institute ZRC SAZU, pp. 169–181.

Huppert, G.N., 1995. Legal protection for caves in the United States. Environmental Geology 26, 121–123.

Huppert, G.N., 2006. Using the law to protect caves: a review of options. In: Hildreth-Werker, V., Werker, J.C. (Eds.), Cave Conservation and Restoration. National Speleological Society, Huntsville, pp. 217–228.

Iannella, M., Fiasca, B., Di Lorenzo, T., Di Cicco, M., Biondi, M., Mammola, S., Galassi, D.M.P., 2021. Getting the 'most out of the hotspot' for practical conservation of groundwater biodiversity. Global Ecology and Conservation 31, e01844.

IUCN, 1948. The International Union for Conservation of Nature. https://www.iucn.org/. (Accessed 12 January 2023).

Juberthie, C., 1995. Underground Habitats and Their Protection. No. 72, Convention on the Conservation of European Wildlife and Natural Habitats. Council of Europe Press.

Keith, D.A., Rodrıguez, J.P., Rodrıguez-Clark, K.M., Nicholson, E., Aapala, K., Alonso, A., Asmussen, M., Bachman, S., Basset, A., Barrow, E.G., Benson, J.S., Bishop, M.J., Bonifacio, R., Brooks, T.M., Burgman, M.A., Comer, P., Comın, F.A., Essl, F., Faber-Langendoen, D., Fairweather, P.G., Holdaway, R.J., Jennings, M., Kingsford, R.T., Lester, R.R., Mac Nally, R., McCarthy, M.A., Moat, J., Oliveira-Miranda, M.A., Pisanu, P., Poulin, B., Regan, T.J., Riecken, U., Spalding, M.D., Zambrano-Martınez, S., 2013. Scientific foundations for an IUCN red list of ecosystems. PLoS One 8, e62111.

Keith, D.A., Rodrıguez, J.P., Brooks, T.M., Burgman, M.A., Barrow, E.G., Bland, L., Comer, P.J., Franklin, J., Link, J., McCarthy, M.A., Miller, R.M., Murray, N.J., Nel, J., Nicholson, E., Oliveira-Miranda, M.A., Regan, T.J., Rodrıguez-Clark, K.M., Rouget, M., Spalding, M.D., 2015. The IUCN red list of ecosystems: motivations, challenges, and applications. Conservation Letters 8, 214–226.

Korbel, K.L., Hose, G.C., 2011. A tiered framework for assessing groundwater ecosystem health. Hydrobiologia 661, 329–349.

Korbel, K.L., Hose, G.C., 2017. The weighted groundwater health index: improving the monitoring and management of groundwater resources. Ecological Indicators 75, 164–181.

Krumholz, L.R., 2000. Microbial communities in the deep subsurfaces. Hydrogeological Journal 8, 41–46.

Lamoreaux, P.E., Powell, W.J., LeGrand, H.E., 1997. Environmental and legal aspects of karst areas. Environmental Geology 29, 23–36.

Linke, S., Turak, E., Asmyhr, M.G., Hose, G., 2019. 3D conservation planning: including aquifer protection in freshwater plans refines priorities without much additional effort. Aquatic Conservation 29, 1063–1072.

Little, K.E., Hayashi, M., Liang, S., 2016. Community-based groundwater monitoring network using a citizen-science approach. Groundwater 54 (3), 317–324.

Mammola, S., Cardoso, P., Culver, D.C., Deharveng, L., Ferreira, R.L., Fišer, C., Galassi, D.M.P., Griebler, C., Halse, S., Humphreys, W.F., Isaia, M., Malard, F., Martinez, A., Moldovan, O.T., Niemiller, M.L., Pavlek, M., Reboleira, A.S.P.S., Souza-Silva, M., Teeling, E.C., Wynne, J.J., Zagmajster, M., 2019. Scientists' warning on the conservation of subterranean ecosystems. BioScience 69, 641–650.

Mammola, S., Leroy, B., 2018. Applying species distribution models to caves and other subterranean habitats. Ecography 41 (7), 1194–1208.

Mammola, S., Lunghi, E., Bilandžija, H., Cardoso, P., Grimm, V., Schmidt, S.I., Martínez, A., 2021. Collecting eco-evolutionary data in the dark: impediments to subterranean research and how to overcome them. Ecology and Evolution 11 (11), 5911–5926.

Mammola, S., Meierhofer, M.B., Borges, P.A.V., Colado, R., Culver, D.C., Deharveng, L., Delić, T., Di Lorenzo, T., Dražina, T., Ferreira, R.L., Fiasca, B., Fišer, C., Galassi, D.M.P., Garzoli, L., Gerovasileiou, V., Griebler, C., Halse, S., Howarth, F.G., Isaia, M., Johnson, J.S., Komerički, A., Martínez, A., Milano, F., Moldovan, O.T., Nanni, V., Nicolosi, G., Niemiller, M.,L.,, Pallarés, S., Pavlek, M., Piano, E., Pipan, T., Sanchez-Fernandez, D., Santangeli, A., Schmidt, S.I., Wynne, J.J., Zagmajster, M., Zakšek, V., Cardoso, P., 2022b. Towards evidence-based conservation of subterranean ecosystems. Biological Reviews of the Cambridge Philosophical Society 97 (4), 1476–1510.

Mammola, S., Frigo, I., Cardoso, P., 2022b. Life in the darkness of caves. Frontiers for Young Minds 10, 657265.

Manenti, R., Barzaghi, B., Tonni, G., Ficetola, G.F., Melotto, A., 2019. Even worms matter: cave habitat restoration for a planarian species increased environmental suitability but not abundance. Oryx 53, 216–221.

Martinez, A., Anicic, N., Calvaruso, S., Sanchez, N., Puppieni, L., Sforzi, T., Fontaneto, D., 2018. A new insight into the Stygofauna Mundi: assembling a global dataset for aquatic fauna in subterranean environments. In: ARPHA Conference Abstracts, vol. 1. Pensoft Publishers, p. e29514.

Megdal, S.B., Gerlak, A.K., Varady, R.G., Huang, L.-Y., 2015. Groundwater governance in the United States: common priorities and challenges. Groundwater 53, 677–684.

Michel, G., Malard, F., Deharveng, L., Di Lorenzo, T., Sket, B., De Broyer, C., 2009. Reserve selection for conserving groundwater biodiversity. Freshwater Biology 54, 861–876.

Moldovan, O.T., 2019. Cave protection in Romania. In: Ponta, G.M., Onac, B.P. (Eds.), Cave and Karst Systems of Romania. Springer, Cham, pp. 537–541.

Niemiller, M.L., Taylor, S.J., 2019. Protecting cave life. In: Culver, D.C., White, W.B., Pipan, T. (Eds.), Encyclopedia of Caves, third ed. Elsevier, pp. 822–829.

Niemiller, M.L., Graening, G.O., Fenolio, D.B., Godwin, J.C., Cooley, J.R., Pearson, W.D., Fitzpatrick, B.M., Near, T.J., 2013. Doomed before they are described? The need for conservation assessments of cryptic species complexes using an amblyopsid cavefish (Amblyopsidae: *Typhlichthys*) as a case study. Biodiversity & Conservation 22 (8), 1799–1820.

Niemiller, M.L., Bichuette, E., Taylor, S.J., 2018. Conservation of cave fauna in Europe and the americas. In: Moldovan, O.T., Kovac, L., Halse, S. (Eds.), Cave Ecology. Dordrecht-Springer, pp. 451–478.

North, L., van Beynen, P., 2016. All in the training: techniques for enhancing karst landscape education through show cave interpretation. Applied Environmental Education and Communication 15, 279–290.

NPSFM, 2020. New Zealand, the National Policy Statement for Freshwater Management. https://environment.govt.nz/acts-and-regulations/national-policy-statements/national-policy-statement-freshwater-management/. (Accessed 12 January 2023).

NWI, 2004. Australian Water National Initiatives. https://www.dcceew.gov.au/water/policy/policy/nwi. (Accessed 12 January 2023).

Overholt, W.A., Trumbore, S., Xu, X., Bornemann, T.L., Probst, A.J., Krüger, M., Küsel, K., 2022. Carbon fixation rates in groundwater similar to those in oligotrophic marine systems. Nature Geoscience 15 (7), 561–567.

Ozimec, R., 2011. Red book of Dinaric cave fauna—an example from Croatia. In: Prelovšek, M., Zupan Hajna, N. (Eds.), Pressures and Protection of the Underground Karst: Cases from Slovenia and Croatia. Inštitut Za Raziskovanje Krasa ZRC SAZU. Postojna/Karst Research Institute ZRC SAZU, Postojna, pp. 182–190.

PASCALIS, 2002. Protocols for the assessment and conservation of aquatic life in the subsurface. https://cordis.europa.eu/project/id/EVK2-CT-2001-00121/de. (Accessed 12 January 2023).

Quevauviller, P., 2005. Groundwater monitoring in the context of EU legislation: reality and integration needs. Journal of Environmental Monitoring 7, 89–102.

Rabelo, L.M., Souza-Silva, M., Lopes Ferreira, R., 2018. Priority caves for biodiversity conservation in a key karst area of Brazil: comparing the applicability of cave conservation indices. Biodiversity & Conservation 27, 2097–2129.

Ramsar Convention, 1975. Convention on Wetlands of International Importance. https://www.dcceew.gov.au/water/wetlands/ramsar. (Accessed 12 January 2023).

Ravbar, N., Šebela, S., 2015. The effectiveness of protection policies and legislative framework with special regard to karst landscapes: insights from Slovenia. Environmental Science & Policy 51, 106–116.

Ravbar, N., Pipan, T., 2022. Karst groundwater dependent ecosystems—typology, vulnerability and protection. In: Mehner, T., Tockner, K. (Eds.), Encyclopedia of Inland Waters, 3, pp. 460–473.

Rodríguez, J.P., Rodríguez-Clark, K.M., Baillie, K.E.M., Ash, N., Benson, J., Boucher, T., Brown, C., Burgess, N.D., Collen, B., Jennings, M., Keith, D.A., Nicholson, E., Revenga, C., Reyers, B., Rouget, M., Smith, T., Spalding, M., Taber, A., Walpole, M., Zager, I., Zamin, T., 2011. Establishing IUCN red list criteria for threatened ecosystems. Conservation Biology 25, 21–29.

Rodríguez, J.P., Rodríguez-Clark, K.M., Keith, D.A., Barrow, E.G., Benson, J., Nicholson, E., Wit, P., 2012. IUCN red list of ecosystems. Sapiens 5, 6–70.

Saccò, M., Blyth, A.J., Douglas, G., Humphreys, W.F., Hose, G.C., Davis, J., Guzik, M.T., Martínez, A., Eberhard, S.M., Halse, S.A., 2022a. Stygofaunal diversity and ecological sustainability of coastal groundwater ecosystems in a changing climate: The Australian paradigm. Freshwater Biology 67, 2007–2023.

Saccò, M., Blyth, A.J., Venarsky, M., Humphreys, W.F., 2022b. Trophic interactions in subterranean environments. In: Mehner, T., Tockner, K. (Eds.), Encyclopedia of Inland Waters, second ed. Elsevier, pp. 537–547.

Saccò, M., Guzik, M.T., van der Heyde, M., Nevill, P., Cooper, S.J., Austin, A.D., White, N.E., 2022c. eDNA in subterranean ecosystems: Applications, technical aspects, and future prospects. Science of the Total Environment 820, 153223.

Sánchez-Fernández, D., Galassi, D.M.P., Wynne, J.J., Cardoso, P., Mammola, S., 2021. Don't forget subterranean ecosystems in climate change agendas. Nature Climate Change 11, 458−459.

SARA, 2002. Canadian Species at Risk Act. https://laws.justice.gc.ca/eng/acts/s-15.3/. (Accessed 12 January 2023).

Soulé, M.E., 1985. What is conservation biology? Bioscience 35, 727−734.

Stein, H., Kellermann, C., Schmidt, S.I., Brielmann, H., Steube, C., Berkhoff, S.E., Fuchs, A., Hahn, H.J., Thulin, B., Griebler, C., 2010. The potential use of fauna and bacteria as ecological indicators for the assessment of groundwater quality. Journal of Environmental Monitoring 12, 242−254.

Tanalgo, K.C., Oliveira, H.F., Hughes, A.C., 2022. Mapping global conservation priorities and habitat vulnerabilities for cave-dwelling bats in a changing world. Science of the Total Environment 843, 156909.

Tercafs, R., 2001. The Protection of the Subterranean Environment: Conservation Principles and Management Tools. P.S. Publishers, Luxembourg.

USFCRPA, 1988. U.S. Federal Cave Ressources Protection Act. https://www.congress.gov/bill/100th-congress/house-bill/1975. (Accessed 12 January 2023).

Weber, D., Flot, J.-F., 2019. Rote Liste und Gesamtartenliste der Grundwasserkrebse (Niphargidae) des Saarlandes. Ministerium für Umwelt und Verbraucherschutz und DELATTINIA 2020.

WCA, 1950. Wildlife Conservation Act of Western Australia. https://www.legislation.wa.gov.au/legislation/statutes.nsf/main_mrtitle_1080_homepage.html. (Accessed 12 January 2023).

WHC, 1975. World Heritage Convention. https://whc.unesco.org/en/convention/. (Accessed 12 January 2023).

WMA, 2000. Water Management Act New South Wales. https://legislation.nsw.gov.au/view/html/inforce/current/act-2000-092. (Accessed 12 January 2023).

Wynne, J.J., Howarth, F.G., Mammola, S., Ferreira, R.L., Cardoso, P., Di Lorenzo, T., Galassi, D.M.P., Medellin, R.A., Miller, B.W., Sánchez-Fernández, D., Bichuette, M.E., Biswas, J., BlackEagle, C.W., Boonyanusith, C., Amorim, I.R., Borges, P.A.V., Boston, P.J., Cal, R.N., Cheeptham, N., Deharveng, L., Eme, D., Faille, A., Fenolio, D., Fišer, F., Fišer, Z., Gon, I.I.I., Goudarzi, S.M.'O., Griebler, F., Halse, C., 27, S., Hoch, H., Kale, E., Katz, A.D., Kováč, L., Lilley, T.M., Manchi, S., Manenti, R., Martínez, A., Meierhofer, M.B., Miller, A.Z., Moldovan, O.T., Niemiller, M.L., Peck, S.B., Giovannini Pellegrini, T., Pipan, T., Phillips- Lander, C.M., Poot, C., Racey, P.A., Sendra, A., Shear, W.A., Souza-Silva, M., Taiti, S., Tian, M., Venarsky, M.P., Yancovic-Pakarati, S., Zagmajster, M., Zhao, Y., 2021. A conservation roadmap for the subterranean biome. Conservation Letters 14 (5), e12834.

# The ecological and evolutionary unity and diversity of groundwater ecosystems—conclusions and perspective

*Florian Malard[1], Christian Griebler[2] and Sylvie Rétaux[3]*

[1]Univ Lyon, Université Claude Bernard Lyon 1, CNRS, ENTPE, UMR 5023 LEHNA, Villeurbanne, France; [2]University of Vienna, Department of Functional & Evolutionary Ecology, Vienna, Austria; [3]Paris-Saclay Institute of Neuroscience, Université Paris-Saclay and CNRS, Saclay, France

Thirty years now separate the first and second edition of *"Groundwater Ecology and Evolution"* (Gibert et al., 1994; Malard et al., 2023). There has been considerable progress in understanding the processes controlling groundwater biodiversity and the functioning of subterranean aquatic ecosystems. Reasons are manifold but two aspects are striking. First, groundwater biologists have since then amassed big datasets at an unprecedented rate and analyzed those using modern tools to test long-standing hypotheses in groundwater ecology and evolution. Second, they have framed these hypotheses within the broader context of theories in ecology and evolution making their findings from groundwater ecosystems more appealing to other disciplines (Mammola and Martinez, 2020; Mammola and Cardoso, 2022). Overall, if the goal of groundwater biological sciences is to provide a general framework of the processes shaping the evolution of groundwater biodiversity and of the functioning of below ground aquatic ecosystems, then, the discipline is doing quite well.

The primary objective for the second edition of *"Groundwater Ecology and Evolution"* was to provide an up-to-date synthesis of scientific achievements, promoting cross-theme collaborations, and providing orientation for future research. We hope this synthesis will play a crucial role in accelerating and advancing scientific knowledge in groundwater ecology and evolution. The second objective, *i.e.,* to emphasize the eco-evolutionary dimension of groundwater, emerged from the editors' feeling that groundwater ecosystems in this respect might not receive the attention they deserve. Ironically, any ecosystem-specific review of state-of-the-art knowledge carries the risk that the synthesis attracts essentially the attention of specialists already convinced of the importance of this ecosystem. We hope we have at least in part avoided that pitfall by attracting the readers' attention to groundwater-based studies that resonate well beyond the frontiers of the ecology and evolution of groundwater organisms (see also Mammola and Martinez, 2020 for a plea about bringing subterranean research outside caves).

In this concluding chapter, we review generalizations that emerge from the contributions in this volume. We find these generalizations fall within an overarching paradigm that flows through many of the chapters in this book. We refer to this paradigm as "the ecological and evolutionary unity and diversity of groundwater ecosystems," in reference to earlier writings by D.C. Culver and T. Pipan on the subject (Culver and Pipan, 2016; Pipan and Culver, 2017). This new emerging paradigm is changing our perception of groundwater ecosystems and the way we use them as model systems for understanding general ecological and evolutionary mechanisms. It also points to pressing research issues that can provide major progress in understanding the diversity of ecological and evolutionary pathways shaping the functioning of groundwater ecosystems.

This concluding chapter is composed of four sections. In the first section entitled *"Pattern detection and description,"* we emphasize the importance of pursuing the exploration of the groundwater biosphere. Its exploration remains a source of unexpected scientific discoveries that continue to shape our perception of groundwater ecosystems from the archetypical view of an environmentally stable and energy-poor ecosystem to a diverse array of groundwater habitats that all have in common the absence of light but differ widely in many other environmental attributes. The second section considers the environmental and biodiversity attributes of groundwater ecosystems that are pushed forward to emphasize the potential of *"Groundwater systems as ecological laboratories."* Despite major advances in groundwater functional ecology, we recognize that our current knowledge and understanding of species response and effect traits does not cover the full range of environmental attributes and selective pressures in groundwater. We may have overemphasized convergent responses to a shared environment at the expense of divergent responses to contrasting selective pressures. We may have overlooked the possibility that adaptation to shared selective pressures may be achieved by different combinations of traits. In the third section entitled *"Groundwater systems as evolutionary laboratories,"* we consider major insights provided by Evo-Devo studies into the molecular, cellular, and developmental basis of eye loss, pigment reduction, and elaboration of extraoptic sensory structures. We point out to the use of shared mechanisms to produce convergent phenotypes but also acknowledge an important level of "molecular tinkering" during evolution to groundwater life. In the last section, we examine why the concept of *"Groundwater's ecosystems supporting services"* struggles to impose itself, thereby delaying the implementation of ecosystem-oriented conservation and management measures for ensuring groundwater sustainability. We identify important research avenues for prioritizing conservation and management efforts in the face of multiple threats to groundwater ecosystems. We ask the reader for understanding that microbial communities in groundwater, while without doubt being key drivers of ecosystem processes and services, are not in focus of our concluding remarks.

## Pattern detection and description

We begin this chapter by urging groundwater biologists to pursue their efforts in describing ecosystem and biodiversity patterns in groundwater. Though this claim may seem at odds with a general trend towards more hypothesis-driven research and a quest for a mechanistic understanding, knowledge of the distribution and extent of groundwater biodiversity and ecosystems remains largely incomplete. To the seven knowledge shortfalls formalized by Hortal and coauthors (2015)—all of which apply to groundwater ecosystems—Ficetola and coauthors (2019) proposed to add an eighth knowledge shortfall, which they named the "Racovitzan impediment." Typically, the biodiversity of an environment that has not been explored cannot be described, nor analyzed and conserved (Ficetola et al., 2019, p. 214). Although this statement sounds like a truism, it underlines a reality that biologists know well: sampling groundwater ecosystems is like exploring unknown land. Dedicated efforts should be made to promote research initiatives in regions of the groundwater biosphere that has not yet been explored especially in Africa, Asia, and South America. However, even in regions that have received a substantial sampling effort, sampling new sites continuously yields species new to science at a rate that largely surpasses the rate of species discovery in most surface environments (Stoch and Galassi, 2010; Ficetola et al., 2019).

Since the 1960s, the scientific exploration of groundwater has continually extended the limits of groundwater ecosystems far beyond the more accessible cave habitats by revealing species communities in phreatic systems (Motas, 1958; Larned, 2012), the hyporheic zone (Orghidan, 1959; Tonina and Buffington, 2023), the hypotelminorheic habitat (Meštrov, 1962; Culver and Pipan, 2014), the deep subsurface (Fredrickson and Hicks, 1987; Borgonie et al., 2011; Kadnikov et al., 2020), chemolithoautotrophically based groundwater ecosystems (Engel, 2019), and groundwater calcretes (Humphreys et al., 2009). These discoveries have revealed an ever-increasing diversity of habitats that extends beyond the narrow contours of a habitat template emphasizing environmental stability and oligotrophy as general features of the groundwater biosphere. Shallow subterranean habitats show pronounced thermal seasonality and high levels of organic-matter compared to deep subterranean habitats (Culver and Pipan, 2011, 2014). Food webs are more complex in groundwater habitats where chemolithoautotrophic production adds up to inputs of detritus from the surface than in habitats where detritus is the only basal source of energy (Herrmann et al., 2020; Venarsky et al., 2023a). The potential for organism dispersal along hyporheic corridors is way higher than within groundwater calcrete archipelago (Cooper et al., 2007; Malard et al., 2017).

The unity of the groundwater biosphere is that all its constituent ecosystems share the absence of light, a characteristic feature that has important evolutionary and ecological consequences. Its diversity resides in numerous ecosystems differing from each other in their physical (e.g. void size), chemical (e.g. redox conditions), and biological features (e.g. predation pressure), thereby resulting in contrasted selective pressures among organisms of different ecosystems and habitats. Admittedly, the absence of light makes groundwater ecosystems unique at least in comparison with most surface environments. However, it does not necessarily make them less diverse, abiotically and biotically. Exploring the diversity of living forms in groundwater is to biodiversity sciences what observational astronomy is to space sciences in that it reveals previously unseen components of the Earth's biodiversity.

## Groundwater systems as ecological laboratories

Interestingly, biologists have pushed forward either the unity or diversity of subterranean ecosystems for testing general hypotheses in ecology (reviewed in Mammola, 2019). For example, Zagmaster and coauthors (2014) used obligate groundwater crustaceans for testing the mechanisms driving the increase in species range size at higher latitudes, a pattern known as the Rapoport effect (Stevens, 1989). The authors argued that since seasonal temperature variation was low in groundwater, any Rapoport effect had to reflect the influence of any other variable but thermal seasonality. In contrast, Mammola and coauthors (2019) used alpine spiders living in cave habitats differing in their thermal seasonality to test Janzen's hypothesis that thermal specialization restricts dispersal along elevational gradients. The test of Janzen's hypothesis by Mammola and coauthors (2019) provides a demonstrative example of using the heterogeneity of subterranean environmental conditions and diversity of species' responses for testing ecological theories. In their desire to emphasize the unique features of groundwater ecosystems, those that make them so different from surface ecosystems, groundwater ecologists may have overlooked, if not unconsciously constrained, their diversity. In the paragraphs below, we rely on evidence and ideas presented in chapters of the second edition of "*Groundwater Ecology and Evolution*" to plead for an increased focus in future research on the study of biological and ecological diversity in groundwater.

In their eloquent plea for minimizing the use of specialized terms in subterranean biology, Culver and coauthors (2023) proposed that the troglomorphy paradigm—the hypothesis that groundwater metazoans show convergent traits because of a convergent selective environment (Christiansen, 1961)—should be replaced by a newly emerging paradigm of a mixture of convergence and divergence (but see also Pipan and Culver, 2012). This proposal is not to refute that groundwater organisms represent fertile ground for the study of mechanisms behind convergent traits. However, the focus on convergent adaptations to shared selective pressures of the groundwater biosphere may have led biologists to overlook the numerous nonconvergent traits that allow species to adapt to contrasted selective pressures. Size of voids, water current, temperature, and the availability of food and dissolved oxygen vary markedly among aquifers (Aquilina et al., 2023; Fišer et al., 2023; Marmonier et al., 2023). And indeed, a number of traits including body size and appendage length (Trontelj et al., 2012; Delić et al., 2016; Pipan and Culver, 2017; Fišer et al., 2023), thermal tolerance breadth (Mermillod-Blondin et al., 2013; Di Lorenzo et al., 2023), resistance to starvation (Salin et al., 2010), and metabolic rate (Simčič and Sket, 2019) vary markedly among groundwater metazoans.

Calls for increased research on species trait diversity in groundwater are apparent in several chapters of the book. Venarsky and coauthors (2023b) recognize that the perspective on life histories of groundwater species has advanced little beyond the generalization that groundwater species exhibit slow life history strategies consistent with a habitat template model characterized by environmental stability and energy limitation. However, they acknowledge that this generalization is from inferences derived from only a few model taxa, with few phylogenetically controlled and replicated comparative studies between related groundwater and surface water organisms and even fewer studies among groundwater organisms belonging to different habitats. Starting from evidence that a high dispersal ability is often associated with fast life history strategies (Beckman et al., 2018), there is also a general perception that the predictability of the groundwater environment might have selected for a low intrinsic dispersal propensity. This perception is reinforced by empirical evidence that groundwater species generally have narrow distribution ranges and that their populations are deeply structured genetically across space. However, narrow species range sizes and restricted gene flow among geographically close populations may reflect strong barriers to dispersal caused by habitat fragmentation, rather than the low organisms' dispersal propensity. Malard and coauthors (2023) propose that groundwater ecologists should take advantage of the diversity of selective pressures in groundwater to test for patterns of covariation between dispersal propensity and phenotypic traits (i.e. dispersal syndromes, Stevens et al., 2014). Debate on the use of surface water standard taxa for environmental risk assessment of groundwater provides another example of a generalization which, to say the least, warrants testing (Di Lorenzo et al., 2019a). This debate has come to focus on the question whether groundwater species are more or less sensitive to chemicals than surface water species. As shown by Di Lorenzo and coauthors (2023), the question is probably too general to admit a single answer, i.e., tolerance of groundwater taxa to chemicals is as variable as that of surface water taxa.

Another important albeit overlooked aspect of the phenotypic diversity of groundwater organisms is that multiple trait solutions can confer the same functional phenotype (Fišer et al., 2023). Natural selection acts on the evolution of traits indirectly through selection on trait functions. However, selection can achieve the same functional outcome via distinct combinations of morphological, behavioral, physiological, and life history traits or any combination of these—a principle known as many-to-one form-to-function mapping (Wainwright et al., 2005; Thompson et al., 2017). Fišer and coauthors (2023) proposed that many-to-one mapping could explain the nonconvergence of morphological traits among populations that colonized groundwater on independent occasions (Konec et al., 2015).

Overall, our current state of knowledge and understanding of the diversity and evolution of biological response traits in groundwater might have been biased by a long-standing but outdated view that groundwater ecosystems would exhibit broadly identical abiotic and biotic features, and hence populations be subject to broadly similar selective pressures (Di Lorenzo et al., 2023; Venarsky et al., 2023b; Fišer et al., 2023). There is today a crucial need for in-depth research to address the role of the heterogeneity of the groundwater biosphere in driving the evolution and diversity of phenotypes. A comprehensive research agenda on phenotypic diversity of groundwater organisms should consider divergent phenotypic changes in response to different selective regimes, convergent phenotypic changes in response to a shared selective regime, as well as non—convergent phenotypic changes that nevertheless allow populations to adapt in unique ways to the same selective regime. This will require developing additional model organisms at population and species levels and setting up comparative experiments for measuring the diverse phenotypic traits contributing to an organism's fitness, trade-offs between these traits, as well as their plasticity. We can only quote Fišer and coauthors writings (2023) to illustrate the difficulty, importance, and hopes of such a research agenda: *"These proposed research steps seem little rewarding in the contemporary molecular '—omics' era that stampedes over much of scientific efforts that are not lightningly fast in producing big data. We plead not to get discouraged, though. After all, we can benefit from the —omics studies only if such big data can be linked to well-documented, measurable and observable biological functions. On the more positive note, technical advances, such as those in computer vision and machine learning, as well as their accessibility via cheap or open-source software and hardware, will eventually find its way underground and transform the traditionally slow quantification of the subterranean phenotype into high throughput phenotyping as well."*

Since the 2000s, the structure and function of food webs and the role of metazoans on ecosystem processes have become prominent topics of groundwater ecology (Boulton et al., 2008; Saccò et al., 2019, 2022). Research has largely contributed to a better understanding of food web complexity by examining the role of habitat attributes (e.g. void size, surface-subsurface connectivity) on food web structure and dynamics, the relative importance of heterotrophic *versus* chemolithoautotrophic processes and top-down *versus* bottom-up limitations, and the role of resource availability and quality on trophic diversification, specialization, and selectivity (Venarsky et al., 2023a). However, there is an urgent need for better understanding the microbial loop and fully integrating it into the function of groundwater food webs (Griebler et al., 2022; Fillinger et al., 2023). The microbial compartment is composed of an interacting set of microorganisms including bacteria, fungi, protozoans (flagellates, ciliates, amoeba), and viruses. Yet, little attention has been paid to the role of mechanisms such as bacterial grazing by microeukaryotes and bacterial lysis via viral infection on microbial dynamics and fluxes of materials and energy through groundwater ecosystems (Karwautz et al., 2022).

Laboratory experiments mimicking the hydrological functioning of water—sediment interfaces and groundwater systems provided clear evidence for a significant influence of bioturbation and biodeposition by facultative and obligate groundwater species on water movement and activity of microorganisms (Mermillod-Blondin, 2011; Hose and Stumpp, 2019). However, the evidence that trophic activity of microeukaryotes and metazoans controls the density and activity of groundwater microorganisms is equivocal (Karwautz et al., 2022; Mermillod-Blondin et al., 2023). Data are lacking for assessing the effect of shredding and grazing activities on the rate of ecosystem processes, such as nutrient cycling. In their conceptual model of the role of metazoans on ecosystem processes in groundwater, Mermillod-Blondin et al. (2023) propose that their role would largely depend on the functional traits of species making up communities, more specifically their feeding mode and rate, sediment mixing rate, burrowing depth, type of biogenic structures produced, and bio-irrigation rate. They further predict that the functional significance of trophic activity of metazoans would increase and that of their engineering activity would decrease with increasing hydrological connectivity of groundwater to surface water. Testing these predictions already provides a full research agenda that requires designing innovative laboratory and field experiments with functionally diverse taxa. Scaling up the results of experiments to ecosystems is an even more challenging yet crucial task if we have to understand how biodiversity changes would affect groundwater ecosystem services delivered to humans.

## Groundwater systems as evolutionary laboratories

The shift in environmental conditions upon colonization of groundwater habitats is considered drastic and colonization events of subterranean aquatic habitats have occurred repeatedly. Thus, groundwater biological systems are a "dream come true" for evolutionary biologists. It is probably why life in groundwater habitats, perhaps more than in any other habitat, has stimulated research in evolutionary biology at large, from population genetics to Evo-Devo, from molecular evolution to ethology, from quantitative genetics to physiology

(Rétaux and Jeffery, 2023). In fact, groundwater organisms offer an infinite collection of solutions to the challenge of living in total and permanent darkness. They have fascinated humans for centuries, including species such as the "olm," the flagship blind salamander species, *Proteus anguinus*, discussed in Kostanjšek and coauthors (2023). Moreover, new 'strange' animals are discovered every day. For example, the first true millipede and leggiest animal on the planet, *Eumillipes persephone*, with a record-breaking 1306 legs, is an eyeless and depigmented diplopod recently discovered in a 60-m deep drill hole in Western Australia (Marek et al., 2021).

How many evolutionary paths lead to a blind and depigmented creature? Since the sixties, the often-remarkable morphology of cave animals has stimulated evolutionary research investigating the causes of reduced features, *i.e.*, the lack of eyes and pigmentation. These traits arose independently in many taxa, thereby providing an excellent case of convergence. In this section, we consider convergence as a phenotypic pattern, i.e., "*the independent evolution of similar phenotypes*" (Rosenblum et al., 2014), without any specifications to the degree of phylogenetic relatedness between taxa or to the genetic mechanisms underpinning that pattern. Of note, some constructive traits or elaborated features can also show striking convergent patterns, as Christiansen (1961) previously noted for elongated appendages among distantly related subterranean taxa. In discussing the similar anatomical modifications to cave life in springtail species of Entomobryinae, he suggested, "*a great diversity of evolutionary paths occurs.*"

Now, 60 years after Christiansen's suggestion, we have learnt a great deal about the molecular, cellular, and developmental processes that convergently shape the anatomy, physiology, behavior and life history traits of groundwater animals. Let us examine their most striking phenotype: eye loss. Cave vertebrates such as *P. anguinus* (Kostanjšek et al., 2023) and *Astyanax mexicanus* (Gross et al., 2023) did not lose their eyes through the same mechanisms and genetic pathways as cave arthropods such as *Asellus aquaticus* (Protas et al., 2023) and *Gammarus minus* (Fong and Carlini, 2023). This is because the developmental processes to construct an eye are markedly different and subject to distinct developmental constraints that restrict morphological evolution. Furthermore, the convergence goes beyond distinct mechanisms in different phyla, i.e., the mechanisms underlying eye loss are not the same between different groundwater vertebrates (Rétaux and Jeffery, 2023) and even between different populations of a given subterranean species (e.g., *A. mexicanus*). This diversity culminates in *Asellus* where there are multiple ways to reduce both pigments and eyes within a single subterranean population (Protas et al., 2011). Even though the exact cellular and molecular pathways involved and the exact cell signaling and cell—cell interactions at play are not the same in all these organisms, Rétaux and Jeffery (2023) have discussed that developmental pleiotropy might be a shared feature underlying eye loss and the appearance of constructive trait(s) in groundwater animals. This means that, due to developmental genes and signaling pathways that act repeatedly in different cell types, tissues, and organs during ontogeny, a modification in one gene can affect negatively the formation of the eye but positively the formation of, to give one example, the olfactory system (Hinaux et al., 2016; Ramaekers et al., 2019).

Groundwater animals also have evolved multiple ways of pigment reduction. The genes *mc1r* (Gross et al., 2009) and *oca2* (Protas et al., 2006) are recurrently mutated and responsible for the *brown* phenotype and albinism in several populations of *Astyanax* cavefish, respectively. These findings provide a striking example of the use of a shared mechanism to produce convergent phenotypes. However, in both cases, different genetic lesions in the same gene occurred in different cavefish populations. Hence, we suggest that some important level of "molecular tinkering" (Jacob, 1977) occurs during evolution that leads to similar troglomorphic phenotypes in originally dissimilar animals. In other words, according to François Jacob's own words, variations and *"novelties come from previously unseen association of old material,"* and *"to create is to recombine"* (Jacob, 1977).

How fast can groundwater creatures lose their eyes? According to our current knowledge, the fastest case of troglomorphic evolution is *A. mexicanus*, which became eyeless and depigmented in less than 20.000 years (Fumey et al., 2018; Policarpo et al., 2021) (or less than 150,000 years depending on authors; Herman et al., 2018). In Chapter 17 of this book, Casane and coauthors (2023) discuss the tempo and extent of the nonfunctionalization of vision genes in cave fish of different genera, based on the hypotheses that these genes evolve under relaxed selection after subterranean colonization and that their nonfunctionalization can serve as a proxy for the time spent in groundwater. The highly variable number of vision pseudogenes in different species indicates that eye loss is very recent in some lineages (e.g., *A. mexicanus*) but much more ancient in others (e.g., *Typhlichthys subterraneus*). Unfortunately, the exact time it actually takes for a species to lose their eyes is unknown to molecular evolutionists because we are looking at the endpoint. Moreover, Policarpo and coauthors (2021) have demonstrated that the extent of eye regression cannot be taken as a proxy for the time spent in the dark. The first European cavefish recently discovered in Southern Germany, a depigmented loach showing highly reduced eyes and which must have colonized groundwater after the last glaciation (20,000–16,000 years ago), is another hint that the troglomorphic phenotype can be acquired at a very rapid evolutionary speed (Behrmann-Godel et al., 2017).

The goal of evolutionary biologists is to understand both the proximal causes (e.g., genetic mutations and molecular mechanisms) and the distal causes (e.g., the evolutionary forces, selection and drift) of phenotypic evolution. Major progress has been made in both directions in past decades. In particular, the species *A. mexicanus* and *A. aquaticus*, which both enable unique and powerful genetic and Evo-Devo analyses, have delivered unprecedented knowledge on the evolution of troglomorphy (see Gross et al., 2023; Protas et al., 2023; Rétaux and Jeffery, 2023). In order to obtain a broader and more generalized picture, there is a crucial need to develop new models for comparative studies, from different phyla, which inhabit various types of groundwater habitats or microhabitats (Rétaux and Jeffery, 2023; see also Fig. 3 in Mammola et al. (2021) for some candidate models). As knowledge on groundwater ecology progresses, a strong link with studies at organismal level should be established *via* the rising Eco-Evo-Devo discipline to better understand the constraints and to better decipher truly adaptive from derived traits. Combining ecology, developmental biology and evolutionary biology (Eco-Evo-Devo) will also benefit from renewed efforts in natural history and zoology research to interpret future big 'omics' data in the framework of hypothesis-driven research and explicit evolutionary models.

As yet, our understanding of molecular processes shaping phenotypes during the transition of organisms from surface water to groundwater is essentially from intraspecific comparative studies, in which the traits of surface aquatic and groundwater populations of the same species are compared (but see Casane et al., 2023). Interspecific comparative studies that consider the outcomes of evolution processes over long time periods, at least longer than the lifespan of natural populations, can complement intraspecific comparative studies. If as suggested above, evolution behaves as a tinker, many tinkering attempts documented at the early stages of subterranean colonization may prove unsuccessful over time. Looking into the genomes of organisms that have colonized the subterranean environment over different time scales may provide insights into those molecular tinkering attempts that are more likely to be successful over the long term. Several species-rich clades comprised of surface and subterranean species provide excellent candidates for interspecific comparative studies among the aquatic fauna (e.g. the asellid isopods, Morvan et al., 2013) or terrestrial fauna (e.g. springtails, carabid beetles, and spiders; Mammola and Isaia, 2017; White et al., 2019). Phylogenetic, genomic, and ecological resources are increasingly available for a certain number of these clades, providing future knowledge insights into the way evolution to subterranean life proceeds (Mammola and Isaia, 2017; Saclier et al., 2018; Mammola et al., 2022a).

## Groundwater ecosystems' supporting services

Groundwater ecosystems are systematically overlooked in general and in water-related legislations, environmental management plans, conservation agendas, and biodiversity targets (Sánchez-Fernández et al., 2021; Fišer et al., 2022; Griebler et al., 2023). Protection of groundwater biodiversity and assessment of the health of groundwater ecosystems are often dashed, if not absent, in legislation acting at scales ranging from national to global (Mammola et al., 2019). In the absence of an overarching groundwater ecosystem governance, management—that is the implementation of rules outlined by governance—of groundwater ecosystems is essentially based on local and regional initiatives (Mammola et al., 2022b). This situation does not promote equitable groundwater ecosystem management across regions. Similarly, for conservation, there are no areas specifically dedicated for the protection of groundwater biodiversity, except at local scales (e.g. regulated access to species-rich caves). Globally, Sánchez-Fernández and coauthors (2021) estimated that only 6.9% of the surficial extent of the subterranean biosphere occasionally overlaps with surface-protected areas. This is, however, a gross overestimate because the authors retained only karst areas and lava fields (i.e. 19.09% of the world continental surface) for calculating the extent of the subterranean biosphere.

There are several reasons why the implementation of groundwater ecosystem—oriented conservation and management measures is so long to come. First, there is still poor awareness about the existence and importance of groundwater ecosystems across the public, policymakers, and managers. This may be due in part because they are "out of sight, out of mind." Yet, the public is potentially more aware of the potential for extraterrestrial life than it is of the occurrence of diversified living forms hidden in groundwater. From our personal experience, the best way to badly begin a discussion with groundwater managers and civil servants is to ask them to cite the name of a single groundwater species. We admit the phrasing is overly provocative, i.e., groundwater managers often know more about the importance of groundwater ecosystem services than they let on, although they do not call them that way. Second, there is often the perception among scientists, managers, and policy-makers that knowledge and understanding of groundwater biodiversity and groundwater ecosystem functioning are still too incomplete to implement cost-effective conservation, assessment, and monitoring schemes. Are species distributions truly so narrow or do they only reflect a too low sampling effort? Are bioindication tools sensitive enough to distinguish natural from human-induced perturbations? We believe this questioning will remain even though mechanistic understanding of groundwater biodiversity and ecosystems will continue growing. If conservation and management interventions have to be considered as integral components of scientific work, then Karl R. Popper' famous quote may help us to decide what to do (Popper, 1959): *"The game of science is, in principle, without end. He who decides one day that scientific statements do not call for any further test, and that they can be regarded as finally verified, retires from the game."* Bearing in mind the importance of groundwater ecosystems to humans and the speed at which they alter them, we shall rather make use of what we already know to act while accepting serious knowledge gaps, rather than continue to wait for some more clues to come. Third, setting up governance and management schemes for groundwater ecosystems not only requires groundwater scientists and practitioners to embrace ignorance but also to resist ignorance production. Embracing ignorance requires taking informed decisions using the best information available whenever accumulating evidence indicates action cannot longer be postponed. Resisting ignorance production requires identifying nonenvironmental considerations that motivate endless calls for more knowledge with the sole objective to postpone actions that would however benefit the majority. Ignorance production may become widespread if critical environmental, economic, and social issues accumulate at a rate faster than they can be addressed.

In the three decades since this book's first edition, the role of groundwater biodiversity in delivering important ecosystem services to humans has gained increased research attention (Griebler and Avramov, 2015). Even at low invertebrate densities, bioturbation activity appears to influence hydrodynamic properties of sediment (e.g. effective porosity; Hose and Stumpp, 2019) that underpin groundwater flow and hence groundwater self-purification processes and finally drinking water provision to humans. Basic data and tools to support the conservation and management of groundwater biodiversity and ecosystems have become increasingly available (Boulton et al., 2023; Hose et al., 2023). Large species occurrence data sets and habitat maps have paved the way for identifying hotspots of groundwater biodiversity in Europe and proposing conservation strategies (De Broyer et al., 2004; Iannella et al., 2021). The Groundwater Health Index, as one example, and its regional refinements have provided an efficient framework for measuring and identifying the ecological status of groundwater ecosystems using biotic and abiotic indicators (Korbel and Hose, 2017; Di Lorenzo et al., 2020; Hose et al., 2023).

Groundwater sustainability requires healthy ecosystems and groundwater biologists have repeatedly generalized unique features of groundwater species, communities, and ecosystems that strongly argue for their management and protection (Mammola et al., 2019; Hose et al., 2022). Many groundwater species have a narrow geographic distribution, small-sized and isolated populations, low dispersal ability, and low reproduction rates, making them extremely sensitive to the many anthropogenic pressures. Groundwater communities potentially show low resistance and resilience to human-induced perturbations because they naturally experience stable environmental conditions. However, such a level of generalization may suggest that all subterranean species and ecosystems require increased conservation and management efforts, an ideal goal that can nip any effort in the bud. Given limited resources for conservation and management, there is a pressing need to prioritize efforts and promote conservation and management actions whose cost-effectiveness can be tested and supported quantitatively (Mammola et al., 2022b). Here again, this need inevitably leads to considering the full environmental and biological diversity of the groundwater biosphere for strategically identifying conservation and management priorities. We foresee several important research avenues in the paragraphs below.

Further developing groundwater ecosystem classifications at a range of spatial scales is essential to support conservation and management decisions (Robertson et al., 2023). Classification units, for example, groundwater ecosystem types, can be used among other conservation targets to fix conservation goals in systematic conservation planning (Boulton et al., 2023). Conservation prioritization also requires developing a multifaceted knowledge of groundwater biodiversity patterns to integrate taxonomic, phylogenetic, and functional diversities into conservation goals (Pollock et al., 2020; Marmonier et al., 2023; Zagmajster et al., 2023). Similarly, research dedicated to the assessment of groundwater ecosystem health should continue promoting methods that account for the structure and composition of the biota (from microbes through to invertebrates), the physicochemical attributes of ecosystems, and functional processes (Hose et al., 2023).

Once conservation goals are set, ecosystem protection measures can readily be oriented using hydrogeology-based vulnerability methods because groundwater ecological classifications heavily rely upon hydrogeological attributes (Boulton et al., 2023). However, we critically need more data on the tolerance of groundwater organisms and sensitivity of groundwater ecosystem processes to stressors for developing environmental risk assessment of groundwater and targeted protection measures (Di Lorenzo et al., 2023). Bearing in mind the strong spatial turnover in groundwater species composition, biological trait-based approaches for assessing the sensitivity of local communities to stressors may perform better than taxonomic approaches (Di Lorenzo et al., 2019b; Hose et al., 2022). Moreover, there is a crucial need to identify the most pressing threats to groundwater ecosystems among climate change, habitat loss and fragmentation, hydrological impacts, and groundwater pollution (Kretschmer et al., 2023). This requires implementing models that predict changes in biodiversity and ecosystem function in response to these various threats.

Better-informed decision making on conservation and management of groundwater ecosystems also requires a better understanding of the ecological and evolutionary processes that generate biodiversity patterns. We foresee two main research challenges for understanding these processes. First, jointly analyzing patterns of taxonomic, functional, and phylogenetic diversity across large geographic areas would reveal geographical variation in the relative influence of speciation, extinction, and dispersal on the diversity and composition of the regional species pool (Borko et al., 2022; Zagmajster et al., 2023). A second challenge is to tease apart the relative contribution of speciation from surface ancestors *versus* subterranean ancestors to groundwater species diversity (Cooper et al., 2023). Disentangling the two types of speciation requires combining multiple approaches, among which a promising one consists in analyzing the evolution of genes specifically involved in traits that regress during groundwater colonization (e.g. vision, pigmentation) (Lefébure et al., 2017; Langille et al., 2021).

## Epilogue

We finish this concluding chapter with another Karl R. Popper's famous quote (Popper, 1962):

> No book can ever be finished. While working on it we learn just enough to find it immature the moment we turn away from it.

At the very moment we write the last lines of this book, we fully appreciate the accuracy and depth of Popper's words. While we measure considerable progress made since the release of *"Groundwater Ecology"* (Gibert et al., 1994), we believe we may have reached a time when groundwater biologists feel cramped in a traditional garb carved on a unity model of the groundwater biosphere that needs revision. In acknowledging this, we turn away from this book with a smile: we are confident that the next generation of scientists will leverage new approaches, methods, and viewpoints to continue pushing the boundaries of groundwater biological sciences further into the unknown.

## Acknowledgments

We thank David C. Culver, Cene Fišer, and Stefano Mammola for constructive comments that improved this chapter. We like to express our gratitude to all contributors and reviewers who dedicated their time and effort to the production of the second edition of *"Groundwater Ecology and Evolution."*

## References

Aquilina, L., Stumpp, C., Tonina, D., Buffington, J.M., 2023. Hydrodynamics and Geomorphology of Groundwater Environments. Chapter 1, this volume.

Beckman, N.G., Bullock, J.M., Salguero-Gómez, R., 2018. High dispersal ability is related to fast life-history strategies. Journal of Ecology 106, 1349–1362.

Behrmann-Godel, J., Nolte, A.W., Kreiselmaier, J., Berka, R., Freyhof, J., 2017. The first European cave fish. Current Biology 27 (7), R257–258.

Borgonie, G., Garcia-Moyano, A., Litthauer, D., Bert, W., Bester, A., van Heerden, E., Möller, C., Erasmus, M., Onstott, T.C., 2011. Nematoda from the terrestrial deep subsurface of South Africa. Nature 474, 79–82.

Borko, Š., Altermatt, F., Zagmajster, M., Fišer, C., 2022. A hotspot of groundwater amphipod diversity on a crossroad of evolutionary radiations. Diversity and Distributions 28, 2765–2777.

Boulton, A.J., Bichuette, M.E., Korbel, K., Stoch, F., Niemiller, M.L., Hose, G.C., Linke, S., 2023. Recent Concepts and Approaches for Conserving Groundwater Biodiversity. Chapter 23, this volume.

Boulton, A.J., Fenwick, G.D., Hancock, P.J., Harvey, M.S., 2008. Biodiversity, functional roles and ecosystem services of groundwater invertebrates. Invertebrate Systematics 22, 103–116.

Casane, D., Saclier, N., Policarpo, M., François, C., Lefébure, T., 2023. Evolutionary Genomics and Transcriptomics in Groundwater Animals. Chapter 17, this volume.

Christiansen, K.A., 1961. Convergence and parallelism in cave Entomobryinae. Evolution 15, 288–301.

Cooper, S., Fišer, C., Zakšek, V., Delić, T., Borko, Š., Faille, A., Humphreys, B.W., 2023. Phylogenies Reveal Speciation Dynamics: Case Studies from Groundwater. Chapter 7, this volume.

Cooper, S.J.B., Bradbury, J.H., Saint, K.M., Leys, R., Austin, A.D., Humphreys, W.F., 2007. Subterranean archipelago in the Australian arid zone: mitochondrial DNA phylogeography of amphipods from central Western Australia. Molecular Ecology 16, 1533–1544.

Culver, D.C., Pipan, T., 2011. Redefining the extent of the aquatic subterranean biotope—shallow subterranean habitats. Ecohydrology 4, 721—730.

Culver, D.C., Pipan, T., 2014. Shallow Subterranean Habitats. Ecology, Evolution and Conservation. Oxford University Press, Oxford.

Culver, D.C., Pipan, T., 2016. Shifting paradigms of the evolution of cave life. Acta Carsologica 44 (3), 415—425.

Culver, D.C., Pipan, T., Fišer, Ž., 2023. Ecological and Evolutionary Jargon in Subterranean Biology. Chapter 4, this volume.

De Broyer, C., Michel, G., Malard, F., Martin, P., Gibert, J., 2004. Action Plan for Conservation of Groundwater Biodiversity. In: Internal Report, Project Contract No EVK2-CT-2001-00121 "Protocols for the Assessment and Conservation of Aquatic Life in the Subsurface". Fifth EU Framework Programme, European Commission.

Delić, T., Trontelj, P., Zakšek, V., Fišer, C., 2016. Biotic and abiotic determinants of appendage length evolution in a cave amphipod. Journal of Zoology 299, 42—50.

Di Lorenzo, T., Avramov, M., Galassi, D.M.P., Iepure, S., Mammola, S., Reboleira, A.S.P.S., Hervant, F., 2023. Physiological Tolerance and Ecotoxicological Constraints of Groundwater Fauna. Chapter 20, this volume.

Di Lorenzo, T., Di Marzio, W.D., Fiasca, B., Galassi, D.M.P., Korbel, K., Iepure, S., Pereira, J.L., Reboleira, A.S.P.S., Schmidt, S.I., Hose, G.C., 2019a. Recommendations for ecotoxicity testing with stygobiotic species in the framework of groundwater environmental risk assessment. Science of the Total Environment 681 (1), 292—304.

Di Lorenzo, T., Fiasca, B., Di Camillo Tabilio, A., Murolo, A., Di Cicco, M., Galassi, D.M.P., 2020. The weighted Groundwater Health Index (wGHI) by Korbel and Hose (2017) in European groundwater bodies in nitrate vulnerable zones. Ecological Indicators 116, 106525.

Di Lorenzo, T., Murolo, A., Fiasca, B., Di Camillo Tabilio, A., Di Cicco, M., Galassi, D.M.P., 2019b. Potential of a trait-based approach in the characterization of an N-contaminated alluvial aquifer. Water 11 (12), 2553.

Engel, A.S., 2019. Chemolithoautotrophy. In: White, W.B., Culver, D.C., Pipan, T. (Eds.), Encyclopedia of Caves. Elsevier, pp. 267—276.

Ficetola, G.F., Canedoli, C., Stoch, F., 2019. The Racovitzan impediment and the hidden biodiversity of unexplored environments. Conservation Biology 33, 214—216.

Fillinger, L., Griebler, C., Hellal, J., Joulian, C., Weaver, L., 2023. Microbial Diversity and Processes in Groundwater. Chapter 9, this volume.

Fišer, C., Borko, Š., Delić, T., Kos, A., Premate, E., Zagmajster, M., Zakšek, V., Altermatt, F., 2022. The European green deal misses Europe's subterranean biodiversity hotspots. Nature Ecology and Evolution 6 (10), 1403—1404.

Fišer, C., Brancelj, A., Yoshizawa, M., Mammola, S., Fišer, Ž., 2023. Dissolving Morphological and Behavioral Traits of Groundwater Animals into a Functional Phenotype. Chapter 18, this volume.

Fong, D.W., Carlini, D.B., 2023. Ecological and Evolutionary Perspectives on Groundwater Colonization by the Amphipod Crustacean *Gammarus minus*. Chapter 16, this volume.

Fredrickson, J.K., Hicks, R.J., 1987. Probing reveals many microbes beneath earth's surface. A.S.M. News 53, 78—79.

Fumey, J., Hinaux, H., Noirot, C., Thermes, C., Rétaux, S., Casane, D., 2018. Evidence for late Pleistocene origin of *Astyanax mexicanus* cavefish. BMC Evolutionary Biology 18, 43.

Gibert, J., Danielopol, D.L., Stanford, J.A., 1994. Groundwater Ecology. Academic Press, San Diego, USA.

Griebler, C., Avramov, M., 2015. Groundwater ecosystem services: a review. Freshwater Science 34 (1), 355—367.

Griebler, C., Fillinger, L., Karwautz, C., Hose, G., 2022. Knowledge gaps, obstacles and research frontiers in microbial ecology. In: Mehner, T., Tockner, K. (Eds.), Encyclopedia of Inland Waters, 2nd edition. Elsevier, pp. 611—624.

Griebler, C., Hahn, H.J., Mammola, S., Niemiller, M.L., Weaver, L., Saccò, M., Bichuette, M.E., Hose, G.C., 2023. Legal Frameworks for the Conservation and Sustainable Management of Groundwater Ecosystems. Chapter 24, this volume.

Gross, J.B., Boggs, T.E., Rétaux, S., Torres-Paz, J., 2023. Developmental and Genetic Basis of Troglomorphic Traits in the Teleost Fish *Astyanax mexicanus*. Chapter 15, this volume.

Gross, J.B., Borowsky, R., Tabin, C.J., 2009. A novel role for Mc1r in the parallel evolution of depigmentation in independent populations of the cavefish *Astyanax mexicanus*. PLoS Genetics 5, e1000326.

Herman, A., Brandvain, Y., Weagley, J., Jeffery, W.R., Keene, A.C., Kono, T.J.Y., Bilandzija, H., Borowsky, R., Espinasa, L., O'Quin, Ornelas-García, C.P., Yoshizawa, M., Carlson, B., Maldonado, E., Gross, J.B., Cartwright, R.A., Rohner, N., Warren, W.C., McGaugh, S.E., 2018. The role of gene flow in rapid and repeated evolution of cave-related traits in Mexican tetra, *Astyanax mexicanus*. Molecular Ecology 27 (22), 4397—4416.

Herrmann, M., Geesink, P., Yan, L., Lehmann, R., Totsche, K., Küsel, K., 2020. Complex food webs coincide with high genetic potential for chemolithoautotrophy in fractured bedrock groundwater. Water Research 170, 115306.

Hinaux, H., Devos, L., Bibliowicz, J., Elipot, Y., Alié, A., Blin, M., Rétaux, S., 2016. Sensory evolution in blind cavefish is driven by early events during gastrulation and neurulation. Development 143, 4521–4532.

Hortal, J., de Bello, F., Diniz, J.A.F., Lewinsohn, T.M., Lobo, J.M., Ladle, R.J., 2015. Seven shortfalls that beset large-scale knowledge of biodiversity. Annual Review of Ecology Evolution and Systematics 46, 523–549.

Hose, G.C., Stumpp, C., 2019. Architects of the underworld: bioturbation by groundwater invertebrates influences aquifer hydraulic properties. Aquatic Sciences 81, 20.

Hose, G.C., Chariton, A.A., Daam, M.A., Di Lorenzo, T., Galassi, D.M.P., Halse, S.A., Reboleira, A.S.P., Robertson, A.L., Schmidt, S.I., Korbel, K.L., 2022. Invertebrate traits, diversity and the vulnerability of groundwater ecosystems. Functional Ecology 36, 2200–2214.

Hose, G.C., Di Lorenzo, T., Fillinger, L., Galassi, D.M.P., Griebler, C., Hahn, H.J., Handley, K.M., Korbel, K., Reboleira, A.S., Siemensmeyer, T., Spengler, C., Weaver, L., Weigand, A., 2023. Assessing Groundwater Ecosystem Health, Status, and Services. Chapter 22, this volume.

Humphreys, W.F., Watts, C.H.S., Cooper, S.J.B., Leijs, R., 2009. Groundwater estuaries of salt lakes: buried pools of endemic biodiversity on the western plateau, Australia. Hydrobiologia 626 (1), 79–95.

Iannella, M., Fiasca, B., Di Lorenzo, T., Di Cicco, M., Biondi, M., Mammola, S., Galassi, D.M.P., 2021. Getting the 'most out of the hotspot' for practical conservation of groundwater biodiversity. Global Ecology and Conservation 31, e01844.

Jacob, F., 1977. Evolution and tinkering. Science 196, 1161–1166.

Kadnikov, V.V., Mardanov, A.V., Beletsky, A.V., Karnachuk, O.V., Ravin, N.V., 2020. Microbial life in the deep subsurface aquifer illuminated by metagenomics. Frontiers in Microbiology 11, 572252.

Karwautz, C., Zhou, Y., Kerros, M.-E., Weinbauer, M.G., Griebler, C., 2022. Bottom-up control of the groundwater microbial food-web in an Alpine aquifer. Frontiers in Ecology and Evolution 10, 854228.

Konec, M., Prevorčnik, S., Sarbu, S.M., Verovnik, R., Trontelj, P., 2015. Parallels between two geographically and ecologically disparate cave invasions by the same species, *Asellus aquaticus* (Isopoda, Crustacea). Journal of Evolutionary Biology 28, 864–875.

Korbel, K.L., Hose, G.C., 2017. The weighted groundwater health index: improving the monitoring and management of groundwater resources. Ecological Indicators 75, 164–181.

Kostanjšek, R., Zakšek, V., Bizjak-Mali, L., Trontelj, P., 2023. The Olm (*Proteus anguinus*), a Flagship Groundwater Species. Chapter 13, this volume.

Kretschmer, D., Wachholz, A., Reinecke, R., 2023. Global Groundwater in the Anthropocene. Chapter 21, this volume.

Langille, B.L., Hyde, J., Saint, K.M., Bradford, T.M., Stringer, D.N., Tierney, S.M., Humphreys, W.F., Austin, A.D., Cooper, S.J.B., 2021. Evidence for speciation underground in diving beetles (Dytiscidae) from a subterranean archipelago. Evolution 75 (1), 166–175.

Larned, Z.T., 2012. Phreatic groundwater ecosystems: research frontiers for freshwater ecology. Freshwater Biology 57, 885–906.

Lefébure, T., Morvan, C., Malard, F., François, C., Konecny-Dupré, L., Guéguen, L., Weiss-Gayet, M., Seguin-Orlando, A., Ermini, L., Der Sarkissian, C., Charrier, N.P., Eme, D., Mermillod-Blondin, F., Duret, L., Vieira, C., Orlando, L., Douady, C.J., 2017. Less effective selection leads to larger genomes. Genome Research 27, 1016–1028.

Malard, F., Capderrey, C., Churcheward, B., Eme, D., Kaufmann, B., Konecny-Dupré, L., Léna, J.-P., Liébault, F., Douady, C.J., 2017. Geomorphic influence on intraspecific genetic differentiation and diversity along hyporheic corridors. Freshwater Biology 62, 1955–1970.

Malard, F., Griebler, C., Rétaux, S., 2023. Groundwater Ecology and Evolution. Elsevier.

Malard, F., Machado, E.G., Casane, D., Cooper, S., Fišer, C., Eme, D., 2023. Dispersal and Geographic Range Size in Groundwater. Chapter 8, this volume.

Mammola, S., 2019. Finding answers in the dark: caves as models in ecology fifty years after Poulson and White. Ecography 42, 1331–1351.

Mammola, S., Cardoso, P., 2022. Caves as simplified settings for testing ecological theory. Karstologia Mémoires 21 (1), 305–308.

Mammola, S., Cardoso, P., Culver, D.C., Deharveng, L., Ferreira, R.L., Fišer, C., Galassi, D.M.P., Griebler, C., Halse, S., Humphreys, W.F., Isaia, M., Malard, F., Martinez, A., Moldovan, O.T., Niemiller, M.L., Pavlek, M.,

Reboleira, A.S.P.S., Souza-Silva, M., Teeling, E.C., Wynne, J.J., Zagmajster, M., 2019. Scientists' warning on the conservation of subterranean ecosystems. BioScience 69 (8), 641–650.

Mammola, S., Isaia, M., 2017. Spiders in caves. Proceedings of the Royal Society B 284, 20170193.

Mammola, S., Lunghi, E., Bilandžija, H., Cardoso, P., Grimm, V., Schmidt, S.I., Hesselberg, T., Martínez, A., 2021. Collecting eco-evolutionary data in the dark: Impediments to subterranean research and how to overcome them. Ecology and evolution 11 (11), 5911–5926.

Mammola, S., Martínez, A., 2020. Let research on subterranean habitats resonate. Subterranean Biology 36, 63–71.

Mammola, S., Meierhofer, M.B., Borges, P.A.V., Colado, R., Culver, D.C., Deharveng, L., Delić, T., Di Lorenzo, T., Dražina, T., Ferreira, R.L., Fiasca, B., Fišer, C., Galassi, D.M.P., Garzoli, L., Gerovasileiou, V., Griebler, C., Halse, S., Howarth, F.G., Isaia, M., Johnson, J.S., Komerički, A., Martínez, A., Milano, F., Moldovan, O.T., Nanni, V., Nicolosi, G., Niemiller, M.L., Pallarés, S., Pavlek, M., Piano, E., Pipan, T., Sanchez-Fernandez, D., Santangeli, A., Schmidt, S.I., Wynne, J.J., Zagmajster, M., Zakšek, V., Cardoso, P., 2022b. Towards evidence-based conservation of subterranean ecosystems. Biological reviews of the Cambridge Philosophical Society 97 (4), 1476–1510.

Mammola, S., Pavlek, M., Huber, B.A., Isaia, M., Ballarin, F., Tolve, M., Čupić, I., Hesselberg, T., Lunghi, E., Mouron, S., Graco-Roza, C., Cardoso, P., 2022a. A trait database and updated checklist for European subterranean spiders. Scientific Data 9, 236.

Marek, P.E., Buzatto, B.A., Shear, W.A., Means, J.C., Black, D.G., Harvey, M.S., Rodriguez, J., 2021. The first true millipede-1306 legs long. Scientific Reports 11, 23126.

Marmonier, P., Galassi, D.M.P., Korbel, K., Close, M., Datry, T., Karwautz, C., 2023. Groundwater Biodiversity and Constraints to Biological Distribution. Chapter 5, this volume.

Mermillod-Blondin, F., 2011. The functional significance of bioturbation and biodeposition on biogeochemical processes at the water–sediment interface in freshwater and marine ecosystems. Journal of the North American Benthological Society 30 (3), 770–778.

Mermillod-Blondin, F., Lefour, C., Lalouette, L., Renault, D., Malard, F., Simon, L., Douady, C.J., 2013. Thermal tolerance breadths among groundwater crustaceans living in a thermally constant environment. Journal of Experimental Biology 216 (9), 1683–1694.

Mermillod-Blondin, F., Hose, G.C., Simon, K.S., Korbel, K., Avramov, M., Vorste, R.V., 2023. Role of Invertebrates in Groundwater Ecosystem Processes and Services. Chapter 11, this volume.

Meštrov, M., 1962. Un nouveau milieu aquatique souterrain : le biotope hypothelminorhéique. Comptes Rendus de l'Académie des Sciences, Paris 254, 2677–2679.

Morvan, C., Malard, F., Paradis, E., Lefébure, T., Konecny-Dupré, L., Douady, C.J., 2013. Timetree of aselloidea reveals species diversification dynamics in groundwater. Systematic Biology 62, 512–522.

Motas, C., 1958. Freatobiologia, o noura ramura a limnologiei (Phreatobiology, a new field of limnology). Natura (Bucharest) 10, 95–105.

Orghidan, T., 1959. Ein neuer lebensraum des unterirdischen wassers: der hyporheische biotop. Archiv für Hydrobiologie 55, 392–414.

Pipan, T., Culver, D.C., 2012. Convergence and divergence in the subterranean realm: a reassessment. Biological Journal of the Linnean Society 107, 1–14.

Pipan, T., Culver, D.C., 2017. The unity and diversity of the subterranean realm with respect to invertebrate body size. Journal of Cave and Karst Studies 79 (1), 1–9.

Policarpo, M., Fumey, J., Lafargeas, P., Naquin, D., Thermes, C., Naville, M., Dechaud, C., Volff, J.N., Cabau, C., Klopp, C., Møller, P.R., Bernatchez, L., García-Machado, E., Rétaux, S., Casane, D., 2021. Contrasting gene decay in subterranean vertebrates: insights from cavefishes and fossorial mammals. Molecular Biology and Evolution 38 (2), 589–605.

Pollock, L.J., O'Connor, L.M.J., Mokany, K., Rosauer, D.F., Talluto, M.V., Thuiller, W., 2020. Protecting biodiversity (in all its complexity): new models and methods. Trends in Ecology and Evolution 35 (12), 1119–1128.

Popper, K., 1959. The Logic of Scientific Discovery. Hutchinson & Co., Ltd., London.

Popper, K., 1962. The Open Society and Its Enemies. Routledge, UK.

Protas, M., Trontelj, P., Prevorčnik, S., Fišer, Z., 2023. Developmental and Genetic Basis of Troglomorphic Traits in the Isopod Crustacean *Asellus aquaticus*. Chapter 14, this volume.

Protas, M.E., Hersey, C., Kochanek, D., Zhou, Y., Wilkens, H., Jeffery, W.R., Zon, L.I., Borowsky, R., Tabin, C.J., 2006. Genetic analysis of cavefish reveals molecular convergence in the evolution of albinism. Nature Genetics 38, 107–111.

Protas, M.E., Trontelj, P., Patel, N.H., 2011. Genetic basis of eye and pigment loss in the cave crustacean, *Asellus aquaticus*. Proceedings of the National Academy of Sciences 108, 5702–5707.

Ramaekers, A., Claeys, A., Kapun, M., Mouchel-Vielh, E., Potier, D., Weinberger, S., Grillenzoni, N., Dardalhon-Cuménal, D., Yan, J., Wolf, Flatt, T., Buchner, E., Hassan, B.A., 2019. Altering the temporal regulation of one transcription factor drives evolutionary trade-offs between head sensory organs. Developmental Cell 50 (6), 780–792.

Rétaux, S., Jeffery, W.R., 2023. Voices from the Underground: Animal Models for the Study of Trait Evolution During Groundwater Colonization and Adaptation. Chapter 12, this volume.

Robertson, A., Brancelj, A., Stein, H., Hahn, H.J., 2023. Classifying Groundwater Ecosystems. Chapter 2, this volume.

Rosenblum, E.B., Parent, C.E., Brandt, E.E., 2014. The molecular basis of phenotypic convergence. Annual Review of Ecology, Evolution, and Systematics 45, 203–226.

Saccò, M., Blyth, A., Bateman, P.W., Hua, Q., Mazumder, D., White, N., Humphreys, W.F., Laini, A., Griebler, C., Grice, K., 2019. New light in the dark-a proposed multidisciplinary framework for studying functional ecology of groundwater fauna. Science of the Total Environment 662, 963–977.

Saccò, M., Blyth, A.J., Venarsky, M., Humphreys, W.F., 2022. Trophic interactions in subterranean environments. In: Mehner, T., Tockner, K. (Eds.), Encyclopedia of Inland Waters, 2nd edition. Elsevier, pp. 537–547.

Saclier, N., François, C.M., Konecny-Dupré, L., Lartillot, N., Guéguen, L., Duret, L., Malard, F., Douady, C.J., Lefébure, T., 2018. Life history traits impact the nuclear rate of substitution but not the mitochondrial rate in isopods. Molecular Biology and Evolution 35, 2900–2912.

Salin, K., Voituron, Y., Mourin, J., Hervant, F., 2010. Cave colonization without fasting capacities: an example with the fish *Astyanax fasciatus mexicanus*. Comparative Biochemistry and Physiology Part A: Molecular & Integrative Physiology 156, 451–457.

Sánchez-Fernández, D., Galassi, D.M.P., Wynne, J.J., Cardoso, P., Mammola, S., 2021. Don't forget subterranean ecosystems in climate change agendas. Nature Climate Change 11, 458–459.

Simčič, T., Sket, B., 2019. Comparison of some epigean and troglobitic animals regarding their metabolism intensity. Examination of a classical assertion. International Journal of Speleology 48, 133–144.

Stevens, G.C., 1989. The latitudinal gradient in geographical range: how so many species coexist in the tropics. The American Naturalist 133, 240–256.

Stevens, V.M., Whitmee, S., Le Galliard, J.-F., Clobert, J., Böhning-Gaese, K., Bonte, D., Brändle, M., Dehling, D.M., Hof, C., Trochet, A., Baguette, M., 2014. A comparative analysis of dispersal syndromes in terrestrial and semi-terrestrial animals. Ecology Letters 17, 1039–1052.

Stoch, F., Galassi, D.M.P., 2010. Stygobiotic crustacean species richness: a question of numbers, a matter of scale. Hydrobiologia 653, 217–234.

Thompson, C.J., Ahmed, N.I., Veen, T., Peichel, C.L., Hendry, A.P., Bolnick, D.I., Stuart, Y.E., 2017. Many-to-one form-to-function mapping weakens parallel morphological evolution. Evolution 71 (11), 2738–2749.

Tonina, D., Buffington, J.M., 2023. Physical and Biogeochemical Processes of Hyporheic Exchange in Alluvial Rivers. Chapter 3, this volume.

Trontelj, P., Blejec, A., Fišer, C., 2012. Ecomorphological convergence of cave communities. Evolution 66 (12), 3852–3865.

Venarsky, M., Simon, K., Saccò, M., François, C., Simon, L., Griebler, C., 2023a. Groundwater Food Webs. Chapter 10, this volume.

Venarsky, M., Niemiller, M.L., Fišer, C., Saclier, N., Moldovan, O.T., 2023b. Life Histories in Groundwater Organisms. Chapter 19, this volume.

Wainwright, P.C., Alfaro, M.E., Bolnick, D.I., Hulsey, C.D., 2005. Many-to-one mapping of form to function: a general principle in organismal design? Integrative and Comparative Biology 45 (2), 256–262.

White, W.B., Culver, D.C., Pipan, T., 2019. Encylopedia of Caves, Third edition. Elsevier.

Zagmajster, M., Eme, D., Fišer, C., Galassi, D., Marmonier, P., Stoch, F., Cornu, J.F., Malard, F., 2014. Geographic variation in range size and beta diversity of groundwater crustaceans: insights from habitats with low thermal seasonality. Global Ecology and Biogeography 23 (10), 1135–1145.

Zagmajster, M., Ferreira, R.L., Humphreys, W.F., Niemiller, M.L., Malard, F., 2023. Patterns and Determinants of Richness and Composition of the Groundwater Fauna. Chapter 6, this volume.

# Index

*Note*: 'Page numbers followed by *"f"* indicate figures, *"t"* indicate tables and *"b"* indicate boxes.'

## A

Abiotic drivers, 39
*Acantocephalus anguillae balcanicus*, 317
Acid-base reactions, 23
*Acidobacteria*, 212, 218–219
*Actinobacteria*, 212, 218–219
Actual evapotranspiration (AET), 128
Adaptation to specific habitats, habitat choice and, 428–429
Adaptationist hypothesis, 400–401
Adaptive radiation, 176
Adaptive Shift Hypothesis (ASH), 95–96, 166.
        *See also* Climatic Relict Hypothesis (CRH)
    critique of, 97–98
    retire as formal categories, 99
Adipose tissue, 408
Adult cave-dwelling morph, 358
Adult cavefish, 292
Adult specimens, 345
Advection, 17
Advection-dispersion-reaction equations (ADRE), 71–72
*Aeromonas hydrophila*, 319
Aesthetascs, 422
Air-filled lungs, 314–315
Albinism, 355
Albino cavefish, 290–291
Algal toxins, 26–27
Aliphatic hydrocarbons, 467
Alkali, 22
Alkaline water, 374
*Allobathynella bangokensis*, 408–409
Allozyme variation, 381
Alluvial aquifers, 463–464, 467–468
    Flathead River, 199
    systems, 555
Alluvial rivers
    hyporheic zone, 64–65
    predicting hyporheic exchange, 65–73
    role of hyporheic flow on water quality, 73–77
Alluvial systems, 122

ecosystem engineering activities by invertebrates, 269–270
trophic actions of invertebrates, 266–269
Alluvial valleys, 14–15
*Alnus glutinosa*, 339–340
Alpine-Dinaric orogenesis, 176
Alps, The, 44–45
Ambimorphs, 100
*Amblyopsis spelea*, 286–287
Ameba, 250
Amensalism, 129
Amino acid changes, 399b
Amphibians, 314–315, 317–318
Amphipoda, 120–122
Amphipods, 269–270, 285–286, 379. *See also* Isopods
    ecological setting and morphological variation, 374–377
    evolutionary perspectives, 383–387
        changes in gene expression associated with adaptation to cave habitat, 384–387
        eye development gene expression, 383
        opsins, 383–384
    family, 158
    fauna, 426
    impetus for colonizing cave streams, 378–380
    melanin pigment loss and innate immunity, 387–388
    multiple independent colonization of cave streams, 380–382
    species, 461–462
    upstream colonization of subterranean waters by *Gammarus minus*, 377–378
Anaerobic hydrocarbon-degrading Peptococcaceae, 220
Anaerobic metabolism, 471
Ancestral population, 380
Ancestral selection, 427–428
Anchialine caves, 51–52
Animal circadian physiology, 417–420
Animal groups, 351
Animal models for study of trait evolution
    brief historical timeline, 286–287

Animal models for study of trait evolution (*Continued*)
    evolutionary developmental biology of groundwater
        organisms, 293–296
    evolutionary genomics of groundwater organisms,
        296–298
    groundwater model systems, 287–289
    timeline of troglomorphic trait evolution, 293
    troglomorphic traits, 289–292
Animal sources, 377
*Anoptichthys*, 352
Antennae, 420
Anterior end of neural plate (ANP), 358
Anthropocene
    anthropogenic threats to groundwater, 490–494
    frameworks for sustainable use of groundwater in
        Anthropocene, 489–490
    groundwater availability and distribution, 484–489
Anthropocene Epoch, 373–374
Anthropogenic abstraction, 492
Anthropogenic action, 489
Anthropogenic pressures, 273–274
Anthropogenic threats, 553–554
    to groundwater, 490–494
        depletion, 491–492
            global groundwater quality, 493–494
            seawater intrusion, 492–493
Anthropogenic water consumption, 489
Antipredation mechanisms, 429
Appalachian cave streams, 129, 190
AQUALIFE project, 40, 42
Aquatic animals, 470–471
Aquatic ecosystems, 265, 552–554
    assessments, 510
Aquatic subterranean habitats, 124–125
Aquatic surface ecosystems, 245
Aquatic surface organisms, 153
Aquatic surface systems, 250
Aquatic vegetation, 16
Aquifers, 13, 211–212, 227, 251
    chemical and nutrient fluxes in, 24–27
    concept, 5–12
    cross-section of aquifer, 6f
    drivers of groundwater flow, 5–7
    function, 17–20
        flow and transport in aquifers, 17–18
        groundwater age, 18–19
        modeling aquifers, 19–20
    geologic types of aquifers, 10–12
        heterogeneity in aquifers, 12f
    hydrodynamics, 5–7
        permeability and hydraulic conductivity, 9–10
        porosity, 7–9
    links to surface hydrology, 13–16

    five scales of nested flow and exchange between
        surface and subsurface water for longitudinal
        profile along river valley, 14f
    productivity, 510
    reservoir, 487
    systems, 536
    type, 153
        classification system, 48
    water–rock interactions within, 22
Aquitards. *See* Compact aquifers
Archaea, 117–118
Aridification, 147–148
Aromatic hydrocarbons, 222
Arrestin, 384–385
Arsenic (As), 225–226
*Arthrobacter aurescens*, 227
Arthropod transcriptomics, 408–409
Artificial food, 472–473
Artificial intelligence, 20
Artificially enriched tracer compounds, 252–253
Asellid isopods, 377–378, 404
*Asellus*, 287–288, 313, 338, 342–343
    *A. kosswigi*, 197–198
    evo-devo research, 344
*Asellus aquaticus*, 99, 102–103, 168, 197–198, 286–287,
        295, 297, 306, 329–330, 409, 422, 430
    comparative transcriptomics, 344–345
    evo-devo, 342–344
    genetic basis of subterranean-related traits, 340–342
    phylogeography and population structure, 330–334
        phenotypic evolution of subterranean populations,
            334–339
        subterranean populations, 331–334, 333t
        widespread surface species with local
            subterranean populations budding off, 330–331
    raising and breeding in laboratory, 339–340
Assessing groundwater ecosystem
    assessing ecosystem health and condition, 503–505
    combining indicators into summary indices, 515–516
    indicators of ecosystem health and condition,
        508–513
    predicting ecosystem health and condition, 516–517
    reference condition for groundwater ecosystems,
        513–514
*Astroblepus pholeter*, 290–291
*Astyanax*, 287–291, 293–294, 313, 353–354
    *A. fasciatus*, 472
    community, 356
    genes, 353–354
    larval eyes, 298
*Astyanax mexicanus*, 96–97, 99, 286–287, 297, 306, 313,
        329–330, 337, 340–343, 345, 358, 394–396, 420,
        449, 472

developmental basis of troglomorphy in *Astyanax*, 357–365
    developmental basis of constructive traits, 360–362
    developmental basis of eye regression, 358–360
    developmental constraints for cavefish evolution, 362–363
    embryonic origins of cavefish evolution, 363–365
history of genetic and genomic studies of troglomorphy in *Astyanax*, 351–357
    candidate gene studies, 353–354
    classical genetic studies, 351–353
    genomics, 357
    QTL mapping, 354–355
    transcriptomics, 356–357
Asynchrony, 191
Australian groundwater calcretes, 169–170
Australian River Assessment System (AUSRIVAS), 502
Axial mesoderm, 363, 365

**B**

Bacterial antibiotic resistance genes, 511–512
Bacterivorous flagellates, 119
*Bacteroidetes*, 212, 218–219
Balkan Clades, 331
*Barbatula barbatula*, 122
Bardet-Biel syndrome protein, 384–385
Basal energy dynamics in groundwater food webs, 242–245
    chemolithoautotrophic food webs, 243–245
    detrital food webs, 242–243
*Batrachochytrium*, 320
    *B. salamandrivorans*, 320
Bayesian Mixing Models (BMMs), 252
Bayesian models, 503–504
Beaver dams, 16
Bedform scale, 66–71
Behavior and interaction with morphology, 421–422
Benthic zone, 73–75
Benzene, toluene, ethylbenzene and xylene (BTEX), 467, 511–512
Bern Convention. *See* Conservation of European Wildlife and Natural Habitats
Biochemical reactions, 493
Biodiversity, 305, 459, 553–554
    hotspots, 534
Biodiversity–Ecosystem Functioning (BEF), 263
Biofilm, 377
Biogeochemical attenuation processes, 535
Biogeochemical cycles, 217–222
Biogeochemical fluxes, toward predictive models of, 221–222

Biogeochemical processes, 73, 75, 216–217, 264, 532–533
Biogeochemical reactions, 24–25
Biogeographic distribution of microorganisms, 132
Biogeographic regions, 154–156
Biogeographical modeling, 198–199
Biogeography, 53–54
BIOLOG Ecoplates, 511
Biological assessment, 513
Biological components, 383
Biological distribution, chemical constraints to, 125–128
Biologists, 157
Biology systems, 450
Biomass, 552
Bioregions, 42–43
Biosphere Reserves (BR), 555
Biota, 512–513
Biotic diversity, 515
Biotic interactions, 128, 130, 528
Biotic Ligand Models (BLM), 468–469
Biotite, 22
Bioturbation processes, 270
Black box models, 19–20
Black morph, 308–310, 313
Blind cavefish, 286–287
Bomb-pulse radiocarbon, 252
Bottlenecks, microbial attenuation of groundwater contaminants and, 222–227
Bottom-up limitation, 241, 247
Brain saves energy, 425
*Brancelia*, 426–427
Breeding dispersal, 185
*Brown* melanophores, 355
Bulk stable isotope analyses, 253
Burrowing worms, 269

**C**

$^{13}$C-actetate, 265
*Caecidotea holsingeri*, 128–129, 190, 265
*Caenorhabditis*, 285–286
Calcite ($CaCO_3$), 21
Calcium, 126
    bicarbonate types, 22–23
Calibration, 176–177
*Cambarus aculabrum*, 536–537
*Cand*. Microgenomates, 219
*Cand*. Parcubacteria, 219
Candidate barriers, 198
Candidate phyla radiation (CPR), 117–118, 212, 218–219
Cannibalism, 192
Capture mark–recapture (CMR), 187

Carbohydrates, 118–119
Carbon dioxide ($CO_2$), 10
    $CO_2$-fixation, 245
Catecholamines, 460–461
Categorical heterogeneity, 10
Cation exchange capacity (CEC), 21
Cave:surface (C:S), 385
Cave(s), 305
    animals, 351
    cave-adapted *Astyanax*, 361–362
    cave-dwelling organisms, 355
    colonization, 379–380, 394
    species, 269
    streams, 269
        multiple independent colonization of, 380–382
    teleosts, 394
Cavefish, 351–352
    evolution
        developmental constraints for, 362–363
        embryonic origins of, 363–365
    transcriptomics, 406–408
Cellulose, 218–219, 511
Central Europe Clade, 331
    subterranean populations of, 331–334
Central nervous system, 415–416
*Ceriodaphnia dubia*, 467–468
Channel-spanning logs, 71
Channel-unit circulation, 16
*Chara* beds, 330
Chemical composition of groundwater, 21–24
    major water quality types, 22–23
    reduced environments, 23–24
    water–rock interactions within aquifers, 22
Chemical compounds in surface waters, origin of, 21
Chemical constraints to biological distribution,
        125–128
Chemical contamination, 458
Chemical fluxes in aquifers, 24–27
Chemical parameters, 378–379
Chemical stress
    physiological tolerance of groundwater organisms
        to, 464–470
        insights and future perspectives, 469–470
        pattern of environmental risk, 465–469
        pattern of tolerance to chemicals, 464–465
Chemically polluted groundwater systems, 272
Chemicals, pattern of tolerance to, 464–465
Chemoautotrophic primary production, 115–116
Chemolithoautotrophic bacteria, 116
Chemolithoautotrophic food webs, 243–245
Chemolithoautotrophic prokaryotes, 243
Chemolithoautotrophs, 220, 225–226
Chemolithoautotrophy in groundwater, 219–220

Chinese cavefish, 289
Chinese hyaline fish (*Sinocyclocheilus hyalinus*), 113
Chlamydiosis, 320
Chloride (Cl), 21
Chlorinated ethenes, 223–224
Chlorinated organic compounds, 223–225
*Chloroflexi*, 212, 218–219
Chlorofluorocarbons (CFC), 19–20
*Chloromyxum protei*, 317
Chytrids, 118–119
Ciliates, 250
Circadian clock, 394–396
*cis*-1, 2-dichlorethene (cis-DCE), 224
Classical genetics, 351–352
Classification
    schemes, 89–90
    systems, 40–41
        of groundwater ecosystems, 41–42
Clay, 8–9
    minerals, 21
    sediments, 122–123
Claybound gravels, 123
Climate, 487–488
    groundwater recharge ratio, 488f
Climate change, 458, 487–488
Climatic oscillations, 175
Climatic Relict Hypothesis (CRH), 95–96, 166.
        *See also* Adaptive Shift Hypothesis (ASH)
    critique of, 97–98
    retire as formal categories, 99
Clogging processes, 274
Coarse particulate organic matter (CPOM), 265–267
Coarse sediments, 214
Coastal anchialine systems, 525–526
Cold, tolerance to, 459
Colonization, 186
    dynamics of stygobionts, 538–540
    impetus for colonizing cave streams, 378–380
        size-selective predation, sexual selection and cave
            colonization, 379–380
    of subterranean biology, 95–99
Community functions, 216
Compact aquifers, 49
Complex aquifers, 19–20
Complex heterogeneous groundwater flow paths, 214
Compound specific methods, 253
Conceptual model of role of invertebrates on
        ecosystem processes and consequences for
        ecosystem services, 270–273
Confined aquifers, 7
*Congeria kusceri*, 320–321
Connectivity of karstic ecosystems, 265–266
Conservation, 320–321

biogeography, 538
    of subterranean species, 538—540
biology, 526
    conserving hotspots of groundwater biodiversity
        and endemism, 537—538
    of groundwater ecosystems and species at risk,
        552—553
    *Proteus* lineages and/or ESU, 322t
Conservation of European Wildlife and Natural
        Habitats, 556
Consolidated rocks, 9—10, 45—46
Constructive traits, 291—292, 355
    developmental basis of, 360—362
Consumer—resource interactions, 241—242, 246
Contaminant plumes, 223
Contaminant-degrading microorganisms,
        222—223
Continental scale, 42—43
    of groundwater ecosystems, 42—43
Continuous heterogeneity, 10
Continuous time random walk (CTRW), 72
Convention on Biological Diversity (CBD),
        554—555
Convention on International Trade in Endangered
        Species (CITES), 554—555
Convergence, 102—103
Convergence evolutionary concept, 101
Copepoda, 424
*Cottus carolinae*, 379
*Crangonyx*, 377—378
Crayfish, 170—171
Cricket clade, 101
*Croatobranchus*, 120, 429
Crustacea, 120—122
Crustacean groups, 424
*cryaa* gene, 360
*crybb1a* gene, 360
*crybb1c* gene, 360
*crybg3* gene, 360
*crybgx* gene, 360
*crygm5* gene, 360
Cryptic diversity, 194—195
Cryptic species, 172, 530
Cryptomycota, 118—119
*Crystallin* gene, 354
Crystallization temperature, 459
Cuban *Lucifuga* species, 394—396
Cyclopoida, 127—128
Cystathionine ß-synthase a (cbsa), 287—288
Cytochrome oxidase subunit I gene (COI), 176—177,
        194—195, 330—331, 378
    sequences, 381—382
Cytophagales, 212

**D**
Damköhler number, 75
*Daphnia*, 516
    *D. magna*, 467—468
Darcy equation, 15
Darcy-based "pumping" model, 66—67
Darcy's law, 6—7
Dark cave environments, 416
*Dechloromonas*, 219—220
Deep groundwater aquifers, 117
Deep-learning models, 504
Defense mechanisms of groundwater species, 462
Degeneration, 293—294
Degradation of petroleum hydrocarbons, 224—225
*Dehalobacter*, 224
*Dehalococcoides*, 224—225, 511—512
Deltaproteobacteria, 218—219
Demographic information, 451
*Dendrocoelum*, 120
Denitrification, 11—12, 269—270
Dense nonaqueous phase liquids (DNAPL), 223—224
Density—Activity—Carbon index (D-A-C index),
        504—505, 514
Depletion, 483, 491—492
*Desulfitobacterium*, 224
Desulfobulbaceae, 223
Detrital food webs, 242—243. *See also*
        Chemolithoautotrophic food webs
    transfer to higher trophic levels, 243
    transport, 242—243
Detritus, 242
Developmental evolution biology of groundwater
        organisms, 293—296
*Diacyclops belgicus*, 464—465
Diapherotrites, Parvarchaeota, Aenigmarchaeota,
        Nanoarchaeota, and Nanohaloarchaeota
        archaea (DPANN archaea), 117—118
Dichloroethene (DCE), 223—224
Differential gene expression, 406
Differentiation with respect to type of food, 426—427
Dikarya, 118—119
*Dina*, 120
Dispersal, 185, 191—192
    cost, 190—191
    dispersal-ecological specialization trade-off, 189—190
    evolution of, 188—192
    extrinsic factors, 188—190
        environmental cues to disperse, 189
        environmental heterogeneity in space and time,
            188—189
        interspecific interactions, 190
    groundwater landscape connectivity modulates
        dispersal, 197—200

Dispersal (*Continued*)
   hypothesis, 186
   importance of, 213—215
   intrinsic factors, 190—192
      body size, 190—191
      inbreeding, 191—192
      intraspecific interactions, 192
      reproductive strategies, 191
   propensity, 186
   range size, 193—196
Dispersivity, 17
Dissolved organic carbon (DOC), 115, 126—127
Dissolved organic matter (DOM), 211, 242, 251, 265
   processing, 245—246
   role of microbial networks in cycling of, 217—219
Dissolved oxygen (DO), 126—127
Diversification process, 165—166, 173, 175
Diversity metrics, 510
DNA, 564
   metabarcoding, 502—503
   repair protein photolyase, 384—385
   sequences, 399b, 451
      data, 534b
   sequencing, 165—166, 538
   technology, 351—352
Dollo's law, 308—310
Domestic wastewater, 493
DRASTIC, 40
Drinking water, 305, 484
   distribution systems, 511
*Drosophila*, 285—286, 295
Drought, 4
Drying process, 273—274
Dune morphology, 68
Dytiscidae, 409

**E**
Earth system, 490
Earth's climate system, 485
Ecological generalist, 104—105
Ecological indices, 40
Ecological opportunity, 176
Ecological processes, 531
   determining microbial community diversity and
      composition, 213—217
Ecological risk assessment (ERA), 516
Ecological specialist, 104—105
Ecological specialization, 189—190
Ecological studies, 241
Ecomorph, 338
Ecoregions, 42—43
Ecosystem

conceptual model of role of invertebrates on
      ecosystem processes and consequences for
      ecosystem services, 270—273
   engineering activities by invertebrates, 269—270
      alluvial systems, 269—270
      Karst systems, 269
   engineers, 269
   environmental impacts on surface water
      —groundwater interfaces and consequences for
      provision of ecosystem services by
      invertebrates, 273—275
   goods and services, 502
   indicators of ecosystem health and condition,
      508—513
      definitions of ecosystem health and indicators
         types, 503b
      functional indicators for assessing health, 510—513
      organizational indicators for assessing health,
         509—510
      stress indicators for assessing health, 513
   services, 270—273, 503b
Ecotonal "windows", 53
Ecotonal groundwater habitats, 52—53
Ecotones, 52
Ecotoxicogenomics, 502—503
Ecotoxicological constraints of groundwater fauna
   physiological tolerance of groundwater invertebrates
      to changing thermal conditions, 458—464
   physiological tolerance of groundwater organisms to
      chemical stress, 464—470
   physiological tolerance of groundwater organisms to
      light, food and oxygen variations, 470—473
Ecotoxicological protocols, 458
   indications for, 470—473
Effective number of codons (ENC), 381—382
*Eigenmannia*, 420
Electron donors, 23
Electron transport system, 428
Element, role of microbial networks in cycling of,
      217—219
Elliott and Brooks model, 68
Emerging contaminants in groundwater, nitrogen
      and, 26—27
Emerging micropollutants, 226—227
Emerging organic contaminants, 226—227
Empirical tests, 403
*emx*3 gene, 362—363
Endangered Species Act of 1973 (ESA), 530, 558
Endemic species, 529—530, 552—553
Endemism, conserving hotspots of groundwater
      biodiversity and, 537—538
Endocrine-disrupting compounds, 26—27

Energy flow pathways, 248–249, 251
Energy-saving mechanisms, 425
Environment and life history plasticity, 448–450
Environment quality, 442–443
Environmental conditions, 443b
  importance of, 213–215
Environmental constraints, 130–131
Environmental cues, 189
Environmental DNA (eDNA), 307–308, 531–532
  and metabarcoding, 532–535
  techniques in groundwater conservation, pitfalls of
    using, 534b
Environmental filtering, 213–214
Environmental heterogeneity in space and time,
    188–189
Environmental mutagenic factors, 404
Environmental parameters, 117
Environmental perturbations, 553–554
Environmental predictability, 188
Environmental Protection of Biodiversity and
    Conservation Act 1999 (EPBC), 557
Environmental radiation, 404
Environmental risk, pattern of, 465–469
Environmental risk assessment (ERA), 465–466
Environmental shift, 393
Enzymes, 511
Epigeomorphs, 100
Epikarstic habitats, 50
Epiphreatic habitats, 51
Eubacteria, 117–118
Eukaryotes, 532
Eunapius subterraneus, 120
Euproctus asper, 422
Europe, RSR patterns, 148–149
Europe-wide analysis, 172
European cave salamander (Proteus anguinus), 113
European Medicines Agency (EMA), 558
European Union Groundwater Directive (EC-GWD),
    272–273, 504
European Union Habitats Directive (EUHD), 556
European Union Water Framework Directive
    (EC-WFD), 504–505
Eurycea spelaea, 95
Eutroglophiles, 93
Evapotranspiration (ETP), 21
EvoDevo toolbox, 293–294
Evolution of gene expression in groundwater,
    405–409
  arthropod transcriptomics, 408–409
  cavefish transcriptomics, 406–408
Evolutionarily significant units (ESUs), 310
Evolutionary development model (evo-devo model),
    342–344

biology, 357–358
  tools commonly needed for evo-devo research, 344t
Evolutionary dynamic processes, 538
Evolutionary genomics, 393
Evolutionary processes, 526
Evolutionary theory, 416
Exophiala salmonis, 319
Experimental protocols, 470
Extra-optical sensory organs, 314–315
Extrapolymeric substances (EPS), 250
Eye(s), 340
  degeneration process, 358–359
  development, 358
    gene expression, 383
  eye-field, 362
  loss, 289–290, 352
    in cavefish, 360
  pigment, 342
  regression, 360
    developmental basis of, 358–360

**F**
Facultative surface taxa, 247
Facultative taxa, 247
Fauna, 119, 509–510
  of epikarst aquifer, 124
Fecundity, 440b
Feeding, 425–427
  differentiation with respect to type of food, 426–427
  energy-saving mechanisms, 425
  on microorganisms, 268
  posture behaviour, 292
Fertilization, 376–377
Fertilizers, 464–465
Fiber cells, 360
Fine particulate organic matter (FPOM), 265
Fine-grained sediments, 214
Firmicutes, 212, 218–219
Fish, 10–11
  predators, 375
Fixation of LoF mutations, 394–396
Flagellates, 250
Flavobacteriosis, 320
Floodplain hyporheic zone, 61–63
Fluorescein, 65
Fluvial hyporheic zone, 61–63
Flux balance analysis (FBA), 221
Focal organisms, 532
Food
  differentiation with respect to type of, 426–427
  food-rich habitats, 425
  physiological tolerance of groundwater organisms
    to, 470–473

Food (*Continued*)
  production, 491
  webs, 241
    clarifying and quantifying food web structure,
        251–253
    dynamics, 249–250
    processes role in groundwater community
        dynamics, 247–248
    studies, 251
Forces, 186
Fossil groundwater, 483
Foussoubie karst system, 124–125
Fractured aquifers, 11, 50
Fractured bedrock aquifers, 11
Fractured rocks, 48
Fractured systems, 11–12
French National Center for Scientific Research
        (CNRS), 307
Freshwater, 483. *See also* Groundwaters (GWs)
  habitats, 329–330
  planetary boundary of freshwater use, 490
  resource, 484–485
  springs, 530
Functional groups, 263
Functional indicators for assessing health,
        510–513
Functional phenotype
  habitat template, 417
  morphological-behavioral functional phenotype,
        417–429
Fungal dark matter, 118–119
Fungi, 118
  in groundwater habitats, 118–119

**G**

Gammaridae, 376–377
*Gammarus*, 287–291, 295
  *G. fossarum*, 128–130, 459
  *G. locusta*, 381
  *G. minus*, 96–97, 99, 101–102, 128–129, 168,
        197–198, 265, 286–287, 295–296, 345, 374, 377,
        448–449
Garden transplantation experiment, 443–445
Gas solubility, 25
Gastropods, 401
Gastrulation process, 365
*Gelyella*, 424
Gemmatimonadetes, 212
Gene expression, 384–385, 393, 460–461
  changes in gene expression associated with
        adaptation to cave habitat, 384–387
    evidence for positive selection, 385
    photolyase, 386–387

evolution of gene expression in groundwater,
        405–409
Gene ontology (GO), 406
Generalized Dissimilarity Modeling (GDM), 540
Generation time (GT), 403
Genes and genome architecture
  evolution of, 394–400
  isopods genomics, 400–405
Genetically divergent species, 194–195
Genetics, 286–287
  basis of subterranean-related traits, 340–342
  diversification, 537
  diversity, 451
  drift, 398b, 400–401
  mapping approach, 340
  variation, 381
Genomes, 393
  elemental composition of genomes in subterranean
        habitats, 404–405
  genome-scale constrained-based models, 222
  size evolution, 400–401
Genomics, *Astyanax mexicanus*, 357
Genotypes, 443–445
*Geobacter*, 221, 224, 512
  *G. metallireducens*, 221
  *G. sulfurreducens*, 221
Geochemical method, 65
Geogenic contamination, 493
Geographic distribution of numerous taxa, 173
Geographic regions, 41–42, 45, 48
Geographical distance, 197
Geologic maps, 3–4
Geothermal energy, 463–464
Geothermal systems, 458–459
German Environment Agency (UBA), 560–561
German Ministry for Education and Research
        (BMBF), 560–561
Global climate change, 273–274
Global conservation, 552
Global distribution, 485
Global groundwater quality, 493–494
Global hotspots, 177
Global pattern of species richness, 152
Global scale of groundwater ecosystems, 42
Global warming, 458–459, 463–464
  insights and future perspectives for global warming
        scenarios, 462–464
Global water cycle, 3, 490
Glycogen, 471
Gnathopods, 426
Governance priorities, 561
GPS technology, 142
Granites, 21

Gravitational gradients, 485
Grazing, 215
Green revolution, 26
Groundwater animals, 286–287
    evolution of gene expression in groundwater,
        405–409
        arthropod transcriptomics, 408–409
        cavefish transcriptomics, 406–408
    evolution of genes and genome architecture,
        394–400
        isopods genomics, 400–405
    measures of selection on coding sequences, 399b
    selection intensity, 398b
Groundwater biodiversity, 113, 115–122, 305, 552
    chemical constraints to biological distribution,
        125–128
        groundwater chemistry, 125–126
        oxygen content and organic matter, 126–128
    metazoans, 119–122
    paleogeographic events and historical climates,
        130–132
    past concepts and approaches in groundwater
        biodiversity conservation, 527–531
    initial impediments to groundwater conservation,
        530–531
        past approaches to groundwater conservation,
            528–530
        past concepts of groundwater biodiversity,
            527–528
    past concepts of, 527–528
    physical constraints to biological distribution,
        122–125
        hydrological connection to surface environment,
            124–125
        size of voids and interconnectedness, 122–124
        temperature, 125
    recent concepts and approaches in groundwater
        biodiversity conservation, 531–543
        conservation biogeography of subterranean
            species, 538–540
        conserving hotspots of groundwater biodiversity
            and endemism, 537–538
        environmental DNA and metabarcoding, 532–535
        systematic conservation planning approaches,
            540–543
        vulnerability mapping, 535–537
    species interactions, 128–130
    toward multifaceted approach to groundwater
        biodiversity patterns, 156–158
    viruses, 116
Groundwater dependent ecosystems (GWDE), 41
Groundwater ecosystems, 4, 115–116, 219–220, 248,
    525

classification systems, 41–42
compact aquifers, 49
conservation of groundwater ecosystems and species
    at risk, 552–553
continental scale, 42–43
current challenges and groundwater conservation,
    563–566
dynamics, 242
ecotonal groundwater habitats, 52–53
fractured aquifers, 50
global scale, 42
habitat/local scale, 48–53
health, 515
karst aquifers, 50–52
landscape scale, 44–48
legal frameworks related to groundwater
    ecosystems, 554–563
protection by water laws, 560–563
    groundwater laws in Australia, 562
    groundwater laws in Central and South America,
        561–562
    groundwater laws in Europe, 560–561
    groundwater laws in United States, 561
    international and national groundwater laws,
        562–563
reference condition for, 513–514
study, assess, and protect groundwater ecosystems,
    553–554
trophic niche diversification in, 248–249
unconsolidated sediment aquifers, 49–50
Groundwater environments, 405
    aquifer concept, 5–12
    aquifer function, 17–20
    chemical and nutrient fluxes in aquifers, 24–27
        biogeochemical reactions, 24–25
        nitrogen and emerging contaminants in
            groundwater, 26–27
        oxygen in groundwater, 25–26
    chemical composition of groundwater, 21–24
Groundwater Fauna Index, 514
Groundwater food webs, 115–116, 248–249, 251
    basal energy dynamics in groundwater food webs,
        242–245
    clarifying and quantifying food web structure,
        251–253
        artificially enriched tracer compounds,
            252–253
        compound specific methods, 253
        natural abundances of stable isotopes and mixing
            models, 251–252
        radiocarbon natural abundances, 252
    role of food web processes in groundwater
        community dynamics, 247–248

Groundwater food webs (*Continued*)
  role of habitat in groundwater food web dynamics, 245–246
    network structure, 245–246
    surface-subsurface connectivity, 246
  trophic niche diversification in groundwater ecosystems, 248–249
  unpacking microbial compartment of groundwater food webs, 250
Groundwater Health Index (GHI), 504
Groundwater invertebrates, 264, 267–268, 464–465
  for bio-indication, 272–273
  physiological tolerance of groundwater invertebrates to changing thermal conditions, 458–464
    insights and future perspectives for global warming scenarios, 462–464
    tolerance to cold, 459
    tolerance to heat, 459–462
Groundwater organisms, 114–115, 189, 448
  current conceptual model of life history evolution in groundwater species, 445–446
    environment and life history plasticity, 448–450
    expanding conceptual model of life history evolution in groundwater species, 448
    life history evolution and population demography, 450–451
    linking life history traits to molecular evolution, 451
    trade-offs among life history, morphological, and physiological traits, 450
  evolutionary developmental biology of groundwater organisms, 293–296
    exchanging eyes for senses during development, 295–296
    loosing eyes, 293–294
    lose eyes during development, 294–295
  evolutionary genomics of, 296–298
    coding mutations, 296–297
    regulatory changes, 297–298
  life history evolution, life history traits, and life table variables, 442–445
  physiological tolerance of groundwater organisms to chemical stress, 464–470
  physiological tolerance of groundwater organisms to light, food and oxygen variations, 470–473
    insights and future perspectives, 472–473
    tolerance to light, 470
    tolerance to oxygen depletion, 470–471
    tolerance to starvation, 471–472
  support for current conceptual model of life history evolution in groundwater species, 446–451
Groundwater temperatures (GWT), 458–459

Groundwater-dependent ecosystems (GDEs), 483–484, 527
Groundwaters (GWs), 3–4, 22–23, 39–40, 330–331, 486
  amphipods, 447–448, 471
  anthropogenic threats to, 490–494
  aquifers, 197–198
  assessment methods, 502–503
  availability and distribution, 484–489
    climate, 487–488
    global distribution, 485
    global groundwater abundance and distribution, 484f
    interactions, 486–487
    sustainable groundwater use and management, 488–489
  biology, 339–340
  chemistry, 125–126
  chemolithoautotrophy in, 219–220
  cnidarians, 120
  communities, 247
    role of food web processes in groundwater community dynamics, 247–248
  conclusive remarks on flagship species in, 322–324
  conservation
    current challenges and, 563–566
    initial impediments to, 530–531
    past approaches to, 528–530
    pitfalls of using eDNA techniques in, 534b
  contamination, 213
  crustaceans, 158
    species richness in Europe, 148–149
  ecology, 141–142
    overlooked part of, 317–320
  ecotones, 52, 128–129
  evolution of gene expression in, 405–409
  exploitation, 274
  extraction, 487, 530
  fauna, 8–11, 448, 470
    patterns of species composition, 152–156
    patterns of species richness, 143–152
    toward multifaceted approach to groundwater biodiversity patterns, 156–158
  flagship species, 305–306
  flow
    in aquifers, 19–20
    drivers of, 5–7
  groundwater–atmosphere interface, 25
  habitats, 40–41, 446, 449, 462–463
    and inhabitants, 305
  hydrochemistry, 214
  isopods, 451
  landscape connectivity modulates dispersal, 197–200

laws in
    Australia, 561–562
    Central and South America, 561–562
    Europe, 560–561
    United States, 561
metazoan communities, 114
microbial attenuation of groundwater contaminants
    and bottlenecks, 222–227
microbial communities, 212
    resistance and resilience of groundwater microbial
    communities to perturbations, 227–230
model
    organisms, 306
    systems, 287–289
morphology and sensory systems of groundwater
    top predator, 313–315
nitrogen and emerging contaminants in, 26–27
oxygen in, 25–26
panda, 324
past concepts and approaches in groundwater
    biodiversity conservation, 527–531
    initial impediments to groundwater conservation,
    530–531
    past approaches to groundwater conservation,
    528–530
    past concepts of groundwater biodiversity,
    527–528
pollution, 464, 493
pool, 3
populations, 191
quality, 484
recharge, 483–484, 487
resources, 4
species, 188–189, 193–194, 457–459
    current conceptual model of life history evolution
    in, 445–446
    groundwater species-rich regions, 151
    richness, 149
species interactions in, 129–130
systems, 245
table, 492–493
vertebrates, 152
withdrawal, 491
Growth rate, 440b
*Gyrinophilus porphyriticus*, 421–422

# H

Haber-Bosch process, 26
Habitat
    choice and adaptation to specific habitats, 428–429
    habitat/local scale of groundwater ecosystems,
    48–53
    role in groundwater food web dynamics, 245–246

"Habitat template" approach, 417, 445–446
*Haemopis*, 120
Hainich oligotrophic aquifer, 127–128
*Haloniscus*, 168
Hard rock geologies, 11
Hard-rock aquifers, 22
Hazardous Concentration 5% (HC5), 517
Health assessments, 508
Healthy ecosystem, 501–502
Heart regenerative capacity, loss of, 291
Heat, tolerance to, 459–462
Heat shock proteins (HSP), 460–461
*Hedgehog* gene, 354, 383
Henry's law, 25
Herbicide S-Metolachlor, 466–467
Heterogeneity, 11–12
    of streambed sediment, 73–75
Heterogeneous environments, 415–416
Heterotrophic nanoflagellates, 250
Heterotrophic production, 115–116
Hierarchic behaviour, 421
High species turnover, 152–156
High-resolution electron microscopy, 117–118
Higher-latitude species, 196
Historical climates, 130–132
Homogenous hydraulic conductivities, 71
Hot moments, 17
Hot spots, 17, 222–223
Hybrid embryos, 406–407
Hybrid models, 536
Hydraulic conductivity, aquifer hydrodynamics,
    9–10
Hydraulic gradient, 487
Hydraulic method, 65
Hydrazine synthase genes, 511–512
Hydrocarbons, 272, 511
    contamination, 228–229
Hydrochemistry, 214–215
Hydrodynamic dispersion, 17–18
Hydrofacies, 123
Hydrogen ($^2$H), 251
Hydrogen sulfide ($H_2S$), 24–25
*Hydrogenophaga*, 219–220
Hydrogeological classification schemes, 40
Hydrogeological indices, 40
Hydrogeological units, 46–47
Hydrogeology, 123
HydroGeoSphere, 19–20
Hydrologic cycle, 485
Hydrological connection to surface environment,
    124–125
Hydrological exchanges, 272–275
    in sedimentary systems, 270

Hypertrophied lateral line system, 425
Hyporheic circulation, 15—16
Hyporheic exchange, 13—14, 61—63
    bedform scale, 66—71
    predicting, 65—73
    reach and landscape scales, 71—73
Hyporheic flow, 61—63
    role on water quality, 73—77
Hyporheic sediments, 272, 527
Hyporheic zone, 25, 52, 61—63
    of alluvial rivers, 64—65
Hyporheos, 64—65
Hypoxia, 428

**I**

Igneous rocks, 21
Immune response, 462
Inbreeding, 191—192
    depression, 191—192
Indicators of ecosystem health and condition, 508—513
Ingolfiellid amphipods, 173
Innate immunity, 387—388
Inorganic pollutants, 329—330
Insecticides, 466—467
Interactions, 486—487
Intergovernmental Panel on Climate Change, 458—459
Internal transcribed spacer (ITS), 331
    ITS-1, 381
    sequences, 381—382
International and national groundwater laws, 562—563
International conventions for protection of groundwater ecosystems, 554—558
International Hydrogeological Map of Europe (IHME), 45
International Union for Conservation of Nature (IUCN), 554
Interspecific interactions, 190
Interstitial habitat, 53
Interstitial movement of riverine water, 61—63
Intraspecific competition, 192
Intraspecific interactions, 192
Intrasystem variability, 449
Intrinsic vulnerability, 40, 535
Invertebrates, 373—374, 425, 504
    conceptual model of role of invertebrates on ecosystem processes and consequences for ecosystem services, 270—273
    ecosystem engineering activities by invertebrates, 269—270

environmental impacts on surface water—groundwater interfaces and consequences for provision of ecosystem services by invertebrates, 273—275
    trophic actions of invertebrates, 265—269
        alluvial systems, 266—269
        Karst systems, 265—266
    trophic and ecosystem engineering activities, 264—265
Ion, 126
Ionic mercury ($Hg^{2+}$), 225
Iron, 23
    iron-rich minerals, 22
Irrigation systems, 491
*Isocapnia* stoneflies, 93—95
Isolation, 551—552
Isopods, 285—286, 343, 377
    genomics, 400—405
        elemental composition of genomes, transcriptomes and proteomes in subterranean habitats, 404—405
        environmental radiation, 404
        genome size evolution, 400—401
        life history traits, 403—404
        molecular evolution rate in subterranean habitats, 401—403

**J**

Jura Mountains, 191—192

**K**

Karst aquifers, 50—52
Karst areas, 176
Karst landscapes, 10
Karst springs, 374
Karst systems, 124
    ecosystem engineering activities by invertebrates, 269
    trophic actions of invertebrates, 265—266
Karstic aquifers, 10—11
Karstic phreatic habitats, 51—52
Karstic regions, 176
Karstic rocks, 48
Karstic systems, 11—12
*Kumonga exleyi*, 559—560

**L**

Lagrangian reference system, 73
Lamarckian processes, 103—104
*Lamprologus lethops*, 289, 396
Land-cover patterns, 246
Landscape ecology, 53—54
Landscape network approach, 531—532

Landscape scales
    of groundwater ecosystems, 44–48
    hyporheic exchange, 71–73
Landscape water regime, 45
Large-bodied species, 190–191
Large-scale toxicity testing, 468–469
Legacy pollutants, 213
Legal frameworks related to groundwater
        ecosystems, 554–563
    groundwater ecosystem protection by water laws,
        560–563
    international and national conventions for protection
        of groundwater ecosystems, 554–558
    legislation focused on protection of endangered
        species, 558–560
*Leptodirus hochenwartii*, 320–321
Lez Karst system, 124
*lhx7* gene, 363
*lhx9* gene, 363
Life history
    environment and life history plasticity, 448–450
    evolution, 442–445
        in groundwater species, current conceptual model
            of, 445–446
        and population demography, 450–451
Life history traits (LHT), 403–404, 442–445
    interactions between life history traits and life table
            variables under different environmental
            conditions, 443b
    and life table variables in studies of groundwater
            fauna, 440b
    linking life history traits to molecular evolution, 451
Life tables, 440b
    variables, 442–445
Lifespan, 440b
Light, physiological tolerance of groundwater
        organisms to, 470–473
Light, tolerance to, 470
Limb bud segment, 409
Limestones, 22
Lipido-proteic-dominant catabolism, 472
Local species richness (LSR), 143. *See also* Regional
        species richness (RSR)
    patterns, 143–146
Locomotion, 424–425
Locomotor appendages, 421–422
Loess aquifers of Southern Germany, 49
Log jams, 16
Longevity, 292
Loosing eyes, 293–294
Lose eyes during development, 294–295
Loss of function (LoF), 394–396
*Lucifuga dentata*, 396

*Lucifuga gibarensis*, 396
Lumped-parameter models, 19–20

**M**

Machine learning approach, 20, 396–400
Macroconsumer biodiversity, 248
Macroecology, 141–142
Macrophytes, 509
Magnesium carbonates (MgCO$_3$), 22
Magnetoreception, 420
Malformation, 293–294
Mammoth Cave region, 150
*Marifugia cavatica*, 120, 289
Marine regressions, 147–148
    marine regression-transgression cycles, 173–174
Marine transgressions and regressions, 42
Marsupia, 376–377
*Mastacembelus*, 289
Mating system type, 191
*Mc1r* gene, 355
Measured environmental concentrations (MEC), 516
*Megadrilus pelagicus*, 426–427
Melanin
    biosynthesis pathway, 290–291
    pigment loss, 387–388
    synthesis, 388
Melanocortin 1 receptor (mc1R), 290–291
Mercury (Hg), 225
    methylation, 225
Metabolic homeostasis, 461
Metabolic interactions, 211–212
Metallic Hg$^o$, 225
Metallic sulfur, 23
Metalloids, 225–226
Metapopulation, 93–95
Metaproteomics, 512
Metazoans, 119–122
    groups, 114
    species, 552
*Methanosaeta*, 212
*Methanosarcina*, 212
Mexican blind cavefish, 197
Mexican cavefish (*Astyanax mexicanus*), 92, 167
Mexican cavefish, 351–352
Microbes, 114, 502
    microbe–invertebrate interactions in groundwater,
        268–269
Microbial activities, 223, 272
Microbial attenuation of groundwater contaminants
        and bottlenecks, 222–227
    chlorinated organic compounds, 223–225
    petroleum hydrocarbons, 222–223
    toxic metals and metalloids, 225–226

Microbial cells
  abundance, 510
  density, 215
  in groundwater environments, 215
Microbial changes, 511
Microbial chemolithoautotrophy, 512
Microbial communities, 217–222
  assembly, 216–217
  composition, 213–214, 223
    ecological processes determining, 213–217
  ecological processes determining microbial
      community diversity, 213–217
  in groundwater, 211–212
  toward predictive models of, 221–222
Microbial diversity and processes in groundwater
  ecological processes determining microbial
      community diversity and composition, 213–217
    differences between planktonic and
        surface-attached microbial communities,
        215–216
    Implications of assembly processes for community
        functioning and knowledge gaps, 216–217
    importance of environmental conditions, dispersal,
        and species interactions, 213–215
  emerging organic contaminants and micropollutants,
      226–227
  microbial attenuation of groundwater contaminants
      and bottlenecks, 222–227
  microbial communities and biogeochemical cycles,
      217–222
    chemolithoautotrophy in groundwater, 219–220
    role of microbial networks in cycling of DOM,
        nitrogen, and element, 217–219
    toward predictive models of microbial
        communities and biogeochemical fluxes,
        221–222
  resistance and resilience of groundwater microbial
      communities to perturbations, 227–230
Microbial fauna, 377
Microbial lineages, 212–213
Microbial networks role in cycling of DOM, nitrogen,
    and element, 217–219
Microbial transformations of inorganic arsenic,
    225–226
Microbiota, 510
Microcrustaceans, 250
Microeukaryotes, 118–119
Micrograzers, 250
Microorganisms, 24–25, 113–114, 122, 264, 272, 320,
    532–533, 552
Micropollutants. *See* Emerging organic contaminants
Microsporidia, 118–119
Migration, 380

*Milyeringa veritas*, 559–560
Minerals, 22
*mitf*1 gene, 356–357
Mitochondria, 428
Mitochondrial DNA, 308–310
Mitochondrial rates, 451
Mixing models, natural abundances of stable isotopes
    and, 251–252
Model organism, 330
Modeling aquifers, 19–20
  lumped parameter models of aquifers, 20f
Modeling of hyporheic exchange in
    gravel-bed rivers, 69
Molecular ecology and conservation genetics,
    310–313
Molecular evolution
  linking life history traits to, 451
  rate in subterranean habitats, 401–403
Molecular methods, 531, 535
Molecular operational taxonomic units (MOTUs), 172
Molecular techniques, 117–118
Molecular tools, 288–289
Mollusca, 120
Molnár János Cave population, 331–334
*Monolistra*, 154–156, 168
Monte Carlo approach, 19–20
Moors, 25–26
Morphological modification for subterranean life,
    99–103
  case studies of troglomorphy, 101–103
  critique and alternative classifications, 103
  troglomorphy and nature of selection, 99–101
Morphological variation, ecological setting and,
    374–377
Morphological-behavioral functional phenotype,
    417–429
  antipredation mechanisms, 429
  feeding, 425–427
  habitat choice and adaptation to specific habitats,
      428–429
  locomotion, 424–425
  reproduction, 427–428
  sensory input, 417–423
Morphology, 330, 416, 420–421
  behavior and interaction with, 421–422
  phenotypic evolution of subterranean populations,
      334–337
  systems of groundwater top predator, 313–315
Morphotypes, 394–396
Motor response, 415–416
Mudstones of Southwestern England, 49
Multi-causality, 142
Multidimensional scaling (MDS), 407–408

Multifaceted approach to groundwater biodiversity patterns, toward, 156–158
Multiple colonizations from surface ancestors, 168–169
Multiple independent colonization of cave streams, 380–382
  allozyme variation, 381
  COI and ITS sequences, 381–382
Mutational-hazard hypothesis (MH hypothesis), 400–401
Mycobacteriosis, 320
*Myotis grisescens*, 95

**N**
*Namanereis*, 120
Nanoarchaeota, 212
Nanoplastics, 26–27
Natal dispersal, 185
National Biodiversity Strategie and Action Plan (NBSAP), 554–555
National Cave Protection Act, 556
National conventions for protection of groundwater ecosystems, 554–558
National Policy Statement for Freshwater Management (NPSFM), 562–563
Natural environment, 501–502
Natural mutant, 365
Natural organic matter, 23
Natural selection, 189, 400–401, 427, 439, 446–447
Natural systems, 490–491
Nature conservation, 305
*Necturus*, 287, 307–308, 313, 316
*Nematoideum protei anguinii*, 317
Neotropics, 537–538
Network structure, 245–246
Networks with EXchange and Subsurface Storage (NEXSS), 72–73
Neural networks, 20
Neutral mutations, 342, 394, 401–403
Neutrality index (NI), 385
Next generation sequencing methods, 165–166
Niphargidae, 158, 171, 429
*Niphargus*, 53, 102, 129, 157, 172, 176, 189, 246, 286–287, 338–339, 424–425
  *N. croaticus*, 129, 157
  *N. hrabei*, 198–199
  *N. hrabeihas*, 409
  *N. inopinatus*, 459
  *N. rhenorhodanensis*, 267–268, 459
  *N. schellenbergi*, 130
  *N. stygius*, 196, 470
  *N. subtypicus*, 129, 157
  *N. timavi*, 128–129, 192

*N. virei*, 191–192, 197–198, 295
*Nitocrella achaiae*, 464–465
Nitrate, 23
  requirements, 504–505
Nitrogen
  cycling, 264
  and emerging contaminants in groundwater, 26–27
  role of microbial networks in cycling of, 217–219
*Nitrosomonas*, 219–220
*Nitrospira*, 212
*Nitrospirae*, 219–220
Noneffective dispersal, 185
Nonrenewable groundwater, 488
Nontroglomorphic species, 285–286
Nonvisual sensory systems, 355
North America, RSR patterns, 149–151
Northern Dinaric Karst Clade, 331, 334
Nuclear mutation rates, 404
Nucleotide excision repair (NER), 387
Nucleotides, 404–405
Numerical models at bedform, 75
Numerical models of hyporheic exchange, 72–73
Nutrients, 487, 525–526
  fluxes in aquifers, 24–27

**O**
Obligate groundwater organisms, 457–458
Obligate groundwater species, 446
Obligate groundwater taxa, 247
*Oca2* gene, 355
*Oculocutaneous albinism* type 2 (*oca2* gene), 287–288
Olfacto-reception, 421
Olfactory sensory neurons (OSNs), 361–362
Olfactory system, 291–292
Oligotrophic environments, 424
Olm (*Proteus anguinus*), 152, 305–306, 316
  conclusive remarks on flagship species in groundwater, 322–324
    evaluation of suitability of, 323t
  conservation, 320–321
  digestive system, 317
  historical rise to fame, 306–307
  molecular ecology and conservation genetics, 310–313
  morphology and sensory systems of groundwater top predator, 313–315
  overlooked part of groundwater ecology, 317–320
  reproductive peculiarities, 315–316
  systematics and evolution, 307–310
Omics methods, 517
One-dimensional Transport with Inflow and Storage program (OTIS program), 71–72
Ontogenetic stages, 405–406

Open framework gravels (OFG), 123
Operational taxonomic units (OTUs), 118–119
*Ophisternon candidum*, 559–560
*Opsin* genes, 353–354, 384–385
    no loss of functional constraints in cave populations, 383–384
Orange eye pigment, 342
*Orconectes*
    *O. australis packardi*, 470
    *O. daqikongensis*, 406, 408
    *O. inermis*, 447–448
    *O. jiarongensis*, 406, 408
Organ Cave (OC), 377–378, 384–385
Organic contaminants, 272, 464
Organic matter (OM), 115–116, 126–128, 242–243
    degradation, 269–270
    processing, 264
Organic pollutants, 329–330
Organic substrates, 118–119
Organizational indicators, 503b
    for assessing health, 509–510
Organohalide-respiring bacteria, 224–225
Orthologous opsin nucleotide sequences, 383–384
*otx2* gene, 356–357
Oxygen (O$_2$), 22–24
    concentration, 470–471, 514
    consumption, 24–25, 353, 461–462
    content, 126–128
    in groundwater, 25–26
    metabolism, 403
    saturation, 551–552
    tolerance to oxygen depletion, 470–471
    variations, physiological tolerance of groundwater organisms, 470–473

**P**

Paleoclimatic events, role of, 173–176
Paleogeographic events, 130–132, 147–148
Paleogeological events, role of, 173–176
*Palmorchestia hypogea*, 168
Parafluvial hyporheic zone, 61–63
Parasites, 317–320
Parental care, 440b
Parental fitness, 447
*Parhyale hawaiensis*, 343
*Paroster*, 172
    *P. microsturtensis*, 198
Particulate organic matter (POM), 242, 251
PASCALIS (European project), 153
Patescibacteria, 117–118
Pathogens, 317–320
*Pax6* gene, 354, 383
*Pde6b* gene, 360

Peatlands, 24–26
Perchloroethylene (PCE), 223–224
Permeability, aquifer hydrodynamics, 9–10
Perturbations, 213
    resistance and resilience of groundwater microbial communities to, 227–230
Pesticides, 514
Petroleum hydrocarbons, 222–223
Phages, 116
Phenoloxidase (PO), 387
Phenotype-genotype association tests, 340, 342
Phenotypic change, 345–346
Phenotypic evolution of subterranean populations, 334–339
Phenotypic plasticity, 449
Phenotypic response, 443–445
Phenotypic variation, 417
Phosphagen, 471
Photolyase, 386–387
    gene expression, 386–387
Photoreactivation, 387
Photoreceptor cells, 313
Phototransduction, 383–384
    genes, 409
*Phreatalona*, 424
*Phreatichthys andruzzii*, 289–290, 294–295, 387
Phreatobites, 93
Phylogenetic analyses, 168, 176–177
Phylogenetic approaches, 167
Phylogenetic diversity, 537
Phylogenetic-informed annotation (PIA), 344–345
Phylogenetically diverging lineages, 118–119
Phylogeny of crayfishes, 193–194
Physiological tolerance
    of groundwater invertebrates to changing thermal conditions, 458–464
    of groundwater organisms to chemical stress, 464–470
    of groundwater organisms to light, food and oxygen variations, 470–473
Pigmentation, 290–291
    genes, 394–396
    genetic basis for, 355
Pigments, 420
    cells, 290–291
Piper ternary diagrams, 22–23
Piston-flow model, 19–20
Plagioclase feldspars, 22
*Plagioporus protei*, 317
Planctomycetes, 212, 219–220
Planetary boundary, 490
    of freshwater use, 490
Planktonic crustaceans, 426–427

Planktonic-attached microbial communities, differences between surface-attached microbial communities and, 215–216
Plasticity, 425
Plate tectonics, 173–174
Platyhelminthes, 120, 127–128
Pleiotropy, 383–384
Pleistocene glacial maxima, 151
*Pmela* gene, 355
*Poecilia Mexicana*, 99, 294–295, 408, 422, 448–449
Polychaeta, 120
Polychlorinated biphenyls (PCBs), 223–224, 321
Polymerase Chain Reaction amplification (PCR amplification), 165–166
Polymorphic allozyme loci, 381
Pool-riffle sequences, 70–71
Population demographics, 450–451
Population demography, 450–451
Population genetic theory, 400–401
*Porcellio scaber*, 343
Porosity, aquifer hydrodynamics, 7–9
Porous media, 5
Postojna-Planina Cave System (PPCS), 307, 311, 331
Postojna-Planina system, 529–530
Precipitation, 22
    anomalies, 195–196
    patterns, 246
Precopulary mate-guarding behavior, 338–339
Precursor compounds, 253
Predation in groundwater, 129–130
Predicted No Effect Concentration (PNEC), 516
Predicting ecosystem health and condition, 516–517
Predictive models of microbial communities and biogeochemical fluxes, toward, 221–222
Premature termination codon (PTC), 386–387
Principal component analyses (PCA), 406–407
Prioritization methods, 541–543
*Proasellus*, 168
    *P. cavaticus*, 191–192, 198–199, 459
    *P. hercegovinensis*, 401
    *P. karamani*, 401
    *P. merdianus*, 243
    *P. slavus*, 457–458
    *P. valdensis*, 189–190, 462
    *P. walteri*, 199–200
Process-oriented approach, 99
Prokaryotes, 117–118, 532
Prokaryotic communities in shallow aquifers, 117
Prokaryotic groundwater communities, 117
Proteidae, 316
Proteids, 307–308
Protein concentrations, 510
Proteobacteria, 212

Proteomes, 405
    elemental composition of proteomes in subterranean habitats, 404–405
*Proteus*, 129–130, 287–288, 290–291, 305–308, 316, 321
    *P. anguinus*, 93, 122, 286–287, 294–295, 305–307, 320–321, 416, 429, 447–448, 471, 532, 556
        *P. anguinus anguinus*, 308–310
        *P. anguinus parkelj*, 310
Protista, 119
Protons (H$^+$), 22
Protozoa, 118
Protozoans, 250
    grazing, 215
*Pseudoanophthalmus*, 96
*Pseudoniphargus* amphipods, 174
Pumping
    exchange, 65–66
    fresh groundwater, 487
    tests, 10
Purification of water, 553–554
Pyrite, 23

## Q

Quality assessment, 513
Quantitative polymerase chain reaction (qPCR), 511–512
Quantitative trait loci (QTL), 287–288, 351–352
    mapping, 354–355
Quartz, 22

## R

*Racekiela cavernicola*, 120
Racovitza's scheme, 93
Radiocarbon (14C) natural abundances, 252
Ramsar Sites (RS), 555
Random amplification of polymorphic DNA (RAPD), 354–355
Range dynamics
    of groundwater species, 196
    of *Niphargus stygius*, 196
Range size of groundwater species, 193–196
Rapoport's effect, 194–195
Reach scales, hyporheic exchange, 71–73
Reach-scale hyporheic circulation, 15–16
Reach-scale hyporheic models, 71–72
Reactive oxygen species (ROS), 403, 471
Reciprocal F1 hybrids, 297
Red eye pigment, 342
Red leg syndrome, 319
Red List of Ecosystems (RLE), 559
Red-sensitive cones, 313
Redds, 64–65

Redox processes, 222–223
Redox reactions in groundwater environments, 23
Reducing environments, 23–24
Reductive traits, 289–291
Reference condition for groundwater ecosystems, 513–514
Regional obligate groundwater fauna, 154
Regional species richness (RSR), 143. *See also* Local species richness (LSR)
    patterns, 146–151
        Europe, 148–149
        North America, 149–151
        South America, 151
Regressive trait, 353
Reliable global assessments, 485
Remote sensing, 531
Renewable groundwater, 488
Representative elementary volume (REV), 10
Reproduction, 427–428, 446–447
Reproductive biology, 427
Reproductive isolation, swinging pendulum from geography to processes generating, 98–99
Reproductive strategies, 191
Residence time, 18–19, 61–63
Resilience, 512–513
    of groundwater microbial communities to perturbations, 227–230
Resistance of groundwater microbial communities to perturbations, 227–230
Resolution grid cells, 492
Respiratory process, 219
Retardation process, 18
Retinal pigmented epithelium (RPE), 359–360
*Rhamdia zongolicensis*, 294–295
Rhodamine, 65
*Rhodoferax ferrireducens*, 221
Rhodopsin, 353–354
Ribulose-1, 5-bisphosphate carboxylase-oxygenase (RuBisCO), 512
Ridge and Valley ecoregions (RV ecoregions), 143–144
Rigorous test, 446
Riparian vegetation, 16
River catchments, 245
River Invertebrate Prediction and Classification System (RIVPACS), 502
River networks, 245–246
Riverine water flows, 61–63
RNA, 511–512
    composition, 404–405
Robust model system, 351–352
Rod cells, 313
*rx3* gene, 362–363

**S**
28S rRNA gene (28S rDNA), 330–331
Saline coastal groundwater, 492
Salinization, 490–491
Saltwater intrusion, 490–491
Sampling
    methods, 565
    strategies, 217
Sand lenses, 123
Sandy gravel matrices, 123
Sanger sequencing, 165–166
*Saprolegnia* sp., 319
Saturated soils, 525–526
Schiødte's classification
    and explanations, 90–92
    scheme, 92
Sculpin predation, 379–380
Seasonality, 427
    of rainfall, 487
Seawater intrusion, 492–493
Semiempirical relationships, 70
Sensory input, 417–423
    behavior and interaction with morphology, 421–422
    beyond darkness, 422–423
    morphology, 420–421
Sensory rearrangements, 422
Sensory systems, 417
    of groundwater top predator, 313–315
Sequencing of cDNA, 406
Sexual dimorphism, 427
Sexual selection, 379–380
Shallow subterranean habitats, 115–116
*Shewanella oneidensis*, 221
*shh* gene, 295
Silicon dioxide ($SiO_2$), 22
Sine oculus, 383
Single colonization from surface ancestors, 168–169
Single-cell eukaryotic microorganisms, 118
Sinking streams, 10
*Sinocycloceilus*, 286–287, 289, 345, 394–396, 406
    *S. anophthalmus*, 408
"Sit-and-wait" strategy, 189
Size–selective predation, 375, 379–380
Skin bacterial communities, 317
Slovenian Cave Conservation Act, 556
Sodium chloride types, 22–23
Soil alteration, 225
Solute Transport and Multirate Mass Transfer-Linear Coordinates model (STAMMT-L model), 72
Solute Transport in Rivers (STIR), 72
Sonic hedgehog genes (*shh* genes), 421
Sorption, 18
South America, RSR patterns, 151

Space, environmental heterogeneity in, 188–189
Spatial heterogeneity, 213
Spatial patterns in population demographics, 241
Spatial scales, 63, 65
Spatially explicit classification system for
  GW fauna, 42
Special Areas of Conservation (SACs), 556
Speciation, 194
  dynamics
    climatic oscillations, 175
    drivers of subterranean diversity, 173–176
    new karst areas, 176
    single colonization *vs.* multiple colonizations from
      surface ancestors, 168–169
    speciation from subterranean ancestors, 169–173
    synthesis and future prospects, 176–177
    Tethys Ocean, plate tectonics and marine
      regression-transgression cycles, 173–174
  events, 166
  process, 338–339
Species accumulation curves, 145–146
Species at Risk Act of 2002 (SARA), 558
Species composition
  patterns of, 152–156
    variation in species composition among localities
      within region, 152–154
    variation in species composition among regions,
      154–156
Species distribution models (SDMs), 538, 541–543
Species interactions, 128–130
  in groundwater, 129–130
  importance of, 213–215
  in surface water, 128–129
Species richness
  global pattern of species richness, 152
  local and regional species richness, 143
    LSR patterns, 143–146
    RSR patterns, 146–151
    patterns of, 143–152
Species sorting, 213–214, 216, 227
Species Survival Commission (SSC), 558–559
Specific sensitivity distribution (SSD), 517
Specific vulnerability, 535
Specimens, 376–377
Speleobiologists, 98–99
Speleogriphacea, 173
Spemann-Mangold organizer, 365
Spirochetes, 218–219
Springs, 17, 53
  hydrology, 53
Stable food resources, 378–379
Stable isotopes
  analyses, 251

natural abundances of stable isotopes and mixing
  models, 251–252
techniques, 187
Stalactites, 50–51
Stalagmites, 50–51
Starvation, tolerance to, 471–472
Statistical methods, 514, 535
Statistical tools, 22–23
*Stenasellus virei*, 338–339
Stenotherm, 378–379
Streambed nests, 64–65
Streamlines in fluid flow, 17
Stress indicators for assessing health, 513
Stressors, 504
*Stygepactophanes*, 420
Stygobionts, 95, 99–100, 114, 556
Stygobite, 93
*Stygobromus* sp, 190–191, 199
  *S. pecki*, 472
  *S. spinatus*, 190
Stygofauna, 48, 530
  protection, 541
Stygophiles, 93, 114
Stygoregions, 43, 513
Stygoxenes, 93, 114, 509–510
Submarine groundwater discharge (SGD), 484
Subsurface dispersal, 196
Subsurface habitats, 125
Subterranean ancestors, speciation from, 169–173
Subterranean animals, 416
Subterranean aquatic habitats, 417
Subterranean biodiversity, 527
  patterns, 156–157
Subterranean biogeographers, 147–148
Subterranean biology
  colonization and speciation, 95–99
    alternative terminology, 98–99
    climatic relict and adaptive shift hypotheses and
      historical background, 95–97
    critique of CRH and ASH, 97–98
    retire CRH and ASH as formal categories, 99
  ecological classifications, 90–95
    critique and alternative classifications, 95
    modern ecological classifications, 92–95
    Schiødte's classification and two possible
      explanations, 90–92
  morphological modification for subterranean life,
    99–103
  overall recommendations, 103–104
Subterranean clades, 168
Subterranean diversity, drivers of, 173–176
Subterranean drainage basin, 378–379
Subterranean dytiscid beetles, 169–170

Subterranean ecomorph, 330
Subterranean ecosystems, 439
Subterranean environments, 175, 375–376, 469–470
Subterranean food chains, 129–130
Subterranean habitats, 93–95, 373–374, 401–403
   elemental composition of genomes, transcriptomes
      and proteomes in, 404–405
   molecular evolution rate in, 401–403
Subterranean life, morphological modification for,
      99–103
Subterranean populations
   of Central Europe Clade, 331–334
   phenotypic evolution of, 334–339
     behavior, 337–339
     changes in quantitative traits of subterranean
       asellus populations, 336t
     life history, 339
     morphology, 334–337
     physiology, 339
Subterranean Reka River, 334
Subterranean speciation, 166–167
Subterranean species, 176, 373–374, 428, 557
   conservation biogeography of, 538–540
Subterranean waters by *Gammarus minus*, upstream
      colonization of, 377–378
Subterranean-related traits, genetic basis of, 340–342
Subtroglophiles, 93
Sulfate, 23
Sulfidic cave habitats, 449
Sulfur ($^{34}$S), 23, 251
Sulfur-oxidizing *Thiothrix* bacteria, 130
*Sulfuricella*, 219–220
*Sulfurospirillum*, 224
Surface ancestors, single colonization *vs.* multiple
      colonizations from, 168–169
Surface aquatic cyprinid fish (*Delminichthys*
      *adspersus*), 199
Surface dispersal of groundwater species, 196
Surface ecomorph, 330, 338
Surface ecosystems, 248, 445–446
Surface environment, hydrological connection to,
      124–125
Surface freshwaters, 39
Surface hydrology, links to, 13–16
Surface streams, 269
Surface waters, 22–23
   deterioration, 492–493
   origin of chemical compounds in, 21
   species interactions in, 128–129
   surface water–groundwater interfaces,
      264–265
     on consequences for provision of ecosystem
       services by invertebrates, 273–275

Surface-attached microbial communities, differences
      between planktonic and, 215–216
Surface-derived plant material, 218–219
Surface-subsurface connectivity, 246
Survival, 440b, 447–448
Sustainability, 543–544
   of groundwater, 489
Sustainable development, 489–490
Sustainable Development Goals (SDGs), 489
Sustainable groundwater, 485
   frameworks for sustainable of groundwater in
      Anthropocene, 489–490
   groundwater and link to United Nations sustainable
      development goals, 489–490
   planetary boundary of freshwater use, 490
   use and management, 488–489
Symbioses, 317–320
Synonymous sites, 381–382
Systematic conservation planning (SCP), 531–532,
      540
   approaches, 540–543
Systematic reserve planning methods, 531–532

**T**
Taxon, 330
Taxonomic assignment, 534b
Taxonomic composition of groundwater species
      communities, 152
Taxonomic-based assessments, 510
Teleost (*Astyanax*), 285–286
   fish, 365
Temperature, 125, 195–196
Temporal patterns in population demographics, 241
Tethys Ocean, 173–174
Tetrapods, 307–308
Thaumarchaeota, 212, 216, 219–220
Thermal conditions, physiological tolerance of
      groundwater invertebrates to changing,
      458–464
Thermal refuges, 373–374
Thermal stress, 460–461
Thermal tolerance, 462
*Thiobacillus*, 219–220
Third suborbital bone (SO3), 360–361
Three-dimensional modeling of groundwater-surface
      water interaction, 15–16
Time, environmental heterogeneity in, 188–189
Time tree approaches, 167
Top-down forces, 248
Top-down pressure, 241, 246
Topography (T), 535
Toxic metals, 225–226
Trade-offs, 293–294

Traditional genetic models, 285–286, 293
*Trans*-acting regulatory changes, 297
Trans-Alpine clade, 331
Transcriptomes, 393
    elemental composition of transcriptomes in
        subterranean habitats, 404–405
    sequencing method, 344
Transcriptomics, *Astyanax mexicanus*, 356–357
Transfer function, 72
Transient storage models, 71–72
Transmitters, 187
Transverse dispersion, 222–223
Tricarboxylic acid (TCA), 245
Trichloroethene (TCE), 223–224
*Trichodina*, 317
*Triplophysa rosa*, 406
*Trocheta*, 120
*Troglichthys rosae*, 536–537
Troglobionts, 92
Troglobites, 92
*Troglocaris*
    populations, 129–130
    *T. anophthalmus*, 198–199
*Troglochaetus beranecki*, 120
Troglomorphic cave populations, 374
Troglomorphic species, 285–286
Troglomorphic syndrome, 188–189
Troglomorphic trait evolution, timeline of, 293
Troglomorphic traits, 285–286, 289–292, 335
    constructive traits, 291–292
    other traits, 292
    reductive traits, 289–291
    in teleost fish *Astyanax mexicanus*
        developmental basis of troglomorphy in *Astyanax*,
            357–365
        history of genetic and genomic studies of
            troglomorphy in *Astyanax*, 351–357
Troglomorphs, 100
Troglomorphy, 100–101, 381
    in *Astyanax*
        developmental basis of, 357–365
        history of genetic and genomic studies of,
            351–357
    case studies of, 101–103
    and nature of selection, 99–101
Troglophiles, 92–93
Trogloxenes, 92
Trophic niche diversification in groundwater
        ecosystems, 248–249
Truncated biodiversity, 509
Tubificid worms, 274–275
Turn-over, 16
    exchange, 61–63

time, 18–19
*Typhlichthys subterraneus*, 394–396
*Typhlocaris*, 174
Typhlogammaridae, 429
Tyrosine, 388
*Tyrp1b* gene, 355

**U**

U.S. Clean Water Act, 561
Ultra-violet (UV), 404
    radiation, 404
    rays, 470
    UV-induced DNA damage, 409
Unconfined aquifer, 6–7
Unconsolidated rocks, 45–46
Unconsolidated sediment aquifers, 49–50
Underground hydrologic system, 4
UNESCO Global Geoparks (UGGp), 555
United Nations (U.N.), 489
United Nations Sustainable Development Goals, 489
    groundwater and link to, 489–490
Unpacking microbial compartment of groundwater
        food webs, 250
Urban development, 373–374

**V**

Vadose habitats, 50–51
"Vector of fluxes" in systems, 270
*Velkovrhia enigmatica*, 120
Ventilation, 474
Verrucomicrobia, 212
Vertebrates, 526
    nervous system, 363
Vertical hydraulic gradient (VHG), 124–125
Vibration attraction behavior (VAB), 361, 421
Vicariance hypothesis, 186
Vinyl chloride (VC), 223–224
Viral infections, 215
Viral lysis, 215
Virus-to-prokaryote ratio (VPR), 116
Viruses, 116, 250
    microeukaryotes, 118–119
    prokaryotes, 117–118
Vision genes, 296, 396
Vision loss, 289–290
Vision pseudogenes, 396
Visual information, 417–420
Visual opsin genes, 409
Void size, 246
Volatile organic compounds (VOCs), 464–465
Volcanic eruptions, 225
*Vorticella*, 317
Vulnerability mapping, 535–537

## W

Ward Spring (WS), 384—385
Wasserhaushaltsgesetz (WHG), 560—561
Wastewater treatment plant, 493
Water cycle components, 485
Water framework Directive (WFD), 42—43
Water management, 490
Water purification, 272
Water purification processes, 269—270
Water quality, 329—330, 531, 560
    hyporheic flow role on, 73—77
    types, 22—23
Water resources, 494
    management, 562
Water scarcity, 487—488
Water table, 5
Water-related ecosystems, 489—490

Waterlouse species, 401
Water—rock interactions within aquifers, 22
Weathering front, 11
Weighting systems, 516
Western Australia (WA), 562
Wetlands, 17
White morph, 308—310, 313
Widespread surface species with local subterranean
    populations budding off, 330—331
World Heritage Convention (WHC), 554—555
World Heritage Properties (WHP), 555

## Y

Yeasts, 118—119

## Z

Zone of tension saturation, 5

Printed in the United States
by Baker & Taylor Publisher Services